High Performance Computing in Science and Engineering '16

Wolfgang E. Nagel • Dietmar H. Kröner •
Michael M. Resch
Editors

High Performance Computing in Science and Engineering '16

Transactions of the High Performance Computing Center, Stuttgart (HLRS) 2016

Springer

Editors
Wolfgang E. Nagel
Zentrum für Informationsdienste
und Hochleistungsrechnen (ZIH)
Technische Universität Dresden
Dresden
Germany

Dietmar H. Kröner
Abteilung für Angewandte Mathematik
Universität Freiburg
Freiburg
Germany

Michael M. Resch
Höchstleistungsrechenzentrum
Stuttgart (HLRS)
Universität Stuttgart
Stuttgart
Germany

Front cover figure: Bag breakup event during the air-assisted atomization of a liquid fuel. The air flow field is colored by particle IDs which depend on the creation time and their respective release position at the inlet. Details can be found in "Smoothed Particle Hydrodynamics for Numerical Predictions of Primary Atomization", by S. Braun, R. Koch and H.-J. Bauer, Institut für Thermische Strömungsmaschinen (ITS), Karlsruher Institut für Technologie (KIT), Karlsruhe, Germany on page 321ff.

ISBN 978-3-319-47065-8 ISBN 978-3-319-47066-5 (eBook)
DOI 10.1007/978-3-319-47066-5

Library of Congress Control Number: 2016963434

Mathematics Subject Classification (2010): 65Cxx, 65C99, 68U20

Printed on acid-free paper

This Springer imprint is published by Springer Nature
The registered company is Springer International Publishing AG
The registered company address is: Gewerbestrasse 11, 6330 Cham, Switzerland

Contents

Part I
Physics

Peter Nielaba

In this section, eight physics projects are described, which achieved important scientific results by using the CRAY XC40 (Hornet and Hazel Hen) of the HLRS. Fascinating new results are being presented in the following pages on astrophysical systems (simulations of galaxy formation, of massive galaxy clusters, and of photodissociation), soft matter systems (simulations of nucleation in colloidal systems and of dynamics and adsorption of fibrinogen), many body quantum systems (simulations of ultracold quantum systems) and elementary particle systems (simulations of nucleon observables and of the anomalous magnetic moment of the muon).

The studies of the astrophysical systems have focused on the galaxy formation, massive galaxy clusters, and on photodissociation of certain molecules.

V. Springel, A. Pillepich, R. Weinberger, R. Pakmor, L. Hernquist, J. Naiman, D. Nelson, M. Vogelsberger, and F. Marinacci from Heidelberg (V.S., A.P., R.W., R.P.), Cambridge USA (L.H., J.N., M.V., F.M.) and Garching (D.N.), in their project GCS-ILLU present results from a new generation of hydrodynamical simulations ("Illustris++" project, AREPO code), including new black hole physics and chemical enrichment models, using more accurate techniques and an enlarged dynamical range. The authors reproduced the appearance of a red sequence of galaxies, quenched by accreting supermassive black holes and computed disk galaxies populations with properties closely matching observational data. In addition, the authors predicted magnetic field amplifications through small-scale dynamo processes for galaxies of different sizes and types.

Yannick M. Bahé and the C-EAGLE collaboration from Garching used in their HLRS project GCS-HYDA the GADGET-3 code to simulate 25 galaxy clusters with high resolution ("Hydrangea" project). By the ongoing data analysis new insights

P. Nielaba (✉)
Fachbereich Physik, Universität Konstanz, 78457 Konstanz, Germany
e-mail: peter.nielaba@uni-konstanz.de

into the physics of galaxy formation in an extreme environment and on the growth of the massive haloes, in which cluster galaxies are embedded, are achieved.

B M McLaughlin, C P Ballance, M S Pindzola, P C Stancil, S Schippers and A Müller from the Universities of Belfast (B.M.M., C.P.B.), Auburn (M.S.P.), Georgia (P.C.S.), and Giessen (S.S., A.M.) investigated in their project *PAMOP* atomic, molecular and optical collisions on petaflop machines in order to support measurements at synchrotron radiation facilities and to study photodissociation effects for astrophysical applications. The Schrödinger and Dirac equations have been solved with the R-matrix or R-matrix with pseudo-states approach, and the time dependent close-coupling method has been used. Various systems and phenomena have been investigated, ranging from X-ray and inner-shell photoionization in atomic oxygen and argon ions, as well as in tungsten ions, the single-photon double ionization in helium, to the photodissociation in SH^+.

The studies of the soft matter systems have focused on nucleation barriers in colloidal systems and on the dynamics and adsorption of fibrinogen.

A. Statt, P. Koß, P. Virnau and K. Binder from the University of Mainz present in their project *colloid* a method to study the free energy barriers for homogeneous nucleation of crystals from a fluid phase, which is not hampered by the fact that the fluid-crystal interface tension in general is anisotropic. By Monte Carlo simulations in the *NpT* ensemble, using the softEAO model for colloidal systems, and by analyzing the equilibrium of a crystal nucleus surrounded by fluid in a small simulation box in thermal equilibrium, the fluid pressure, chemical potential and the volume of the nucleus have been computed to obtain the nucleation barrier. Interesting deviations from the classical nucleation theory with spherical nucleus assumptions have been discovered and analysed.

S. Köhler, F. Schmid and G. Settanni from the University of Mainz investigated in their project *Flexadfg* dynamical properties of fibrinogen and of the initial adsorption stages of fibrinogen on mica and graphite surfaces by atomistic Molecular Dynamics simulations. The adsorption simulations on mica showed a preferred adsorption orientation in a reversible process without large deformations of the protein, and the adsorption simulations on graphite showed an irreversible character and a formation of a large quantity of protein-surface contacts which eventually lead to deformations of the protein and the onset of denaturation.

In the last granting period, quantum mechanical properties of elementary particle systems have been investigated as well as the quantum many body dynamics of trapped bosonic systems.

O.E. Alon, R. Beinke, L.S. Cederbaum, M.J. Edmonds, E. Fasshauer, M.A. Kasevich, S. Klaiman, A.U.J. Lode, N.G. Parker, K. Sakman, M.C. Tsatsos, A.I. Streltsov from the Universities of Haifa (O.E.A.), Heidelberg (R.B., L.S.C., S.K., A.I.S.), Newcastle (M.J.E., N.G.P.), Tromso (E.F.), Stanford (M.A.K.), Basel (A.U.J.L.), Wien (K.S.), Sao Paulo (M.C.T.) studied in their project *MCTDHB* ultracold atomic systems by their method termed multiconfigurational time-dependent Hartree for bosons (MCTDHB). The principal investigators have focused on seven topics: (a) single shots imaging of dynamically created quantum many-body vortices, (b) many-body tunneling dynamics of Bose-Einstein condensates and vortex states in 2D, (c) transition from vortices to solitonic vortices in 2D trapped Bose-Einstein condensates, (d) variance as a sensitive probe of correlations and uncertainty product of an out-of-equilibrium many-particle system, (e) development of a multiconfigurational time-dependent Hartree method for fermions ("MCTDH-X") (f) trapped fermions escape, (g) composite fragmentation of multi-component Bose-Einstein condensates.

C. Alexandrou, K. Jansen, G. Koutsou and C. Urbach from Nicosia (C.A., G.K.), Zeuthen (K.J.) and Bonn (C.U.) investigated in their project GCS-Nops the inner structure of the proton and other hadrons by lattice chromodynamics. By generating the ensemble using directly the physical value of the pion and nucleon masses, the principal investigators were able to compute the hadron spectrum, the axial and tensor charges moments of parton distribution functions and the quark contents of the nucleons.

P. Marquard and M. Steinhauser from Zeuthen (P.M.) and Karlsruhe (M.S.) computed in their project NumFeyn multi-loop Feynman integrals for the electron contribution to the anomalous magnetic moment of the muon, using the FIESTA package.

The Illustris++ Project: The Next Generation of Cosmological Hydrodynamical Simulations of Galaxy Formation

Volker Springel, Annalisa Pillepich, Rainer Weinberger, Rüdiger Pakmor, Lars Hernquist, Dylan Nelson, Shy Genel, Mark Vogelsberger, Federico Marinacci, Jill Naiman, and Paul Torrey

Abstract Cosmological simulations of galaxy formation provide the most powerful technique for calculating the non-linear evolution of cosmic structure formation. This approach starts from initial conditions determined during the Big Bang – which are precisely specified in the cosmological standard model – and evolves them forward in time to the present epoch, thereby providing detailed predictions that test the cosmological paradigm. Here we report first preliminary results from a new

V. Springel (✉)
Astronomisches Recheninstitut, Zentrum für Astronomie der Universität Heidelberg, Mönchhofstr. 12–14, 69120, Heidelberg, Germany

Heidelberg Institute for Theoretical Studies, Schloss-Wolfsbrunnenweg 35, 69118, Heidelberg, Germany
e-mail: volker.springel@h-its.org

A. Pillepich
Max-Planck Institute for Astronomy, Königstuhl 17, 69117, Heidelberg, Germany
e-mail: pillepich@mpia-hd.mpg.de

R. Weinberger • R. Pakmor
Heidelberg Institute for Theoretical Studies, Schloss-Wolfsbrunnenweg 35, 69118, Heidelberg, Germany
e-mail: rainer.weinberger@h-its.org; ruediger.pakmor@h-its.org

L. Hernquist • J. Naiman
Center for Astrophysics, Harvard University, 60 Garden Street, 02138, Cambridge, MA, USA
e-mail: lars.hernquist@cfa.harvard.edu; jill.naiman@cfa.harvard.edu

D. Nelson
Max-Planck Institute for Astrophysics, Karl-Schwarzschild-Str. 1, 85740, Garching, Germany
e-mail: dnelson@mpa-garching.mpg.de

S. Genel
Department of Astronomy, Columbia University, 550 W. 120th St., 10027, New York, NY, USA
e-mail: shygenelastro@gmail.com

M. Vogelsberger • F. Marinacci • P. Torrey
Kavli Institute for Astrophysics and Space Research, MIT, 02139, Cambridge, MA, USA
e-mail: mvogelsb@mit.edu; fmarinac@mit.edu; ptorrey@mit.edu

W.E. Nagel et al. (eds.), *High Performance Computing in Science and Engineering '16*, DOI 10.1007/978-3-319-47066-5_1

generation of hydrodynamical simulations that excel with new physics, enlarged dynamic range and more accurate numerical techniques. The simulations of our ongoing Illustris++ project on HazelHen successfully reproduce the appearance of a red sequence of galaxies that are quenched by accreting supermassive black holes, while at the same time yielding a population of disk galaxies with properties that closely match observational data. Also, we are able to predict the amplification of magnetic fields through small-scale dynamo processes in realistic simulations of large galaxy populations, thereby providing novel predictions for the field strength and topology expected for galaxies of different size and type.

1 Introduction

In principle, simulations of cosmic structure formation are well-specified initial value problems that ought to be able to predict galaxy formation in an ab-initio manner. However, the enormous dynamic range and the complex baryonic processes in galaxy formation make this an extremely challenging multi-scale, multi-physics problem whose full understanding is still a distant goal. Nevertheless, earlier simulations have already proven instrumental for developing our current understanding of structure formation, even given their underlying simplifications. Indeed, cosmological simulations have played a crucial role in establishing ΛCDM as the leading cosmological theory, despite our present ignorance of the true physical nature of dark matter and dark energy.

In particular, dark matter only simulations such as the Millennium simulations [1–3] have led to significant physical insight and reached a high degree of maturity and accuracy. However, such DM-only simulations do not provide predictions regarding the galaxies themselves, and an extra step is required in order to bridge the gap with observations. Primarily two approaches have been used to establish this link: (1) the technique of "semi-analytical modeling", whereby baryonic physics is modeled coarsely at the scale of an entire galaxy and applied in post-processing on top of DM simulations [4, 5], and (2) hydrodynamic simulations, where the evolution of the gaseous component of the Universe is treated using the methods of computational fluid dynamics. The latter approach, together with subgrid prescriptions that provide numerical closure and that take into account astrophysical processes related to star formation, enables the complex interaction of the different baryonic components (gas, stars, black holes) to be treated on a much smaller scale, ideally yielding a self-consistent and powerfully predictive calculation.

Our group has published in 2014 the presently largest hydrodynamic simulations of galaxy formation [6–8]. This simulation suite – dubbed "Illustris" – used a different approach than the ones so far commonly adopted in astrophysics to simulate gas on a computer ("smoothed particle hydrodynamics", SPH, and "Eulerian" mesh-based methods, typically utilizing adaptive mesh refinement, AMR). Illustris employed a moving, unstructured mesh as it has been implemented in our code AREPO [9]: like in AMR, the volume of space is discretized into many individual cells, but as in SPH, these cells move with time, adapting to the flow of gas in their vicinity. As a result, the mesh itself, constructed through a Voronoi tessellation of space, has no preferred directions or regular grid-like structure. Over the past few years, we have shown that this new type of approach for simulating gas has significant advantages over the other two methods, particularly for large cosmological simulations like Illustris [10–13].

One of the major achievements of the Illustris simulation is its ability to track the small-scale evolution of gas and stars within a representative portion of the Universe. The calculation yielded a population of thousands of well-resolved elliptical and spiral galaxies, reproduced the observed radial distribution of galaxies in clusters and the characteristics of hydrogen on large scales, and at the same time it matches the metal and hydrogen content of galaxies on small scales. However, the analysis of Illustris has also revealed a number of tensions with observational data. In particular, it has become clear that the physical model used for the so-called radio-mode feedback [14] of accreting supermassive black holes has been too strong and violent at the scale of galaxy groups and low-mass clusters, causing a depletion of their baryon content. At the same time, this physical model still proved insufficient to quench the central galaxies in these systems to the required degree, causing these galaxies to become too massive. Other problems we identified were the normalization of the faint-end of the galaxy luminosity function, too large galaxy sizes, and a lack of a pronounced bimodality in the galaxy color distribution. In addition, important physical ingredients such as magnetic fields were still missing.

This provides the motivation for the 'Illustris++' project that we currently carry out as a GCS large-scale project on HazelHen at HLRS. The primary scientific goal of our project is to calculate new, unprecedentedly large hydrodynamic simulation models of the universe that improve upon the earlier Illustris project in several important respects. Most importantly, we aim to improve the feedback models by using a newly developed model for black hole accretion and its associated energy release, by employing a considerably improved multi-species chemical enrichment model, by adjusting the treatment of galaxy winds, and last but not least, by adding magnetic fields to our simulations, opening up a rich new area of predictions that are still poorly explored, given that the body of cosmological magneto-hydrodynamic (MHD) simulations is still very small [15–22]. In particular, we aim to study the strength of magnetic field amplification through structure formation as a function of halo and galaxy size. We will also be able to quantify for the first time the expected distribution of magnetic field properties for galaxies of a given size, and to explore the role of winds and strong feedback events in "polluting" the intergalactic-medium with magnetic fields. In addition, we aim for a larger number of resolution elements, and a larger simulation volume than realized previously. This is necessary to study the regime of galaxy clusters better (which are rare and can only be found in a sufficiently large volume), and to allow a sampling of the massive end of the galaxy and black hole mass functions.

At the time of this writing, our project is still running, and only a subset of the production calculations have finished, with some of the main runs presently being computed. In this article, we describe some of the developments undertaken for the project, detail practical and technical aspects of our work, and the status of our runs. We also describe a few preliminary results in an exemplary fashion.

2 Physics and Code Developments for Illustris++

2.1 New Blackhole Physics Model

As discussed above, we replaced the so-called 'radio-mode' of supermassive black hole accretion and feedback in our AREPO code with a novel implementation. The physics of active galactic nuclei (AGN) is crucial for quenching star formation in large galaxies, particularly the central galaxies in groups and clusters of galaxies. While our previous model for blackhole growth in Illustris worked well in certain respects [23], it also showed significant deficits, in particular, it reduced the Sunyaev-Zeldovich decrement in galaxy groups as a result of excessive gas loss, and led to a still insufficient reduction of the star formation rate in the largest clusters, making the central galaxies not red enough.

We have therefore adopted a new kinetic feedback model for AGN driven winds, motivated by recent theoretical suggestions that conjecture advection dominated inflow-outflow solutions for the accretion flows onto the black holes in this regime [24]. For a detailed description of the new model we refer to our recent preprint [25]. In brief, our approach estimates the gas accretion rate through the Bondi-Hoyle-Lyttleton model. However, unlike in previous work, we have eliminated any artificial boost factor to the accretion rate in favor of starting with a slightly higher seed mass of $5 \times 10^5\,M_\odot$. In terms of feedback, we distinguish between a quasar mode for high accretion rates where the feedback is purely thermal, and a kinetic mode for low accretion rate states where the feedback is purely kinetic. The latter replaces the old radio-mode. Instead, we now inject kinetic energy directly at the position of the black hole, in random directions, so that the time-averaged momentum injection vanishes. The distinction between the two feedback modes is based on the Eddington ratio of the black hole accretion. For Eddington ratios above 0.1, the black hole is assumed to be always in the quasar mode, but for lower Eddington ratios we make the threshold dependent on the black hole mass, such that it becomes progressively easier for low mass black holes to stay in the quasar mode. Or expressed differently, large black hole masses will eventually transition to the kinetic mode, and as this has a higher impact on the host system than the thermal feedback of the quasar mode, this will tend to reduce the accretion rate further, such that the system will then typically stay in the kinetic mode. In the kinetic feedback state, strong quenching of cooling flows and star formation in the halo is possible, such that the corresponding galaxy quickly reddens.

2.2 Hierarchical Time Integration

The simulations carried out in Illustris++ represent a significant challenge not only in terms of size and spatial dynamic range, but also in terms of the dynamic range in timescales. In particular, the strong kinetic AGN feedback, which couples to the densest gas in galaxies, induces very small timesteps for a small fraction of the mass. Over the course of 13 billion years of cosmic evolution, we need up to $\sim 10^7$ timesteps in total. This would be completely infeasible with time integration schemes that employ global timesteps, but even for the individual timestepping we use in AREPO, this represents a formidable problem. It can only be tackled if the computation of sparsely populated timesteps can be made extremely fast so that they do not dominate the total CPU time budget. This in turn requires elimination of overheads that touch the full particle/cell system on such timesteps.

For Illustris++, we have developed a novel hierarchical timestepping scheme in our AREPO code that solves this in a mathematically clean fashion. This is done by recursively splitting the Hamiltonian describing the dynamics into a 'slow' and a 'fast' system (similar to [26]). One important feature of this time integration scheme is that the split-off fast system is self-contained, i.e. its evolution does not rely on any residual coupling with the "slow" part. This means that our goal, namely that poorly populated short timesteps can be computed without touching any parts of the system living on longer timesteps, can be realized.

2.3 Chemical Enrichment Model

For Illustris++, we have also improved our modelling of chemical enrichment, both by using updated yield tables that account for the most recent results of stellar evolution calculations, and by making the tracking of different chemical elements more accurate and informative. The most important technical measure to achieve this has been the introduction of a fiducial "other chemical elements" mass bin, such that together with the explicit tracking of 9 chemical elements (H, He, C, N, O, Ne, Mg, Si, Fe), the metal abundance vector accounts for the full mass content of every cell or star. Since we do a spatial reconstruction for every element individually, the previous code could arrive at extrapolated flux vectors at cell interfaces with an unphysical sum of the 9 explicitly tracked elements, leaving for the other elements a negative contribution. In our new treatment, the density of these other elements is reconstructed as well, and the abundance pattern is renormalized after extrapolation, thereby always leading to physically viable chemical compositions at flux exchanges.

The other significant change we made is that we now account with special chemical tagging fields separately for metals produced by asymptotic giant branch (AGB) stars, type-II supernovae, and type-Ia supernovae. This has not been done before in such hydrodynamical simulations, and opens up a rich additional set of analysis possibilities which are largely unexplored thus far. Given that the metallicity patterns in the circumgalactic medium are emerging as a critical observational diagnostic and constraint for the feedback physics, this is a very timely extension of our modelling capabilities.

2.4 Hydrodynamical Accuracy Improvements

For certain problems, the original implementation of AREPO reached only first-order convergence in the L_1 norm. In [27] we have shown that this can be rectified by simple modifications in the time integration scheme and the spatial gradient estimates of the code, both acting together to improve the accuracy of the code. As a result, the new formulation used for Illustris++ is now second-order accurate under the L_1 norm even in unfavorable situations. As a welcome side effect, conservation of angular momentum is substantially improved, too. We have found that these changes can significantly improve the results of smooth test problems. On the other hand, we also showed that cosmological simulations of galaxy formation are unaffected for well resolved galaxies, demonstrating that the numerical errors eliminated by the new formulation do not impact these simulations significantly. Nevertheless, the improved accuracy of the new formulation is clearly to be preferred, and we expect that small, poorly resolved galaxies are rendered more accurately in Illustris++ than before, corroborating the advantage of our moving-mesh technique compared to SPH or AMR codes in this regime.

We have also made important improvements to the ideal magnetodynamics (MHD) solver in our code [28], primarily in the form of an additional timestep criterion that controls the size of the Powell source terms applied for divergence control. Previously, this was not checked explicitly, instead the timestep of a cell was determined only by the Courant condition and a kinematical timestep constraint. It could thus happen under rare conditions that the source terms would apply order unity corrections to the magnetic field over the course of a timestep, leading to a relatively large local error. In our new code used for Illustris++, this is now safely prevented, increasing both the accuracy and robustness of our MHD implementation.

2.5 Elimination of All-to-All Communication Steps

In the AREPO code, we need to carry out, at multiple places, parallel, distributed tree walks that serve to calculate, e.g., the short-range gravitational field, the local enrichment region of stellar populations, or the zone of accretion around a supermassive black hole. Algorithmically, each MPI rank first does a range-search on its local domain, during which it also detects if the search region overlaps with other domains. In the latter case, a search request to the foreign domain is registered. These are exchanged to the corresponding target rank and processed in a second phase of the distributed tree walk. The number of these search requests for each of the other possible MPI ranks is stored in a table. After the first tree walk phase, the table is communicated with an MPI_Alltoall such that each rank knows how many items it needs to import from each other processor. As the domain layout is highly irregular as a result of the work-load and memory-balancing, the detailed communication pattern arising here is irregular and sparse, and cannot be predicted ahead of time.

The sparseness however implies that for a large number of MPI ranks mostly zero entries are communicated in the MPI_Alltoall. This has not been a critical source of overhead thus far if the dynamic range of the simulation is limited, but becomes an issue in simulations with 10^4 MPI ranks and beyond, where the smallest timesteps (which are carried out most frequently) need to be very fast so as to not start dominating the total CPU budget. Illustris++ runs on HazelHen are our first large production simulations where this source of overhead plays a sizable role.

To mitigate this, we have developed during the first project phase a relatively involved rewrite of our communication patterns in the parallel tree walks, which can completely eliminate the need for an MPI_Alltoall. Instead, we can now employ a sparse communication pattern that makes use of an asynchronous collective barrier. The latter is in principle available as part of MPI-3.0, but for compatibility reasons with older systems that do not yet support MPI-3.0 we have implemented an efficient sparse asynchronous communication pattern for the distributed tree walks ourselves, relying only on MPI-2 features.

3 Simulation Set and Production Runs

After obtaining access to HORNET/HazelHen, we have first carried out a limited number of test simulations plus a number of science runs of zooms into the Illustris volume targeting individual galaxies (these also served to test our new physics implementations, and several publications about them are currently in preparation). These had confirmed that our simulation code AREPO runs effectively and without technical issues on the new Cray XC40 machine. This also helped us to establish the precise performance for our simulation work-load, both with respect to computational throughput, communication bandwidth and I/O speed. In all three areas, the high expectations we had for the XC40 were met (module some I/O issues that initially surfaced, but which could be resolved by switching our project to a different filesystem). Also, our tests did not reveal any technical obstacles against carrying out our simulations, aside from a surprisingly low memory ceiling for the application code on the compute nodes. We could initially not use more than ~3900 MB per core in large partition runs without sometimes falling victim to OOMs, caused by memory needs of the I/O subsystem and MPI buffers. How to work around this reliably required a lot of experimenting and testing on our end.

After finalizing our modification of the physical model of Illustris++, we adjusted our plans for the primary science runs in the project by first carrying out "IllustrisPrime", a very demanding simulation with 18 billion resolution elements in a $75\,h^{-1}$Mpc box similar to the original Illustris run, but now using the new full physics model of Illustris++ with all its improvements, as well as including magneto-hydrodynamics (MHD). We also now adopted the newest cosmological models as determined by the PLANCK Satellite. This cosmological simulation is the first that includes MHD and resolves galaxy formation at high resolution, opening up many possibilities for novel predictions. Also, IllustrisPrime will be ideal to convincingly demonstrate that we can solve the problems at the bright end of the galaxy luminosity functions that have troubled all previous simulations in the field, including our older Illustris simulation that presently defines the state-of-the-art in this area.

In Table 1, we give an overview of our primary production simulations, omitting smaller test calculations. We are currently still in the process of finalizing one of

Table 1 Overview of primary production runs carried out by the Illustris++ project. All simulations use PLANCK cosmological parameters and are carried out with a tracer particle method that is faithful with respect to the mass flux in the system between all baryonic components. We typically use two Monte Carlo tracers per Voronoi cell, i.e. $N_{\mathrm{tracers}} = 2 \times N_{\mathrm{cells}}$. All simulations follow more than 13 billion years of cosmic evolution, with smallest timesteps of order a few thousand years

Symbolic name	Boxsize	N_{dm}	N_{cells}	MPI ranks	Physics	Run status
L75n1820TNG	$75\,h^{-1}$Mpc	1820^3	1820^3	10,752	Final full physics model	Advanced
L75n1820MF	$75\,h^{-1}$Mpc	1820^3	1820^3	12,000	Alternative AGN feedback	Finished
L75n910TNG	$75\,h^{-1}$Mpc	910^3	910^3	2688	Final full physics model	Finished
L75n455TNG	$75\,h^{-1}$Mpc	455^3	455^3	336	Final full physics model	Finished
L205n2500TNG	$205\,h^{-1}$Mpc	2500^3	2500^3	24,000	Final full physics model	Started
L35n2160TNG	$35\,h^{-1}$Mpc	2160^3	2160^3	16,320	Final full physics model	Started
L25n512TNG	$25\,h^{-1}$Mpc	512^3	512^3	1200	Final full physics model	Finished
L12.5n512TNG	$12.5\,h^{-1}$Mpc	512^3	512^3	1200	Final full physics model	Finished

our main production runs using 10,752 cores on HazelHen, while IllustrisPrime has already finished. In addition, we are carrying out two further large calculations which just have been started. They are substantially larger and either excel in volume or mass resolution, respectively. We have already transferred more than 240 TB of production data to the Heidelberg Institute of Theoretical Studies, in part by using fast gridftp services offered by HLRS. From the ongoing runs, we expect about 200 TB of additional data, which we will also transfer to Heidelberg for the scientific analysis in the coming years.

4 Selected Preliminary Results

In Fig. 1, we illustrate the large-scale distribution of different quantities in the IllustrisPrime simulation (L75n1820MF). From top to bottom, we show projections of the gas density field, the mean mass-weighted metallicity, the mean magnetic field strength (field energy weighted), the dark matter density, and the stellar density. The displayed regions are $75\,h^{-1}$Mpc from left to right, and $3.75\,h^{-1}$Mpc deep. We can nicely see the cosmic web on large-scales, formed both in the dark matter and the diffuse gas. The color hue in the gas distribution shown on top encodes the mass-weighted temperature across the slice. We see that the largest halos are filled with hot plasma, and in addition, there is clearly evidence for very strong outflows in the largest halos impinging on the intergalactic medium, leading to relatively widespread heating.

The bottom panel in Fig. 1 displays the stellar mass density. Clearly, on the scales shown in this image, the individual galaxies appear as very small dots, illustrating that the stellar component fills only a tiny fraction of the volume. However, our simulations have enough resolution and dynamic range to actually resolve the internal structure of these galaxies in remarkable detail. This is shown in Fig. 2, which zooms in on two disk galaxies formed in our simulations. The one on the right hand panel is in a more massive halo and has a more massive black hole. This in fact has made it start to transition into the quenched regime, which here begins by a reduced star formation in the center as a result of kinetic AGN feedback. The outskirts of the galaxy still support some level of star formation, causing blue spiral arms.

Of particular interest in our new simulations is the magnetic field that builds up in halos and galaxies. In Fig. 3, we show a typical disk galaxy in a face-on orientation, plotting the magnetic vector field overlaid on a rendering of the gas density in the background. We see that the field is ordered in the plane of the disk, where it has been amplified by shearing motions to sizable strength. Interestingly, there are multiple field reversals and a complicated topology of the field surrounding the disk. The magnetic field not only provides additional pressure for the gas, but also is of critical importance for transport processes of heat energy and cosmic rays. Our realistic field topologies should be very useful for studying the propagation of

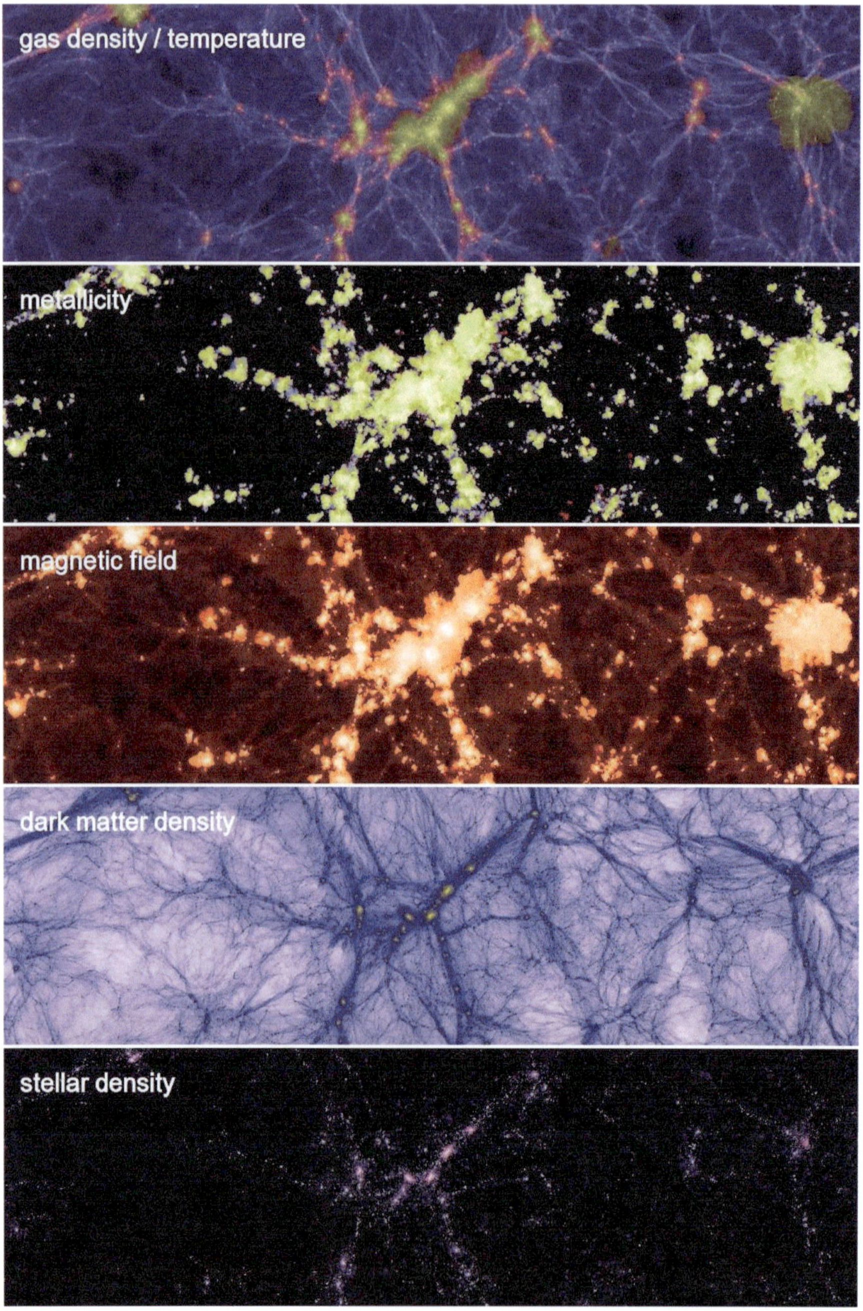

Fig. 1 Thin projections through the L75n1820MF simulation, showing the gas density field, the metallicity, the magnetic field strength, the dark matter density, and the stellar density

Fig. 2 Stellar mass distributions of two disk galaxies in halos of mass $8.3 \times 10^{11}\,M_\odot$ (*left*) and $2.0 \times 10^{12}\,M_\odot$ (*right*), respectively, in face-on (*top*) and edge-on projections (*bottom*). The stellar colors are assigned according to their age and metallicity

cosmic rays in the Milky Way, and for analyzing the deflections of ultra-high energy cosmic rays of extra-galactic origin.

In Fig. 4, we show an analysis of the typical magnetic field strengths reached in halos of different size. We here plot radial profiles for four different halo masses, stacking up to 50 halos in a narrow mass range around the virial masses 10^{10}, 10^{11}, 10^{12}, and $10^{13}\,M_\odot$. We see that field strengths of several μG are reached in the centres of galaxies in halos of masses $10^{11} - 10^{12}\,M_\odot$, in good agreement with typical observed fields. In smaller halos, the fields are still notably weaker, presumably because here they have not yet been amplified as efficiently. In larger halos, $10^{13}\,M_\odot$ and beyond, they are also weaker in the centers, but for a different reason. Here some of the magnetic flux is expelled by strong nuclear outflows driven by AGN feedback. In any case, the strength of the simulated fields implies a remarkable amplification relative to the primordial fields that we seeded in the initial conditions. This initial field strength is empirically largely unconstrained, but our results reached for the field strength in galactic centres do not depend on the value we used (which in our case was 10^{-11} Gauss) over a wide dynamic range, because the magnetic amplification processes stop once the dynamo processes responsible for the exponential amplification saturate. This happens when the magnetic pressure becomes comparable to the thermal pressure.

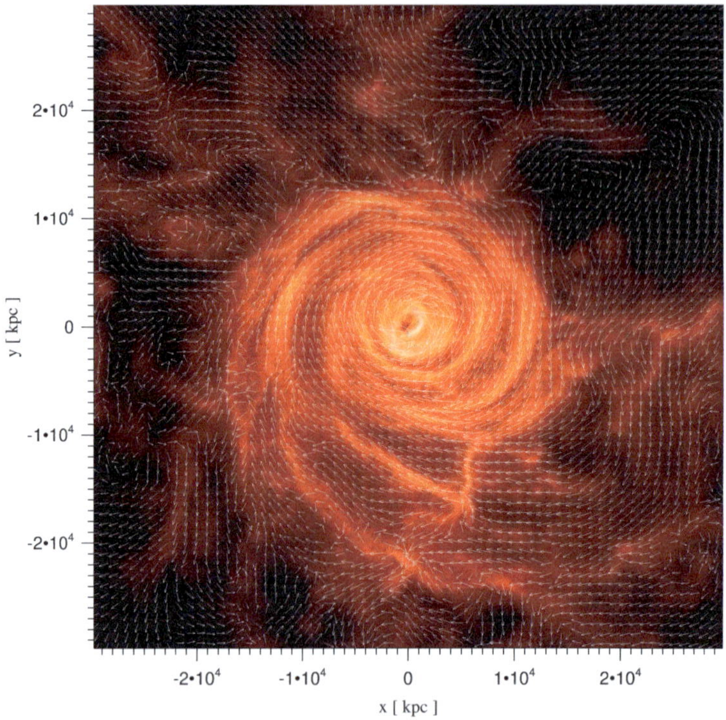

Fig. 3 Magnetic field structure in a disk galaxy (the one displayed in the left-hand panel of Fig. 2), overlaid over a rendering of the gas density structure (color-scale in the background). The length of the drawn vectors is made only weakly dependent on the magnetic field strength (as $\propto |\mathbf{B}|^{1/4}$) in order to see more of the field structure in the regions with weaker fields

On large scales, however, the magnetic field strengths reached in voids still reflect the initial field. This is clearly seen in Fig. 5, which gives a phase-space diagram of gas density versus magnetic field strength. The correlation $B \propto \rho^{2/3}$ (indicated as a solid line) reflects adiabatic expansion/compression of the initial field set at the starting redshift $z = 127$. At baryonic overdensities of around 100, we see that much larger fields are created. This is in part due to the amplification of the field through large-scale shearing flows and in part due to a small-scale dynamo driven by star formation and galactic wind feedback on small scales.

In sum, our calculations demonstrate that already an extremely tiny magnetic field left behind by the Big Bang is sufficient to explain orders of magnitude larger field strengths observed today. Interestingly, the magnetic field strength found in the simulation agrees very well with the values measured for the Milky Way and neighboring galaxies. This is remarkable given that there are no free parameters influencing the magnetic field amplification that could be tuned to modify the final field strength reached in our simulated galaxies.

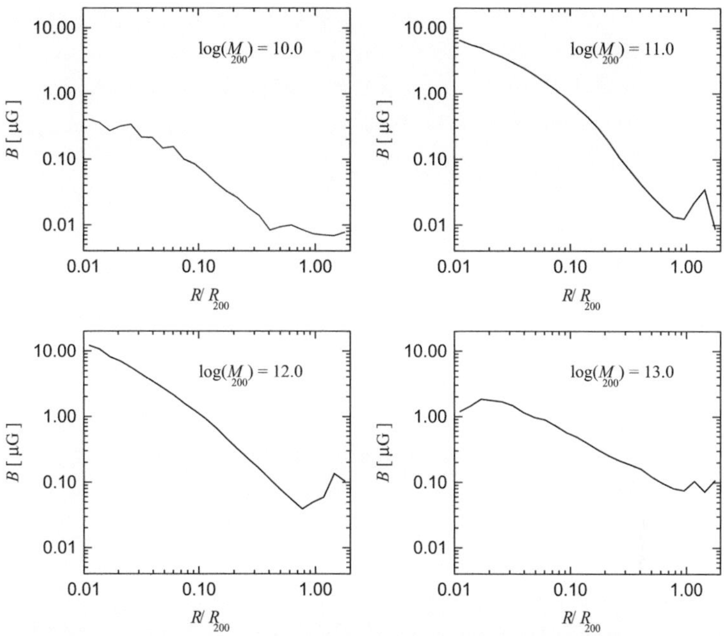

Fig. 4 Spherically averaged profiles of the mean magnetic field strength in halos of different mass. Each panel shows a stacked set of up to 50 halos in a narrow mass bin around a different virial mass, as labeled in each panel

Another powerful application of our simulations lies in studies of the metal enrichment in the universe. We track 9 chemical elements explicitly, and all other elements are lumped together in a 10-th fiducial component so that the advected metallicity vectors always correspond to physically meaningful abundance patterns in all situations. In addition, we use a newly developed metal tagging technique, allowing us to characterize which fraction of metals in every cell or star originated from AGB stars, supernova type II explosions, or type-Ia explosions.

In Fig. 6, we show a break down of the total metal content in the gas phase of the simulated universe at different times as a function of gas density. The individual histograms are normalized to the total metal content in the gas at the corresponding epoch. The distributions can hence answer the question at which gas densities the majority of the metals can be found. Interestingly, we see that most of the metals are actually stored at gas densities that correspond to the circumgalactic medium, whereas only a smaller fraction is contained in the star-forming interstellar medium, and very little in the low-density intergalactic medium. The relative shares between these phases are time-dependent, with more metals found in low density gas towards late times. This is most likely a result of the feedback that expelled these metals from the galaxies.

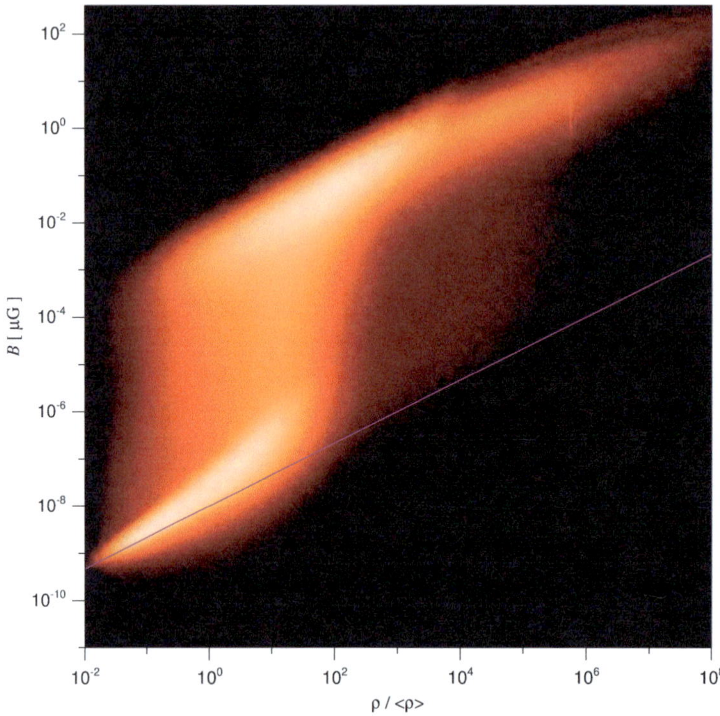

Fig. 5 Phase-space diagram of the magnetic field strength versus gas overdensity at $z = 0$ in one of our Illustris++ simulations. The *line* shows the locus corresponding to adiabatic compression or expansion of the initial field strength

5 Conclusions

Understanding the feedback processes in galaxy formation and evolution is the principal challenge in theoretical extragalactic astronomy. This question is also of critical relevance for cosmology, as baryonic processes impact the distribution of dark matter, and hence in turn affect cosmological probes that aim to constrain, for example, the physical properties of dark energy. Solving the feedback conundrum is unthinkable without further refining the simulation models and the employed numerical methods. This is due to the multi-scale and multi-physics nature of the problem, which tends to limit analytic approaches for studying the problem to highly schematic and correspondingly uncertain models.

The Illustris++ project aims to redefine the state-of-the-art of cosmological hydrodynamical simulations of galaxy formation. In particular, our new simulations make significant progress on predicting the bright end of the galaxy luminosity function through the use of a new model for AGN feedback. Also, realistic galaxy sizes, morphologies and colors are obtained at the same time. The scientific analysis of the simulation promises a rich harvest and will primarily focus on testing the

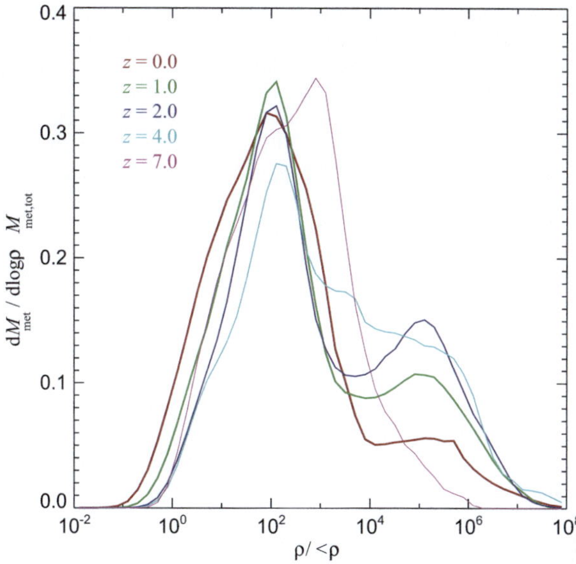

Fig. 6 Distribution of the gas phase metal content with respect to baryonic overdensity at different epochs, as labelled. Each distribution is normalized to the total metal mass in the gas at the corresponding time

models further. Our simulation predictions for the gas around galaxies, the so-called circum-galactic medium (CGM) are particular timely, as the Cosmic Origins Spectrograph (COS) on board of the Hubble Space Telescope (HST) has provided a wealth of absorption line data probing this phase. In addition, our simulations allow us to make novel, testable predictions for the magnetic field strength in different environments, and its correlation with other galaxy properties.

Acknowledgements The authors gratefully acknowledge computer time through the project GCS-ILLU on Hornet/HazelHen at HLRS. We acknowledge financial support through subproject EXAMAG of the Priority Programme 1648 'SPPEXA' of the German Science Foundation, and through the European Research Council through ERC-StG grant EXAGAL-308037, and we would like to thank the Klaus Tschira Foundation.

References

1. Springel, V., White, S.D.M., Jenkins, A., Frenk, C.S., Yoshida, N., Gao, L., Navarro, J., Thacker, R., Croton, D., Helly, J., Peacock, J.A., Cole, S., Thomas, P., Couchman, H., Evrard, A., Colberg, J., Pearce, F.: Nature **435**, 629 (2005). doi:10.1038/nature03597
2. Boylan-Kolchin, M., Springel, V., White, S.D.M., Jenkins, A., Lemson, G.: MNRAS **398**, 1150 (2009). doi:10.1111/j.1365-2966.2009.15191.x
3. Angulo, R.E., Springel, V., White, S.D.M., Jenkins, A., Baugh, C.M., Frenk, C.S.: MNRAS **426**, 2046 (2012). doi:10.1111/j.1365-2966.2012.21830.x

4. Kauffmann, G., Colberg, J.M., Diaferio, A., White, S.D.M.: MNRAS **303**, 188 (1999). doi:10.1046/j.1365-8711.1999.02202.x
5. Cole, S., Lacey, C.G., Baugh, C.M., Frenk, C.S.: MNRAS **319**, 168 (2000). doi:10.1046/j.1365-8711.2000.03879.x
6. Vogelsberger, M., Genel, S., Springel, V., Torrey, P., Sijacki, D., Xu, D., Snyder, G., Bird, S., Nelson, D., Hernquist, L.: Nature **509**, 177 (2014). doi:10.1038/nature13316
7. Vogelsberger, M., Genel, S., Springel, V., Torrey, P., Sijacki, D., Xu, D., Snyder, G., Nelson, D., Hernquist, L.: MNRAS **444**, 1518 (2014). doi:10.1093/mnras/stu1536
8. Genel, S., Vogelsberger, M., Springel, V., Sijacki, D., Nelson, D., Snyder, G., Rodriguez-Gomez, V., Torrey, P., Hernquist, L.: MNRAS **445**, 175 (2014). doi:10.1093/mnras/stu1654
9. Springel, V.: MNRAS **401**, 791 (2010). doi:10.1111/j.1365-2966.2009.15715.x
10. Vogelsberger, M., Sijacki, D., Kereš, D., Springel, V., Hernquist, L.: MNRAS **425**, 3024 (2012). doi:10.1111/j.1365-2966.2012.21590.x
11. Sijacki, D., Vogelsberger, M., Kereš, D., Springel, V., Hernquist, L.: MNRAS **424**, 2999 (2012). doi:10.1111/j.1365-2966.2012.21466.x
12. Torrey, P., Vogelsberger, M., Sijacki, D., Springel, V., Hernquist, L.: MNRAS **427**, 2224 (2012). doi:10.1111/j.1365-2966.2012.22082.x
13. Bauer, A., Springel, V.: MNRAS **423**, 2558 (2012). doi:10.1111/j.1365-2966.2012.21058.x
14. Sijacki, D., Springel, V., Di Matteo, T., Hernquist, L.: MNRAS **380**, 877 (2007). doi:10.1111/j.1365-2966.2007.12153.x
15. Dolag, K., Bartelmann, M., Lesch, M.: A&A **348**, 351 (1999)
16. Dolag, K., Bartelmann, M., Lesch, H.: A&A **387**, 383 (2002). doi:10.1051/0004-6361:20020241
17. Dolag, K., Grasso, D., Springel, V., Tkachev, I.: J. Cosmol. Astropart. Phys. **1**, 009 (2005). doi:10.1088/1475-7516/2005/01/009
18. Donnert, J., Dolag, K., Lesch, H., Müller, E.: MNRAS **392**, 1008 (2009). doi:10.1111/j.1365-2966.2008.14132.x
19. Bonafede, A., Dolag, K., Stasyszyn, F., Murante, G., Borgani, S.: MNRAS **418**, 2234 (2011). doi:10.1111/j.1365-2966.2011.19523.x
20. Kotarba, H., Lesch, H., Dolag, K., Naab, T., Johansson, P.H., Donnert, J., Stasyszyn, F.A.: MNRAS **415**, 3189 (2011). doi:10.1111/j.1365-2966.2011.18932.x
21. Beck, A.M., Lesch, H., Dolag, K., Kotarba, H., Geng, A., Stasyszyn, F.A.: MNRAS **422**, 2152 (2012). doi:10.1111/j.1365-2966.2012.20759.x
22. Marinacci, F., Vogelsberger, M., Mocz, P., Pakmor, R.: MNRAS **453**, 3999 (2015). doi:10.1093/mnras/stv1692
23. Sijacki, D., Vogelsberger, M., Genel, S., Springel, V., Torrey, P., Snyder, G.F., Nelson, D., Hernquist, L.: MNRAS **452**, 575 (2015). doi:10.1093/mnras/stv1340
24. Yuan, F., Narayan, R.: ARA&A **52**, 529 (2014). doi:10.1146/annurev-astro-082812-141003
25. Weinberger, R., Springel, V., Hernquist, L., Pillepich, A., Marinacci, F., Pakmor, R., Nelson, D., Genel, S., Vogelsberger, M., Naiman, J., Torrey, P.: ArXiv e-prints (2016)
26. Pelupessy, F.I., Jänes, J., Portegies Zwart, S.: New A **17**, 711 (2012). doi:10.1016/j.newast.2012.05.009
27. Pakmor, R., Springel, V., Bauer, A., Mocz, P., Munoz, D.J., Ohlmann, S.T., Schaal, K., Zhu, C.: MNRAS **455**, 1134 (2016). doi:10.1093/mnras/stv2380
28. Pakmor, R., Bauer, A., Springel, V.: MNRAS **418**, 1392 (2011). doi:10.1111/j.1365-2966.2011.19591.x

Hydrangea: Simulating a Representative Population of Massive Galaxy Clusters

Yannick M. Bahé, for the C-EAGLE collaboration

Abstract Galaxy clusters are the most massive bound structures in the Universe, and contain not only up to several thousand galaxies, but also extended haloes of dark matter and hot gas. Observations show that galaxies in clusters differ from those living in more isolated parts of the Universe, but the physics of how clusters shape their galaxies is at present not well understood. Not only does this constitute a major gap in our understanding of galaxy formation, but it also limits the use of galaxy clusters as cosmological probes. In the *Hydrangea* project, we have created a suite of 24 simulated galaxy clusters at unprecedented resolution, using a state of the art galaxy formation model developed for the EAGLE project. Detailed scientific analysis of the simulation outputs, which has only just begun, is expected to lead to major new insight into the physics of both galaxy formation in an extreme environment and the growth of the massive haloes in which cluster galaxies are embedded.

1 Introduction

Galaxy clusters are collections of large numbers of galaxies – up to several thousands in the most extreme cases – occupying a region in our Universe that is typically a few megaparsec (Mpc[1]) in size. Moreover, observations have firmly established that the galaxies which are seen in optical light are in fact only a minor component of these objects: the dominant constituents in terms of mass are instead extended, diffuse 'haloes' of dark matter (DM) and very hot gas (the 'intra-cluster medium' or ICM) that together account for typically more than 90 % of the mass of a galaxy cluster [1]. Including these optically invisible components, the mass of the largest such objects exceeds 10^{15} times the mass of our Sun ($M_\odot$), making

[1]The *parsec* (pc) is the standard unit of length in astronomy, with 1 pc = 3.08×10^{16} m.

Y.M. Bahé (✉)
Max Planck Institut für Astrophysik, Karl-Schwarzschild-Str. 1, 85748, Garching, Germany
e-mail: ybahe@mpa-garching.mpg.de

W.E. Nagel et al. (eds.), *High Performance Computing in Science and Engineering '16*, DOI 10.1007/978-3-319-47066-5_2

galaxy clusters the most massive gravitationally bound structures in the present-day Universe.

Scientifically, galaxy clusters are of interest in contemporary astrophysics for at least three reasons. The first is that the close proximity of their member galaxies to each other, as well as the presence of the DM and ICM haloes, constitute an extreme environment for galaxy formation, a detailed understanding of which is an integral part of the wider quest to understand how galaxies and larger-scale structures formed and evolved in our Universe over the last 13 billion years. Secondly, it has become clear that galaxy clusters are also one of the most promising probes to investigate the composition and expansion history of the Universe itself, including the as-yet poorly understood nature of Dark Energy (DE). Finally, the concentration of vast amounts of DM in galaxy clusters also makes them interesting astrophysical laboratories that can help us to better understand the nature of this dominant gravitating constituent of our Universe.

There is ample observational evidence that galaxies in groups and clusters are different from "field" galaxies that formed in more isolated regions of the Universe, such as our Milky Way. Their colours are typically red – indicating a lack of recent and ongoing star formation – whereas field galaxies tend to be blue due to recently formed young, massive stars. A second key difference is their morphology: cluster galaxies are biased towards "early" (elliptical) types, whereas many isolated galaxies show a pronounced "late-type" morphology with a prominent spiral disk [2]. A multitude of physical mechanisms have been suggested that could explain these trends as a result of transformations that occur when galaxies fall into a group or cluster: these include ram pressure stripping of cold [3] and hot gas [4], tidal stripping by the group/cluster potential [5] and galaxy-galaxy interactions [6]. However, understanding to what extent, and on which timescales, each of these processes actually affect galaxies has so far proved elusive [7, 8].

Making progress here is one of the most important goals in extragalactic astrophysics, not only because groups and clusters harbour a significant fraction (approximately one third) of all galaxies in the local Universe – so that understanding their evolution constitutes a key part of understanding galaxy formation in general – but also because the usefulness of galaxy clusters as precision probes of DE and cosmology is compromised by systematic effects that include the influence of cluster galaxies and the hot gaseous intra-cluster medium (ICM) [9], unless these effects are well understood and accounted for. Several large-scale surveys are currently under way, or planned for the near future, to study DE with weak gravitational lensing measurements of galaxy clusters, including the Dark Energy Survey (DES), the Large Synoptic Survey Telescope (LSST), and the European Space Agency's Euclid mission, so that a better understanding of baryonic processes in galaxy clusters is urgently required.

Major new observational insight is expected in the near future from a number of large integral-field unit surveys such as KMOS-Cluster, SAMI and MaNGA, as well as the eRosita X-ray telescope. However, the non-linear nature of galaxy evolution – several of the above-mentioned transformation mechanisms will likely amplify each other's effect – makes it impossible to accurately model these observations without recourse to detailed numerical calculations.

Cosmological hydrodynamical simulations self-consistently include the majority of the mechanisms described above, and their impact on galaxies can be studied in much more detail than in semi-analytic models (the currently most widely used tool for this purpose). Furthermore, as long as the simulation parameters are only adjusted to reproduce realistic field galaxies (i.e. those which have not been subject to the environmental processes we wish to study), such simulations also possess considerable predictive power. Our previous work with the GIMIC simulations [10] has already shown, for example, that ram pressure stripping plays a much bigger role in shaping cluster satellite galaxies than naively indicated by semi-analytic models [8], and that the direct influence of the cluster environment may persist out to at least $5r_{200}$, much further than previously thought [11]. However, these simulations did not produce realistic field galaxies, which put significant limits on their predictive power.

The Virgo consortium's EAGLE project [12] has developed a model of a representative population of field galaxies which matches a wide range of observed scaling relations, as well as the galaxy stellar mass function. However, galaxy clusters ($M_{200} > 10^{14}\,\mathrm{M}_{\odot}$)[2] occupy only a tiny fraction of the Universe by volume, and are therefore poorly sampled in EAGLE: The "AGNdT9" model – which gives much more realistic properties of massive galaxies and galaxy groups compared to the standard "Ref" model, owing to a refined parameterisation of AGN feedback – was only run in a $(50\,\mathrm{Mpc})^3$ box and therefore includes only one cluster with M_{200} just above $10^{14}\,\mathrm{M}_{\odot}$ at z = 0 [12]. This is an order of magnitude below the mass of e.g. the well-studied nearby Coma cluster. At higher redshift ($z > 1$), clusters are absent even in the largest $(100\,\mathrm{Mpc})^3$ EAGLE run, preventing any kind of evolutionary study or comparison to the rapidly growing number of observations in this field [13]. Conversely, simulations of very large cosmological volumes (such as COSMO-OWLS [14] or BAHAMAS [15]) include many galaxy clusters, but cannot resolve details within individual galaxies. Neither of these existing simulations are therefore well-suited to studying the evolution of cluster galaxies.

Overcoming this problem is the objective of the C-EAGLE[3] simulations, a family of related projects aiming to make progress through high resolution cosmological hydrodynamical simulations of galaxy clusters based on the "zoomed initial conditions" [16] technique in combination with the simulation code used successfully for the EAGLE simulations. The *Hydrangea* project, a core part of this effort, has simulated a sample of 24 galaxy clusters in the mass range $M_{200} = 10^{14} - 3 \times 10^{15}\,\mathrm{M}_{\odot}$. Motivated by our prior work with GIMIC, the simulations are set up with a high-resolution zoom region extending out to $10r_{200}$ from the cluster

[2]M_{200} is defined as the mass within r_{200}, the radius inside which the mean density equals 200 times the critical density of the Universe (ρ_{crit}).

[3]Abbreviation for "Cluster-EAGLE", which also refers to the sea eagle (*Haliaeetus pelagicus*) as the most massive member of the eagle family.

centre, to capture the large-scale environmental influence. C-EAGLE also features an additional set of galaxy clusters of similar mass as those in *Hydrangea* but simulated only out to $5r_{200}$ – thus reducing the simulated volume by a factor of 8 – for improved statistical power in studying the properties of the ICM, a small suite of ultra-high resolution simulations to study the formation of dwarf galaxies in clusters, and a set of simulated galaxy groups. In combination with the existing EAGLE runs, these simulations will allow extensive new insight not only into the physics of galaxy formation in an extreme environment, but also the formation and evolution of the cluster haloes themselves.

2 Simulation Code

Our simulations are run with a heavily modified version of the cosmological TreePM/SPH code GADGET-3, last described by [17], that was developed, optimized, and extensively tested for the EAGLE project [12]. The code is fully MPI-parallelized, with a sophisticated domain decomposition scheme that assigns to each MPI-task the particles in a large number of disjoint cells; this significantly improves load-balancing in highly clustered systems such as those we have simulated.

Gravitational forces are calculated on large scales by Fourier-transforming in parallel a periodic mesh covering the entire simulation volume (implemented through the FFTW library). On smaller scales, a Barnes-Hut tree algorithm is used, together with direct summation on the smallest scales. The code also uses an additional isolated mesh covering only the high-resolution region of zoom simulations, which results in an orders-of-magnitude speed-up. Particles are integrated in time on variable time-steps nested hierarchically on up to 20 levels. Different from "standard" GADGET-3, our code uses a time-step limiter [18] which ensures that timesteps are kept short after particles experience significant changes in their internal energy. This significantly improves the accuracy in the treatment of feedback [19].

Hydrodynamical forces are evaluated with the Smoothed Particle Hydrodynamics (SPH) approach, which is implemented in the entropy-conserving formulation [20]. The SPH implementation has been modified significantly for the EAGLE project through a series of measures collectively referred to as "Anarchy" (Dalla Vecchia, in prep.) which include the conservative pressure-entropy formalism of [21], the artificial viscosity switch of [22], an artificial conduction switch similar to that of [23], and the C^2 Wendland kernel [24]. These modifications largely eliminate the inaccuracies related to contact discontinuities and spurious fragmentation present in older versions of SPH [19].

The most significant modifications of the GADGET-3 code relate to the implementation of relevant physical processes on unresolved scales ("sub-grid physics"). We only summarized these briefly; for details the reader is referred to the description of the "AGNdT9" model in [12]. Gas cooling and chemical enrichment are implemented following [25, 26] by explicitly tracking the abundance of the 11 most important chemical species (H, He, C, N, O, Ne, Mg, Si, S, Ca, and Fe) and interpolating tabulated cooling curves [27] on an element-by-element basis. Gas particles dense enough to form stars [28] are converted to star particles in a probabilistic way normalised to the observed Kennicutt-Schmidt relation [29, 30]. Energy feedback from star formation is implemented stochastically in a single thermal mode [31]. Similarly, feedback from accreting supermassive black holes is modelled in a single thermal mode, by heating a small number of gas particles by 10^9 K [12, 32].

3 Galaxy Cluster Simulations

Like all C-EAGLE runs, the *Hydrangea* simulations are based upon a very large, low resolution (particle masses of $m_{\rm DM} \approx 8 \times 10^{10}\,{\rm M}_\odot$) dark matter only simulation realised in a cubic box of side length 3200 Mpc (comoving). This simulation contains more than 300,000 dark matter haloes with a mass in excess of $10^{14}\,{\rm M}_\odot$, including almost 3000 extremely massive objects with $M_{200} > 10^{15}\,{\rm M}_\odot$. We discarded those with a relatively close more massive neighbour (within 20 r_{200} or 30 Mpc, whichever is larger), and those within 200 Mpc from the simulation box edge. Out of the remaining haloes, 29 objects in the mass range $14.0 < \log_{10}(M_{200}/{\rm M}_\odot) < 15.25$ and distributed uniformly in M_{200} were selected at random for re-simulation as our 'core' sample. Furthermore, one even more massive object ($M_{200} \approx 10^{15.4}{\rm M}_\odot$) was selected for comparison to the most massive observed clusters.

For each object selected for re-simulation, high-resolution zoomed initial conditions (ICs) were generated with the `IC_2LPT_GEN` code [33], using second order Lagrangian perturbation theory. The dark matter particle mass $m_{\rm DM} = 9.7 \times 10^6{\rm M}_\odot$ in the high-resolution region of these ICs is almost 10,000 times smaller than in the parent simulation; the mass of baryon (star and gas) particles is lower still by a factor of $f_b \equiv \Omega_b/\Omega_m = 6.36$, the cosmic baryon fraction, so that the simulation contains (initially) equal numbers of DM and baryon particles.[4] In order to correctly model the tidal forces acting on the high-resolution region, the remaining volume of the 3200 Mpc simulation box is filled with a relatively small number of very massive 'boundary' particles.

To test the quality of the high-resolution ICs, each simulation was first run in N-body only mode, i.e. without hydrodynamics or subgrid models. The motivation for doing this is twofold: first, such a simulation incurs only a small fraction of the computational cost of a full hydrodynamical run and therefore constitutes an economical way to test the quality of the high-resolution ICs by comparing

[4]During the course of a simulation, some baryon particles are 'swallowed' by black holes, so that the final number of baryon particles is typically slightly lower.

the masses of the simulated cluster haloes in the high-resolution run to the low-resolution counterparts in the parent simulation. On the other hand, a lot of insight into the effect of baryonic physics can be gained from comparing dark matter only and hydrodynamical simulations started from the same ICs, as any differences can be ascribed solely to the presence of gas in the latter [9]. We have verified that for all objects in our sample, the final cluster masses in the high-and low resolution runs are virtually identical (differing by less than 10 %), and that no low-resolution 'boundary' particles are present within $>12r_{200}$ from the cluster centre at $z = 0$.

However, simulating all 30 haloes with full hydrodynamics in a high-resolution region of $10r_{200}$ would have incurred a very high computational cost. To accommodate the project within the constraints of available resources, we therefore selected a subsample of 24 objects for the *Hydrangea* runs, including the very massive cluster with $M_{200} \approx 10^{15.4}\,\mathrm{M}_{\odot}$ but otherwise biased towards lower mass haloes which are individually cheaper to simulate, but also contain a smaller number of galaxies and therefore benefit especially from an enlarged sample size. Hydrodynamical simulations of the remaining six clusters, which are not part of *Hydrangea*, were performed only out to $5r_{200}$. Eleven *Hydrangea* haloes have masses M_{200} in the range $10^{14.0}$ to $10^{14.5}\,\mathrm{M}_{\odot}$, eight between $10^{14.5}$ and $10^{15.0}\,\mathrm{M}_{\odot}$, and five between $10^{15.0}$ and $10^{15.5}\,\mathrm{M}_{\odot}$.

3.1 Simulations Performed at HLRS

Simulating a galaxy cluster at the high resolution required to adequately resolve individual galaxies (i.e. with baryon particle masses of $m_b \approx 10^6\,\mathrm{M}_{\odot}$) is particularly challenging in case of the most massive objects with $M_{200} \gtrsim 10^{15}\,\mathrm{M}_{\odot}$. At this mass scale, each cluster is resolved into $>2 \times 10^9$ particles, requiring at least 4 TB of memory, while the most massive object we are simulating contains almost five billion particles and requires nearly 10 TB of memory. The Hornet/HazelHen system at HLRS provides 5 GB of memory per core, and is therefore ideally suited to run these extremely memory intensive massive clusters. We have therefore concentrated the computing time allocated to our project at HLRS on 12 of the 25 objects, biased towards those with the highest mass, while the remaining 13 clusters were run on other machines (*Odin* and *Hydra* at the MPCDF, Garching, as well as *Cosma-5* in Durham/UK). In addition, six of the DM only runs were performed on Hornet/HazelHen.

In addition to the simulation run itself – i.e. the advancement in time of the simulated system from the initial conditions at redshift $z = 127$ to the present epoch, $z = 0$ – a second crucial part is to catalogue the outputs in order to identify bound systems representing galaxies and groups/clusters. For consistency with the original EAGLE project, this is achieved with the `SUBFIND` algorithm [34]. Although not as computationally expensive as the simulation itself, the cost is nevertheless non-negligible, both in terms of computing time – up to 1.5 million CPU-hrs for the most massive objects – and memory, which translates into a requirement for up to

Table 1 The three largest simulations performed at HLRS. Note that the wallclock time includes queueing between individual jobs

Halo ID	Mass $[10^{15}\,M_\odot]$	N_{core}	Wallclock time [days]	CPU time $[10^6$ CPU-hr$]$
22	1.05	2048	187	4.8
28	1.70	2048	164	4.2
40	2.19	4096	281	10.3

200 nodes on Hornet/HazelHen. Most of the SUBFIND analysis for the massive clusters has therefore been performed on Hornet/HazelHen as well.

Production runs at HLRS began on 16 June 2015, following a brief period of testing our code on the Cray XC40 system after access was granted on June 3. The longest running simulation performed at HLRS, of the cluster with $M_{200} \approx 10^{15.4}\,M_\odot$, finished on April 29, 2016. The post-processing of simulation outputs with SUBFIND was completed on May 12, 2016, thus concluding our calculations at HLRS slightly ahead of the scheduled project completion date (May 15).

For each simulation, 30 full snapshots were stored between redshift $z = 14$ and $z = 0$, with a constant gap of 500 Myr between them. In addition, we saved a larger number of 'snipshots' that only contain the most essential and most rapidly time-varying quantities calculated by the simulation, such as particle positions, velocities, and SPH-interpolated density. For each simulation, we have stored at least 178 such snipshots, resulting in a maximum time between any two outputs of 125 Myr. In total, the simulations run at HLRS have so far produced 350 TB of data, which has been continuously transferred to the Virgo Data Archive at the Max Planck Computing and Data Facility (MPCDF) in Garching for scientific analysis. Table 1 lists the cluster mass, run time, and number of cores used for the three most demanding simulations performed at HLRS. The entire *Hydrangea* project has produced more than 500 TB of raw data, and simulated the formation of more than 20,000 galaxies with stellar mass $M_{star} \geq 10^9\,M_\odot$.

4 Simulation Results

In Fig. 1 we show a visualisation of one of the most massive clusters simulated as part of *Hydrangea* at HLRS, halo 28 with a mass at $z = 0$ of $10^{15.23}\,M_\odot$. The left column shows the density of dark matter, gas and stars in a cubic 30 Mpc region centered on the cluster (projected along the simulation z-axis). Clearly visible is the massive central cluster, which is surrounded by a number of smaller clusters and connected to the surrounding Universe by filaments of gas and dark matter. In the right column, we show from top to bottom the mass-weighted metallicity of the simulated gas (i.e. the fraction of mass in elements heavier than Helium), its temperature, and velocity. In these panels, the central cluster shows a strong

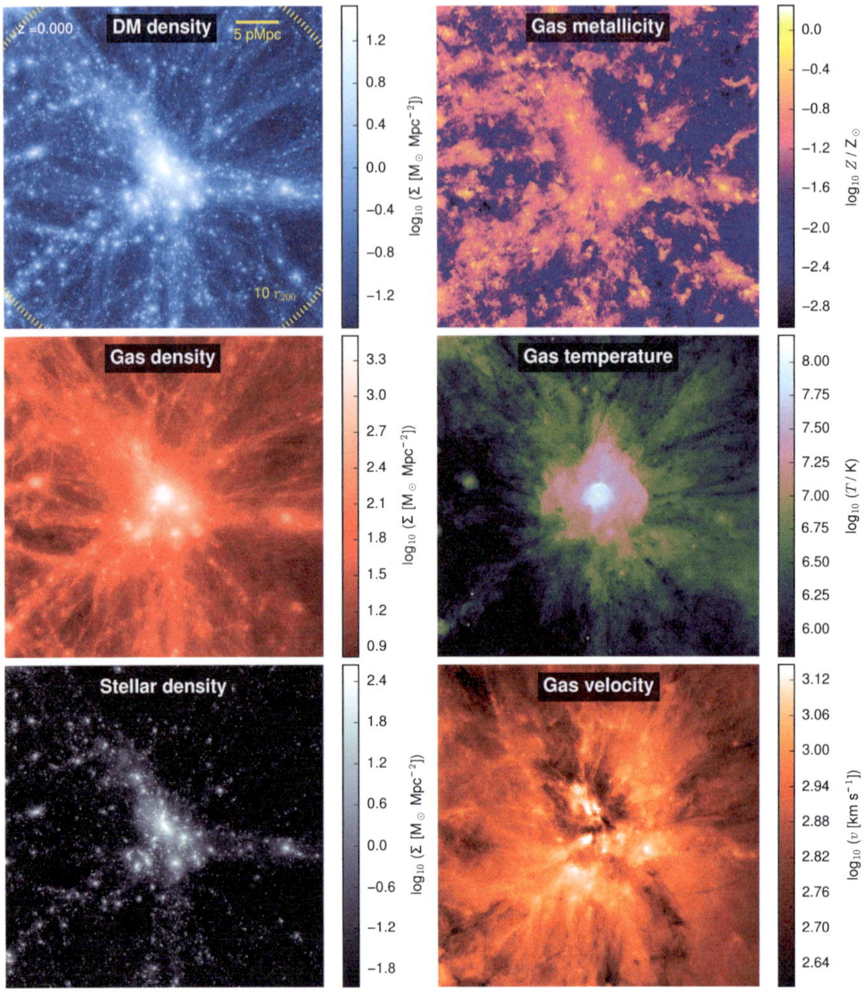

Fig. 1 Visualisation of halo 28 at redshift $z = 0.0$, one of the most massive *Hydrangea* clusters simulated at HLRS. The *left column* shows, from top to bottom, the projected density of dark matter, gas, and stars in a cubic box of 40 Mpc side length, centered on the cluster. The nominal edge of the high-resolution region, at a distance of $10r_{200} = 25\,\text{Mpc}$ is indicated with the *dotted yellow lines* visible in the corners of the *top-left panel*. In the *right column*, we show the projected mass-weighted metallicity of the gas, its temperature, and (bulk) velocity. In qualitative agreement with observations, the central cluster is filled with very hot, metal-enriched gas that shows a complex dynamical structure. Each point in the stellar density map (*bottom left*) represents a simulated galaxy

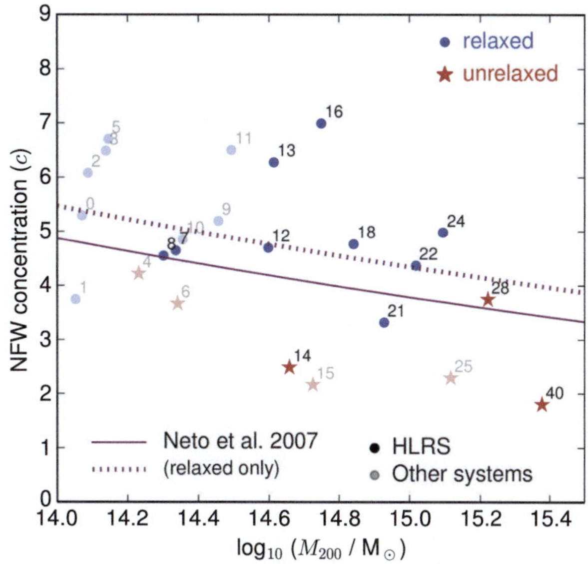

Fig. 2 Distribution of the *Hydrangea* clusters in mass and concentration. As expected, more massive systems are overall slightly less concentrated, but at any given mass, our simulations contain clusters of very different concentration. Investigating the extent to which the galaxy population mirrors this diversity amongst the cluster sample is one key question to be addressed in the upcoming scientific analysis

overabundance of metals, in qualitative agreement with observations [35]. The gas in the simulated cluster, the 'intra-cluster medium' (ICM) is extremely hot ($T > 10^8$ K), and relatively hot ($T \gtrsim 10^6$ K) almost to the edge of the simulation volume. On close inspection, individual galaxies that are visible as small points in the stellar density map (top left) also show up as spots of relatively cool ($T \lesssim 10^6$ K) gas against the hot ICM. The velocity map (bottom right) shows the complex interplay of high-bulk-velocity gas that is inflowing at speeds exceeding 1000 km s^{-1}, and the hot gas in the virialised haloes whose bulk velocity is much lower (dark red).

Figure 2 presents an overview of the entire *Hydrangea* cluster sample, in terms of the mass and concentration of the cluster halo at the centre of each simulation volume. Masses are defined as total mass within r_{200}, while the concentration is derived by fitting an NFW profile [36] to the DM mass distribution as described by [19, 37]. Galaxies classified as 'relaxed' according to the substructure and offset criteria of [37] are marked as blue circles, while 'unrelaxed' systems violating at least one of these criteria are shown as red stars. Dark shaded symbols indicate runs performed at HLRS; as can be seen, this includes the majority of objects with $M_{200} \gtrsim 10^{14.6}\,M_{\odot}$. Our sample includes galaxy clusters with widely differing concentrations at approximately the same mass.

Finally, we show in Fig. 3 the star formation activity in our simulated galaxies, quantified as the fraction of galaxies that is not forming stars at a significant rate, i.e. whose specific star formation rate sSFR $\equiv$ SFR/$M_{\rm star} < 10^{-11}\,{\rm yr}^{-1}$. Observations have shown that the fraction of such 'passive' satellite galaxies in

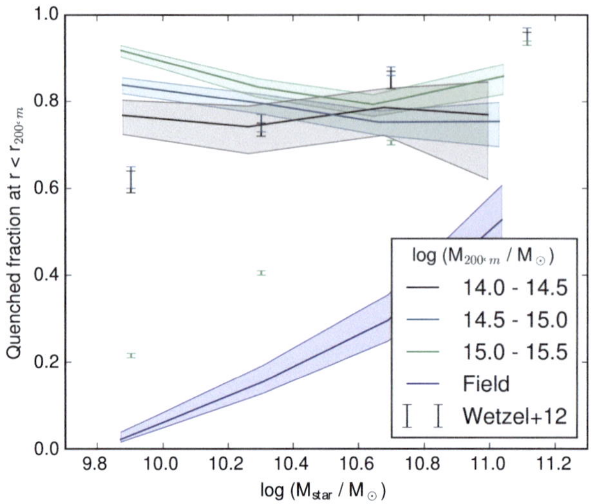

Fig. 3 Fraction of simulated galaxies within $r_{200,m}$ that are passive (i.e. have a specific star formation rate sSFR $< 10^{-11}\,\mathrm{yr}^{-1}$). Differently coloured bands denote different ranges of halo mass, as indicated in the bottom right corner. In approximate agreement with observations, cluster galaxies have a much higher passive fraction than field galaxies, the difference being greatest in the most massive clusters (*green*)

clusters increases with both stellar mass and the mass of the cluster [38].[5] As Fig. 3 shows, our simulations reproduce the latter observation well, especially for galaxies at the lower end of the mass range considered, $M_{\mathrm{star}} \approx 10^{10}\,\mathrm{M}_{\odot}$.

The full scientific exploitation of the rich dataset produced by the *Hydrangea* simulations and related C-EAGLE projects has only just begun, and is expected to take several years. This analysis includes quantitative comparisons of the simulated galaxies and the intra-cluster medium to observational data (e.g. [35, 38–40]), as well as projects aiming to obtain a detailed understanding of the physical processes operating in galaxy clusters that lead to e.g. the lack of star formation in cluster galaxies, the change in galaxy morphology from disk-dominated to elliptical, and the formation of structures in the ICM. The results of these studies will be reported in the astrophysical literature.

5 Summary

We have produced the *Hydrangea* simulations of two dozen massive galaxy clusters, a ground-breaking new tool to study the formation of galaxies in the most extreme environment in our Universe. The simulations use a well-tested code developed

[5]Given that the age of the Universe is approximately 10^{10} yr, such galaxies must have formed stars at a much higher rate in the past in order to build up their current stellar mass.

for the EAGLE project and are in large part run on the HLRS Cray XC40 Hornet/HazelHen system, each using up to 4096 cores. Memory is a key limiting factor in our simulations, which require up to 10 TB of RAM, making the high-memory machine at HLRS an ideal system to run them. All simulations have been completed within the allocated time frame, and the same is true for cataloging of outputs with the SUBFIND code. We are now beginning the scientific analysis of the simulation data, which is expected to lead to more than a dozen publications over the coming years.

References

1. Voit, G.M.: Rev. Modern Phys. **77**, 207 (2005). doi:10.1103/RevModPhys.77.207
2. Hogg, D.W., Blanton, M.R., Brinchmann, J., Eisenstein, D.J., Schlegel, D.J., Gunn, J.E., McKay, T.A., Rix, H.W., Bahcall, N.A., Brinkmann, J., Meiksin, A.: ApJ Lett. **601**, L29 (2004). doi:10.1086/381749
3. Gunn, J.E., Gott, J.R., III.: ApJ **176**, 1 (1972). doi:10.1086/151605
4. Larson, R.B., Tinsley, B.M., Caldwell, C.N.: ApJ **237**, 692 (1980). doi:10.1086/157917
5. Merritt, D.: ApJ **264**, 24 (1983). doi:10.1086/160571
6. Moore, B., Lake, G., Quinn, T., Stadel, J.: MNRAS **304**, 465 (1999). doi:10.1046/j.1365-8711.1999.02345.x
7. Wetzel, A.R., Tinker, J.L., Conroy, C., van den Bosch, F.C.: MNRAS (2013, in press). doi:10.1093/mnras/stt469
8. Bahé, Y.M., McCarthy, I.G.: MNRAS **447**, 969 (2015). doi:10.1093/mnras/stu2293
9. Velliscig, M., van Daalen, M.P., Schaye, J., McCarthy, I.G., Cacciato, M., Le Brun, A.M.C., Dalla Vecchia, C.: MNRAS **442**, 2641 (2014). doi:10.1093/mnras/stu1044
10. Crain, R.A., Theuns, T., Dalla Vecchia, C., Eke, V.R., Frenk, C.S., Jenkins, A., Kay, S.T., Peacock, J.A., Pearce, F.R., Schaye, J., Springel, V., Thomas, P.A., White, S.D.M., Wiersma, R.P.C.: MNRAS **399**, 1773 (2009). doi:10.1111/j.1365-2966.2009.15402.x
11. Bahé, Y.M., McCarthy, I.G., Balogh, M.L., Font, A.S.: MNRAS **430**, 3017 (2013). doi:10.1093/mnras/stt109
12. Schaye, J., Crain, R.A., Bower, R.G., Furlong, M., Schaller, M., Theuns, T., Dalla Vecchia, C., Frenk, C.S., McCarthy, I.G., Helly, J.C., Jenkins, A., Rosas-Guevara, Y.M., White, S.D.M., Baes, M., Booth, C.M., Camps, P., Navarro, J.F., Qu, Y., Rahmati, A., Sawala, T., Thomas, P.A., Trayford, J.: MNRAS **446**, 521 (2015). doi:10.1093/mnras/stu2058
13. Muzzin, A., Wilson, G., Demarco, R., Lidman, C., Nantais, J., Hoekstra, H., Yee, H.K.C., Rettura, A.: ApJ **767**, 39 (2013). doi:10.1088/0004-637X/767/1/39
14. Le Brun, A.M.C., McCarthy, I.G., Schaye, J., Ponman, T.J.: MNRAS **441**, 1270 (2014). doi:10.1093/mnras/stu608
15. McCarthy, I.G., Schaye, J., Bird, S., Le Brun, A.M.C.: ArXiv e-prints (2016)
16. Katz, N., Quinn, T., Bertschinger, E., Gelb, J.M.: MNRAS **270**, L71 (1994)
17. Springel, V.: MNRAS **364**, 1105 (2005). doi:10.1111/j.1365-2966.2005.09655.x
18. Durier, F., Dalla Vecchia, C.: MNRAS **419**, 465 (2012). doi:10.1111/j.1365-2966.2011.19712.x
19. Schaller, M., Dalla Vecchia, C., Schaye, J., Bower, R.G., Theuns, T., Crain, R.A., Furlong, M., McCarthy, I.G.: MNRAS **454**, 2277 (2015). doi:10.1093/mnras/stv2169
20. Springel, V., Hernquist, L.: MNRAS **333**, 649 (2002). doi:10.1046/j.1365-8711.2002.05445.x
21. Hopkins, P.F.: MNRAS **428**, 2840 (2013). doi:10.1093/mnras/sts210
22. Cullen, L., Dehnen, W.: MNRAS **408**, 669 (2010). doi:10.1111/j.1365-2966.2010.17158.x
23. Price, D.J.: J. Comput. Phys. **227**, 10040 (2008). doi:10.1016/j.jcp.2008.08.011

24. Wendland, H.: Adv. Comput. Math. **4**, 389 (1995)
25. Wiersma, R.P.C., Schaye, J., Smith, B.D.: MNRAS **393**, 99 (2009). doi:10.1111/j.1365-2966.2008.14191.x
26. Wiersma, R.P.C., Schaye, J., Theuns, T., Dalla Vecchia, C., Tornatore, L.: MNRAS **399**, 574 (2009). doi:10.1111/j.1365-2966.2009.15331.x
27. Ferland, G.J., Korista, K.T., Verner, D.A., Ferguson, J.W., Kingdon, J.B., Verner, E.M.: PASP **110**, 761 (1998). doi:10.1086/316190
28. Schaye, J.: ApJ **609**, 667 (2004). doi:10.1086/421232
29. Kennicutt, R.C., Jr.: ApJ **498**, 541 (1998). doi:10.1086/305588
30. Schaye, J., Dalla Vecchia, C.: MNRAS **383**, 1210 (2008). doi:10.1111/j.1365-2966.2007.12639.x
31. Dalla Vecchia, C., Schaye, J.: MNRAS **426**, 140 (2012). doi:10.1111/j.1365-2966.2012.21704.x
32. Rosas-Guevara, Y.M., Bower, R.G., Schaye, J., Furlong, M., Frenk, C.S., Booth, C.M., Crain, R.A., Dalla Vecchia, C., Schaller, M., Theuns, T.: MNRAS **454**, 1038 (2015). doi:10.1093/mnras/stv2056
33. Jenkins, A.: MNRAS **403**, 1859 (2010). doi:10.1111/j.1365-2966.2010.16259.x
34. Springel, V., White, S.D.M., Tormen, G., Kauffmann, G.: MNRAS **328**, 726 (2001). doi:10.1046/j.1365-8711.2001.04912.x
35. Yates, R.M., Thomas, P.A., Henriques, B.M.B.: ArXiv e-prints (2016)
36. Navarro, J.F., Frenk, C.S., White, S.D.M.: ApJ **462**, 563 (1996). doi:10.1086/177173
37. Neto, A.F., Gao, L., Bett, P., Cole, S., Navarro, J.F., Frenk, C.S., White, S.D.M., Springel, V., Jenkins, A.: MNRAS **381**, 1450 (2007). doi:10.1111/j.1365-2966.2007.12381.x
38. Wetzel, A.R., Tinker, J.L., Conroy, C.: MNRAS **424**, 232 (2012). doi:10.1111/j.1365-2966.2012.21188.x
39. Dressler, A.: ApJ **236**, 351 (1980). doi:10.1086/157753
40. Sun, M., Jones, C., Forman, W., Vikhlinin, A., Donahue, M., Voit, M.: ApJ **657**, 197 (2007). doi:10.1086/510895

PAMOP Project: Computations in Support of Experiments and Astrophysical Applications

B.M. McLaughlin, C.P. Ballance, M.S. Pindzola, P.C. Stancil, S. Schippers, and A. Müller

Abstract Our computation effort is primarily concentrated on support of current and future measurements being carried out at various synchrotron radiation facilities around the globe, and photodissociation computations for astrophysical applications. In our work we solve the Schrödinger or Dirac equation for the appropriate collision problem using the R-matrix or R-matrix with pseudo-states approach from first principles. The time dependent close-coupling (TDCC) method is also used in our work. A brief summary of the methodology and ongoing developments implemented in the R-matrix suite of Breit-Pauli and Dirac-Atomic R-matrix codes (DARC) is presented.

B.M. McLaughlin (✉) • C.P. Ballance
Centre for Theoretical Atomic Molecular and Optical Physics (CTAMOP), School of Mathematics & Physics, The David Bates Building, Queen's University, 7 College Park, Belfast BT7 1NN, UK
e-mail: bmclaughlin899@btinternet.com; c.ballance@qub.ac.uk

M.S. Pindzola
Department of Physics, 206 Allison Laboratory, Auburn University, 36849, Auburn, AL, USA
e-mail: pindzola@physics.auburn.edu

P.C. Stancil
Department of Physics and Astronomy and the Center for Simulational Physics, University of Georgia, 30602-2451, Athens, GA, USA
e-mail: stancil@physast.uga.edu

S. Schippers
I. Physikalisches Institut, Justus-Liebig-Universität Giessen, 35392, Giessen, Germany
e-mail: Stefan.Schippers@physik.uni-giessen.de

A. Müller
Institut für Atom- und Molekülphysik, Justus-Liebig-Universität Giessen, 35392, Giessen, Germany
e-mail: Alfred.Mueller@iamp.physik.uni-giessen.de

W.E. Nagel et al. (eds.), *High Performance Computing in Science and Engineering '16*, DOI 10.1007/978-3-319-47066-5_3

1 Introduction

Our research efforts continue to focus on the development of computational methods to solve the Schrödinger and Dirac equations for atomic and molecular collision processes. Access to leadership-class computers such as the Cray XC40 at HLRS allows us to benchmark our theoretical solutions against dedicated collision experiments at synchrotron facilities such as the Advanced Light Source (ALS), Astrid II, BESSY II, SOLEIL and PETRA III and to provide atomic and molecular data for ongoing research in laboratory and astrophysical plasma science. In order to have direct comparisons with experiment, semi-relativistic, or fully relativistic computations, involving a large number of target-coupled states are required to achieve spectroscopic accuracy. These computations could not be even attempted without access to high performance computing (HPC) resources such as those available at leadership computational centers in Europe (HLRS) and the USA (NERSC, NICS and ORNL). We use the R-matrix and R-matrix with pseudo-states (RMPS) methods to solve the Schrödinger and Dirac equations for atomic and molecular collision processes.

Satellites such as *Chandra* and *XMM-Newton* are currently providing a wealth of x-ray spectra on many astronomical objects, but a serious lack of adequate atomic data, particularly in the K-shell energy range, impedes the interpretation of these spectra. With the break-up and demise of the recently launched Astro-H satellite in the spring of 2016, it has left a void in x-ray observational data for a variety of atomic species of prominent astrophysical interest of paramount importance (Kallman T, Private communication, 2015). In the intervening period before the next x-ray satellite mission, we shall continue to benchmark laboratory photoionization cross section measurements against sophisticated theoretical methods.

The motivation for our work is multi-fold; (a) Astrophysical Applications [1–4], (b) Fusion and plasma modelling, (c) Fundamental interest and (d) Support of experimental measurements and Satellite observations. For heavy atomic systems [5, 6], little atomic data exists and our work provides results for new frontiers on the application of the R-matrix; Breit-Pauli and DARC parallel suite of codes. Our highly efficient R-matrix codes are widely applicable to the support of present experiments being performed at synchrotron radiation facilities. Examples of our results are presented below in order to illustrate the predictive nature of the methods employed compared to experiment.

The main question asked of any method is, how do we deal with the many body problem? In our case we use first principle methods (ab initio) to solve our dynamical equations of motion. Ab initio methods provide highly accurate, reliable atomic and molecular data (using state-of-the-art techniques) for solving the Schrödinger and Dirac equation. The R-matrix non-perturbative method is used to model accurately a wide variety of atomic, molecular and optical processes such as; electron impact ionization (EII), electron impact excitation (EIE), single and double photoionization and inner-shell x-ray processes. The R-matrix method provides cross sections and rates used as input for astrophysical modeling codes

Table 1 Photoionization cross section calculations: timings for the $J = 1$ scattering symmetry of W^{2+} ions. The scattering model used included 392-states, 1728 coupled channels, and 800,000 energy points. The R-matrix outer region module PSTGBF0DAMP performance on Hazel Hen, the Cray XC40 at HLRS, is presented for a different number of cores

R-matrix (Module)	Number of runs	Speed-up (factor)	Cray XC40 (Number of cores)	Total core time (minutes)
PSTGBF0DAMP	1	1.00	1000	451.525
PSTGBF0DAMP	1	2.01	2000	224.588
PSTGBF0DAMP	1	3.93	4000	114.866
PSTGBF0DAMP	1	5.82	8000	77.523
PSTGBF0DAMP	1	9.77	10,000	46.193

such as; CLOUDY, CHIANTI, AtomDB, XSTAR necessary for interpreting experiment/satellite observations of astrophysical objects as well as fusion and plasma modeling for JET and ITER.

2 R-Matrix Code Performance: Photoionization

The use of massively parallel architectures allows one to do calculations which previously could not have been addressed. This approach enables large scale relativistic calculations for trans-iron elements of Kr-ions, Xe-ions, Se-ions [5, 6] and W-ions [10, 11]. It allows one to provide atomic data in the absence of experiment, and for that purpose takes advantage of the linear algebra libraries available on most architectures. Further developments of the dipole codes benefit from similar developments made to the existing excitation R-matrix codes [6–9]. In Table 1 we show typical timings required in the determination of the photoionization cross section results for W^{2+} ions, for the $J = 1$ even scattering symmetry. Timings and speed up factors are given for the outer region module PSTGBF0DAMP used to determine photoionization cross sections. One clearly sees that using between 1000 to 10,000 cores, a speed up of nearly a factor of 10 is obtained with almost perfect scaling of this outer region module.

3 X-Ray and Inner-Shell Processes

3.1 K-Shell Photoionization of Atomic Oxygen Ions: O^{4+} *and* O^{5+}

The launch of the satellite Astro-H (re-named Hitomi) on February 17, 2016, was expected to provide x-ray spectra of unprecedented quality and would have required a wealth of atomic and molecular data on a range of collision processes to assist with the analysis of spectra from a variety of astrophysical objects. The subsequent break-up 40 days later on March 28, 2016 of Hitomi leaves a void in observational x-ray spectroscopy. Measurements of cross sections for photoionization of atoms and ions are essential data for testing theoretical methods in fundamental atomic physics and for modeling of many physical systems, for example, terrestrial plasmas, the upper atmosphere, and a broad range of astrophysical objects (quasar stellar objects, the atmosphere of hot stars, proto-planetary nebulae, H II regions, novae, and supernovae) [12, 13].

Limited wavelength observations for x-ray transitions were recently made on atomic oxygen, neon, magnesium and their ions with the High Energy Transmission Grating (HETG) on board the *CHANDRA* satellite [14]. Strong absorption K-shell lines of atomic oxygen, in its various forms of ionization, have been observed by the XMM-Newton satellite in the interstellar medium, through x-ray spectroscopy of low-mass x-ray binaries [15]. The Chandra and XMM-Newton satellite observations may be used to identify absorption features in astrophysical sources, such as active galactic nuclei (AGN), x-ray binaries, and for assistance in benchmarking theoretical and experimental work [16–21].

Absolute cross sections for the K-shell photoionization of Be-like (O^{4+}) and Li-like (O^{5+}) atomic oxygen ions were measured (in their respective K-shell regions) by employing the ion-photon merged-beam technique at the SOLEIL synchrotron-radiation facility in Saint-Aubin, France. High-resolution spectroscopy with $E/\Delta E \approx 4000$ ($\approx$140 meV, FWHM) was achieved with photon energy from 550 eV up to 675 eV. Rich resonance structure observed in the experimental spectra is analyzed using the R-matrix with pseudosates (RMPS) method.

Detailed spectra for Be-like [O^{4+}] and Li-like [O^{5+}] atomic oxygen ions in the vicinity of the K-edge were measured. This work is the culmination of photoionization cross section measurements on the atomic oxygen isonuclear sequence. Previous studies on this sequence, focused on obtaining photoionization cross sections for the O^{+} and O^{2+} ions [17] and the O^{3+} ion [16], where differences of 0.5 eV in the positions of the K_α resonance lines with prior satellite observations were found. This will have major implications for astrophysical modelling.

Figure 1 shows the spectra for Be-like atomic oxygen in the region of the strong 1s $\rightarrow$ 2p resonance. To compare directly with the SOLEIL measurements, the theoretical R-matrix cross sections have been convoluted with a Gaussian profile width of 220 meV at FWHM. For O^{4+} as the $1s^2 2s2p\ ^3P^o$ metastable state is present in the photon beam, an admixture of 70 % of the ground state and 30 % of the metastable state, of the respective cross sections, appears to simulate experiment suitably well. The theoretical cross section results presented in Fig. 1 indicate excellent agreement with the SOLEIL experimental measurements. Similarly in Fig. 2, the SOLEIL spectra for Li-like atomic oxygen in the region of the strong 1s $\rightarrow$ 2p resonance are illustrated. To compare with the SOLEIL measurements, the theoretical cross sections have been convoluted with a Gaussian profile width of 350 meV at FWHM. We note that for both ions, the theoretical results from the R-matrix with pseudostates method (RMPS) show suitable agreement with the SOLEIL measurements [22].

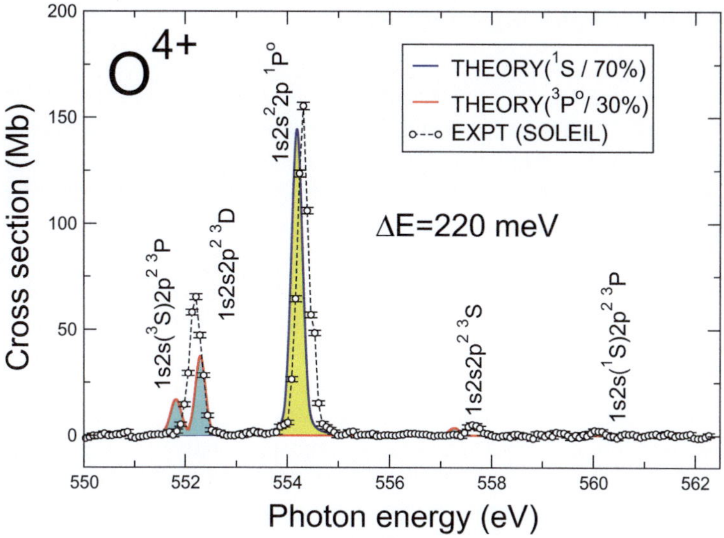

Fig. 1 SOLEIL experimental *K*-shell photoionization cross section of O^{4+} ions in the 550–560 eV photon energy range. Measurements were taken with a 220 meV band-pass at FWHM [22]. *Solid points* (experiment): the error bars give the statistical uncertainty. *Solid line* (R-matrix with pseudostates 526-levels) assuming an admixture of 70 % ($1s^2 2s^2\ ^1S$) and 30 % ($1s^2 2s2p\ ^3P^o$). The strong 1s → 2p resonances are clearly visible in the spectra

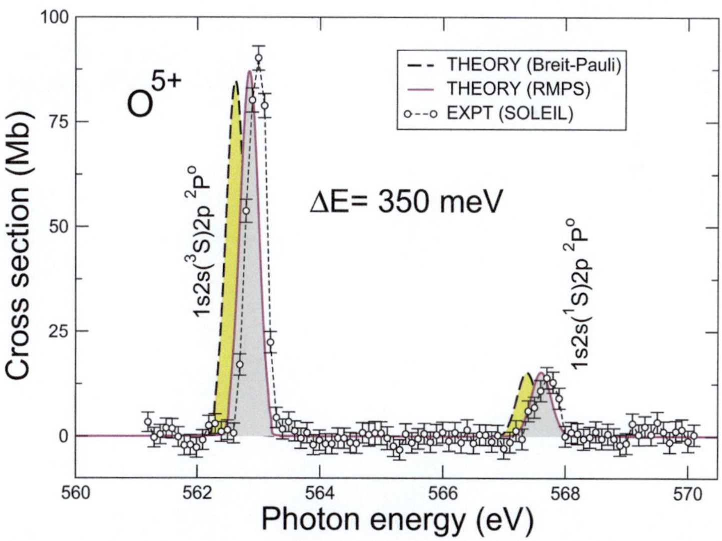

Fig. 2 SOLEIL experimental *K*-shell photoionization cross section of O^{5+} ions in the 560–570 eV photon energy range. Measurements were taken with a 350 meV band-pass at FWHM [22]. *Solid points* (experiment): the error bars give the statistical uncertainty. *Solid (magenta) line* R-matrix with pseudostates, 120-levels for the $1s^2 2s\ ^2S$ ground state. *Dashed (black) line* Breit-Pauli approximation. The strong 1s → 2p resonances are clearly visible in the spectra

3.2 *L-Shell Photoionization:* Ar^+

Photoionization cross-sections were obtained using the relativistic Dirac Atomic R-matrix Codes (DARC) for valence and *L*-shell energy ranges between 27 and 270 eV. A total of 557 levels arising from the dominant configurations $3s^2 3p^4$, $3s3p^5$, $3p^6$, $3s^2 3p^3[3d, 4s, 4p]$, $3p^5 3d$, $3s^2 3p^2 3d^2$, $3s3p^4 3d$, $3s3p^3 3d^2$, $2s^2 2p^5$ and $3s^2 3p^5$ have been included in the target wavefunction representation of the residual Ar^{2+} ion, including up to 4p in the orbital basis. The target wavefunctions were obtained using the GRASP code [23, 24], and the collision calculations were performed using a parallel version of the DARC codes [7–9, 26]. Direct comparisons of the photoionization cross sections in the valence region showed excellent agreement with previous R-matrix results and ALS measurements [27].

Photoionization cross section calculations were performed in the *L*-shell energy region between 250 and 280 eV in order to compare directly with the measurements made by Bizau and co-workers at the SOLEIL radiation facility in France [28]. To compare directly with the SOLEIL measurements, theory was convoluted with a 140 meV Gaussian profile width at FWHM to match the experiment.

Figure 3 illustrates the photoionization cross-section, as a function of the incident photon energy in eV across the *L*-shell threshold region from 250 to

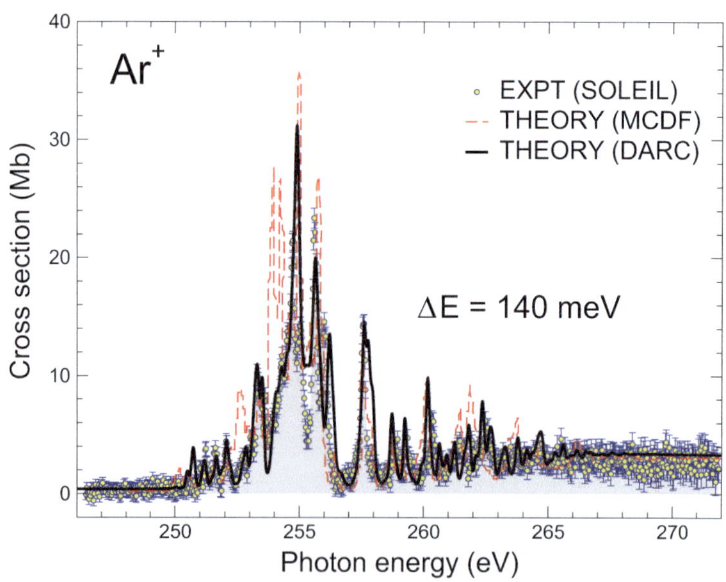

Fig. 3 Photoionization cross sections (Mb) as a function of the photon energy (eV) in the Ar^+ *L*-shell region between 250 and 270 eV. The *(blue) circles* are the experimental measurements from SOLEIL taken at a band pass of 140 meV at FWHM. The *dashed (red) line* are the MCDF theoretical results and the *solid (black) line* are the DARC (model DARC3) results. The theoretical results were statistically weighted for the initial ground state and convoluted with a Gaussian profile width of 140 meV at FWHM [29]

270 eV. Comparisons are made between the experimental results from SOLEIL, and theoretical work, MCDF and DARC. In order to match the SOLEIL experimental spectrum an energy shift of 7.5 eV to the DARC calculations was necessary [29].

3.3 Photoionization of Tungsten (W) Ions: W^{2+} *and* W^{3+}

Although not directly relevant to fusion, photoionization of tungsten atoms and ions is interesting because it can provide details about spectroscopic aspects and, as time-reversed photorecombination, provides access to the understanding of one of the most important atomic collision processes in a fusion plasma, electron-ion recombination. R-matrix theory is a tool to obtain information about electron-ion and photon-ion interactions in general. Electron-impact ionization and recombination of tungsten ions have been studied experimentally [30–37] while there are no detailed measurements on electron-impact excitation of tungsten atoms in any charge state. Thus, the present study on photoionization of these complex systems and comparison of the experimental data with R-matrix calculations provides benchmarks and guidance for future theoretical work on electron-impact excitation.

For comparison with the measurements made at the ALS, state-of-the-art theoretical methods using highly correlated wavefunctions were applied that include relativistic effects. An efficient parallel version [10, 11] of the DARC [24–26] suite of codes continues to be developed and applied to address electron and photon interactions with atomic systems, providing for hundreds of levels and thousands of scattering channels. These codes are presently running on a variety of parallel high performance computing architectures world wide [7–9]. DARC calculations on photoionization of heavy ions carried out for Se^+ [5], Xe^+ [6], Fe^+ [38], Xe^{7+} [39], W^+ [40, 41, 44], Se^{2+} [42], and Kr^+[48], ions showed suitable agreement with high resolution ALS measurements. Large-scale DARC photoionization cross section calculations on neutral sulfur compared to photolysis experiments, made in Berlin [49], and measurements performed at SOLEIL for 2p removal in Si^+ ions by photons [50] both showed suitable agreement.

Experimental and theoretical results are reported for single-photon single ionization of W^{2+} and W^{3+} tungsten ions. Experiments were performed at the photon-ion merged-beam setup of the Advanced Light Source in Berkeley. Absolute cross sections and detailed energy scans were measured over an energy range from about 20 to 90 eV at a bandwidth of 100 meV. Broad peak features with widths typically around 5 eV have been observed with almost no narrow resonances present in the investigated energy range. Theoretical results were obtained from a Dirac-Coulomb *R*-matrix approach. The calculations were carried out for the lowest-energy terms

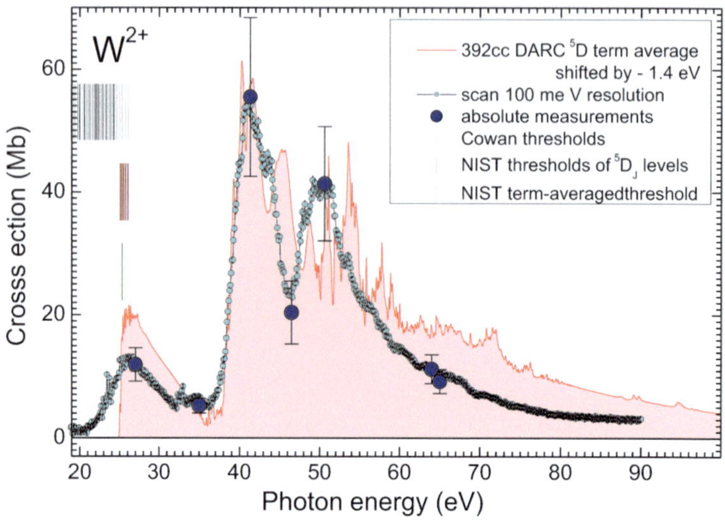

Fig. 4 Photoionization of W^{2+} ions measured at energy resolution 100 meV. Energy-scan measurements (*small circles* with statistical error bars) were normalized to absolute cross-section data represented by *large circles* with total error bars. The *black vertical bars* at energies below 26 eV represent ionization thresholds of all $5d^4$, $5d^3 6s$, and $5d^2 6s^2$ levels with excitation energies lower than the excitation energy of the lowest level ($^5G_2$) within the $5d^3 6p$ configuration. These thresholds were calculated by using the Cowan code [45] as implemented by Fontes and coworkers [46] and were shifted by about 0.5 eV to match the ground level ionization threshold from the NIST tables [47]. The *(brown) vertical bars* between 25 and 26 eV indicate the NIST ionization potentials of the levels within the $5d^4\,{}^5D$ ground-term. The lowest *(green) vertical bar* which matches the cross-section onset shows the NIST ground-term-averaged ionization potential. The *solid (red) line* with (*light red*) *shading* represents the result of the present 392-level DARC calculation (125 μeV step size) of the ground-term-averaged photoionization cross section, convoluted with a Gaussian of 100 meV width. The theoretical cross sections are shifted by −1.4 eV to match experiment [43]

of the investigated tungsten ions with levels $5s^2 5p^6 5d^4\,{}^5D_J$ $J = 0, 1, 2, 3, 4$ for W^{2+} and $5s^2 5p^6 5d^3\,{}^4F_{J'}$ $J' = 3/2, 5/2, 7/2, 9/2$ for W^{3+}. As illustrated in Fig. 4 for W^{2+} ions, suitable agreement is achieved below 60 eV, but at higher energies there is a factor of approximately two difference between experiment and theory. In Fig. 5, assuming a statistically weighted distribution of ions in the initial ground-term levels, over the energy range investigated, good agreement between theory and experiment for W^{3+} ions is achieved [43].

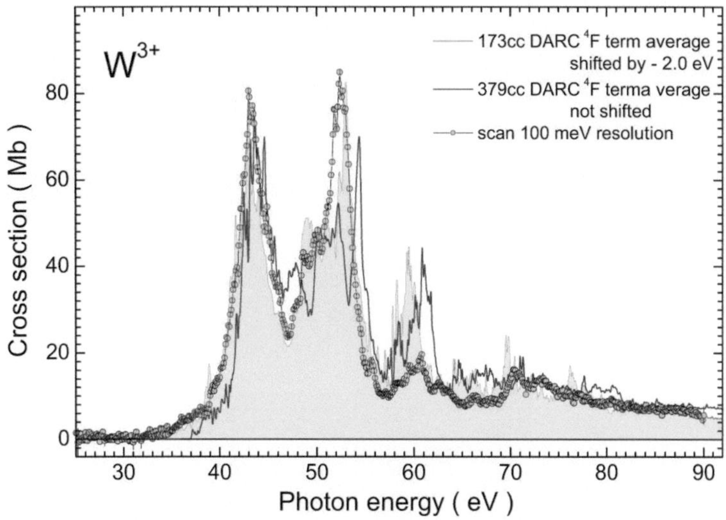

Fig. 5 Comparison of the measured photoionization cross section of W^{3+} with the present 173-level DARC calculation (87 μeV step size; *thin red line* with shading) and the present 379-level DARC result (109 μeV step size; *solid blue line* without shading). The theory curves were obtained by convolution of the original spectra with a Gaussian of 100 meV width. Only the 173-level calculations are shifted down in energy by 2.0 eV so that the steep rise of the experimental cross section function at about 40 eV is matched

4 Single-Photon Double Ionization: He

The time-dependent close-coupling (TDCC) method [51] was used to perform single-photon double ionization cross section calculations of He in the 1s2p ^{3}P^o excited state. Total and energy differential cross sections for the 1s2p ^{3}P^o excited state are presented for the TDCC (ℓ_1, ℓ_2, L) and TDCC (ℓ_1 j_1, ℓ_2 j_2, J) representations. Figure 6 illustrates the total TDCC total cross sections, and Fig. 7 that for the differential cross section, as a function of the ejected electron energy in eV, for each initial He(1s2p $^3P^o_{0,1,2}$) fine-structure level. Differences found between the level resolved single-photon double ionization cross sections are due to varying degrees of continuum correlation found in the outgoing two electrons [52].

5 Photodissociation: SH$^+$

Photodissociation cross sections for the SH$^+$ radical are computed from all rovibrational (RV) levels of the ground electronic state X $^3\Sigma^-$ for wavelengths from threshold to 500 Å. The five electronic transitions, 2 $^3\Sigma^-$ $\leftarrow$ X $^3\Sigma^-$, 3 $^3\Sigma^-$ $\leftarrow$ X $^3\Sigma^-$, A $^3\Pi$ $\leftarrow$ X $^3\Sigma^-$, 2 $^3\Pi$ $\leftarrow$ X $^3\Sigma^-$, and 3 $^3\Pi$ $\leftarrow$ X $^3\Sigma^-$,

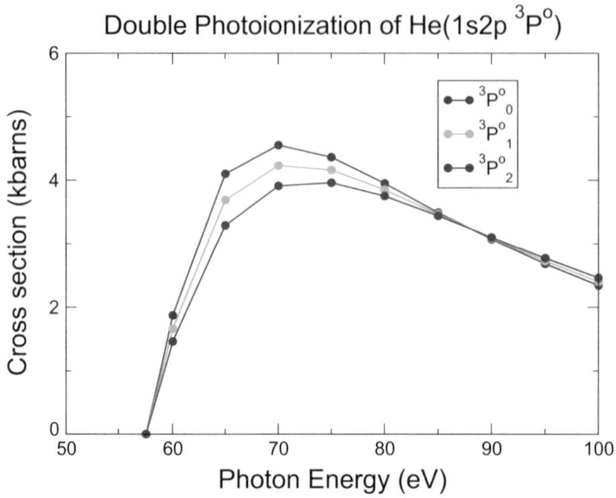

Fig. 6 Total cross sections (kbarns) as a function of photon energy using the time dependent close-coupling (TDCC) method. Results are shown for the initial individual fine-structure states of He(1s2p $^3P^o_J$), where $J = 0$, 1 and 2 [52]

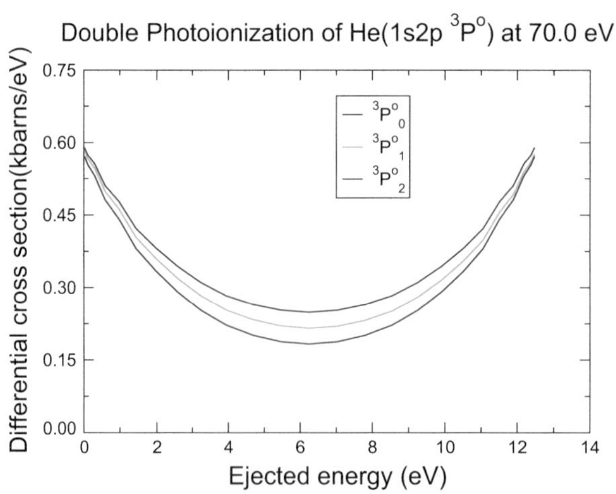

Fig. 7 Differential cross sections (kilobarns/eV) as a function of the ejected electron energy in eV using the time dependent close-coupling (TDCC) method at a photon energy of 70 eV. Results are shown for the initial individual fine-structure states of He(1s2p $^3P^o_J$), where $J = 0$, 1 and 2 [52]

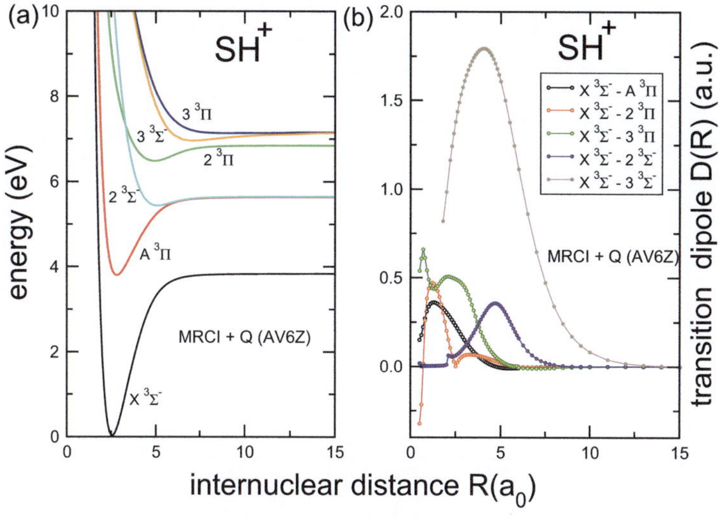

Fig. 8 (**a**) Relative electronic energies (eV) for the SH^+ molecular cation, as a function of bond separation at the MRCI+Q level of approximation with an AV6Z basis. Energies are relative to the ground state near equilibrium (2.6 a_0). The states shown are for the transitions connecting the $X^3\Sigma^- \rightarrow 2\,^3\Sigma^-, 3\,^3\Sigma^-, A\,^3\Pi, 2\,^3\Pi, 3\,^3\Pi$ states involved in the photodissociation process. (**b**) Dipole transition moments $D(R)$ (a.u.) for the $X\,^3\Sigma^- \rightarrow A\,^3\Pi, 2\,^3\Sigma^-, 3\,^3\Sigma^-, 2\,^3\Pi, 2\,^3\Pi$, transitions. The MRCI + Q approximation with an AV6Z basis set was used to calculate the transition dipole moments

are treated with a fully quantum-mechanical two-state model, (i.e. no non-adiabatic coupling between excited states was included in our work). The photodissociation calculations incorporate adiabatic potential energy curves (PEC) and transition dipole moment (TDM) functions computed in the multi-reference configuration interaction approach [53] with the Davidson correction (MRCI+Q) [54], using an augmented-correlation-consistent polarized valence sextuplet basis set, designated as aug-cc-pV6Z or AV6Z, as illustrated in Fig. 8. We have adjusted our ab initio data to match available experimental molecular data and asymptotic atomic limits. Local thermodynamic equilibrium (LTE) photodissociation cross sections were computed which assume a Boltzmann distribution of RV levels in the $X\,^3\Sigma^-$ molecular state of the SH^+ cation. The LTE cross sections are presented for temperatures in the range 1000–10,000 K.

As far as we are aware, the current work is the first explicit photodissociation calculations for the SH^+ radical ion. An estimate was made in van Dishoeck et al. [55] of the SH^+ cross section by scaling that of CH^+. As illustrated in Fig. 9, there is suitable agreement, however the current results are about a factor of 3 larger, therefore we would expect the photodissociation rate to be enhanced by a similar amount.

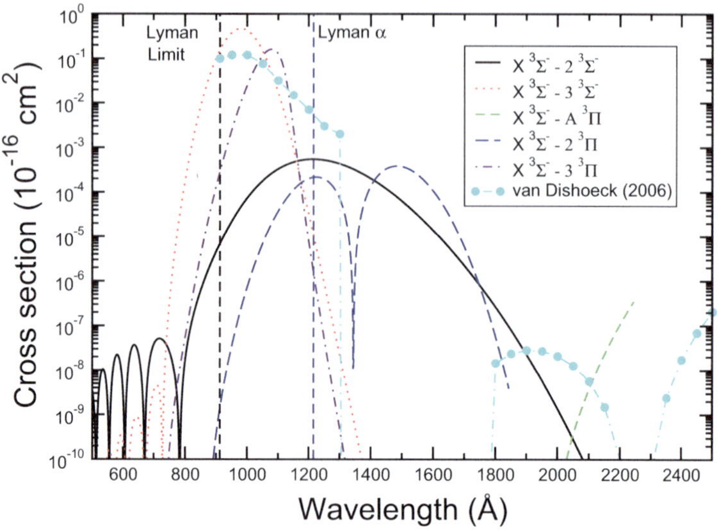

Fig. 9 Comparison of SH^+ photodissociation cross sections for $v'' = 0$ and $J'' = 0$ with estimates from Ref. [55]

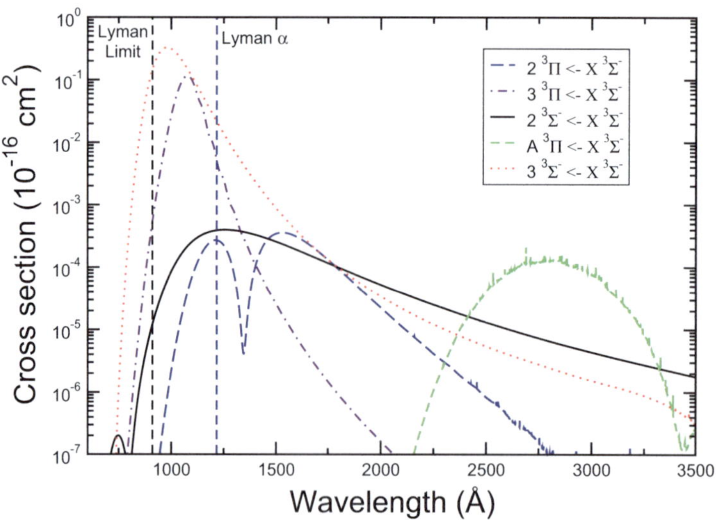

Fig. 10 Total SH^+ LTE photodissociation cross section at 3000 K for all electronic transitions

In Fig.10, the LTE cross sections for all five transitions are compared at 3,000 K. This should be compared to Fig. 9 for $v'' = 0, J'' = 0$ where it is seen that the cross sections are larger in the LTE case for wavelengths longer than $\sim$1500 Å.

The SH^+ radical ion, sulfanylium, was not detected in the interstellar medium (ISM) until as late as 2010 [56]. It is however, an important tracer of gas condensations in dense regions and also probes the warm surface layers of photo-dominated regions (PDR) [57]. Furthermore, its abundance is expected to be enhanced in x-ray dominated regions (XDR) [58]. In their model of the Orion Bar PDR, Nagy et al. [57] find that photodissociation accounts for a maximum of about 4.4 % of the total destruction rate of SH^+, since reactive collisions with H and dissociative recombination by electrons are more efficient. However, they adopted the estimated cross section of [55] for $v'' = 0$, $J'' = 0$. We point out that the adoption of the current cross sections would enhance the photodissociation contribution to greater than 10 %. We note that the photodissociation rates are not given here as they are sensitive to the local radiation field and dust properties. The latter is quite different in the Orion Bar from the average ISM of the galaxy. The densities and temperatures (10^5–10^6 cm^{-3} and $\sim$1000 K) of the Orion Bar PDR begin to approach the regime where photodissociation from excited states might contribute which is currently neglected in all models. Furthermore, LTE conditions are almost satisfied, but at 1000 K there is not a significant difference between the LTE and $v'' = 0$, $J'' = 0$ cross sections [59].

6 Summary

The power of the predictive nature of the R-matrix approach within a non-relativistic or a fully relativistic approach for photoionization cross sections, valence or inner-shell, resonance energy positions, Auger widths and strengths is illustrated. Quantal calculation of photodissociation cross sections and rates for astrophysical applications require as input accurate potential energy curves and transition dipole moments. Access to leadership architectures is essential to our research work such as the Cray XC40 at HLRS which provides an integral contribution to our computational effort in atomic, molecular and optical collision processes.

Acknowledgements A Müller and S Schippers acknowledge support by Deutsche Forschungsgemeinschaft under project numbers Mu-1068/10, Mu-1068/20 and through NATO Collaborative Linkage grant 976362. B M McLaughlin acknowledges support from the US National Science Foundation through a grant to ITAMP at the Harvard-Smithsonian Center for Astrophysics, under the visitor's program, the RTRA network *Triangle de le Physique* and a visiting research fellowship (VRF) from Queen's University Belfast. M S Pindzola acknowledges support by NSF and NASA grants through Auburn University. P C Stancil acknowledge support by NASA grants through University of Goergia at Athens. This research used computational resources at the National Energy Research Scientific Computing Center in Berkeley, CA, USA, and at the High Performance Computing Center Stuttgart (HLRS) of the University of Stuttgart, Stuttgart, Germany. The Oak Ridge Leadership Computing Facility at the Oak Ridge National Laboratory, provided additional computational resources, which is supported by the Office of Science of the U.S. Department of Energy under Contract No. DE-AC05-00OR22725. The Advanced Light Source is supported by the Director, Office of Science, Office of Basic Energy Sciences, of the US Department of Energy under Contract No. DE-AC02-05CH11231.

References

1. Hasoglu, M.F., Abdel Naby, S.A., Gorczyca, T.W., Drake, J.J., McLaughlin, B.M.: K-shell photoabsorption studies of the carbon isonuclear sequence. Astrophys. J. **724**, 1296 (2010)
2. McLaughlin, B.M.: Inner-shell photoionization, fluorescence and Auger yields. In: Ferland, G., Savin, D.W. (eds.) Spectroscopic Challenges of Photoionized Plasma, Astronomical Society of the Pacific. ASP Conference Series, vol. 247, p. 87. Astronomical Society of the Pacific, San Francisco (2001)
3. Kallman, T.R.: Challenges of plasma modelling: current status and future plansa. Space Sci. Rev. **157**, 177 (2010)
4. McLaughlin, B.M., Ballance, C.P.: Photoionization, fluorescence and inner-shell processes. In: McGraw-Hill (ed.) McGraw-Hill Yearbook of Science and Technology, p. 281. McGraw Hill, New York (2013)
5. McLaughlin, B.M., Ballance, C.P.: Photoionization cross section calculations for the halogen-like ions Kr^+ and Xe^+. J. Phys. B: At. Mol. Opt. Phys. **45**, 085701 (2012)
6. McLaughlin, B.M., Ballance, C.P.: Photoionization cross-sections for the trans-iron element Se^+ from 18 eV to 31 eV. J. Phys. B: At. Mol. Opt. Phys. **45**, 095202 (2012)
7. McLaughlin, B.M., Ballance, C.P.: Petascale computations for large-scale atomic and molecular collisions, ch 15. In: Resch, M.M., Kovalenko, Y., Fotch, E., Bez, W., Kobaysahi, H. (eds.) Sustained Simulated Performance 2014. Springer, New York (2014)
8. McLaughlin, B.M., Ballance, C.P., Pindzola, M.S., Müller, A.: PAMOP: petascale atomic, molecular and optical collisions, ch 4. In: Nagel, W.E., Kröner, D.H., Resch M.M. (eds.) High Performance Computing in Science and Engineering'14. Springer, New York (2015)
9. McLaughlin, B.M., Ballance, C.P., Pindzola, M.S., Schipprs, S., Müller, A.: PAMOP: petascale computations in support of experiments, ch 4. In: Nagel, W.E., Kröner, D.H., Resch, M.M. (eds.) High Performance Computing in Science and Engineering'15. Springer, New York (2016)
10. Ballance, C.P., Griffin, D.C.: Relativistic radiatively damped R-matrix calculation of the electron-impact excitation of W^{46+}. J. Phys. B: At. Mol. Opt. Phys. **39**, 3617 (2006)
11. Ballance, C.P., Loch, S.D., Pindzola, M.S., Griffin, D.C.: Electron-impact excitation and ionization of W^{3+} for the determination of tungsten influx in a fusion plasma. J. Phys. B: At. Mol. Opt. Phys. **46**, 055202 (2013)
12. Kjeldsen, H., Kristensen, B., Brooks, R.L., Folkman, H., Knudsen, H., Andersen, T.: Absolute state-slected measurements of the photoionization cross section of N^+ and O^+ ions. Astrophys. J. Suppl. Ser. **138**, 219 (2002)
13. Garcia, J., Mendoza, C., Bautista, M.A., Gorczyca, T.W., Kallman, T.R., Palmeri, P.: K-shell photoabsorption of oxygen ions. Astrophys. J. Suppl. Ser. **158**, 68 (2005)
14. Liao, J.-Y., Zhang, S.-N., Yao, Y.: Wavelength measurements of K transitions of oxygen, neon, and magnesium with X-ray absorption lines. Astrophys. J. **774**, 116 (2013)
15. Pinto, C., Kaastra, J.S., Costantini, E., de Vries, C.: Interstellar medium composition through X-ray spectroscopy of low-mass X-ray binaries. Astron. Astrophys. **551**, 25 (2013)
16. McLaughlin, B.M., Bizau, J.M., Cubaynes, D., Al Shorman, M.M., Guilbaud, S., Sakho, I., Blancard, C., Gharaibeh, M.F.: K-shell photoionization of B-like (O^{3+}) oxygen ions: experiment and theory. J. Phys. B: At. Mol. Opt. Phys. **47**, 115201 (2014)
17. Bizau, J.M., Cubaynes, D., Guilbaud, S., Al Shorman, M.M., Gharaibeh, M.F., Ababneh, I.Q., Blancard, C., McLaughlin, B.M.: K-shell photoionization of O^+ and O^{2+} ions: experiment and theory. Phys. Rev. A **92**, 023401 (2015)
18. Gorczyca, T.W., Bautista, M.A., Hasoglu, M.F., Garcia, J., Gatuzz, E., Kasstra, J.S., Kallman, T.R., Manson, S.T., Mendoza, C., Raasen, A.J.J., de Vries, C.P., Zatsarinny, O.: A comprehensive X-ray absorption model for atomic oxgen. Astrophys. J. **779**, 78 (2013)
19. Gatuzz, E., Garcia, J. Mendoza, C., Kallman, T.R., Witthoeft, M., Lohfink, A., Bauitista, M.A., Palmeri, P., Quinet, P.: Photoionization modeling of oxygen K absorption in the interstellar medium: the Chandra grating spectra of XTE J1817–330. Astrophys. J. **768**, 60 (2013)

20. Gatuzz, E., Garcia, J. Mendoza, C., Kallman, T.R., Witthoeft, M., Lohfink, A., Bauitista, M.A., Palmeri, P., Quinet, P.: Erratum: photoionization modeling of oxygen K absorption in the interstellar medium: the Chandra grating spectra of XTE J1817–330. Astrophys. J. **778**, 83 (2013)
21. Gatuzz, E., Garcia, J., Mendoza, C., Kallman, T.R., Bautista, M.A., Gorczyca, T.W: Physical properties of the interstellar medium using high-resolution Chandra spectra: O K-edge absorption. Astrophys. J. **790**, 131 (2014)
22. Bizau, J.M., Cubaynes, D., Guilbaud, S., Al Shorman, M.M., El Ghazaly, M.O.A., Gharaibeh, M.F., Sakho, I., McLaughlin, B.M.: K-shell photoionization of O^{4+} and O^{5+} ions: experiment and theory. Mon. Not. R. Astro. Soc. (MNRAS) (2016, in press)
23. Dyall, K.G., Grant, I.P., Johnson, C.T., Plummer, E.P.: GRASP: a general-purpose relativistic atomic structure program. Comput. Phys. Commun. **55**, 425 (1989)
24. Grant, I.P.: Quantum Theory of Atoms and Molecules: Theory and Computation. Springer, New York (2007)
25. Norrington, P.H., Grant, I.P.: Low-energy electron scattering by Fe XXIII and Fe VII using the dirac R-matrix method. J. Phys. B: At. Mol. Opt. Phys. **20**, 4869 (1987)
26. R-matrix DARC and BP codes. http://connorb.freeshell.org (2016)
27. Covington, A.M., Aguilar, A., Covington, I.R., Hinojosa, G., Shirley, C.A., Phaneuf, R.A., Álvarez, I., Cisneros, C., Dominguez-Lopez, I., Sant'Anna, M.M., Schlachter, A.S., Ballance, C.P., McLaughlin, B.M.: Valence-shell photoionization of chlorinelike Ar^{+} ions. Phys. Rev. A **84**, 013413 (2011)
28. Blancard, C., Cossé, Ph., Faussurier, G., Bizau, J.-M., Cubaynes, D., El Hassan, N., Guilbaud, S., Al Shorman, M.M., Robert, E., Liu, X.-J., Nicolas, C., Miron, C.: L-shell photoionization of Ar^{+} to Ar^{3+} ions. Phys. Rev. A **85**, 043408 (2012)
29. Tyndall, N.B., Ramsbottom, C.A., Ballance, C.P., Hibbert, A.: Valence and *L*-shell photoionization of Cl-like argon using *R*-matrix techniques. Mon. Not. Roy. Astro. Soc. (MNRAS) **456**, 366 (2016)
30. Müller, A.: Fusion-related ionization and recombination data for tungsten ions in low to moderately high charge states. Atoms **3**, 120 (2015)
31. Rausch, J., Becker, A., Spruck, K., Hellhund, J., Borovik Jr, A., Huber, K., Schippers S., Müller, A.: Electron-impact single and double ionization of W^{17+}. J. Phys. B: At. Mol. Opt. Phys. **44**, 165202 (2011)
32. Stenke, M., Aichele, K., Harthiramani, D., Hofmann, G., Steidl, M., Völpel, R., Salzborn, E.: Electron-impact single-ionization of singly and multiply charged tungsten ions. J. Phys. B: At. Mol. Opt. Phys. **28**, 2711 (1995)
33. Schippers, S., Bernhardt, D., Müller, A., Krantz, C., Grieser, M., Repnow, R., Wolf, A., Lestinsky, M., Hahn, M., Novotný, O., Savin, D.W.: Dielectronic recombination of xenonlike tungsten ions. Phys. Rev. A **83**, 012711 (2011)
34. Krantz, C., Spruck, K., Badnell, N.R., Becker, A., Bernhardt, D., Grieser, M., Hahn, M., Novotný, O., Repnow, R., Savin, D.W., Wolf, A., Müller, A., Schippers S.: Absolute rate coefficients for the recombination of open *f*-shell tungsten ions. J. Phys. Conf. Ser. **488**, 012051 (2014)
35. Spruck, K., Badnell, N.R., Krantz, C., Novotný, O., Becker, A., Bernhardt, D., Grieser, M., Hahn, M., Repnow, R., Savin, D.W., Wolf, A., Müller, A., Schippers, S.: Recombination of W^{18+} ions with electrons: absolute rate coefficients from a storage-ring experiment and from theoretical calculations. Phys. Rev. A **90**, 032715 (2014)
36. Borovik, A. Jr., Ebinger, B., Schury, D., Schippers, S., Müller, A.: Electron-impact single ionization of W^{19+} ions. Phys. Rev. A **93**, 012708 (2016)
37. Badnell, N.R., Spruck, K., Krantz, C., Novotný, O., Becker, A., Bernhardt, D., Grieser, M., Hahn, M., Repnow, R., Savin, D.W., Wolf, A., Müller, A., Schippers, S.: Recombination of W^{19+} ions with electrons: absolute rate coefficients from a storage-ring experiment and from theoretical calculations. Phys. Rev. A **93**, 052703 (2016)
38. Fivet, V., Bautista, M.A., Ballance, C.P.: Fine-structure photoionization cross sections of Fe II. J. Phys. B: At. Mol. Opt. Phys. **45**, 035201 (2012)

39. Müller, A., Schippers, S., Esteves-Macaluso, D., Habibi, M., Aguilar, A., Kilcoyne, A.L.D., Phaneuf, R.A., Ballance, C.P., McLaughlin, B.M.: High resolution valence shell photoionization of Ag-like (Xe^{7+}) Xenon ions: experiment and theory. J. Phys. B: At. Mol. Opt. Phys. **47**, 215202 (2014)
40. Müller, A., Schippers, S., Hellhund, J., Holosto, K., Kilcoyne, A.L.D., Phaneuf, R.A., Ballance, C.P., McLaughlin, B.M.: Single-photon single ionization of W^+ ions: experiment and theory. J. Phys. B: At. Mol. Opt. Phys. **48**, 2352033 (2015)
41. Müller, A.: Precision studies of deep-inner-shell photoabsorption by atomic ions. Phys. Scr. **90**, 054004 (2015)
42. Macaluso, D.A., Aguilar, A., Kilcoyne, A.L.D., Red, E.C., Bilodeau, R.C., Phaneuf, R.A., Sterling, N.C., McLaughlin, B.M.: Absolute single-photoionization cross sections of Se^{2+}: experiment and theory. Phys. Rev. A **92**, 063424 (2015)
43. McLaughlin, B.M., Ballance, C.P., Schippers, S., Hellhund, J., Kilcoyne, A.L.D., Phaneuf, R.A., Müller, A.: Photoionization of tungsten ions: experiment and theory for W^{2+} and W^{3+}. J. Phys. B: At. Mol. Opt. Phys. **49**, 065201 (2016)
44. Müller, A., Schippers, S., Hellhund, J., Kilcoyne, A.L.D., Phaneuf, R.A., Ballance, C.P., McLaughlin, B.M.: Single and multiple photoionization of W^{q+} tungsten ions in charged states $q = 1, 2, .., 5$: experiment and theory. J. Phys. Conf. Ser. **488**, 022032 (2014)
45. Cowan, R.D.: The Theory of Atomic Structure and Spectra. University of California Press, Berkeley (1981)
46. Fontes, C.J., Zhang, H.L., Abdallah, J. Jr., Clark, R.E.H., Kilcrease, D.P., Colgan, J.P., Cunningham, R.T., Hakel, P., Magee, N.H., Sherrill, M.E.: The Los Alamos suite of relativistic atomic physics codes. J. Phys. B: At. Mol. Opt. Phys. **48**, 144014 (2015)
47. Kramida, A.E., Ralchenko, Y., Reader, J., NIST ASD Team: NIST Atomic Spectra Database (version 5.2). National Institute of Standards and Technology, Gaithersburg (2014)
48. Hinojosa, G., Covington, A.M., Alna'Washi, G.A., Lu, M., Phaneuf, R.A., Sant'Anna, M.M., Cisneros, C., Álvarez, I., Aguilar, A., Kilcoyne, A.L.D., Schlachter, A.S., Ballance, C.P., McLaughlin, B.M.: Valence-shell single photoionization of Kr^+ ions: experiment and theory. Phys. Rev. A **86**, 063402 (2012)
49. Barthel, M., Flesch, R., Rühl, E., McLaughlin, B.M.: Photoionization of the $3s^2 3p^4\ ^3$P and the $3s^2 3p^4\ ^1$D,^{1} S states of sulfur: experiment and theory. Phys. Rev. A **91**, 013406 (2015)
50. Kennedy, E.T., Mosnier, J.-P., Van Kampen, P., Cubaynes, D., Guilbaud, S., Blancard, C., McLaughlin, B.M., Bizau, J.-M.: Photoionization cross sections of the aluminumlike Si^+ ion in the region of the $2p$ threshold (94–137 eV). Phys. Rev. A **90**, 063409 (2014)
51. Pindzola, M.S., Robicheaux, F., Loch, S.D., Berengut, J.C., Topcu, T., Colgan, J., Foster, M., Griffin, D.C., Ballance, C.P., Schultz, D.R., Minami, T., Badnell, N.R., Witthoeft, M.C., Plante, D.R., Mitnik, D.M., Ludlow, J.A., Kleiman, U.: The time-dependent close-coupling method for atomic and molecular collision processes. J. Phys. B: At. Mol. Opt. Phys. **40**, R39 (2007)
52. Li, Y., Pindzola, M.S., Colgan, J.P.: Double photoionization of He from the 1s2p ^{3}P^o excitated state. J. Phys. B: At. Mol. Opt. Phys. **49**, 19205 (2016)
53. Helgaker, T., Jorgesen, P., Oslen, J.: Molecular Electronic-Structure Theory. Wiley, New York (2000)
54. Langhoff, S., Davidson, E.R.: Configuration interaction calculations on the nitrogen molecule. Int. J. Quantum Chem. **8**, 61 (1974)
55. van Dishoeck, E.F., Jonkheid, B., van Hemert, M.C.: Photoprocesses in protoplanetary disks. Faraday Discuss. **133**, 855 (2006)
56. Benz, A.O., et al.: Hydrides in young stellar objects: radiation tracers in a protostar-disk-outflow system. Astron. Astrophys. **521**, A35 (2010)
57. Nagy, Z., et al.: The chemistry of ions in the orion Bar I. – CH^+,SH^+, and CF^+. Astron. Astrophys. **550**, A96 (2013)
58. Abel, N.P., Federman, S.R., Stancil, P.C.: The effects of doubly ionized chemistry on SH^+ and S^{2+} abundances in X-ray-dominated regions. Astrophys. J. **675**, L81 (2008)
59. McMillan, E.C., Shen, G., McCann, J.F., McLaughlin, B.M., Stancil, P.C.: Rovibrationally resolved photodissociation of SH^+. J. Phys. B: At. Mol. Opt. Phys. **49**, 084001 (2016)

Estimation of Nucleation Barriers from Simulations of Crystal Nuclei Surrounded by Fluid in Equilibrium

Antonia Statt, Peter Koß, Peter Virnau, and Kurt Binder

Abstract Nucleation rates for homogeneous nucleation are commonly estimated in terms of an Arrhenius law involving the nucleation barrier, written in terms of a competition of the contribution in surface free energy of the nucleus and the free energy gain proportional to the nucleus volume. For crystal nuclei this "classical nucleation theory" is hampered by the problem that the nucleus in general is non spherical, since the interfacial excess free energy depends on the orientation of the interface relative to the crystal axes. This problem can be avoided by analyzing the equilibrium of a crystal nucleus surrounded by fluid in a small simulation box in thermal equilibrium. Estimating the fluid pressure and the chemical potential, as well as the volume of the nucleus, suffices to obtain the nucleation barrier, if the equation of state of the pure phases as well as the coexistence pressure are known. This method is demonstrated to work using a coarse-grained model for colloids with an effective attraction due to small polymers, comparing two choices of the attraction strength.

Keywords Colloids • Nucleation • Crystallization • Asakura-Oosawa model

1 Introduction and Overview

When a material is brought out of equilibrium by a sudden change of thermodynamic parameters (e.g. temperature T, pressure p, etc.), such that a phase boundary of a phase transition (e.g. the melting/crystallization line $T_m(p)$ in the p, T plane) is crossed, the old phase (e.g. the fluid) is metastable, and the new phase (e.g. the crystal) forms by nucleation [1, 2]. Homogeneous nucleation (due to statistical fluctuations) requires to overcome a free energy barrier (Fig. 1), ΔF^*,

A. Statt • P. Koß
Graduate School Materials Science in Mainz, Staudinger Weg 9, D-55099, Mainz, Germany
e-mail: statt@uni-mainz.de

P. Virnau (✉) • K. Binder
Institut für Physik, Johannes Gutenberg-Universität, Staudinger Weg 7, D-55099, Mainz, Germany
e-mail: virnau@uni-mainz.de

W.E. Nagel et al. (eds.), *High Performance Computing in Science and Engineering '16*, DOI 10.1007/978-3-319-47066-5_4

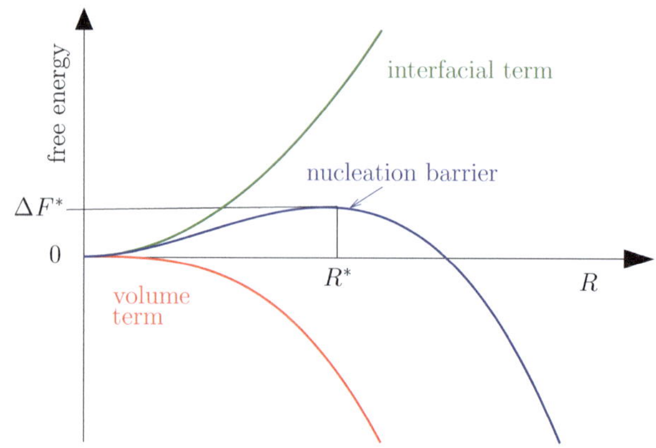

Fig. 1 Formation free energy contribution of a nucleus ΔF as function of its linear dimension R. In $d = 3$ dimensions, the volume term is negative and scales like R^3, but the interfacial term is positive and scales like R^2. Thus a nucleation barrier ΔF^* for a "critical droplet" with linear dimension R^* results

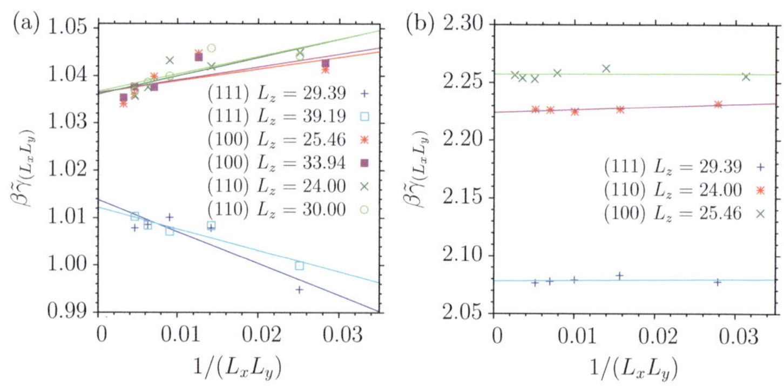

Fig. 2 Finite size scaling for the reduced interfacial tension of the soft effective Asakura-Oosawa (softEffAO) model at two reduced interaction strengths, $\eta_p^r = 0.1$ (*left part*) and $\eta_p^r = 0.2$ (*right part*), plotted vs inverse interfacial area, using $L_x \times L_y \times L_z$ geometry and periodic boundary conditions. Three orientations of the interface are shown, (111) [i.e. a closed packed interface in the face-centered cubic crystal lattice], (110) and (100) (Part (**a**) is taken from Ref. [3], Part (**b**) from Ref. [4])

due to the unfavorable surface free energy contribution. Classical nucleation theory estimates this barrier making two assumptions: (i) The critical nucleus can be described by a spherical droplet, R^* being its radius. (ii) The interfacial free energy just is $4\pi R^{*2}\gamma$, γ being the interfacial tension of a flat planar interface. However, while these assumptions look rather reasonable for the nucleation of liquid droplets from supersaturated vapor, they make little sense for crystal nucleation: the spherical shape of the nucleus is not consistent with its regular crystal structure, and furthermore γ is not isotropic, but rather depends somewhat on the orientation of the interface relative to the crystal axes. This is demonstrated in Fig. 2 for the model of attractive colloidal particles studied in the present work [3, 4]. This model will be explained in the following section. Here it suffices to know that for weak attraction between the colloidal particles this anisotropy is rather weak (left part of Fig. 2), and hence an almost spherical droplet shape may be expected, while for stronger attraction (right part of Fig. 2) the anisotropy is more noticeable. Then the crystal shape will deviate from a sphere. Each point in Fig. 2a, b took around 4×24 h on $\sim$1000 CPUs in parallel. If the interfacial free energy were known for arbitrary interface orientation (and not just for the three choices (111), (110) and (100) displayed in Fig. 2), one could find the equilibrium crystal shape from the Wulff construction [5]. First of all, this procedure is cumbersome, and knowing $\gamma(\mathbf{n})$ for only three orientations $\mathbf{n}$ of the interface, this is only possible approximately. If we could do that, the surface term in Fig. 1 could be written as function of the nucleus volume V as

$$F_{\text{surf}}(V) = V^{2/3} \int_{A_w} \gamma(\mathbf{n}) d\mathbf{s} \equiv A_w \bar{\gamma} V^{2/3} \quad , \tag{1}$$

where the surface integral $\int d\mathbf{s}$ is extended over a crystal having the Wulff shape and unit volume. The corresponding surface area is A_w, and $\bar{\gamma}$ is then an average interfacial tension. Then ΔF in Fig. 1 becomes (p_c is the pressure in the crystal and p_l is the pressure in the liquid)

$$\Delta F = -(p_c - p_\ell)V + F_{\text{surf}}(V) = -(p_c - p_\ell)V + A_w \bar{\gamma} V^{2/3} \quad , \tag{2}$$

and the barrier ΔF^* occurs for $V = V^*$ with $\partial(\Delta F(V))/\partial V \,|_{V^*} = 0$. This yields

$$V^{*1/3} = \frac{2A_w\bar{\gamma}}{3(p_c - p_\ell)} \quad , \quad \Delta F^* = \frac{1}{3} A_w \bar{\gamma} V^{*2/3} \quad . \tag{3}$$

Now the present study simply exploits the idea [6, 7] to combine both Eqs. (3) as follows and expand the pressures at the coexistence conditions as

$$\Delta F^* = \frac{1}{2}\Big(p_c - p_\ell\Big)V^* \quad , \tag{4}$$

$$p_c \approx p_{\text{coex}} + (6/\pi)\eta_m(\mu_c(p_c) - \mu_{\text{coex}}),$$
$$p_l \approx p_{\text{coex}} + (6/\pi)\eta_f(\mu_l(p_l) - \mu_{\text{coex}}), \tag{5}$$

with $\eta_m(\eta_f)$ being the packing fractions of the (spherical) colloidal particles where the melting (freezing) sets in. Since in equilibrium the chemical potential for the nucleus coexisting with fluid is homogeneous,

$$\mu_c(p_c) = \mu_l(p_l) = \mu \quad , \tag{6}$$

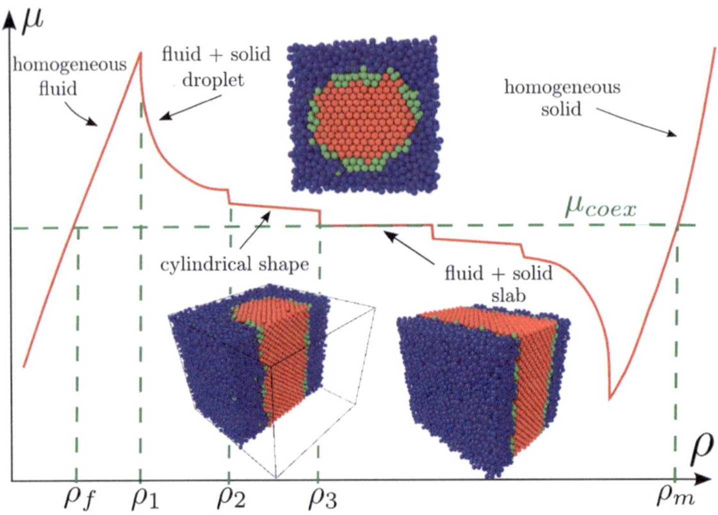

Fig. 3 Schematic plot of the chemical potential μ vs. density ρ (or packing fraction $\eta = (\pi\sigma_c^3/6)\rho$, σ_c being the colloid diameter, respectively), for a system undergoing a phase transition from liquid (at density ρ_f) to solid (at density ρ_m) at $\mu = \mu_{\text{coex}}$ in the thermodynamic limit (broken horizontal straight line) and in a box of finite volume V_{box} (full curve). Due to finite size effects, the homogeneous liquid is stable until the density ρ_1 where the droplet evaporation/condensation transition occurs. For $\rho_1 < \rho < \rho_2$ a nucleus surrounded by liquid is stable: this is the region of interest, where μ_l, p_l, and V^* need to be extracted. At ρ_2, a transition occurs to a cylinder-like nucleus, stabilized by the periodic boundary conditions that are applied throughout. For $\rho = \rho_3$ a transition to a slab-like crystal with two planar interfaces occurs. In the slab region, theory requires $\mu = \mu_{\text{coex}}$, if the periodic boundary condition is commensurable with the crystal periodicity. The different states are illustrated with snapshot pictures of configurations of the model with $\eta_p^r = 0.1$ (particles in the crystal are shown in *red*, in the fluid in *blue*, in the interface region in *green*) (From [6])

Eqs. 4, 5 can be rewritten as

$$\Delta F^* = \frac{1}{2}(\eta_m - \eta_f)(\mu - \mu_{\text{coex}})V^* \quad . \tag{7}$$

Thus the task of the simulation is to locate coexistence conditions (p_{coex}, μ_{coex}, η_m, η_f) in the bulk, "measure" the chemical potential (or the pressure p_l) of the liquid surrounding the nucleus, and "measure" the volume V^* of the nucleus. The latter can simply be done by a finite-size generalization of the lever rule, when we carry out a "measurement" of p_l in a simulation box of finite volume V_{box} at a chosen constant packing fraction η

$$\eta V_{\text{box}} = \eta_l(p_l)(V_{\text{box}} - V^*) + \eta_c(p_c)V^* \quad . \tag{8}$$

As a caveat, we mention that V_{box} has to be chosen large enough so that fluctuations of μ and p_l are relatively small, and one can only work in a restricted range of packing fractions (avoiding both the "droplet-evaporation/condensation" transition [8] and the appearance of cylinder-like droplets or slab structures [9], see Fig. 3).

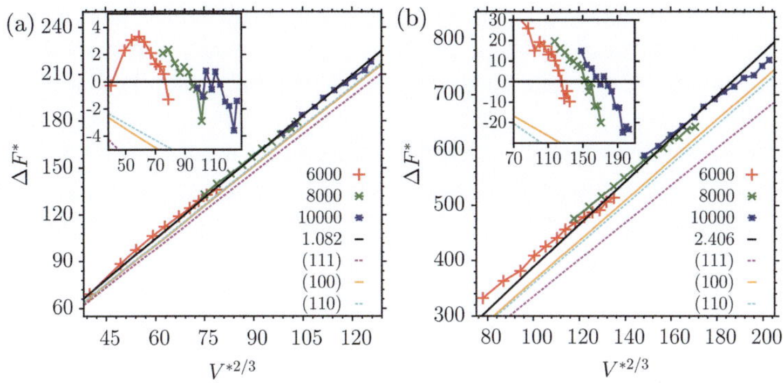

Fig. 4 Nucleation barriers $\Delta F^*(V^*)$ plotted vs $V^{*2/3}$ for $\eta_p^r = 0.1$ (**a**) and $\eta_p^r = 0.2$ (**b**). Here units $k_BT = 1$ and $\sigma_c = 1$ are used. Three system sizes are included in each case, containing $N = 6000$, 8000 or 10,000 particles, respectively. *Full straight lines* show fits assuming a spherical surface (replacing A_w by $A_{\text{iso}} = (36\pi)^{1/3}$ in Eq.(3)) and then fitting $\bar{\gamma}$, with result $\bar{\gamma} = 1.082$ (**a**) and $\bar{\gamma} = 2.406$ (**b**). The *dotted lines* indicate the predictions when one would take, in case (**a**) $\gamma_{111} = 1.013$, $\gamma_{110} = 1.044$ and $\gamma_{100} = 1.039$, and in case (**b**) $\gamma_{111} = 2.078$, $\gamma_{110} = 2.224$ and $\gamma_{100} = 2.256$ rather than $\bar{\gamma}$. The *inset* in the figure shows the differences of the data to the fits (From [4])

Before presenting any details on our procedures, we show central results to show that the strategy outlined above works (Fig. 4). Indeed, apart from small deviations, for all choices of N used there is a broad regime where the proportionality of ΔF^* to $V^{*2/3}$ holds, and the important feature is that these data superimpose to a common straight line irrespective of N in each case. This property is crucial, because we want to be able to describe nucleation phenomena in bulk materials, not in nanoscopically small boxes with periodic boundary conditions. The use of such boxes is needed to be able to study nuclei in thermal equilibrium – a nucleus on top of the barrier in Fig. 1 is unstable against thermal fluctuations, of course, and cannot be straightforwardly studied.

However, as the insets in Fig. 4a, b show, there occur minor deviations from the fit to a common straight line, but these deviations are of the order of a few percent only. These deviations are to a fewer extent statistical errors, but to a larger extent systematic. We attribute the systematic errors due to the fact that in our geometry the chemical potential in the system is not strictly constant (as tacitly assumed in Fig. 1), but fluctuates. This fluctuation is larger the smaller N is. A second systematic effect comes from the translational entropy of the nucleus in the simulation box, which scales proportional to $\ln(V_{\text{box}})$ and hence $\ln(N)$. More research will be needed to clarify the nature of these systematic corrections quantitatively.

In any case, the deviations due to the choice of a spherical droplet shape and use of any of the interface tensions of planar interfaces (γ_{111}, γ_{110} or γ_{100}, respectively) are distinctly larger than these systematic errors, and would lead to a significant underestimation of the nucleation barrier, in particular for $\eta_p^r = 0.2$. We expect that this discrepancy will increase further for still larger η_p^r, where ultimately faceted crystals [5] will result. We recall that in the simplistic lattice gas model at low temperatures T the nucleus shape tends to a simple cube, and then the ratio of the actual barrier ΔF^* to the spherical approximation gradually tends to $6/\pi$ as $T \to 0$ [10].

In the next section, we shall give some details on the model that we have used for our study, and in the third section, some details of the actual analysis that yielded Fig. 4 will be given. Section 4 summarizes our conclusions, and gives an outlook to future work.

2 The Model and Its Bulk Properties

Our choice of model is motivated by colloidal suspensions, for which nucleation rates have been studied extensively both by experiment and simulations (see review [11]), focusing on the limit of hard-sphere like colloids. However, it must be noted that experimentally it is not possible to determine the packing fraction η better than with 1 % accuracy, moreover it is not possible to manufacture colloids which are perfectly uniform in size, and also some additional other interactions (in addition to the hard-sphere-like repulsion) are always present [12]. Hence we do not focus on hard spheres here, but rather focus on colloidal suspensions of hard-sphere-like particles where small polymers added to the suspension provide an attractive interaction, whose range can be controlled by the polymer radius and whose strength can be controlled by the polymer fugacity [13]. The standard model for such systems is the well-known Asakura-Oosawa (AO) model [14, 15]. Since the colloid-colloid interaction for this model is singular for distances equal to the colloid diameter σ_c, it is inconvenient from a simulator's perspective, and since it is also an idealization of reality anyway, it is advisable to work with the so called "softEffAO model" [4, 6, 7], where the singular potential is replaced by an almost equivalent smooth potential, which leads to a very similar phase separation as the original AO model. For $r \leq \sigma_c (= 1)$ the repulsive potential hence is not infinite, but replaced by $[k_B T = 1]$

$$U_{\text{rep}}(r) = 4\left(\left[\frac{b\sigma_c}{r-\varepsilon\sigma_c}\right]^{12} + \left[\frac{b\sigma_c}{r-\varepsilon\sigma_c}\right]^{6} - \left[\frac{b\sigma_c}{b\sigma_c+q-\varepsilon\sigma_c}\right]^{12} - \left[\frac{b\sigma_c}{b\sigma_c+q-\varepsilon\sigma_c}\right]^{6}\right). \tag{9}$$

where $b = 0.01$, ϵ is specified below, $q = 0.15$ is a convenient choice of constants. The attractive part of the potential is described by $\sigma_c < r \leq \sigma_c(1+q)$

$$U_{\text{att}}(r) = \eta_p^r \left(1+q^{-1}\right)^3 \left[1 - \frac{3r/\sigma_c}{2(1+q)} + \frac{(r/\sigma_c)^3}{2(1+q)^3}\right] , \tag{10}$$

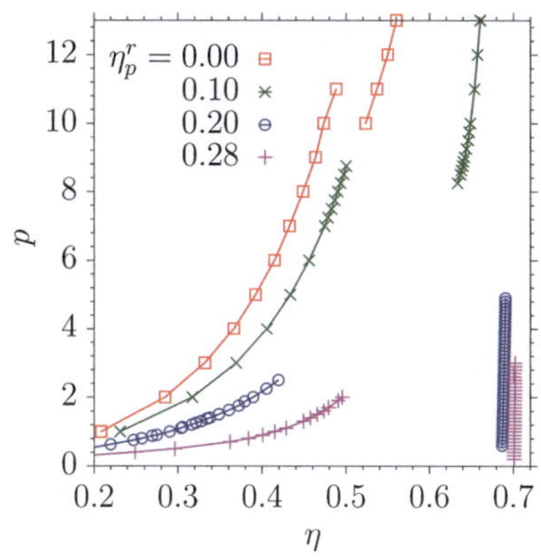

Fig. 5 Normalized pressure p versus packing fraction η for several choices of attraction strength η_p^r, as indicated. The branches at the left side represent the liquid phase and the branches at the right side the crystal

while $U_{\rm att}(r >= \sigma_c(1+q)) = 0$. The parameter ϵ is chosen such that the total potential is smoothly differentiable at $r = \sigma_c$, which yields $\epsilon = 0.967118$ ($\eta_p^r = 0.1$) or $\epsilon = 0.9892$ ($\eta_p^r = 0.2$), respectively. For this potential it is straightforwardly possible to compute the pressure applying the Virial formula, unlike for the original AO model.

Figure 5 shows then the phase diagram of this model for different choices of η_p^r. These data were taken by sampling the packing fraction η by Monte Carlo runs in the NpT-ensemble, using $N = 4000$ colloidal particles. The data for the crystal were obtained using a perfect fcc lattice as initial condition, of course. Ideally, the liquid branch should only occur for pressures $p \le p_{\rm coex}$. However, as usually observed for NpT simulations of first order transitions, this is not the case: there is a regime of pressures where both phases are stable or metastable, respectively, and from the data of Fig. 5 a straightforward estimation of $p_{\rm coex}$ is not possible. We have determined $p_{\rm coex}$ by the method proposed by Zykova-Timan et al. [16]. In this method, one studies slab configurations where in the initial state a crystal domain (of volume $V_c = L \times L \times L_c$) and a liquid domain (of volume $V_l = L \times L \times (5L - L_c)$) are present. Periodic boundary conditions are used, and thus the domains are separated by two planar $L \times L$ interfaces (L is chosen such, that at the chosen pressure the crystal is not distorted). If the chosen pressure exceeds $p_{\rm coex}$ and we let the system evolve in the Monte Carlo run, we expect that the crystal grows on expense of the liquid, while the opposite behavior occurs for $p < p_{\rm coex}$ (see [16, 17] for more details). Plotting the volume change of the total system versus Monte Carlo "time" for various pressures we identify $p_{\rm coex}$ as the pressure where no volume change occurs (Fig. 6). As discussed in [4, 7, 17], this method is not as straightforward as it looks, since there is both the need to take averages over many equivalent runs to reduce statistical noise in the curves such as shown in Fig. 6, and there is the need to study several choices of L (or n, respectively) to extrapolate the resulting estimates

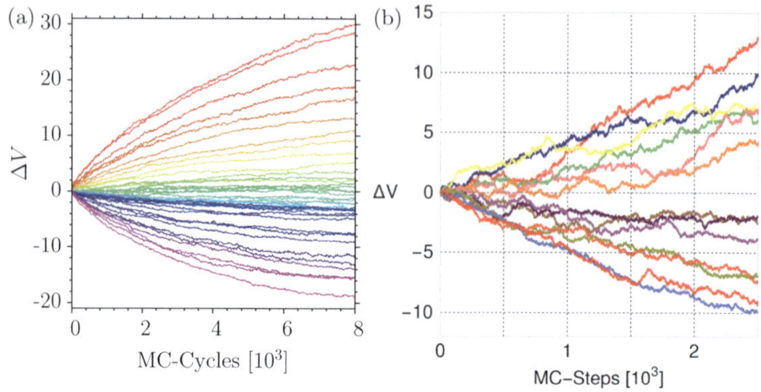

Fig. 6 (**a**) Volume change as a function of the number of Monte Carlo steps for $\eta_p^r = 0.2$, choosing $n = 10$ lattice planes in x and y directions, and pressures from $p = 0.6$ (*red, top*) to $p = 3.0$ (*magenta, bottom*). (**b**) Same plot as (**a**) for $\eta_p^r = 0.28$

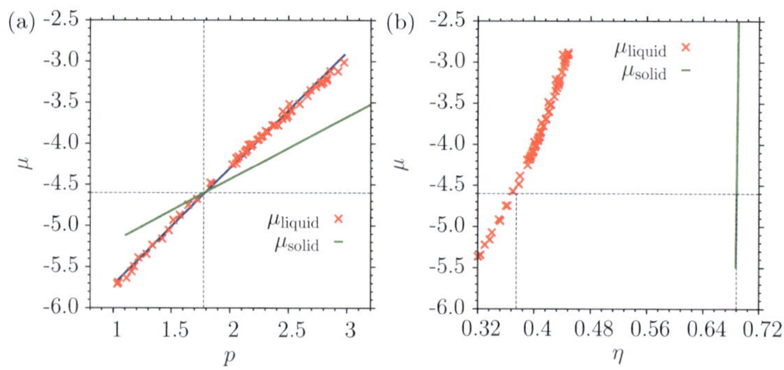

Fig. 7 Chemical potential as a function of pressure (**a**) and packing fraction (**b**), for the softEffAO model with $\eta_p^r = 0.2$

of $p_{\text{coex}}(n)$ vs n^{-2} in order to obtain an estimate for the true coexistence pressure that applies in the thermodynamic limit. When p_{coex} is known and the liquid and solid branches $\eta_l(p)$ and $\eta_c(p)$ are known (Fig. 5), we immediately can read off $\eta_f = \eta_l(p_{\text{coex}})$, $\eta_m = \eta_c(p_{\text{coex}})$, and from the estimation of the pressure p_l of the liquid coexisting with the nucleus we can infer μ {Eq. 6} from the linear expansion of p_l {Eq.5}, and using also the expansion for p_c {Eq. 5} we find $\eta_c(p_c)$ and V^* can then be inferred from Eq. 8. However, it is advisable to check that one works close enough to coexistence conditions such that the linear expansions, Eq 5, are actually valid. For this purpose, a new method to estimate the chemical potential has been developed [4, 6, 7, 17], since in many cases of interest the standard Widom particle insertion method [18] cannot be applied. Figure 7 shows, as an example, the chemical potential μ for $\eta_p^r = 0.2$ plotted against both pressure and packing fraction, using the estimate of p_{coex} as estimated above from the method explained in Fig. 6. Indeed one finds that the curves μ vs. p for both phases are straight lines in the regime of interest. Using $p_{\text{coex}} = 1.78 \pm 0.02$ we found $\mu_{\text{coex}} = -4.60 \pm 0.04$ in this case, leading to $\eta_f = 0.374 \pm 0.002$, $\eta_m - 0.688 \pm 0.001$.

3 Simulation Analysis of the Nucleus-Fluid Equilibrium

The nucleus-fluid equilibrium was studied by putting part of the particles in a subsystem with perfect crystal structure at $\eta = \eta_m$ and filling the rest of the box with particles having the expected density of the fluid, and then equilibrating the system. It is important to verify that the resulting crystal nucleus does not depend in any significant way on this rather arbitrary initial state. This fact is illustrated in Fig. 8, where it is illustrated for $\eta_p^r = 0.2$ that different shapes of the initial crystal lead to rather similar shapes of the resulting equilibrated nucleus, and the corresponding distributions of pressures and density in the fluid region surrounding the nucleus are within error identical. Note that in this case $p_l \approx 2.454$ and $\eta_l = 0.4184 \pm 0.0001$.

The actual data for $p_l(\eta)$ for three different choices of N are then shown in Fig. 9. One can see that the variation of the pressure as function of the total packing fraction η decreases, in accordance with the schematic variation of μ vs. η in the region where the nucleus coexists with surrounding fluid (Fig. 3). The larger N the smaller the pressure (for $N \to \infty$ these curves converge towards the horizontal variation $p = p_{\text{coex}}$). From this study we find the packing fraction $\eta_l(p)$ of the surrounding fluid, at the same time, and these data are also shown in Fig. 9: they coincide at a common curve and this curve is nothing but the pressure vs. packing fraction curve of the corresponding homogeneous system (Fig. 5).

Thus, our actual numerical results validate the assumptions made in our finite-size generalization of the lever rule, Eq. 8. From the knowledge of η, $\eta_l(p)$ and $\eta_c(p)$ (using Fig. 7) we have obtained all the necessary input to deduce V^* via Eq. 8, and using then Eqs. 4 or Eq. 7, respectively, the data shown in Fig. 4 result.

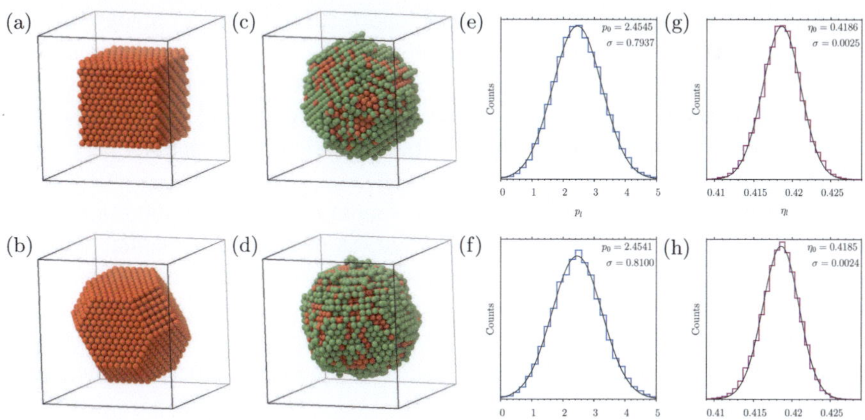

Fig. 8 Different crystalline seeds (left column, part (**a**), (**b**)) lead to very similar shapes of the equilibrated crystalline nuclei (next column, part (**c**), (**d**)) and almost identical distributions of pressure (part (**e**), (**f**)) and density (part (**g**), (**h**)) of the surrounding fluid. All data refer to the case $N = 10{,}000$, $\eta = 0.48$, $\eta_p^r = 0.2$. The equilibrated nuclei shapes where obtained after about 10^{10} Monte Carlo cycles, with each cycle comprised of N Monte Carlo trial moves

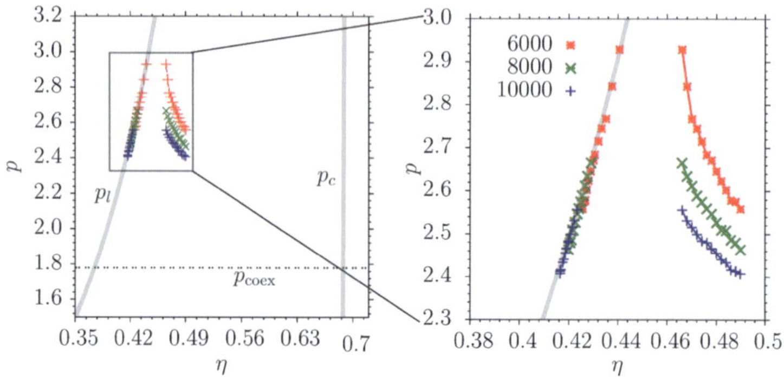

Fig. 9 Pressure p_l in the liquid surrounding a crystalline nucleus, as shown in Fig. 8, plotted vs. the total packing fraction η for the softEffAO model choosing $\eta_p^r = 0.2$. Three choices of N are shown, $N = 6000$, 8000, and 10,000, respectively. The region of interest is shown on the right with strongly magnified scales. Values for the packing fraction of the fluid η_l are included and lie on top of the bulk equation of state for the liquid branch. From [4]

4 Conclusions

A method to study the free energy barrier for homogeneous nucleation of crystals from a fluid phase has been presented, which is not hampered by the fact that the fluid-crystal interface tension in general is anisotropic. In the softEffAO model, variation of the parameter η_p^r that controls the strength of the effective attraction between the colloidal particles allows to control this anisotropy (Fig. 2), and indeed deviations from the standard (inappropriate) assumption of spherical nucleus shape were found (Fig. 4). In the present report, several steps of analysis of the simulation data have been explained. We also emphasize the need for accessing a fast supercomputer such as HORNET at the HLRS Stuttgart for the research: typical system sizes involve systems with 10^4 colloidal particles, and for obtaining data such as shown in Fig. 6 averages over 100 runs carried out in parallel need to be taken.

While the present work has addressed a simple model system, appropriate for colloidal suspensions, future work should address interparticle potentials that are relevant for materials science, since nucleation of crystals is a very relevant problem there. Also an application to study the formation of ice nuclei in the atmosphere is planned, since this problem is of central importance in the context of climate modeling. In all cases, complementary studies of the kinetic aspects of nucleation phenomena will be needed.

Acknowledgements We would like to thank the DFG for funding in the framework of the priority program on heterogeneous nucleation (SPP 1296, grant N° VI 237/4-3). P. K. is a recipient of a DFG-fellowship/DFG-funded position through the Excellence Initiative by the Graduate School Materials Science in Mainz (GSC 266). We thank the HLRS Stuttgart for generous grants of computer time at the HORNET supercomputer. The authors gratefully acknowledge the computing time granted on the supercomputer Mogon at Johannes Gutenberg University Mainz (www.hpc.uni-mainz.de).

References

1. Binder, K., Stauffer, D.: Adv. Phys. **25**, 343 (1976)
2. Kashchiev, D.: Nucleation: Basic Theory with Applications. Butterworth-Heinemann, Oxford (2000)
3. Schmitz, F., Virnau, P.: J. Chem. Phys. **142**, 354110 (2015)
4. Statt, A.: Dissertation, Johannes Gutenberg Universität Mainz (2015). URN: urn:nbn:de:hebis:77-41750
5. Herring, C.: Phys. Rev. **82**, 87 (1951)
6. Statt, A., Virnau, P., Binder, K.: Phys. Rev. Lett. **114**, 026101 (2015)
7. Statt, A., Virnau, P., Binder, K.: Mol. Phys. **113**, 2556–2570 (2015)
8. Binder, K.: Physica A **319**, 99 (2003)
9. MacDowell, L.G., Shen, V.K., Errington, J.R.: J. Chem. Phys. **125**, 034705 (2006)
10. Schmitz, F., Virnau, P., Binder, K.: Phys. Rev. E **81**, 70533 (2013)
11. Palberg, T.: J. Phys. Condens. Matter **26**, 333101 (2014)
12. Royall, C.P., Poon, W.C.K., Weeks, E.R.: Soft Matter **9**, 17 (2013)
13. Poon, W.C.K.: J. Phys. Condens. Matter **14**, R859 (2002)
14. Asakura, S., Oosawa, F.: J. Chem. Phys. **22**, 1255 (1954)
15. Binder, K., Virnau, P., Statt, A.: J. Chem. Phys. **141**, 140901 (2014)
16. Zykova-Timan, T., Horbach, J., Binder, K.: J. Chem. Phys. **133**, 014705 (2010)
17. Statt, A., Schmitz, A., Virnau, P., Binder, K.: Monte Carlo simulation of crystal-liquid phase coexistence. In: Nagel, W.E., et al. (eds.) High Performance Computing in Science and Engineering'15: Part I: Physics. Springer, Berlin (2016)
18. Widom, B.: J. Chem. Phys. **39**, 2808 (1963)

The Internal Dynamics and Early Adsorption Stages of Fibrinogen Investigated by Molecular Dynamics Simulations

Stephan Köhler, Friederike Schmid, and Giovanni Settanni

Abstract Fibrinogen, a plasma glycoprotein of vertebrates, plays an essential role in blood clotting by polymerizing into fibrin upon activation. It also contributes, upon adsorption on material surfaces, to determine their biocompatibility and has been implicated as a cause of thrombosis and inflammation at medical implants. Here we present the first fully atomistic simulations of the initial stages of the adsorption process of fibrinogen on mica and graphite surfaces. The simulations reveal a weak adsorption on mica that allows frequent desorption and reorientation events. This adsorption is driven by electrostatic interactions between the protein and the silicate surface as well as the counter ion layer. Preferred adsorption orientations for the globular regions of the protein are identified. The adsorption on graphite is found to be stronger with fewer reorientation and desorption events, and showing the onset of denaturation of the protein.

1 Introduction

Fibrinogen (Fg) is a 340 kD multi-chain glyco-protein which can polymerize into fibrin, one of the main component of blood clots. Fibrin formation and lysis (fibrinolysis) are tightly controlled processes along the pathway leading to coagulation [1]. Fg, once activated by thrombin, which cleaves the fibrinopeptide A and B (FpA, FpB), exposes specific A- and B-knobs which bind to the corresponding a- and b-holes of neighbor Fg molecules and initiate the fibrin polymerization process. Fibrin is later stabilized by additional non-covalent and covalent interactions. By further interacting with other blood components through its integrin binding sites, fibrin plays an important role in regulating coagulation and immune response. Fibrinolysis on the other hand is effected by plasmin, which cleaves fibrin on specific cleavage points in a well defined temporal sequence [2–4].

The elongated structure of human Fg, as shown by the crystallographic data [5], is formed by two symmetric units which dimerize through a central globular E region. Each symmetric unit (protomer) is constituted by three peptide chains Aα,

S. Köhler • F. Schmid • G. Settanni (✉)
Institut für Physik, Johannes Gutenberg University, Mainz, Germany
e-mail: settanni@uni-mainz.de

W.E. Nagel et al. (eds.), *High Performance Computing in Science and Engineering '16*, DOI 10.1007/978-3-319-47066-5_5

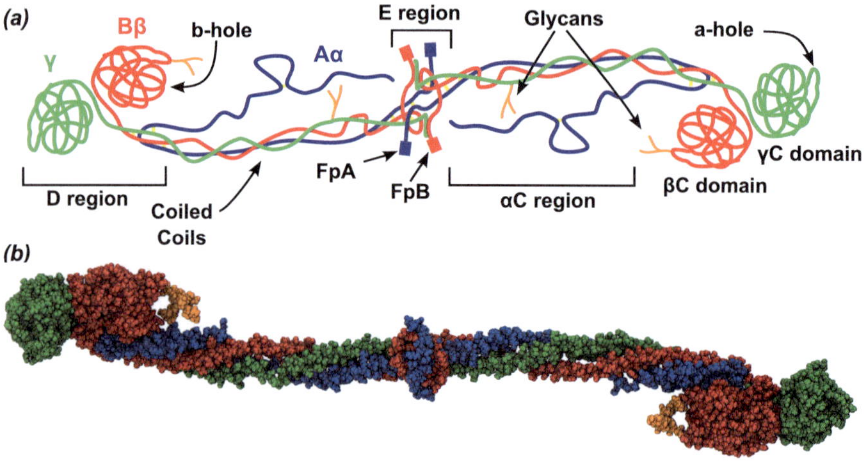

Fig. 1 The fibrinogen molecule. (**a**) Schematic representation of the fibrinogen molecule. The three chains of Fg, Aα, Bβ and γ are shown in *blue*, *red* and *green*, respectively. (**b**) Van der Waals representation of the crystallographic structure (pdb 3GHG) of Fg, color coded as in (**a**). Carbohydrates are in *orange*. The αC region and the FpA and FpB peptides were not resolved in the crystal structure (Reprinted from Ref. [6]. Copyright (2015) Köhler et al. under the terms of the Creative Commons Attribution License)

Bβ and γ which depart from their N-terminal region (E region), form an elongated coiled-coil region, and end into two globular domains forming the D region (Fig. 1). The C terminal segment of the Aα chain, i.e. the αC region, as well as the N-terminal parts of chain Aα and Bβ, including FpA and FpB, are mostly disordered (thus, not resolved in the crystal).

The D region contains several integrin binding sites, including the P1 and P2 sites (residues γ190–202 and γ377–395, respectively) which are known to bind leukocyte integrin $\alpha_M\beta_2$ [7, 8], and site H12 (residues γ392–411), which binds to the platelet integrin receptor $\alpha_{IIb}\beta_3$ [9]. In particular, P1 is partly located in a cleft between the γC and βC domain (binding cleft). Additionally, the D region contains the a- and b-"holes" which are the binding sites of the "knobs" at the end of the Fp tethers of the E region and play a major role in fibrin formation.

Although the available crystallographic structures of Fg show a relatively limited variability, atomic force microscopy images of adsorbed Fg on several surfaces reveal a large degree of conformational flexibility. Indeed, the typical tri-nodular structure of Fg, as observed in adsorption studies, where the three nodules correspond to the two D regions and the central E region, is very variable [10], and the angle formed by the three nodules has a wide distribution [11, 12]. The source of this conformational flexibility at the molecular level is not well understood. Early sequence analysis [13] and comparison of several crystallographic structures of Fg [5, 14, 15] suggested the presence of a hinge point in the middle of the coiled-coil regions connecting the E and D regions. With the help of the simulations described below, we have suggested a possible role of this hinge point and the extent of flexibility that it confers to the Fg molecule.

Two surfaces often used in Fg adsorption experiments are mica and graphite [10, 11, 16–20]. They represent an ideal charged/hydrophilic and a non-polar surface, respectively, and their sheet structure allows for the production of atomically flat surfaces. The techniques used for these experiments, however, do not allow to spatially resolve the mechanism behind the flexibility of Fg or the atomic scale details of the adsorbed state. Simulations, which have been used to study protein adsorption on mica [21–23] and graphite/graphene [24–28], can help to fill the spatial resolution gap. The adsorption of the γC domain of Fg on various self assembled monolayer surfaces has been investigated using atomistic molecular dynamics (MD) simulations, which showed rolling motions but no deformations [29]. The adsorption of Fg on graphene has also been investigated using atomistic MD simulations [28], which showed slow equilibration possibly driven by the formation of hydrophobic contacts and conformational rearrangements. Further simulations of Fg explored its mechanical response to external forces [30, 31], as well as its flexibility in solution [6]. Fg adsorption has been also studied using simplified models [32–34], where Fg is replaced by one or a small number of interacting objects that represent the whole molecule or the globular regions. In these models, as well as in models of fibrin polymerization [35], the internal flexibility of Fg is either ignored or treated approximately, although it may play a very important role especially in the characterization of its hydrodynamic properties.

After presenting the methodological tools used in our work we show how we have addressed two key aspects of Fg behavior, namely its interanal dynamics and its adsorption properties.

In Sect. 3, we report the results of extensive molecular dynamics (MD) simulations performed on Fg in solution. The simulations allow for the identification of large bending motions centered at a hinge point on the coiled-coil region of Fg. We also show how these bending motions may provide a conserved mechanism facilitating the action of plasmin in fibrinolysis.

In Sect. 4 we report on atomistic molecular dynamics simulations of the initial stages of Fg adsorption on mica and graphite. In these simulations we address the speed, strength and reversibility of the adsorption process on both surfaces, as well as the emergence of preferential adsorption orientations for the Fg protomer. We also address the change in the flexibility of Fg upon adsorption as well as the possible onset of deformation/denaturation.

2 Simulation Methods

The simulations are based on the crystal structure of human Fg (PDB ID: 3GHG) [5]. The carbohydrate groups that are only partly resolved in the crystal have been modelled using VMD and introduced in some of the simulations. The unresolved parts of the protein structure (the αC domain and the N terminal segments of all the chains) have not been included in the calculations. Several molecular constructs have been prepared to assess the role of the different components of the Fg molecule. Rectangular periodic simulation boxes with explicit TIP3P water [36] and physiological ion concentration (150 mMol [NaCl]) were prepared using VMD [37] (see Tables 1 and 2 for box sizes of the solution and adsorption simulations, respectively).

Isobaric-isothermal simulations were set up at a temperature of 310 K and pressure of 1atm using NAMD [38] with a Langevin thermostat and a Langevin piston barostat [39, 40] using 200 and 100 ps^{-1} as decay time, respectively. The covalent bonds involving hydrogen atoms were fixed in length and a 2fs timestep was used. The CHARMM22 force field with CMAP corrections [41] was used with its recent extension to carbohydrates [42] in combination with ParamChem (http:/ www.paramchem.org) and the CHARMM generalized force field (CGenFF) [43]. The van der Waals forces were cut off at 1.2 nm while PME was used for long range electrostatic interactions with a grid spacing of 1 Å. After energy minimization (NAMD's conjugate gradient algorithm, 15,000 steps) of hydrogen atoms and water molecules, the systems were heated and equilibrated for 10 ns. Production runs statistics are given in Tables 1 and 2 for the solution and adsorption simulations, respectively.

In the case of the adsorption simulations, mica and graphite were chosen as model solid surfaces for adsorption. The graphite surface was built as a six layer graphene sheet using the Carbon Nanostructure Builder within the program VMD

Table 1 List of the simulations of Fg in solution

System	Initial box size [nm]	N[a]	Simulation time[b][ns]
Dimer			
Mono-glycosilated	13.27 × 48.59 × 12.70	788,173	77, 88
Unglycosilated	13.27 × 48.59 × 12.70	786,811	25, 20
Protomer			
Di-glycosilated	11.02 × 35.07 × 10.42	381,304	45, 21, 21
Mono-glycosilated	12.28 × 27.89 × 11.57	381,397	199, 188
Unglycosilated	12.28 × 27.89 × 11.57	380,169	135, 109, 100, 82, 51, 43, 30, 20, 20, 14

[a]Total number of particles in the system
[b]Each number indicates the time length of an independent simulation

Table 2 List of the simulations of Fg adsorption

System	Initial box size [nm]	N	Simulation time [ns]
Mica, 0		681,900	120, 120, 95, 94, 71, 70, 48, 48, 45, 43
Mica, 120	18.47 × 29.91 × 14.30	681,937	120, 96, 91, 90, 85, 70, 45, 45, 44, 43
Mica, 240		681,909	120, 100, 70, 70, 70, 46, 45, 41, 38, 30
Graphite, 0		440,893	51,48, 45, 20, 20, 18
Graphite, 120	12.18× 27.36 × 13.03	440,875	52, 50, 49, 49, 26, 19
Graphite, 240		440,698	51, 49, 48, 44, 26, 23

[37] and modeled using standard CHARMM aromatic carbon parameters. The mica surface was constructed according to a recently published model [44] which was successfully adopted to simulate peptide adsorption [45, 46]. The mica surface consists of a two-layer sheet with realistic surface defects. The defects are point defects where an aluminum atom substitutes a silicon atom. Potassium ions are evenly distributed on the two sides of the mica slab. The solid surfaces were constructed as being continuous in the xy-plane by defining covalent bonds that wrap around the periodic boundaries. Then, a 12 nm high water box was constructed on top of the solid surface using VMD. After the equilibration of the solvated surface box, the first protomer and residues α27–65, β58–95 and γ14–40 of the second protomer (see Fig. 2) were added to the simulation box. The protein was added in such a way that the minimal distance to the solid surface was at least 0.8 nm. Three different initial orientations (labeled 0, 120, 240) were constructed by rotating Fg around its long axis by 120°. This procedure limits the sampled space to so called side-on adsorption which is known to be the dominant adsorption mode, at least in the dilute regime [11, 33]. After this step, the surface systems were further minimized and equilibrated for 0.75 ns before starting the production runs listed in Table 2).

To identify the collective motions of the whole Fg molecule and of its subdomains we performed several principal component analyses (PCA) [47] using wordom [48] and GROMACS utilities [49]. DynDom [50] was used to identify rigid domains and hinges of motion. The overlap between spaces spanned by the dominant PCA modes of different simulations was used to quantify the similarity of the observed dynamics [51].

In the adsorption simulations the orintations of the D and E domain are investigated separately. To characterize the different adsorption orientations we

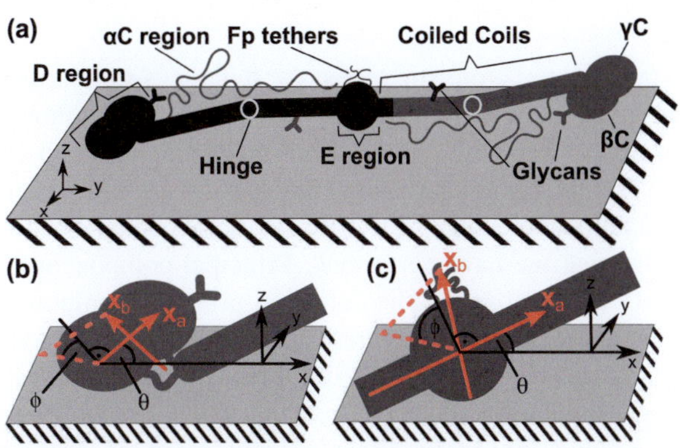

Fig. 2 (**a**) Schematic representation of Fg near a solid surface. The simulated part of the protein is colored in *black*. In (**b**) and (**c**) a close-up view of the D and E region, respectively, where the vectors used to characterize the orientation of regions with respect to the surface are indicated with *red arrows*. See main text for a detailed description (Reprinted with permission from Köhler et al. [60]. Copyright 2015 American Chemical Society)

defined an angle θ describing the tilting of a relevant axis of the region with respect to the surface and an angle ϕ describing the rotation around the identified axis (Fig. 2b, c). Both θ and ϕ are defined separately for the D and E regions. A contact between the protein and the solid surface is defined when a heavy protein atom comes closer than 0.5 nm to the surface. If such a contact is formed in the globular D and E regions and persists longer than 1 ns, we call this an adsorbed state. A contact between a given residue and the solid surface is called persistent if it forms in all sets of simulations, independent of the starting configuration. As a reference for an unbiased adsorption process, we also measured the fraction of heavy atoms on the protein surface contributed by each residue type: charged residues contributed 52 % of the surface atoms, polar uncharged 33 %, hydrophobic 11.5 % and the carbohydrate groups 3.5 %. Here protein atoms were considered as being on the protein surface if they were within 0.2 nm of water atoms. To detect spreading of the globular regions of the protein during an adsorption event, we monitored the "domain height", which we define as the z-component of the distance between the center of mass and that protein atom which is closest to the surface of the material.

The simulations have been carried out, in part, on Hornet/Hazelhen at the High performance computing center Stuttgart. The simulations were carried out using NAMD [38], which has been parallelized using MPI and can be specifically compiled for the XC40 architecture. The large classical MD systems studied here are particularly fit to the Cray XC40 architecture, as, thanks to the high performance interconnect, they scale well up to 4000 cores. The typical job used for these calculations involved 50–100 nodes and lasted for less than 3 h allowing to collect about 1 ns of trajectory per job, depending on system size.

3 Simulations of Fibrinogen Internal Dynamics

We have performed several atomistic molecular dynamics simulations of Fg, either in its full dimeric state or considering only one of the two symmetric protomers. In either case we have simulated glycosilated and unglycosilated constructs. The cumulative time length of the simulations reaches 1.3 μs with several continuous stretches of simulation reaching 0.2 μs. Fg undergoes large bending motions in all the simulations that we have performed. Principal component analysis (PCA) is used to quantify these motions. The dominant principal components of motions (PCA modes, Fig. 3a–c) of the Fg protomer are the same in all sampled trajectories as revealed by a large overlap between the three dominant modes ranging from 0.6 to 0.9 between simulation subsets. This means that neither glycosilzation, nor dimerization state play a fundamental role in determining the large scale motion of Fg. Because of the overlap of the largest PCA modes in the different simulations sets, the analysis presented here is done using all the available data merged together in a single set, which improves the statistical significance of the results. The first three PCA modes span the degrees of freedom associated with bending at a hinge point in the coiled-coil region (Fig. 3a–c), while the 4th PCA mode is related to a

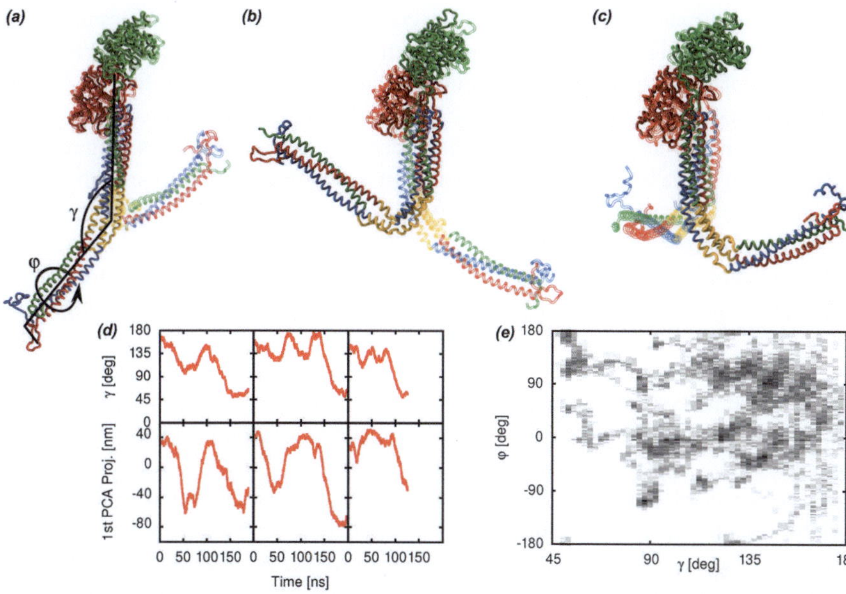

Fig. 3 Characterization of the large bending motions of fibrinogen. (**a**)–(**c**) Dominant PCA modes of the Fg protomer with the hinge region highlighted in *yellow* (chains colored according to Fig. 1). For each PCA mode, the two structures with the largest (*solid*) and smallest (*transparent*) projection along the PCA mode are represented. An illustration of the bending angle γ and the torsion angle φ is superimposed to the first PCA mode. The three groups of atoms used to define the γ angle are the E region (α50–58, β82–90, γ23–31), the hinge region (α99–110, β130–155, γ70–100) and the D region (β200–458, γ140–394). The four groups of atoms used to define the φ are one part of the E region (α50–58, γ23–31), another part of the E region (β82–90, γ23–31), the hinge region (α99–110, β130–155, γ70–100) and the D region (β200–458, γ140–394) (**d**) Time series of the γ angle and of the projection of the trajectories along the first PCA component from selected simulation runs. The plots show both the reversibility of the motion and the time scale along which it occurs. (**e**) Distribution of the bending angle γ and dihedral angle φ around the hinge of the Fg protomer as observed in the present simulations. The elongated conformation of Fg observed in the crystals correspond to a γ angle close to 160° (Reprinted from Ref. [6]. Copyright (2015) Köhler et al. under the terms of the Creative Commons Attribution License)

pure torsion of the coiled coil along its axis (not shown). The motions are reversible as shown by the time series of the PCA projections (Fig. 3d) Lower ranking PCA modes provide smaller contributions to the overall variance so they will not be analyzed further.

The program DynDom [50], applied to the extremal structures observed along the first PCA mode (Fig. 3a) of the Fg protomer, has been used to identify the regions of the molecule which are more rigid in our simulations, as well as the connecting hinge regions. DynDom reports the presence of two relatively rigid regions, separated by a hinge. The E region and the N-terminal part of the coiled-coil region represent one of the two rigid domains, while the C-terminal part of the coiled-coil region along with the D region represent the second. The hinge region is located approximately in the middle of the coiled-coil region and includes the break in the α-helical structure of the γ chain, which gives rise to a flexible loop (residues γ70–78), along with the neighbor residues on the Aα and Bβ chains (Fig. 3a–c). The break in the α-helical structure of the γ-chain is facilitated by two proline residues.

The bending around the identified hinge can be described by a bending angle γ and a torsion angle φ defined using groups of atoms from the E region, the hinge region and the D region (Fig. 3a). The γ and φ angles strongly correlate with the projections along the dominant PCA modes. Our simulation data show a consistent and significant bending occurring at the hinge region and reaching bending angles below 90° (Fig. 3e). The time it takes for the Fg structure to reach a bending angle below 110° from conformations similar to the crystal structure (bending angle above 150°) is 19 ± 1 ns along the trajectories, averaged over the 12 observed events (see Fig. 3d). The reverse process occurs twice in the simulations, taking 20 and 26 ns. The simulations of the full Fg dimer do not show significant correlations between the angle values observed at the two hinges.

Comparison of the crystallographic structures of Fg coiled-coil regions from various organisms already hinted at the presence of a flexible hinge [5]. This hypothesis is also supported by hydrogen-deuterium exchange experiments [52]. The latter are in good agreement with our simulations: amino acids from the coiled-coil region with lower helical probability in the simulations (Fig. 4b) correspond to amino acids with low protection factors in the experiments. The hinge is positioned on the non-helical segment of the γ chain (γ70–78), most probably due to the resulting reduction in the stiffness of the coiled coil. This segment is non-helical also in the other crystallized Fg structures [14, 15]. In addition, this segment has markedly helix-breaking features in most of the available Fg sequences from vertebrates that we have analyzed, showing a large density of proline and glycine residues as well as high probability to be a disordered/hot loop as revealed by the program DisEMBL [53] (Fig. 4a). This analysis supports the idea that the non-helical segment of the γ chain provides a function that is strongly conserved across vertebrates possibly linked to the bending motion of the coiled-coil region. Besides providing flexibility to the individual Fg molecules as well as the fibrin fibers [5], the bending at the hinge may help expose the plasmin cleavage sites located nearby on the coiled-coil region [13]. Our simulations strongly support this hypothesis showing that the α-helical structure around the plasmin cleavage sites Aα104–105 and Bβ133–134 is partly disrupted by the bending motions, and the exposure to the solvent of the involved peptide bonds increases (Fig. 4c, d). The twisting of fibrin fibers [54] compresses molecules in the center of the fiber and stretches them on the perimeter. If the hinge bending is necessary to accommodate such deformation, it is reasonable to believe that the bending motions at the hinge may actually be reduced by tension applied along the fiber axis. Thus, fibrinolysis assisted by the bending motions at the hinge may selectively take place on fibrin molecules subject to reduced tension. This hypothesis is supported by experimental evidence indicating reduced plasmin fibrinolytic effectiveness on fibrin fibers subject to mechanical tension [55].

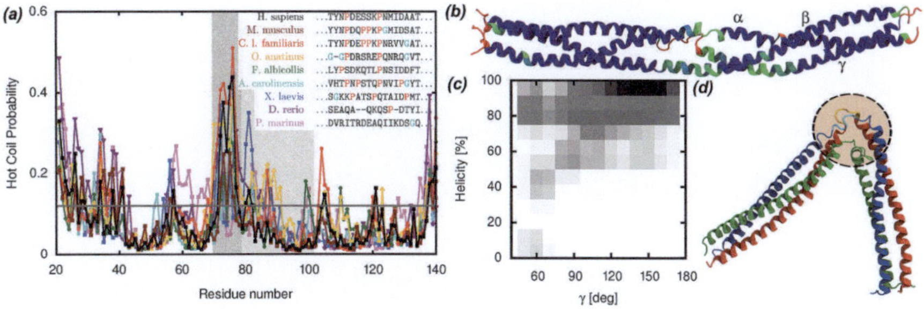

Fig. 4 Functional role of the bending motions in the coiled-coil region of fibrinogen. (**a**) DisEMBL "hot coil" predictions for the sequences of the γ chain from several vertebrates, highlighting the fact that the flexibility of the non-helical loop is a conserved feature. The hinge region is shaded and, within that region, the non-helical loop segment is *dark shaded*. The *inset* legend reports the sequence alignment of the non-helical loop region across the same vertebrates, highlighting the content of glycine and proline residues. (**b**) Cartoon representation of the coiled-coil region of Fg colored according to the fraction of the simulation time spent in an α-helical conformation (*red* = 0, *green* = 0.85, *blue* = 1). The N-termini of the segments are on the left. The regions with lower helical fraction are in good agreement with regions with lower protection factors as determined in H/D exchange experiments [52]. (**c**) Probability distribution of the fraction of helical residues around the Aα104–105 and Bβ133–134 plasmin cleavage sites as a function of the bending angle γ. *Dark shades* correspond to high probability. The three residues preceding and following the cleavage sites (i.e., Aα102–107 and Bβ131–136) have been included in the calculation of the helicity. Larger bending (lower γ angle) correlates with lower helical content. (**d**) Snapshot of the conformation of the bent coiled-coil region (chains colored as in Fig. 1) showing the disrupted secondary structure around the plasmin cleavage sites (rendered *yellow* and *cyan* inside the *dashed circle*) (Reprinted from Ref. [6]. Copyright (2015) Köhler et al. under the terms of the Creative Commons Attribution License)

4 Simulations of Fibrinogen Adsorption

Mica The dominant large scale motions of Fg on the mica surface (as identified using PCA), are bending motions at the hinge, which closely resemble those previously described for Fg in Sect. 3. A large essential dynamics (ED) overlap (0.69–0.71) is observed between the three largest PCA modes measured in the different sets of simulations. The overlap with previously reported solution simulations is also high (0.63). Furthermore the sampled distribution of hinge conformations (Fig. 5a), as well as the observed bending time of (16 $\pm$ 6)ns are in reasonable agreement with the corresponding results in solution. In some instances hinge bending coupled with protein-surface contact formation can lead to sliding and rolling motions on the surface. In a previous simulation of the γC domain [29] of Fg a rolling motion has been observed. To our knowledge the sliding motion has not been observed previously on this system.

In simulations on mica, the total number of contacts formed with the surface reaches a plateau at $\approx$15 contacts (average over the simulations) after about 30 ns (Fig. 6a). The fraction of contacts formed by the different types of residues

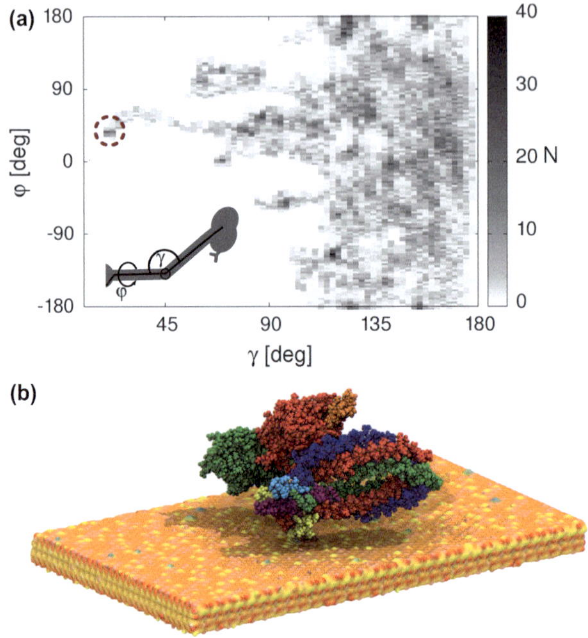

Fig. 5 (**a**) Distribution of the hinge bending and dihedral angle at a mica surface. The definition of the angles is shown in the *inset*. (**b**) Maximally bent state of Fg at the mica surface (highlighted in (**a**)). A collision of the D and E regions prevents further bending at the hinge. The Aα, Bβ and γ chains from the whole protomer are rendered in *blue*, *red* and *green*, respectively, the carbohydrates in orange. For clarity the solvent is not shown (Reprinted with permission from Köhler et al. [60]. Copyright 2015 American Chemical Society)

resembles essentially the one expected from the surface distribution of residues – the only exception being that positively charged residues contribute more than would be expected while the contribution of negatively charged and polar residues is slightly lower than expected. This phenomenon is explained by the negatively charged nature of the mica surface. Another significant feature of the simulations is the observation of frequent desorption and reorientation events. On average a globular region was only adsorbed for (14 ± 3)ns before leaving the surface again. We observed a total of 51 adsorption events in the D region and 45 in the E region. The great flexibility provided by the hinge allow us to treat these events separately. Similarly, the adsorption orientation of the D- and E-regions can be analyzed separately. Several different adsorption orientations have been identified for both globular regions by dividing the space of the adsorption angles into a small number of adsorption orientation states (Fig. 7a–b). For all simulated systems, the adsorption orientation states of the D- and E-regions overlap significantly (Table 3), although a bias towards the initial orientation state is certainly visible. These data clearly indicate that transitions from one orientation state to another occur frequently. We observed 68 reorientation events (changes in the adsorption orientation state) for the

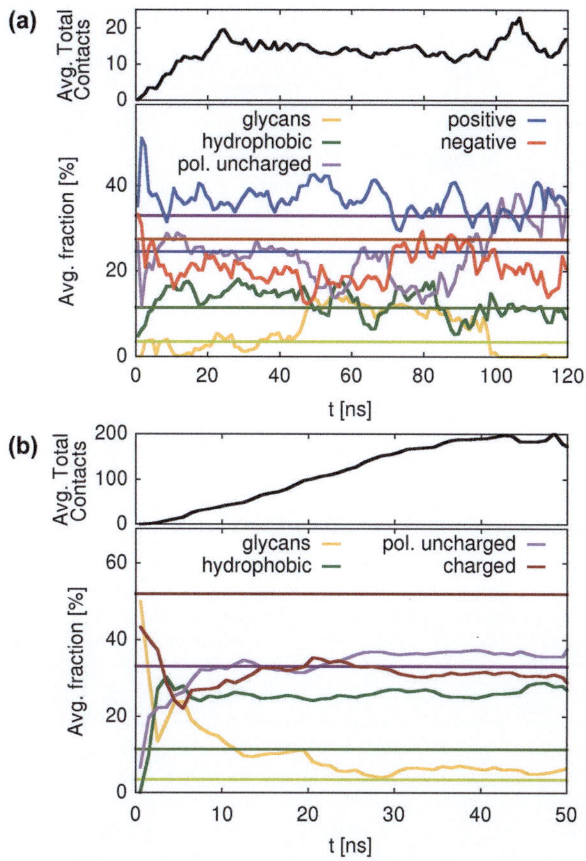

Fig. 6 Average total number of contacts and fraction contributed by the different types of residues during adsorption on (**a**) mica and (**b**) graphite. The straight *horizontal darkened lines* represent the expected fractions according to the exposed surface area in the crystal structure (after equilibration) (Reprinted with permission from Köhler et al. [60]. Copyright 2015 American Chemical Society)

D-region and 22 for the E-region. Neglecting trajectories where the D- (E-) region never contacted the surface this gives an average time of 27 ns (14 ns) between reorientation events.

More specifically, the E region shows three distinct adsorption orientation states. The orientation E1 is significantly populated in all sets of simulations. In cases where the simulation starts with this orientation it never leaves it, while simulations starting with the other orientations often show reorientation towards E1. These data support the idea that E1 is a preferred adsorption orientation. In the E1 orientation the flexible Fp tethers point toward the surface. The presence of many charged residues in this region likely explains the preference for this adsorption orientation. In simulations starting from orientation E1 (Mica/240), 60 % of the Fg-surface contacts are provided by the E region. In Mica/0 and Mica/120 simulations, these

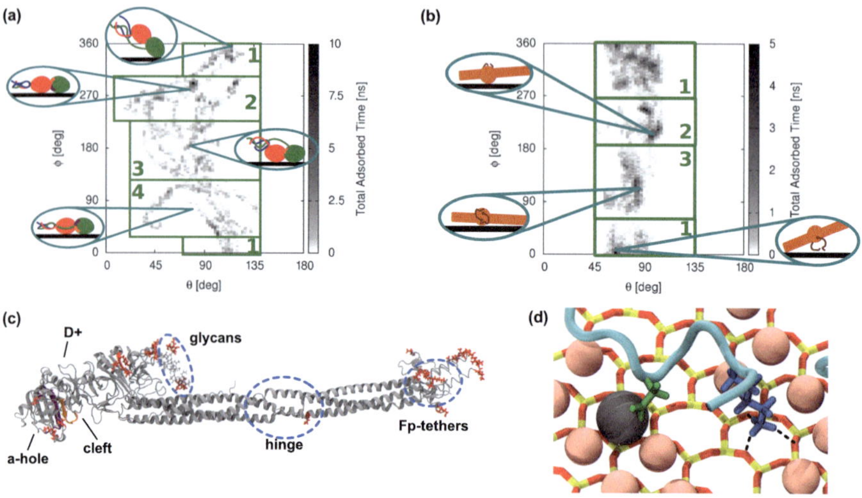

Fig. 7 (**a**), (**b**) Distribution of the adsorption angles for the D- and E-region, respectively, and schematic representation of the corresponding adsorption orientation states. Chains are color coded as in Fig. 5. The *orange patch* between the βC and γC domain indicates the binding cleft while the *purple region* identifies the P2 and H12 binding sites. In orientation D4 the P2 and H12 binding sites face away from the surface and are available for binding. (**c**) The residues α27–28,α38,α92,β345–348, β361–363, β365, γ323, γ356–357), γ361 and carbohydrates 479–480 from the first whole protomer and α27–30, α37–38, α63–65, β91 and γ38–40 from the second truncated protomer form persistent contacts on mica regardless of initial orientation (*red licorice*). The carbohydrate cluster (glycans) attached to the βC domain is shown in *grey licorice*, the P1 site is rendered in *orange* and the P2 and H12 sites in *purple*. (**d**) Example snapshot of a pair of oppositely charged amino acids (lysine in *blue*, aspartic acid in *green*) anchoring the E region in an Fp-down orientation. The aspartic acid interacts with a sodium ion (*gray*) that has replaced potassium (*pink*) from the counter ion layer. Hydrogen bonds between the lysine and the silicate ring are indicated by the *black dotted lines*. For clarity only the topmost atoms of the mica surface are shown (Reprinted with permission from Köhler et al. [60]. Copyright 2015 American Chemical Society)

Table 3 Fraction of the time spent in each of the adsorption orientation states defined in Fig. 7a ,b. States containing the initial orientation are in bold

System	D1	D2	D3	D4	E1	E2	E3
Mica, 0	**33 %**	28 %	5 %	34 %	46 %	**44 %**	10 %
Mica, 120	0 %	**60 %**	37 %	3 %	13 %	28 %	**59 %**
Mica, 240	8 %	0 %	**42 %**	50 %	**100 %**	0 %	0 %
All	16 %	32 %	25 %	27 %	56 %	21 %	23 %

numbers are significantly lower (30 % and 19 %), further supporting that E1 is indeed the preferred adsorption state.

The D-region shows several populated adsorption orientation states. In this case, a preference for the orientation D4 is detectable. D4 is observed in all sets of simulations, although no set of simulation starts from there. This orientation corresponds to an adsorbed state where the binding cleft between γC- and βC-domain (containing part of the P1 site) is facing the surface, while the sites P2 and H12 face the solution. This may have consequences on the accessibility of the integrin binding sites upon adsorption on mica. Indeed, the preferential exposure of the P2 and H12 binding sites may explain the integrin-mediated adhesion forces measured in recent AFM experiments of leukocytes [56] and platelets [10, 57] on fibrinogen-coated mica surfaces.

Further support for the presence of preferred adsorption orientations comes from the identification of persistent contacts, i.e., contacts that are observed in all sets of simulations (Fig. 7c). The position of the residues involved in persistent contacts clearly highlights a preferred adsorption orientation of the D region. All persistent contacts in the D region are located on the side expected to contact the surface in orientation D4. Two persistent contacts are found in the carbohydrate cluster. This cluster is mobile enough to form these contacts in the orientations D2–4. The persistent contacts in the E region are mostly located in the Fp-tethers, thus corresponding to a E1 adsorption orientation. The truncated ends of the second protomer also form persistent contacts, which are likely artefacts due to the truncation of the sequence. The charged patches on the D region contribute some of the persistent contacts with the mica surface, however, most of the persistent contacts are formed in regions where the side chains of positively and negatively charged residues are in close proximity. Further investigation shows that opposite-charge pairs provide an ideal balance for the negatively charged mica surface with its positive counter ion layer. The negative side chain can trap ions from the counter ion layer, while the positive side chain interacts directly with the surface. The interactions of lysine with mica are particularly favorable as the positive charge of the amine is attracted to the negative charge of silicate rings while it can also form hydrogen bonds with the oxygen atoms (see Fig. 7d). Visual inspection of the simulations shows that a single lysine contact can be enough to anchor the E-region. The strong interaction of lysine with silica surfaces has already been reported in previous simulations [58]. The same mechanism has also been observed during the adsorption of the Fg γC domain on charged self assembled monolayers [29]. Given the interaction pattern involving pairs of oppositely charged residues identified above, the preference for the D4 orientation could also be related to the number of such pairs facing the surface in orientation D4, which is larger than on the opposite side (34 pairs versus 26).

Graphite As for mica, the hinge behavior of Fg at a graphite surface is consistent with the behavior in the solution simulations. The overlap between the PCA modes in the three orientations is slightly lower than for mica (0.65 overlap). Comparing the behavior at graphite with the one observed in the solution simulations, we find that the bending has similar overlap (0.65 overlap). The same holds for the angle distribution (not shown) and the bending time of (17 ± 5)ns.

The adsorbed orientations can be grouped into the same classes as on mica. However, almost no trajectory samples multiple orientations. Only one trajectory shows evidence of a reorientation event. The orientations of adsorbed molecules are mostly determined by the starting orientations at the beginning of the simulations. All observed deviations are due to reorientations occurring before adsorption. Thus, we cannot discuss desorption or reorientation times in this case.

We have observed 13 adsorption events of the D region and 11 for the E region. In only three of these cases was a desorption observed and in only one case did the desorption occur after adsorption events lasting more than 1 ns. Very few reorientations were observed on graphite. This supports the idea that graphite is "stickier" than mica for Fg. The hinge bending can provide the impetus to bring the globular regions of Fg close to the surface. In the case of graphite this leads to almost irreversible adsorption events. Thus a bending event not occurring in a plane parallel to the surface will lead to irreversible contact formation and hinder further bending motions, which eventually influences the sampling of the hinge angles and the bending modes. The persistent contacts are limited to an isolated residue in the disordered part of the hinge region, the carbohydrate group and the flexible loops of the a-hole (not shown). Since the a-hole is located at the tip of the D-region, it can be brought into contact with the surface from several initial orientations through hinge motions. The existence of persistent contacts in this region is thus more likely a geometric effect than the result of specific interactions. No persistent contacts are formed in the E-region.

The sticky nature of the graphite surface is also supported by the large number of contacts formed between Fg and the graphite surface (Fig. 6b). In fact, the number of heavy atoms in contact with the surface is almost an order of magnitude larger than in the case of mica. It is also interesting to note that the distribution of contacts markedly differs from that on mica. As expected, the fraction of hydrophobic contacts between the protein and the graphite surface is larger than their fraction on the solvent exposed surface area. In turn, the contribution of the charged residues to contacts with the graphite surface is reduced. In fact, the overall distribution of the contacting residues is closer to the distribution of residues in the whole protein (41 % polar, 26 % charged, 33 % hydrophobic), not just the protein surface. About half of the aromatic residues involved in contacts with the surface form $\pi - \pi$ interactions, in agreement with previous simulation data for the adsorption of proteins on graphene [27, 28].

A pronounced flattening of the adsorbed domains is observed upon adsorption on graphite (Fig. 8). This is likely the onset of denaturation, which is known to occur if Fg adsorbs to hydrophobic surfaces [59]. The denaturation of globular domains leads to their spreading on the surface. As an indication of this we monitored the change in domain height during individual adsorption events. In mica, this difference is centered around zero, indicating that domains do not spread (Fig. 8a). On graphite, however, changes are more pronounced and lead systematically to a reduction in domain height (Fig. 8b). This trend is in line with experimental findings according to which the domain height changes very little if Fg is adsorbed on mica, but is reduced on graphite [20]. It should be mentioned that changes in the domain height on mica can also occur as a result of rolling of the non spherical domains. The electric field on the charged mica surface is unable to induce unfolding during the simulation time, mostly because it is neutralized by the counterion layers [22].

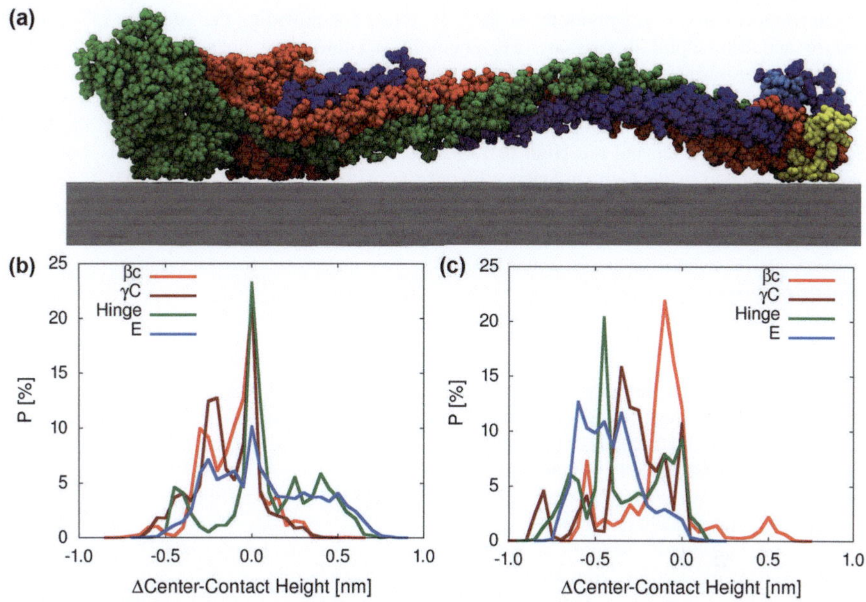

Fig. 8 (**a**) The adsorbed conformation of Fg on graphite shows a noticeable flattening of the domains. Coloring as in Fig. 5b. Histogram of the change in domain height for (**b**) mica and (**c**) graphite (Reprinted with permission from Köhler et al. [60]. Copyright 2015 American Chemical Society)

5 Conclusions

We have used molecular dynamics simulations to investigate the dynamics of Fg and the initial adsorption stages of Fg on mica and graphite surfaces. All sets of simulations, show a hinge bending mechanism which is possibly conserved across vertebrates and may facilitate fibrinolysis. The adsorption simulations on mica show a reversible process which does not encompass large deformations of the protein. They also reveal the presence of a preferred adsorption orientation, in agreement with our proposed model of Fg adsorption [6] which was fitted to experimental data. The adsorption simulations on graphite have an irrevesible character and show the formation of a large quantity of protein-surface contacts which eventually lead to deformations of the protein and the initiation of spreading in agreement with experiments [20].

Note Some parts of this report have been reprinted (adapted) with permission from Ref. [6]. Copyright (2015) Köhler et al. under the terms of the Creative Commons Attribution License. Other parts of this report have been reprinted (adapted) with permission from Ref. [60]. Copyright (2015) American Chemical Society.

Acknowledgements The authors thank Prof. H. Heinz for providing the structure of the mica surface and for helpful discussions. SK gratefully acknowledges financial support from the Graduate School Materials Science in Mainz. GS gratefully acknowledges financial support from the Max-Planck Graduate Center with the University of Mainz. We gratefully acknowledge support with computing time from the HPC facility Mogon at the university of Mainz, the Jülich Supercomputing Center and the High performance computing center Stuttgart. This work was partially supported by the German Science Foundation within SFB 1066 (project Q1).

References

1. Weisel, J.W.: J. Thromb. Haemost. **5**(Suppl 1), 116 (2007). doi:10.1111/j.1538-7836.2007.02504.x. http://dx.doi.org/10.1111/j.1538-7836.2007.02504.x
2. Takagi, T., Doolittle, R.F.: Biochemistry **14**(23), 5149 (1975). doi:10.1021/bi00694a020. http://dx.doi.org/10.1021/bi00694a020
3. Takagi, T., Doolittle, R.F.: Biochemistry **14**(5), 940 (1975)
4. Mihalyi, E.: Ann. N. Y. Acad. Sci. **408**(1), 60 (1983). doi:10.1111/j.1749-6632.1983.tb23234.x. http://dx.doi.org/10.1111/j.1749-6632.1983.tb23234.x
5. Kollman, J., Pandi, L., Sawaya, M., Riley, M., Doolittle, R.: Biochemistry **48**(18), 3877 (2009). doi:10.1021/bi802205g. http://pubs.acs.org/doi/abs/10.1021/bi802205g
6. Köhler, S., Schmid, F., Settanni, G.: PLoS Comput. Biol. **11**(9), e1004346 (2015). doi:10.1111/j.1538-7836.2007.02504.x. doi:10.1111/j.1538-7836.2007.02504.x
7. Altieri, D.C., Plescia, J., Plow, E.F.: J. Biol. Chem. **268**(3), 1847 (1993)
8. Ugarova, T.P., Solovjov, D.A., Zhang, L., Loukinov, D.I., Yee, V.C., Medved, L.V., Plow, E.F.: J. Biol. Chem. **273**(35), 22519 (1998)
9. Kloczewiak, M., Timmons, S., Hawiger, J.: Biochem. Biophys. Res. Commun. **107**(1), 181 (1982)
10. Soman, P., Rice, Z., Siedlecki, C.A.: Langmuir **24**(16), 8801 (2008). doi:10.1021/la801227e. http://pubs.acs.org/doi/abs/10.1021/la801227e
11. Yermolenko, I.S., Lishko, V.K., Ugarova, T.P., Magonov, S.N.: Biomacromolecules **12**(2), 370 (2011). doi:10.1021/bm101122g. http://pubs.acs.org/doi/abs/10.1021/bm101122g
12. Protopopova, A.D., Barinov, N.A., Zavyalova, E.G., Kopylov, A.M., Sergienko, V.I., Klinov, D.V.: J. Thromb. Haemost. **13**(4), 570 (2015). doi:10.1111/jth.12785. http://dx.doi.org/10.1111/jth.12785
13. Doolittle, R.F., Goldbaum, D.M., Doolittle, L.R.: J. Mol. Biol. **120**(2), 311 (1978). doi:http://dx.doi.org/10.1111/j.1749-6632.1983.tb23234.x. http://www.sciencedirect.com/science/article/pii/0022283678900700
14. Brown, J.H., Volkmann, N., Jun, G., Henschen-Edman, A.H., Cohen, C.: Proc. Natl. Acad. Sci. U. S. A. **97**(1), 85 (2000). doi:10.1073/pnas.97.1.85. http://www.pnas.org/content/97/1/85.abstract
15. Yang, Z., Kollman, J.M., Pandi, L., Doolittle, R.F.: Biochemistry **40**(42), 12515 (2001)
16. Beijbom, L., Larsson, U., Kaveus, U., Hebert, H.: J. Ultrastruct. Mol. Struct. Res. **98**(3), 312 (1988). doi:10.1016/S0889-1605(88)80923-6. http://www.sciencedirect.com/science/article/pii/S0889160588809236
17. Marchin, K.L., Berrie, C.L.: Langmuir **19**(23), 9883 (2003). doi:10.1021/la035127r. http://pubs.acs.org/doi/abs/10.1021/la035127r
18. Tunc, S., Maitz, M.F., Steiner, G., Vazquez, L., Pham, M.T., Salzer, R.: Colloids Surf. B **42**(3–4), 219 (2005). doi:10.1016/j.colsurfb.2005.03.004. http://www.sciencedirect.com/science/article/pii/S0927776505000986
19. Sit, P., Marchant, R.E.: Surf. Sci. **491**(3), 421 (2001). doi:10.1016/S0039-6028(01)01308-5. http://www.sciencedirect.com/science/article/pii/S0039602801013085

20. Agnihotri, A., Siedlecki, C.A.: Langmuir **20**(20), 8846 (2004). doi:10.1021/la049239+. http://pubs.acs.org/doi/abs/10.1021/la049239%2B
21. Heinz, H.: J. Comput. Chem. **31**(7), 1564 (2010). doi:10.1002/jcc.21421. http://dx.doi.org/10.1002/jcc.21421
22. Starzyk, A., Cieplak, M.: J. Chem. Phys. **139**(4), 045102 (2013). doi:10.1063/1.4813854. http://dx.doi.org/10.1063/1.4813854
23. Kubiak-Ossowska, K., Burley, G., Patwardhan, S.V., Mulheran, P.A.: J. Phys. Chem. B **117**(47), 14666 (2013). doi:10.1021/jp409130s. http://dx.doi.org/10.1021/jp409130s
24. Raffaini, G., Ganazzoli, F.: Langmuir **19**(8), 3403 (2003). doi:10.1021/la026853h. http://pubs.acs.org/doi/abs/10.1021/la026853h
25. Utesch, T., Daminelli, G., Mroginski, M.A.: Langmuir **27**(21), 13144 (2011). doi:10.1021/la202489w. http://pubs.acs.org/doi/abs/10.1021/la202489w
26. Kang, S.G., Huynh, T., Xia, Z., Zhang, Y., Fang, H., Wei, G., Zhou, R.: J. Am. Chem. Soc. **135**(8), 3150 (2013). doi:10.1021/ja310989u. http://pubs.acs.org/doi/abs/10.1021/ja310989u
27. Baweja, L., Balamurugan, K., Subramanian, V., Dhawan, A.: Langmuir **29**(46), 14230 (2013). doi:10.1021/la4033805. http://dx.doi.org/10.1021/la4033805
28. Chong, Y., Ge, C., Yang, Z., Garate, J.A., Gu, Z., Weber, J.K., Liu, J., Zhou, R.: ACS Nano **9**(6), 5713 (2015). doi:10.1021/nn5066606. http://dx.doi.org/10.1021/nn5066606
29. Agashe, M., Raut, V., Stuart, S.J., Latour, R.A.: Langmuir **21**(3), 1103 (2005). doi:10.1021/la0478346. http://pubs.acs.org/doi/abs/10.1021/la0478346
30. Lim, B.B., Lee, E.H., Sotomayor, M., Schulten, K.: Structure **16**(3), 449 (2008). doi:10.1016/j.str.2007.12.019. http://www.sciencedirect.com/science/article/pii/S0969212608000476
31. Zhmurov, A., Brown, A.E., Litvinov, R.I., Dima, R.I., Weisel, J.W., Barsegov, V.: Structure **19**(11), 1615 (2011). doi:10.1016/j.str.2011.08.013
32. Adamczyk, Z., Barbasz, J., Cieśla, M.: Langmuir **26**(14), 11934 (2010). doi:10.1021/la101261f. http://pubs.acs.org/doi/abs/10.1021/la101261f
33. Adamczyk, Z., Barbasz, J., Cieśla, M.: Langmuir **27**(11), 6868 (2011). doi:10.1021/la200798d. http://pubs.acs.org/doi/abs/10.1021/la200798d
34. Vilaseca, P., Dawson, K.A., Franzese, G.: Soft Matter **9**, 6978 (2013). doi:10.1039/C3SM50220A. http://dx.doi.org/10.1039/C3SM50220A
35. Rocco, M., Molteni, M., Ponassi, M., Giachi, G., Frediani, M., Koutsioubas, A., Profumo, A., Trevarin, D., Cardinali, B., Vachette, P., Ferri, F., Prez, J.: J. Am. Chem. Soc. **136**(14), 5376 (2014). doi:10.1021/ja5002955. http://dx.doi.org/10.1021/ja5002955
36. Jorgensen, W.L., Chandrasekhar, J., Madura, J.D., Impey, R.W., Klein, M.L.: J. Chem. Phys. **79**(2), 926 (1983). doi:10.1063/1.445869. http://link.aip.org/link/?JCP/79/926/1
37. Humphrey, W., Dalke, A., Schulten, K.: J. Mol. Graph. **14**, 33 (1996)
38. Phillips, J.C., Braun, R., Wang, W., Gumbart, J., Villa, E., Chipot, C., Skeel, R.D., Kale, L., Schulten, K.: J. Comput. Chem. **26**, 1781 (2005)
39. Martyna, G.J., Tobias, D.J., Klein, M.L.: J. Chem. Phys. **101**(5), 4177 (1994). doi:10.1063/1.467468. http://link.aip.org/link/?JCP/101/4177/1
40. Feller, S.E., Zhang, Y., Pastor, R.W., Brooks, B.R.: J. Chem. Phys. **103**(11), 4613 (1995). doi:10.1063/1.470648. http://link.aip.org/link/?JCP/103/4613/1
41. Mackerell, A.D., Feig, M., Brooks, C.L.: J. Comput. Chem. **25**(11), 1400 (2004). doi:10.1002/jcc.20065. http://dx.doi.org/10.1002/jcc.20065
42. Guvench, O., Mallajosyula, S.S., Raman, E.P., Hatcher, E., Vanommeslaeghe, K., Foster, T.J., Jamison, F.W., MacKerell, A.D.: J. Chem. Theory Comput. **7**(10), 3162 (2011). doi:10.1021/ct200328p. http://pubs.acs.org/doi/abs/10.1021/ct200328p
43. Vanommeslaeghe, K., Hatcher, E., Acharya, C., Kundu, S., Zhong, S., Shim, J., Darian, E., Guvench, O., Lopes, P., Vorobyov, I., Mackerell, A.D.: J. Comput. Chem. **31**(4), 671 (2010). doi:10.1002/jcc.21367. http://dx.doi.org/10.1002/jcc.21367
44. Heinz, H., Koerner, H., Anderson, K.L., Vaia, R.A., Farmer, B.L.: Chem. Mater. **17**(23), 5658 (2005). doi:10.1021/cm0509328. http://pubs.acs.org/doi/abs/10.1021/cm0509328
45. Bertran, O., Curcó, D., Zanuy, D., Alemán, C.: Faraday Discuss **166**, 59 (2013)

46. Maity, S., Zanuy, D., Razvag, Y., Das, P., Alemn, C., Reches, M.: Phys. Chem. Chem. Phys. **17**(23), 15305 (2015). doi:10.1039/c5cp00088b. http://dx.doi.org/10.1039/c5cp00088b
47. Kitao, A., Hirata, F., Go, N.: Chem. Phys. **158**, 447 (1991). doi:http://dx.doi.org/10.1016/0301-0104(91)87082-7. http://www.sciencedirect.com/science/article/pii/0301010491870827
48. Seeber, M., Cecchini, M., Rao, F., Settanni, G., Caflisch, A.: Bioinformatics **23**(19), 2625 (2007)
49. Spoel, D.V.D., Lindahl, E., Hess, B., Groenhof, G., Mark, A.E., Berendsen, H.J.C.: J. Comput. Chem. **26**(16), 1701 (2005). doi:10.1002/jcc.20291. http://dx.doi.org/10.1002/jcc.20291
50. Poornam, G.P., Matsumoto, A., Ishida, H., Hayward, S.: Proteins: Struct. Funct. Bioinf. **76**(1), 201 (2009). doi:10.1002/prot.22339. http://dx.doi.org/10.1002/prot.22339
51. Hess, B.: Phys. Rev. E. **62**, 8438 (2000). doi:10.1103/PhysRevE.62.8438. http://link.aps.org/doi/10.1103/PhysRevE.62.8438
52. Marsh, J.J., Guan, H.S., Li, S., Chiles, P.G., Tran, D., Morris, T.A.: Biochemistry **52**(32), 5491 (2013). doi:10.1021/bi4007995. http://pubs.acs.org/doi/abs/10.1021/bi4007995
53. Linding, R., Jensen, L.J., Diella, F., Bork, P., Gibson, T.J., Russell, R.B.: Structure **11**(11), 1453 (2003). doi:http://dx.doi.org/10.1016/j.str.2003.10.002. http://www.sciencedirect.com/science/article/pii/S0969212603002351
54. Weisel, J.W., Nagaswami, C., Makowski, L.: Proc. Natl. Acad. Sci. U. S. A. **84**(24), 8991 (1987)
55. Varj, I., Stonyi, P., Machovich, R., Szab, L., Tenekedjiev, K., Silva, M.M.C.G., Longstaff, C., Kolev, K.: J. Thromb. Haemost. **9**(5), 979 (2011). doi:10.1111/j.1538-7836.2011.04203.x. http://dx.doi.org/10.1111/j.1538-7836.2011.04203.x
56. Yermolenko, I.S., Fuhrmann, A., Magonov, S.N., Lishko, V.K., Oshkadyerov, S.P., Ros, R., Ugarova, T.P.: Langmuir **26**(22), 17269 (2010). doi:10.1021/la101791r. http://pubs.acs.org/doi/abs/10.1021/la101791r
57. Podolnikova, N.P., Yermolenko, I.S., Fuhrmann, A., Lishko, V.K., Magonov, S., Bowen, B., Enderlein, J., Podolnikov, A.V., Ros, R., Ugarova, T.P.: Biochemistry **49**(1), 68 (2010). doi:10.1021/bi9016022. http://pubs.acs.org/doi/abs/10.1021/bi9016022
58. Patwardhan, S.V., Emami, F.S., Berry, R.J., Jones, S.E., Naik, R.R., Deschaume, O., Heinz, H., Perry, C.C.: J. Am. Chem. Soc. **134**(14), 6244 (2012). doi:10.1021/ja211307u. http://pubs.acs.org/doi/abs/10.1021/ja211307u
59. Sivaraman, B., Latour, R.A.: Biomaterials **31**(5), 832 (2010). doi:10.1016/j.biomaterials.2009.10.008. http://dx.doi.org/10.1016/j.biomaterials.2009.10.008
60. Köhler, S., Schmid, F., Settanni, G.: Langmuir **31**(48), 13180–13190 (2015). doi:10.1021/acs.langmuir.5b03371. PMID: 26569042. http://dx.doi.org/10.1021/acs.langmuir.5b03371.

Vorticity, Variance, and the Vigor of Many-Body Phenomena in Ultracold Quantum Systems: MCTDHB and MCTDH-X

Ofir E. Alon, Raphael Beinke, Lorenz S. Cederbaum, Matthew J. Edmonds, Elke Fasshauer, Mark A. Kasevich, Shachar Klaiman, Axel U.J. Lode, Nick G. Parker, Kaspar Sakmann, Marios C. Tsatsos, and Alexej I. Streltsov

Abstract During the past year of the MCTDHB project at the HLRS, we continued to strive and conquest further applications, developments, and expansion of the MultiConfigurational Time-Dependent Hartree for Bosons (MCTDHB) method in the context of ultracold atomic systems. We also announce the MCTDH-X package, the Multiconfigurational Time-Dependent Hartree for Indistinguishable Particles X package, which is able to treat identical bosons and fermions, with or without spin/internal degrees of freedom, alike. Here we report on a plethora of results and versatile applications which include: (i) single-shot imaging of fluctuating vortices in a fragmented Bose-Einstein condensate (BEC); (ii) the many-body

O.E. Alon
Department of Physics, University of Haifa at Oranim, 36006, Tivon, Israel

R. Beinke • L.S. Cederbaum • S. Klaiman • A.I. Streltsov (✉)
Theoretische Chemie, Physikalisch-Chemisches Institut, Universität Heidelberg, Im Neuenheimer Feld 229, D-69120, Heidelberg, Germany
e-mail: Alexej.Streltsov@pci.uni-heidelberg.de

M.J. Edmonds • N.G. Parker
Joint Quantum Centre (JQC) Durham-Newcastle, School of Mathematics and Statistics, Newcastle University, NE1 7RU, Newcastle upon Tyne, England, UK

E. Fasshauer
Department of Chemistry, University of Tromsø – The Arctic University of Norway, Centre for Theoretical and Computational Chemistry, N-9037, Tromsø, Norway

M.A. Kasevich
Department of Physics, Stanford University, 94305, Stanford, CA, USA

A.U.J. Lode
Department of Physics, University of Basel, Klingelbergstrasse 82, CH-4056, Basel, Switzerland

K. Sakmann
Vienna Center for Quantum Science and Technology, Atominstitut TU Wien, Stadionallee 2, 1020, Vienna, Austria

M.C. Tsatsos
Instituto de Física de São Carlos, Universidade de São Paulo, Caixa Postal 369, 13560-970, São Carlos, São Paulo, Brazil

W.E. Nagel et al. (eds.), *High Performance Computing in Science and Engineering '16*, DOI 10.1007/978-3-319-47066-5_6

tunneling and fragmetnation of vortices in 2D trapped BECs; (iii) the transition from vortices to solitonic vortices in 2D trapped BECs; (iv) the variance of a many-particle system being very sensitive to correlations even in the infinite-particle limit; (v) the consequences of the latter on the out-of-equilibrium uncertainty product of an evolving BEC; (vi) the mechanism of tunneling to open space of a few interacting polarized fermions; and (vii) composite fragmentation of multi-components BECs (i.e., with internal degrees of freedom). These are all exciting results made throughout the allocation of computer time by the HLRS to the MCTDHB project. Finally, further perspectives and future research plans are briefly discussed.

1 Introductory Remarks

During the past 10 years, the Multiconfigurational Time-Dependent Hartree for Bosons (MCTDHB) method [1–8], designed to solve efficiently the many-particle time-dependent Shrödinger equation of interacting bosons, has led to many results and developments, primarily in the many-body physics of ultracold trapped Bose-Einstein condensates. There are now a couple of software packages that has implemented the MCTDHB method very efficiently [9–12]. We have been proud to report on these achievements, many of them were made possible with the generous allocation of high-performance computer resources by the HLRS to the MCTDHB project, in the reports of the previous years [13–15]. We aim at continuing the above tradition in the report of the present year. In what follows we summarize our research works reported in Refs. [16–22]. Within our report, we advertise the MCTDH-X package, the Multiconfigurational Time-Dependent Hartree for Indistinguishable Particles X package, which is able to treat the many-body dynamics of identical bosons and fermions alike, whether with or without spin/internal degrees of freedom [12].

2 Single Shots of Dynamically Created Quantum Many-Body Vortices

In the field of many-body quantum physics the theoretically most easily accessible quantities are low order correlation functions, such as the single-particle density, the single-particle momentum distribution as well as the respective two-body correlation functions.

Unlike many other subfields of physics the field of ultracold quantum gases opens the rare possibility to investigate high order correlation functions of many-body quantum systems, essentially at no cost. Usually an absorption image of an atomic cloud is taken at the end of an experimental run, which measures the position of the particles. According to the postulates of quantum mechanics the positions $\mathbf{r}_1, \ldots, \mathbf{r}_N$

of these particles are distributed according to the N-particle probability density

$$P(\mathbf{r}_1, \ldots, \mathbf{r}_N) = |\Psi(\mathbf{r}_1, \ldots, \mathbf{r}_N)|^2, \tag{1}$$

where $\Psi(\mathbf{r}_1, \ldots, \mathbf{r}_N)$ is the many-body wave function of the system. Absorption images of ultracold quantum gases thus sample the N-particle probability density, i.e., a random deviate $(\mathbf{r}'_1, \ldots, \mathbf{r}'_N)$ of $P(\mathbf{r}_1, \ldots, \mathbf{r}_N)$ represents an expected outcome of a single shot of an experiment.

In the special case where the system of bosons is fully condensed [23], the many-body wave function factorizes into an N-fold product of a single function $\Psi(\mathbf{r}_1, \ldots, \mathbf{r}_N) = \phi(\mathbf{r}_1) \times \ldots \phi(\mathbf{r}_N)$. Accordingly, there are no correlations between particles, because the N-particle probability density factorizes $P(\mathbf{r}_1, \ldots, \mathbf{r}_N) = P(\mathbf{r}_1) \times \cdots \times P(\mathbf{r}_N)$ with $P(\mathbf{r}) = |\phi(\mathbf{r})|^2$. The popular Gross-Pitaevskii mean-field approximation assumes the many-body wave function to be of this form.

However, for any other type of many-body wave function the particles are more or less correlated. This becomes apparent in the general decomposition of multivariate probability distributions

$$P(\mathbf{r}_1, \ldots, \mathbf{r}_N) = P(\mathbf{r}_1)P(\mathbf{r}_2|\mathbf{r}_1) \times \cdots \times P(\mathbf{r}_N|\mathbf{r}_{N-1}, \ldots, \mathbf{r}_1), \tag{2}$$

where e.g. $P(\mathbf{r}_2|\mathbf{r}_1)$ denotes the conditional probability of finding a particle at $\mathbf{r}_2$, given another one is at $\mathbf{r}_1$. Using Eq. (2) the authors have recently developed an algorithm which allows the simulation of single shots from arbitrary many-body wave functions [16]. This generalizes previous work in this direction for special cases [24–27]. A powerful algorithm to obtain highly accurate many-body wave functions of dynamic many-boson systems is the MCTDHB method [2, 4, 9, 28], which we use here as a tool to obtain the wave function of a rotating condensate. In the following we review some of the results on fluctuating many-body vortices, see [16] for details.

Quantized vortices are a hallmark of Gross-Pitaevskii mean-field theory and typically display a density node [29]. Their appearance is related to a critical rotation velocity. It was recently discovered that stirring a BEC can lead to many-body vortices below the mean-field critical velocity [27, 30, 31]. In contrast, the single particle density of many-body vortices only has a finite value at the vortex core. However, it is important to distinguish between the single-particle density $\rho(\mathbf{r}) = NP(\mathbf{r})$, which is the average over many single shots and single shots themselves which are random deviates of $P(\mathbf{r}_1, \ldots, \mathbf{r}_N)$.

Consider the ground state of a repulsively interacting BEC of $N = 10{,}000$ bosons in a 2D harmonic trap with $\omega_x = \omega_y = 1$ at an interaction strength $\lambda = 17$. The many-body ground state using $M = 2$ orbitals is practically fully condensed with $\rho_1/N = 99.99\,\%$ and therefore well-described by Gross-Pitaevskii mean-field theory. We then switch on a time-dependent stirring potential $V_s(\mathbf{r}, t) = \frac{1}{2}\eta(t)[x(t)^2 - y(t)^2]$ that imparts angular momentum onto the BEC. Here $x(t)$ and $y(t)$ vary harmonically and the amplitude $\eta(t)$ is linearly ramped up from zero to a finite value until time $t = 80$, kept constant there until $t = 300$ and ramped back down again until $t = 380$.

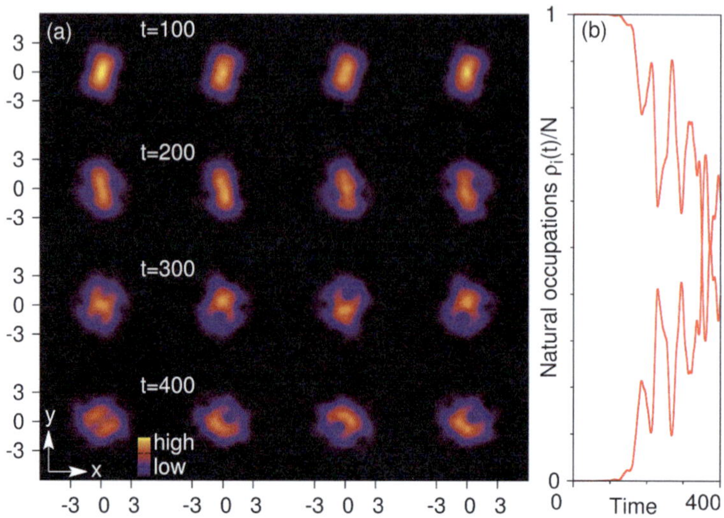

Fig. 1 *Fluctuating many-body vortices.* A repulsive condensate in the ground state of a harmonic trap is stirred by a rotating potential in two spatial dimensions. Over the course of time the system fragments and in single shots vortices appear at random positions. (**a**) First column: single-particle density at different times. Second to fourth column: single shots at the same times. (**b**) Fragmentation of the condensate as a function of time. Starting from a condensed state, the system of bosons fragments as it is stirred. While the system is condensed single shots and the single-particle density look alike. When the system is fragmented vortices appear at random positions. Parameter values: $N = 10{,}000$. Interaction strength: $\lambda = 17$. See text for details. All quantities shown are dimensionless (Figure from Ref. [16])

The first column of Fig. 1a shows the single-particle density at different times. The remaining three columns show single shots taken at the same times as shown for the density. Figure 1b shows the evolution of the natural occupations as a function of time. The system is condensed as long as only a single natural occupation is occupied. As expected from the discussion above single shots merely reproduce the single-particle density for such BECs. However, over the course of time an additional natural orbital becomes occupied, i.e., the BEC becomes fragmented [32]. Each single shot then shows a clear vortex with no particles at its core. These vortices appear randomly at different locations in each shot, in contrast to their mean-field counterparts. The average over many single shots reproduces the single-particle density. The fact that the vortices appear randomly at different locations in each shot explains the finite value of the vortex core in the average over many single shots, i.e., the single-particle density.

3 Many-Body Tunneling Dynamics of Bose-Einstein Condensates and Vortex States in 2D

Tunneling phenomena in two-dimensional (2D) trapped Bose-Einstein condensates (BECs) have attracted high attention in recent years. Especially the tunneling dynamics of trapped vortices on the mean-field level were studied, e.g., in an harmonic potential with a Gaussian potential barrier [33], in 2D superfluids [34], between two Gaussian wells [35], or between two pinning potentials [36]. In three-dimensional double-well potentials, macroscopic superpositions of vortex states during the tunneling dynamics have been found [37].

The motivation of this work is to investigate the full many-body Schrödinger dynamics of a tunneling 2D BEC with definite total angular momentum. To this end, we consider a 2D radial double-well trap, comprised of the trap center and an external rim. Both parts are separated by a ring-shaped potential barrier. We discuss repulsive condensates made of $N = 100$ bosons with zero total angular momentum, $L = 0$, and vortex states with $L = N$. We demonstrate numerically that BECs carrying definite total angular momentum do indeed tunnel through the potential barrier. We find that many-body effects set in at weaker interactions when the tunneling system carries angular momentum. A general conclusion stemming from our results is that the long time tunneling dynamics of 2D BECs cannot be described by a standard mean field, like the Gross-Pitaevskii equation, even in the regime of weak interaction between the bosons (Fig. 2).

The calculations were carried out by using the hybrid MPI and OpenMP implementation [10] of the multiconfigurational time-dependent Hartree for bosons (MCTDHB) method [2, 4]. To study the long time tunneling dynamics in 2D, we were in need of computational resources allowing for parallel multi-core/multi-thread simulations and simulation times up to several weeks.

4 Transition from Vortices to Solitonic Vortices in 2D Trapped Bose-Einstein Condensates

Quantized vortices and dark solitons are the fundamental nonlinear excitations of atomic Bose-Einstein condensates in two/three dimensions and one dimension, respectively [38]. Quantized vortices are defects in the quantum phase about which the superfluid flows with quantized circulation, while dark solitons are non-dispersive waves characterized by a density depression and phase slip. Since the early days of atomic condensates, both vortex structures and dark solitons have been experimentally generated and studied. However, recent experiments have reported intriguing structures called solitonic vortices [39–41]. These excitations, first predicted by Brand and Reinhardt [42], lie at the crossover between one and two/three dimensions, where dark solitons are dimensionally unstable yet the transverse confinement negatives conventional vortices. Motivated by these observations

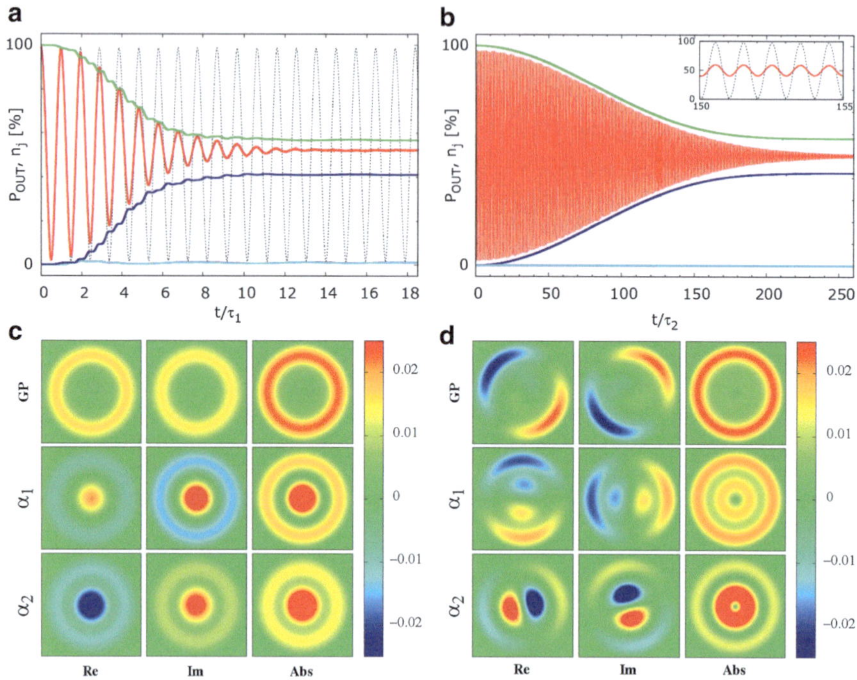

Fig. 2 Panels (**a**)+(**b**): Mean-field ($M = 1$) and many-body ($M = 4$) tunneling dynamics of BECs with $N = 100$ interacting particles held in the radial double well with (**a**) $L = 0$ and $\Lambda = 2$ and (**b**) $L = N$ and $\Lambda = 0.2$. In the Gross-Pitaevskii cases, the occupation probability of the external part, $P_{OUT}(t)$ (*dotted gray curve*), oscillates without damping. In the many-body cases (*solid red*), $P_{OUT}(t)$ is damped and saturates for long times. Only the first two natural orbitals are occupied (*solid green* and *blue*), the other two (*solid magenta* and *light blue*, atop of each other) carry only a tiny fraction of the particles. Panels (**c**)+(**d**): Real, imaginary, and absolute value of the Gross-Pitaevskii mean-field orbital (top rows) and of the first two natural orbitals α_1 and α_2 of the many-body simulation ($M = 4$, middle and bottom rows) for (**c**) $L = 0$ and (**d**) $L = N$ after the density has collapsed. Whereas the mean-field orbital is localized in the external rim, the many-body orbitals are delocalized, covering both the trap's center and rim. See [17] for more details. All quantities are dimensionless (Figure panels adapted from Ref. [17])

we examine the crossover from vortices to solitonic vortices in condensates under harmonic trapping [18] based on extensive numerical simulations of the Gross-Pitaevskii equation, which describes the nonlinear dynamics of the mean-field condensate wavefunction. This is performed using the MCTDH-X package [12] with $N = 100$ and $M = 1$.

We map out the vortex-solitonic vortex crossover in terms of its oscillation frequency as a function of the trap aspect ratio, ω_y/ω_x, for various interaction strengths (Fig. 3a). As the condensate is deformed, the vortex passes from the 2D regime, characterized by slow elliptical trajectories, to the solitonic vortex regime, characterized by linear oscillations with a relatively high frequency, akin to that of

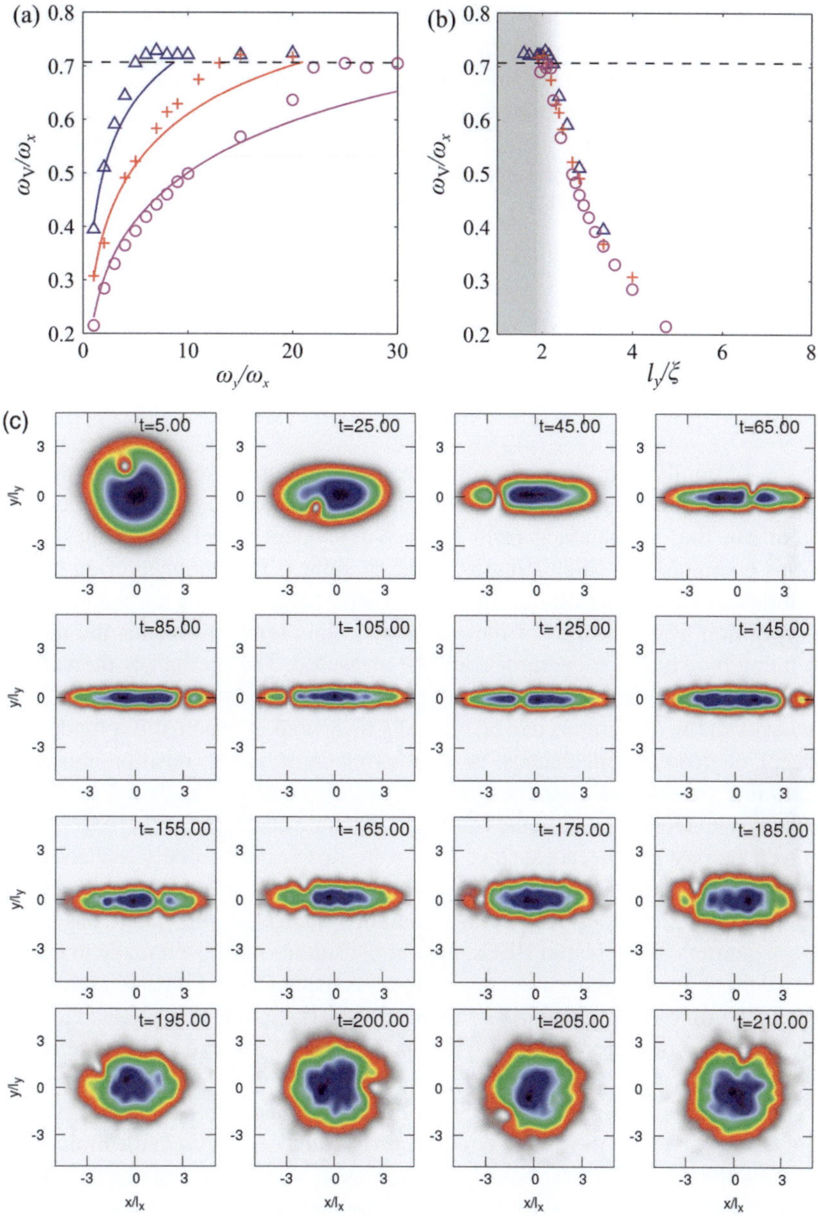

Fig. 3 *Vortex-solitonic vortex transition*. (**a**) The vortex oscillation frequency ω_v increases as the trap ratio ω_y/ω_x is increased, saturating to the dark soliton prediction $\omega_x/\sqrt{2}$, shown here for different interaction strengths (*blue*, *red*, *pink*) from simulations (points) and an analytical prediction (lines). (**b**) When plotted as a function of the condensate width l_y divided by the healing length ξ, the data fall on a common curve. (**c**) Evolution of the condensate density during an example trap deformation cycle (circular → elongated → circular) (Figure panels adapted from Ref. [18])

a dark soliton. When plotted as a function of the condensate width divided by the healing length, the oscillation frequencies fall on a common curve (Fig. 3b).

Next we examine the hysteresis of the system in traps with time-dependent aspect ratio, when the system, starting from a circular trap, is being elongated and then returned back to its initial shape. When the solitonic regime is probed during the hysteresis cycle, angular momentum is lost from the system but, remarkably, the vortex can re-emerge (Fig. 3c).

5 Variance as a Sensitive Probe of Correlations and Uncertainty Product of an Out-of-Equilibrium Many-Particle System

The out-of-equilibrium dynamics of a quantum system is described by the time-dependent Schrödinger equation. All physical information on the evolving quantum system can thus be obtained from its time-dependent wavefunction by applying various operators and calculating expectation values. The variance of an operator quantifies to what extent the system under investigation is in an eigenstate or a superposition of eigenstates of the operator. In this sense it dictates the quantum resolution by which the operator could be measured. The product of the variances of two operators defines an uncertainty product. The uncertainty product quantifies to what extent two operators can be mutually measured. As such, it is a fundamental concept in quantum mechanics. A famous example is the position–momentum uncertainty product of a single quantum particle which is analyzed in quantum mechanics textbooks for both the static and dynamic cases, see, e.g., [43].

Over the past two decades, since they were first experimentally realized, Bose-Einstein condensates (BECs) made of ultracold trapped bosonic atoms have become a popular ground to study interacting quantum systems [44–49]. There has been an intense theoretical interest in BECs, and ample studies have been made to describe their static and particularly dynamic properties using Gross-Pitaevskii, mean-field theory. The time-dependent Gross-Pitaevskii equation governs a mean-field theory which assumes that each and every boson is described by one and the same time-dependent one-particle function throughout the evolution of the BEC in time.

The general paradigm is that Gross-Pitaevskii theory properly describes the ground state as well as the out-of-equilibrium dynamics of BECs in the limit of large particle numbers. To be explicit, Lieb, Seiringer, and Yngvason have rigorously proven for trapped BECs with two-body repulsive interaction in the limit of an infinite particle number and at constant interaction parameter (i.e., when keeping the product of the number of particles times the scattering length fixed), that the ground-state energy and density of the condensate converge to those obtained by minimizing the Gross-Pitaevskii energy functional [50]. Thereafter, Lieb and Seiringer proved in the same limit that the ground state is 100 % condensed [51]. In the case of out-of-equilibrium dynamics, Erdős, Schlein, and Yau have rigorously proven that an expanding initially-trapped BEC, after the trap is released, still exhibits 100 % condensation [52]. Furthermore, the condensate density evolves according to that predicted by the time-dependent Gross-Pitaevskii equation. Whereas these results imply that the fraction of depleted particles vanishes in the infinite particle limit, the absolute number of non condensed particles is always non-zero in the interacting system. The latter is central to the present investigations.

In [19] we analyze the *ground state* of a trapped BEC and demonstrate that, even in the infinite particle limit when the BEC is 100 % condensed, the variance of a many-particle operator can substantially differ from that predicted by the Gross-Pitaevskii theory. The existence of many-body effects beyond those predicted by the mean-field Gross-Pitaevskii theory stems from the necessity of performing the infinite particle limit only after the quantum mechanical observalbe is evaluated and not prior to its evaluation. This is essential since otherwise any trace of many-body correlations is washed out before the quantum mechanical observable can be evaluated. This is explained in length both analytically and numerically in Ref. [19], see Fig. 4 for an example.

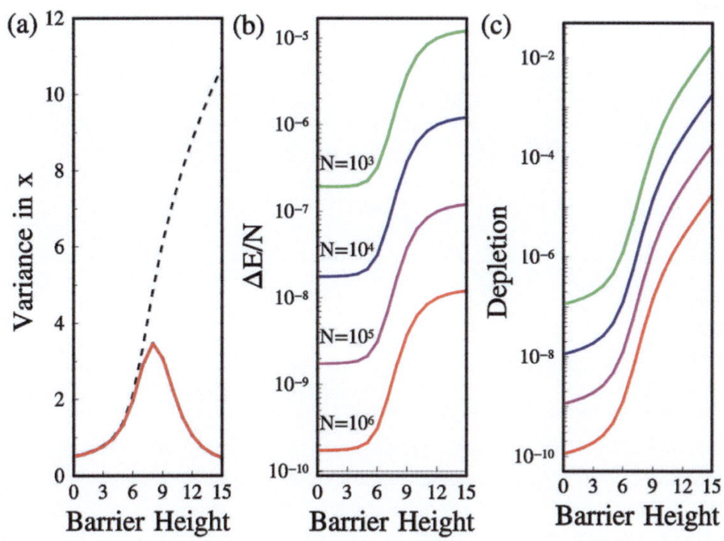

Fig. 4 The variance of the many-particle position operator, $\hat{X} = \sum_{j=1}^{N} \hat{x}_j$, of a weakly-interacting BEC held in a symmetric trap for different barrier heights. Results for $N = 1000$ (in *green*), 10,000 (in *blue*), 100,000 (in *magenta*), and 1,000,000 (in *red*) bosons are shown. The interaction parameter is $\Lambda = \lambda_0(N-1) = 0.1$. (**a**) Shown is the variance $\frac{1}{N}\Delta^2_{\hat{X}}$ (full curves) and the Gross-Pitaevskii variance $\Delta^2_{\hat{x},GP}$ (in *black*; *dashed curve*). Large differences arise from a certain barrier height. All four curves for the different numbers of particles lie atop each other. (**b**) The many-body energy and (**c**) the depletion are seen to approach and coincide with the Gross-Pitaevskii results, as is expected from the literature. Note the small values on the *y*-axes of panels (**b**) and (**c**). In contrast, the variance converges to a value different from the Gross-Pitaevskii results for not too shallow barriers. See [19] for more details. The quantities shown are dimensionless (Figure from Ref. [19])

In [20] we generalize our result for the ground state to the dynamics of an *out-of-equilibrium* BEC. Dynamics is generally more intricate than statics, and involves (sometimes many) excitations. We show, analytically and numerically, that the evolution in time of the uncertainty product of two operators can deviate from that of the Gross-Pitaevskii dynamics, even in the infinite particle limit. We explicitly demonstrate this deviation for the center-of-mass position–momentum uncertainty product of a freely expanding BEC, see Fig. 5, as well as to the dynamics of a trapped BEC [20]. The uncertainty product is an example of an observable of a BEC that, rather than depending on the depleted fraction which vanishes in the infinite particle limit, depends on the depleted total number of particles which always exists in the interacting system. Our results thus advocate that one has to use a many-body propagation theory, such as MCTDHB, to describe the out-of-equilibrium dynamics of observables like the uncertainty product of BECs, even in the limit of an infinite number of particles when the system becomes 100 % condensed.

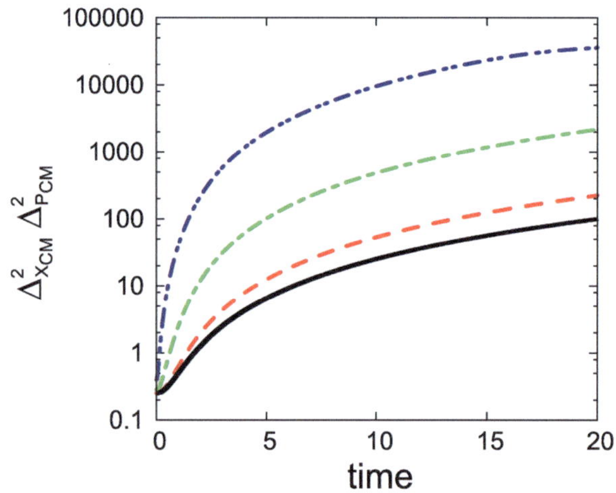

Fig. 5 Time-dependent center-of-mass position–momentum uncertainty product $\Delta^2_{\hat{X}_{CM}}(t)\Delta^2_{\hat{P}_{CM}}(t)$ of a BEC released from an harmonic trap. An illustrative numerical example with 1D bosons. The one-body Hamiltonian is $-\frac{1}{2}\frac{\partial^2}{\partial x^2}+\frac{x^2}{2}$, and the interboson interaction is contact, $\lambda_0\hat{W}(x_1-x_2)=\lambda_0\delta(x_1-x_2)$. Shown and compared as a function of time are the Gross-Pitaevskii results for the interaction parameters $\Lambda=\lambda_0(N-1)=1$ (in *red; dashed*), $\Lambda=10$ (in *green; dashed–dotted*), and $\Lambda=100$ (in *blue; dashed–double-dotted*) and the analytical, many-body result $\frac{1}{4}(1+t^2)$ valid $\forall\Lambda$ (in *black; full curve*). The position–momentum uncertainty product computed at the Gross-Pitaevskii level differs from the analytical, many-body result. The difference increases upon increasing Λ, meaning that the pace of growth of the uncertainty at the mean-field level depends on the interaction parameter Λ (the y-axis is plotted in logarithmic scale). The many-body uncertainty product grows as t^2, and the mean-field uncertainty product is seen to grow in a similar manner in time (see Ref. [53] for the mean-field analysis of the expansion). The uncertainty product constitutes a macroscopic probe of the time-dependent correlations of a BEC, even when the system becomes 100 % condensed in the limit of an infinite number of particles. See [20] for more details. The quantities shown are dimensionless (Figure from Ref. [20])

6 Beyond Structureless Bosons: Results Obtained with the MCTDH-X Package

6.1 *Trapped Fermions Escape*

The tunneling of one trapped fermion like an electron, proton and even classes of atoms or molecules through a barrier can in many cases be solved even analytically. But how do several trapped fermions behave? Do they tunnel through the barrier together or do they go one-by-one? This issue has been investigated experimentally [54] showing a sequential tunneling process, but has not been fully understood theoretically yet. We therefore utilize the MCTDHF approach [3, 55–57] implemented in the MCTDH-X program package [12]. We first confine two fermions in a parabolic potential (see Fig. 6a) and then propagate the wavefunction. As the time evolves, fermions with two different kinetic energies, which are here characterized by their momenta k, are observed. This can be explained by sequential tunneling of the fermions since the first fermion will feel both the external potential and the interaction with the second fermion in the potential. However, the energy of the second fermion is only influenced by the potential and therefore is lower (see Fig. 6b).

For stronger interparticle interactions a third kinetic energy is observed which fits to the collective movement of two fermions. From Fig. 7, it can be seen that the two-body correlation function for k_1 and k_2 is larger for different energies than their own (i.e., maxima on the off-diagonal), while for a momentum of k_{red} the other electron is observed with the same energy, which manifests in a peak on the diagonal in Fig. 7. We therefore conclude that fermions escape over the barrier together while they travel alone when they tunnel through the barrier [21].

6.2 *Composite Fragmentation of Multi-component Bose-Einstein Condensates*

Much like humans in a society may form groups that have different opinions and interests, ultracold bosonic atoms in a Bose-Einstein condensate may exhibit “social behavior”. A system of indistinguishable bosons with two internal degrees of freedom is naturally divided into two groups + and −. As a whole, these two groups may have “the same interests”, i.e., be in a coherent, condensed state or “have different interests”, i.e., be in an incoherent, fragmented state.

Herein, we investigate the ground state of a system of two-component bosons that are trapped in harmonic potentials whose minima are spatially split as a function of the distance Δ between them. To this end, we use the multiconfigurational time-dependent Hartree for indistinguishable particles software [12]. As a first step, we show how the two groups + and − react to the distance Δ between their parabolic

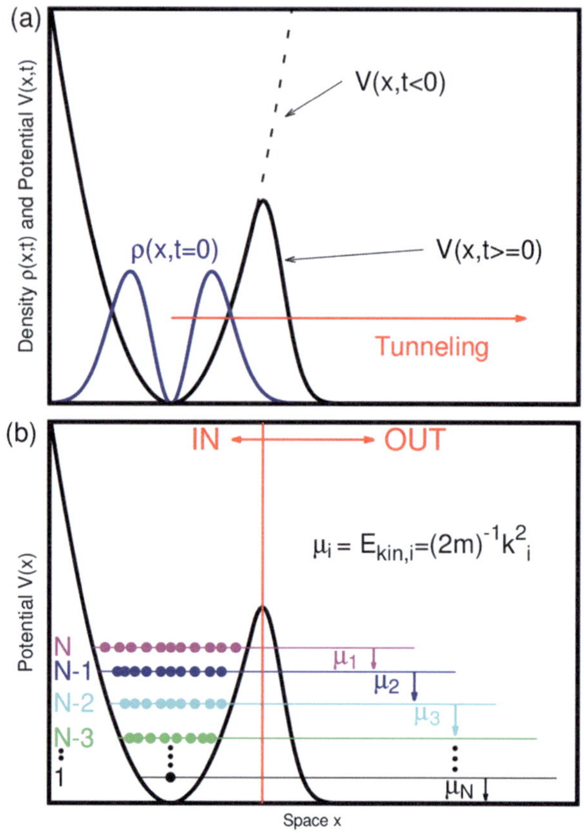

Fig. 6 (**a**) Potential in which the fermions are trapped. Initially, the fermions are confined to the parabolic potential $V(x, t < 0)$ and their wavefunction has the density $\rho(x, t = 0)$ (sketched). The potential $V(x, t \geq 0)$ is then opened to allow for fermions to escape. (**b**) When several fermions are confined in space and repel each other, the energy in the system is higher than for one single fermion. When the first fermion leaves the confined space by tunneling it takes the energy stemming from its interaction with the other fermions and removes it from the other fermions. Hence, their energy is lowered. All quantities are dimensionless (Figure material from Ref. [21])

confinements, by plotting the component densities $\rho^+(x) = \sum_{kq} \rho_{kq}\phi_k^{+,*}(x)\phi_q^+(x)$, $\rho^-(x) = \sum_{kq} \rho_{kq}\phi_k^{-,*}(x)\phi_q^-(x)$ and the composite density $\rho^{+/-}(x) = \rho^+(x) + \rho^-(x)$ in the left part of Fig. 8. The $+$ and $-$ groups' densities $\rho^+(x)$ and $\rho^-(x)$ center themselves around the minima of their respective potentials quite intuitively. When the splitting becomes sufficiently large ($\Delta \sim 4$), fragmentation in the system emerges (bottom panel in left part of Fig. 8): the system of the two groups of atoms loses its coherence – the "society" starts to host different interests.

To understand how the different interests are distributed between the groups, we plot the correlation function of the system in the right part of Fig. 8. Quite intuitively, the coherence within the different components or groups is maintained

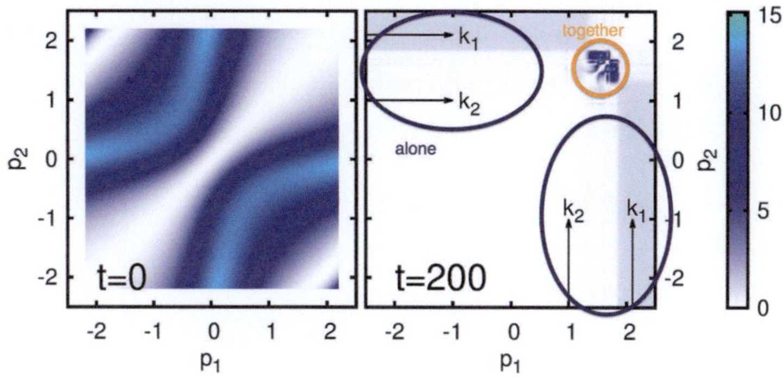

Fig. 7 Two-body momentum correlation function $g^{(2)}(p_1, p_2) = \frac{\sum_{kqsl} \rho_{kqsl} \phi_k^*(p_1)\phi_q^*(p_2)\phi_s(p_1)\phi_l(p_2)}{\sum_{kq} \rho_{kq}\phi_k(p_1)\phi_q(p_1) \sum_{kq} \rho_{kq}\phi_k(p_2)\phi_q(p_2)}$. For sequential tunneling the probability that the two momenta are different is pronounced (i.e., maxima on the off-diagonal), while two fermions traveling together have the same momentum (i.e., maxima on the diagonal). The latter feature increases with the interparticle interaction amongst the fermions and, therefore, this process can be assigned to the joint over the barrier escape. All quantities are dimensionless (Figure material from Ref. [21])

while it is lost between them: when the splitting Δ is large enough, one hence finds

$$|g^{(1),-}|^2 = \left| \frac{\sum_{kq} \rho_{kq}\phi_k^{-,*}(x')\phi_q^-(x)}{\sqrt{\left(\sum_{kq} \rho_{kq}\phi_k^{-,*}(x')\phi_q^-(x')\right)\left(\sum_{kq} \rho_{kq}\phi_k^{-,*}(x)\phi_q^-(x)\right)}} \right|^2 \approx 1 \quad \text{and}$$

$$|g^{(1),+}|^2 = \left| \frac{\sum_{kq} \rho_{kq}\phi_k^{+,*}(x')\phi_q^+(x)}{\sqrt{\left(\sum_{kq} \rho_{kq}\phi_k^{+,*}(x')\phi_q^+(x')\right)\left(\sum_{kq} \rho_{kq}\phi_k^{+,*}(x)\phi_q^+(x)\right)}} \right|^2 \approx 1 \quad \text{while}$$

$$|g^{(1),+/-}|^2 = \left| \frac{\sum_{kq\alpha} \rho_{kq}\phi_k^{\alpha,*}(x')\phi_q^\alpha(x)}{\sqrt{\left(\sum_{kq\alpha} \rho_{kq}\phi_k^{\alpha,*}(x')\phi_q^\alpha(x')\right)\left(\sum_{kq\alpha} \rho_{kq}\phi_k^{\alpha,*}(x)\phi_q^\alpha(x)\right)}} \right|^2 \to 0.$$

Members of the same group in our "society" of atoms share interests while the members of different groups have different interests. We term this distribution of interests among groups "composite fragmentation". Composite as opposed to component fragmentation can only occur in systems of ultracold bosonic atoms with (multiple) internal degrees of freedom [22].

7 Concluding Remarks and Future Plans

We have presented in the above chapters, continuing thereby the tradition set forward in previous years [13–15], our scientific work within the MCTDHB project supported by the HLRS for the last year. Our work span a variety of applications and systems. In Ref. [16] we dealt with single-shot simulations of dynamic quantum many-body systems and their implications on imaging of fluctuating vortices in rotating fragmented BECs; In Ref. [17] we explored the many-body tunneling dynamics of BECs and vortex states in 2D circular traps and the resulting emergence

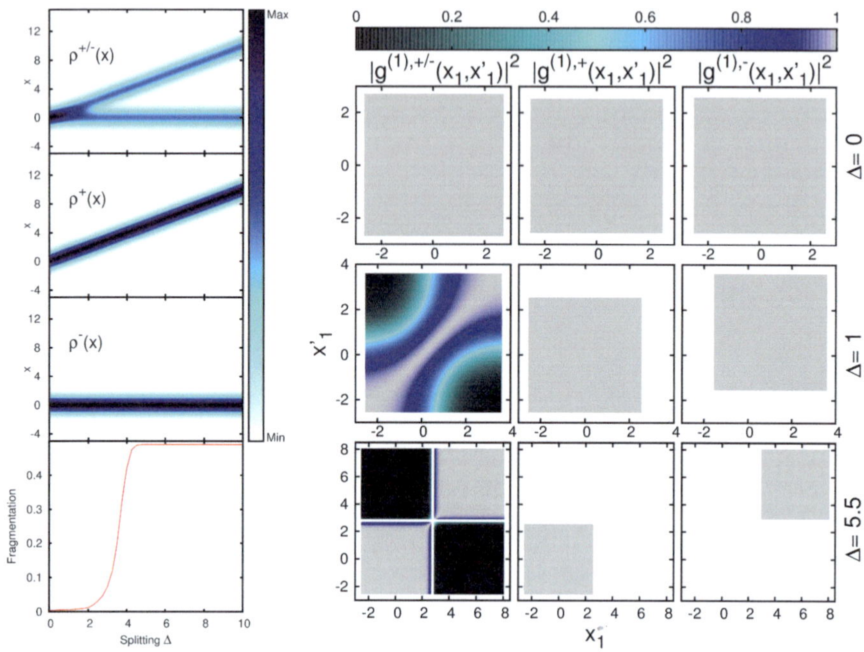

Fig. 8 *Left*: Ground-state density and fragmentation as a function of the splitting Δ. The *top panel* shows the composite density $\rho^{+/-}(x) = \sum_{kq\alpha} \rho_{kq} \phi_k^{\alpha,*}(x)\phi_q^{\alpha}(x)$ of the two components $\alpha = +$ and $\alpha = -$. The *second* and *third panel* depict the component densities $\rho^{\alpha}(x) = \sum_{kq} \rho_{kq} \phi_k^{\alpha,*}(x)\phi_q^{\alpha}(x)$ of the system in the respective internal state $\alpha = +$ and $\alpha = -$. The *bottom panel* shows the fragmentation of the system. Fragmentation is energetically favorable as soon as the overlap of the densities of the internal states becomes small (cf. *top* and *bottom panels*). *Right*: Signatures of composite fragmentation in the one-body correlation function as a function of the splitting Δ. The rows of panels correspond to the splittings $\Delta = 0$, $\Delta = 1$, and $\Delta = 5.5$ from top to bottom. The values of fragmentation are $F = 0.004$, $F = 0.006$, and $F = 0.490$, respectively. The first column shows the composite correlation function of both internal states $|g^{(1),+/-}|^2$, the middle column the correlation function of the $\alpha = +$ state, $|g^{(1),+}|^2$, and the right column the correlation function of the $\alpha = -$ state, $|g^{(1),-}|^2$. The correlations are only plotted for coordinates (x, x') if the component (composite) one-body density at these coordinates is larger than 0.05, to avoid analyzing component (composite) correlations where there are practically no particles. While the component correlations exhibit full coherence, i.e., $|g^{(1),\alpha}|^2 \approx 1$ in the middle and left column, the composite correlation function shows a quick loss of coherence between the components, i.e., $|g^{(1),+/-}|^2 \approx 0$ on the off-diagonals in the left column: the fragmentation in the system is of "composite" type. All quantities shown are dimensionless, see text for further discussion (Figure material from Ref. [22])

of fragmentation; In Ref. [18] we described the transition from vortices to solitonic vortices in 2D trapped BECs and its hysteresis manifestations in traps with a time-dependent aspect ratio; In Ref. [19] we have enjoyed the privilege of using the computational resources of the HLRS to demonstrated that the variance of many-body operators for the ground-state of trapped BECs is a sensitive probe of correlations enduring the infinite-particle limit, and in Ref. [20] we, similarly, extended this result to the out-of-equilibrium dynamics of BECs; In Ref. [21] we disseminated the multiconfigurational time-dependent Hartree method for fermions (within the MCTDH-X package [12]), and reported its implementation, exactness, and application to the mechanism of a few polarized fermions tunneling to open space; and in Ref. [22] we derived the multiconfigurational time-dependent Hartree method for bosons with internal degrees of freedom, and applied it to the prediction of composite fragmentation of multi-component Bose-Einstein condensates. All in all, we have had a fruitful research year within the MCTDHB project supported by the HLRS.

Last but not least, we would like to touch upon some of the theoretical and computational challenges ahead of us in the upcoming research year. We aim at extending our dynamical studies mentioned above on 2D trapped BECs by calculating the many-body excitation spectra of the latter within the framework of the linear-response theory of MCTDHB (LR-MCTDHB, see Refs. [58–60] for details). Especially for dense grids and many orbitals, these computations will need a very large amount of cores and MPI processes in order to find low-lying eigenvalues of the large, sparse, and non-hermitian linear-response matrix. Very large, or dense, grids will also be needed to accurately describe many-body quantum turbulence in BECs [61], and quench/relaxation dynamics [62] for BECs beyond purely 1D scenarios, such as in Ref. [63]. Finally, simulating the out-of-equilibrium dynamics of fermionic systems, e.g., to explore the BCS-BEC crossover in a trapped finite system, see in this context Ref. [64], one would need increasingly large multiconfigurational spaces. These are the tip of the iceberg of computational challenges awaiting the MCTDHB and MCTDH-X packages [10, 12] and the computational resources allocated to us in the upcoming research year by the HLRS.

Acknowledgements Financial support by the Deutsche Forschungsgemeinschaft (DFG) is gratefully acknowledged. OEA acknowledges funding by the Israel Science Foundation (Grant No. 600/15). RB acknowledges support from the Heidelberg Graduate School of Fundamental Physics (HGSFP). MJE and NGP acknowledge support by EPSRC (UK) Grant No. EP/M005127/1. EF gratefully acknowledges funding from the Research Council of Norway (RCN) through CoE Grant No. 179568/V30 (CTCC). AUJL acknowledges financial support by the Swiss SNF and the NCCR Quantum Science and Technology. KS acknowledges support through the Karel Urbanek Postodcoral Research Fellowship from the Applied Physics Department of Stanford University. MCT acknowledges funding by FAPESP and CePOF at IFSC-University of São Paulo.

References

1. Streltsov, A.I., Alon, O.E., Cederbaum, L.S.: General variational many-body theory with complete self-consistency for trapped bosonic systems. Phys. Rev. A **73**, 063626 (2006)
2. Streltsov, A.I., Alon, O.E., Cederbaum, L.S.: Role of excited states in the splitting of a trapped interacting bose-einstein condensate by a time-dependent barrier. Phys. Rev. Lett. **99**, 030402 (2007)
3. Alon, O.E., Streltsov, A.I., Cederbaum, L.S.: Unified view on multiconfigurational time propagation for systems consisting of identical particles. J. Chem. Phys. **127**, 154103 (2007)

4. Alon, O.E., Streltsov, A.I., Cederbaum, L.S.: Multiconfigurational time-dependent Hartree method for bosons: Many-body dynamics of bosonic systems. Phys. Rev. A **77**, 033613 (2008)
5. Sakmann, K., Streltsov, A.I., Alon, O.E., Cederbaum, L.S.: Exact quantum dynamics of a bosonic Josephson junction. Phys. Rev. Lett. **103**, 220601 (2009)
6. Lode, A.U.J., Sakmann, K., Alon, O.E., Cederbaum, L.S., Streltsov, A.I.: Numerically exact quantum dynamics of bosons with time-dependent interactions of harmonic type. Phys. Rev. A **86**, 063606 (2012)
7. Meyer, H.-D., Gatti, F., Worth, G. A. (eds.): Multidimensional Quantum Dynamics: MCTDH Theory and Applications. Wiley-VCH, Weinheim (2009)
8. Proukakis, N.P., Gardiner, S.A., Davis, M.J., Szymanska, M.H. (eds.): Quantum Gases: Finite Temperature and Non-equilibrium Dynamics. Cold Atoms Series, vol. 1. Imperial College Press, London (2013)
9. Streltsov, A.I., Sakmann, K., Lode, A.U.J., Alon, O.E., Cederbaum, L.S.: The Multiconfigurational Time-Dependent Hartree for Bosons Package, version 2.3. Heidelberg (2013)
10. Streltsov, A.I., Cederbaum, L.S., Alon, O.E., Sakmann, K., Lode, A.U.J., Grond, J., Streltsova, O.I., Klaiman, S.: The Multiconfigurational Time-Dependent Hartree for Bosons Package, version 3.x. Heidelberg (2006-Present). http://mctdhb.org
11. Streltsov, A.I., Streltsova, O.I.: The Multiconfigurational Time-Dependent Hartree for Bosons Laboratory, version 1.5 (2015) http://MCTDHB-lab.org; http://QDlab.org
12. Lode, A.U.J., Tsatsos, M.C., Fasshauer, E.: The Multiconfigurational Time-Dependent Hartree for Indistinguishable Particles X Package (2015) http://mctdhx.org; http://ultracold.org; http://schroedinger.org; http://mctdh.bf
13. Lode, A.U.J., Sakmann, K., Doganov, R.A., Grond, J., Alon, O.E., Streltsov, A.I., Cederbaum, L.S.: Numerically-exact schrödinger dynamics of closed and open many-boson systems with the MCTDHB package. In: Nagel, W.E., Kröner, D.H., Resch, M.M. (eds.) High Performance Computing in Science and Engineering '13: Transactions of the High Performance Computing Center, Stuttgart (HLRS) 2013, pp. 81–92. Springer, Heidelberg (2013)
14. Klaiman, S., Lode, A.U.J., Sakmann, K., Streltsova, O.I., Alon, O.E., Cederbaum, L.S., Streltsov, A.I.: Quantum many-body dynamics of trapped bosons with the MCTDHB package: towards new horizons with novel physics. In: Nagel, W.E., Kröner, D.H., Resch, M.M. (eds.) High Performance Computing in Science and Engineering '14: Transactions of the High Performance Computing Center, Stuttgart (HLRS) 2014, pp. 63–86. Springer, Heidelberg (2015)
15. Alon, O.E., Bagnato, V.S., Beinke, R., Brouzos, I., Calarco, T., Caneva, T., Cederbaum, L.S., Kasevich, M.A., Klaiman, S., Lode, A.U.J., Montangero, S., Negretti, A., Said, R.S., Sakmann, K., Streltsova, O.I., Theisen, M., Tsatsos, M.C., Weiner, S.E., Wells, T., Streltsov, A.I.: MCTDHB physics and technologies: excitations and vorticity, single-shot detection,measurement of fragmentation, and optimal control in correlated ultra-cold bosonic many-body systems. In: Nagel, W.E., Kröner, D.H., Resch, M.M. (eds.) High Performance Computing in Science and Engineering '15: Transactions of the High Performance Computing Center, Stuttgart (HLRS) 2015, pp. 23–50. Springer, Heidelberg (2016)
16. Sakmann, K., Kasevich, M.: Single-shot simulations of dynamic quantum many-body systems. Nat. Phys. **12**, 451 (2016)
17. Beinke, R., Klaiman, S., Cederbaum, L.S., Streltsov, A.I., Alon, O.E.: Many-body tunneling dynamics of Bose-Einstein condensates and vortex states in two spatial dimensions. Phys. Rev. A **92**, 043627 (2015)
18. Tsatsos, M.C., Edmonds, M.J., Parker, N.G.: Transition from vortices to solitonic vortices in trapped atomic Bose-Einstein condensates. Phys. Rev. A **94**, 023627 (2016)
19. Klaiman, S., Alon, O.E.: Variance as a sensitive probe of correlations. Phys. Rev. A **91**, 063613 (2015)
20. Klaiman, S., Streltsov, A.I., Alon, O.E.: Uncertainty product of an out-of-equilibrium many-particle system. Phys. Rev. A **93**, 023605 (2016)

21. Fasshauer, E., Lode, A.U.J.: Multiconfigurational time-dependent Hartree method for fermions: implementation, exactness, and few-fermion tunneling to open space. Phys. Rev. A **93**, 033635 (2016)
22. Lode, A.U.J.: The multiconfigurational time-dependent Hartree method for bosons with internal degrees of freedom: theory and composite fragmentation of multi-component Bose-Einstein condensates. Phys. Rev. A **93**, 063601 (2016)
23. Penrose, O., Onsager, L.: Bose-Einstein condensation and liquid helium. Phys. Rev. **104**, 576 (1956)
24. Javanainen, J., Yoo, S.M.: Quantum phase of a Bose-Einstein condensate with an arbitrary number of atoms. Phys. Rev. Lett. **76**, 161 (1996)
25. Castin, Y., Dalibard, J.: Relative phase of two Bose-Einstein condensates. Phys. Rev. A **55**, 4330 (1997)
26. Dziarmaga, J., Karkuszewski, Z.P., Sacha, K.: Images of the dark soliton in a depleted condensate. J. Phys. B **36**, 1217 (2003)
27. Dagnino, D., Barberán, N., Lewenstein, M.: Vortex nucleation in a mesoscopic Bose superfluid and breaking of the parity symmetry. Phys. Rev. A **80**, 053611 (2009)
28. Streltsov, A.I., Alon, O.E., Cederbaum, L.S.: General mapping for bosonic and fermionic operators in fock space. Phys. Rev. A **81**, 022124 (2010)
29. Fetter, A.L.: Rotating trapped Bose-Einstein condensates. Rev. Mod. Phys. **81**, 647 (2009)
30. Dagnino, D., Barberán, N., Lewenstein, M., Dalibard, J.: Vortex nucleation as a case study of symmetry breaking in quantum systems. Nat. Phys. **5**, 431 (2009)
31. Weiner, S.E., Tsatsos, M.C., Cederbaum, L.S., Lode, A.U.J.: Angular momentum in interacting many-body systems hides in phantom vortices. arXiv:1409.7670
32. Nozières, P., James, D.S.: Particle vs. pair condensation in attractive Bose liquids. J. Phys. (Fr.) **43**, 1133 (1982)
33. Martin, A.M., Scott, R.G., Fromhold, T.M.: Transmission and reflection of Bose-Einstein condensates incident on a Gaussian tunnel barrier. Phys. Rev. A **75**, 065602 (2007)
34. Arovas, D.P., Auerbach, A.: Quantum tunneling of vortices in two-dimensional superfluids. Phys. Rev. B **78**, 094508 (2008)
35. Salgueiro, J.R., Zacarés, M., Michinel, H., Ferrando, A.: Vortex replication in Bose-Einstein condensates trapped in double-well potentials. Phys. Rev. A **79**, 033625 (2009)
36. Fialko, O., Bradley, A.S., Brand, J.: Quantum tunneling of a vortex between two pinning potentials. Phys. Rev. Lett. **108**, 015301 (2012)
37. Garcia-March, M.A., Carr, L.D.: Vortex macroscopic superpositions in ultracold bosons in a double-well potential. Phys. Rev. A **91**, 033626 (2015)
38. Kevrekidis, P.G., Frantzeskakis, D.J., Carretero-González, R. (eds.): Emergent Nonlinear Phenomena in Bose-Einstein Condensates. Springer, Berlin (2008)
39. Becker, C., Sengstock, K., Schmelcher, P., Kevrekidis, P.G., Carretero-González, R.: Inelastic collisions of solitary waves in anisotropic Bose-Einstein condensates: sling-shot events and expanding collision bubbles. New J. Phys. **15**, 113028 (2013)
40. Donadello, S., Serafini, S., Tylutki, M., Pitaevskii, L.P., Dalfovo, F., Lamporesi, G., Ferrari, G.: Observation of solitonic vortices in Bose-Einstein condensates. Phys. Rev. Lett. **113**, 065302 (2014)
41. Ku, M.J.H., Ji, W., Mukherjee, B., Guardado-Sanchez, E., Cheuk, L.W., Yefsah, T., Zwierlein, M.W.: Motion of a solitonic vortex in the BEC-BCS crossover. Phys. Rev. Lett. **113**, 065301 (2014)
42. Brand, J., Reinhardt, W.P.: Solitonic vortices and the fundamental modes of the snake instability: possibility of observation in the gaseous Bose-Einstein condensate. Phys. Rev. A **65**, 043612 (2002)
43. Cohen-Tannoudji, C., Diu, B., Laloë, F.: Quantum Mechanics, vol. 1. Wiley, New York (1977)
44. Dalfovo, F., Giorgini, S., Pitaevskii, L.P., Stringari, S.: Theory of Bose-Einstein condensation in trapped gases. Rev. Mod. Phys. **71**, 463 (1999)
45. Leggett, A.J.: Bose-Einstein condensation in the alkali gases: some fundamental concepts. Rev. Mod. Phys. **73**, 307 (2001)

46. Bloch, I., Dalibard, J., Zwerger, W.: Many-body physics with ultracold gases. Rev. Mod. Phys. **80**, 885 (2008)
47. Pitaevskii, L., Stringari, S.: Bose-Einstein Condensation. Oxford University Press, Oxford (2003)
48. Leggett, A.J.: Quantum Liquids: Bose Condensation and Cooper Pairing in Condensed Matter Systems. Oxford University Press, Oxford (2006)
49. Pethick, C.J., Smith, H.: Bose-Einstein Condensation in Dilute Gases, 2nd edn. Cambridge University Press, Cambridge (2008)
50. Lieb, E.H., Seiringer, R., Yngvason, J.: Bosons in a trap: a rigorous derivation of the Gross-Pitaevskii energy functional. Phys. Rev. A **61**, 043602 (2000)
51. Lieb, E.H., Seiringer, R.: Proof of Bose-Einstein condensation for dilute trapped gases. Phys. Rev. Lett. **88**, 170409 (2002)
52. Erdős, L., Schlein, B., Yau, H.-T.: Rigorous derivation of the Gross-Pitaevskii equation. Phys. Rev. Lett. **98**, 040404 (2007)
53. Brazhnyi, V.A., Kamchatnov, A.M., Konotop, V.V.: Hydrodynamic flow of expanding Bose-Einstein condensates. Phys. Rev. A **68**, 035603 (2003)
54. Serwane, F., Zürn, G., Lompe, T., Ottenstein, T.B., Wenz, A.N., Jochim, S.: Deterministic preparation of a tunable few-fermion system. Science **332**, 6027 (2011)
55. Caillat, J., Zanghellini, J., Kitzler, M., Koch, O., Kreuzer, W., Scrinzi, A.: Correlated multielectron systems in strong laser fields: a multiconfiguration time-dependent Hartree-Fock approach. Phys. Rev. A **71**, 012712 (2005); Zanghellini, J., Kitzler, M., Fabian, C., Brabec, T., Scrinzi, A.: An MCTDHF approach to multielectron dynamics in laser fields. Laser Phys. **13**, 1064 (2003)
56. Kato, T., Kono, H.: Time-dependent multiconfiguration theory for electronic dynamics of molecules in an intense laser field. Chem. Phys. Lett. **392**, 533 (2004)
57. Nest, M., Klamroth, T., Saalfrank, P.: The multiconfiguration time-dependent Hartree-Fock method for quantum chemical calculations. J. Chem. Phys. **122**, 124102 (2005)
58. Grond, J., Streltsov, A.I., Lode, A.U.J., Sakmann, K., Cederbaum, L.S., Alon, O.E.: Excitation spectra of many-body systems by linear response: general theory and applications to trapped condensates. Phys. Rev. A **88**, 023606 (2013)
59. Alon, O.E., Streltsov, A.I., Cederbaum, L.S.: Unified view on linear response of interacting identical and distinguishableparticles from multiconfigurational time-dependent Hartree methods. J. Chem. Phys. **140**, 034108 (2014)
60. Alon, O.E.: Many-body excitation spectra of trapped bosons with general interaction by linear response. J. Phys. Conf. Ser. **594**, 012039 (2015)
61. Tsatsos, M.C., Tavares, P.E.S., Cidrim, A., Fritsch, A.R., Caracanhas, M.A., dos Santos, F.E.A., Barenghi, C.F., Bagnato, V.S.: Quantum turbulence in trapped atomic Bose-Einstein condensates. Phys. Rep. **622**, 1 (2016)
62. Lode, A.U.J., Chakrabarti, B., Kota, V.K.B.: Many-body entropies, correlations, and emergence of statistical relaxation in interaction quench dynamics of ultracold bosons. Phys. Rev. A **92**, 033622 (2015)
63. Gring, M., Kuhnert, M., Langen, T., Kitagawa, T., Rauer, B., Schreitl, M., Mazets, I., Adu Smith, D., Demler, E., Schmiedmayer, J.: Relaxation and prethermalization in an isolated quantum system. Science **337**, 1318 (2012)
64. von Stecher, J., Greene, C.H.: Spectrum and dynamics of the BCS-BEC crossover from a few-body perspective. Phys. Rev. Lett. **99**, 090402 (2007)

Nucleon Observables as Probes for Physics Beyond the Standard Model

Constantia Alexandrou, Karl Jansen, Giannis Koutsou, and Carsten Urbach

Abstract We discuss the results of our ongoing project on Hazel Hen at HLRS concerning observables that shed light on the inner structure of the proton and other hadrons. We use techniques from lattice quantum chromodynamics to evaluate these observables on a gluon field ensemble of a $48^3 \times 96$ lattice at a lattice spacing of $a = 0.094(1)$ fm. The novelty of this ensemble is that it is generated directly at the physical value of the pion mass such that any extrapolation from heavier pion masses can be avoided, eliminating thus this systematic uncertainty. By employing state of the art lattice QCD algorithms we were able to compute the hadron spectrum, the axial and tensor charges moments of parton distribution functions and the quark contents of the nucleons.

1 Introduction

The project is embedded in the field of high energy particle physics. In particular, it addresses the strong interaction which binds together quarks and gluons. This leads to the formation of the observed hadronic matter, e.g. the proton and neutron as the most prominent examples.

C. Alexandrou
Department of Physics, University of Cyprus, P.O. Box, 20537, 1678, Nicosia, Cyprus
e-mail: alexandrou@cyi.ac.cy

K. Jansen (✉)
NIC, DESY, Platanenallee 6, 17538, Zeuthen, Germany
e-mail: karl.jansen@desy.de

G. Koutsou
Computation-Based Science and Technology Research Center, The Cyprus Institute, 20 Kavafi Str., 2121, Nicosia, Cyprus
e-mail: g.koutsou@cyi.ac.cy

C. Urbach
Helmholtz-Institut für Strahlen-und Kernphysik (Theorie) and Bethe Center for Theoretical Physics, Universität Bonn, 53115, Bonn, Germany
e-mail: urbach@hiskp.uni-bonn.de

W.E. Nagel et al. (eds.), *High Performance Computing in Science and Engineering '16*, DOI 10.1007/978-3-319-47066-5_7

The theory that is to explain the interaction of quarks and gluons is quantum chromodynamics. It is a local quantum field theory which can be written down in a most elegant and compact way. Still, we demand from this theory that it describes the phenomena of the strong interaction from very small to large O(1fm) scales. For distances much below 1fm the quarks behave almost freely, we speak of *asymptotic freedom*. When the distance becomes at O(1fm) the interaction between quarks and gluons becomes strong, they bind strongly together and form the observed hadrons. In fact, the force between quarks and gluons becomes so strong that they cannot be detected in experiment as asymptotic particles. We speak of *the confinement of quarks*.

The fact that the interaction becomes very strong hinders that the theory can be evaluated in perturbation theory since no small parameter occurs. Thus, in order to test QCD as the correct theory of the strong interaction, non-perturbative methods have to be employed. The way to use such non-perturbative approaches is to formulate the theory on a discrete 4-dimensional space time grid. This leads to the notion of *Lattice QCD* (LQCD). By rotating the standard Minkowski time to Euclidean time, LQCD can be interpreted as statistical mechanics model (in the sense of the Ising model). This makes it possible to perform numerical simulations of the system and hence in turn to evaluate the theory from first principles and non-perturbatively.

Following this approach, the present project has concentrated on the computation of important hadronic observables. Starting with the benchmark calculation of the hadronic spectrum, in particular the baryon masses, a number of quantities that characterize the properties of hadrons, their spin, their angular momentum and their quark content has been computed. The novelty of the project is that all calculation have been performed directly at the physical value of the nucleon and pion masses (the physical point), see below. This is a most substantial progress for such lattice calculations since it avoids the demanding and often not very controlled extrapolation from heavier than physical pion masses to the physical one. Working directly at the physical point eliminates therefore an – often dominating – systematic error and opens a new road for understanding the strong force. Besides understanding better nuclear matter, the observables computed in this project, the charges, the neutron electric dipole moment and the quark contents of the nucleon play a most significant role as input to the interpretation of worldwide ongoing and planned experiments to detect new physics beyond the standard model.

2 Lattice QCD Setup

These main results of our last project period are summarized in two recent publications, Refs. [1, 2]. There is, in addition, a recent work on the σ-terms in QCD [3] related to the quark content of the nucleon and there are a number of papers in preparation. These publications are listed in the bibliography section from where the links can be followed to their arXiv entry on which they have been posted. The last project period allowed us to substantially progress our analysis on the $48^3 \cdot 96$ lattice with $N_f = 2$ flavors of mass-degenerate quarks employing maximally twisted mass fermions [4–6]. As one – somewhat unexpected – outcome of our project, with our present statistics we do still observe tensions with experimental/phenomenological results. It is therefore necessary to obtain as precise as possible values for the considered physical quantities in order to discriminate a statistical fluctuation as the cause for this tension and a real discrepancy. Thus it is necessary to reach full statistics for the $48^3{\cdot}96$ lattice. In addition, and most importantly, it will be necessary to analyze a $64^3 \cdot 128$ lattice which we want to generate in the near future. We have already thermalized gluon field configurations on the $64^3 \cdot 128$ lattice at exactly the same parameters as the previous run on a $48^3 \cdot 96$ lattice. Our first results for this large lattice with still rather low statistics and spatial length of about 6 fm, are very promising in that we can reproduce the physical values of the nucleon and pion masses and several decay constants.

More specifically, the resources granted to us on Hazel Hen during the last computer allocation period led to obtaining first results for several important nucleon structure quantities directly at the physical values of the up- and down-quark mass. In addition, we investigated and implemented new algorithms for speeding up our calculations in order to make best practice on the used architecture. By complementing GPU resources used for the calculation of disconnected diagrams, we have calculated to unprecedented accuracy the nucleon light, strange and charm σ-terms [3]. Furthermore, we have obtained results for all local and one-derivative nucleon matrix elements, including the nucleon axial g_A and tensor g_T charges, momentum fraction $\langle x\rangle_q$ helicity $\langle x\rangle_{\Delta q}$ and transversity $\langle x\rangle_{\delta q}$ [1, 2].

For all calculations we used an ensemble of twisted mass fermion configurations of size $48^3 \times 96$ including a clover term, with degenerate up- and down-quarks with masses tuned to reproduce the physical value of the pion mass.

3 Nucleon Mass and Lattice Spacing

As a first very important step, the determination of the lattice spacing for the here used ensemble is necessary. We calculated the lattice spacing from the nucleon mass itself. This has been achieved by combining older results at unphysical values of the pion mass with the lattice data for the pion and nucleon masses directly at the physical point. The pion mass dependence of the ratio of the nucleon mass to the pion mass as function of the pion mass squared has been fitted to heavy baryon chiral perturbation theory. As can be see in Fig. 1 this dependence is very well described by the theoretical formulae. In particular, we find for the ratio $m_N/m_{\pi^\pm} = 7.6(3)$ which is consistent with the physical value. Thus taking this as being at the physical pion mass we set $m_N = 0.938\,\text{GeV}$ and determine from this physical value the lattice spacing as $a = 0.094(1)\,\text{fm}$. Note that this value is fully consistent with a scale setting procedure using pure gauge quantities, see [1].

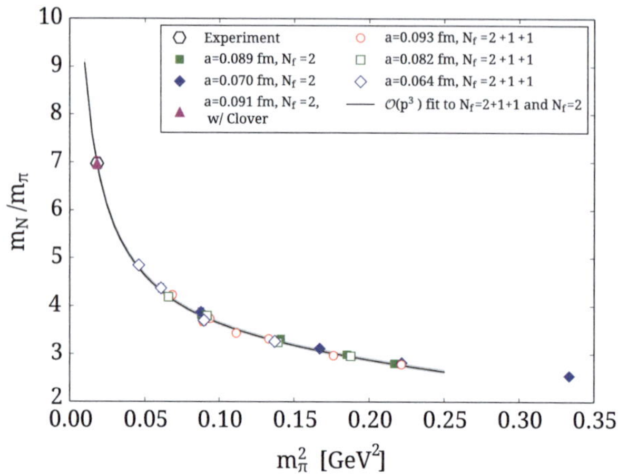

Fig. 1 The ratio of the nucleon mass to the pion mass as a function of the pion mass squared. For determining the pion mass squared the scale is set using the nucleon mass at the physical point as described in the text. The fit includes the points with heavier than physical pion masses (*circles*, *diamonds* and *squares*) but not the ensemble with the clover term (*filled triangle*)

4 Three-Point Functions

After computing the hadronic masses as benchmark quantities we now turn to the calculation of more complicated observables that derive from three-point functions. To this end, on each lattice gauge-configuration we compute three-point functions originating from 16 randomly chosen positions, and for each of these positions we compute the nucleon three-point function with the single unpolarized and three polarized projections, at three sink-source time separations. This means we needed to invert the twisted mass clover operator 4992 times per gauge-field configuration. These calculations are only feasible when using multiple right-hand-side (rhs) methods such as the polynomialy accelerated Arnoldi method used in this project and multi-grid, which we plan to employ in the future. Both methods yield speed-ups of between 30 to 100 times compared to standard Conjugate Gradients (CG) technique.

4.1 Nucleon σ-Terms

An important outcome during the past allocation period was the calculation of the nucleon light, strange and charm σ-terms. The precise values of these quantities are important for Dark Matter searches since they enter cross-sections of nuclei scattering with Weakly Interacting Massive Particles (WIMPs) through Higgs boson exchange. These are examples of quantities where lattice QCD can provide predictions, in particular in the case of the strange and charm σ-terms where direct measurements are not possible.

Following a first calculation at pion mass of around 370 MeV [7], which was used to develop the required methodology, we used Hazel Hen resources to obtain the σ-terms directly from the nucleon matrix elements for the first time at the physical values of the light quark masses. Our results have been reported in Ref. [3], where guided by the analysis at 370 MeV, which showed large contaminations by excited states, we computed multiple sink-source separations in order to reliably identify the ground state matrix element.

Having results at multiple sink-source separations and at measurements of O(10^4), we were able to apply different analyses on our data which carry different systematics in terms of the excited state contaminations. This is shown in Fig. 2, where we use three methods to obtain the σ-terms: fits to the standard plateau which

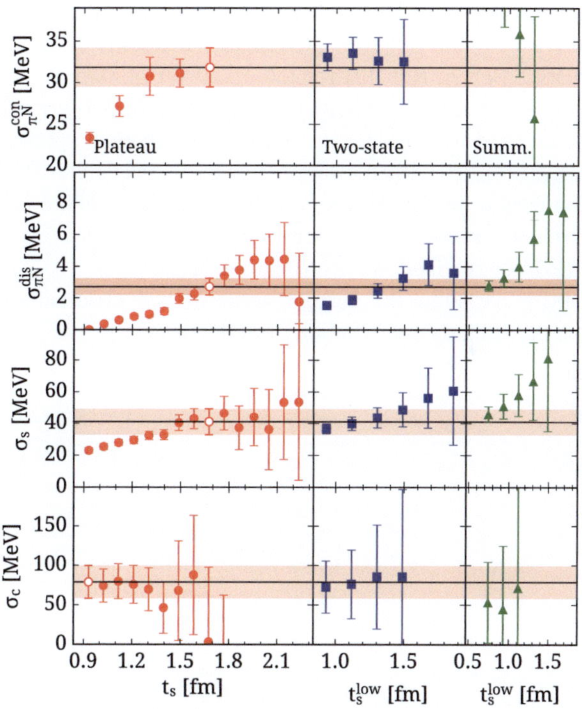

Fig. 2 Connected and disconnected contributions to the light σ-term ($\sigma_{\pi N}$, top two rows), the strange σ-term (σ_s, third row), and the charm σ-terms (σ_c, bottom row). Results obtained using the standard plateau method are shown in the first column (*red circles*) as a function of the sink-source separation, using a two-state fit in the center column (*blue squares*) as a function of the smallest sink-source separation included in the fit and using the summation method in the right column (*green triangles*) as a function of the smallest sink-source separation included in the fit. The *red band* indicates the final value taken, which is the plateau value when all three methods agree

ignores excited states, fits which include an excited state ("two-state") and fits to the ratio summed over the insertion time coordinate ("Summation"). Using these three methods to ensure systematic uncertainties were under control we were able to obtain values for the σ-terms to a precision which can be used in phenomenological studies such as in Ref. [8]. Furthermore this is one of few calculations of the charm σ-terms and one of the most accurate predictions for the strange σ-term.

We note that Hazel Hen resources were used for computing the connected contribution of the $\sigma_{\pi N}$. They were complemented by GPU resources for the calculation of the disconnected contributions to the σ-terms leveraging optimally our available computer resources.

4.2 Nucleon Charges and Moments of Generalized Parton Distributions

Apart from the dedicated calculation of the σ-terms highlighted above, our production runs yield all local and one-derivative matrix elements of the nucleon. These results were reported on in Ref. [2] for the initial set of statistics obtained at the time. Our current goals are a high-statistics evaluation of these matrix elements at multiple sink-source separations in order to reliably investigate excited state effects. Currently we have obtained 6400 measurements and expect about 10,000 by the end of the allocation period. For an error of about 2 % we need about $O(10^5)$ statistics up to sink-source separations of $\sim$1.5 fm, which is made possible with our new multi-grid method as explained in the current proposal.

In Figs. 3 and 4 we show the current status of two observables, the nucleon axial charge and the nucleon quark helicity fraction for the physical point ensemble.

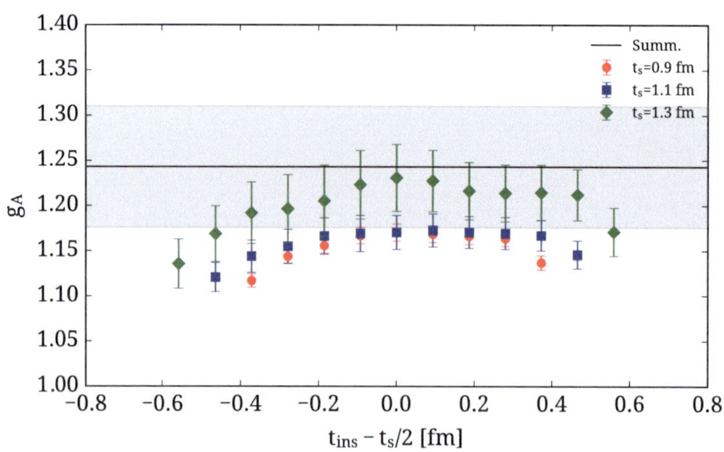

Fig. 3 Nucleon axial charge of the cA2.09.48 ensemble as a function of the distance between and insertion and source time (t_{ins}) for three sink-source separations. The band is the result of the summation method

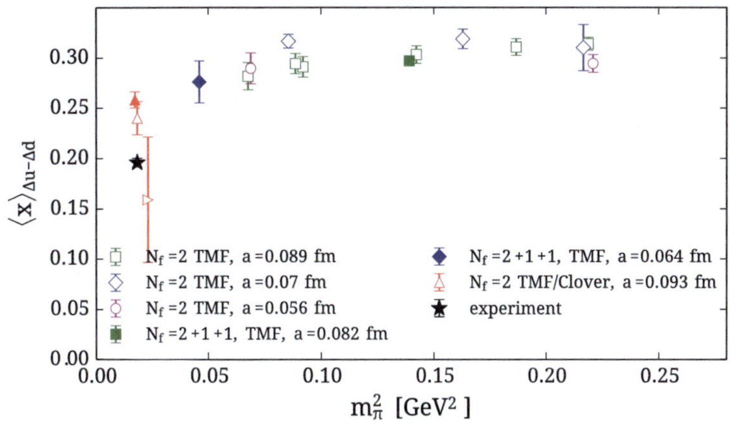

Fig. 4 Nucleon quark helicity fraction calculated on the physical point ensemble (*triangles*) compared to results obtained at heavier than physical pion masses. The *upwards pointing triangle* is the result of a fit to plateau, while the *rightwards pointing triangle* is the result of the summation method

The axial charge is given as a function of the insertion to source time separation for three sink-source separations. Excited states are suppressed exponentially with both these separations. We also show the summation method, in which t_{ins} is summed over and a two-parameter fit is performed of which the slope is g_A, leading however to a larger error. The helicity is shown as a function of the pion mass compared to various lattice results from the literature at heavier than physical pion masses and to experiment. At the physical point we show the result of the plateau method at separation 1.3 fm, while slightly shifted we show the result of the summation method. As can be seen in both cases there is a trend towards the experimentally measured values of both quantities but a tension still remains as mentioned above.

Furthermore the errors at larger sink-source separations and especially for the summation method are large and need to be reduced in order to decide on the excited state effects. These results clearly demonstrate the need for more statistics, especially at larger sink-source separations, in order to allow definite conclusions on such quantities to be drawn and to be able to decide, whether there is a real tension. If it turns out that there is a discrepancy between our lattice data and experimental/phenomenological results, it becomes necessary to investigate which other systematic effect of our lattice calculation could be responsible for this. It is our opinion that then finite volume effects are the cause for the tension and hence a calculation on a larger volume at fixed action parameters becomes mandatory. As mentioned above we have in any case already initiated the generation of gluon field configurations on a $64^3 \times 128$ lattice which will allow us to provide a definite answer to this question.

5 Conclusion

With the computing time allocated to us on the Hazel Hen at HLRS we could achieve substantial progress in the evaluation of physical quantities related to the structure of hadrons and the search for new physics beyond the standard model. With these resources, we were able to analyze gluon field configurations that were generated on a $48^3 \times 96$ lattice at a lattice spacing of about $a \approx 0.09$ fm and, most importantly, the physical value of the pion and nucleon masses.

The latter fact avoids the extrapolation from unphysically large pion masses to the physical one and thus eliminates an often dominating systematic uncertainty. In this way, we could not only compute the spectrum of many hadrons observed in nature and reproduce their physical values but we could also address more complicated observables that are important to shed light on the inner structure of nucleons and serve as input for the interpretation of experiments looking for beyond the standard model physics: by computing the first moment of parton and gluon distribution functions, we can determine the quark and gluon average momentum in the proton; having the axial and tensor charges at our disposal we can also make a complete analysis of the proton spin; the neutron electric dipole moment is important to understand charge and parity violation from the strong interaction; finally having results for the so-called σ-terms, we can provide important input for dark matter search.

Interestingly and somewhat unexpected, we still see tensions of some quantities with experimental or phenomenological analyses. It will be most important to understand this tension by increasing the statistics of the present calculation and by also using a larger lattice of size $64^3 \times 128$ to address finite volume effects. It would be most reassuring to find agreement of these quantities with experimental/phenomenological results, which would open the road for a controlled analysis of further quantities such as generalized form factors [9, 10], the neutron electric dipole moment [11] or even the parton distribution functions themselves [12].

Acknowledgements We would like to thank all members of the European Twisted Mass Collaboration (ETMC) in which this work is embedded for a most enjoyable and fruitful collaboration. Without this common effort it would have not been possible to obtain the interesting and important results described in this report.

References

1. Abdel-Rehim, A., et al.: Simulating QCD at the Physical Point with $N_f = 2$ Wilson Twisted Mass Fermions at Maximal Twist (2015). 1507.05068
2. Abdel-Rehim, A., et al.: Nucleon and pion structure with lattice QCD simulations at physical value of the pion mass. Phys. Rev. **D92**, 114513 (2015). 1507.04936
3. Abdel-Rehim, A., et al. [ETM Collaboration]: Direct evaluation of the quark content of nucleons from lattice QCD at the physical point. Phys. Rev. Lett. **116**(25), 252001 (2016). doi:10.1103/PhysRevLett.116.252001. [arXiv:1601.01624 [hep-lat]]

4. Frezzotti, R., Rossi, G.C.: Chirally improving Wilson fermions. I: O(a) improvement. JHEP **08**, 007 (2004). hep-lat/0306014
5. Frezzotti, R., Rossi, G.C.: Twisted-mass lattice QCD with mass non-degenerate quarks. Nucl. Phys. Proc. Suppl. **128**, 193–202 (2004). hep-lat/0311008
6. Frezzotti, R., Rossi, G.C.: Chirally improving Wilson fermions. II: four-quark operators. JHEP **10**, 070 (2004). hep-lat/0407002
7. Abdel-Rehim, A., et al.: Disconnected quark loop contributions to nucleon observables in lattice QCD. Phys. Rev. **D89**, 034501 (2014). 1310.6339
8. Hoferichter, M., de Elvira, J.R., Kubis, B., Meißner, U.-G.: Remarks on the pion-nucleon sigma-term. arXiv:1602.07688 [hep-lat]
9. Alexandrou, C., et al.: Nucleon electromagnetic form factors in twisted mass lattice QCD. Phys. Rev. D **83**, 094502 (2011). doi:10.1103/PhysRevD.83.094502. [arXiv:1102.2208 [hep-lat]]
10. Alexandrou, C., Constantinou, M., Dinter, S., Drach, V., Jansen, K., Kallidonis, C., Koutsou, G.: Nucleon form factors and moments of generalized parton distributions using $N_f = 2 + 1 + 1$ twisted mass fermions. Phys. Rev. D **88**(1), 014509 (2013). doi:10.1103/PhysRevD.88.014509. [arXiv:1303.5979 [hep-lat]]
11. Alexandrou, C., Athenodorou, A., Constantinou, M., Hadjiyiannakou, K., Jansen, K., Koutsou, G., Ottnad, K., Petschlies, M.: Phys. Rev. D **93**(7), 074503 (2016). doi:10.1103/PhysRevD.93.074503. [arXiv:1510.05823 [hep-lat]]
12. Alexandrou, C., Cichy, K., Drach, V., Garcia-Ramos, E., Hadjiyiannakou, K., Jansen, K., Steffens, F., Wiese, C.: Lattice calculation of parton distributions. Phys. Rev. D **92**, 014502 (2015). doi:10.1103/PhysRevD.92.014502. [arXiv:1504.07455 [hep-lat]]

Numerical Evaluation of Multi-loop Feynman Integrals

Peter Marquard and Matthias Steinhauser

Abstract The main focus of our activities in the period from July 2015 to June 2016 was on the numerical computation of multi-dimensional integrals needed for the electron contribution to the anomalous magnetic moment of the muon.

1 Introduction

Many applications in particle physics require the computation of so-called Feynman integrals which have the following form

$$\int \mathrm{d}^d p_1 \cdots \mathrm{d}^d p_L \prod_i \frac{1}{k_i^2 - m_i^2} \,. \tag{1}$$

Here p_j and k_i are momenta in Minkowski space where k_i are linear combinations of p_j and external momenta q_l. m_i are scalar quantities which can be identified with the masses of the involved particles. The peculiarity of Eq. (1) is the non-integer dimensionality of each of the momentum integrals since $d = 4 - 2\epsilon$. Here ϵ is a regularization parameter which has been introduced since in general the integrals diverge in four space-time dimensions. Eventually, we are interested in the limit $\epsilon \to 0$ where the divergences manifest themselves as poles in ϵ. To obtain a physical quantity these poles are removed with the help of the renormalization procedure.

The main aim of this project is the numerical evaluation of integrals as given in Eq. (1) where $L = 4$, the masses m_i are either 0 or m, and for the only external momentum we have $q^2 = m^2$. Such integrals are the building blocks for the relation of quark masses defined in the $\overline{\mathrm{MS}}$ and on-shell scheme (as reported in the previous report) and the anomalous magnetic moment which was the main focus in the period from July 2015 to June 2016.

P. Marquard
Deutsches Elektronen-Synchrotron DESY, Platanenallee 6, 15738, Zeuthen, Germany

M. Steinhauser (✉)
Institut für Theoretische Teilchenphysik, Karlsruher Institut für Technologie, 76128, Karlsruhe, Germany
e-mail: Matthias.Steinhauser@kit.edu

W.E. Nagel et al. (eds.), *High Performance Computing in Science and Engineering '16*, DOI 10.1007/978-3-319-47066-5_8

The anomalous magnetic moment of the muon, a_μ, is among the most precisely measured quantities in particle physics. It is measured to a precision of 0.54 parts per million which matches the precision of the Standard Model theory prediction [1, 2]. However, since many years one observes a discrepancy of about three to four standard deviations which survives persistently all improvements both on the theory and the experimental side.

The theory prediction can be split into hadronic, electroweak and QED contribution (see Ref. [3] for further references). The non-perturbative hadronic contribution is further subdivided into the vacuum polarization and light-by-light contribution and has reached next-to-next-to-leading order accuracy [4]. It is nevertheless the main source to the uncertainty of the theory prediction. On the other hand, the electroweak part is known up to two-loop order and thus well under control. The numerically largest contribution arises from QED corrections. In this respect the four-loop corrections play a special role since their numerical impact is of the same order of magnitude as the discrepancy between theory and experiment. Note that the four-loop corrections have only been computed by one group using entirely numerical methods [3]. In the approach of our group the calculation proceeds analytically up to a point where a_μ is expressed as a linear combination of about 380 integrals of the type described above. A large fraction of them have been computed at the HLRS.

2 FIESTA

As already mentioned in the previous reports, the workhorse for our calculations performed at the HLRS is the program package `FIESTA` [5–7]. For convenience we repeat the main features of `FIESTA`.

`FIESTA` is developed since 2008 with the participation of the Institute for Theoretical Particle Physics (TTP) at KIT. `FIESTA` stands for Feynman Integral Evaluation by a Sector decompositTion and applies the method of sector decomposition [8] to obtain finite expression for the coefficients of the Laurent series of Eq. (1) in $\epsilon = (4 - d)/2$. These finite expressions are multi-dimensional parameter integrals with in general large integrands of the size of a few hundred MB up to a GB.

In practice the preparation of the integrand is performed within Mathematica on the local cluster. The expressions are transferred in form of a data base to the HLRS where the time-consuming Monte-Carlo integration is performed. `FIESTA` uses a simple master slave model for the parallelization, where the integrands are distributed from the master to the slaves using MPI and each term is integrated by a slave using a single core.

The scaling behavior of `FIESTA` is almost linear with the number of cores as has been demonstrated in the previous report.

Since the calculation of $g-2$ is technically closely related to the calculation of the $\overline{\mathrm{MS}}$–on-shell relation, which we reported on previously, we only needed about 13,000 node hours to complete the calculation of $g-2$. For some parts of the calculation it would be helpful to have a queue with a longer wall-time limit available to circumvent the considerable waiting time in the queue.

For our calculation we used up to 1536 cores on `Hornet`. For about 20 % of the integrals we used more than 240 cores. In this way we could respect the CPU time limit of 24 h and at the same time obtain the required precision.

All integrals were calculated multiple times with different precision to ensure the convergence of the Monte-Carlo integration. We observe the expected $1/\sqrt{N}$ behavior in the reduction of the statistical uncertainty where N is the number of sample point.

3 Anomalous Magnetic Moment of the Muon

The numerically most important four-loop contribution comes from the Feynman diagrams which contain a closed electron loop since in that case numerically large logarithms $\log(m_\mu/m_e) \approx \log(206) \approx 5.3$ are present which in some cases are even raised to fourth power. Let us in the following concentrate on these kind of corrections. Sample Feynman diagrams are shown in Fig. 1

The underlying Feynman integrals to the Feynman diagrams in Fig. 1 contain two mass scales: m_e and m_μ. Since $m_e \ll m_\mu$ it is natural to perform an expansion in the mass ratio. From the appearance of logarithmic contributions mentioned above it is obvious that a naive Taylor expansion is not appropriate and would lead to wrong results. Instead we apply a so-called asymptotic expansion which is an algorithmic prescription to factorize the original expressions into integrals which depend either on m_e or on m_μ.

As a example let us present our results for the diagram class III (cf. Fig. 1). Leaving out the overall factor $(\alpha/\pi)^4$ the contribution to a_μ reads [9]

$$\begin{aligned}
A_2^{(8),\mathrm{III}} &= 1.15444 \pm 0.00446 - 1.80996\ell_x \\
&\quad + x[-0.849197] \\
&\quad + x^2[-1.95556 \pm 0.00400 - 1.25333\ell_x] \\
&\quad + x^3[-20.2365 - 15.3527\ell_x] \\
&= [1.1544 \pm 0.0045 + 9.6500] \\
&\quad + [-0.004107] \\
&\quad + [-0.00004574 \pm 0.00000009 + 0.00015630] \\
&\quad + [-0.000002289 + 0.000009260]
\end{aligned}$$

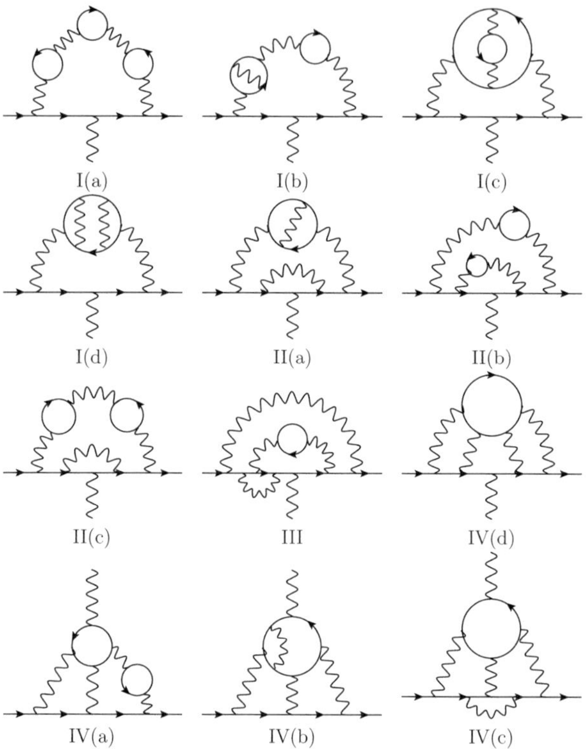

Fig. 1 Four-loop example Feynman diagrams contributing to a_μ containing at least one closed electron loop. The external *solid lines* represent muons, the *solid loops* denote electrons or muons and the *wavy lines* represent photons

$$
\begin{aligned}
&= [10.8044 \pm 0.0045] \\
&\quad + [-0.004107] \\
&\quad + [0.00011056 \pm 0.00000009] \\
&\quad + [0.000006970] \\
&= 10.8004 \pm 0.0045\,,
\end{aligned}
\tag{2}
$$

where in the first row the expansion in x and its dependence on $\ell_x = \log(x)$ is shown. After the first equality sign the numerical values of x and $\ell_x \approx -5.33\ldots$ are inserted, but the resulting summands are kept separated, which indicates the relative behavior between the constant and the logarithmic terms. Afterwards the sums for every order in x are evaluated to demonstrate the convergence of the asymptotic series. At the end the final contribution of the diagram class is shown.

Note that our result in Eq. (2) is in perfect agreement to the result of Ref. [3] which reads 10.7934 ± 0.0027.

Our final result for $A_2^{(8)}(m_\mu/m_e)$ is given by [9–11]

$$A_2^{(8)} = A_2^{(8),\text{lbl}} + A_2^{(8),\text{rem}} = 126.34(38) + 6.53(30) = 132.86(48)\,. \quad (3)$$

Our numerical uncertainty amounts to approximately $0.5 \times (\alpha/\pi)^4 \approx 1.5 \times 10^{-11}$. It is larger than the uncertainty in Ref. [3]. Nevertheless it is sufficiently accurate as can be seen by the comparison to the difference between the experimental result and theory prediction which is given by[1]

$$a_\mu(\text{exp}) - a_\mu(\text{SM}) \approx 249(87) \times 10^{-11}\,. \quad (4)$$

Note that the uncertainty in Eq. (4) receives approximately the same amount from experiment and theory (i.e. the hadronic contribution). Even after a projected reduction of the uncertainty by a factor four both in $a_\mu(\text{exp})$ and $a_\mu(\text{SM})$ our numerical precision is a factor ten below the uncertainty of the difference.

4 Outlook

In the following years we plan to consider the universal contribution to the leptonic anomalous dimension where only one lepton flavor is present. Furthermore, we plan to extend the results of Ref. [12] and to present the relation between the $\overline{\text{MS}}$ and on-shell quark mass for generic number of colors and massless quark flavors. Both applications require more precise master integrals which we plan to compute on the `Hazel Hen` cluster at the HLRS.

References

1. Bennett, G.W., et al., [Muon G-2 Collaboration]: Final report of the Muon E821 anomalous magnetic moment measurement at BNL. Phys. Rev. D **73**, 072003 (2006). [hep-ex/0602035]
2. Roberts, B.L.: Status of the Fermilab Muon $(g-2)$ experiment. Chin. Phys. C **34**, 741 (2010). [arXiv:1001.2898 [hep-ex]]
3. Aoyama, T., Hayakawa, M., Kinoshita, T., Nio, M.: Complete tenth-order QED contribution to the Muon g-2. Phys. Rev. Lett. **109**, 111808 (2012). [arXiv:1205.5370 [hep-ph]]
4. Kurz, A., Liu, T., Marquard, P., Steinhauser, M.: Hadronic contribution to the Muon anomalous magnetic moment to next-to-next-to-leading order. Phys. Lett. B **734**, 144 (2014). doi:10.1016/j.physletb.2014.05.043. [arXiv:1403.6400 [hep-ph]]
5. Smirnov, A.V., Tentyukov, M.N.: Feynman integral evaluation by a sector decomposition approach (FIESTA). Comput. Phys. Commun. **180**, 735 (2009). [arXiv:0807.4129 [hep-ph]]. Preprint No. TTP08-32

[1] This result is taken from Ref. [3].

6. Smirnov, A.V., Smirnov, V.A., Tentyukov, M.: FIESTA 2: parallelizeable multiloop numerical calculations. Comput. Phys. Commun. **182**, 790 (2011). [arXiv:0912.0158 [hep-ph]]. Preprint No. TTP09-39
7. Smirnov, A.V.: FIESTA 3: cluster-parallelizable multiloop numerical calculations in physical regions. Comput. Phys. Commun. **185**, 2090 (2014). [arXiv:1312.3186 [hep-ph]]
8. Heinrich, G.: Sector decomposition. Int. J. Mod. Phys. A **23**, 1457 (2008). [arXiv:0803.4177 [hep-ph]]
9. Kurz, A., Liu, T., Marquard, P., Smirnov, A., Smirnov, V., Steinhauser, M.: Electron contribution to the Muon anomalous magnetic moment at four loops. Phys. Rev. D **93**(5), 053017 (2016). doi:10.1103/PhysRevD.93.053017. [arXiv:1602.02785 [hep-ph]]
10. Kurz, A., Liu, T., Marquard, P., Smirnov, A.V., Smirnov, V.A., Steinhauser, M.: Light-by-light-type corrections to the muon anomalous magnetic moment at four-loop order. Phys. Rev. D **92**(7), 073019 (2015). doi:10.1103/PhysRevD.92.073019. [arXiv:1508.00901 [hep-ph]]
11. Kurz, A., Liu, T., Marquard, P., Smirnov, A.V., Smirnov, V.A., Steinhauser, M.: Higher order hadronic and leptonic contributions to the Muon $g - 2$. arXiv:1511.08222 [hep-ph]
12. Marquard, P., Smirnov, A.V., Smirnov, V.A., Steinhauser, M.: Quark mass relations to four-loop order. Phys. Rev. Lett. **114**(14), 142002 (2015). [arXiv:1502.01030 [hep-ph]]

Part II
Molecules, Interfaces, and Solids

Christoph van Wüllen and Holger Fehske

The following chapter reveals that chemistry, material science and solid state physics have profited substantially from the computational resources provided by both the High Performance Computing Center Stuttgart and the Steinbuch Centre for Computing Karlsruhe. A particular challenge was the multi-scale character of the problems the projects were concerned with in the simulations.

From the broad range of the field, seven contributions have been selected for presentation. First-principle DFT and MD calculations have been used to study molecular systems such as molecules bound to gold surfaces and water structures at the water-mineral interface, and to study processes that could be termed as 'molecular engineering' of semiconductors. As to the modelling of structural, electronic and transport properties of solids, such methods have been used to study rare earth silicide thin films, lithium ion diffusion through NZP structures, and laser ablation in covalently bonded materials. The last contribution presents a time-dependent DMRG treatment of quantum transport in nano-devices attached to metallic leads.

The work by D. Marx, M. Wollenhaupt and M. Z. Michoff from the University of Bochum applies first-principles molecular dynamics methods to study in detail what happens if mechanical forces act on a molecular system, how the molecules stretch and finally how bond rupture takes place. They compared aliphatic and aromatic

C. van Wüllen (✉)
Fachbereich Chemie, Technische Universität Kaiserslautern
Erwin-Schrödinger-Str. 52, 67663 Kaiserslautern, Germany
e-mail: vanwullen@chemie.uni-kl.de

H. Fehske
Institut für Physik, Lehrstuhl Komplexe Quantensysteme,
Ernst-Moritz-Arndt-Universität Greifswald

Felix-Hausdorff-Str. 6, 17489 Greifswald, Germany
e-mail: fehske@physik.uni-greifswald.de

thiolates bound to a gold surface, and the interesting observation was, that although aliphatic thiolates bind more strongly to the gold surface (higher thermal desorption energy), the mechanical force required to pull the molecule off the surface is lower than for aromatic thiolates. The reason is a different mechanochemical pathway, namely a Au-Au bond rupture for aliphatic thiloates vs. Au-S bond rupture for aromatic thiolates. In a related investigation, it was investigated how thiotic acid with a PEG chain stretches under mechanical force, and how chemical its reactivity is modulated by the mechanical force. It was found that stretching molecule facilitates nucleophilic attac at one of the sulfur atoms. The QuantumESPRESSO and CPMD programs have been used to carry out these calculations. Using several levels of parallelization, these codes offer good scaling up to several thousands of processors.

K. Remi and M. Sulpizi from the University of Mainz investigate the interface between water and fluorite (CaF_2) at a microscopic level. Such water-mineral interactions are of great importance in materials science but there are also environmental and medical issues. Experimentally, vibrational sum frequency generation (VSFG) is used to obtain spectroscopic signatures from the interface area, since only an assembly of *oriented* water molecules and not bulk (isotropic) water generates a VSFG signal. In the simulations, such a signal is obtained from a molecular dynamics run using quantum mechanical forces (Born-Oppenheimer molecular dynamics using density functional methods, as implemented in the CP2K program), and calculating the contribution of (only) the interfacial water molecules to the time-dependent dipole moment. This has been done for different setups (low, neutral and high pH), and the molecular origin of signatures observed in experimental VSFG spectra could be clarified.

The group of R. Tonner at the University of Marburg studies 'molecular engineering' of functional semiconductors. Experimentally, thin films are grown on materials such als silicon using metal-organic vapour phase epitaxy. The calculations assess the thermodynamic stability of such phases. For example, it is demonstrated that G(NAsP) is stabilized (with respect to segregation into GaN, GaAs and GaP islands) if grown on Si(001) because of the strain imposed by the epitaxial growth process. It has been further found out that bismuth atoms in dilute Ga(AsBi) are not distributed evenly but tend to cluster, and this has some effect on how the band gap of the host GaAs is modified by Bi doping. Finally, a new mechanism of how in-plane vibrations of planar molecules adsorbed on a metal surface aquire infrared intensity has been studied. These calculations used the VASP program which allows making efficient use of modern supercomputers in highly parallel calculations.

The material physics group at the University Paderborn, conducted by W. G. Schmidt, combines energy density functional (DFT) calculations with ab initio thermodynamics to provide a microscopic structural model for the silicon thin film 5×2 phase observed in the sub-monolayer regime, where the lack of theoretical investigations is particularly severe. Within their DFT framework the generalised gradient approximation is used, in Perdew-Burke-Ernzerhof formulation, as implemented in the VASP package. The authors demonstrate that the 5×2 structure is characterised by alternating Si honeycomb and Seiwatz chains, with rare earth atoms located in the interjacent channels. The simulated data impressively agrees with measured STM images of the sub-monolayer structure on a Si(111) surface.

Materials crystallising in the NZP structure, named after $NaZr_2(PO_4)_3$, with Na exchanged by Li, are promising candidates for solid-state electrolytes, primarily because they form a 3D diffusion network for Li ions. C. Elsässer and collaborators from the University Freiburg and the Fraunhofer Institute for Mechanics of Materials analysed the diffusion of Li through various NZP compounds by combining DFT simulations based on *Quantum ESPRESSO PWscf* code and static energy calculations with bond valence potentials. In this way important structure-property relationships could be identified, which allowed, e.g., to predict the migration barrier heights directly from crystal structure characteristics. For LISICON, which possesses high ionic Li conductivities under special conditions, the Ti and P were substituted by a variety of isovalent elements, in order to discuss the influence on Li-ion diffusion. Calculating the activation energies for the migration of a Li vacancy, the authors found out that the Li ion can escape the cage through the bottleneck much easier when the coordination polyhedron around Li-ion is larger.

An outstanding example for the predictive power of large-scale (multi-million particle) molecular dynamics (MD) simulations is the work by A. Kiselev, J. Roth, and H.-R. Trebin from the Institute for Functional Matter and Quantum Technologies at the University Stuttgart. Within the framework of a self-consistent continuum-atomistic two-temperature (TTM) approach the authors model carrier-lattice interaction and electron-hole recombination processes in covalently bound materials subject to strong laser radiation fields. To this end the hitherto existing MD simulations for metals with high charge carrier concentration are developed further to treat semiconductors where charge carriers have to created first by the laser pulse. The focus is on laser ablation in silicon. Here, dynamical interactions, depending on the electron temperature, have to be taken into account. Compared to a simple rescale model (with laser energy introduced by a rescaled kinetic energy of the particles), the ablation thresholds are much higher and the material is completely vaporised. The results demonstrate the importance of the combined MD-TTM algorithm with temperature adapted potentials. Beyond doubt such a approach paves the way for the numerical treatment of general non-equilibrium phenomena in highly excited covalent systems.

The challenging problems of non-linear quantum transport and transient dynamics of currents in strongly interacting nano-structures were addressed by B. Schoenauer from the Center for Extreme Matter and Emergent Phenomena at Utrecht University and P. Schmitteckert from the Institute for Theoretical Physics and Astrophysics at the University Würzburg. By means of an elaborate time-dependent density matrix renormalisation group (td-DMRG) technique, the authors studied a paradigmatic model system, where an interacting ring-structure is sandwiched between two metallic leads. To prepare the system in a state of non-equilibrium it is quenched by applying an external potential to the leads (which is switched of for all times $t > 0$). Krylov subspace methods are used to facilitate the calculation of the matrix exponential function needed to describe the time evolution. This way, the authors try to answer the question of whether local observables, such as the currents through the links, always relax to a steady state. However, it turned out that oscillating ring-currents, which are orders of magnitude larger than transmitted currents, dominate the transport, in particular for strong interactions. Results obtained for different voltages show that the frequency of this oscillation is independent of the bias, and does not decay on the time scales accessible to the calculations (which, however, not completely rules out a relaxation of the currents inside the ring structure in the very long-time limit). A detailed (td-DMRG) analysis of the time dependence of the reduced density matrix sheds light on the states that contribute to the current oscillation.

We finally like to emphasise that all projects introduced by the reviewers have in common, besides a high scientific quality, the need for powerful computers to achieve their results. That is why the leading-edge systems at the HLRS and KIT SCC are a prerequisite for such ambitious research.

Mechanochemistry of Ring-Opening Reactions: From Cyclopropane in the Gas Phase to Thiotic Acid on Gold in the Liquid Phase

Martin Zoloff Michoff, Miriam Wollenhaupt, and Dominik Marx

Abstract We have studied the mechanochemistry of a system consisting of a PEG molecular model with a bidentate thiolate anchor covalently attached to a metallic surface and subject to solvation. Functionalized gold nanoparticles show increasing potential in real-life applications. In these situations, they are exposed to different chemical environments and to mechanical tension due to contact and friction with other molecules. We have used accelerated ab initio molecular dynamics to study the effect of mechanical forces on the chemical reactivity of such a system. Although the results reported here can only be considered as preliminary, a very interesting mechanochemical behavior is displayed by the system under study. We have focused on the attack of OH^- on either anchoring S atom. We have determined the free energy of activation for each reaction at two values of a constant external force acting on the molecule and, thus, stressing the gold–thiolate interface. We have also studied the thermal and mechanical detachment of aliphatic and aromatic thiolates adsorbed on gold surfaces. In this case, we have found an interesting change in mechanism of the mechanical detachment that depends on the nature of electronic properties of the carbon skeleton, but that remains unchanged for substituents of widely different nature on the aromatic ring.

1 Scientific Background

The conventional way of starting chemical reactions – for example in an industrial process but also in a flask in the laboratory – is to use thermal, photochemical, or electrochemical activation [1]. These techniques have been applied for hundreds of

M. Zoloff Michoff (✉) • M. Wollenhaupt • D. Marx
Lehrstuhl für Theoretische Chemie, Ruhr-Universität Bochum, 44801, Bochum, Germany
e-mail: martin.zoloff@theochem.ruhr-uni-bochum.de;
miriam.wollenhaupt@theochem.ruhr-uni-bochum.de;
dominik.marx@theochem.ruhr-uni-bochum.de

W.E. Nagel et al. (eds.), *High Performance Computing in Science and Engineering '16*, DOI 10.1007/978-3-319-47066-5_9

years; taking for example the procedure of cooking, the effect of adding heat was efficiently used even though no one knew what was happening on a molecular level.

Nowadays recently developed methods such as atomic force microscopy [2] or sonochemical processes in an ultrasonic bath [3] provide the possibility of applying mechanical forces to single molecules while monitoring the stress and the resulting change in the structure of molecules.

In the field of theoretical chemistry, codes for calculating geometries, energies, and even reaction pathways as a function of constant force rather than upon imposing structural constraints (i.e. isotensional versus isometric stretching, see Ref. [1] for a comprehensive discussion) have been implemented and successfully used. All this research is related to the field of (covalent) mechanochemistry, which deals with the influence of external mechanical forces on molecules and, in particular, on their reactions, see Refs. [1, 4, 5] for reviews.

Thiolate–gold interfaces have been intensely studied for many decades using a wide array of experimental and computational methodologies [6, 7]. The pronounced interest in these particular hybrid molecule/metal junctions and interfaces is due to a multitude of potential applications such as tailoring the properties of surfaces [8–12], chemical anchors for molecular electronics applications [13–15] or coating agents for the stabilization of gold nanoparticles [16, 17]. Thus, the results obtained by this study will be of great relevance to diverse research fields, such as tailored surfaces for electrochemical processes, molecular electronics and in medicinal applications, just to mention a few. For instance, the potential applications of gold nanoparticles, AuNPs, in several medical applications, such as X-ray imaging, photothermal therapy, radiotherapy and targeted drug delivery just to mention a few, has attracted an enormous attention in the last few years [18–28].

To prevent agglomeration and to increase their circulation time in the blood stream, nanoparticles are typically coated with a biocompatible polymer such as polyethylene glycol, PEG [29]. The chemical modification of AuNPs with PEG ligands has been usually performed by covalently attaching PEG chains to the metallic surface through a thiolate linkage. In the first studies on PEG-AuNPs, the PEG ligands employed were terminated by a monothiol functional group [30]. However, in most recent works, multivalent thiol linkers are being used. One of the most common examples are PEG ligands that are appended to thioctic acid, TA (or dyhydrolipoic acid, DHLA, in its reduced form) [31–36]. This functional group provides with a bidentate anchoring to gold surfaces. Experimental studies have shown that TA terminated PEG ligands provide an enhanced colloidal stability to AuNPs under a wide range of experimental conditions with respect to their monothiolated counterparts [33–36].

Despite the steeply increasing interest in exploiting multidentate thiols for functionalization, all experimental as well as theoretical studies performed so far have only dealt with the structure and thermodynamic stability of these adsorbates. Information about how such multidentate anchoring functionalities would respond to external stress is scarce whereas monothiol-based interfaces and point contacts have been intensely studied [37–47]. To the best of our knowledge, only one experimental study dealing with the mechanical rupture of a dithiolate linkage to gold has been reported [48]. Interestingly, the measured force required to remove a single TA molecule from the gold substrate resulted in an about 3.4 times *smaller* rupture force compared to that of a simple Au–S bond. This suggests that SAMs of multidentate thiols may be *less* stable under tensile stress than anticipated from their thermal properties via thermal desorption experiments and computed binding energies.

To shed some light into this open topic, we have carried out a comprehensive computational study on the thermal and mechanical detachment of a series of bidentate thiolates adsorbed on a gold surface [49]. In this work, the effect of the chain length separating sulfur atoms has been studied. It was found that thermal desorption always yields cyclic disulfides. In contrast, mechanochemical desorption leads to cyclic gold complexes, where metal atoms are extracted from the surface and kept in tweezer-like arrangements by the sulfur atoms. Interestingly, the flexibility of the chain is shown to crucially impact on the mechanical strength of the junction. Given these insights, what remained to be explored is to what extent solvent effects might affect the rupture scenario and thus the mechanical strength of nanojunctions.

We have carried out a systematic computational study, comparing mono and bidentate thiolate ligands, that expands our knowledge and provides with key information in order to understand in a very detailed level the mechanochemical behavior of the thiolate–gold interface.

On one hand, we have investigated the mechanical and thermal desorption of prototypical aliphatic alkanethiolates, such as ethyl (**Et-S**) and butylthiolate (**Bu-S**), and a series of substituted *p*-methyl thiophenolates adsorbed on gold surfaces (see Fig. 1). Since the detailed adsorption structure thiolate ligands on gold is still being debated [6, 7, 50], we have considered two different models for the adsorption of the molecules: a perfect flat Au(111) surface and a surface with a vacancy defect on the adsorption site.

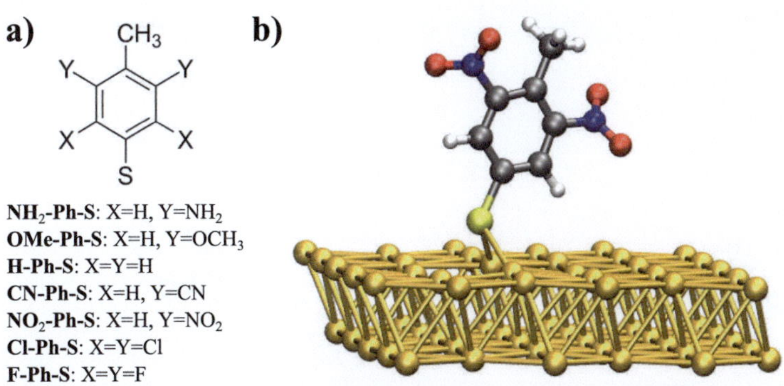

Fig. 1 (**a**) Molecular structure of the substituted thiophenols studied. (**b**) Adsorbate of **Ar–NO_2–S** on a perfect Au(111) surface, being shown as an illustrative example of the initial structure for the thermal and mechanical desorption. Note that for clarity purposes, only the two top layers of the Au slab are shown

The surface has been modeled using 4 layers of a 5×6 slab, using ca. 15 Å of vacuum in the direction perpendicular to the surface to avoid spurious interactions between the periodic images. This approach has been well–established in the Marx group also within mechanochemistry. To explore the mechanical desorption of the proposed systems, we have used the "*isometric*" approach, in which the carbon of the methyl group that is common to all molecular species, was constrained to move in a plane parallel to the bottom gold layer of the slab. The atoms in this layer were kept fixed at their bulk positions throughout the simulations. The distance between these two planes can be termed as the "*stretching parameter*" (*D*), which was increased stepwise by increments of 0.2 Å until final breakage of the molecule–metal junction was observed.

On the other hand, we have also explored the mechanochemistry of such hybrid interfaces in more realistic conditions, that is to say including the effects of finite temperature and those can arise from a fully explicit solvation environment.

In this study we have focus on the PEG ligands attached to a gold surface by means of a bidentate thiolate linking functionality such as the thioctic acid (**TA**, see Fig. 2).

As it was already mentioned, the **PEG-TA** system is widely used in experiments to coat gold surfaces. Thus, there is a high interest in this system, but little is yet known about its response to an external mechanical stress.

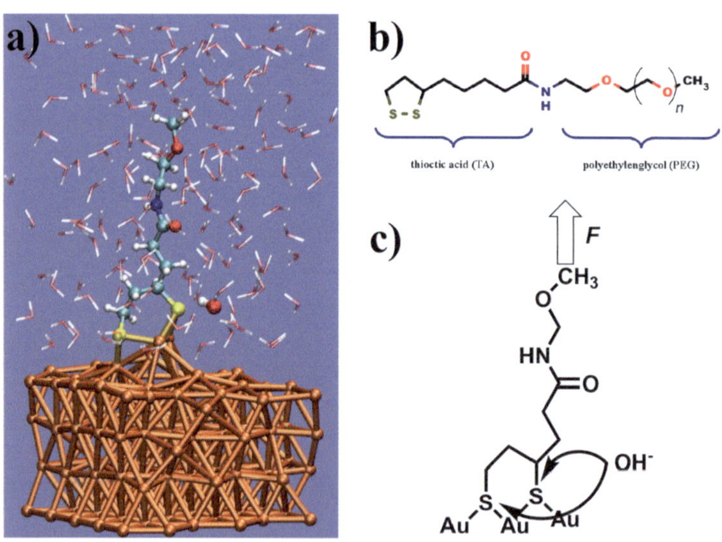

Fig. 2 (**a**) Illustration of the system studied using AIMD simulations, showing the Au slab representing the metallic surface, a model of the **PEG-TA** conjugate adsorbed on it, and the solvation environment described explicitly at an atomistic level. An OH^- molecule has been highlighted using larger spheres. (**b**) Molecular structure of the **PEG-TA** conjugate. (**c**) Scheme showing possible reaction pathways to be explored in the presence of OH^-. The *thick arrow* indicates the external force, explicitly included in the AIMD simulations and located on the carbon of the terminal methyl group of the molecule and directed perpendicular to the metallic surface

To study this system, we have used ab initio molecular dynamics (AIMD) simulations, in which the solvent molecules were explicitly included.

We have focused on the effect of the external force on the reactivity of this particular thiolate–gold interface in the presence of OH^-. The reaction pathway for attack of the nucleophilic species on either anchoring S atom has been explored using an advanced enhanced sampling technique, such as thermodynamic integration.

The effect of the an external force was explicitly included in the AIMD simulations. This "*isotensional*" approach involves the concept of the "force transformed free energy surface" [51]. This approach has been successfully used to tackle the covalent mechanochemistry of molecular systems such as cyclopropane rings [51, 52] and diethyl disulfide [53, 54]. This constitutes, as far as we know, the first attempt to study the reactivity of a fully solvated molecule–metal junction under the effect of constant external tensile stress.

2 Results and Discussion

2.1 *Mechanochemistry of Aliphatic and Aromatic Thiolates on Gold Surfaces*

Thermal desorption was determined by energy computing the energy difference between the optimized structure of the molecule adsorbed on the surface and the corresponding surface and molecule separated at infinite distance. For the mechanical desorption, by using the "isometric approach" described above, we obtained the energy profile as a function of the *D*, from which the force profile as a function of *D* can be obtained straightforward by simple derivation. The energy profiles typically show a series of elastic and plastic deformations. The rupture force, F_{rup} is obtained as the maximum value along the stretching pathway for the last elastic portion of the curve, ie. leading to final breakage of the gold–thiolate interface. These results are summarized in Table 1.

Table 1 Thermal desorption energies (E_{des}), and mechanical rupture force (F_{rup}). The coloring indicates if the mechanical desorption occurs through a Au–Au (**blue**) or a S–Au (**red**) bond rupture

Molecule	Au(111) Flat		Au(111) Vacancy	
	E_{des}, eV	F_{rup}, nN	E_{des}, eV	F_{rup}, nN
Bu-S	2.15	1.68	2.64	1.75
Et-S	2.15	1.46	2.63	1.72
NH_2-Ph-S	1.74	2.25	2.22	1.82
OCH_3-Ph-S	1.69	2.20	2.15	1.82
H-Ph-S	1.69	2.21	2.06	1.74
CN-Ph-S	1.69	2.11	2.20	1.80
NO_2-Ph-S	1.68	2.32	2.06	1.77
Cl-Ph-S	1.55	1.73	1.98	2.24
F-Ph-S	1.54	1.69	2.01	2.19

On one hand, it can be seen that aliphatic thiolates have a higher desorption energy than the thiophenolates. Within the aromatic derivatives, there is no definite trend regarding the nature of the substituent. This was also observed for *p*-substituted thiophenolates adsorbed on gold surfaces [55], and can be explained in terms of the nature of the S–Au bonding interactions. Although the **Cl-Ph-S** and **F-Ph-S** derivatives display a mild decrease in the E_{des}, this could also be attributed to steric effects from the *ortho* substituents.

Interestingly, for the flat Au(111) all aromatic thiophenolates display a similar mechanical desorption mechanism regardless of the nature of the substituent, and which notably differs from the one displayed by the aliphatic derivatives. For the latter, the breakage always occurs at a Au–Au bond, with a F_{rup} of $\sim$1.6 nN, whereas for the aromatic disubstituted molecules the final rupture takes place at a S–Au, with F_{rup} values of $\sim$2.2 nN. Thus, the aliphatic thiolates display a *higher* thermal desorption energy, but are mechanically detached with a F_{rup} that is $\sim$37 % *lower* than that of the aromatic thiolates.

To determine whether this detachment scenario depends on the detailed structure of the S–Au junction, additional pulling computational experiments were carried out in which **Bu-S** and **H-Ph-S** were scrambled at an early stage in the opposite mechanical desorption pathway. This is illustrated in Fig. 3. As it can be appreciated,

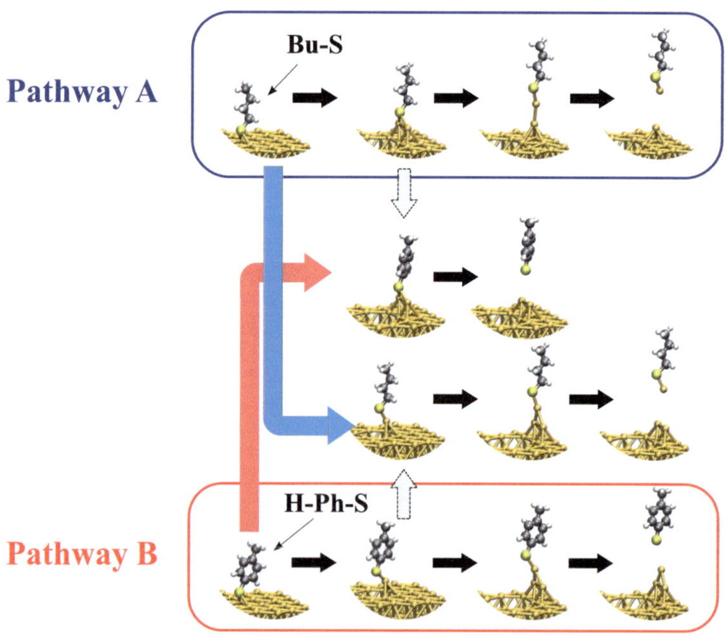

Fig. 3 Illustrative scheme of the two different mechanical detachment pathways observed for aliphatic (Pathway A, top) and aromatic (Pathway B, bottom) thiolates on flat Au(111). Only some key structures along the pathway are being sequentially shown. In each case, the S–Au junction taken from the second relevant structure along the pathway is kept and the carbon skeleton of the molecule exchanged. Illustrative structures of the mechanical detachment pathways originated from these starting points are shown

the aliphatic molecule detaches via a Au–Au rupture, whereas the aromatic thiolate does so through the breakage of a S–Au bond, regardless of the detailed structure of the Au–S junction. Several parameters derived from the electronic structure, such as charges and bond orders correlate well with these observations.

Finally, the presence of a vacancy defect on the adsorption site provides with less coordinated Au atoms to which the molecule attach more strongly, as it is reflected in the E_{des} values in Table 1. This does not change the mechanical detachment scenario for the aliphatic thiolates, but it does for the thiophenolates. The aromatic derivatives display now a Au–Au breakage with a F_{rup} of 1.8 nN, which means that a higher adsorption interaction leads to a lower mechanical stability.

We are now performing more analysis to have a deeper understanding of this phenomena, and we foresee that these results will have great impact in the design of such hybrid molecule/metal interfaces.

2.2 Mechanochemical Activation of Hydroxide Attack on the Anchoring Moiety of PEG-Thioctic Acid Adsorbed on a Gold Surface

To define the minimum model for the **PEG-TA** conjugate, we have studied the properties under stress of a molecule composed of thioctic acid with two ethyleneglycol units appended (**LONG**). This is the largest molecular system we consider it could be feasible to treat in AIMD in a full solvated simulation box.

We have also determined the mechanical properties of two shorter models of the **PEG-TA** conjugate: one with one less ethyleneglycol unit (**SHORT-1**), an another one using a modified version of the thioctic acid with two CH_2 units less (**SHORT-2**). The comparison of the mechanical properties of the three proposed molecular models is shown in Fig. 4.

As it can be observed, most of the relevant mechanical features displayed by the largest model considered are retained by the model labeled as **SHORT-2**. This will be the molecular model chosen to represent the **PEG-TA** conjugate in our simulations.

In order to study the possible adsorption structures of **PEG-TA** on gold surfaces, many different adsorption sites and geometries have been probed using as a model the cyclic portion of the thioctic acid molecule. To account for different bonding scenarios, three types of surfaces were considered: flat Au(111), and two types of point defects, a vacancy and an adatom. The most stable structures found for each type of surface are shown in Fig. 5. The corresponding desorption energies to give the cyclic disulfide are 1.27 eV, for the flat Au(111) surface, 1.76 eV for the surface with a vacancy and 1.46 eV for the surface with a gold adatom.

Then we studied the mechanical detachment of the minimum **PEG-TA** model previously determined adsorbed on the Au(111) surface with an adatom. The results are shown in Fig. 6a. For comparison purposes, we have also included the results

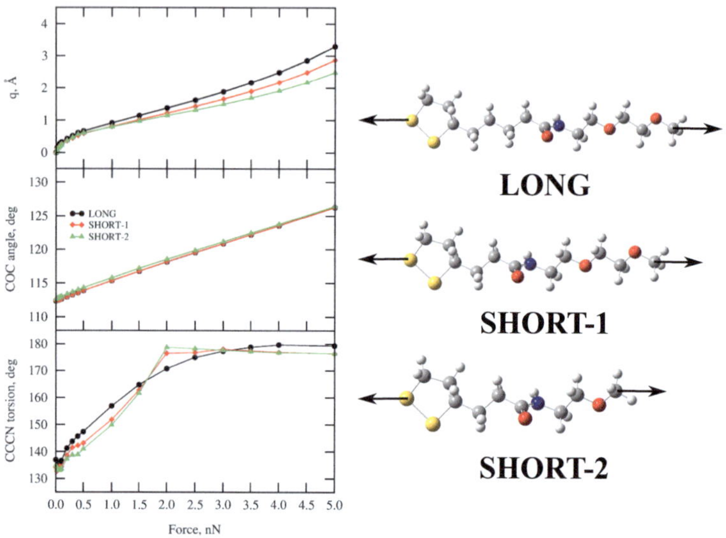

Fig. 4 Mechanical properties of the proposed molecular models for **PEG-TA** as a function of the external force applied, *left panel* from top to bottom: stretching coordinate q (distance between the atoms to which the force is applied); $C(H_2)$–O–$C(H_3)$ angle; and C–C–C(=O)–N torsion angle. The molecular structures are illustrated in the *right panel*. The *arrows* indicate the atoms on which the external force is applied to

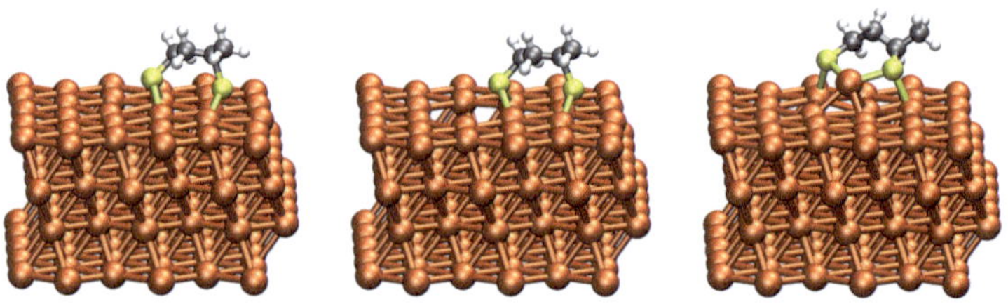

Fig. 5 Most stable structures found for the adsorption of the cyclic portion of thioctic acid on different gold surfaces. From left right, adsorbates on flat Au(111), surface with a vacancy and surface with an adatom

from the mechanical desorption of the cyclic moiety in **TA** adsorbed on a defective Au(111) surface with one vacancy (see Fig. 6b) [49]. Both simulations were carried out in vacuum.

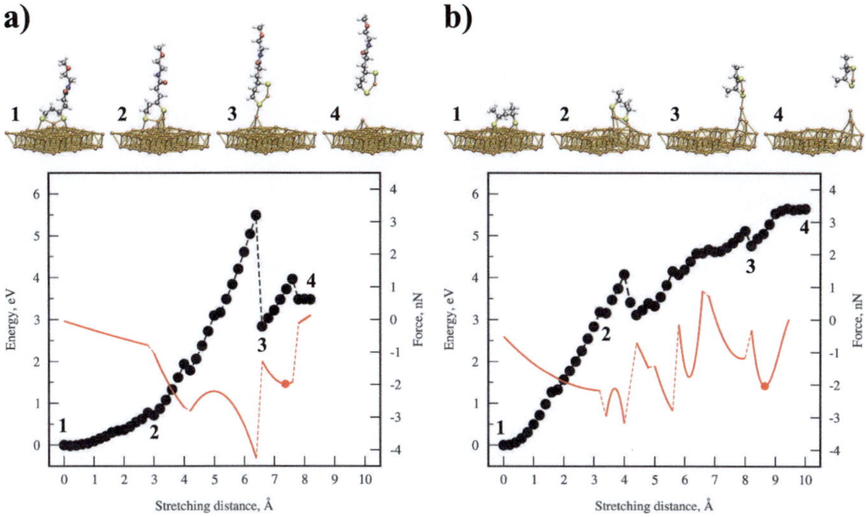

Fig. 6 Pathway of mechanical desorption of (**a**) **PEG-TA** model on a gold surface with an adatom, and (**b**) Cyclic moiety of **TA** on a gold surface with a vacancy. In both cases the following plots are shown: Total electronic energy along the mechanical desorption pathway (*filled black circles*, left axis), force *versus* distance curves for regions of elastic deformation (*solid red lines*, right axis); the connecting *broken red lines* are merely guides to the eye through discontinuous plastic deformation events, as a function of the stretching parameter *D*. The *filled red circle* indicates the F_{rup} value. Some relevant structures along each stretching pathway are shown on the top of each plot

Two points should be specially noted: on one hand, the mechanical stretching pathways does not differ much when these two defective surfaces are considered. On the other, as expected, the PEG chain does not greatly influence the mechanical stability of the molecule–metal junction. The main differences are noted at the initial stages of the stretching of the **PEG-TA** model, which corresponds to the unfolding of its soft dihedral degrees of freedom. Most noticeably, at the last stage just before the final breakage the detailed geometry of the molecule–metal contact is very similar in both cases. It corresponds to the molecule being bonded to the surface by one sulfur atom and with one metallic atom complexed by both sulfur atoms in a tweezer-like arrangement. Further stretching leads to the detachment of the final product, a cyclic complex with one gold being extracted from the surface, by means of a S–Au bond rupture with a very similar rupture force, F_{rup}, value: 2.05 nN for the cyclic moiety initially adsorbed on the surface with a vacancy, and 1.99 nN for the **PEG-TA** model on the surface with an adatom.

From the atomic charges evolution of the stretching of the **PEG-TA** model, we could determine that the effect of the external force has the largest impact on the sulfur atoms. These atoms become more positive as the molecule–metal junction is stretched during the first stages, making them prone to an attack by a nuclephilic species such as OH^- (aq). Because the attachment point of the chain to which the force is applied is not symmetrical with respect to the S–Au bonds, then it can be foreseen that the external stress will not be equally transduced to both thiolate–gold linkages. Therefore, we considered of interest to examine the effect of the external force on the free energy pathway for the attack of OH^- (aq) on both sulfur sites. For this purpose, our starting point was a pre-stretched **PEG-TA** structure on gold, obtained from our preliminary "in vacuo" study. After solvation and equilibration with water and one OH^- impurity of the selected structure, we then proceeded with the AIMD simulations at constant force. We started from a somewhat lower force than the one used to pre-stretch **TA-PEG**, and then the constant force was increased in a stepwise manner. Using this procedure, we covered a range of 1.2–2.6 nN, up to now. We are still running simulations at higher forces, since we have not yet observed the detachment of the molecule from the surface.

Notably, at a value of 2.2 nN a structural change at the gold–thiolate interface is observed. This is illustrated in Fig. 7. At low forces, both anchoring sulfur atoms are attached to two gold atoms, with one common atom to which both S atoms are attached to. As the force increases, for the S atom one labeled as S_1 in Fig. 7, the S–Au bond with central Au atom is notably elongated.

This observation suggests that a value of 2.2 nN could be a threshold beyond which the relative reactivity of the anchoring S atoms may dramatically change. We then proceeded to assess this hypothesis by means of determining the free

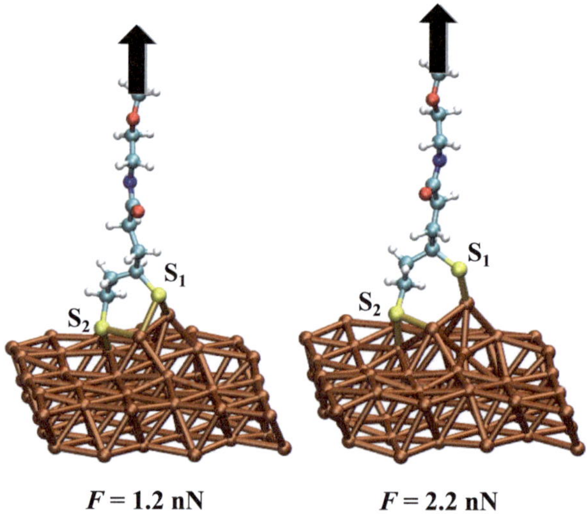

Fig. 7 Illustrative structures at F = 1.2 nN and F = 2.2 nN for **TA-PEG** in a solvation environment. The *arrows* are placed on the carbon atom to which the external force is applied to in the direction indicated. Water molecules and the two bottom Au layers have been removed for clarity

energy of reaction of OH^- acting as a nucleophilic species on either S atom at two different values of constant external force, i.e. 1.2 nN and 2.0 nN, that is to say the two extreme values in the elastic regime of the S–Au bond stretching. As was already mentioned, this was done using the enhanced sampling technique of thermodynamic integration. The distance between the oxygen atom in OH^- and either S atom was used as the reaction coordinate. In the "Reactant State (RS)", the nucleophilic species is at a distance of ~3.5 Å from either S atom. In the "Product State (PS)", OH is attached to the S atom at a bonding distance of 1.65 Å. The value of the reaction coordinate was then decreased stepwise from the RS situation to the PS in steps of 0.2 Å. Between increments of the reaction coordinate, we took special care to run molecular dynamics long enough to allow for the equilibration of the Lagrange multiplier value. This is typically converged within 1–1.5 ps.

The energy profile at each value of the external force and for the attacking of OH^- at either anchoring S were then obtained by integrating the mean force at each value of the reaction coordinate between the RS and the PS. The free energy of activation, $\Delta A^{\ddagger}$, was then computed as the difference between the RS and the maximum value along the curve. These values are summarized in Table 2.

Notably, the effect of the external force on the reactivity is very mild, contrary to our expectation. A preliminary analysis showed that this seems to be related to the fact that the desolvation process of the nucleophile is the major contributor to the activation energy of the reaction, and that this process is ahead of bond formation, ie. they do not develop synchronously when going from the reactant to the transition state.

Interestingly, there is ~30 % decrease in the activation energy of OH^- attack on S_2 with respect to S_1. In principle, this is counterintuitive, since S_1 is directly attached to the carbon chain to which the force is applied to. It would be expected that as a consequence the corresponding S–Au bonds would be more stressed and, therefore, it would require *less* energy to break them. Although it can be seen that upon stretching the gold–thiolate interface there is indeed a larger proportion of the external force being transduced to the S_1–Au bonds, the analysis of the electronic structure shows that there is also a charge transfer from the Au atoms to S_1, thus resulting in this S atom becoming *more* negatively charged and, as a result, less prone to an attack from a nucleophilic species.

These preliminary results show that this system displays an interesting mechanochemical behavior, that still needs to be properly understood. To fulfill that aim, we need to prolong the molecular dynamics simulations to have a better statistical sampling. This will be achieved by the use of metadynamics simulations in its multiple walker implementation, which will allow us to exploit its massive parallelization in order to greatly accelerate the calculations in real time.

Table 2 Activation free energies ($\Delta A^{\ddagger}$) for the attack of OH^- on S_1 and S_2 at F = 1.2 nN and F = 2.0 nN

Force	$\Delta A^{\ddagger}$ (S_1), kcal/mol	$\Delta A^{\ddagger}$ (S_2), kcal/mol
1.2	29.5	19.7
2.0	27.2	18.0

3 Software and Computational Resources

For the study of the aliphatic/aromatic thiolate–gold junctions in vacuum, calculations were performed using the Quantum ESPRESSO suite [56]. This code implements a variety of methods and algorithms based on the solution of the density-functional theory (DFT) problem, using a plane waves basis set and pseudopotentials to represent electron-ion interactions. It is designed to have a high performance on massively parallel architectures by implementing several levels of parallelization.

The mechanochemical study of the solvated bidentate thiolate–gold interface was carried out using ab initio molecular dynamics. AIMD will be performed using the cost-effective Car–Parrinello algorithm [57] as implemented in the CPMD code [58]. This is also a plane-wave/pseudopotential implementation of Density Functional Theory (DFT). Parallelization of the code has been done on different levels, allowing for a shared memory parallelization (OpenMP), a distributed memory scheme (MPI) and a hybrid MPI/OpenMP parallelization scheme. This typically allows to scale to several thousand processors, depending of the size and nature of the specific system.

Applying Car–Parrinello propagation to small gap or metallic systems requires some care [58]. First of all, separate thermostats must be applied to the nuclear and electronic subsystems. Secondly, Nosé–Hoover chain of thermostats establish ergodic sampling, in particular when independent such chains are coupled to each cartesian degree of freedom of the moving nuclei in the so-called "massive thermostating" approach. It turned out that unusually high tight criteria were required to integrate the equations of motion of the thermostats using a high-order Suzuki–Yoshida algorithm. Moreover, a very short time step of ca. 0.05 fs is required for a good energy conservation.

Typical calculations were run using the distributed memory (MPI) parallelization scheme. For the runs using Quantum Espresso, typically 192 cores (8 nodes in the Hazelhen system) were used, with an average wall time of 5.8 h (equivalent to ca. 1100 core-hours per run). A typical run required approximately 1.5 GB of disk space for permanent storage and additional 4.5 GB for scratch data. For CPMD, most runs were performed using so far ca. 200 processors in average.

In the future, we will be using metadynamics for an efficient sampling of the force-transformed free energy surface. Drawing on our implementation of the "multiple walkers technique" [59] in CPMD, we will be able to use several thousands processors for a single simulation while maintaining a linear scaling. We will typically use 10 walkers to sample the force-transformed free energy surface, thus requiring in the order of ca. 2000 cores per run.

Acknowledgements Partial financial support is provided by the DFG Koselleck Grant "Understanding Mechanochemistry" to D.M. We wish to thank Przemyslaw Dopieralski and Martin Krupička for their contributions to this work.

References

1. Ribas-Arino, J., Marx, D.: Chem. Rev. **2012**(112), 5412–5487
2. Grandbois, M., Beyer, M., Rief, M., Clausen-Schaumann, H., Gaub, H.E.: Science **1999**(283), 1727–1730
3. Hickenboth, C.R., Moore, J.S., White, S.R., Sottos, N.R., Baudry, J., Wilson, S.R.: Nature **2007**(446), 423–427
4. Beyer, M.K., Clausen-Schaumann, H.: Chem. Rev. **2005**(105), 2921–2948
5. Caruso, M.M., Davis, D.A., Shen, Q., Odom, S.A., Sottos, N.R., White, S.R., Moore, J.S.: Chem. Rev. **2009**(109), 5755–5798 (Anfang)
6. Pensa, E., Cortés, E., Corthey, G., Carro, P., Vericat, C., Fonticelli, M.H., Benítez, G., Rubert, A.A., Salvarezza, R.C.: Acc. Chem. Res. **2012**(45), 1183–1192
7. Häkkinen, H.: Nat. Chem. **2012**(4), 443–455
8. Uysal, A., Stripe, B., Lin, B., Meron, M., Dutta, P.: Phys. Rev. Lett. **2011**(107), 115503
9. Hamoudi, H., Neppl, S., Kao, P., Schüpbach, B., Feulner, P., Terfort, A., Allara, D., Zharnikov, M.: Phys. Rev. Lett. **2011**(107), 027801
10. Zayak, A.T., Hu, Y.S., Choo, H., Bokor, J., Cabrini, S., Schuck, P.J., Neaton, J.B.: Phys. Rev. Lett. **2011**(106), 083003
11. Vericat, C., Vela, M.E., Benitez, G., Carro, P., Salvarezza, R.C.: Chem. Soc. Rev. **2010**(39), 1805–1834
12. Li, F.-S., Zhou, W., Guo, Q.: Phys. Rev. B **2009**(79), 113412
13. Saffarzadeh, A., Demir, F., Kirczenow, G.: Phys. Rev. B **2014**(89), 045431
14. Batista, R.J.C., Ordejón, P., Chacham, H., Artacho, E.: Phys. Rev. B **2007**(75), 041402
15. Tao, N.J.: Nat. Nanotechnol. **2006**(1), 173–181
16. Zhao, P., Li, N., Astruc, D.: Coord. Chem. Rev. **2013**(257), 638–665
17. Chen, X., Strange, M., Häkkinen, H.: Phys. Rev. B **2012**(85), 085422
18. Pathak, R.K., Kolishetti, N., Dhar, S.: WIREs Nanomed. Nanobiotechnol. **2015**(7), 315–329
19. Majdalawieh, A., Kanan, M.C., El-Kadri, O., Kanan, S.M.: J. Nanosci. Nanotechno. **2014**(14), 4757–4780
20. Liu, X., Li, H., Jin, Q., Ji, J.: Small **2014**(10), 4230–4242
21. Howes, P.D., Chandrawati, R., Stevens, M.M.: Science **2014**(346), 1247390
22. Vigderman, L., Zubarev, E.R.: Adv. Drug Deliv. Rev. **2013**(65), 663–676
23. Mieszawska, A.J., Mulder, W.J.M.; Fayad, Z.A., Cormode, D.P.: Mol. Pharm. **2013**(10), 831–847
24. Kumar, D., Saini, N., Jain, N., Sareen, R., Pandit, V.: Expert Opin. Drug Deliv. **2013**(10), 397–409
25. Rana, S., Bajaj, A., Mout, R., Rotello, V.M.: Adv. Drug Deliv. Rev. **2012**(64), 200–216
26. Parveen, S., Misra, R., Sahoo, S.K.: Nanomed. Nanotechnol. Biol. Med. **2012**(8), 147–166
27. Papasani, M.R., Wang, G., Hill, R.A.: Nanomed. Nanotechnol. Biol. Med. **2012**(8), 804–814
28. Dykman, L., Khlebtsov, N.: Chem. Soc. Rev. **2012**(41), 2256–2282
29. Otsuka, H., Nagasaki, Y., Kataoka, K.: Adv. Drug Deliv. Rev. **2003**(55), 403–419
30. Wuelfing, W.P., Gross, S.M., Miles, D.T., Murray, R.W.: J. Am. Chem. Soc. **1998**(120), 12696–12697
31. Sebby, K.B., Mansfield, E.: Anal. Bioanal. Chem. **2015**(407), 2913–2922
32. Gao, J., Huang, X., Liu, H., Zan, F., Ren, J.: Langmuir **2012**(28), 4464–4471
33. Oh, E., Susumu, K., Jain, V., Kim, M., Huston, A.: J. Colloid Interf. Sci. **2012**(376), 107–111
34. Oh, E., Susumu, K., Goswami, R., Mattoussi, H.: Langmuir **2010**(26), 7604–7613
35. Zhang, G., Yang, Z., Lu, W., Zhang, R., Huang, Q., Tian, M., Li, L., Liang, D., Li, C.: Biomaterials **2009**(30), 1928–1936
36. Mei, B.C., Oh, E., Susumu, K., Farrell, D., Mountziaris, T.J., Mattoussi, H.: Langmuir **2009**(25), 10604–10611
37. Gorman, C.B., Carroll, R.L., He, Y., Tian, F., Fuierer, R.: Langmuir **2000**(16), 6312–6316

38. Krüger, D., Fuchs, H., Rousseau, R., Marx, D., Parrinello, M.: J. Chem. Phys. **2001**(115), 4776–4786
39. Keel, J.M., Yin, J., Guo, Q., Palmer, R.E.: J. Chem. Phys. **2002**(116), 7151–7157
40. Krüger, D., Fuchs, H., Rousseau, R., Marx, D., Parrinello, M.: Phys. Rev. Lett. **2002**(89), 186402
41. Krüger, D., Rousseau, R., Fuchs, H., Marx, D.: Angew. Chem. Int. Ed. **2003**(42), 2251–2253
42. Xu, B., Tao, N.J.: Science **2003**(301), 1221–1223
43. Konôpka, M., Rousseau, R., Štich, I., Marx, D.: J. Am. Chem. Soc. **2004**(126), 12103–12111
44. Chen, F., Zhou, A., Yang, H.: Appl. Surface Sci. **2009**(255), 6832–6839
45. Seema, P., Behler, J., Marx, D.: Phys. Chem. Chem. Phys. **2013**(15), 16001–16011
46. Xue, Y., Li, X., Li, H., Zhang, W.: Nat. Commun. **2014**, 5 (2014)
47. Seema, P., Behler, J., Marx, D.: Phys. Rev. Lett. **2015**(115), 036102
48. Langry, K.C., Ratto, T.V., Rudd, R.E., McElfresh, M.W.: Langmuir **2005**(21), 12064–12067
49. Zoloff Michoff, M.E., Ribas-Arino, J., Marx, D.: Phys. Rev. Lett. **2015**(114), 075501
50. Pei, Y., Zeng, X.C.: Nanoscale **2012**(4), 4054–4072
51. Dopieralski, P., Ribas-Arino, J., Marx, D.: Angew. Chem. Int. Ed. **2011**(50), 7105–7108
52. Wollenhaupt, M., Krupička, M., Marx, D.: ChemPhysChem **2015**(16), 1565–1565
53. Dopieralski, P., Ribas-Arino, J., Anjukandi, P., Krupička, M., Kiss, J., Marx, D.: Nat. Chem. **2013**(5), 685–691
54. Dopieralski, P., Ribas-Arino, J., Anjukandi, P., Krupička, M., Marx, D.: Angew. Chem. Int. Ed. **2015**(55), 1304–1308
55. Miranda-Rojas, S., Muñoz Castro, A., Arratia-Pérez, R., Mendizábal, F.: Phys. Chem. Chem. Phys. **2013**(15), 20363–20370
56. Giannozzi, P., et al.: J. Phys. Condens. Matter **2009**(21), 395502
57. Car, R., Parrinello, M.: Phys. Rev. Lett. **1985**(55), 2471–2474
58. Marx, D., Hutter, J.: Ab Initio Molecular Dynamics. Cambridge University Press, Cambridge (2009)
59. Raiteri, P., Laio, A., Gervasio, F.L., Micheletti, C., Parrinello, M.: J. Phys. Chem. B **2006**(110), 3533–3539

Microscopic Insights into the Fluorite/Water Interfaces from Vibrational Sum Frequency Generation Spectroscopy

Rémi Khatib and Marialore Sulpizi

Abstract Water/mineral interfaces are central to a wide range of environmental and technological processes. In this report we provide a quantitative, molecular-level understanding of the CaF_2/water interface using Density Functional Theory-based molecular dynamics simulations.

In particular through the comparison of calculated Vibrational Sum Frequency Generation spectra to the experimental ones, we give a structural characterisation of the interface at different pH. At low pH, the surface is positively charged, causing a substantial degree of water ordering. Our results suggest that the surface charge originates from the dissolution of fluoride ions of the topmost layer, rather than from proton adsorption to the surface.

At high pH we observe the presence of Ca-OH species pointing into the water. Such OH groups do not establish hydrogen bonds with the surrounding water, and are therefore responsible for the "free OH" signature which is recorded in the Vibrational Sum Frequency Generation spectrum.

1 Introduction

Water-mineral interactions are of general importance for a wide range of environmental, chemical, metallurgical, and ceramic processes [18, 19]. The interaction of fluorite (CaF_2) with water is of specific relevance for industrial, environmental and medical applications, e.g. for understanding fluorine dissolution in drinking water [22]. Recently, there has been a proposal to use CaF_2 as an analogue of UO_2 in dissolution experiments in order to understand the long term dissolution behaviour of spent nuclear fuel. This has accordingly raised the interest in the interaction of CaF_2 with water [5].

Despite the apparent importance of the fluorite/water interface, it has been challenging to obtain detailed insights into this interface at the molecular-scale. Recently, Frequency Modulation Atomic Force Microscopy (FM-AFM) [12] has provided important new information on molecular length scales by analysing the

R. Khatib • M. Sulpizi (✉)
Johannes Gutenberg University Mainz, Staudinger Weg 7, 55099, Mainz, Germany
e-mail: sulpizi@uni-mainz.de

W.E. Nagel et al. (eds.), *High Performance Computing in Science and Engineering '16*, DOI 10.1007/978-3-319-47066-5_10

fluorite/water interface, not only as function of the pH, but also as function of the concentration of ions in the solution and addressing fluorite/water interfaces with saturated and supersaturated solutions. At high pH, the presence of surface adsorbates is detected and attributed to calcium hydroxo complexes [12]. At low pH atomic scale disorder was observed, which could be attributed to either partial dissolution of the topmost layer by the creation of F^- vacancies, or to proton adsorption at the interface. Still experiments seem not to be able to distinguish between the two possible scenarios [12].

As another surface sensitive technique, Vibrational Sum Frequency Generation Spectroscopy (VSFG) has the ability to selectively address the nanometric interfacial water layer, and indeed has contributed substantially to our understanding of the physical and chemical properties of the CaF_2/water interface [2, 3]. VSFG is rather unique in its ability to provide the vibrational spectrum of water molecules specifically at the interface, as the selection rule of VSFG requires symmetry to be broken, i.e. no VSFG signal can be generated from the adjacent centrosymmetric bulk. Previous VSFG investigations of water at the CaF_2/water interface by the Richmond group [2, 3] have revealed dramatic changes in the interfacial hydrogen bonding structure upon changing the pH of the aqueous phase. In particular at low pH, the VSFG experiments have suggested that positive charge develops on the surface, causing orientation of water molecules into highly ordered, tetrahedrally coordinated states. At near-neutral pH, the VSFG signal vanishes and this has been interpreted as the result of a more random orientation of the interfacial water molecules at a near-neutral surface. Finally in the basic pH regime dissociative adsorption was hypothesised to take place on the solid surface resulting in the formation of Ca-OH species. Open questions are still: how do these OH groups contribute to the VSFG spectrum? What type of order is established in the interfacial water region?

Here we review a recent simulation study aimed at answering these questions and to provide a new microscopic understanding of the CaF_2/water interface as function of pH [11]. We explore the effect of surface termination on the interfacial water arrangement and we show the importance of the local electrical field due to ions in solution in the near-surface region on water orientation. Such a detailed analysis is now possible thanks to recent advances in the computational techniques. In particular, we use Density Functional Theory (DFT)-based molecular dynamics (MD) simulations, which allow an accurate description of the structure and dynamics of hydrogen bonding in highly heterogeneous environments, also including electronic polarisation. A newly developed approach is used for the calculation of the VSFG spectra [11] which only requires the atomic positions and velocities without the cost of the additional calculation of molecular dipoles and polarizabilities. At the same time appropriate selection rules for the VSFG are also taken into account. The spectra are calculated using velocity-velocity correlation functions (VVCF) over several 100 ps time scale. This is possible thanks to the use of massively parallel architectures, such as the Cray XE6 (Hermit, HRLS) and the Cray XC40 (Hazel Hen, HRLS) used in the present work. This permits us to build several models, which include about 500 atoms each and span e.g. different surface charges.

2 Methodology

2.1 Simulation Setup

Several models are used to describe the fluorite/water interface over a wide range of pH. The reference system – an interface between CaF_2 (111) and water at neutral pH – is composed of 88 water molecules and 60 formula units of CaF_2 contained in a $11.59 \times 13.38 \times 34.0$ Å cell periodically repeated in the (*X*, *Y*, *Z*) directions. All the other models have close compositions and size to allow inter-system comparisons. The thickness of water slabs is around 20 Å along the *z*-axis, which is reasonable compromise between the need to achieve bulk-like properties far from the surface and the computational cost. Simulations were carried out with the package CP2K/Quickstep [25], consisting in Born-Oppenheimer MD (BOMD) BLYP [1, 13], electronic representation including Grimme (D3) correction for dispersion [7], GTH pseudopotentials [6, 8], a combined Plane-Wave (280 Ry density cutoff) and TZV2P basis sets. All the BOMD are performed using the NVT ensemble. The Nosé-Hoover thermostat is used to control the average temperature at 330 K. Trajectories are accumulated for at least 50 ps (whom 10 ps of equilibration) with a time step of 0.5 fs.

2.2 Method for VSFG

The starting equation to calculate the VSFG response function from molecular dynamics simulations have been introduced by Morita [9, 15–17]:

$$\chi_{PQR}^{(2),R} = \frac{-i}{k_B T \omega} \int_0^{+\infty} e^{i\omega t} \left\langle \dot{A}_{PQ}(t) \dot{M}_R(0) \right\rangle dt \tag{1}$$

Here $\chi^{(2),R}$ is the resonant part of second-order susceptibility tensor, (P, Q, R) are any directions of the laboratory frame $(\hat{\mathbf{X}}, \hat{\mathbf{Y}}, \hat{\mathbf{Z}})$, ω is the frequency of the IR beam, A_{PQ} and M_R are respectively the components of the total polarizability tensor and the total dipole moment and the dot stands for the time derivative.

If we suppose that at the frequencies of interest only the O-H stretching has an impact on the spectra, the total polarizability and dipole moment of the system (A_{PQ}, M_R) can be decomposed into individual (OH) bond contributions ($\alpha_{mn,PQ}$, $\mu_{mn,R}$), where the sum is done over all the N_m bonds of the M molecules:

$$\begin{cases} \dot{A}_{PQ}(t) = \sum_{m=1}^{M}\sum_{n=1}^{N_m} \dot{\alpha}_{mn,PQ}(t) \\ \dot{M}_R(t) = \sum_{m=1}^{M}\sum_{n=1}^{N_m} \dot{\mu}_{mn,R}(t) \end{cases} \tag{2}$$

Moreover, thanks to basic geometry considerations, one can express the dipole moment of the A-B bond from the molecular frame ($\boldsymbol{\mu}_b$) to the laboratory frame ($\boldsymbol{\mu}_l$):

$$\boldsymbol{\mu}_l = \mathbf{D}\boldsymbol{\mu}_b \tag{3}$$

where **D** is the direction cosine matrix projecting the bond frame onto the laboratory frame. In the following, we will assume that (1) the bond elongations are small enough to make Taylor expansion at the first order and (2) the stretching mode of the bond is much faster than the modes involving a bond reorientation – for example the libration. The second assumption means that $\dot{D}_{Ri} \approx 0$ and that $\frac{dr_z}{dt} \gg \frac{dr_x}{dt} \approx \frac{dr_y}{dt}$ Therefore $\dot{\mu}_R$ can be simplified into:

$$\begin{aligned} \dot{\mu}_R(0) &\approx \sum_i^{x,y,z} D_{Ri}(0)\dot{\mu}_i(0) \\ &\approx \sum_i^{x,y,z} D_{Ri}(0)\left(\sum_j^{x,y,z} \frac{\partial \mu_i}{\partial r_j}\left.\frac{dr_j}{dt}\right|_{t=0}\right) \\ &\approx \sum_i^{x,y,z} D_{Ri}(0)\frac{\partial \mu_i}{\partial r_z}v_z(0) \end{aligned} \tag{4}$$

where $v_z(0) = \left.\frac{dr_z}{dt}\right|_{t=0}$ corresponds to the projection of the velocity on the bond axis.

With the same methodology for the polarizability, one deduces that:

$$\dot{\alpha}_{PQ}(t) \approx \sum_i^{x,y,z}\left[D_{Pi}(t)\sum_j^{x,y,z}\left(\frac{\partial \alpha_{ij}}{\partial r_z}D_{Qj}(t)\right)\right] \times v_z(t) \tag{5}$$

Table 1 Calculated derivatives of the dipole moment (D.Å^{-1}) and polarizability (Å^2) of the O-H bond in a bulk of water and in CaFOH monomer. The results are given within the bond frame

	$\frac{\partial \mu_x}{\partial r}$	$\frac{\partial \mu_y}{\partial r}$	$\frac{\partial \mu_z}{\partial r}$	$\frac{\partial \alpha_{xx}}{\partial r}$	$\frac{\partial \alpha_{yy}}{\partial r}$	$\frac{\partial \alpha_{zz}}{\partial r}$	$\frac{\partial \alpha_{xy}}{\partial r}$	$\frac{\partial \alpha_{xz}}{\partial r}$	$\frac{\partial \alpha_{yz}}{\partial r}$
H_2O	−0.15	0.0	2.1	0.40	0.53	1.56	0.0	0.02	0.0
H_3O^+	−0.11	0.0	1.7	0.47	0.40	1.50	0.0	0.0	0.0
HO^-	0.0	0.0	1.6	0.5	0.5	2.3	0.0	0.0	0.0

The use of equation (4) and (5) into equation (2) brings important computational advantages. Indeed the velocities and the direction cosine matrix (v_z, $\mathbf{D}$) can be readily obtained from the DFT-MD trajectories while $\frac{\partial \alpha_{ij}}{\partial r_z}, \frac{\partial \mu_i}{\partial r_z}$ can be parametrized [4]. Our approach avoid the additional direct calculation of the bond dipole moment and polarizabilities which, at an *ab initio* level certainly requires a considerable additional computational cost, e.g. the cost of the Wannier centres localisation [23]. Finally, with the splitting of the dipole moment and polarizability into their bond contributions, it is easy to decompose the signal into its auto-, intramolecular and intermolecular parts.

The parametrization of $\frac{\partial \alpha_{ij}}{\partial r_z}$ and $\frac{\partial \mu_i}{\partial r_z}$ is based on the calculation of the maximally localised Wannier functions (MLWF) [14] and has been done through the methodology developed by Salanne et al. [21]. The values are obtained by a 2-point numerical differentiation: a single O-H bond is elongated by ±0.02 Å. For the O-H bond of water molecules, a trajectory of 128 H_2O inside a cubic box (c = 15.6404 Å) has been simulated and an average involving more than 4000 bonds distributed over a dynamic of 40 ps has been done. One formula unit of HCl has been added to the previous box in order to do the same kind of sampling about the O-H bond of the hydronium. Finally, for the O-H bond of the grafted hydroxide ions, the derivatives are those obtained on a linear monomer of CaFOH. All these values are resumed in the Table 1.

3 Results and Discussion

We describe here atomistic models for the fluorite/water interfaces at different pH conditions. The models are used to calculate the interface vibrational spectra and to provide their molecular interpretation.

At low pH positive charge is expected to accumulate at the fluorite/water interface. In particular the following reaction is expected to take place:

$$(CaF_2)_{surf} + H^+_{aq} \rightleftarrows (CaF^+)_{surf} + HF_{aq}. \tag{6}$$

Fluoride ions dissolving into the water solution leave positive vacancies on the surfaces, which are responsible for the aligning of the water molecules. As the VSFG signal increases with increasing interfacial order in the system, a large VSFG

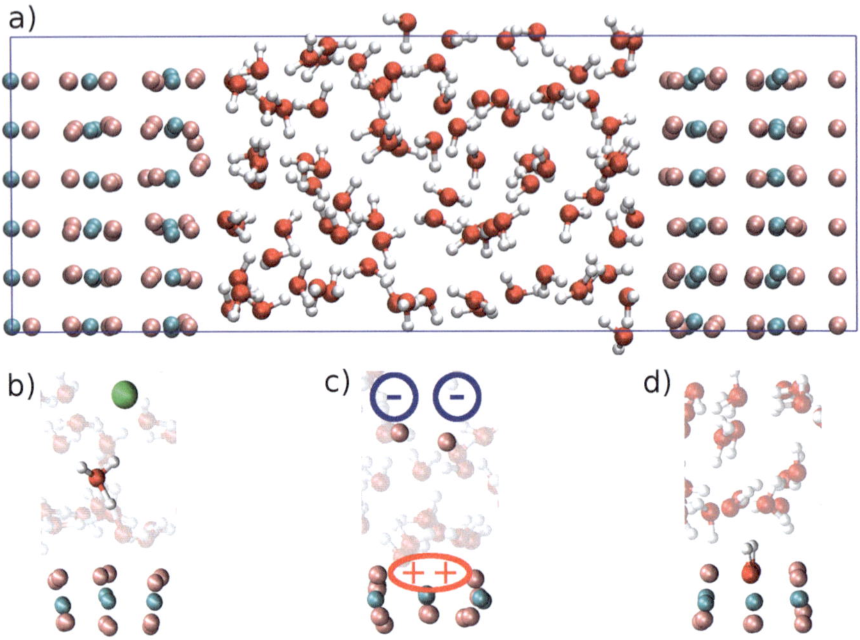

Fig. 1 (**a**) Random snapshot of the system used to describe the CaF_2/H_2O interface for the neutral pH. Miniatures highlighting the differences between the neutral pH and the (**b**) low pH with an excess of proton in the form of dissociated HCl, (**c**) low pH system with partial dissolution of fluoride ions, (**d**) high pH with 6 substitutions of fluorides by hydroxides per surface. For (**b**–**d**), the water molecules are transparent in order to highlight the ions position. The hydrogens are coloured in *white*, the oxygens in *red*, the fluorines in *pink*, the clorines in *green* and the calciums in *turquoise*

signal is detected [2, 3]. For low pH, model systems which resemble the final equilibrium state can be built with various concentrations of fluorite vacancies on the surface, which correspond to different extents of positive charge on the surface (Fig. 1). In particular our model consists of a CaF_2 slab in contact with water where two equivalent interfaces are present. Fluoride counterions are added to the solution to compensate the positive surface charge, i.e. to get an overall neutral system. We find that the F^- ions tend to prefer to be solvated by water, and form a diffuse layer in the near-surface region. Overall, the surface-localised positive charge and the near-surface negative counterions generate a double layer, giving rise to a rather strong electrical field at the solid/liquid interface. We have considered more extreme conditions with 2.58 vacancies.nm^{-2} (4 vacancies on each surface) and milder conditions with 1.29 vacancies.nm^{-2} (2 vacancies) or with 0.64 vacancies.nm^{-2} (1 vacancy), respectively.

At high pH, the hydroxide ions in excess are expected to react with the CaF_2 surface leading to the following substitution:

$$(CaF_2)_{surf} + HO^-_{aq} \rightleftarrows (CaFOH)_{surf} + F^-_{aq} \qquad (7)$$

The Ca-OH groups on the surface have been suggested as the responsible for the narrow band signal at 3645 cm^{-1} [2, 3]. For high pH, we have constructed a model where a surface modification of the CaF_2 has taken place in response to the increased concentration of OH groups in the solution. In the topmost fluorite layer, F^- were partially or totally replaced by HO^- (Fig. 1). Different concentrations of OH have been considered in order to establish a relation between the VSFG signal intensity and the pH: 1, 6 and 12 substitution over the 12 available sites per surface.

Using the described models we have calculated the spectral responses from the surface sensitive vibrational density of states using surface specific VVCF (see method sections for details) for the XXZ polarization (the indexes of $\chi^{(2)}$ will be omitted).

In the case of low pH the spectra for the different vacancy densities are reported in the top row of Fig. 2. The common feature for all the different concentrations of surface vacancies is the presence of a broad negative band in the $\text{Im}\chi^{(2)}$ spectrum, which, for the 1 and 2 vacancies systems, is located around 3300 cm^{-1}. As the charge concentration increases to 4 positive charges, the intensity of the band increases and the band position moves towards lower frequencies, with a maximum located at 3100 cm^{-1}. If we compare the calculated spectra to the experimental ones [11], we can see that such strong red shift for the 4 vacancies system is not consistent with the experiment. Better agreement is found for the 1 and 2 vacancies

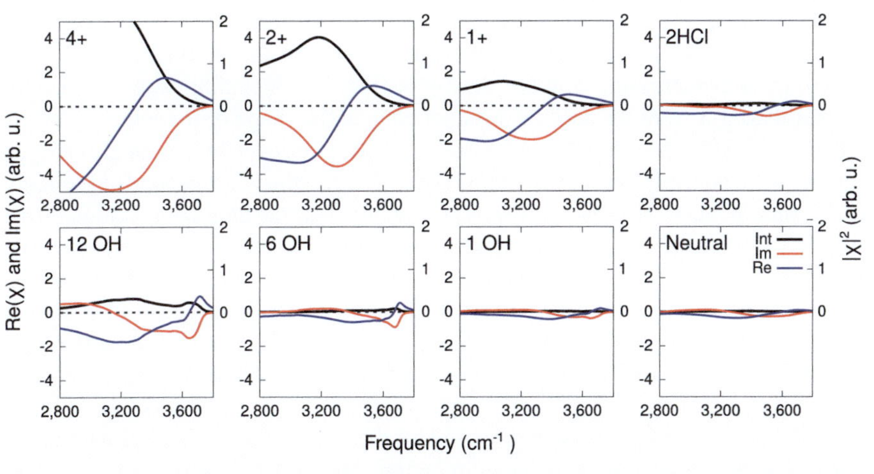

Fig. 2 Comparison of the $\text{Im}\chi^{(2)}$, $\text{Re}\chi^{(2)}$ and $|\chi^{(2)}|^2$ for different values of the surface defect concentration (*plain lines*). *Top panels*: low pH. *Bottom panels*: high pH. In order to facilitate the comparison, the spectra with 2 HCl per surface have been plotted in *dotted lines* on the spectra with 2 vacancies per surface

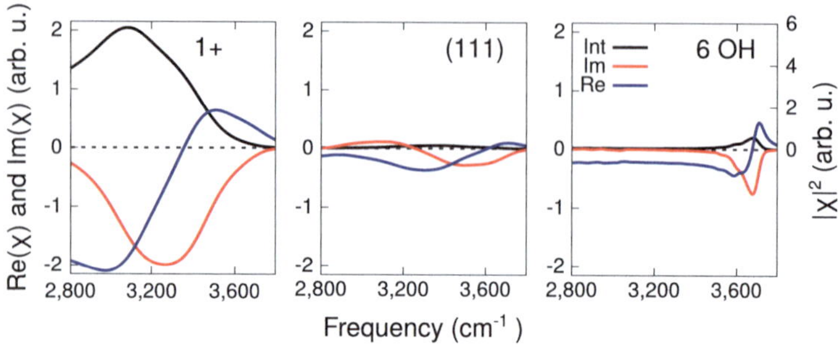

Fig. 3 Re$\chi^{(2)}$, Im$\chi^{(2)}$ and $|\chi^{(2)}|^2$ obtained from simulations (*blue*, *red* and *black* respectively). Low pH (1+), neutral (111) and high pH (6 OH) systems are considered

systems. Additional information can also be extracted from a comparison between the calculated and experimentally measured Re$\chi^{(2)}$. The computed Re$\chi^{(2)}$ (Fig. 2, blue lines) shows two main peaks, a positive peak at higher frequencies and a negative one at lower frequencies. In the case of 1 or 2 positive charges on the surface the peak position and the crossing from positive to negative values are in good agreement with the experimental spectra (Fig. 3). However as the defect number increases to 4, we notice a very strong shift of the negative band to lower frequencies which also shifts the zero crossing toward 3200 cm^{-1}. Moreover, also for the intensity spectrum the best match between theory and experiment is found for 1 or 2 vacancies per site. Overall these considerations suggest that the vacancy density is around 0.65 per nm^2 for the experimental condition of pH = 2.

What is the molecular origin of the strong negative band in Im$\chi^{(2)}$? A detailed molecular analysis unveils that such a band is due to an ordered layer of water which builds up at the interface, with water dipoles oriented toward the bulk. The water order extends over 4–5 Å, as it can be deduced from the convergence of the Im$\chi^{(2)}$ spectrum with increasing probing thickness (Fig. 4a). Including water molecules further than 5 Å from the surface does not change the shape or the intensity of the calculated VSFG spectrum. It is interesting to notice that even for a strongly charged interface the aqueous order only extends over 4–5 Å, which corresponds to roughly 2–3 layers of water. However, we should note here that the high computational cost of electronic structure based methods imposes severe limitations on the size of the accessible models. In this respect our model is expected to capture the contribution to the spectra of the Stern layer (possibly the major contribution here), but cannot account for the full diffuse layer, which is expected to extend over a few nanometers thickness. In the case of CaF_2 the experimentally estimated Debye length is around 30 Å [10].

Alternative models have been proposed for the low pH fluorite/water interface. In particular, as also mentioned in the introduction, one of the suggested interpretation for the atomic scale disorder observed at low pH in the FM-AFM experiments, is proton adsorption at the interface [12]. In order to investigate the spectral response of such a system and to compare it to the experimental one, we build an additional

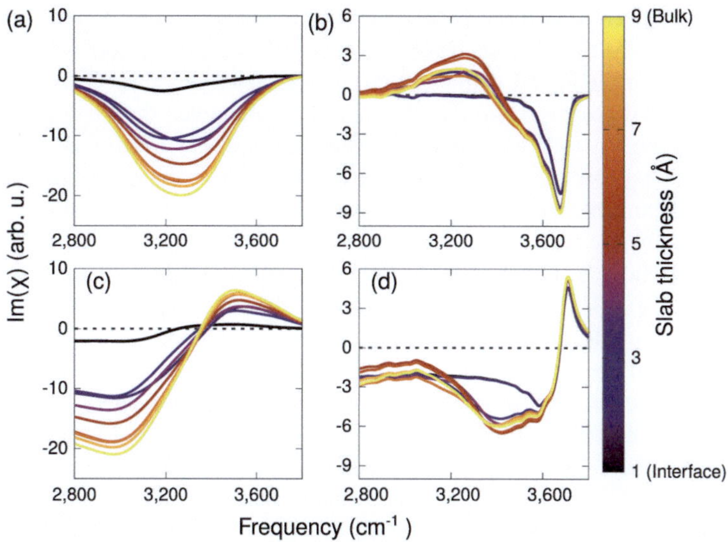

Fig. 4 Im$\chi^{(2)}$ (*top*) and Re$\chi^{(2)}$ (*bottom*) as function of the layer thickness included in the calculation. *Left panels*: low pH (1 defect per surface); *Right panels*: high pH (6 substitutions per surface)

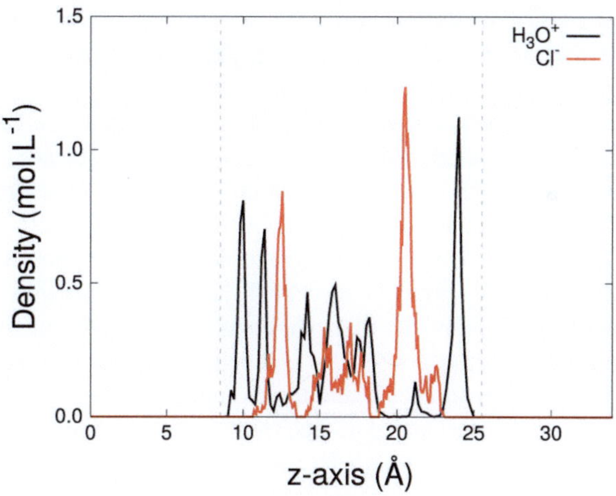

Fig. 5 Density profile of H_3O^+ and Cl^- along the *Z*-axis. As a guide for the eyes, the position of the CaF_2 interface is represented by a *dashed grey line*

model without fluoride vacancy, but instead with an excess of protons in the form of dissociated HCl is present (4 HCl, 2.5 M solution). Such a system is reported in Fig. 1b and would eventually corresponds to 2 excess protons per surface. The proton distribution at the interface is reported in Fig. 5.

The calculated VSFG spectra for this system are shown in Fig. 2 in the last panel of the top row. The first striking result is that overall the signal is much weaker than that obtained for the model with two fluorine vacancies per surface, which exhibit the same overall positive charge at the interface. Moreover, the main peak in the $\mathrm{Im}\chi^{(2)}$ is located at $\approx$3500–3600 cm^{-1}, which is quite far from peak location in the experimental spectra. This analysis would suggest that the excess proton alone cannot be responsible for the measured spectra, which instead originates from the water aligned by the positive fluorine vacancies.

Let's now move to the analysis of the high pH conditions. The imaginary and real part of the VSFG spectrum together with the intensity spectrum calculated from the surface selective VVCF analysis are presented in the bottom row of Fig. 2 for the three different values of OH concentration on the surface. For the 1 and 6 substitutions two main features can be observed in the imaginary part: the first is a positive band between 3280 and 3400 cm^{-1}, the second is a negative feature between 3400 and 3700 cm^{-1}. In the case of 12 OH substitutions, the overall profile of $\mathrm{Im}\chi^{(2)}$ is very different, with a broad negative band extending up to 3200 cm^{-1} where a crossing to positive values is finally observed. The real part and the intensity spectrum have a very high intensity below 3600 cm^{-1} (Fig. 3), which is not present in the experiment [11]. The best agreement between calculated and experimental spectra is found for the models with 1 or 6 OH substitutions. From this we can set, for the experimental pH = 13, an upper limit of 6 OH substitutions per surface corresponding to 3.87 substitutions.nm^{-2}.

As done for the low pH, also for the high pH conditions, we can decompose the overall signal in molecular contributions, thus providing a microscopic interpretation of the experimental spectra. In particular, the peak between 3600 and 3700 cm^{-1} is only associated with the OH groups on the surface, namely those OH groups which replace F^- in the topmost layer, which is clear from the purple spectrum in the bottom panel of Fig. 4. This frequency is very close to that of "free OH" [20, 26], indeed such an OH group does not form any hydrogen bond with water. This is clearly shown in the radial distribution function of the Ca-OH hydrogen with water oxygens: the distance between the proton of the Ca-OH and the oxygen from water (red curve, Fig. 6) is much larger than the distance between the proton from one water molecule and the oxygen from the next water molecule (black curve Fig. 6). The presence of a "free OH" signal at the solid/liquid interfaces is not so uncommon. A similar high frequency peak has also been observed for the alumina/water interface [24], where no hydrogen bond is formed between the surface OH groups and the water molecules.

In addition to the "free OH" peak, the high pH spectra, also present a band between 3280 and 3400 cm^{-1}, which is instead associated with hydrogen bonded water molecules at the interface. These hydrogen bonded waters have an opposite orientation with respect to that of the OH groups, as evident from the opposite sign of $\mathrm{Im}\chi^{(2)}$ for the two different peaks. The water ordering is not very pronounced and saturates with a distance of 2 Å (Fig. 4).

Finally, let's briefly comment on the neutral pH conditions. The neutral pH model is given by a fluorine terminated surface in contact with neutral water (no excess of

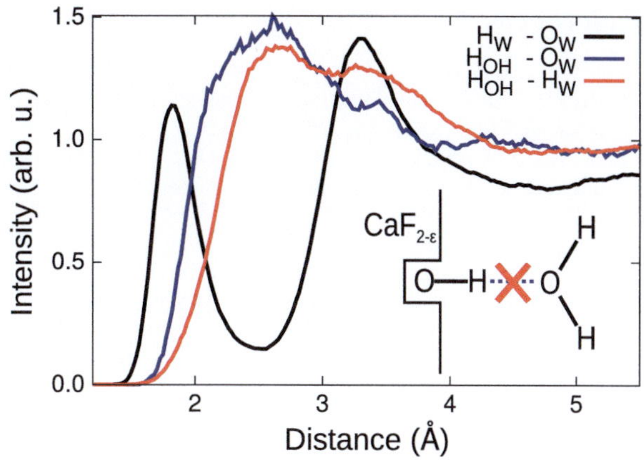

Fig. 6 HO and HH radial distribution functions. The subscript "W" stands for water, while the subscript "OH" the grafted hydroxide

hydronium or hydroxide). The calculated Im$\chi^{(2)}$ is reported in Fig. 3 (blue line), along with the overall signal intensity (Fig. 3, black line). The signal for the Im$\chi^{(2)}$ is very weak and presents a negative sign in the higher frequency region (3400–3500 cm^{-1}) and a positive band in the lower frequency range (3000–3200 cm^{-1}). A molecular analysis shows that, there is a strongly adsorbed layer of water at the interface with little or no preferential orientation at the interface.

4 Conclusions

We reviewed an *ab initio* molecular dynamics study which permitted to elucidate the details of the fluorite/water interface. The calculated VSFG spectra using surface selective VVCFs provide a molecular assignment of the different features observed in the experimental spectra. We find that at low pH the strong band in the hydrogen bond region is due to the highly ordered water as the surfaces is positively charged, due to the F^- dissolution. We also show that an eventual excess proton at the interface can only have a minor impact on the spectra. At high pH the "free OH" signal is due to the surface Ca-OH groups, which do not hydrogen bond strongly to water. The very good agreement between theory and experiments in both the Re$\chi^{(2)}$ and Im$\chi^{(2)}$ permits to pin down the atomistic details of the CaF_2 interface with water and to provide a first molecular interpretation of the spectra.

Acknowledgements This work was supported by the DFG Research Grant SU 752/2-1. All the dynamics were simulated on the supercomputers of the High Performance Computing Center (HLRS) of Stuttgart (Grant 2DSFG).

References

1. Becke, A.D.: Density-functional exchange-energy approximation with correct asymptotic behavior. Phys. Rev. A **38**, 3098–3100 (1988)
2. Becraft, K.A., Moore, F.G., Richmond, G.L.: Charge reversal behavior at the CaF_2/H_2O/SDS interface as studied by vibrational sum frequency spectroscopy. J. Phys. Chem. B **107**(16), 3675–3678 (2003)
3. Becraft, K.A., Richmond, G.L.: In situ vibrational spectroscopic studies of the CaF_2/H_2O interface. Langmuir **17**(25), 7721–7724 (2001)
4. Corcelli, S.A., Skinner, J.L.: Infrared and Raman line shapes of dilute HOD in liquid H_2O and D_2O from 10 to 90 degree. J. Phys. Chem. A **109**(28), 6154–6165 (2005)
5. Godinho, J., Piazolo, S., Evins, L.: Effect of surface orientation on dissolution rates and topography of CaF_2. Geochim. Cosmochim. Acta **86**, 392–403 (2012)
6. Goedecker, S., Teter, M., Hutter, J.: Separable dual-space gaussian pseudopotentials. Phys. Rev. B **54**, 1703–1710 (1996)
7. Grimme, S., Antony, J., Ehrlich, S., Krieg, H.: A consistent and accurate ab initio parametrization of density functional dispersion correction (DFT-D) for the 94 elements H-Pu. J. Chem. Phys. **132**(15), 154104 (2010)
8. Hartwigsen, C., Goedecker, S., Hutter, J.: Relativistic separable dual-space gaussian pseudopotentials from H to Rn. Phys. Rev. B **58**, 3641–3662 (1998)
9. Ishiyama, T., Takahashi, H., Morita, A.: Vibrational spectrum at a water surface: a hybrid quantum mechanics/molecular mechanics molecular dynamics approach. J. Phys. Condens. Matter **24**(12), 124107 (2012)
10. Jena, K.C., Covert, P.A., Hore, D.K.: The effect of salt on the water structure at a charged solid surface: differentiating second- and third-order nonlinear contributions. J. Phys. Chem. Lett. **2**(9), 1056–1061 (2011)
11. Khatib, R., Backus, E.H.G., Bonn, M., Perez-Haro, M.-J., Gaigeot, M.-P., Sulpizi, M.: Water orientation and hydrogen-bond structure at the fluorite/water interface. Sci. Rep. **6**, 24287 (2016)
12. Kobayashi, N., Itakura, S., Asakawa, H., Fukuma, T.: Atomic-scale processes at the fluorite-water interface visualized by frequency modulation atomic force microscopy. J. Phys. Chem. C **117**(46), 24388–24396 (2013)
13. Lee, C., Yang, W., Parr, R.G.: Development of the Colle-Salvetti correlation-energy formula into a functional of the electron density. Phys. Rev. B **37**, 785–789 (1988)
14. Marzari, N., Vanderbilt, D.: Maximally localized generalized Wannier functions for composite energy bands. Phys. Rev. B **56**, 12847–12865 (1997)
15. Morita, A., Hynes, J.T.: A theoretical analysis of the sum frequency generation spectrum of the water surface. II. Time-dependent approach. J. Phys. Chem. B **106**(3), 673–685 (2002)
16. Morita, A., Ishiyama, T.: Recent progress in theoretical analysis of vibrational sum frequency generation spectroscopy. Phys. Chem. Chem. Phys. **10**, 5801–5816 (2008)
17. Nihonyanagi, S., Ishiyama, T., Lee, T.-k., Yamaguchi, S., Bonn, M., Morita, A., Tahara, T.: Unified molecular view of the air/water interface based on experimental and theoretical $\chi^{(2)}$ spectra of an isotopically diluted water surface. J. Am. Chem. Soc. **133**(42), 16875–16880 (2011)
18. Putnis, A.: Why mineral interfaces matter. Science **343**, 1441–1442 (2014)
19. Putnis, C.V., Ruiz-Agud, E.: The mineral-water interface: Where minerals react with the environment. Elements **9**(3), 177–182 (2013)
20. Roberts, S.T., Petersen, P.B., Ramasesha, K., Tokmakoff, A., Ufimtsev, I.S., Martinez, T.J.: Observation of a zundel-like transition state during proton transfer in aqueous hydroxide solutions. PNAS **106**(36), 15154–15159 (2009)
21. Salanne, M., Vuilleumier, R., Madden, P.A., Simon, C., Turq, P., Guillot, B.: Polarizabilities of individual molecules and ions in liquids from first principles. J. Phys. Condens. Matter **20**(49), 494207 (2008)

22. Saxena, V., Ahmed, S.: Dissolution of fluoride in groundwater: a water-rock interaction study. Environ. Geol. **40**(9), 1084–1087 (2001)
23. Sulpizi, M., Salanne, M., Sprik, M., Gaigeot, M.-P.: Vibrational sum frequency generation spectroscopy of the water liquid-vapor interface from density functional theory-based molecular dynamics simulations. J. Phys. Chem. Lett. **4**(1), 83–87 (2013)
24. Tong, Y., Wirth, J., Kirsch, H., Wolf, M., Saalfrank, P., Campen, R.K.: Optically probing Al-O and O-H vibrations to characterize water adsorption and surface reconstruction on α-alumina: an experimental and theoretical study. J. Chem. Phys. **142**(5), 054704 (2015)
25. VandeVondele, J., Krack, M., Mohamed, F., Parrinello, M., Chassaing, T., Hutter, J.: Quickstep: fast and accurate density functional calculations using a mixed gaussian and plane waves approach. Comput. Phys. Commun. **167**(2), 103–128 (2005)
26. Walrafen, G.E., Douglas, R.T.W.: Raman spectra from very concentrated aqueous NaOH and from wet and dry, solid, and anhydrous molten, LiOH, NaOH, and KOH. J. Chem. Phys. **124**(11), 114504 (2006)

Growth, Structural and Electronic Properties of Functional Semiconductors Studied by First Principles

Andreas Stegmüller, Phil Rosenow, Lisa Pecher, Nikolay Zaitsev, and Ralf Tonner

Abstract Ab initio calculations of thermodynamic, electronic and vibrational properties of functional semiconductor materials relevant for silicon photonics were performed. The thermodynamics of hydrogen coverage on Si(001) was investigated as important chemical growth processes depend on surface structure and govern the structural quality of the materials deposited. Exemplarily, the influence of strain and chemical effects on the band gaps was studied by the two optically active alloys Ga(NAsP) and dilute Ga(AsBi). Furthermore, electron-vibron coupling of NTCDA molecules in an interface with the Ag(111) surface was investigated, identifying interfacial dynamical charge transfer to occur and enable the detection of in-plane IR modes. The DFT calculations on electronic structure and dynamics have been found in good agreement with experimental observations.

1 Introduction

Many semiconductor solar cells or logical devices, transistors, are based on silicon (Si) crystal substrates. One potential pathway increasing device efficiencies beyond the so-called red brick wall – the physical limits of miniaturization and device fabrication – is to employ optically active compound semiconductors within Si-based devices. Then, the device can make use of specifically designed optoelectronic properties that enable optical telecommunication or even nonlinear optical effects. Due to the indirect band gap of pure Si, functionality of conventional devices is limited due to excitation inefficiency and various loss mechanisms [1, 2].

The project reported is part of a research program that investigates fabrication and properties of new semiconductor materials. One class is III/V materials that comprise chemical elements from groups 13 and 15 at various relative concentrations which allows the adjustment of electronic band gaps and atomic structure for integration in Si-based devices [3, 4].

For defect-free growth and integration highly specific deposition techniques have been developed [5]. The growth of thin films in the nanometer scale is possible

A. Stegmüller • P. Rosenow • L. Pecher • N. Zaitsev • R. Tonner (✉)
Philipps-Universität Marburg, Marburg, Germany
e-mail: tonner@chemie.uni-marburg.de

W.E. Nagel et al. (eds.), *High Performance Computing in Science and Engineering '16*, DOI 10.1007/978-3-319-47066-5_11

with metal-organic vapour phase epitaxy (MOVPE). However, multiple challenges arise that, up to now, limit certain film-substrate combinations, stacking of multiple materials and the efficiency of laser devices producible [6]. Conditions during material deposition and film formation from metal-organic sources determine the structural quality. Due to high complexity of the physical and chemical processes involved fundamental understanding on their interplay is hard to achieve [7]. Some insight could already be gained in the course of the research conducted. It has been found, that chemical decomposition and diffusion kinetics as well as surface and interface thermodynamics significantly affect the substrate-film interface structure and, thus, its electronic properties [8]. Further research activity will be presented in the following.

The thermodynamic properties of the initial Si(001) substrate surface with respect to hydrogen coverage are adressed in the first section of this report. In the second section, the electronic and optical properties of the functional material Ga(NAsP) is investigated exploring composition-dependent strain and growth effects. Furthermore, the local ordering of Bi clusters in a dilute Ga(AsBi) material is studied uncovering the influence of chemical ordering effects on the band gap.

Another example of the relationship between atomic and electronic structure at substrate-film interfaces is presented in the final section where the electron-phonon coupling of 1,4,5,8-naphthalenetetracarboxylic dianhydride (NTCDA) on a silver Ag(111) substrate is adressed.

For reliable results on such properties, methods based on density functional theory (DFT) were employed. In order to achieve realistic models, large, periodic structural models and basis sets had to be used. The combination of large system sizes and accurate ab initio methods lead to computationally demanding calculations and is thus only feasible with supercomputing capacities.

2 Thermodynamic Properties of Hydrogen on Si(001) Under Chemical Vapor Deposition Conditions from Ab Initio Approaches

As outlined in the introduction, surface processes play a decisive role in epitaxy growth of integrated semiconductor heterostructures. In MOVPE, metal-organic molecular compounds interact with the substrate and decompose prior to, upon or after adsorption to the surface [5, 6].

As many of the III/V materials deposited are metastable, defect-free growth relies on an understanding of the growth-determining processes. The extent of hydrogen coverage on the common semiconductor substrate Si(001) is of particular importance as nucleation of functional materials on Si substrates is largely determined by the microscopic state of the surface [1, 10].

Different ab initio approaches to determine Gibbs free reaction energies and hydrogen coverage on the Si(001) surface were applied [9].

2.1 Methods

Total energies were determined by DFT methods applying the VASP 5.3.5 [11–14] software with a plane-wave basis set and the projector-augmented wave procedure (PAW) [15, 16]. The expansion of plane-wave basis functions was stopped at 350 eV while electronic energies and atomic forces were converged to 10^{-5} eV and 10^{-2} eV/Å in electronic and structural relaxation, respectively.

The lattice constant of Si was optimized to $a = 5.421$ Å applying the exchange-correlation functional by Perdew, Burke and Ernzerhof (PBE) [17, 18] and the D3 [19, 20] correction for long-range, attractive van der Waals interactions. Further calculations were performed with the HSE06 hybrid functional [21, 22] with D3 as indicated in the sections below.

The D3 parameters for the PBE0 hybrid functional [20] were used for HSE06 calculations while 50.2 and 21.2 Å were used as cutoffs for the interaction radius and for the determination of coordination numbers, respectively, for all calculations.

Momentum space was expanded in a Γ-centered grid derived via the Monkhorst-Pack method [23] with a (4 2 1)-division of the Si(001)c(4x2) surface cell containing four dimers. For other cell sizes the k-mesh division was scaled according to inverse lattice vectors. The asymmetric slabs contain eight Si layers with the two bottom layers frozen to bulk positions and hydrogen saturation at the bottom.

Model cells for Si(001)c(4x2) at different coverages are presented in Fig. 1.

The coverage is defined as $\theta = 1$ hydrogen atom per surface Si atom, i.e. the monohydride configuration H/Si(001). For lower coverages it was assumed that Si-Si dimers will either be both hydrogenated or both pristine due to stabilization as Si(001)c(4x2) reconstruction with buckled dimers for $\theta = 0$. Thus, it can be easily understood that fully covered $\theta = 1$ as well as uncovered $\theta = 0$ Si(001) are stable. A coverage of $\theta = 0.5$ seems to be stabilized with adjacent, fully and uncovered dimer rows.

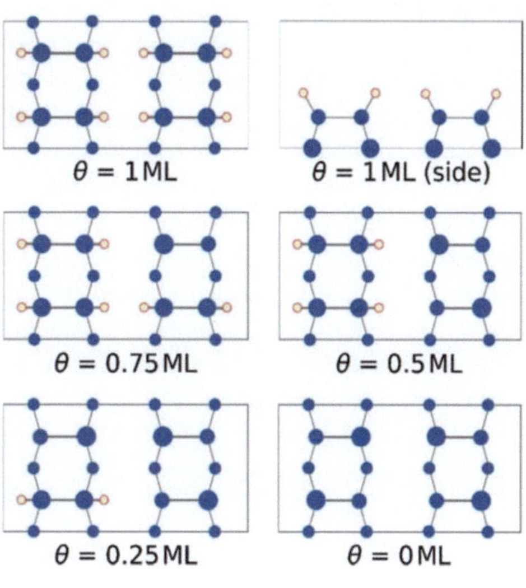

Fig. 1 Supercells of the Si(001) surface with different coverages θ used in this study (Figure reprinted with permission from Ref. [9])

For vibration calculations the Phonopy 1.8.2 code [24, 25] was used and a 2×2 supercell containing 16 dimers lead to converged results (PBE-D3) and used as standard for phonon density of state calculations with a q-mesh of $8 \times 4 \times 1$ points in reciprocal space, centered at the Γ-point.

2.2 Results

The temperature dependence of the hydrogen coverage in thermodynamic equilibrium was studied computing the phonon spectrum in a supercell approach. As an approximation to these demanding computations, an interpolated phonon (IP) approach was found to give comparable accuracy.

Following the *ab initio thermodynamics* (AITD) ansatz Gibbs reaction energies were calculated [26] as

$$\Delta G = \frac{N_H}{A} \times \left(\Delta E + \Delta\mu + \frac{1}{2}\mathrm{R}T \times ln(\frac{p}{p^0}) \right) \tag{1}$$

neglecting the volume change of the surface (with respect to [26]) and applying the reaction energies ΔE of the desorption reaction

$$[\mathrm{Si} - \mathrm{H}] \rightarrow [\mathrm{Si}_\theta] + \frac{1}{2}\mathrm{H}_2, \tag{2}$$

$$\Delta E = \frac{E_\theta + \frac{N_H}{2} \times E_{H_2} - E_{1ML}}{N_H} \tag{3}$$

as calculated by PBE-D3 and HSE06-D3. The hybrid functional resulted in roughly $15\,\mathrm{kJ\,mol^{-1}}$ higher desorption energies and found to be closer to the results of other studies [27].

N_H is the number of desorbing hydrogen atoms per unit cell, A is the surface area, R is the universal gas constant, T is the temperature, p is the hydrogen partial pressure and p^0 is the standard pressure (1013 mbar); the chemical potential of hydrogen $\Delta\mu$ can be derived from the literature (NIST) and the Shomate equations [28]. For intermediate coverages, also configurational entropy was accounted for [9]. Furthermore, the Gibbs energies for those configurations can be linearly interpolated between full and zero coverage. This computationally efficient approach is referred to as *interpolated phonon* (IP) in the following and was found in good agreement with explicit phonon (EP) calculations [9]. Following the explicit phonon approach, force constants from supercell calculations were used for thermodynamic corrections as described in [24, 25].

The phonon spectrum of H/Si(001) and Si(001) as well as a comparison of AITD and IP approaches are presented in Fig. 2.

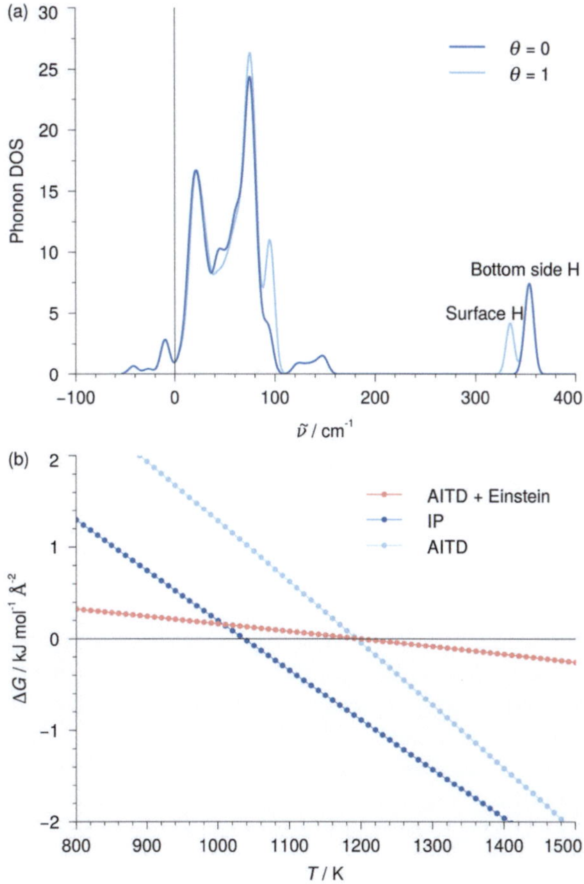

Fig. 2 (**a**) Phonon density of states (DOS) of hydrogenated and pristine Si surface. The negative frequencies stem from the frozen bottom layers. (**b**) Free energy of complete desorption for Einstein model [29, 30], interpolated phonon (IP) and AITD approach in comparison. The electronic energies were computed with HSE06-D3 (Figure reprinted with permission from Ref. [9])

In Fig. 3 temperature dependence of the hydrogen coverage is drawn for the IP and the AITD approach, respectively. Furthermore, the performance of the two exchange-correlation functionals applied can be retraced.

Regarding the onset temperature and the temperature range of desorption, calculation of interpolated phonons in combination with electronic desorption energies by HSE06+D3 gives best agreement with the experimental data available.

From about 800 K, a temperature relevant in chemical vapour deposition (CVD) techniques, desorption of hydrogen is significant and leads to coverages $\theta < 0.95$ monolayer. The bare Si(001) surface is present at about 1400 K and above. Strong changes in hydrogen coverage are found between 1000 and 1200 K in good

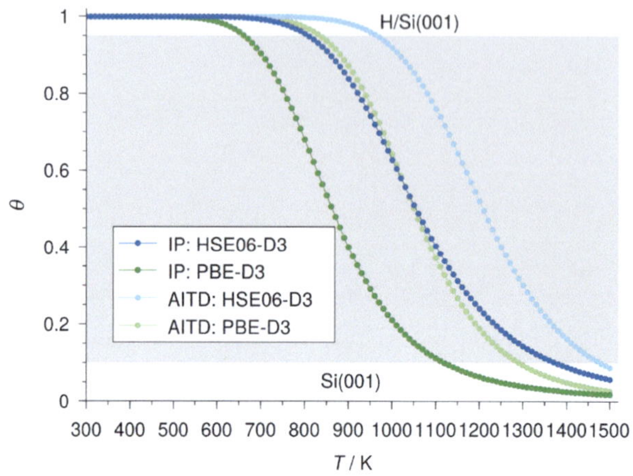

Fig. 3 Temperature dependence of coverage θ for IP and AITD approach. Binding energies computed with PBE-D3 and HSE06-D3. The *grey-shaded area* indicates the range of the graph with partially hydrogenated surface ($0.95 > \theta > 0.1$ ML) (Figure reprinted with permission from Ref. [9])

agreement with previous reflectance anisotropy spectroscopy experiments [31, 32]. In summary, the *ab initio thermodynamics* AITD approach shows less accurate behaviour in comparison with experiment, even with Einstein corrections.

These findings allow a rational choice for the surface state in the computational treatment of chemical reactions under typical metal organic vapor phase epitaxy conditions on Si(001).

3 Growth-Dependent Electronic and Optical Properties of Active Ga(NAsP) and Dilute Ga(AsBi) Materials

Functional semiconductors based on the compound materials Ga(NAsP) and Ga(AsBi) are being developed for integration into light-emitting, logical devices based on a Si(001) substrate. The variation of group 15 element's concentration is used for band structure engineering and tuning of emission wavelengths.

As discussed previously, next to mechanical strain effects, the growth behaviour and structural quality of Ga(NAsP) determine the device performance. The strain energy is thus investigated in the first part of this section for Ga(NAsP) grown on Si and GaP substrates.

In the second section, local ordering of Bi atoms in dilute bismide Ga(AsBi) films is investigated. While the formation of those clusters is likely a growth phenomenon (bulk diffusion of Bi is expected to be low), its consequences on the electronic structure of the alloy are evaluated by DFT calculations on bulk-type supercell models of the film formed.

3.1 *Strain Energy of Ga(NAsP) on Si and GaP [33]*

Ga(NAsP) exhibits a direct band gap in the Vis/IR range promoting the material for efficient light emitting devices performing efficiently even at room temperature [34–36]. It is almost lattice-matched to Si(001) and GaP(001), and can thus be epitaxially grown.

A minimal strain energy decreases the probability of dislocation defects formed during growth. Defect-free samples of Ga(NAsP)/GaP/Si(001) were realized at moderate and high growth temperatures between 575 and 700 °C [33]. While the material was found homogeneous at high temperatures (decreasing compositional disorder inside the bulk film) the roughness at the QW interfaces was smallest at low temperatures. The QW roughness also increases with the thickness of the film grown [37–41].

These results point towards thermodynamic stability of Ga(NAsP) grown on GaP(001) although dilute nitride III/V materials were expected to be metastable [33].

In the following, the composition $Ga(N_xAs_{0.85-x}P_{0.15})$ is investigated with an N concentration x between 0 and 0.25 and its stability with respect to different substrates is presented.

3.1.1 Methods

The calculations were performed according to Sect. 2.1 unless the following modifications. The PBE-D3 functional was used with a cut-off energy of 500 eV for the plane-wave basis expansion and a Γ-centered (6 6 6) k-grid for primitive cells. For $(2 \times 4 \times 5)$ supercells, (4 3 2) intersections were used. Mimicing the epitactic nature of Ga(NAsP) growth on Si(001) and GaP(001), the x/y cell parameters were constrained and the cells were relaxed stepwise in z towards hypothetical bulk-like Ga(NAsP) (*theoretical epitaxy*).

The strain relaxation energy (SRE) and the phase separation energy (PSE) are defined as

$$SRE = E^{strained}_{\mathrm{Ga(NAsP)}} - E^{bulk}_{\mathrm{Ga(NAsP)}} \tag{4}$$

and

$$PSE = E^{strained}_{\mathrm{Ga(N}_x\mathrm{As}_y\mathrm{P}_z)} - \mathrm{N}\left(xE^{strained}_{\mathrm{GaN}} + yE^{strained}_{\mathrm{GaAs}} + zE^{strained}_{\mathrm{GaP}}\right). \tag{5}$$

The former is the energy difference between the strained and the relaxed bulk film of the compound material. The latter is given by the energy difference between the strained compound material and the strained binary materials with respect to the substrate. All materials exhibit zinc blende structure with the respective substrate's lateral lattice constant.

3.1.2 Results

Figure 4 shows the SRE of Ga(NAsP) on Si and Ga(AsP) substrates as well as the PSE of Ga(NAsP) on a Si lattice and its equilibrium lattice constant.

The SRE of Ga($N_xAs_{0.85-x}P_{0.15}$) decreases from 0 % to 20 % N content as the material is decreasingly compressively strained with respect to a Si substrate (Fig. 4a) black, filled dots). Above an N ratio of 20 %, the material becomes tensilely strained and the SRE rises. In contrast, on hypothetical Ga(AsP) substrate, a deposited film becomes increasingly tensilely strained when adding nitrogen. Low SRE values support the hypothesis of thermodynamic stability for a given compound material-substrate system.

The tendency for phase separation into the binary materials in different strained environments was studied by the PSE for Ga($N_xAs_{0.85-x}P_{0.15}$) (Fig. 4b). The PSE of the quarternary material with silicon's lattice constant (black dots) is negative for low N contents and stabilizes further for concentrations up to 15 %. For higher concentrations it drastically increases and becomes positive for 25 % N incorporation. Then, there is a thermodynamic drive to separate into the binary components GaN, GaP and GaAs, presumably dominated by the contribution of GaN which is highly strained in the respective environments. This can be followed by the behaviour of the PSE on Ga($N_xAs_{0.85-x}P_{0.15}$) with equilibrium lattice constant which increases monotonously with N concentration (yellow dots, Fig. 4 b).

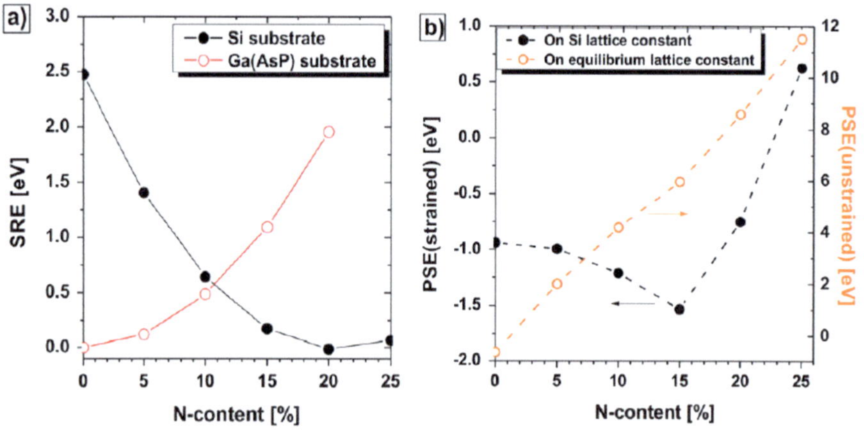

Fig. 4 Computational results for (**a**) strain relaxation energy (SRE) of Ga($N_xAs_{0.85-x}P_{0.15}$) on Si (*filled black symbols*) and virtual Ga($As_{0.85}P_{0.15}$) (*open red symbols*) substrates. In (**b**) the values of the phase separation energy (PSE) of Ga($N_xAs_{0.85-x}P_{0.15}$) strained to Si lattice constant (*filled black symbols*, left hand side axis) and on its equilibrium lattice constant (*open orange symbols*, right hand side axis) are plotted. Energy values refer to the (2 × 4 × 5) supercell (Figure reprinted with permission from Ref. [33])

These results support experimental findings and the hypothesis drawn: Depending on composition and the substrate, Ga(NAsP) is thermodynamically stabilized at certain N concentrations (15–20 %). In this composition range, the lattice-match to Si is optimal. QW layers of Ga(NAsP) can be grown at higher temperatures resulting in high quality films measured by roughness and compositional disorder.

3.2 Local Bi Ordering in Dilute Ga(AsBi) and Its Effect on Electronic Structure

Dilute Ga(AsBi) materials are applied for band gap engineering of III/V semiconductors. Starting from GaAs, the band gap can linearly be reduced by adding bismuth (Bi) at low concentrations (<1.9 %) so that light-emitting devices in the IR region were realized. This band gap reduction originates in an up-shift of the valence bands accompanied by a down-shift of the spin-orbit split-off band [43–49].

At higher Bi concentrations, the band gap narrowing becomes non-linear and, furthermore, a dependence of the band gap with respect to internal Bi bonding arrangements arises. Bismuth may form local clusters with multiple Bi atoms in close vicinity which makes a chemical perspective on the stability and bonding nature in those materials worthwhile. The homogeneity of Bi distribution in $Ga(As_{1-x}Bi_x)$ was studied with a periodic DFT approach and the results are presented in the following [42, 43].

3.2.1 Methods

Various local configurations at different concentrations x were modeled by $3 \times 3 \times 3$ supercells as sketched in Fig. 5 applying the general settings mentioned in Sect. 2.1. The cut-off energy of the basis set was 510 eV and the atomic structures were optimized with PBE-D3. Electronic structure (band gaps) was calculated by the meta-GGA functional by Tran and Blaha, TB09 (modified Becke-Johnson LDA), [50, 51] with spin-orbit coupling [52, 53] and (3 3 3) k-points.

The nature of different Ga_nBi_m bonding patterns was investigated by the Crystal Orbital Hamilton Population analysis [54–56] using the Lobster software (v 2.0.0, Koga local basis [57, 58]). This analysis evaluates the (anti-)bonding character of an atom pair based on the Hamilton matrix elements contributing to the overlap of the atoms' projected local orbitals.

The results of the COHP analysis for different Bi configurations inside Ga(AsBi) with concentrations of x = 0.031, 0.047 and 0.063 are presented in Fig. 6. The configurations can be classified with respect to the local Bi arrangements as

1. dispersed, Bi atoms at maximum separation, or
2. clustered, Bi atoms are bonded to a shared Ga atom.

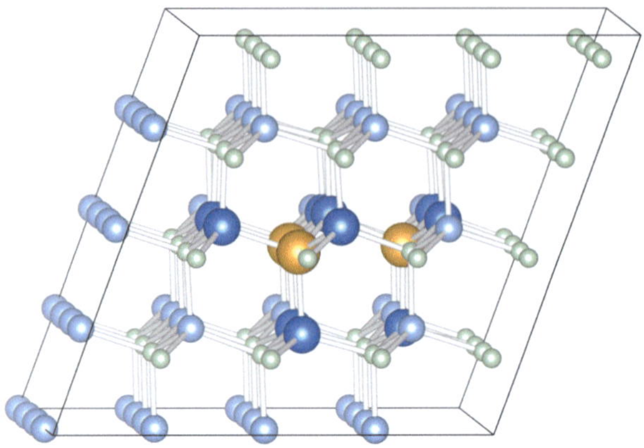

Fig. 5 3 × 3 × 3 supercell of a clustered arrangement of Bi atoms inside dilute Ga(AsBi) with highlighted cluster atoms (Ga: *blue*, As: *green*, Bi: *orange*) (Figure from Ref. [42])

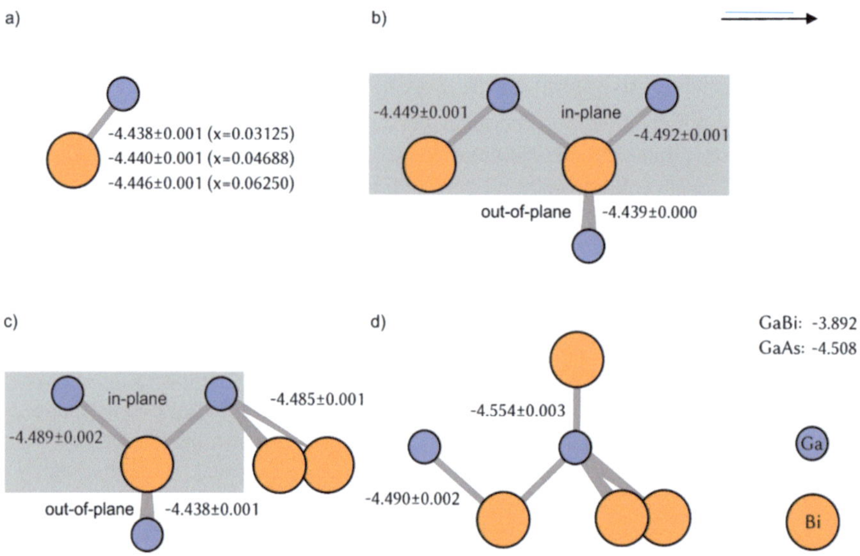

Fig. 6 Local Ga_nBi_m configurations classified as (**a**) dispersed with concentrations of x = 0.03125, 0.04688 and 0.0625, and (**b**) clustered containing (**b**) Bi_2, (**c**) Bi_3 and (**d**) Bi_4 units. Averaged iCOHP values are fiven for equivalent bonds in eV per bond together with the standard deviation. For the clustered arrangements bonds to Ga in the same (in-plane, shaded) and in other crystallographic planes (out-of-plane) can be distinguished (Figure from Ref. [42])

3.2.2 Results: Local Ordering

The total energies of the different configurations studied do not differ significantly [42]. However, the stabilities of local Bi arrangements can clearly be distinguished based on the COHP analyis as evaluated by the energies gained by intergration over the COHP elements (iCOHP). The iCOHP values for Ga-Bi bonds for several local configurations next to the bonding types in bulk GaAs and hypothetical GaBi are presented in Fig. 6.

In pure GaAs (−4.508 eV/bond) and GaBi (−3.892 eV/bond) the iCOHP bond strengths serve as a reference for the maximum and minimum bonding interaction, respectively. It becomes obvious that the III-V bonds in GaAs are much stronger than in binary GaBi, which is not known experimentally. All configurations of dilute bismide materials considered, the bond strengths are close to the iCOHP of ideal GaAs, in agreement with the structures found in experiment and negligible total energy differences.

For the dispersed configurations, the Ga-Bi bond strength (iCOHP = −4.438, −4.440, −4.446 eV/bond) increases with Bi concentration x, however, for the clustered arrangements, stronger bonds were found. In line with the dispersed situations, the iCOHP values also increase with Bi concentration in the clustered arrangements, where a Ga-Bi-Ga unit forms a plane as indicated by the shaded area in Fig. 6a–d. In-plane Ga-Bi bonds tend to be more favourable (iCOHP = −4.489, −4.490, −4.492 eV/bond) than out-of-plane bonds (iCOHP = −4.439, −4.438 eV/bond). The strongest Ga-Bi bond was found for a $GaBi_4$ tetrahedral arrangement (at the highest x studied, clustered arrangement) with an iCOHP of −4.554 eV/bond. Remarkably, this value is a larger than the absolute iCOHP of the GaAs bond in the ideal binary material.

Thus, it was concluded that clustered, i.e. heterogeneously dispersed, arrangements of Bi atoms forming Ga-Bi bonds in a dilute bismide GaAs have stronger bonds and are more likely to occur than homogeneously dispersed Bi atoms. This is, of course, only a thermodynamic view and growth processes (kinetics, defect formation etc.) might influence the formation of certain configurations. This conclusion was, however, in good agreement with quantitative analysis by high-resolution high angle annular dark field (transmission electron microscopy) images of Ga(AsBi) samples at similar Bi concentrations grown under metal-organic vapour phase epitaxy conditions for photonics applications [42].

3.2.3 Results: Effect on Electronic Structure

The effect of band gap narrowing due this localized character of clustered Bi atom arrangements in dilute $Ga(As_{1-x}Bi_x)$ alloys was investigated and the results are presented in the following [43].

In dilute Ga(NAs) a band gap reduction was found which was explained as anticrossing of localized, empty s(N) orbitals with the conduction band of GaAs. It was shown that the electron mobility was affected by the N concentration which hints towards effects on the conduction band in dilute Ga(NAs).

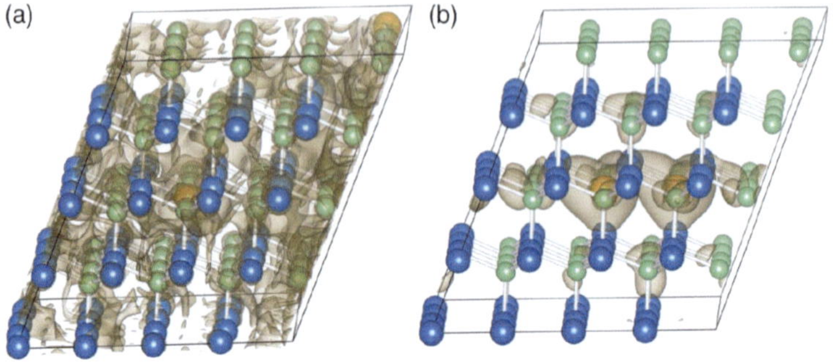

Fig. 7 Band decomposed charge density of the heavy hole band for two atom [111] chain (**a**) and cluster (**b**) arrangements. The charge density results from integration over the whole Brillouin zone. Every isovalue is set to 10 % of the respective maximum (Figure reprinted with permission from Ref. [43])

In contrast, for dilute Ga(AsBi) materials the hole mobility is decreased in the bismide compared to pure GaAs (effect on valence band) [59–62]. This is an effect in the valence band – more precisely, as will be shown, a hybridization of p(Bi) orbitals which depends on the Bi concentration and local configuration in Ga(AsBi).

Figure 7 shows the partial charge density of the heavy hole band for a dispersed (left) and a clustered (right) arrangement at a Bi concentration of $x = 0.047$ (2 Bi atoms per supercell). The band decomposition clearly displays the delocalized (a) (dispersed) and localized (b) (clustered) character of the valence band for the two arrangements.

The band structures of pristine GaAs and four dilute Ga(AsBi) supercell models were calculated by DFT and a band unfolding technique [63]. The formation of a tail in the valence band is highly dependent on the relative local arrangement of the Bi atoms measured by the Bi-Bi separation inside the supercell.

This behaviour can be explained by the tendency of the p(Bi) orbitals to hybridize with decreasing separation along the [111] axis (the supercell's diagonal) [43]. In Fig. 8 the band gap of the host GaAs material is plotted next to the localized Bi levels for four different Bi concentrations in dilute Ga(AsBi).

Furthermore, the band gap narrowing effect was found to increase with Bi concentration (compressive strain). Compared to experimental photoluminescence measurements, the bowing rates determined indicate a two-scale disorder effect for high Bi concentrations.

Thus, three effects determine the band gap narrowing of dilute Ga(AsBi): chemical arrangements of the dopant atoms, strain and macroscopic disorder.

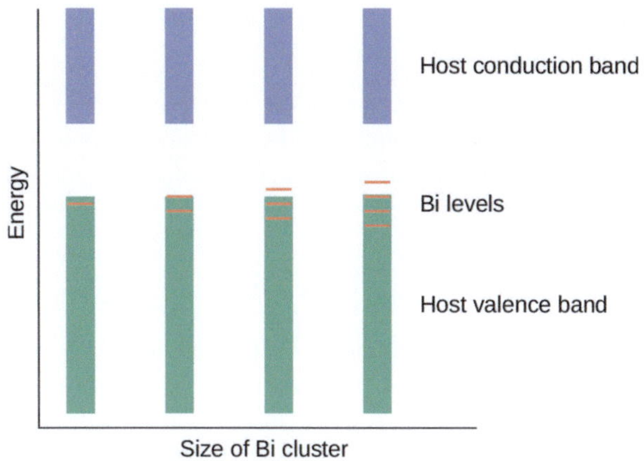

Fig. 8 Band gaps of pristine GaAs and dilute Ga(AsBi) with clustered arangements of Bi atoms in a 8 × 8 × 8 supercell. The evolution of Bi defect levels is shown as a function of Bi cluster size

4 Electron-Phonon Coupling of NTCDA on Ag(111)

The following section expands the research on the interplay of chemical and electronic properties of semiconductor materials conducted. The behaviour of wide-gap aromatic molecules on metal substrates at finite temperatures was studied by DFT methods and the electronic response of vibrational modes was investigated. As the character of semiconductor interfaces determines device performance the insights on electron-vibron coupling gained from this study will guide further investigations on III/V-Si systems.

Infrared spectroscopy experiments of NTCDA adsorbates on Ag(111) show active in-plane molecular modes which should be inactive following selection rules for molecules [64]. These modes do not lead to a change in the molecule's dipole moment orthogonal to the surface. A dynamical dipole moment emerges at the interface between the adsorbate layer and the substrate. For this quantity, no theoretical or experimental proof beyond heuristic models has been provided [65].

Here, the *interfacial dynamical charge transfer* (IDCT) was studied for a sub-monolayer NTCDA on Ag(111) system and the relative importance compared to nuclear motion at the interface was quantified. A rationale for the amount of IDCT for specific vibrational modes at the interface is provided going beyond an empirical evaluation presented earlier [66] without applying time-dependent treatment [67], which is unfeasible for systems as large as the one studied. Electron-vibron coupling was investigated for similar systems before with the conclusion that all the prerequisites for IDCT (strong adsorption, dynamical partial occupation of the lowest unoccupied molecular orbital) are fulfilled for the system under investigation [68, 69]. Schöll et al. found that the dipole moment μ_{IDCT} of the NTCDA/Ag(111) system is affected by a strong electron-vibron coupling [70].

4.1 Methods

IR intensity is proportional to the square of the change in dipole moment μ_{dyn}^2 for a given mode. The dipole moment is affected by nuclear motion as well as dynamic charge transfer from the metal substrate to the molecule across the adsorbate-substrate interface.

$$\mu_{dyn} = \mu_{\text{nucl}} + \mu_{IDCT} \tag{6}$$

Both components will be estimated by DFT calculations applying the settings described in Sect. 2.1 with the PBE-D3 functional, a 350 eV cut-off energy and a Γ-centered (3 3 1) k-grid within VASP 5.2.12. The lattice parameter was determined as $a = 4.073$ Å for a four-layer Ag(111) slab which was optimized in a 4×4 supercell (equilibrium adsorption geometry shown in Fig. 9).

The charge transferred Δq according to μ_{IDCT} will be estimated by the Natural Population Analysis scheme [71, 72] (which shows the same qualitative trends as the Atoms In Molecules (AIM) scheme [73]).

IR spectra and vibrational modes were calculated by the finite-differences ($\pm\Delta Q$) method of a partial Hessian matrix, displacing the adsorbate atoms and computing intensities based on the z-component of μ_{dyn}. A negligible influence of the substrate atom displacements on the spectra was found and the substrate was thus frozen. The IR spectra produced were found to be in good agreement with experimental measurements [64, 65]. The vibrational modes directly provide the nuclear motions and thus μ_{nucl}.

4.2 Results

The NTCDA molecules are bound to Ag(111) via attractive van der Waals interactions. However, the bent geometry of the adsorbate molecule suggests a more

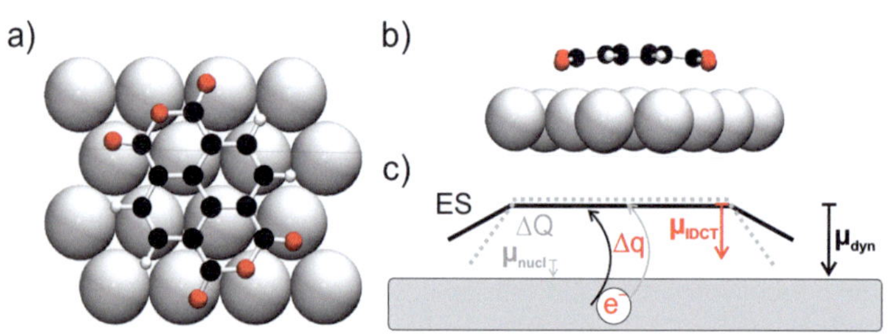

Fig. 9 Schematic of most stable adsorption geometry of NTCDA on a bridge position on Ag(111) in top (**a**) and side (**b**) view; (**c**) contributions to dynamic dipole moment (Figure reprinted with permission from Ref. [64])

covalent character of the Ag-O bonds. The molecule is distorted propagating shorter Ag-O distances than Ag-C which lifts the planarity of the free molecule (type-averaged bond lengths d(O-Ag) = 2.577 ± 0.004 Å, d(C-Ag) = 2.905 ± 0.145 Å). This reduces the symmetry from D_{2h} to approximately C_{2v} [65]. The adsorption energy is large ($\Delta E_{ads} = 2.09\,\text{eV}$) and the molecule's LUMO is partially filled. This fulfills important prerequisits for IDCT [74, 75].

The calculated as well as the experimentally measured IR spectra agree very well in intensities and mode energies. As can be seen in Table 1, the in-plane molecular modes with symmetry a_g become IR active [64, 65] in the adsorbate-substrate system due to IDCT which has a large contribution to μ_{dyn}.

The IDCT was quantified by the mode-specific charge transfer from the substrate to the adsorbate, Δq, which was found to correlate to μ_{dyn} (Fig. 10a). Depending on the mode symmetry, nuclear motions contribute less to μ_{dyn} (e.g. a_g modes). Those modes are dominated by the IDCT dipole moment, which can be derived as μ_{IDCT} by a linear correlation with Δq, Fig. 10a. μ_{nucl} as given in Table 1 is the difference between μ_{dyn} and μ_{IDCT}.

For this value, on the other hand, good correlation was found for $\mu_{\Delta Q}$ computed by finite-difference displacements as shown in Fig. 10b. The dipole moment from nuclear motion μ'_{nucl} can then be directly derived from $\mu_{\delta Q}$ as given in Table 1.

Table 1 Computed properties of vibrational modes. See Fig. 9c for definition of terms. Table reused with permission from Ref. [64]

No.	$\tilde{\nu}$[a]	sym.[b]	int.[c]	μ_{dyn}[d]	μ_{IDCT}[d]	μ'_{nucl}[d]	Δq[e]	$\mu_{\Delta \mathrm{Q}}$[d]
1	648.1	a_g	0.004	0.18	0.08	+0.11	0.01	+0.14
2	665.4	b_{2g}	0.000	0.00	–	–	0.00	0.00
3	717.9	b_{3u}	0.058	−0.59	0.52	−1.08	0.11	−1.14
4	720.8	a_u	0.000	0.00	–	–	0.00	+0.01
5	813.0	b_{3u}	0.002	0.11	0.22	−0.18	0.04	−0.17
6	987.0	a_g	0.002	0.12	0.13	+0.06	0.02	+0.09
7	1104.2	a_g	0.021	0.37	0.22	+0.02	0.04	+0.04
8	1256.9	a_g	0.038	0.53	0.35	+0.05	0.07	+0.08
9	1345.5	a_g	0.581	1.95	1.88	+0.07	0.42	+0.10
10	1404.8	a_g	0.133	0.94	0.92	+0.09	0.20	+0.12
11	1435.8	b_{1u}	0.000	0.02	–	–	0.00	0.00
12	1509.8	b_{2u}	0.000	0.00	–	–	0.00	0.00
13	1565.6	a_g	1.000	2.51	2.44	+0.17	0.55	+0.21
14	1625.7	a_g	0.264	1.27	1.84	−0.62	0.41	−0.64
15	1628.9	b_{1u}	0.024	0.37	0.57	−0.19	0.12	−0.18

[a]Vibrational modes in cm^{-1}
[b]Mode symmetries
[c]IR intensities normalized to highest value
[d]Dipole moments in Debye
[e]Charges in *e*

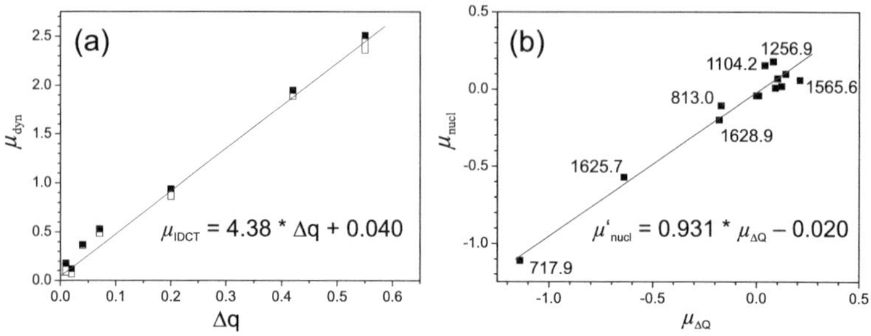

Fig. 10 (**a**) Correlation of charge transfer (Δq) and μ_{dyn} to determine μ_{IDCT}; the *open rectangles* refer to first order correction of μ_{dyn} due to nonzero μ_{nucl}. (**b**) Correlation of $\mu_{\Delta Q}$ with μ_{nucl} to determine μ'_{nucl} (Figure reprinted with permission from Ref. [64])

As can be seen comparing the dipole contributions μ_{nucl} and μ_{IDCT} in Table 1, the role of IDCT dominates the dynamic dipole moments μ_{dyn} and associated infrared activities. The contribution of nuclear motion (out-of-plane bending) is less important throughout the 15 modes investigated. For a_g symmetric modes, the magnitude of IDCT can be derived from partial charges as shown by the NPA-derived atomic charges Δq. According to Eq. 6, contributions from nuclear motion can be determined reliably if IDCT is the main contribution.

We derived unequivocal evidence for the dominating role of IDCT for dynamic dipole moments, associated IR activities and thus electron-vibron coupling.

Acknowledgements The authors acknowledge the research training group (Graduiertenkolleg GRK 1782) “Functionalization of Semiconductors”, the collaborative research center (Sonderforschungsbereich SFB 1083) “Structure and Dynamics of Internal Interfaces” (both funded by the German Research Foundation DFG). AS thanks the Beilstein Institut, Frankfurt am Main, for support.

References

1. Liang, D., Bowers, J.E.: Nat. Photonics **4**(8), 511 (2010)
2. Carrère, H., Marie, X.: chap. 6. In: Balkan, N., Xavier, M. (eds.) Semiconductor Modeling Techniques. Springer Series in Materials Science, vol. 159, pp. 153–195. Springer, Berlin/Heidelberg (2012). doi:10.1007/978-3-642-27512-8. http://www.springerlink.com/index/10.1007/978-3-642-27512-8
3. Liebich, S., Zimprich, M., Beyer, A., Lange, C., Franzbach, D.J., Chatterjee, S., Hossain, N., Sweeney, S.J., Volz, K., Kunert, B., Stolz, W.: Appl. phys. Lett. **99**(7), 071109 (2011)
4. Kunert, B., Volz, K., Nemeth, I., Stolz, W.: J. Lumin. **121**(2), 361 (2006)
5. Stringfellow, G.B.: Mater. Sci. Eng. B **87**(2), 97 (2001)
6. Beyer, A., Ohlmann, J., Liebich, S., Heim, H., Witte, G., Stolz, W., Volz, K.: J. Appl. Phys. **111**(8), 0835341 (2012)
7. Stegmüller, A., Rosenow, P., Tonner, R.: Phys. Chem. Chem. Phys. **16**(32), 17018 (2014)
8. Beyer, A., Stegmüller, A., Oelerich, J.O., Jandieri, K., Werner, K., Mette, G., Stolz, W., Baranovskii, S.D., Tonner, R., Volz, K.: Chem. Mater. **28**(10), 3265 (2016)

9. Rosenow, P., Tonner, R.: J. Chem. Phys. **144**, 204706 (2016)
10. Brauers, A.: J. Cryst. Growth **107**, 281 (1991)
11. Kresse, G., Hafner, J.: Phys. Rev. B **47**(1), 558 (1993)
12. Kresse, G., Hafner, J.: Phys. Rev. B **49**(20), 14251 (1994)
13. Kresse, G., Furthmüller, J.: Comput. Mater. Sci. **6**(1), 15 (1996)
14. Kresse, G., Furthmüller, J.: Phys. Rev. B Condens. Matter Mater. Phys. Condens. Matter **54**(16), 11169 (1996)
15. Blöchl, P.: Phys. Rev. B **50**(24), 17953 (1994)
16. Kresse, G., Joubert, D.: Phys. Rev. B **59**(3), 1758 (1999)
17. Perdew, J.P., Burke, K., Ernzerhof, M.: Phys. Rev. Lett. **77**(18), 3865 (1996)
18. Perdew, J.P., Burke, K., Ernzerhof, M.: Phys. Rev. Lett. **78**(7), 1396 (1997)
19. Grimme, S., Antony, J., Ehrlich, S., Krieg, H.: J. Chem. Phys. **132**(15), 154104 (2010)
20. Grimme, S., Ehrlich, S., Goerigk, L.: J. Comput. Chem. **32**, 1456 (2011)
21. Vydrov, O.A., Heyd, J., Krukau, A.V., Scuseria, G.E.: J. Chem. Phys. **125**(7), 074106 (2006)
22. Heyd, J., Scuseria, GE., Ernzerhof, M.: J. Chem. Phys. **124**(21), 219906 (2006)
23. Pack, J.D., Monkhorst, H.J.: Phys. Rev. B **13**(12), 5188 (1976)
24. Togo, A., Tanaka, I.: Scr. Mater. **108**, 1 (2015)
25. Togo, A., Oba, F., Tanaka, I.: Phys. Rev. B Condens. Matter Mater. Phys. **78**(13), 1 (2008)
26. Kaxiras, E., Bar-Yam, Y., Joannopoulos, J.D., Pandey, K.C.: Phys. Rev. B **35**(18), 9625 (1987)
27. Dürr, M., Höfer, U.: Surf. Sci. Rep. **61**(12), 465 (2006)
28. Chase, M.W.: J. Phys. Chem. Ref. Data **25**(5), 1297 (1996)
29. Einstein, A.: Ann. der Phys. **22**, 180 (1906)
30. Hill, T.: An Introduction to Statistical Thermodynamics. Dover Publications, New York (1986)
31. Shi, J., Kang, H.C., Tok, E.S., Zhang, J.: J. Chem. Phys. **123**(3), 034701/1-034701/8 (2005)
32. Puzder, A., Williamson, A.J., Grossman, J.C., Galli, G.: Phys. Rev. Lett. **88**(9), 097401 (2002)
33. Wegele, T., Beyer, A., Ludewig, P., Rosenow, P., Duschek, L., Jandieri, K., Tonner, R., Stolz, W., Volz, K.: J. Phys. D Appl. Phys. **49**(7), 075108 (2016)
34. Rosemann, N.W., Metzger, B., Kunert, B., Volz, K., Stolz, W., Chatterjee, S.: Appl. Phys. Lett. **103**(25) (2013)
35. Hossain, N., Sweeney, S.J., Rogowsky, S., Ostendorf, R., Wagner, J., Liebich, S., Zimprich, M., Volz, K., Kunert, B., Stolz, W.: Electron. Lett. **47**(16), 931 (2011)
36. Liebich, S., Zimprich, M., Beyer, A., Lange, C., Franzbach, D.J., Chatterjee, S., Hossain, N., Sweeney, S.J., Volz, K., Kunert, B., Stolz, W.: Appl. Phys. Lett. **99**(7), 17 (2011)
37. Sakaki, H., Noda, T., Hirakawa, K., Tanaka, M., Matsusue, T.: Appl. Phys. Lett. **51**(23), 1934 (1987)
38. Dura, J.A., Pellegrino, J.G., Richter, C.A.: Appl. Phys. Lett. **69**(8), 1134 (1996)
39. Penner, U., Rücker, H., Yassievich, I.N.: Semicond. Sci. Technol. **13**, 709 (1999)
40. Erol, A.: Dilute III-V Nitride Semiconductors and Material Systems. Springer, Berlin (2008)
41. Dargam, T.G., Koiller, B.: Solid State Commun. **105**(4), 211 (1998)
42. Knaub, N., Rosenow, P., Beyer, A., Jandieri, K., Ludewig, P., Tonner, R., Volz, K.: (2016, submitted)
43. Bannow, L.C., Rubel, O., Rosenow, P., Badescu, S.C., Hader, J., Moloney, J.V., Tonner, R., Koch, S.W.: Phys. Rev. B **93**(20), 205202 (2016)
44. Alberi, K., Wu, J., Walukiewicz, W., Yu, K.M., Dubon, O.D., Watkins, S.P., Wang, C.X., Liu, X., Cho, Y.J., Furdyna, J.: Phys. Rev. B Condens. Matter Mater. Phys. **75**(4), 1 (2007)
45. Francoeur, S., Seong, M.J., Mascarenhas, A., Tixier, S., Adamcyk, M., Tiedje, T.: Appl. Phys. Lett. **82**, 3874 (2003)
46. Ludewig, P., Knaub, N., Hossain, N., Reinhard, S., Nattermann, L., Marko, I.P., Jin, S.R., Hild, K., Chatterjee, S., Stolz, W., Sweeney, S.J., Volz, K.: Appl. Phys. Lett. **102**(24), 100 (2013)
47. Marko, I.P., Ludewig, P., Bushell, Z.L., Jin, S.R., Hild, K., Batool, Z., Reinhard, S., Nattermann, L., Stolz, W., Volz, K., Sweeney, S.J.: J. Phys. D Appl. Phys. **47**(34), 345103 (2014)
48. Fuyuki, T., Yoshioka, R., Yoshida, K., Yoshimoto, M.: Appl. Phys. Lett. **103**(20), 3 (2013)

49. Lewis, R., Beaton, D., Lu, X., Tiedje, T.: J. Cryst. Growth **311**(7), 1872 (2009)
50. Becke, A.D., Johnson, E.R.: J. Chem. Phys. **124**(22) (2006)
51. Tran, F., Blaha, P.: Phys. Rev. Lett. **102**(22), 5 (2009)
52. Kim, Y.S., Hummer, K., Kresse, G.: Phys. Rev. B Condens. Matter Mater. Phys **80**(3), 1 (2009)
53. Kim, Y.S., Marsman, M., Kresse, G., Tran, F., Blaha, P.: Phys. Rev. B Condens. Matter Mater. Phys. **82**(20), 1 (2010)
54. Dronskowski, R., Blöchl, P.E.: J. Phys. Chem. **97**(33), 8617 (1993)
55. Deringer, V.L., Tchougréeff, A.L., Dronskowski, R.: J. Phys. Chem. A **115**(21), 5461 (2011)
56. Maintz, S., Deringer, V.L., Tchougréeff, A.L., Dronskowski, R.: J. Comput. Chem. **34**(29), 2557 (2013)
57. Koga, T., Kanayama, K., Watanabe, S., Thakkar, A.J.: Int. J. Quantum Chem. **71**(6), 491 (1999)
58. Koga, T., Kanayama, K., Watanabe, T., Imai, T., Thakkar, A.J.: Theor. Chem. Acc. **104**(5), 411 (2000)
59. Kini, R.N., Ptak, A.J., Fluegel, B., France, R., Reedy, R.C., Mascarenhas, A.: Phys. Rev. B Condens. Matter Mater. Phys. **83**(7), 1 (2011)
60. Nargelas, S., Jarasiunas, K., Bertulis, K., Pacebutas, V.: Appl. Phys. Lett. **98**(8), 1 (2011)
61. Beaton, D.A., Lewis, R.B., Masnadi-Shirazi, M., Tiedje, T.: J. Appl. Phys. **108**(8), 2 (2010)
62. Cooke, D.G., Hegmann, F.A., Young, E.C., Tiedje, T.: Appl. Phys. Lett. **89**(12), 83 (2006)
63. Rubel, O., Bokhanchuk, A., Ahmed, S.J., Assmann, E.: Phys. Rev. B Condens. Matter Mater. Phys. **90**(11), 1 (2014)
64. Rosenow, P., Jakob, P., Tonner, R.: J. Phys. Chem. Lett. **7**, 1422 (2016)
65. Tonner, R., Rosenow, P., Jakob, P.: Phys. Chem. Chem. Phys. **18**, 6316 (2016)
66. Braatz, C.R., Öhl, G., Jakob, P.: J. Chem. Phys. **136**(13), 134706/1-134706/8 (2012)
67. Weigel, A., Dobryakov, A., Klaumünzer, B., Sajadi, M., Saalfrank, P., Ernsting, N.P.: J. Phys. Chem. B **115**(13), 3656 (2011)
68. Langreth, D.C.: Phys. Rev. Lett. **54**(2), 126 (1985)
69. Chabal, Y.J.: Phys. Rev. Lett. **55**(8), 845 (1985)
70. Schöll, A., Zou, Y., Kilian, L., Hübner, D., Gador, D., Jung, C., Urquhart, S.G., Schmidt, T., Fink, R., Umbach, E.: Phys. Rev. Lett. **93**(14), 93 (2004)
71. Reed, A.E., Weinstock, R.B., Weinhold, F.: J. Chem. Phys. **83**(2), 735 (1985)
72. Dunnington, B.D., Schmidt, J.R.: J. Chem. Theory Comput. **8**(6), 1902 (2012)
73. Bader, R.: Atoms in Molecules – A Quantum Theory. Oxford University Press (1990)
74. Tautz, F.S.: Prog. Surf. Sci. **82**(9–12), 479 (2007)
75. Bendounan, A., Forster, F., Schöll, A., Batchelor, D., Ziroff, J., Umbach, E., Reinert, F.: Surf. Sci. **601**(18), 4013 (2007)

Submonolayer Rare Earth Silicide Thin Films on the Si(111) Surface

S. Sanna, C. Dues, U. Gerstmann, E. Rauls, D. Nozaki, A. Riefer, M. Landmann, M. Rohrmüller, N.J. Vollmers, R. Hölscher, A. Lücke, C. Braun, S. Neufeld, K. Holtgrewe, and W.G. Schmidt

Abstract Rare earth induced silicide phases of submonolayer height and 5×2 periodicity on the Si(111) surface are investigated by density functional theory and ab initio thermodynamics. The most stable silicide thin film consists of alternating Si Seiwatz and honeycomb chains aligned along the $[1\bar{1}0]$ direction, with rare earth atoms in between. This thermodynamically favored model is characterized by a minor band gap reduction compared to bulk Si and explains nicely the measured scanning tunneling microscopy images.

1 Introduction

Metallic rare earth (RE) silicides can be grown epitaxially as thin films on the Si(111) substrate by rare earth deposition and thermal treatment [1–3]. The resulting metal/semiconductor interface is characterized by an extraordinarily low Schottky barrier height of 0.3–0.4 eV on n-type substrates. Due to the marginal lattice mismatch [4] between substrate and thin film, the interface furthermore has a low defect concentration and is very stable. Therefore rare earth silicides on n-type silicon are considered ideal candidates for Ohmic contacts [5, 6]. The relatively high barrier height on p-type substrates makes them interesting for infrared detectors and photovoltaic applications [7]. For submonolayer coverage, a variety of structures with different periodicities was found [8–12].

Despite the large and growing interest in silicide thin films on Si(111), our knowledge of these systems is still fragmentary. A multitude of surface reconstructions or nanostructures with different periodicities has been observed, depending on the rare earth species and rare earth coverage [8–19]. The observed structures are characterized by different stoichiometries and heights. For the case of dysprosium silicide, e.g., a full monolayer results in a film with 1×1 periodicity and $DySi_2$ hexagonal structure, multilayer silicides grow in a film with $\sqrt{3} \times \sqrt{3}$ periodicity

S. Sanna (✉) • C. Dues • U. Gerstmann • E. Rauls • D. Nozaki • A. Riefer • M. Landmann • M. Rohrmüller • N.J. Vollmers • R. Hölscher • A. Lücke • C. Braun • S. Neufeld • K. Holtgrewe • W.G. Schmidt
Lehrstuhl für Theoretische Physik, Universität Paderborn, 33095 Paderborn, Germany
e-mail: simone.sanna@uni-paderborn.de

W.E. Nagel et al. (eds.), *High Performance Computing in Science and Engineering '16*, DOI 10.1007/978-3-319-47066-5_12

and Dy_3Si_5 composition, while submonolayer coverage results in structures with $2\sqrt{3} \times 2\sqrt{3}$ or 5×2 periodicity [10].

Computational studies of two-dimensional rare-earth silicides are rare and limited to the simplest yttrium and erbium silicide structures [15, 18–23]. The lack of theoretical investigations is particularly severe in the submonolayer range, where, to the best of our knowledge, no theoretical investigations are available.

The present paper aims at providing microscopic structural models for the silicide thin film 5×2 phase observed in the submonolayer regime. To this end, we combine total energy density functional theory (DFT) calculations with ab initio thermodynamics. Calculations are performed using Tb (atomic number 65) and Dy (atomic number 66) as prototypical trivalent rare earths.

2 Methodology

Total-energy density functional theory (DFT) calculations are performed within the generalized gradient approximation [24] (GGA) in the Perdew-Burke-Ernzerhof formulation [25] as implemented in the Vienna ab initio simulation package (VASP) [26, 27]. Projector augmented wave [28, 29] (PAW) potentials with projectors up to $l = 1$ for H, $l = 2$ for Si and $l = 3$ for the rare earth atoms, as well as a plane wave cutoff of 400 eV have been used. As no other valence state than RE^{3+} has been observed for the rare earth ions in the silicide structures, we constrain the valence state of the investigated rare earth ions treating f electrons as core states. This approach, commonly referred to as frozen-core method, allows for a proper treatment of the lanthanides within DFT [30–32].

Six Si bilayers stacked along the [111] crystallographic direction model the substrate. The periodic supercell contains in addition the silicide thin film of variable structure and height, and a vacuum region of at least 15 Å. The dangling bonds at the bottom of the slab are saturated by H atoms. The atomic positions are relaxed until the residual Hellmann-Feynman forces are lower than 0.001 eV/Å. Thereby three Si bilayers and the hydrogen atoms are kept frozen. Test calculations show that adding further substrate layers does not result in noticeable changes of the calculated geometries and band structures. Dipole-correction algorithms have been used to correct the spurious interactions of the slabs with their periodic images [33, 34].

Simulated constant-current STM images are calculated within the Tersoff-Hamann approach [35, 36] on the basis of the partial densities of states (LDOS).

In order to compare the formation energy of silicide films with different composition, we use the Landau potential Ω, approximated as [37, 38]

$$\Omega(\mu_{\mathrm{Si}}, \mu_{\mathrm{RE}}) \approx E^{DFT}(N_{\mathrm{Si}}, N_{\mathrm{RE}}) - \sum_{i}^{\mathrm{Si,RE}} \mu_i N_i. \tag{1}$$

In this equation, $E^{DFT}(N_{\mathrm{Si}}, N_{\mathrm{RE}})$ is the DFT total energy of a slab containing N_{Si} silicon atoms and N_{RE} rare earth atoms. μ_{Si} and μ_{RE} are the corresponding chemical potentials and represent the experimental growth conditions. The sum in Eq. 1 also extends to the H atoms employed to saturate the Si dangling bonds at the bottom side of the slabs. The usage of the total rather than the free energy for the calculation of Ω is an approximation. It is justified as long as the entropic contributions are of similar magnitude for the different silicide films.

The Landau potential Ω in Eq. 1 is expressed as a function of the chemical potentials μ_{Si} and μ_{RE}. Their thermodynamically allowed range is constrained by several conditions. The upper limits are given by the bulk phases,

$$\mu_{\mathrm{i}} \leq \mu_{\mathrm{i}}^{\mathrm{bulk}} \quad i = \mathrm{Si}, \mathrm{RE}. \tag{2}$$

Furthermore the silicide films are in equilibrium with the Si substrate, which represents an infinite reservoir of Si atoms. This pins the value of μ_{Si} to $\mu_{\mathrm{Si}}^{\mathrm{bulk}}$ and allows to express the Landau potential as $\Omega = \Omega(\mu_{\mathrm{RE}})$. The RE chemical potential can be controlled experimentally with the amount of rare earth deposited on the Si substrate before annealing. If we restrict our investigation to silicide phases with a given stoichiometry $\mathrm{RE}_\alpha \mathrm{Si}_\beta$, the lower limit of μ_{RE} is given by

$$\alpha \cdot \mu_{\mathrm{Si}}^{\mathrm{bulk}} + \beta \cdot \mu_{\mathrm{RE}} = \mu_{\mathrm{Si}_\alpha \mathrm{RE}_\beta}^{\mathrm{bulk}} \tag{3}$$

where we use $\mu_{\mathrm{Si}} = \mu_{\mathrm{Si}}^{\mathrm{bulk}}$ as we consider Si rich conditions. However, we also consider lower values of μ_{RE}, representing non-stoichiometric silicides with dilute rare earth concentrations.

The values of the bulk chemical potentials are estimated by the total energy per formula unit calculated within DFT using hexagonal stoichiometric RESi_2 phases with AlB_2 structure (space group D_{3d}^3) for the silicides, hexagonal close-packed structures (space group D_{6h}^4) for the metallic elemental rare earths and the gas phase of hydrogen.

Highly customizable parallelization schemes are implemented in VASP. In particular, parallelization (and data distribution) over bands, parallelization (and data distribution) over plane wave coefficients, as well as parallelization over k-points (no data distribution) can to be used at the same time on massively parallel systems such as the Cray XC40 (Hazel Hen), in order to obtain high computational efficiency.

The performance of the Cray XC40 in combination with the available parallelization routines has been tested and optimized with VASP 5.3.5 and a test system consisting of 1620 electrons distributed over 1120 orbitals of different symmetry.

The first step of the parallelization procedure is the distribution of the workload related to each orbital on a certain number of cores. It turns out that a number of cores corresponding approximately either to the square root of all available cores or to the number of cores per node works best for the HRLS Hazel Hen. In addition, this method significantly improves the stability due to reduced memory

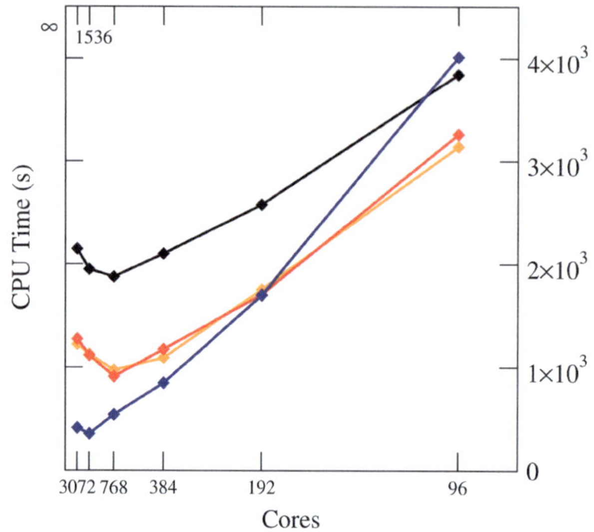

Fig. 1 CPU time on the HRLS CRAY XE6 for the self consistent calculation of the electronic structure of a $LiNbO_3$ slab within different parallelization schemes. See text for details

requirements. Numerical results for the test configuration are shown in Fig. 1 (red line). This setup results in a roughly linear scaling up to 768 cores.

Within the described approach for parallel computing, it is possible to steer the data distribution mode. In particular, the plane wise data distribution in real space can be activated. This allows for a much reduced communication during the Fourier transforms (FFTs). Unfortunately, the resulting load balancing is worsened. Therefore, the suitability of the plane wise data distribution must be tested for the particular computational architecture and in dependence of the number of processors. The results of our tests are shown in Fig. 1 (orange line). As expected, advantages of the plane wise data distribution occur for a relatively small numbers of processors, but are outweighted by load-balancing problems for calculations that employ 768 cores.

The computational techniques for the modeling of atomic system discussed here rely on mathematical libraries, in particular the Linear Algebra Package LAPACK or its distributed-memory implementation ScaLAPACK. The calculations discussed above have been performed with ScaLAPACK. This speeds up the calculations by up to a factor of two compared to LAPACK calculations, shown by the black line in Fig. 1. Even if only 96 cores are used, there is a noticeable speed up achieved by using the scalable linear algebra package.

Starting with VASP Version 5.3.2 it is possible to use additionally a parallelization over the k points used to sample the Brillouin zone in the reciprocal space calculations. Thereby it is possible to specify the number of k points that are to be treated in parallel. Within the group of cores that share the work for an individual k point also the electronic states and/or plane wave coefficients are treated in parallel.

The results of the corresponding tests are shown in Fig. 1 (blue line). It can be seen that the *k* point parallelization leads to an additional saving of computer time, provided more than 192 cores are used. The speed up with respect to calculations without *k* point parallelization amounts up to a factor two. Moreover, this approach allows for the extension of the roughly linear scaling to 1536 cores.

3 Results

The Si(111) surface reconstructs 2×1 if cleaved at room temperature. Annealing at 400 °C leads to a superstructure with a 7×7 periodicity [39]. However, since rare earth adsorbates typically prevent the formation of these surface reconstructions, we focus on the unreconstructed Si(111) surface. It is characterized by a hexagonal surface unit cell. Cutting bulk Si perpendicularly to the [111] crystallographic directions results in broken sp^3 bonds at the topmost Si layer, creating one dangling bond per surface unit cell. Indeed, the corresponding surface band structure shows a single surface state localized within the electronic band gap. The band is half filled and crosses the Fermi level.

During the silicide growth process, rare earth ions are deposited at the Si(111) surface. In order to identify the energetically favorable adsorption sites, we have calculated the potential energy surface for the adsorption of isolated atoms at the Si(111) 1×1 surface. It is calculated constraining the lateral coordinates of the rare earth atom and allowing its height as well as the remaining degrees of freedom of the uppermost three Si substrate bilayers to relax. We determined the energy for 56 lateral positions on a rectangular grid (average spacing 0.5 Å). The energy between the grid points is then evaluated by bicubic interpolation of the calculated data. The outcome of our calculation in the case of Dy is shown in Fig. 2. The rare earth ion prefers adsorption at the hcp site (T_4, global minimum) or at the fcc site (H_3, local minimum). The energy difference between the two sites amounts to 225 meV, and the low energy barrier between the two minima (about 300 meV) indicates a relatively high mobility of the adsorbates.

5×2 silicide phases of Gd, [11, 40] Tb, [12] Dy, [10] Ho, [41] and Er [42] have been recently observed by STM. This structure consists of a silicide submonolayer termination and is – exactly as the $2\sqrt{3} \times 2\sqrt{3}$ phase – metastable [42]. The two structures are in competition at low rare earth coverage. However, both are transformed into more stable silicides upon annealing. Available STM images reveal chain-like structures in the $[1\bar{1}0]$ direction (three equivalent chain orientations are thus possible).

However, without precise knowledge of the rare earth content, it is hard to extract structural information from the STM images. Consequently, very few attempts to assign a structural model to the rare earth film are available in the literature [11, 12].

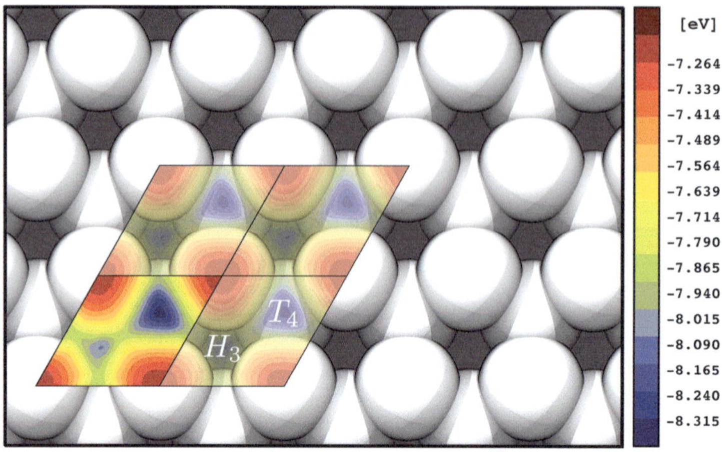

Fig. 2 Calculated potential energy surface for a single Dy atom at Si(111)

In order to develop a model for the 5×2 superstructure, several factors can be considered. First of all, the stable silicides depend on the lanthanide coverage. In detail, $2\sqrt{3} \times 2\sqrt{3}$, 5×2 and finally the 1×1 surface periodicity is observed with increasing rare earth availability. Considering the size of the respective surface unit cells, this limits the number of rare earth atoms to a maximum of 10 atoms per surface unit. Furthermore, available STM images reveal the presence of ordered chains along the $[1\bar{1}0]$ direction. Thus the rare earth atoms are placed in our models in the stable positions determined in above, so that oriented chain-like structures are formed. As the rare earth atoms in silicide films with 1×1 periodicity are completely covered by a silicon double layer that is absent in the films with $2\sqrt{3} \times 2\sqrt{3}$ periodicity, it is plausible that the lanthanide ions in the 5×2 phase are at least partially covered by Si atoms. Therefore, rare earth layers covered to a different extent are simulated. In order to capture possible Si dimerization effects as known from other silicon surfaces – e.g the Si(001) – doubled 5×1 unit cells with artificial dimerization in the $[1\bar{1}0]$ directions are employed. Models consisting of alternating Si Seiwatz and honeycomb chains with rare earth atoms in between, as originally proposed by Battaglia et al. [11] and then by Franz et al. [12], have been tested as well.

Following the criteria above, we have developed 14 structural models, which are shown in Figs. 3 and 5. Besides the labels *a* to *n*, corresponding to Fig. 3a–n, the number of rare earth atoms per 5×2 unit cell is indicated in the picture. While certainly further models are conceivable, the models described above are expected to allow for the derivation of general trends and conclusions. As the slabs modeling the surface structures contain a different number of Si and rare earth atoms, their DFT total energy cannot be directly compared. In order to determine the thermodynamically stable structures, we calculate the Landau potential as described in Sect. 2. As neither the contribution to the free energy of the hydrogen atoms

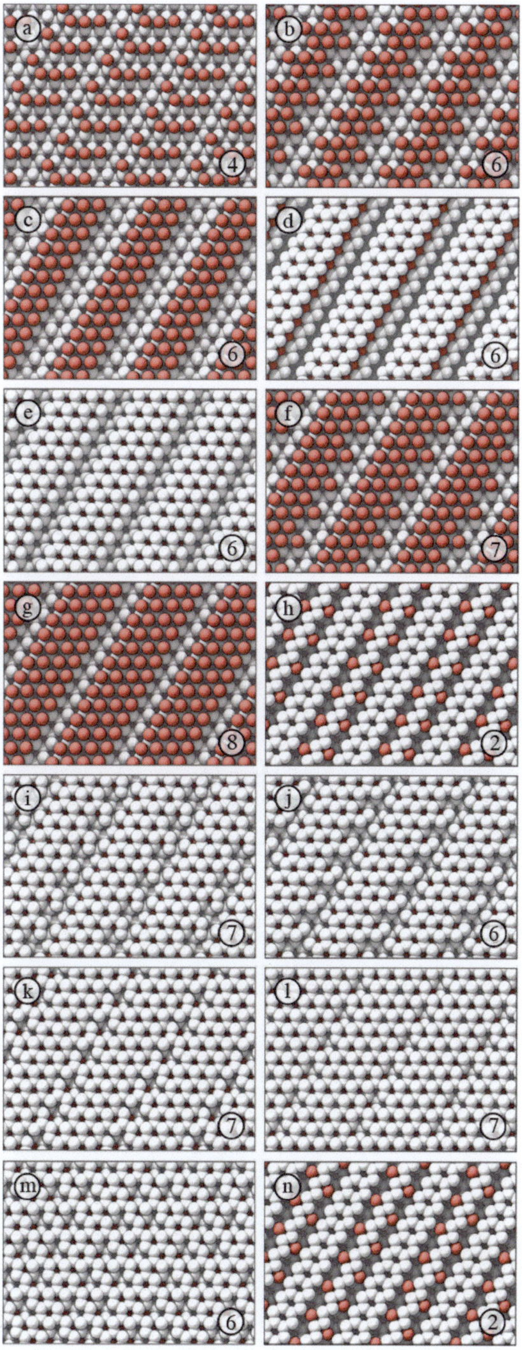

Fig. 3 Structural models for RE induced Si(111)(5 × 2) surface reconstructions. Numbers in the *lower right corner* indicate the number of RE atoms per unit cell

nor the size of the surface unit cell are considered for this particular calculation, the calculated values do not correspond to the absolute formation energies of the structures and the Landau potential is labeled by Ω'. However, as both the surface unit cell as well as the number of hydrogen atoms used to passivate the dangling bonds at the bottom side of the slabs are the same for all configurations, a relative comparison of the different structures is possible. Even if the 5×2 phase has been observed for different rare earth silicides, we limit our investigation to Dy silicide due to the high demand of computational resources. However, based on our experience with the other silicide structures discussed above, the results may again be extrapolated to all trivalent rare earths.

The phase diagram in Fig. 4 shows that structures with rare earth atoms in the channels between honeycomb and Seiwatz chains (*h*, *n*) are favored. Indeed, for most values of the chemical potentials, which are relevant for submonolayer coverage, the structures labeled by *h* and *n* are the most stable configurations, while for strongly Dy rich conditions the models *m* and *i* can be formed. These structures are less relevant, however, since at these values of the rare earth chemical potential monolayer or multilayer silicides are formed.

Thus, the energetically almost degenerate models *h* and *n* (energy difference 18 meV per 5×2 unit cell) with two rare earth atoms per unit cell describe the observed phase. Considering that four and eight Si atoms per 5×2 unit cell form the Seiwatz and honeycomb chains, respectively, the stoichiometry of the silicide layer at the Si(111) surface can be expressed as $RESi_6$. The difference between the two

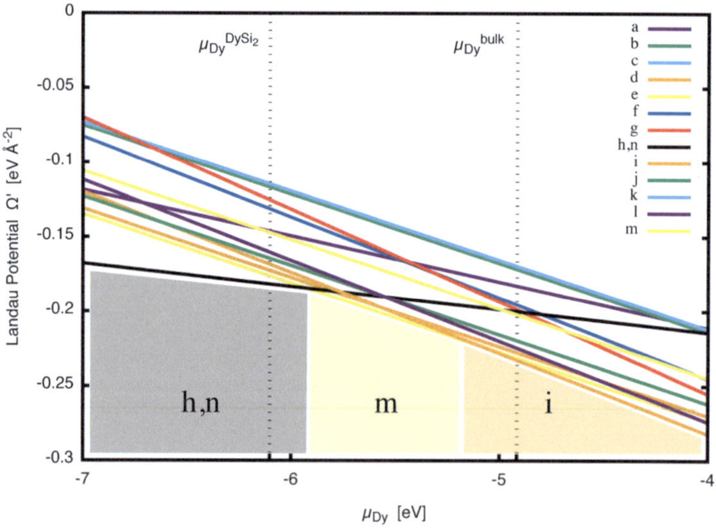

Fig. 4 Calculated phase diagram for the dysprosium adsorbed Si(111) surface with 5×2 periodicity as a function of the dysprosium chemical potential μ^{Dy}. Two representative values of μ^{Dy}, corresponding to Dy in its metallic hcp bulk phase and to Dy in hexagonal $DySi_2$ state are indicated. Si-rich conditions are assumed

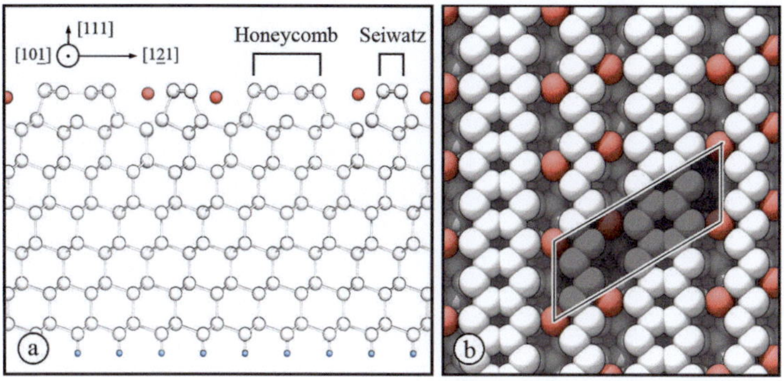

Fig. 5 Side (**a**) and top view (**b**) of the thermodynamically stable rare earth induced surface reconstruction with 5 × 2 periodicity on the Si(111) surface. The termination corresponds to structure *h* in Figs. 3 and 4 and consists of alternating Si Seiwatz and honeycomb chains. The surface unit cell is highlighted

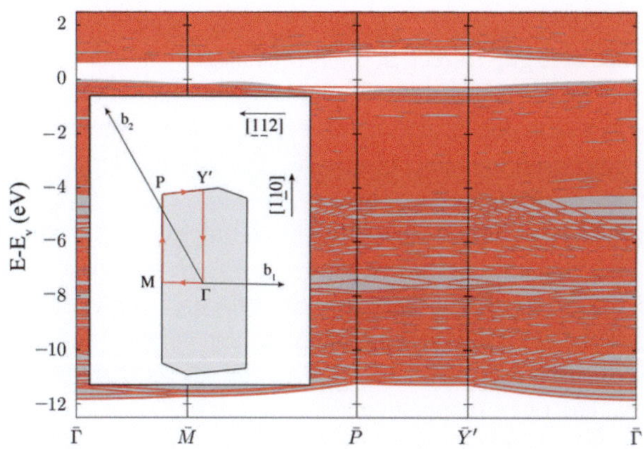

Fig. 6 Calculated surface band structure for model *h* in Figs. 3 and 4. Projected bulk bands are shown in *grey*. The *inset* shows the surface Brillouin zone of the 5 × 2 structure

models *h* and *n* consists in the different alignments of neighboring rare earth atom rows on both sides of the honeycomb chains. The slightly more stable model *h*, in which the rare earth atoms are aligned in-phase, is shown in more detail in Fig. 5. It is known that on Si(111) a honeycomb chain is stabilized by one electron per 3 × 1 unit cell, while a zigzag Seiwatz chain requires two electrons per 2 × 1 unit cell [11]. Thus, the 5 × 2 phase is built of two 3 × 1 surface units with honeycomb chains and two 2 × 1 surface units containing Seiwatz chains. The structure is stabilized by two trivalent rare earth atoms, which provide six electrons per unit cell.

Figure 6 shows the calculated electronic band structure of model *h*. We mention that the Si bulk band gap calculated here is about 0.67 eV i.e. slightly smaller than measured, due to the underestimation of the band gaps in DFT calculations [37]. Almost no surface localized electronic states are present in the bulk gap region. The

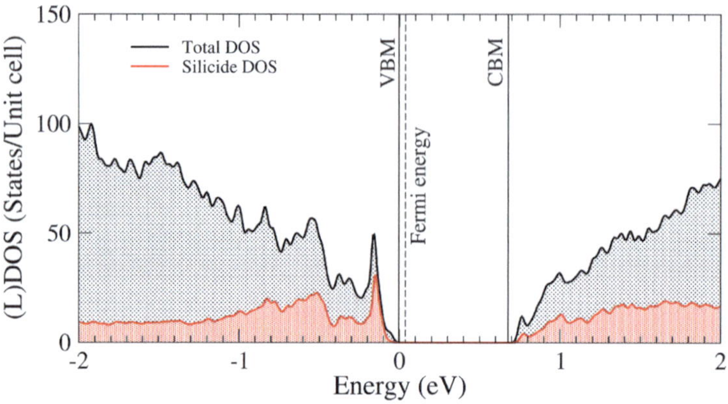

Fig. 7 Calculated density of states for model *h* in Figs. 3 and 4. The total density of states is shown *black*, while the silicide contribution is shown *red*

fundamental electronic gap (direct, at $\overline{\Gamma}$) is only slightly smaller than the calculated Si bulk gap. This confirms that the submonolayer silicide with 5×2 periodicity on the Si(111) surface is semiconducting. The inset of Fig. 6 shows the 5×2 surface Brillouin zone employed for the calculations. The atomic chains are parallel to the long sides of the surface Brillouin zone.

The (local) density of states of the slab modeling the $DySi_6$ silicide with 5×2 periodicity on the Si(111) surface is shown in Fig. 7. The total density of states is represented by the black curve, while the red curve represents the local DOS of the silicide layer. The dotted lines indicate the valence and conduction band edges of bulk Si. The calculated (L)DOS again shows that the silicide layer is semiconducting. The overall appearance of the total DOS is very similar to the Si bulk DOS, with the exception of a minor reduction of the fundamental bandgap. This effect is due to the electronic states close to the conduction band minimum. Otherwise the presence of the $DySi_6$ layer does not strongly affect the band gap region of the substrate.

The knowledge of the thermodynamically stable structural model now also allows for the interpretation of the STM images and the identification of the observed features. In the filled state images (Fig. 8a, c) the bright spots are assigned to the honeycomb (broad rows) and Seiwatz (thin rows) chains, which capture the electrons from the rare earth atoms. The latter are thus not visible at this bias. In contrast, in the empty state images (Fig. 8b, d), the rare earth atoms donating their electrons are visible, while the dark rows show the location of the honeycomb chains. As between the chains different equivalent lattice sites are available for the rare-earth atoms, different STM patterns are possible. These correspond to an in-phase alignment between neighboring rare earth rows (model *h*) or a zigzag alignment (model *n*).

In contrast to monovalent and divalent ions, for which also other $n \times 2$ phases with odd $n \neq 5$ have been observed, 5×2 is the only possible $n \times 2$ periodicity for

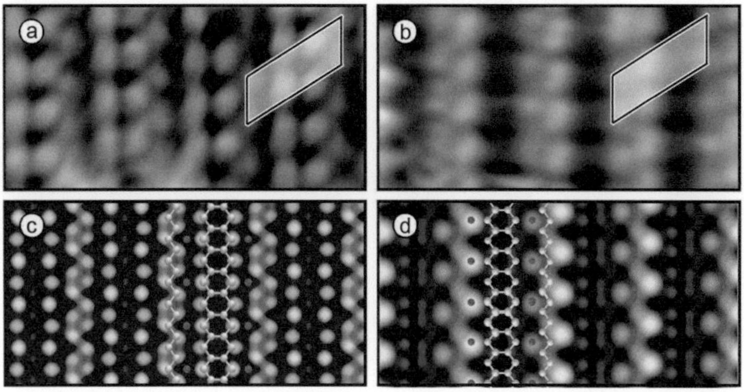

Fig. 8 (**a**), (**b**) Measured STM images of the 5 × 2 Tb silicide submonolayer structure on the Si(111) surface [12] in comparison with simulated data in (**c**), (**d**). Experimental STM images refer to voltages of −1.5 V [(**a**), (**c**) filled states] and 1.5 V [(**b**), (**d**) empty states], and tunneling currents of 100 pA. The 5 × 2 surface unit cell is indicated

trivalent rare earths.[1] Sticking to the models consisting of alternating honeycomb and Seiwatz chains, phases of 7 × 2 or 9 × 2 periodicity could in principle be built by one honeycomb chain and two or three Seiwatz chains, respectively. These structures, however, would have to be stabilized by 10 and 14 electrons per unit cell, respectively. This condition cannot be satisfied by an integer number of trivalent donors, which explains why no other $n \times 2$ phase than the 5 × 2 has been observed for trivalent rare earths.

4 Conclusions

Two-dimensional silicide structures of 5×2 periodicity formed at the Si(111) surface upon rare earth deposition in the submonolayer regime have been investigated theoretically. The DFT calculations allow for the identification of a structural model that is compatible with the experimental data.

According to this model, the 5 × 2 structure is characterized by alternating Si honeycomb and Seiwatz chains oriented along the $[10\overline{1}]$ crystallographic direction. The rare earth atoms are located in the channels between the chains. The formation

[1]It is also important to notice that the adsorption of divalent metals at the Si(111) typically leads to a $n \times 2$ surface reconstruction, with n an odd integer. Thus, as suggested by Battaglia et al. [11], the 5 × 2 phase might be induced by divalent lanthanides such as Yb, Eu, Sm or Tm. In this case, they would give rise to completely different structures, similar to the reconstructions formed by deposition of divalent alkaline-metal earths (Mg, Ca, Sr, Ba). These are not investigated in this work, as we only consider lanthanides in the trivalent state (Dy^{3+},Tb^{3+}).

of the silicide structures with 5×2 periodicity does not strongly affect the electronic properties of the substrate, but slightly reduces the band gap.

Acknowledgements The Deutsche Forschungsgemeinschaft (DFG) is acknowledged for financial support (FOR1700, SCHM 1361/21). The calculations were performed at the High Performance Computing Center Stuttgart (HLRS) and the Paderborn Center for Parallel Computing (PC^2).

References

1. Paki, P., Kafader, U., Wetzel, P., Pirri, C., Peruchetti, J.C., Bolmont, D., Gewinner, G.: Phys. Rev. B **45**, 8490 (1992)
2. d'Avitaya, F.A., Perio, A., Oberlin, J.C., Campidelli, Y., Chroboczek, J.A.: Appl. Phys. Lett. **54**(22), 2198 (1989)
3. Knapp, J.A., Picraux, S.T.: Appl. Phys. Lett. **48**(7), 466 (1986)
4. Wetzel, P., Pirri, C., Paki, P., Peruchetti, J., Bolmont, D., Gewinner, G.: Solid State Commun. **82**(4), 235 (1992)
5. Tu, K.N., Thompson, R.D., Tsaur, B.Y.: Appl. Phys. Lett. **38**(8), 626 (1981)
6. Vandré, S., Preinesberger, C., Busse, W., Dähne, M.: Appl. Phys. Lett. **78**(14), 2012 (2001)
7. Vandré, S., Kalka, T., Preinesberger, C., Dähne-Prietsch, M.: Phys. Rev. Lett. **82**, 1927 (1999)
8. Lohmeier, M., Huisman, W.J., ter Horst, G., Zagwijn, P.M., Vlieg, E., Nicklin, C.L., Turner, T.S.: Phys. Rev. B **54**, 2004 (1996)
9. Roge, T., Palmino, F., Savall, C., Labrune, J., Pirri, C.: Surf. Sci. **383**(2–3), 350 (1997)
10. Engelhardt, I., Preinesberger, C., Becker, S., Eisele, H., Dähne, M.: Surf. Sci. **600**(3), 755 (2006)
11. Battaglia, C., Cercellier, H., Monney, C., Garnier, M.G., Aebi, P.: EPL (Europhys. Lett.) **77**(3), 36003 (2007)
12. Franz, M., Große, J., Kohlhaas, R., Dähne, M.: Surf. Sci. **637–638**, 149 (2015)
13. Wanke, M., Franz, M., Vetterlein, M., Pruskil, G., Höpfner, B., Prohl, C., Engelhardt, I., Stojanov, P., Huwald, E., Riley, J., Dähne, M.: Surf. Sci. **603**(17), 2808 (2009)
14. Wanke, M., Franz, M., Vetterlein, M., Pruskil, G., Prohl, C., Höpfner, B., Stojanov, P., Huwald, E., Riley, J.D., Dähne, M.: J. Appl. Phys. **108**(6), 064304 (2010)
15. Stauffer, L., Mharchi, A., Pirri, C., Wetzel, P., Bolmont, D., Gewinner, G., Minot, C.: Phys. Rev. B **47**, 10555 (1993)
16. Kitayama, H., Tear, S., Spence, D., Urano, T.: Surf. Sci. **482–485**(Part 2), 1481 (2001)
17. Bonet, C., Spence, D., Tear, S.: Surf. Sci. **504**, 183 (2002)
18. Rogero, C., Koitzsch, C., González, M.E., Aebi, P., Cerdá, J., Martín-Gago, J.A.: Phys. Rev. B **69**, 045312 (2004)
19. Rogero, C., Martín-Gago, J.A., Cerdá, J.I.: Phys. Rev. B **74**, 121404 (2006)
20. Koitzsch, C., Bovet, M., Garnier, M., Aebi, P., Rogero, C., Martín-Gago, J.: Surf. Sci. **566–568**(Part 2), 1047 (2004) (Proceedings of the 22nd European Conference on Surface Science)
21. Magaud, L., Reinisch, G., Pasturel, A., Mallet, P., E. Dupont-Ferrier, Veuillen, J.Y.: EPL (Europhys. Lett.) **69**(5), 784 (2005)
22. Wetzel, P., Saintenoy, S., Pirri, C., Bolmont, D., Gewinner, G.: Phys. Rev. B **50**, 10886 (1994)
23. Cocoletzi, G.H., de la Cruz, M.R., Takeuchi, N.: Surf. Sci. **602**(2), 644 (2008)
24. Perdew, P., Chevary, J.A., Vosko, S.H., Jackson, K.A., Pederson, M.R., Singh, D.J., Fiolhais, C.: Phys. Rev. B **46**, 6671 (1992)
25. Perdew, J.P., Burke, K., Ernzerhof, M.: Phys. Rev. Lett. **77**, 3865 (1996)
26. Kresse, G., Furthmüller, J.: Comput. Mater. Sci. **6**, 15 (1996)
27. Kresse, G., Furthmüller, J.: Phys. Rev. B **54**, 11169 (1996)
28. Bloechl, P.E.: Phys. Rev. B **50**(24), 17953 (1994)

29. Kresse, G., Joubert, D.: Phys. Rev. B **59**, 1758 (1999)
30. Anisimov, V.I., Aryasetiawan, F., Lichtenstein, A.I.: J. Phys. Condensed Matter **9**(4), 767 (1999)
31. Sanna, S., Schmidt, W.G., Frauenheim, T., Gerstmann, U.: Phys. Rev. B **80**, 104120 (2009)
32. Sanna, S., Frauenheim, T., Gerstmann, U.: Phys. Rev. B **78**, 085201 (2008)
33. Neugebauer, J., Scheffler, M.: Phys. Rev. B **46**(24), 16067 (1992)
34. Bengtsson, L.: Phys. Rev. B **59**(19), 12301 (1999)
35. Tersoff, J., Hamann, D.R.: Phys. Rev. Lett. **50**, 1998 (1983)
36. Tersoff, J., Hamann, D.R.: Phys. Rev. B **31**, 805 (1985)
37. Bechstedt, F.: Principles of Surface Physics. Advanced Texts in Physics. Springer, Berlin/Heidelberg (2003)
38. Sanna, S., Schmidt, W.G.: Phys. Rev. B **81**(21), 214116 (2010)
39. Lüth, c: Surfaces and Interfaces of Solid Materials. Springer Study Edition. Springer, Berlin/Heidelberg (1995)
40. Kirakosian, A., McChesney, J., Bennewitz, R., Crain, J., Lin, J.L., Himpsel, F.: Surf. Sci. **498**(3), L109 (2002)
41. Perkins, E., Scott, I., Tear, S.: Surf. Sci. **578**(1–3), 80 (2005)
42. Wetzel, P., Pirri, C., Gewinner, G., Pelletier, S., Roge, P., Palmino, F., Labrune, J.C.: Phys. Rev. B **56**, 9819 (1997)

Computational Analysis of Li Diffusion in NZP-Type Materials by Atomistic Simulation and Compositional Screening

Daniel Mutter, Britta Lang, Benedikt Ziebarth, Daniel Urban, and Christian Elsässer

Abstract Solid state electrolytes (SSEs) can become a key component for the development of novel reliable, safe, and highly efficient Li-ion batteries. This work focuses on the vacancy-mediated diffusion of Li ions through solid compounds with NZP crystal structures [e.g. $LiTi_2(PO_4)_3$ (LTP); NZP stands for $NaZr_2(PO_4)_3$], which is a promising class of materials for the application as SSEs. Since this crystal structure is known to be stable for many combinations of elements on the cation positions, the activation energies for vacancy jumps were calculated in this work for a variety of NZP-type compounds with different compositions. First-principles calculations based on density functional theory were performed to determine the migration barrier heights, and to correlate their values to structural characteristics. In addition, the bond valence method was applied to the NZP-type compounds, which not only helps to identify diffusion networks and transition points, but which can also be valuable for predicting qualitative trends by systematic compositional screening.

1 Introduction

In recent years, there has been a rapidly growing industrial demand for energy storage materials combining large specific energy and power densities with high safety, harmlessness for health and abundant availability of the processed elements. An increase of the safety of current Li-ion batteries could be achieved by replacing the likely flammable and toxic liquid electrolytes by ion conducting solid compounds. Materials crystallizing in the NASICON [2, 3, 6, 8] or NZP crystal

D. Mutter (✉) • C. Elsässer
Freiburger Materialforschungszentrum (FMF), Albert-Ludwigs-Universität Freiburg, Stefan-Meier-Straße 21, 79104 Freiburg, Germany

Fraunhofer Institute for Mechanics of Materials IWM, Wöhlerstraße 11, 79108 Freiburg, Germany
e-mail: daniel.mutter@iwm-extern.fraunhofer.de

B. Lang • B. Ziebarth • D. Urban
Fraunhofer Institute for Mechanics of Materials IWM, Wöhlerstraße 11, 79108 Freiburg, Germany

W.E. Nagel et al. (eds.), *High Performance Computing in Science and Engineering '16*, DOI 10.1007/978-3-319-47066-5_13

structure, named after the compound $NaZr_2(PO_4)_3$, with Na being exchanged by Li, can be regarded as promising candidates for solid-state electrolytes (SSEs), mainly due to their three-dimensional diffusion network for Li ions [10] and their capability of accommodating many different combinations of elements on the Zr and P sublattices by maintaining the NZP crystal structure [18, 19]. The variety of possible elemental substitutions allows for a systematic screening of many different compositions with the goal of finding novel materials with desired properties such as high ionic conductivity and low thermal expansion. In this work, the diffusion of Li through various NZP-type compounds is analyzed and screened by combining quantum-mechanical ab-initio simulations with static energy calculations based on bond valence potentials. Structure–property relationships were identified [12] leading to the possibility of qualitatively predicting migration barriers directly from crystal structure characteristics. After describing the NZP crystal structure in detail, the employed computational methods are explained concisely. Results for vacancy-mediated Li-ion migration in $LiX_2(LO_4)_3$, where X and L denote ions substituting Ti and P, respectively in $LiTi_2(PO_4)_3$ (LTP or LISICON [11]) are presented and discussed. Finally, a summary of the usage of computational resources on the ForHLR I supercomputer is given.

2 Structure of NZP-Type Materials

The general structure of NZP or LTP compounds can be described by the formula $(M1)^{[6]}\,(M2)_3^{[8]}\,X_2^{[6]}(L^{[4]}O_4)_3$ [15, 16, 18, 19]. M1 and M2 denote interstitial positions which are fully or partly occupied by Li, X and L are the positions of Zr/Ti and P, respectively. The oxygen coordination numbers of the cations are given by superscripts in square brackets. NZP compounds crystallize in a rhombohedral structure with the space group $R\bar{3}c$. Two XO_6 octahedra and three LO_4 tetrahedra being connected by oxygen atoms form the basic $X_2(LO_4)_3$ units of the structural framework, which are called 'lanterns' due to their characteristic shape (see Fig. 1). In between the connected lanterns, three-dimensional migration paths exist for the Li ions from M1 to M2 positions and vice versa. From the site multiplicities of M1 (Wyckoff position 6b) and M2 (18e) it is obvious that there are three times as much M2 than M1 positions. Each M1 is surrounded by six M2 sites, and each M2 connects two M1 sites, leading to the three-dimensional network. The occupation of M1 and M2 positions with Li ions depends on the oxidation states of X and L ions such that charge neutrality is ensured, e.g. $LiTi_2(PO_4)_3$ with $Ti^{(+IV)}$ and $P^{(+V)}$, $Li_4Zr_2(SiO_4)_3$ with $Zr^{(+IV)}$ and $Si^{(+IV)}$, or $Li_3Al_2(VO_4)_3$ with $Al^{(+III)}$ and $V^{(+V)}$. In the case of mixed occupation of X sites with atoms of valencies +IV and +V, some of the M1 sites have to be empty, leading to the possibility of vacancy-mediated diffusion without having to incorporate additional vacancies.

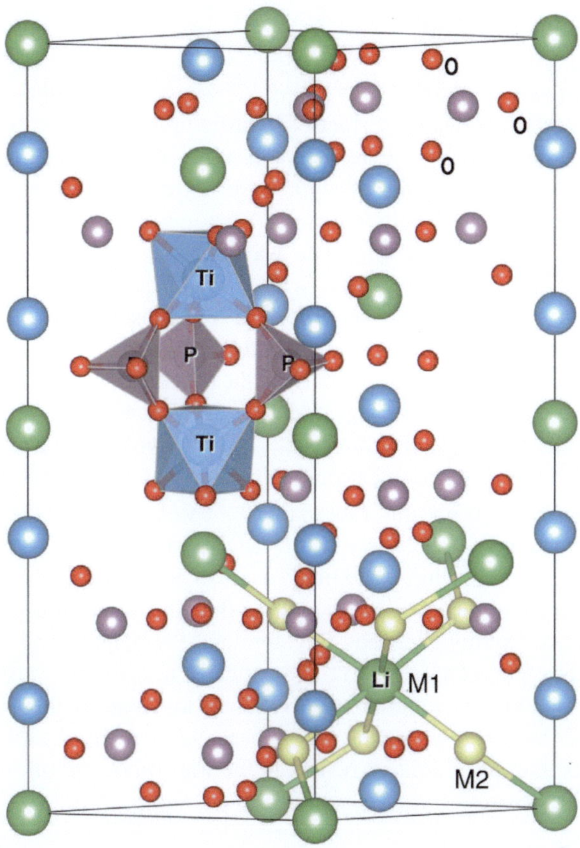

Fig. 1 Hexagonal supercell of the rhombohedral NZP structure, shown for $LiTi_2(PO_4)_3$ (LTP). One 'lantern' configuration is visualized by coordination tetrahedra of oxygen (*red spheres*) around P atoms (*violet spheres*) and by octahedra around Ti atoms (*blue spheres*). In LTP, Li ions occupy the M1 positions (*green spheres*), and the M2 positions are empty (*yellow spheres*). Only M2 sites surrounding one Li ion are shown for clarity

3 Computational Methods

Transition paths and activation energies (migration barrier heights) in the material class described above were calculated by means of ab-initio methods based on the density functional theory (DFT) [7] and the nudged elastic band (NEB) method [9], and by static energy calculations using bond valence (BV) potentials [1].

3.1 Ab-Initio Calculations

The DFT code *Quantum ESPRESSO PWscf* [7] was applied to obtain ground-state configurations and energies of perfect NZP structures, structures with vacancies, and structures with migrating Li atoms at transition points. Since compounds with more than one Li atom per formula unit were not considered (no occupation of M2 sites), the perfect supercells contain 108 atoms (6 Li, 12 X, 18 L, and 72 O), arranged as depicted in Fig. 1. The wavefunctions of the valence electrons were expanded in a plane-wave basis with a cutoff energy of 476 eV, and their interaction with the ionic cores was described by ultrasoft pseudopotentials [20]. The exchange-correlation contribution to the total energy was taken into account by the general gradient approximation (GGA) in the formulation of Perdew, Burke, and Ernzerhof [17]. Brillouin-zone integration was performed on a $3 \times 3 \times 1$ grid, set up by the scheme of Monkhorst and Pack [14]. Cell-volume relaxation was done by total-energy minimization, and atomic relaxation was stopped when the minimal force acting on an atom became less than 10^{-3} eV/Å. For the NEB calculations, 5 intermediate images were chosen along an initially straight path between the previously relaxed and the initial and final states. In general, at each ionic relaxation step, the component of the force on an atom in the direction of this path is replaced by forces of elastic springs between the atom and the neighboring images of this atom [9], thereby impeding atoms at energetically unfavorable positions along a path between two energy minima from relaxing into one of these minima. The components of the forces perpendicular to the path are not changed, and so the correlated relaxation of all images results in the transition path of minimal energy across the saddle point.

3.2 The Bond Valence Method

In a simplified picture of localized chemical bonds in a system consisting of anions and cations, the *atomic valence* $V(A)$, which denotes the total number of electrons an atom A contributes to bonding, is exactly subdivided into portions v_i belonging to the individual bonds i with the N surrounding counterions [4]. It is further intuitive to assume that these *bond valences* describe the strength of the bond, which generally decreases with the atomic distance. This dependency can be expressed by an exponential function:

$$V(A) = \sum_{i=1}^{N} v_i = \sum_{i=1}^{N} \exp[(R_0 - R_i)/b] \,, \tag{1}$$

with distances R_i between the bonded atoms. The parameters R_0 and b can be adjusted to achieve minimal mismatch of $V(A)$ with the oxidation state of atom A in known stable configurations. Higher sum mismatches should therefore correspond to energetically less favorable and less stable structures [5]. The total energy is linked to the bond valences by:

$$E_{\text{bv}} = D_0 \left[\sum_{i=1}^{N} \left(\frac{v_i - v_{\min}}{v_{\min}} \right)^2 - N \right] , \tag{2}$$

which is formally derived by assuming a Morse potential with the dissociation energy D_0 for the interaction between oppositely charged ions [1]. $v_{\min} = \exp[(R_0 - R_{\min})/b]$ describes the bond valence at the equilibrium distance $R_{\min}$. Optimized parameters R_0, b, D_0, $R_{\min}$, and the cutoff distance R_{cut} are available for a large variety of oxides [1]. In addition to the attractive energy expressed by (2), a repulsive Coulomb interaction between equally charged ions A_1 and A_2 at a distance $R_{A_1-A_2}$ is taken into account by:

$$E_{\text{Coulomb}} = \frac{1}{4\pi\varepsilon_0} \frac{q_{A_1} q_{A_2}}{R_{A_1-A_2}} \text{erfc} \left(\frac{R_{A_1-A_2}}{\rho_{A_1-A_2}} \right) , \tag{3}$$

with effective, screened charges q and a screening factor ρ [1].

4 Results and Discussion

4.1 Energy Landscape Calculations

In order to identify the paths for vacancy-mediated Li-ion diffusion, the energy of a Li ion at each position in the cell was calculated with the BV approach in the LTP structure. The lowest energy values are found, as expected, for Li at M1 positions. At a certain value of higher energy, which can be regarded as the activation energy within this model, the isosurface becomes interconnected and forms a continuous network throughout the whole system, as depicted in Fig. 2. A vacancy at a M1 site, visualized by the empty circle in the upper Li layer, can now move in the crystal by successive jumps along e.g. the dashed line, thereby effectively enabling Li-ion movement in the opposite direction. Since the energies result from a static calculation with an effective interaction potential, they cannot be considered as quantitatively accurate, but can be valuable for predicting qualitative trends of activation energies when e.g. incorporating defects or screening a large variety of elemental substitutions.

4.2 Activation Energies for NZP Materials

Activation energies for the migration of a Li vacancy were calculated with DFT for several substitutions at the Ti and P positions of LTP. To this end, defect-free systems

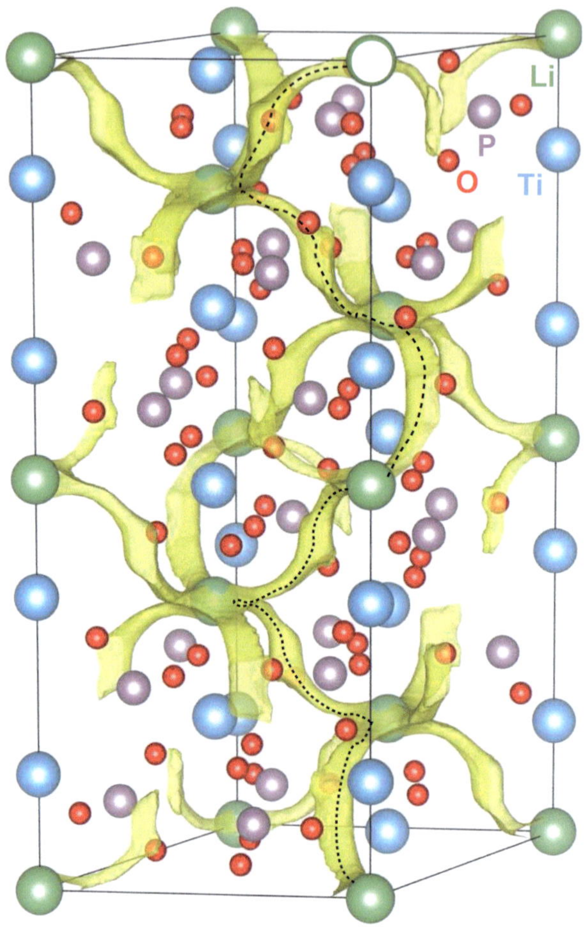

Fig. 2 Hexagonal supercell of rhombohedral LTP with an energy isosurface of constant $E(\text{Li})$ as calculated with the BV method. The energy value was chosen as the minimum value at which the isosurface formed an interconnected network (+0.8 eV relative to the energy of Li at the M1 positions). The Li vacancy in the supercell is visualized by the empty sphere in the uppermost Li layer

were set up followed by a relaxation of the cell volume and atomic positions. The volume was then kept constant for the simulations of the structures containing one vacancy at a M1 site (initial state), and for those structures with two vacancies at adjacent M1 sites and an interstitial Li at the intermediate M2 site (i.e., the transition state). The activation energy was obtained as the energy difference between these two configurations. In order to calculate the energy along the migration paths, NEB runs were performed for a few of the considered systems (see Fig. 3), leading to mirror-symmetric barriers.

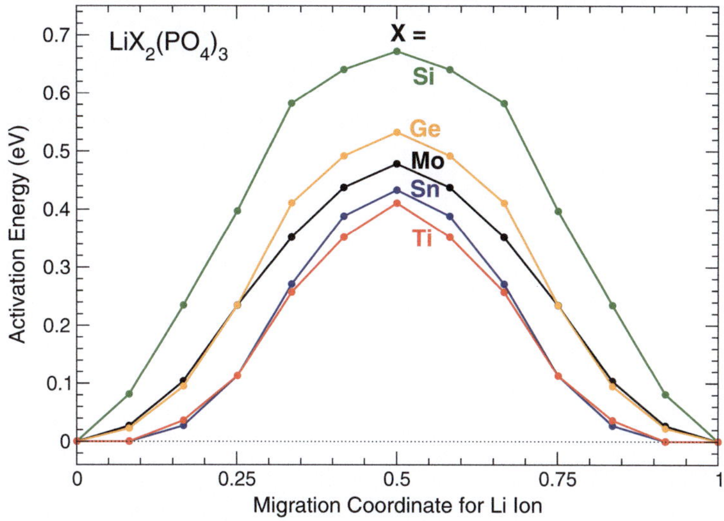

Fig. 3 Minimum energy paths for the migration of a Li ion from a M1 to an adjacent M1 position across the transition point (M2) in $LiX_2(PO_4)_3$ for 5 different tetravalent elements X [12]

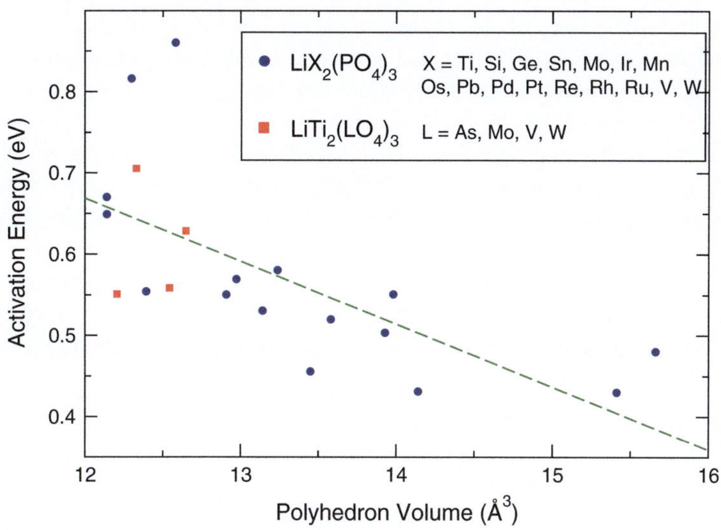

Fig. 4 Dependence of the activation energy for vacancy-mediated Li-ion migration on the volume of the LiO_6 octahedron for a variety of NZP compounds

In order to analyze the influence of different elements on the Ti and P sites of LTP, the activation energy was plotted against the volume of the oxygen octahedron around the Li ion (Fig. 4). There is a clear trend that larger polyhedron volumes correspond to lower migration energies. When a Li ion moves from a M1 to a M2

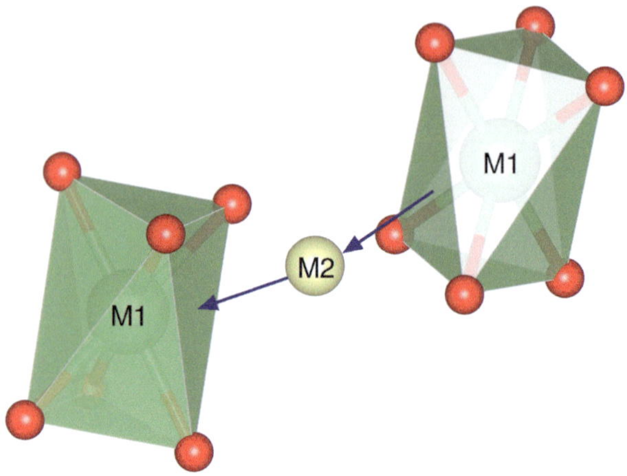

Fig. 5 Oxygen octahedra around adjacent M1 positions. A possible diffusion direction, which is indicated by *arrows*, crosses the bottleneck formed by three oxygen atoms at the face of the octahedron

position, it has to cross one of the triangular faces of the octahedron spanned by three O atoms, and so the energy barrier is higher the smaller this bottleneck is (see Fig. 5). For systems containing Ge, Ti, Sn, and Hf, a similar relationship between activation energy and bottleneck size in NZP structures was experimentally observed by Martinez et al. using X-ray diffraction and electrical impedance spectroscopy [13].

The trend is also maintained when, instead of a total replacement of Ti by other elements in LTP, only a partial substitution of just one Ti atom in the cell is considered, leading to $LiX_{0.2}Ti_{1.8}(PO_4)_3$. This was shown earlier in this project for a large set of tri-, tetra-, and pentavalent elements X [12]. In the case of $LiSi_{0.2}Ti_{1.8}(PO_4)_3$ it was found that the migration barriers along a complete path through the cell, as indicated for example by the dashed line in Fig. 2, vary by up to 0.2 eV depending on the distance of the moving ion from the substituted atom.

5 Computer Resources

For the DFT calculations, the MPI parallelized pwscf code of the software *Quantum ESPRESSO* was run on the ForHLR I *singlenode* queues using 20 processors. K-point parallelization was not taken into account (specified by the flag *-npool = 1* of the *mpirun* command), since this was found to lead to the shortest execution times for 20 cores. In Fig. 6, the decrease of computation time with the number of employed cores is shown for two representative systems of LTP with 107 atoms: system S_1 containing one vacancy on a M1 position leading to 5 irreducible k-points (resulting from symmetry operations on the original 3×3×1 grid of k-points),

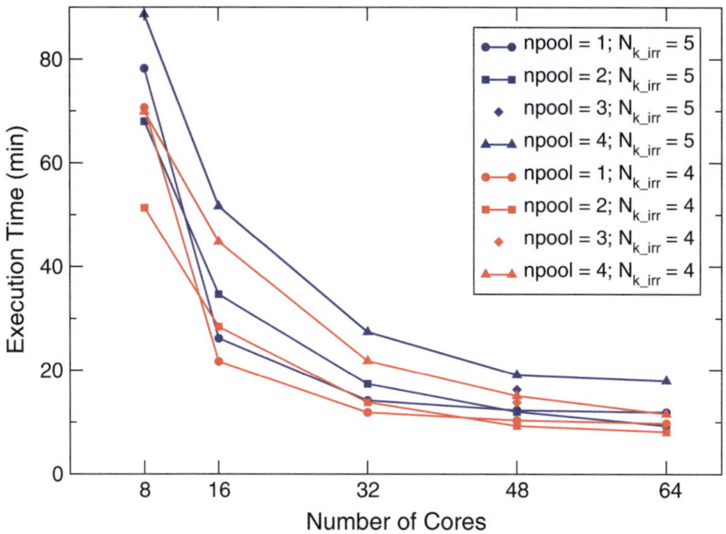

Fig. 6 Dependence of execution time of a single electronic self-consistency run on the number of cores. Different settings for the *mpirun* option *npool* were applied, which corresponds to k-point parallelization, and two systems with different numbers of irreducible k-points ($N_{\text{k\_irr}}$) were considered. The calculations using 8 and 16 cores were performed on the *singlenode* queue, and those using 32, 48, and 64 cores on the *multinode* queue with 2, 3, and 4 nodes, respectively, and 16 cores in each case

and system S_2 with two vacancies on adjacent M1 positions and an interstitial on a M2 position (saddle point), resulting in 4 irreducible k-points. One electronic self-consistency loop (QE setting *calculation*= "*scf*") was calculated, which resulted in convergence after 16 steps for both systems. The good scaling behavior of *Quantum ESPRESSO* is visible in this plot, based on which the use of 20 cores and *npool* = *1* in the simulations presented in this work was considered a reasonable setting.

Figure 7 shows the dependence of computation time on the total number of calculation steps, i.e., all the electronic self-consistency steps plus ionic relaxation steps, for three different systems of LTP [the perfect system with 108 atoms (S_0), and S_1 and S_2 as described above] with the same input parameters for the QE run. It is clearly visible that the saddle-point configurations need more steps to relax than the perfect crystals. On average over all points, one step takes about 110 s on one node and 20 processing units on the machines of the ForHLR I *singlenode* queue.

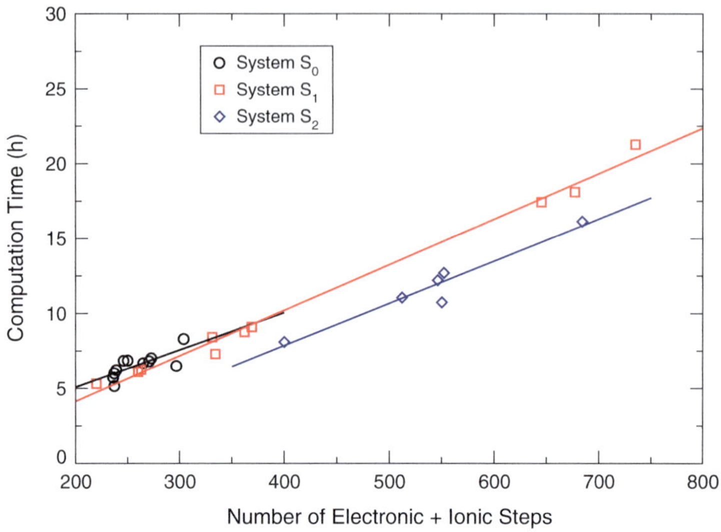

Fig. 7 Dependence of the computation time on the sum of electronic self-consistency and ionic relaxation steps for the three characteristic systems considered in this work: a defect-free cell with 108 atoms leading 4 irreducible k-points (S_0), a cell with one Li vacancy at M1 (107 atoms) leading to 5 irreducible k-points (S_1), and a cell with two vacancies at M1 sites and one interstitial Li atom at the transition point M2 (107 atoms) leading to 4 irreducible k-points (S_2). The different points belong to different substitutions of Ti in $LiTi_2(PO_4)_3$. Lines are linear fits

6 Summary and Outlook

In this work static energy calculations based on bond valence potentials and density functional theory calculations were applied to study migration paths and activation energies for Li ions in ion-conducting NZP-type SSE materials. Based on the compound $LiTi_2(PO_4)_3$, Ti and P atoms were substituted by a variety of isovalent elements to analyze their influence on Li-ion diffusion. The larger the coordination polyhedron around a Li-ion is, the easier the Li ion can escape the cage through the bottleneck. In the next step of this project, a systematic compositional screening approach will be applied combining qualitative results of bond valence calculations with molecular dynamics and ab-initio simulations in order to discover hitherto unknown combinations of elements in NZP compounds leading to stable crystal structures with low migration barriers and therefore high conductivities for Li ions.

Acknowledgements This work was funded by the German Research Foundation (DFG Grant no. El 155/26-1). The DFT calculations were performed on the computational resource ForHLR Phase I funded by the Ministry of Science, Research, and Arts Baden-Württemberg and DFG ("Deutsche Forschungsgemeinschaft").

References

1. Adams, S., Prasada Rao, R.: High power lithium ion battery materials by computational design. Phys. Stat. Solidi A **208**, 1746 (2011)
2. Alamo, J.: Chemistry and properties of solids with the [NZP] skeleton. Solid State Ion. **63**, 547 (1993)
3. Anantharamulu, N., Rao, K.K., Rambabu, G., Kumar, B.V., Radha, V., Vithal, M.: A wide-ranging review of NASICON type materials. J. Mater. Sci. **46**, 2821 (2011)
4. Brown, I.D.: Chemical and steric constraints in inorganic solids. Acta Crys. **B48**, 553 (1992)
5. Brown, I.D., Poeppelmeier, R. (eds.): Bond Valences. Springer, Berlin (2014)
6. Delmas, C., Nadiri, A., Soubeyroux, J.: The NASICON-type titanium phosphates $ATi_2(PO_4)_3$ (A = Li, Na) as electrode materials. Solid State Ion. **28**, 419 (1988)
7. Giannozzi, P., Baroni, S., Bonini, N., Calandra, M., Car, R., Cavazzoni, C., Ceresoli, D., Chiarotti, G.L., Cococcioni, M., Dabo, I.: Quantum ESPRESSO: a modular and open-source software project for quantum simulations of materials. J. Phys. Condens. Matter **21**, 395502 (2009)
8. Hagman, L.O., Kierkegaard, P.: The crystal structure of $NaMe_2^{IV}(PO_4)_3$; Me^{IV} = Ge, Ti, Zr. Acta Chem. Scand. **22**, 1822 (1968)
9. Henkelman, G., Uberuaga, B.P., Jonsson, H.: A climbing image nudged elastic band method for finding saddle points and minimum energy paths. J. Chem. Phys. **113**, 9901 (2000)
10. Kamaya, N., Homma, K., Yamakawa, Y., Hirayama, M., Kanno, R., Yonemura, M., Kamiyama, T., Kato, Y., Hama, S., Kawamoto, K.: A lithium superionic conductor. Nat. Mater. **10**, 682 (2011)
11. Knauth, P.: Inorganic solid Li ion conductors: an overview. Solid State Ion. **180**, 911 (2009)
12. Lang, B., Ziebarth, B., Elsässer, C.: Lithium ion conduction in $LiTi_2(PO_4)_3$ and related compounds based on the NASICON structure: a first-principles study. Chem. Mater. **27**, 5040 (2015)
13. Martinez, A., Pecharroman, C., Iglesias, J.E., Rojo, J.M.: Relationship between activation energy and bottleneck size for Li^+ ion conduction in NASICON materials of composition $LiMM'(PO_4)_3$; M,M' = Ge, Ti, Sn, Hf. J. Phys. Chem. B **102**, 372 (1998)
14. Monkhorst, H.J., Pack, J.D.: Special points for Brillouin-zone integrations. Phys. Rev. B **13**, 5188 (1976)
15. Orlova, A.I.: Isomorphism in phosphates of the $NaZr_2(PO_4)_3$ structural type and radiochemical properties. Radiochemistry **44**, 423 (2002)
16. Orlova, A.I., Koryttseva, A.K.: Phosphates of pentavalent elements: structure and properties. Crystallogr. Rep. **49**, 724 (2004)
17. Perdew, J.P., Burke, K., Ernzerhof, M.: Generalized gradient approximation made simple. Phys. Rev. Lett. **77**, 3865 (1996)
18. Pet'kov, V.I., Asabina, E.A., Markin, A.V., Smirnova, N.N.: Synthesis, characterization and thermodynamic data of compounds with NZP structure. J. Therm. Anal. Calorim. **91**, 155 (2008)
19. Pet'kov, V.I., Orlova, A.I.: Crystal-chemical approach to predicting the thermal expansion of compounds in the NZP family. Inorg. Mater. **39**, 1013 (2003)
20. Vanderbilt, D.: Soft self-consistent pseudopotentials in a generalized eigenvalue formalism. Phys. Rev. B **41**, 7892 (1990)

Molecular Dynamics Simulations of Silicon: The Influence of Electron-Temperature Dependent Interactions

Alexander Kiselev, Johannes Roth, and Hans-Rainer Trebin

Abstract The well-known two-temperature model for solids with highly excited electrons is extended from metals to semiconductors. It is combined with classical molecular dynamics simulations to study laser ablation in semiconductors where charge carriers are created by the absorption of the laser light. The model is improved by extending the static modified Tersoff potential to a dynamical interaction which depends on the electron temperature of the material. Results are presented for single and double pulses in silicon and are compared to a simple rescale model where the laser energy is added as kinetic energy to the atoms.

1 Introduction

The non-equilibrium phenomena in highly excited covalent systems induced by strong laser radiation fields have received much attention in recent years. These ultrafast processes are still not well understood despite many theoretical and computational investigations.

Here we use multi-million particle molecular dynamics (MD) simulations to study the laser ablation in covalently bonded materials. A combined self-consistent continuum-atomistic model (TTM) [1] was applied for the carrier-lattice interaction and the electron-hole recombination processes. In addition, the temporal and spatial dependence of the interaction on exited carrier density was taken into account by fitting the interatomic forces to finite-temperature density functional theory calculations. The influence of the pulse shape on the ablation has been investigated by studying the behavior of single and double pulses.

The results are compared to laser ablation studied with a simplified molecular dynamics simulations approach where the laser energy is applied directly to the atoms by scaling the kinetic energy. Here the interaction is modeled by fixed

A. Kiselev • J. Roth (✉) • H.-R. Trebin
Institut für Funktionelle Materie und Quantentechnologien, Universität Stuttgart, Stuttgart, Germany
e-mail: alexander.kiselev@fmq.uni-stuttgart.de; johannes.roth@fmq.uni-stuttgart.de; trebin@itap.uni-stuttgart.de

W.E. Nagel et al. (eds.), *High Performance Computing in Science and Engineering '16*, DOI 10.1007/978-3-319-47066-5_14

Tersoff potentials. The results demonstrate the importance of the combined MD-TTM approach and electron-temperature adapted potentials.

We first present the continuum two-temperature model, then we add molecular dynamics simulations. We introduce electron-temperature dependent interactions and show how to determine them. We end up with results obtained by the new method.

2 Continuum-Atomistic Modeling

We begin with the modeling of laser pulse propagation. The power density of a laser pulse on the surface of the sample can be given as a Gaussian function in space and time:

$$I(t) = \sqrt{\frac{4\ln 2}{\pi}}\frac{(1-R)\phi}{\pi b^2 t_p} e^{-(r^2/b^2)} e^{-4\ln 2(t-t_0)^2/t_p^2}, \tag{1}$$

where R is the reflectivity, ϕ is the laser fluence, b is the laser pulse width, t_p is the laser pulse duration and t_0 is the time at the maximal laser intensity.

For spatially homogeneous distributed laser power density this equation can be simplified to

$$I(t) = \sqrt{\frac{4\ln 2}{\pi}}\frac{(1-R)\phi}{t_p} \exp\left[-4\ln 2\frac{(t-t_0)^2}{t_p^2}\right]. \tag{2}$$

The propagation of the laser pulse is given by the rate equation [1]

$$\frac{\partial I(x,t)}{\partial x} = -(\alpha + \Theta n)I - \beta I^2 \tag{3}$$

if the direction of the laser beam is the x-axis. Here α and β are one- and two-photon absorption coefficients, respectively, and Θ is the free-carrier absorption cross section. This model extends the Lambert-Beer law widely used for metals and takes into account a variable number density of free carriers n in covalent materials. The dynamics of electron-hole pairs density is given by

$$\frac{\partial n}{\partial t} = \frac{\alpha I}{h\nu} + \frac{\beta I^2}{2h\nu} - \gamma n^3 + \theta n, \tag{4}$$

where ν is the photon frequency and the last two terms correspond to Auger recombination and impact ionization processes, respectively.

The total energy density of the electron-hole pairs can be treated as a sum of potential and kinetic energy densities:

$$U = nE_g + 3nk_B T_c \tag{5}$$

with band-gap energy E_g and carrier temperature T_c. Based on this equation the following model for the energy transport process have been suggested [1, 2]

$$3nk_B\frac{\partial T_c}{\partial t} = \nabla\cdot(k_c\nabla T_c) - \frac{3nk_B}{\tau_c}(T_c - T_l) + (\alpha + \Theta n)I - \frac{\partial n}{\partial t}(E_g + 3k_BT_c) - n\frac{\partial E_g}{\partial t}. \tag{6}$$

The corresponding energy transport equation for the lattice (index l) can also be derived:

$$C_l\frac{\partial T_l}{\partial t} = \nabla\cdot(k_l\nabla T_l) + \frac{3nk_B}{\tau_c}(T_c - T_l). \tag{7}$$

These two coupled partial differential equations, well known as the *T*wo-*T*emperature *M*odel (TTM) have been established for continuum modeling of electronic and lattice sub-systems after ultrashort laser irradiation.

For a more realistic description of the non-equilibrium lattice dynamics we replace Eq. 7 by the *M*olecular *D*ynamics (MD) equation of motion:

$$m_i\frac{d^2r_i}{dt^2} = F_i - \xi m_i v_i^T, \tag{8}$$

where ξ is the TTM-MD coupling constant defined by

$$\xi = \frac{\frac{1}{N_{FD}}V\sum_{m=1}^{N_{FD}}\frac{3nk_B}{\tau_c}(T_l - T_c^m)}{\sum_{k=1}^{N_v}m_k(v_k^T)^2}. \tag{9}$$

Here m_k and v_k^T are the mass and the thermal velocity of the k-th atom, N_{FD} is the number of electronic iterations within a single MD step and N_v is the number of atoms in a volume V.

In this work the heat transport equation for the carriers has been solved by a *F*inite-*D*ifference (FD) method and the molecular dynamics simulations have been carried out with IMD, the *I*TAP *M*olecular *D*ynamics simulation package [3, 4].

For modeling the reflectivity and laser field absorption process we use the Drude formula for the dielectric function [1, 5]:

$$\varepsilon = \varepsilon_r - \left(\frac{\omega_p}{\omega_L}\right)^2\frac{1}{1 + i\nu/\omega_L}, \tag{10}$$

which leads to the reflectivity

$$R = \frac{(\Re(\tilde{n}) - 1)^2 + \Im(\tilde{n})^2}{(\Re(\tilde{n}) + 1)^2 + \Im(\tilde{n})^2} \tag{11}$$

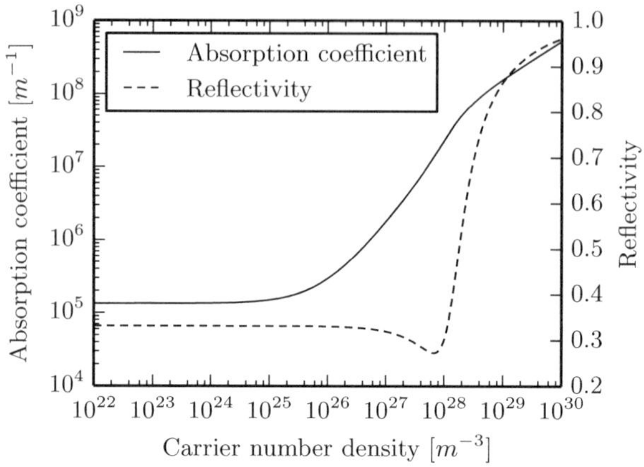

Fig. 1 Absorption coefficient α and reflectivity R for a wavelength of 775 nm

and the one-photon absorption coefficient

$$\alpha = \frac{2\omega_L \Im(\tilde{n})}{c}, \tag{12}$$

where ω_L is the laser frequency, ν is the collision frequency parameter, c is the speed of light, ω_p is the plasma frequency, $\tilde{n}$ is the complex refractive index and ε_r is the intrinsic dielectric constant. Figure 1 shows the dependency of the absorption coefficient and reflectivity on the carrier number density for a laser wavelength of 775 nm.

3 Interatomic Potentials

During the last decades a significant number of interatomic potentials for silicon have been developed and widely used in molecular dynamics simulations:

- Modified Embedding Atom Method (MEAM)
- Stillinger-Weber Potential (SW)
- Tersoff Potential (T3)
- Modified Tersoff Potential (MOD)
- Environment-Dependent Interatomic Potential (EDIP).

A comparison of modeled physical properties with experimental data is presented in Table 1. Except for the melting temperature value, all of the listed physical properties are in good agreement with experiment. However, since the melting temperature is a major thermodynamical material property for modeling the laser ablation, the molecular dynamics simulations in this work were performed using the MOD potential.

Table 1 Elastic constants, bulk modulus and melting temperatures of silicon using different interatomic potentials compared with experimental data [6]

Property	Exp	MEAM	SW	T3	MOD	EDIP
C_{11}, GPa	166	**167**	**162**	143	**166**	175
C_{12}, GPa	64	**65**	82	75	**65**	**62**
C_{44} relaxed, GPa	80	**80**	60	69	**77**	71
B, GPa	99	**99**	108	**98**	**99**	**100**
T_m, K	1683	2990	**1691**	2547	**1681**	**1520**

In the original Tersoff interaction the total potential energy V is modeled as a sum of pair-like repulsive V_R and attractive V_A interactions with environment-dependent coefficient b:

$$V = \frac{1}{2}\sum_{i\neq j} f_C(r^{ij})[V_R(r^{ij}) - b^{ij}V_A(r^{ij})] \tag{13}$$

$$V_R = A\exp(-\lambda r^{ij}), \quad V_A = B\exp(-\mu r^{ij}) \tag{14}$$

$$b^{ij} = \left(1 + (\zeta^{ij})^{\eta}\right)^{-\delta} \tag{15}$$

$$\zeta^{ij} = \sum_{k(\neq i,j)} f_C(r^{ik}) \cdot g(\cos\theta) \cdot e^{\alpha(r^{ij}-r^{ik})^{\beta}}. \tag{16}$$

The modified angular-dependent term

$$g(\cos\theta) = c_1 + \frac{c_2(h-\cos\theta)^2}{c_3 + (h-\cos\theta)^2}\left[1 + c_4 e^{-c_5(h-\cos\theta)^2}\right]$$

and cutoff function

$$f_c(r) = \frac{1}{2} + \frac{9}{16}\cos\left(\pi\frac{r-R_1}{R_2-R_1}\right) - \frac{1}{16}\cos\left(3\pi\frac{r-R_1}{R_2-R_1}\right)$$

were introduced in the MOD potential to improve the melting temperature value. The parameters for this potential are listed in Table 2.

Under strong laser irradiation the anti-bonding states of covalent materials are occupied. This has the consequence that the potential energy surface and thus the interatomic interactions change nearly instantaneously. The resulting interatomic forces can induce non-thermal processes in the lattice such as melting or phase transformation, equivalently to ordinary thermal processes. To take these effects into account, the MOD potentials for silicon, called MOD*, were parameterized dependent on the electronic temperature by using finite-temperature density functional theory (FTDFT) calculations.

First we prepared a set of silicon configurations: simple cubic (sc), body-centered cubic (bcc), face-centered cubic (fcc) and cubic diamond crystal structures

Table 2 MOD* parameters at zero temperature

$A[eV]$	$B[eV]$	$\lambda_1[1/\AA]$	$\lambda_2[1/\AA]$	η	δ	α	β	c_1	c_2	c_3	c_4	c_5	h	R_1	R_2
3173.17	104.446	3.25	1.29407	1.0	0.577175	2.3	1.0	0.1889	688746.3	10^6	1.0	26.0	−0.365	2.8	3.5

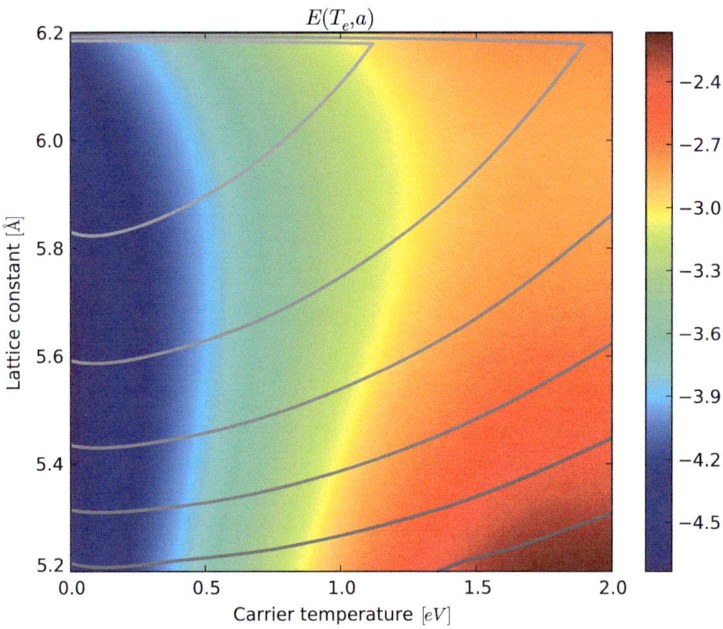

Fig. 2 Dependency of cohesive energy on carrier temperature and the lattice constant for silicon

containing 8 primitive unit cells at 20 different lattice constants. Additional shear deformations were applied to obtain data sets for the fitting of the elastic constants.

For each of these configurations DFT calculations were performed using VASP (Vienna ab initio simulation program [7]) with the projector augmented wave (PAW [8]) method and the local density approximation (LDA). A cut-off energy for the plane-wave basis set of 450 eV and a regular k-point mesh of $9 \times 9 \times 9$ were used. In Fig. 2 the dependency of cohesive energy on carrier temperature and the lattice constant for the cubic diamond structures are plotted. The resulting cohesive energies, interatomic forces and stress tensor components were applied for evaluation of the MOD* potential parameters at 21 different electronic temperatures in range of 0 and 2.0 eV. The parameter sets were fitted by using the `potfit` program [9, 10] developed and available at our institute.

A polynomial fit was carried out to achieve a smooth dependency on the carrier temperature T_c [11] for each potential parameter P (Table 3):

$$P(T_c) = \sum_{n=0}^{N} a_n T_c^n.$$

Table 3 Temperature dependent MOD* parameters

P	a_0	a_1	a_2	a_3	a_4	a_5	a_6	a_7	a_8
A	3173.17	0.	425.514	−4735.05	7639.62	−7356.67	4517.85	−1545.23	217.302
B	104.446	0.	37.3199	−295.457	173.421	169.029	−218.979	84.8027	−11.354
λ_2	1.29407	0.	0.104408	−0.887133	0.548067	−0.275339	0.284062	−0.15779	0.0295958
δ	0.577175	−0.016636	0.033080	−0.026383	0.016919	−0.003495	−0.000214	0.	0.

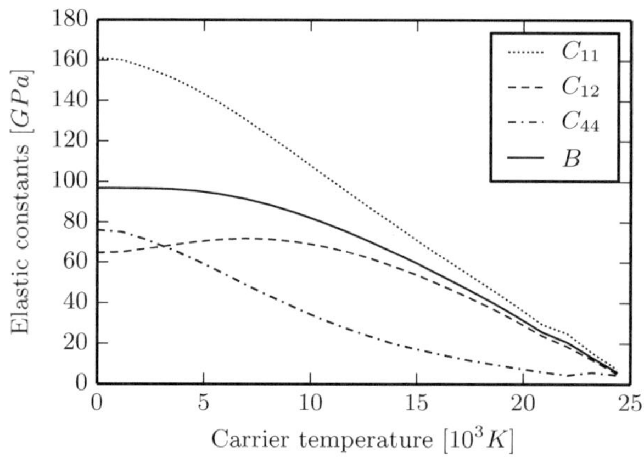

Fig. 3 Dependence of elastic constants C_{ij} and bulk modulus B on carrier temperature T_c

Here the a_0 coefficients, which correspond to the MOD* parameters for silicon at zero temperature (Table 2), are differ from the original MOD parameters since we used the DFT data instead of experimental data of elastic constants for the fitting (Fig. 3).

For the electron-temperature dependent potentials used in molecular dynamics simulations the force calculations have to be extended:

$$F = -\left(\sum_k \frac{\partial V}{\partial P_k}\frac{\partial P_k}{\partial T_c}\right)\nabla T_c - \nabla V,$$

where P_k are the temperature dependent potential parameters, which were calculated at each atomic position r_i by using a trilinear interpolation method.

4 Results

The molecular dynamics simulations of laser ablation were performed for a box of constant size $1124 \times 4.34 \times 4.34\,\text{nm}^3$ with approximately 10^6 silicon atoms and a time step of 0.101806 fs. This represents a 1-μm thick silicon film. The simulation domain was divided into 750 FD cells along the x-axis, which corresponds to the [1 0 0] crystallographic direction. In this case each of these FD cells contains nearly 1333 atoms. Periodic boundary conditions were applied in y- and z-directions whereas open boundaries were assumed along the x-direction. The electronic-temperature dependent MOD* potential was used throughout this work.

The two-temperature-model for the electronic system was solved on a regular finite difference grid with a time step of 0.0051 fs i.e., 200 electronic iterations

Table 4 Laser pulse and material properties of silicon [1]

Parameter	Value
Laser wavelength λ	775 nm
Laser pulse duration t_p	100 fs
Time at the laser peak intensity t_0	300 fs
Carriers thermal conductivity k_c	$(-3.47 \times 10^8 + 4.45 \times 10^6 T_c)\,\text{eV} \cdot \text{s}^{-1}\text{Å}^{-1}\text{K}^{-1}$
Energy relaxation time τ_c	$240 \cdot (1 + \frac{n}{6\times10^{20}\text{cm}^3})$ fs
One-photon absorption coefficient α	$648.585 \cdot \exp(T_l/430)\,\text{cm}^{-1}$
Free-carrier absorption cross section Θ	$5.1 \times 10^{-18} \cdot \frac{T_l}{T_{room}}\,\text{cm}^2$
Two-photon absorption coefficient β	0
Band gap energy E_g	$(1.16 - 7.02 \times 10^{-4} \frac{T_l^2}{T_l+1080} - 1.5 \times 10^{-8} n^{1/3})\,\text{eV}$
Auger recombination coefficient γ	$3.8 \times 10^{-31}\,\text{cm}^6\text{s}^{-1}$
Impact ionization coefficient θ	$3.6 \times 10^{10} \cdot \exp(-\frac{1.5E_g}{k_B T_c})\,\text{s}^{-1}$

within a single MD step. The material parameters for silicon were chosen according to [1] and are listed in Table 4.

The laser pulses were modeled with a Gaussian temporal profile with the full width at half-maximum of 100 fs and a wavelength of 775 nm. This wavelength corresponds to a photon energy of 1.6 eV which is higher than the band gap, thus the two-photon absorption can be neglected. The second pulse delay for double-pulse simulations a value of 0.25 ps was chosen and for the laser field absorption mechanism and reflectivity the Drude model was applied.

First we performed molecular dynamics simulations at constant room temperature (300 K) and zero pressure for a few thousand steps in order to reach an equilibrium atomic configuration, while the electronic temperature was also kept constant according to the Fermi-Dirac distribution.

Then we investigated the evolution of carrier and lattice temperatures after laser irradiation with single and double pulses for the fluences between 0.02 and $0.15\,\text{J/cm}^2$.

Figure 4 shows the carrier density at the front film surface for single pulses with laser fluences of 0.075, 0.1 and $0.12\,\text{J/cm}^2$, and double pulses with laser fluences of 0.05, 0.075 and $0.1\,\text{J/cm}^2$. As expected, we observe a linear dependency of the maxima on laser fluences according to the carrier number rate equation (4) for both laser pulse sequences. We can also see increasing of maxima at the second laser pulse due to the rise of carrier number densities after the first peaks. The Auger recombination, on the other hand, decreases the carriers number density.

The temporal evolution of the carrier temperatures for single and double pulses is plotted in Fig. 5. A nearly linear dependency of the maxima can be observed here also. The carrier temperatures maxima are shifted. The rapid increase of carriers temperature during decreasing of carrier number density is a consequence of the fifth term on the RHS of energy balance equation (6), which is proportional to the negative time derivative of density n. Here the potential energy of carriers is

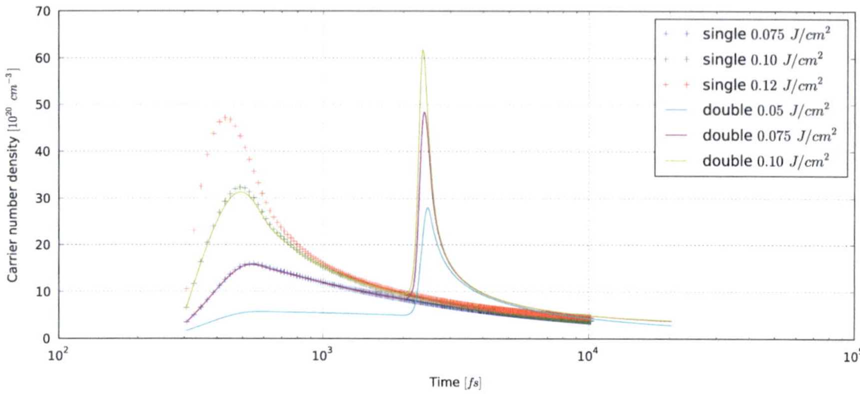

Fig. 4 Carrier density at the front film surface for single and double pulses at several laser fluences

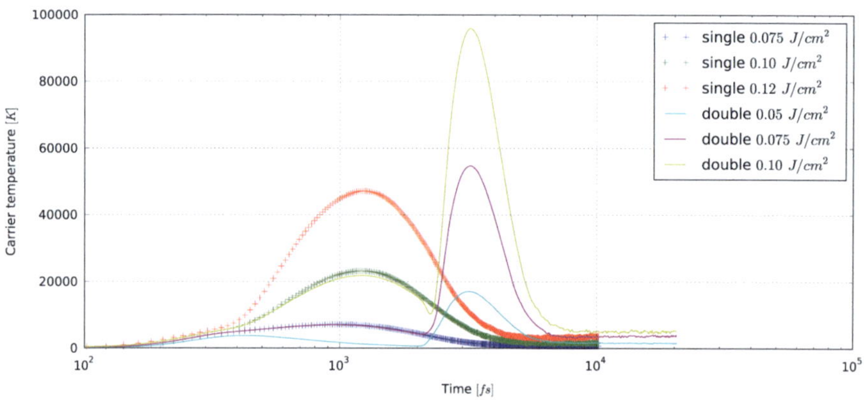

Fig. 5 Carrier temperature at the front film surface for single and double pulses at several laser fluences

converted to the kinetic energy of carriers, characterized by their temperature. The higher peak positions of the second pulse maxima indicate an increasing role of impact ionization, which is proportional to the number of electrons. Furthermore in highly excited covalent systems a significant decrease of the absorption length and reflectivity can be observed, which also leads to higher energy absorbing capacity.

Figure 6 shows the lattice temperature at the front film surface for single and double pulses. The energy exchange of carriers and lattice occur on a time scale of 1 to 10 ps. After equilibration the plotted temperatures at the front of the sample correspond to the average temperatures of the whole simulation domain. Increasing fluctuations at higher laser fluences indicate a phase transition on the material surface. The ablation thresholds were observed at $\sigma_E = 0.1$ J/cm^2 and $t = 0.09$ ps for single pulses and at $\sigma_E = 0.07$ J/cm^2 and $t = 3.2$ ps for double pulse sequences. These can be clearly seen in Figs. 7 and 8, where spatial and temporal evolution

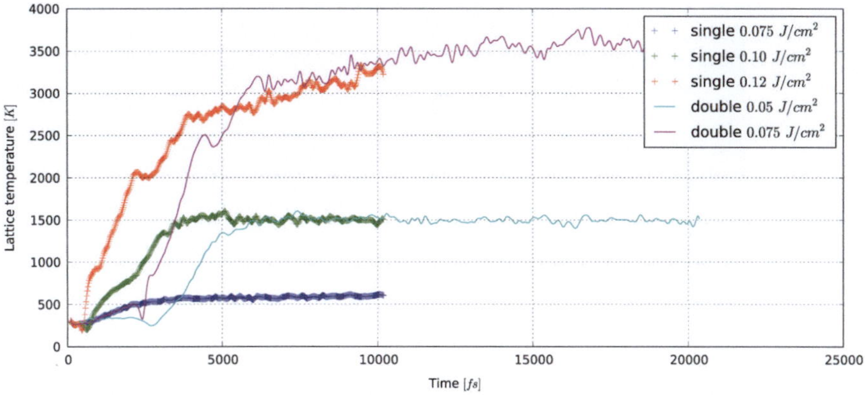

Fig. 6 Lattice temperature at the front film surface for single and double pulses at several laser fluences

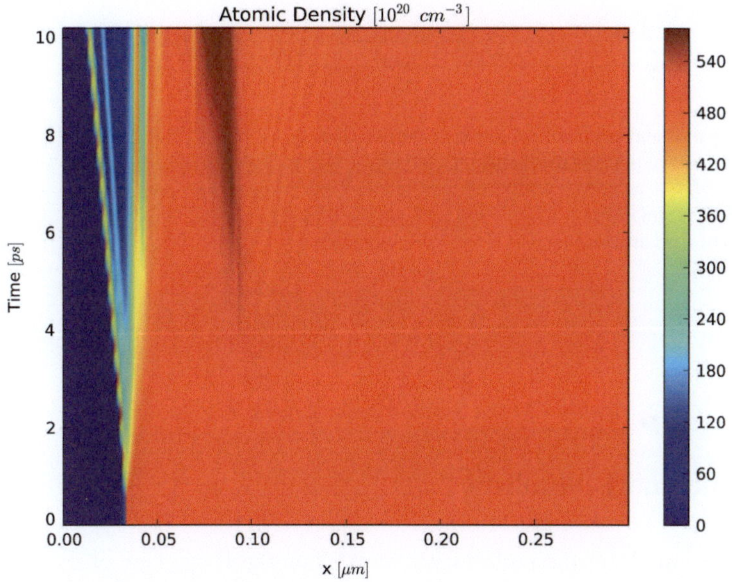

Fig. 7 Spatial and temporal evolution of atomic density for single pulses with laser fluence of $0.12\,\mathrm{J/cm^2}$ above the damage threshold

of atomic densities from single pulse and double pulse simulations, respectively, is plotted. The calculated results for single pulses are comparable to experimental values for the ablation threshold in silicon as reported by Pronko et al. [12] $\sigma_E = 0.17\,\mathrm{J/cm^2}$ ($\lambda = 800\,\mathrm{nm}$, $t_p = 100\,\mathrm{fs}$) and $\sigma_E = 0.108\,\mathrm{J/cm^2}$ ($\lambda = 786\,\mathrm{nm}$, $t_p = 90\,\mathrm{fs}$).

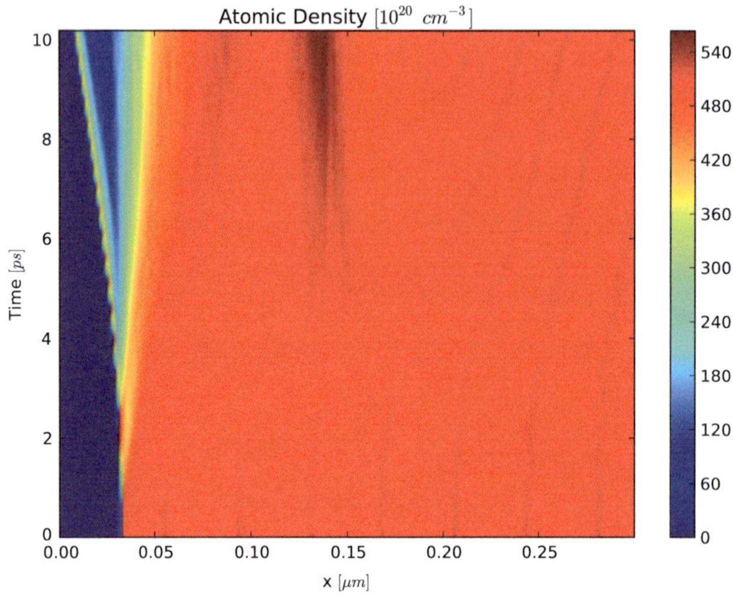

Fig. 8 Spatial and temporal evolution of atomic density for double pulses with laser fluence of 0.10 J/cm^2 above the damage threshold

Table 5 Performance numbers of a shear rate simulation

Shear rate [s^{-}1]	Node × hours	MD steps	Quotient
$3{\cdot}10^4$	57,600	30,990,000	538
$3{\cdot}10^5$	8400	6,584,050	784
$3{\cdot}10^7$	1200	1,211,490	1010
$3{\cdot}10^8$	1050	877,300	836
minim	3600	490,000	136

5 Benchmark Numbers

As a benchmark for the performance of IMD on the Hazel Hen of the HLRS we report the results of a shear rate simulation since numbers of the overall performance of IMD and the performance in ablation simulations have been given in previous reports. The system studied was a block of Ag-Cu alloy with the size 52×52×70 nm containing 16 million atoms. Several simulations with different shear rates have been carried out together with a minimized shearing (minim in Table 5). The most interesting column in Table 5 is the last labeled “quotient”. It contains the number of MD steps per node × hour, also called wall time. Obviously this number varies largely and not very systematically. The reason is that the performance depends strongly on the specific type of defects generated at a certain shear rate. This also shows that it is not possible to use one of the cases for optimization.

In the row labeled "minim" in Table 5 the shearing is carried out in such a way that the energy of the probe is minimized at each time steps. This explains the much smaller performance in this case.

6 Conclusions

We have extended the ansatz to combine a continuum two-temperature-model for the electron part with molecular dynamics simulations to treat the atomistic part from simple metals with a high constant charge carrier concentration to semiconductors where charge carriers have to be created first by the laser beam. It turned out to be essential to introduce electron-temperature dependent interactions for the semiconductors. For metals it might be also interesting to test the influence of high electron temperature but to our knowledge no in-depth study exists yet. A comparison of the results to a simple rescale model where the laser energy is introduced by rescaling the kinetic energy of the particles shows a completely different behavior of the ablation process. The ablation thresholds are much higher and the material is completely vaporized. It might be interesting to relate the results of the rescale and the TTM model in the case of ns-pulses, but this cannot be done for fs-pulses. Furthermore, even for the ns-pulses the electron-temperature modified atomic interactions have to be taken into account. The main challenge of the current approach are the electronic parameters. In addition to electron heat capacity, heat conductivity and electron-phonon-coupling parameters there a dozen more parameters that can only be extracted from literature. For germanium this should be no problem and work is under way, but for compound semiconductors it could be impossible to find all parameters. The atomic interaction on the other hand can be fitted in all cases even for very complex compounds with our interaction fitting code `potfit`.

References

1. Gan, Y., Chen, J.K.: Combined continuum-atomistic modeling of ultrashort-pulsed laser irradiation of silicon. Appl. Phys. A **105**, 427–437 (2011)
2. Agassi, D.: Phenomenological model for pisosecond-pulse laser annealing of semiconductors. J. Appl. Phys. **55**, 4376–4383 (1984)
3. Roth, J., Gähler, F., Trebin, H.-R.: A molecular dynamics run with 5.180.116.000 particles. Int. J. Mod. Phys. C **11**, 317–22 (2000)
4. Stadler, J., Mikulla, R., Trebin, H.-R.: IMD: a software package for molecular dynamics studies on parallel computers. Int. J. Mod. Phys. C **8**, 1131–1140 (1997)
5. Sokolowski-Tinten, K., von der Linde, D.: Generation of dense electron-hole plasma in silicon. Phys. Rev. B **61**, 2643–2650 (2000)
6. Timonova, M., Thijsse, B.J.: Thermodynamic properties and phase transitions of silicon using a new MEAM potential. Comput. Mat. Sci.**48**, 609–620 (2010)

7. Kresse,G., Hafner, J.: Ab initio molecular dynamics for liquid metals. Phys. Rev. B **47**, 558–561 (1993)
8. Kresse, G., Joubert, D.: From ultrasoft pseudopotentials to the projector augmented-wave method. Phys. Rev. B **59**, 1758–1775 (1999)
9. Brommer, P., Gähler, F.: Potfit: effective potentials from ab-initio data. Model. Simul. Mat. Sci. Eng. **15**, 295–304 (2007)
10. Brommer, P., Gähler, F.: Effective potentials for quasicrystals from ab-initio data. Philos. Mag. **86**, 753–758 (2006)
11. Shokeen, L., Schelling, P.K.: Thermodynamics and kinetics of silicon under conditions of strong electronic excitation. J. Appl. Phys. **109**, 073503 (2011)
12. Pronko, P., VanRompay, P., Horvth, C., Loesel, F., Juhasz, T., Liu, X., Mourou, G.: Avalanche ionization and dielectric breakdown in silicon with ultrafast laser pulses. Phys. Rev. B **58**, 2387 (1998)

Non-linear Quantum Transport in Interacting Nanostructures

Benedikt Schoenauer and Peter Schmitteckert

Abstract We study numerically whether the transient dynamics of local currents inside interacting regions, coupled to non-interacting leads, can be distinctly different from the transient dynamics of currents which are transmitted through these interacting regions. Here, we present a simple model system incorporating an asymmetric ring structure as an interacting impurity region into a non-interacting tight-binding chain. For this model system, we observe local currents, restricted to the ring structure, that can easily outgrow the net current through the system. The transient oscillations of these currents may also decay on time scales orders of magnitude larger than the decay time of transient features in the transmitted currents, for which a decay time is known (Wingreen et al., Phys Rev B 48(11):8487–8490, 1993) to always exist and to be proportional to the inverse resonance width.

1 Introduction

Transport properties of strongly interacting quantum systems are a major challenge in todays condensed matter theory. In our project we apply the density matrix renormalization group (DMRG) method [13, 16, 17, 23, 26, 27] to study transport properties [4–6, 9, 18, 19, 22, 24] of quantum devices attached to metallic leads. In previous projects [8, 9, 19, 21, 22, 24] we studied the conductance of single impurities coupled to metallic leads. To this end we developed two complementary approaches to obtain the conductance of a structure coupled to left and right leads. First we used the Kubo approach [2] to obtain the linear conductance. Combined with leads described in momentum space [3, 20] we obtained high resolution in energy. The second approach is based on simulating the time evolution of an

B. Schoenauer
Center for Extreme Matter and Emergent Phenomena, Institute for Theoretical Physics, Utrecht University, Princetonplein 5, 3584 CE Utrecht, The Netherlands
e-mail: b.m.schonauer@uu.nl

P. Schmitteckert (✉)
Lehrstuhl für Theoretische Physik I, Physikalisches Institut, Am Hubland, Universität Würzburg, 97074 Würzburg, Germany
e-mail: peter.schmitteckert@physik.uni-wuerzburg.de

W.E. Nagel et al. (eds.), *High Performance Computing in Science and Engineering '16*, DOI 10.1007/978-3-319-47066-5_15

initial state with a charge imbalance and is reviewed in [7]. In cooperation with Edouard Boulat and Hubert Saleur we have been able to show that our approach is in excellent agreement with analytical calculations in the framework of the Bethe ansatz [4]. This agreement is remarkable as the numerics is carried out on a lattice model, while the analytical result is based on field theoretical methods in the continuum limit. Most strikingly, we proved the existence of a negative differential conductance (NDC) regime even in this simple model of a single resonant level with interaction on the contact links. In an extension of this approach we presented results for current-current correlations, including shot noise, based on our real time simulations in an earlier HPC report [8], see also [5, 6]. We then managed to include a counting field into our time dependent simulations which allowed us to obtain the full counting statistics (FCS) via the direct simulation of the cumulant generating function [10, 21]. Finally we have been able to obtain the sub leading corrections to the FCS of charge transport in the inverse measuring time [11, 12]. Despite this success story of obtaining steady state transport properties from quenches in the charge imbalance it is an open question, whether a system will always relax to the true steady state. In classical mechanics it is well known that driven systems can end up in an oscillatory or even chaotic state. In order to extend studies to this issue we expand our impurity systems to interacting structures with an internal degree of freedom.

The initial state of our spin-less fermion systems is given by the lowest eigenstate of the Hamilton matrix. The time evolution as a solution of the linear Schrödinger equation is then given by the action of a matrix exponential of the Hamiltonian governing the time evolution on the initial state. The numerical problem arises from the fact that the dimension of the corresponding Hilbert space is given by 2^M with M the number of lattice sites typically ranging from a few tens to a few hundreds of sites. The DMRG is now an iterative procedure to search for a suitable subspace of the complete Hilbert space that is sufficient to represent the system. Exploiting particle number conservation allows us to implement a block structure for the Hamilton matrix and by employing a dyadic representation of the vector Hilbert space the matrix vector product needed for the sparse matrix diagonalization and matrix exponentials can be represented by a set of matrix multiplications, for details see [22]. Our parallelization strategy consist of a master-worker queue where the work chunks are given BLAS-3 and LAPACK function calls. These matrix operation are then evaluated in a single threaded operation [22]. In addition we enhance the concurrency of the code by an asynchronous evaluation of the scalar products. Specifically in the evaluation of the matrix exponential we can schedule several steps concurrently, i.e. we can fill the worker queue from the main thread while the worker threads are still occupied with a previous scalar product. We can parallelize the sweeping procedure and distribute it over several nodes. However, to obtain a single I/V-characteristic we have to perform a complete DMRG simulation for each voltage value of interest. Since we have to run many DMRG jobs we use an embarrassingly parallel strategy here, as a distributed calculation would lead to a decreased performance. By default we have allocated an entire node consisting of 20 processors to each DMRG calculation. These processors were used by four master threads and sixteen worker threads as described in [22]. Each of the processors was assigned 3200 MB of memory. Using these default resources our DMRG calculations had an average runtime of two weeks. We typically used a few hundred GB of scratch data on the local disk `$TMP` which was written to the `$WORK`-directory for an eventual restart of the jobs.

2 Motivation of the Model System

As pointed out in the introduction, the intention of this work is to establish and investigate a simple model system that comprises a small additional amount of degrees of freedom compared to the commonly studied single impurity systems. The model is suitable to approximate benzene-like molecules and structures, as well. From the model system we expect new insights into the non-equilibrium behavior of interacting structures. From its similarity to benzene we hope to better understand the charge transport behavior of the periodic benzene structure that is graphene. The obvious system that fulfills our prerequisites is a preferably small interacting ring structure. A similar model system has already been used by Bohr and Schmitteckert [1] to study interaction and interference effects in benzene. We extend this model to also incorporate asymmetries, which for example occur for doped benzene or graphene sheets and for which recent density functional theory (DFT) studies by Walz et al. [25] predict that they give rise to large circular currents.

3 Hamiltonian of the Model System

The complete, unperturbed Hamiltonian H_0 of the system reads

$$H_0 = H_{\text{ring}} + H_{\text{leads}} + H_{\text{coupling}} \,, \tag{1}$$

where the three individual Hamiltonians describe the three different regions of the system.

3.1 Interacting Ring Structure

First, we present the ring structure. It is basically a lattice discretized ring, which is quadratic and consists of four localized orbitals. We will subsequently refer to these orbitals as sites to emphasize their localized nature. Between the sites there is a tunneling matrix element J', which is also called hopping element. To decrease the complexity of the models we only allow for spin-less fermions, which is equivalent to a strong external magnetic polarization, that enforces the same spin polarization for each fermion in the system. The electron-electron interaction is modeled as repulsive and involves only particles on neighboring sites. The parameter U specifies the interaction strength. On a single site in the upper half of the ring structure a gate voltage ϵ_T is applied. The gate potential lifts the symmetry between the lower and the upper half of the ring and also the degeneracy of the ground state of the ring structure. It is also meant to act as a restoring force on particles propagating through the ring structure and thereby to enhance non-equilibrium effects. The Hamiltonian of the uncoupled structure can be written as

$$H_{\text{ring}} = \sum_{\langle x,y \rangle} U \left(n_x - \frac{1}{2} \right) \left(n_y - \frac{1}{2} \right) - J' \left(d_x^\dagger d_y + \text{h.c.} \right) + \epsilon_T n_2 \,,$$

where $\langle x, y \rangle$ denotes neighboring sites in the ring, d_x is the annihilation operator for a fermion on site x and $n_x = d_x^\dagger d_x$ is the occupation number operator at a site x. The gate potential is applied only to site $x = 2$. The structure of the interaction term preserves the particle-hole symmetry.

3.2 Non-interacting Tight-Binding Leads

The metallic leads are modeled as a half-infinite chain of sites which is often referred to as tight-binding chain because it describes materials, in which the electrons are tightly bound to the ions of the crystal lattice and can only tunnel from orbital to orbital. The leads are also modeled as sufficiently large, such that the electron-electron interaction is completely screened and one can therefore omit the interaction term. Furthermore, since the ring structure is small, one assumes, that only a single transport channel of the leads couples to it, which makes it possible to model the leads as two one-dimensional chains. This fits well with the preference of the density-matrix renormalization group for one-dimensional systems. The electrons in the leads are also spin-less and the Hamiltonian is given by

$$H_{\text{lead}} = -J \left(\sum_{|x| \geq 2} c_x^\dagger c_{x-1} + \text{h.c.} \right) , \tag{2}$$

where J denotes the tunneling matrix element in the leads and c_x the annihilation operator at a site in the leads. Here, the index x runs from ± 2 to $\pm\infty$.

The non-interacting tight-binding chain can be solved analytically by a plane wave ansatz with wave vector k. This ansatz yields the dispersion relation of the system, which reads

$$E(k) = -2J \cos(k) \ . \tag{3}$$

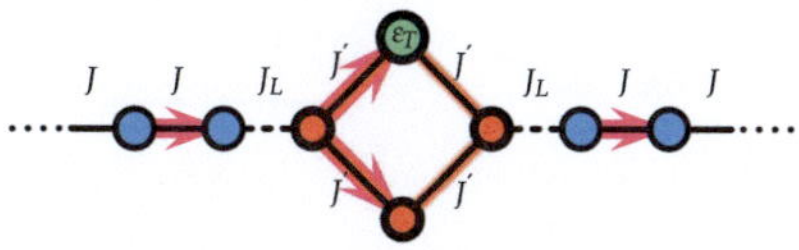

Fig. 1 Schematic representation of the two model system. The four ring sites making up the structure are shown as *red* and *green circles*. The *red bonds* indicate sites with nearest-neighbor interaction. An on site gate potential ϵ_T is applied to the *green* site in the upper half of the ring. *Blue* sites belong to the noninteracting tight-binding leads. The coupling J_L between leads and ring structure is represented by a *dashed line*

The energy band of the system consequently has the shape of a cosine and the bandwidth $D = 4J$.

3.3 Coupling Between Ring Structure and Leads

The ring structures are symmetrically coupled to the leads as shown in Fig. 1. The tunneling element J_L is chosen to be smaller than the tunneling in the ring and the leads respectively to pronounce the features induced by the interacting nanostructure. In the limit of very small coupling, a perturbation theory in the coupling is possible and subject of future work. The coupling Hamiltonian is given by

$$H_{\text{coupling}} = -J_L \left(d_1^\dagger c_{-1} + d_{N_R}^\dagger c_1 + \text{h.c.} \right) , \tag{4}$$

where the operator $d_1^\dagger$ creates a particle on the left site of the ring structure and $d_{N_R}^\dagger$ creates a particle on the right site of the ring.

4 Quenching of the System

To prepare a system in a state of non-equilibrium it is quenched by applying an external potential to the leads. For this, an operator H_{SD} is added to the unperturbed Hamiltonian H_0. The quenched Hamiltonian reads

$$H_{\text{quench}} = H_0 + H_{\text{SD}} = H_0 + \frac{V_{\text{SD}}}{2} \left(\sum_{x \leq -1} n_x - \sum_{x \geq 1} n_x \right) . \tag{5}$$

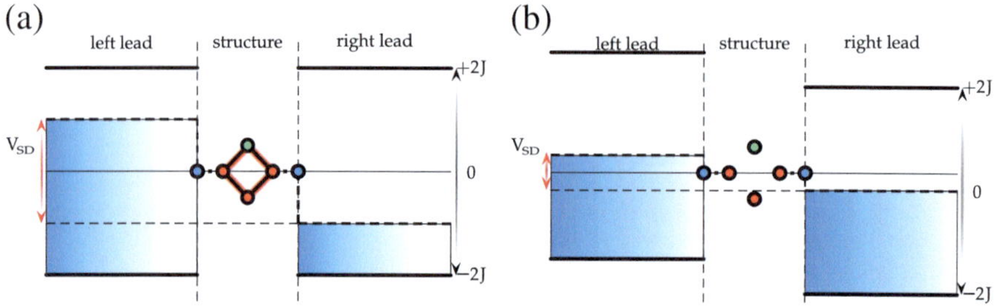

Fig. 2 Different quench setups. (**a**) Schematic representation of the initial conditions corresponding to an Hamiltonian where the bias voltage $\pm V_{\mathrm{SD}}/2$ is applied to left respectively right lead at $t = 0$ and switched off afterwards. The voltage is applied exclusively to the leads and not to the ring structure. The bandwidth 4 J of the leads originates from cosine band of the tight-binding chain. (**b**) Representation of the quench setup where the bias voltage $\pm V_{\mathrm{SD}}/2$ is applied to leads in the Hamiltonian that is used for time evolution. Half of the states in both leads are occupied and the bands are shifted against each other for all times $t > 0$. If $V_{\mathrm{SD}} > 2\,\mathrm{J}/e$ electrons at the Fermi level of the left lead and holes close to the Fermi level in the right lead have no state in the opposite lead to tunnel into. The current decreases as a result, which is unphysical

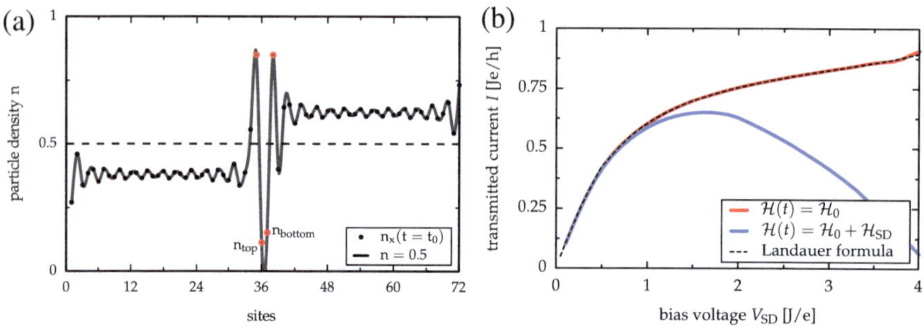

Fig. 3 Initial setup of the system. (**a**) Electron density of the ground state ψ_0 when a bias $V_{\mathrm{SD}} = 0.4\,\mathrm{J}/e$ is applied at $t = 0$. The electron density is below $n = 0.5$ in the left lead and above $n = 0.5$ in the right lead. In both leads we observe Friedel oscillations of the electron density. The particle number on the sites in the ring structure adds up to exactly $N = 2$, because the nearest-neighbor interaction forces half filling in the ring structure. The second result of the interaction is a clear particle density wave in the ring structure, where the left and the right site of the ring are almost completely filled and the top and bottom site are nearly empty. (**b**) I-V characteristics that result from the two different quench setups that are shown in Fig. 2. If the bias voltage is applied only at $t = 0$ the current agrees with the analytical results for all values of V_{SD}. If the bias voltage is added to the time evolution Hamiltonian, the current decreases for $V_{\mathrm{SD}} > 2\,\mathrm{J}/e$ because the two band are shifted against each so far that certain particles or holes close to the Fermi level cannot tunnel into the opposite lead

The ground state ψ_0 of this Hamiltonian H_{quench}, which is calculated using the standard finite lattice DMRG, is characterized by a charge imbalance, that is depicted in Fig. 3a. The energy difference between the highest occupied energy levels in the right lead (source) and the left lead (drain) should be equivalent to the bias or source-drain voltage V_{SD}.

For all times $t > 0$, the external potential is switched off and one calculates the time-resolved expectation values of a chosen set of observables for the state ψ_0 while the dynamics are governed by H_0.

It is equally possible to determine the initial state ψ_0 for the unperturbed system H_0 and to apply the external potential for all times $t > 0$, which results in the time-dependent behavior of ψ_0 being determined by H_{quench}. This is equivalent to an energy shift of the entire band of the leads by $\pm V_{\text{SD}}/2$. As shown in Fig. 3b, this leads to unphysical results for current for bias voltages $V_{\text{SD}} > 2\,\text{J}/\text{e}$. For $V_{\text{SD}} \geq 4\text{J}/\text{e}$, where $D = 4\text{J}/\text{e}$ is the bandwidth, the current vanishes completely because energy conservation prohibits the tunneling of particles respectively holes from one energy band to the other.

We refer to these two options as instantaneous quenches, since the switching of the bias voltage occurs with an infinite velocity. The former of the two quench methods has been our standard procedure for the system quench because of the obvious drawbacks of the latter one. It has been applied whenever nothing else is stated. In both cases the leads also act as a particle reservoir.

5 Numerical Dynamics

For a numerical calculation of the time evolution the time is discretized into time steps $\Delta t = t_j - t_{j-1} = 0.4\,\text{J}^{-1}$. The time evolution is performed within the td-DMRG calculation in the fashion explained in detail in [7]. We apply Krylov subspace methods to facilitate the time evolution via a matrix exponential function as it rigorously assures unitarity. Depending on the choice of the quench procedure the time evolution is either

$$\left|\psi\left(t_j\right)\right\rangle = e^{-iH_0\left(t_j - t_{j-1}\right)}\left|\psi\left(t_{j-1}\right)\right\rangle \tag{6}$$

or

$$\left|\psi\left(t_j\right)\right\rangle = e^{-i(H_0 + H_{\text{SD}})\left(t_j - t_{j-1}\right)}\left|\psi\left(t_{j-1}\right)\right\rangle . \tag{7}$$

The maximum number of time steps is limited by the system size because after a finite time τ_T the fastest wave packets have reached the hard wall boundary of the system and are reflected.

6 Standard Parameters and Observables

To properly explore our model numerically, we vary the parameters of our Hamiltonian. To simultaneously ensure a certain degree of comparability between different calculations, we have defined a set of standard parameters, that is used for most of the calculations. In one set of calculations only a single parameter is varied while the other parameters assume their standard values. The standard parameters help to reduce the overall computational costs.

The standard system size is set to $M = 72$ system sites. The number $M = 72$ is a compromise between a system size large enough to study the dynamics for a reasonable time $t < \tau_T$ and the calculation time that is needed for the system. Using the chosen standard system size a basic DMRG calculation has a reasonable runtime of at most two days depending on the ForHLRI computer cluster.

As a result of the chosen system size, we adopt a bias voltage of $V_{\mathrm{SD}} = 0.4\,\mathrm{J/e}$. The frequency of the oscillation artifacts that result from said bias voltage is large enough to obtain reasonable results from our fitting procedure while the bias voltage is still small enough to avoid the additional errors that occur for bias voltages close to the band width. From early calculation we deduce that $V_{\mathrm{SD}} = 0.4\,\mathrm{J/e}$ is the most suitable choice for the standard parameter since it is equal to the actual effective bias voltage for our chosen standard system size and thus reflects well the behavior of an infinite system to which the same bias voltage is applied.

The standard value of the on-site gate potential $\epsilon_T = 0.5\,\mathrm{J}$ derives from our first calculations where we have found a suitable oscillation frequency of the ring currents in response to the gate potential.

We define three different regimes for the interaction strength. The regime of weak interaction is by default calculated with a value $U = 0.1\,\mathrm{J}$. The strong interaction regime uses a default value $U = 1.0\,\mathrm{J}$ and very strong interaction is commonly calculated using $U = 2.0\,\mathrm{J}$. Calculations have been done for a wide range of interaction strengths to confirm, that the chosen values properly represent the particular regimes.

For the coupling between leads and ring structure we have chosen a standard value of $J_L = 0.5\,\mathrm{J}$ to model that ring and leads consist of different materials with an imperfect coupling.

Early calculations have been done using a time discretization of $\Delta t = 0.25\,\mathrm{J}^{-1}$. For later we have used a more coarse discretization of $\Delta t = 0.4\,\mathrm{J}^{-1}$. Calculations for both discretization widths haven been compared to ensure that they yield the same results.

The DMRG calculations have kept a minimum of $N = 700$ states per block and a maximum of $N = 2800$ states per block. Particular calculations that constantly needed 2800 per block were rerun using a minimum of $N = 900$ and a maximum of $N = 3600$.

By default we have calculated the current through four distinct bonds of the systems. The particular bonds and the direction of positive current are indicated by the purple arrows in Fig. 4. We will subsequently denote the current through the bonds located in the leads as the *transmitted current* and the currents through the indicated bonds in the ring as *upper link current* and *lower link current*.

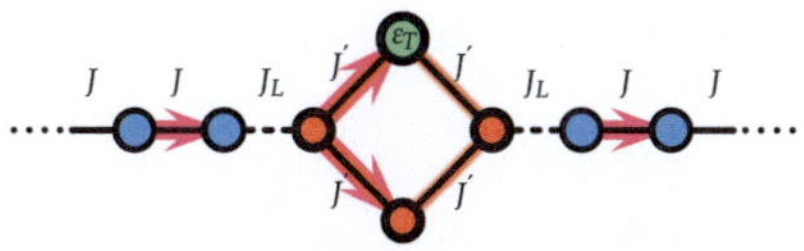

Fig. 4 The model system and the position of the measured currents. The image shows the model system. A gate potential is marked by the green circle and is given the standard value $\epsilon_T = 0.5\,\mathrm{J}$. For the nearest-neighbor interaction between the sites of the ring (marked by the *red* links) we distinguish three regimes. The standard interaction strength for weak interaction is $U = 0.1\,\mathrm{J}$, for strong interaction $U = 1.0\,\mathrm{J}$ and very strong interaction $U = 2.0\,\mathrm{J}$. The hopping elements in the ring are $J' = 1\,\mathrm{J}$ while the couplings between ring and leads are $J_L = 0.5\,\mathrm{J}$. The **transmitted current** is the mean value of the currents measured in the *left* and *right* leads marked by the *purple arrows*. The **upper link current** is measured at the position of the *upper purple arrow* in the respective ring. The **lower link current** is measured at the position of the *lower arrow* in the ring. The *purple arrows* also indicate the direction of positive current

7 Transient Dynamics of Currents in the Interacting System

We now turn our focus to the main objective of this work, which is the investigation of the time-dependent behavior of currents in the model system. We obtain this behavior from our td-DMRG calculations and examine it for local transient and local steady state regimes. In this way we try to answer our initial question of whether the local observables, e.g. the currents through the studied links, always relax to a steady state. Constant non-equilibrium effects of the currents would hereby indicate that *the certain relaxation* is in fact a false assumption. The properties of the currents in the ring structures, which will subsequently be called *ring currents*, for finite interaction are moreover interesting in connection with the work by Walz et al. [25]. They find ring currents orders of magnitude larger than the transmitted current in their DFT studies of hydrogen doped Graphene.

For a thorough analysis of the dynamics of the system, we perform td-DMRG calculations for a wide range of interaction strengths $0\,\mathrm{J} < U \leq 6.0\,\mathrm{J}$ and bias voltages $0.1\,\mathrm{J} \leq V_{\mathrm{SD}} \leq 4\,\mathrm{J/e}$. Several calculations have been done for varying system sizes to detect if particular features of the time-dependent currents are caused by the finite system size. The data is analyzed manually and checked for transient and steady state regimes. If oscillations are found in the currents, they are first fitted with cosine of fixed frequency $\omega = V_{\mathrm{SD}}$. An oscillation with frequency $\omega = V_{\mathrm{SD}}$ would indicate finite size effects. If the oscillations and the fits mismatch, a second cosine fit with variable frequency is performed to obtain a value for the amplitude and the frequency of the oscillation. Figures 5, 6, 7, 8, and 9 display the time-dependent currents (as labelled in Fig. 4) for different interaction strengths U and the standard parameters for $V_{\mathrm{SD}} = 0.4\,\mathrm{J/e}$ and $\epsilon_T = 0.5\,\mathrm{J}$.

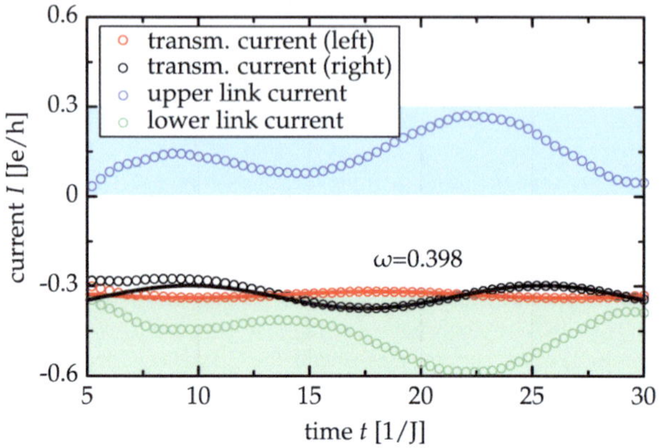

Fig. 5 Time-dependent currents of the quadratic ring system for $U = 0.1\,\mathrm{J}$. The plot displays the measured currents as a function of time t for a system with $M = 72$ sites and parameters $\epsilon_T = 0.5\,\mathrm{J}$ and $V_{\mathrm{SD}} = 0.4\,\mathrm{J/e}$. The currents are labelled according to Fig. 4. The oscillation frequency of the transmitted current for $t > 10\,\mathrm{J}^{-1}$ is $\omega = 0.398 \simeq 0.4 = V_{\mathrm{SD}}$. The oscillation of the currents is therefore regarded as a finite size effect

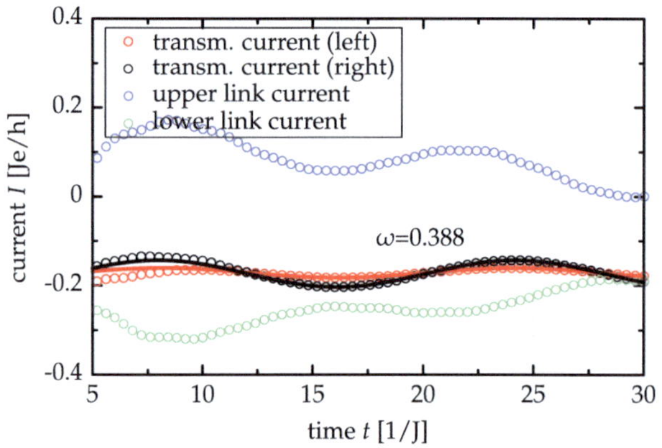

Fig. 6 Time-dependent currents of the quadratic ring system for $U = 0.5\,\mathrm{J}$. The plot displays the measured currents as a function of time t for a system with $M = 72$ sites and parameters $\epsilon_T = 0.5\,\mathrm{J}$ and $V_{\mathrm{SD}} = 0.4\,\mathrm{J/e}$. The currents are labelled according to Fig. 4. The oscillation frequency of the transmitted current for $t > 10\,\mathrm{J}^{-1}$ is $\omega = 0.388 \simeq 0.4 = V_{\mathrm{SD}}$. The oscillation of the currents is therefore regarded as a finite size effect

7.1 Weak Interaction $U \leq 0.1\,\mathrm{J}$

From Fig. 5, we conclude, that weak interaction does not visibly change the time-dependancy of the currents. For times $t > 10\,\mathrm{J}^{-1}$ no more transient effects can be observed. Remaining oscillations have a frequency $\omega = V_{\mathrm{SD}}$ and are only a result of the finite system size. For the oscillations inside the ring we observe a phase difference of $\Delta\phi = \pi/2$ compared to the oscillations of the transmitted current.

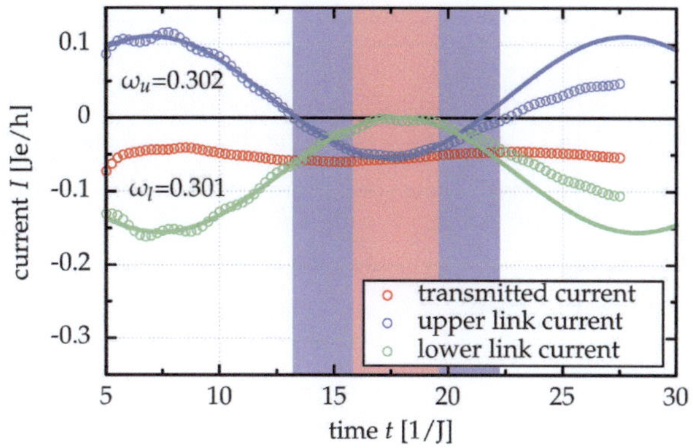

Fig. 7 Time-dependent currents of the quadratic ring system for $U = 1.0\,J$. The image shows the measured currents as a function of time t for a system with $M = 72$ sites and parameters $\epsilon_T = 0.5\,\text{J}$ and $V_{\text{SD}} = 0.4\,\text{J/e}$. The currents are labelled according to Fig. 4. The oscillation frequency of the upper ring current $\omega_u = 0.302$ and the frequency of the lower ring current $\omega_l = 0.301$ at times $t \geq 5\,\text{J}^{-1}$ are noticeably smaller than the bias voltage $V_{\text{SD}} = 0.4\,\text{J/e}$. The oscillation of the ring currents is therefore assumed to be a novel effect, that is not caused by the finite system size. The transmitted current still exhibits an oscillation with frequency $\omega \simeq V_{\text{SD}}$. This remaining oscillation of the transmitted current corresponds to the familiar finite size effect

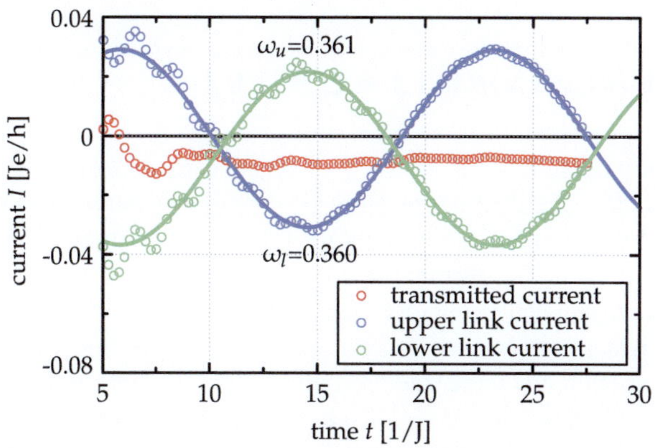

Fig. 8 Time-dependent currents of the quadratic ring system for $U = 2.0\,J$. The plot shows the measured currents as a function of time t for a system with $M = 72$ sites and parameters $\epsilon_T = 0.5\,\text{J}$ and $V_{\text{SD}} = 0.4\,\text{J/e}$. The currents are labelled according to Fig. 4. The oscillation frequency of the upper ring current $\omega_u = 0.361$ and the frequency of the lower ring current $\omega_l = 0.360$ at times $t \geq 5\,\text{J}^{-1}$ are not equal to the bias voltage $V_{\text{SD}} = 0.4\,\text{J/e}$. The oscillation of the ring currents is assumed to be a novel effect, that is not caused by the finite system size. The oscillation of the transmitted current with frequency $\omega \simeq V_{\text{SD}}$ has become barely noticeable

Calculations for different bias voltages yield qualitatively identical results so that it can be concluded that the studied currents in the system reliably relax to the steady state in the regime of weak interaction.

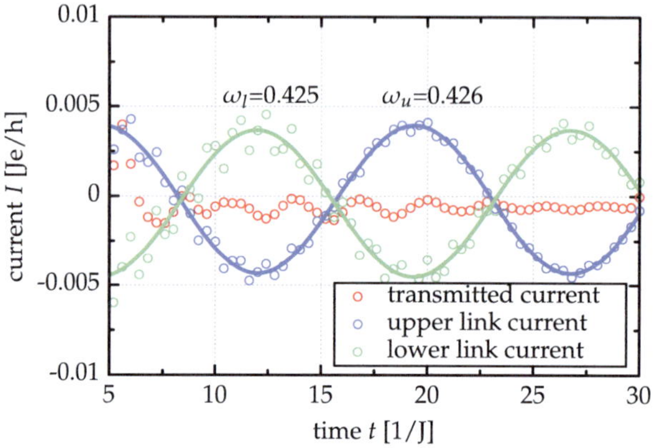

Fig. 9 Time-dependent currents of the quadratic ring system for $U = 4.0\,J$. The figure displays the measured currents as a function of time t for a system with $M = 72$ sites and parameters $\epsilon_T = 0.5\,\mathrm{J}$ and $V_{\mathrm{SD}} = 0.4\,\mathrm{J/e}$. The currents are labelled according to Fig. 4. The oscillation frequency of the upper ring current $\omega_u = 0.426$ and the frequency of the lower ring current $\omega_l = 0.425$ at times $t \geq 5\,\mathrm{J}^{-1}$ have now become larger than to the bias voltage $V_{\mathrm{SD}} = 0.4\,\mathrm{J/e}$. The oscillation of the ring currents is still assumed to be a novel effect, that is not caused by the finite system size. An oscillation of the transmitted current with frequency $\omega \simeq V_{\mathrm{SD}}$ no longer visible. The oscillation artifacts seems to be increasingly suppressed with growing interaction strength U

7.2 *Strong Interaction* $0.1\,\mathrm{J} < U \leq 1.0\,\mathrm{J}$

For interaction strengths $0.1\,\mathrm{J} < U \leq 1.0\,\mathrm{J}$, one recognizes a behavior of the currents that is increasingly different from the case of vanishing interaction. In Fig. 6 we display the time-dependent currents for interaction strength $U = 0.5\,\mathrm{J}$ and in Fig. 7 for $U = 1.0\,\mathrm{J}$. The former shows a significant decrease of the amplitude of the finite size induced oscillations and the appearance of a single oscillation period of deviant frequency for times $t < 15\,\mathrm{J}^{-1}$ in the ring currents only. For increasing interaction, the finite size oscillation becomes more and more suppressed and effectively vanishes for $U = 1.0\,\mathrm{J}$. Meanwhile, another oscillation with a relatively large amplitude and frequency $\omega \neq V_{\mathrm{SD}}$ becomes the dominant feature of the ring currents. Calculations for different bias voltages show that the frequency of this oscillation is completely independent of the applied bias. The amplitude of the oscillation is also large enough for the upper link current to temporarily change direction. From Fig. 7, one can deduce the dynamics of the currents in the ring structure to be as follows:

1. $t < 13\,\mathrm{J}^{-1}$: A relatively large current flows through the lower half of the ring from the right side to left side of the ring structure. At the left ring site only about one third of the electrons leaves the ring structure and tunnels into the lead while the other two thirds flow from the left ring site to the top site of the ring.

2. $13\,\mathrm{J}^{-1} \lesssim t < 16\,\mathrm{J}^{-1}$: Two small currents flow from the top and bottom sites of the ring structure towards the left ring site from where the current flows into the lead.
3. $16\,\mathrm{J}^{-1} \leq t \lesssim 20\,\mathrm{J}^{-1}$: A current is only flowing from the top site to the left site. The current from the left ring site into the leads is equivalent to the upper link current. At $t = 20\,\mathrm{J}^{-1}$ the systems changes back to the behavior of the previous time frame and is subsequently oscillating between the states 1 and 3, with 2 being the intermediate state of the oscillation.

7.3 Very Strong Interaction $U > 1.0\,\mathrm{J}$

In the regime of very strong interaction, we find a continued establishment of the ring current oscillations that we described in the previous subsection. Figures 8 and 9 present example calculations for interaction strengths of $U = 2.0\,\mathrm{J}$ and $U = 4.0\,\mathrm{J}$ where the ring currents show distinct oscillations with amplitudes several times larger than the mean transmitted current. The oscillation that is caused by the finite system size and that could be observed in the transmitted current for weaker interaction is now entirely suppressed.

For the frequency of the ring current oscillation, we observe an increase with increasing interaction strength while it remains independent of the bias voltage. A variation of the gate potential ϵ_T results in a proportional variation of the oscillation frequency. In contrast to the regime of strong interaction we do not find a noticeable decay of the amplitude of the ring current oscillations anymore. The upper and the lower ring current are in anti phase and both now change direction periodically. The dynamics of the currents in the ring structure, described in Sect. 7, is therefore modified to be:

1. A relatively large current flows through the lower half of the ring from the right side to left side of the ring structure. At the left ring site only about one third of the electrons leaves the ring structure and tunnels into lead while the other two thirds flow from the left ring site to the top site of the ring.
2. Two small currents flow from the top and bottom sites of the ring structure towards the left ring site from where the current flows into the lead.
3. A large current is now flowing from the top site of the ring structure to the left site. Here only a fraction of the electrons tunnel into the lead while the majority is flowing from the left ring site to the bottom site. The system periodically oscillates back and forth between states 1 and 3, with state 2 being the intermediate state of the oscillation.

We find that strong repulsive interaction leads to a qualitatively different dynamical behavior of the currents in the system, especially the currents in the ring structure. For the ring currents we find an oscillation that becomes the dominant feature with increasing interaction strength.

At least for very strong interaction, the oscillation of the ring currents does not decay on the time scales accessible to our calculations. Because of this permanent oscillation one could suspect that the currents of the system might not relax to the steady state for interaction strengths $U > 1.0\,\mathrm{J}$.

At this point, we would like to remind the reader that the primary intention of this work is to examine whether the local ring currents in the chosen model system demonstrably remain in a transient state or eventually relax to a steady state. The observation of a permanent transient state would effectively falsify the "steady state relaxation assumption" that is made when the Landauer formula [14, 15] is employed. This Landauer formula is widely used to calculate electronic properties of materials, particularly molecular materials.

The results we have shown so far suggest that the quadratic interacting ring structure might already fit the bill. However, one first has to determine the origin of the seemingly permanent oscillation of the ring currents in order to confirm that the quadratic ring structure has indeed the desired properties.

Since oscillating currents are already known as a consequence of finite system sizes it stands to reason that the finite system size may also be the cause of the newly found ring current oscillation. We have therefore performed additional calculations in which a potential origination from finite system size can be unveiled. In a first attempt we vary the system size and search for potential shifts in frequency, amplitude or phase of the oscillation. A second approach makes use of damped boundary conditions, see [2, 7, 20]. Such boundary conditions of the leads reduce the energy gaps in the vicinity of the Fermi level and should also result in deviating properties of the oscillation, if it appears due to a finite level discretization.

The results, as of yet, indicate that the frequency of the ring current oscillation depends solely on the interaction strength and the on-site gate potential. From this we conclude that the oscillation is not equivalent to the known finite site oscillation artifact. It brings about the question in which way the two parameters actually influence the oscillation. This question is addressed in Sect. 9.

7.4 Transient Dynamics for Damped Boundary Conditions

In order to further rule out the finite system size as the origin of the ring current oscillation, we perform td-DMRG calculations for systems with modified leads. The last ten sites of each lead are coupled by exponentially decreasing tunneling elements. This is known as damped boundary conditions (DBC) and is explained in detail in [2, 7, 20]. The purpose of the damped boundary conditions is a reduction of the energy gap at the Fermi level. This energy gap is responsible for several finite size effects such as the additional oscillations on top of the steady state [7]. The modification of an effect in response to damped boundary conditions is a good indicator for a finite size effect. Damped boundary conditions have a significant drawback though. They drastically reduce the time τ_T, which is the time before reflected wave packages return to the interacting structure. The size of the system

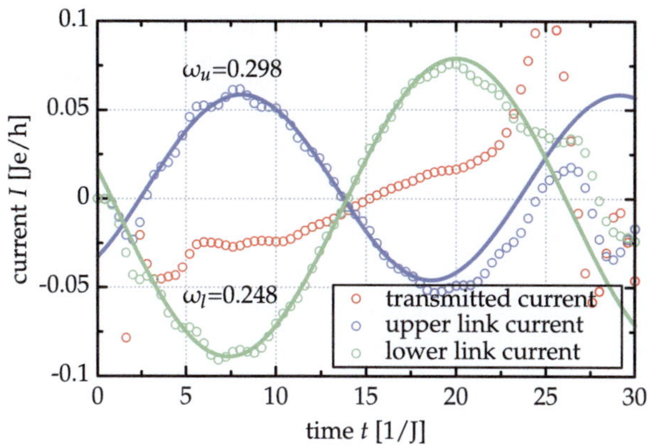

Fig. 10 Time-dependent currents of the quadratic ring system with damped boundary condition for $U = 1.0J$. The image shows the measured currents as a function of time t for a system with parameters $\epsilon_T = 0.5\,\mathrm{J}$ and $V_{\mathrm{SD}} = 0.4\,\mathrm{J/e}$. The size of the system is originally $M = 72$ sites. Because of the damped boundary conditions the effective system size is $M \simeq 52$ sites. Wave packages already get reflected at the first link with a decreased tunnel matrix element $J_{\mathrm{DBC}} = \Lambda^n J$, which result in a significantly smaller transit time $\tau_T \approx 0.25\,\mathrm{J}^{-1}$. The currents are labelled according to Fig. 4. The oscillation frequency of the upper ring current $\omega_u = 0.298$ and the frequency of the lower ring current $\omega_l = 0.248$ at times $5\,\mathrm{J}^{-1} \leq t \leq 20\,\mathrm{J}^{-1}$ are noticeably smaller than the bias voltage $V_{\mathrm{SD}} = 0.4\,\mathrm{J/e}$. The oscillations persist despite the fine energy resolution at the Fermi level due to the damped boundary conditions. This suggests, that the oscillation of the ring currents is not a finite size effect

for which we illustrate the results in Fig. 10 is thus effectively reduced from $M = 72$ sites to $M = 52$ sites resulting in a time $\tau_T = \frac{52}{2\mathrm{J}} = 26\,\mathrm{J}^{-1}$.

Regular td-DMRG calculations were performed for a system of $M = 72$ sites where the tunneling matrix element of the ten outermost sites of both leads was set to $J_{\mathrm{DBC}} = \Lambda^n \cdot J$. We choose $\Lambda = 0.5$ and $n = 1 \ldots 10$, were $n = 10$ for the last site of each lead respectively. For the other parameters of the system a set of values was chosen for which we have observed ring current oscillations in prior calculations.

In Fig. 10 we show the results of a calculation, that exemplifies the set of calculations employing damped boundary conditions. For parameters for which an oscillation of the ring currents can be found in a regular system one also finds these oscillations for a system modified by damped boundary conditions. By comparing Figs. 7 and 10 one however discovers that the amplitude of the oscillation is approximately 40 % smaller for the system with damped boundary conditions. Although this might be a hint for a finite size effect, calculations for regular systems with $M = 96$ and $M = 150$ sites show no reduction in the amplitude, where one would expect a reduced amplitude by 25 % and 52 % for an oscillation caused by the finite system size.

From the calculations employing damped boundary conditions one can again conclude that the oscillation of the ring currents is not related to the familiar finite size oscillation artifacts. They neither depend on the bias voltage V_{SD} nor do they decay proportional to M^{-1}. Considering our intention of finding a system that remains in a permanent transient state, this oscillation of the ring currents proves to be quite promising.

8 Limit of Long Time

So far, we have found the ring structure model to be a promising candidate for a model system, for which the relaxation of local observables to the steady state occurs at times $t \to \infty$. The presented calculations have mainly been aimed at studying a wide range of system parameters, targeting only system sizes of $M \simeq 72$ sites and times $t \leq 30\,\mathrm{J}^{-1}$. Short time frames do not allow to deduce whether the oscillation of the ring currents and transmitted currents is actually permanent or decays after $t = 30\,\mathrm{J}$. We have therefore performed a second series of calculations to specifically target longer simulated times.

A calculation of longer times needs to go hand in hand with the calculation of larger system sizes. This was pointed out in Sect. 5 and means an enormous increase of computer time for a small increase in simulated time. Therefore only few calculations have been done to explore the long time limit in a first series of calculations, that was performed before access to the ForHLRI computer cluster was obtained. The chosen system size for these first calculations has been $M = 96$ system sites, resulting in a transit time $\tau_T \simeq 45\,\mathrm{J}^{-1}$. Since this is a rather small increase in simulated time, some additional measures have been taken to obtain information about even longer times. The system parameters were chosen such that the frequency of the ring current oscillations is large enough to observe several oscillation periods but small enough that a wave contains sufficient data points to properly determine amplitude of the oscillation. In Fig. 11 we show the results of a calculation for quadratic ring structure using the parameters $U = 4.0\,\mathrm{J}$ and $\epsilon_T = 0.5\,\mathrm{J}$. For the chosen parameters, we obtain an oscillation frequency of $\omega \simeq 0.43\,\mathrm{J}$, which meets the requirements.

In conjunction with the fit of a cosine to both ring currents one can estimate from the amplitude of the data points whether and how fast the oscillation decays. A close look at Fig. 11 reveals that all data points lie on the fitted cosine function for times $30\,\mathrm{J}^{-1} \leq t \leq \tau_T$. One cannot recognize a decay in amplitude, neither exponential nor linear. Instead one observes a more clean oscillation with progressive time, suggesting that other transient effects have already decayed on the calculated time scale. The decay of this ring current oscillation is therefore either not taking place at all or extremely slow. More recent calculations for system sizes of $M = 120$ sites confirm this finding.

Our results do not completely rule out an eventual relaxation of the local currents inside the interacting ring structure to a local steady state, but they strongly suggest that a relaxation is at the very least taking place on time scales orders of magnitude larger than the time we can simulate using td-DMRG.

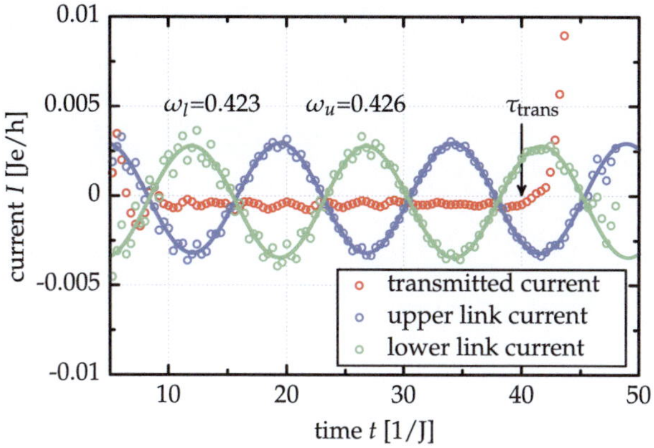

Fig. 11 Limit of long time for the quadratic ring system and $U = 4.0J$. The image displays the measured currents as a function of time t for a system with $M = 72$ sites and parameters $\epsilon_T = 0.5\,\mathrm{J}$ and $V_{\mathrm{SD}} = 0.4\,\mathrm{J/e}$. The currents are labelled according to Fig. 4. The oscillation frequency ring currents $\omega_u \simeq 0.43$ is chosen such that sufficiently many half-waves fit into the time frame $5\,\mathrm{J}^{-1} \leq t \leq 50\,\mathrm{J}^{-1}$ while being small enough that the amplitude of the half-waves can be clearly resolved. The oscillation amplitude of the ring currents is not noticeably decaying for times $t \leq t_{\mathrm{trans}} \equiv \tau_T$. A decay of the amplitudes is therefore supposed to happen at times scale that are orders of magnitude larger than the time, that can be simulated with td-DMRG. Our results do not exclude, that the ring current oscillation is not decaying at all

9 Study of the Uncoupled Interacting Structure

The td-DMRG calculations for the quadratic ring structure show that the frequency of the ring current oscillation is independent of the applied bias voltage and rather depends on the interaction strength U. They are particularly depending on the size of the gate potential ϵ_T. As can be seen from Figs. 6 and 7 one also needs a large interaction strength to observe the ring current oscillation for the quadratic structure in the first place. This motivates the investigation of the isolated interacting ring structure to determine how the interaction influences the eigenstates of the ring structure. A special focus of this investigation is hereby on the spectrum of the ring structure as a function of the interaction strength. Energy differences in the spectrum, that are comparable to the frequency of the oscillation, could hint at states, that are involved in the oscillation process. If such energy differences are found in the spectrum, one can then check in td-DMRG calculations if the oscillation is indeed connected to the corresponding eigenstates of the ring.

In order to obtain the spectrum, we construct the Hamiltonian of the ring structure in the complete many-body Hilbert space basis. A complete diagonalization of the Hamiltonian matrix is performed to obtain each energy eigenstate and eigenvector. We then repeat this calculation for a wide range of interaction strengths and various gate potentials. For selected eigenstates of the system, we calculate the expectation value of n_x, the particle density on the sites in the ring. From this local particle density we get additional insight into the spatial structure of the eigenstates.

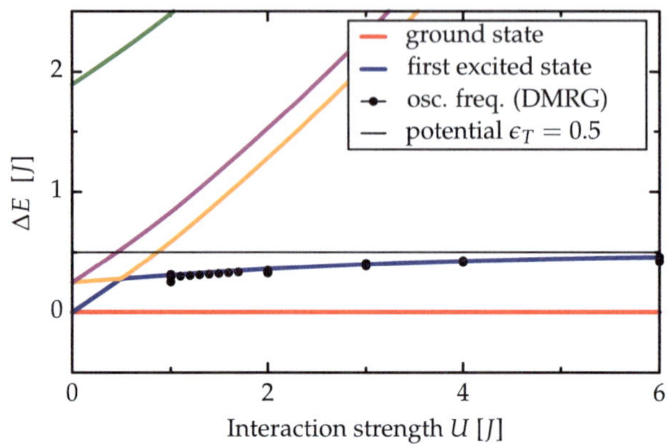

Fig. 12 Low energy spectrum of the uncoupled interacting quadratic ring structure. The plot shows the lowest energy eigenvalues of the uncoupled quadratic ring as a function of the interaction strength U for $\epsilon_T = 0.5$ J. The ground state energy E_0 is subtracted from each energy eigenvalue so that $\Delta E = E - E_0$. The *black data points* indicate the oscillation frequency obtained by td-DMRG for a given value of U. The oscillation frequencies agree with the energy difference between the ground state and the particular excited state. This is also the lowest excited state for interaction strength $U \geq 0.5$ J. The excited state corresponds to the electron density shown in Fig. 13b

By subtracting the value of the ground state energy from each eigenvalue of the quadratic ring structure one arrives at the spectrum that is depicted in Fig. 12. For $U = 0$, one finds a twofold degeneracy of the ground state energy and the first excited level. A finite interaction strength lifts both degeneracies. A further increase in interaction strength leads to a steep growth in energy for respectively one of each of the previously degenerate states. One particular excited state is however only slowly growing in energy as a function of U and becomes the lowest exited state for $U > 0.5$ J. This dependence on interaction strength is reminiscent of the frequency of the ring current oscillation. A comparison of the two does indeed show a good agreement, indicating that this state is the main contributor to the oscillation phenomenon. The energy of the excited state asymptotically approaches a value $\Delta E = 0.5\,\mathrm{J} = \epsilon_T$ for $U \to \infty$ and is $E = 0.25\,\mathrm{J} = \frac{\epsilon_T}{2}$ in the limit of vanishing interaction. From this we conclude that the ring site, to which the gate voltage is applied, is completely empty in the ground state and completely occupied in the particular excited state for $U \to \infty$. In Fig. 13 we show the local electron density on the sites of the ring for an interaction strength $U = 2.0$ J and a gate potential $\epsilon_T = 0.5$ J for four eigenstates of the quadratic ring structure with the lowest energy. One can see that the two lowest energies correspond to states that represent a charge density wave. For the ground state one discovers, that the left and right ring sites are filled while the top and bottom sites are empty. The lowest excited state can be described by the opposite picture. Now the top and bottom ring sites are filled while the other two sites are empty. The other two eigenstates comprise a different number of electrons in the system. With increasing interaction, a half-filling of the system

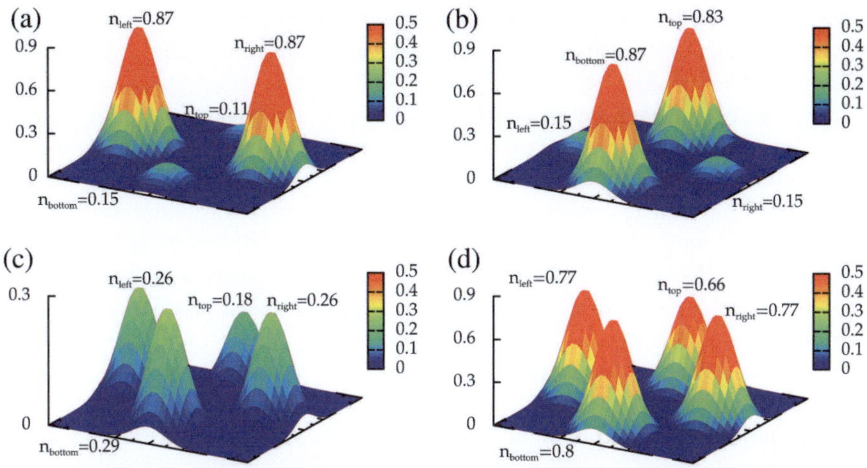

Fig. 13 Local electron density on the sites of the uncoupled interacting quadratic ring structure. The images display the local electron density on the sites of the quadratic ring structure for $U = 2.0\,\mathrm{J}$ and $\epsilon_T = 0.5\,\mathrm{J}$. The local particle density is given as the expectation value $\langle \hat{n}_x \rangle$ of the eigenstates corresponding to the four lowest energy eigenvalues. The two lowest energies are characterized by a particle density wave shown in (**a**) and (**b**). The ground state (**a**) is the particle density wave with small electron density on the site of the gate potential ϵ_T. In the state (**c**) the ring is occupied by a single electron and in state (**d**) by three electrons. States (**c**) and (**d**) become energetically less favorable compared to (**a**) and (**b**) for increasing interaction strength U

becomes increasingly favorable, explaining why these two state are far higher in energy as compared to the ground state of the system for strong interaction.

From the spectrum of the interacting ring structure we have identified the two states that are likely to be involved in the ring current oscillation. The ground state of the system corresponds to a density wave with a small local electron density on several ring sites. It thus seems reasonable that at least one other state is involved in the charge transport through the ring structure. If this state was the particular excited state, an oscillation between the ground state and the excited state might be the cause for the ring current oscillation. We have therefore performed an additional series of td-DMRG calculations, where the time-dependent reduced density matrix of the ring structures has been measured. From these calculations, one can determine whether such an oscillation between the eigenstates takes place.

10 Calculation of the Reduced Density Matrix of the Interacting Structure

Studying the uncoupled rings we have seen that two particular low-lying eigenstates of the ring structures have an energy difference that coincides with the frequency of the ring current oscillation. We are consequently interested in how these two eigenstates contribute to the oscillation phenomenon. To this end we perform td-DMRG calculations in which the time-dependent probability of the eigenstates of the ring is determined.

One can obtain the time-dependent probability of the quantum state in the ring from the reduced density matrix of the ring structure. This reduced density matrix is calculated by tracing out the basis states of the leads from the density matrix $\rho = |\psi_0\rangle\langle\psi_0|$ that describes the pure quantum state of the entire system. By extracting the probability of each eigenstate of the ring structures and analyzing its time-dependency, we try to determine whether the eigenstates, that we have identified in the spectrum of the uncoupled ring, are indeed involved in the ring current oscillation.

The time-dependent probability of the ground state and the one particular excited state is shown in Fig. 14. From it, we find a probability of the ground state, that is close to unity. The contribution of the other eigenstates of the ring structure to the global ground state is consequently orders of magnitude smaller. The particularly interesting excited state has a probability $\sim 10^{-4}$ while the other two eigenstates, whose electron density is shown in Fig. 14, have a mean probability of $\sim 10^{-3}$.

For the study of the time-dependency of the probabilities, we distinguish two cases, namely the case in which the applied bias voltage is $V_{\mathrm{SD}} \leq \epsilon_T/e$ and the case in which $V_{\mathrm{SD}} > \epsilon_T/e$. The former is depicted in Fig. 14a and the latter in Fig. 14b, respectively.

10.1 $V_{\mathrm{SD}} \leq \epsilon_T/e$

In Fig. 14a we observe an oscillation of both the ground state and the first excited state. However both oscillations differ in frequency. The ground state oscillation frequency is equal to the bias voltage. We thus conclude that this oscillation is due to the finite size effect discussed in [7]. The excited state oscillates with the frequency, that we expect from our calculation for the uncoupled ring, and is equal to the frequency of the ring current oscillation. This is another indicator that said excited state contributes to the ring current oscillation. All other eigenstates of the uncoupled ring have been studied in the same fashion. Only few of them exhibit an oscillatory behavior but none has a frequency that matches the frequency of the ring current oscillation. The most notable of the other eigenstates is the one that corresponds to Fig. 13c. It also oscillats with frequency $\omega = V_{\mathrm{SD}}$ having the same amplitude as the ground state and a phase shift of π compared to the ground state.

10.2 $V_{\mathrm{SD}} > \epsilon_T/e$

The probability of the ground state and the particular excited state as a function of time changes significantly for bias voltages larger then the on-site gate potential.

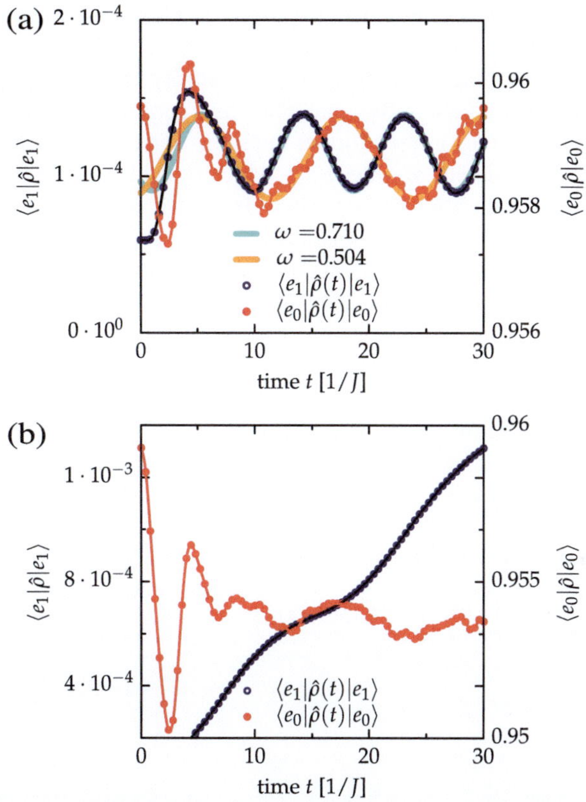

Fig. 14 Time-dependent reduced density matrices of the interacting ring structures. The figures show the probability of the ground state (*right axis*) and the first excited state (*left axis*) as a function of time. Figures (**a**) and (**b**) picture the probabilities for the quadratic ring structure ($M = 72$ sites) in the case (**a**) $V_{SD} \leq \epsilon_T/e$ and (**b**) $V_{SD} > \epsilon_T/e$. In (**a**) we have chosen $\epsilon_T = 1.0\,\mathrm{J}$ and $U = 2.0\,\mathrm{J}$ for which we find an oscillation frequency of the ring currents $\omega \approx 0.7$. The oscillation frequency of the probability of the first excited state has the same frequency. This indicates that the particular excited state is involved in the oscillation effect. The ground state probability oscillates with $\omega = V_{SD}$. In (**b**) we find a similar behavior of the ground state probability and the excited state probability for the hexagonal ring structure and parameters $U = 2.0\,\mathrm{J}$ and $\epsilon_T = 0.5\,\mathrm{J}$. (**b**) shows an increasing probability of the excited state modulated by a frequency ω that does not match the ring current oscillation frequency for $U = 2.0\,\mathrm{J}$ and $\epsilon_T = 0.5\,\mathrm{J}$

One can no longer observe a distinct oscillation of the ground state with a frequency $\omega = V_{SD}$ or any other frequency. The probability of the first excited state as a function of time is now qualitatively different as well. The probability now increases seemingly linearly and an oscillation is solely modulated onto this linear function. The frequency of this oscillation does also not match frequency of the oscillation of the ring currents. Due to its linear growth the probability of the excited states now reaches values of up to 10^{-3} whereas the probability of the ground state is slightly smaller than before. When examining the other eigenstates of the uncoupled quadratic ring structure we find none that oscillate with the same frequency as the ring currents.

The study of the time-dependent reduced density matrix also points to the particular excited state as a substantial state that contributes to the ring current oscillation. However it also raises further questions. The occupation probability of the particular excited state is of order 10^{-4}, which is small considering that the ring currents have oscillation amplitudes $\sim 10^{-2}\,\mathrm{eJ/h}$. A time evolution calculation for the uncoupled ring using exact diagonalization and the occupation probabilities of the ground state and the one excited state from the reduced density matrix yields ring current oscillation that possess the right frequency and phase but only amplitudes of order $10^{-5}\,\mathrm{eJ/h}$. This discrepancy has yet to be understood. A second question concerns the probability of the particular excited for bias voltages larger than the on-site gate potential ϵ_T. The occupation probability increases monotonously while the ring current oscillation retains the behavior seen for smaller voltages. This is also not consistent with an explanation that assumes a switching between ground state and excited state as the cause of the ring current oscillation.

Acknowledgements This work was performed on the computational resource ForHLR I funded by the Ministry of Science, Research and the Arts Baden-Württemberg and DFG ("Deutsche Forschungsgemeinschaft") within project QWHISTLE.

References

1. Bohr, D., Schmitteckert, P.: The dark side of benzene: interference vs. interaction. Ann. Phys. **524**(3–4), 199–204 (2012)
2. Bohr, D., Schmitteckert, P., Wölfle, P.: Dmrg evaluation of the kubo formula – conductance of strongly interacting quantum systems. Europhys. Lett. **73**, 246 (2006)
3. Bohr, D., Schmitteckert, P.: Strong enhancement of transport by interaction on contact links. Phys. Rev. B **75**(24), 241103(R) (2007)
4. Boulat, E., Saleur, H., Schmitteckert, P.: Twofold Advance in the Theoretical Understanding of Far-From-Equilibrium Properties of Interacting Nanostructures. Phys. Rev. Lett. **101**(14), 140601 (2008)
5. Branschädel, A., Boulat, E., Saleur, H., Schmitteckert, P.: Numerical evaluation of shot noise using real-time simulations. Phys. Rev. B **82**, 205414 (2010)
6. Branschädel, A., Boulat, E., Saleur, H., Schmitteckert, P.: Shot noise in the self-dual interacting resonant level model. Phys. Rev. Lett. **105**, 146805 (2010)
7. Branschädel, A., Schneider, G., Schmitteckert, P.: Conductance of inhomogeneous systems: real-time dynamics. Ann. Phys. **522**(9), 657–678 (2010)
8. Branschädel, A., Schmitteckert, P.: Conductance of correlated nanostructures. In: High Performance Computing in Science and Engineering'10. Springer, Berlin (2010)
9. Branschädel, A., Ulbricht, T., Schmitteckert, P.: Conductance of correlated nanostructures. In: Nagel, W.E., Kröner, D.B., Resch, M. (eds.) High Performance Computing in Science and Engineering'09, pp. 123–137. Springer, Berlin (2009)
10. Carr, S.T., Bagrets, D.A., Schmitteckert, P.: Full counting statistics in the self-dual interacting resonant level model. Phys. Rev. Lett. **107**(20), 206801 (2011)
11. Carr, S.T., Schmitteckert, P., Saleur, H.: Transport through nanostructures: finite time vs. finite size. Phys. Rev. B **89**, 081401 (2014)

12. Carr, S.T., Schmitteckert, P., Saleur, H.: Full counting statistics in the not-so-long-time limit. Phys. Scr. T **165**, 014009 (2015)
13. Hallberg, K.A.: New trends in density matrix renormalization. Adv. Phys. **55**(5–6), 477–526 (2006)
14. Landauer, R.: Spatial variation of currents and fields due to localized scatterers in metallic conduction. IBM J. Res. Dev. **1**(3), 223–231 (1957)
15. Meir, Y., Wingreen, N.S.: Landauer formula for the current through an interacting electron region. Phys. Rev. Lett. **68**(16), 2512–2515 (1992)
16. Noack, R.M., Manmana, S.R.: Diagonalization- and numerical renormalization-group-based methods for interacting quantum systems. AIP Conf. Proc. **789**, 93–163. AIP Publishing (2005)
17. Peschel, I., Wang, X., Kaulke, M., Hallberg, K. (eds.): Density Matrix Renormalization – A New Numerical Method in Physics. Springer, Berlin (1999)
18. Schmitteckert, P.: Nonequilibrium electron transport using the density matrix renormalization group method. Phys. Rev. B **70**(12), 121302 (2004)
19. Schmitteckert, P.: Signal transport in and conductance of correlated nanostructures. In: Nagel, W.E., Kröner, D.B., Resch, M. (eds.) High Performance Computing in Science and Engineering'07, pp. 99–106. Springer, Berlin (2007)
20. Schmitteckert, P.: Calculating Green functions from finite systems. J. Phys. Conf. Ser. **220**, 012022 (2010)
21. Schmitteckert, P.: Obtaining the full counting statistics of correlated nanostructures from time dependent simulations. In: High Performance Computing in Science and Engineering'11. Springer, Berlin (2011)
22. Schmitteckert, P., Schneider, G.: Signal transport and finite bias conductance in and through correlated nanostructures. In: Nagel, W.E., Jäger, W., Resch, M. (eds.) High Performance Computing in Science and Engineering'06, pp. 113–126. Springer, Berlin (2006)
23. Schollwöck, U.: The density-matrix renormalization group. Rev. Mod. Phys. **77**(1), 259–315 (2005)
24. Ulbricht, T., Schmitteckert, P.: Signal transport in and conductance of correlated nanostructures. In: Nagel, W.E., Kröner, D.B., Resch, M. (eds.) High Performance Computing in Science and Engineering'08, pp. 71–82. Springer, Berlin (2008)
25. Walz, M., Wilhelm, J., Evers, F.: Current patterns and orbital magnetism in mesoscopic dc transport. Phys. Rev. Lett. **113**(13), 136602 (2014)
26. White, S.R.: Density matrix formulation for quantum renormalization groups. Phys. Rev. Lett. **69**(19), 2863–2866 (1992)
27. White, S.R.: Density-matrix algorithms for quantum renormalization groups. Phys. Rev. B **48**(14), 10345–10356 (1993)

Part III
Reactive Flows

Dietmar Kröner

The four contributions in the section "Reactive Flows" indicate that the numerical simulations of reactive flows could be more improved concerning the accuracy, the efficiency, the parallelization and the scalability. The first two papers and the last one are based on the OpenFOAM software and the third one on the in–house code TASCOM3D. All four projects were supported by the German Research Council (DFG).

In the first contribution about "DNS Analysis of the Correlation of Heat Release Rate with Chemiluminescence Emissions in Turbulent Combustion" by F. Zhang, T. Zirwes, P. Habisreuther and H.Bockhorn the authors perform DNS computations of the methane-air combustion in turbulent flow, modeled by the compressible Navier-Stokes equations with gravity and an additional equation for species transport, diffusion and reaction. The pressure is given by the ideal gas law, dynamic viscosity and heat conductivity seem to be constant. The chemical reaction mechanism consists of 18 species and 69 fundamental reactions, containing the optically active OH radical. The main goal was the investigation of the correlation between heat release rate and the luminescent species in turbulent flames. The implementation uses the open source software OpenFOAM for the CFD part and Cantera for the chemical reaction. Aim of this work is the validation of a correlation between the presence of the OH radical, which can be measured optically, and the heat release in the chemical reaction. Such a correlation is assumed in practice for the technical optimization of combustion chambers.

The underlying grid for the numerical simulations contains 16 million finite volumes and the parallel implementation uses 3,600 processor cores from the Hazel Hen cluster.

D. Kröner (✉)
Abteilung für Angewandte Mathematik, Universität Freiburg, Hermann-Herder-Str. 10, 79104 Freiburg, Germany
e-mail: dietmar.kroener@mathematik.uni-freiburg.de

In the second contribution about "Direct Numerical Simulation of Non-Premixed Syngas Combustion using OpenFOAM" by S. Vo, A. Kronenburg, O.T. Stein and E.R. Hawkes a benchmark setting of turbulent non-premixed syngas combustion is simulated with the OpenFOAM software package with additional functionality provided by another group. Different grid resolutions and different models of the species diffusion are used. The results are compared to benchmark data gained from a dedicated high order DNS solver, to study the effects of the lower order discretization provided by OpenFOAM. The results are provided in a concise manner and the computations performed with OpenFOAM are in good agreement with the benchmark of the more specialized code. The contribution proves that OpenFOAM can be used for direct numerical simulations. However it turns out that OpenFOAM's low order discretization schemes are likely to affect simulations with different set-ups. Weak and strong scaling tests are performed in order to analyse the parallel performance of the OpenFOAM solver on the Hazel Hen architecture.

In the third contribution about "Numerical Simulations of Rocket Combustion Chambers with Supercritical Injection" by M. Seidl, R.Keller, P. Gerlinger, and M. Aigner a number of improvements to the compressible, implicit CFD solver TASCOM3D are described and numerical results for the simulation of rocket combustion chamber are presented. The code is validated by nonreactive and reactive benchmark tests at high pressures. The authors consider two simulations, a non-reactive cryogenic nitrogen jet dissolving into a warm nitrogen surrounding and a reactive simulation in a model rocket combustor. The simulation results matched experimental observations very well in a qualitative and quantitative manner.

In the fourth contribution about "Two-Zone Fluidized Bed Reactors for Butadiene Production: A Multiphysical Approach with Solver Coupling for Supercomputing Application" by M. Hettel, J. A. Denev and O. Deutschmann the numerical modelling of a two-zone fluidized bed reactor in a laboratory-scale for production of the basic chemical 1,3-butadiene from n-butane is considered.

The final aim of the project is to model the complex interaction of all relevant processes including the gas-phase flow field, the movement of solid particles, the heterogeneously catalyzed reactions on the inner particle surface and the intra-particle transport phenomena. The main parts of the mathematical model are the compressible Navier Stokes, transport equations for the species and Newton's equations for the particles.

The authors used the CFDEM coupling software which couples the DEM engine LIGGGHTS to the open source CFD code within OpenFoam. The computations were performed on the research cluster of the state of Baden-Württemberg JUSTUS, which is located in Ulm and on ForHLR-I, ForHLR-II.

The limitations conserning the physical modelling, the software implementation, and the architecture or combinations of the supercomputers are discussed.

A DNS Analysis of the Correlation of Heat Release Rate with Chemiluminescence Emissions in Turbulent Combustion

Feichi Zhang, Thorsten Zirwes, Peter Habisreuther, and Henning Bockhorn

Abstract The essential correlation of heat release rate and chemiluminescence emission from turbulent combustion is quantitatively analyzed by means of direct numerical simulation (DNS) of premixed methane/air flames, employing a detailed reaction mechanism with 18 species and 69 elementary reactions, and the mixture-averaged transport method. One-dimensional freely propagating laminar flames have first been studied for different stoichiometries varying from fuel-lean to fuel-rich conditions. There, the local generation of the chemiluminescent OH* species correlates strongly with the heat released by the combustion reaction, especially in the fuel-lean range. Three-dimensional DNS have then been applied to calculate a synthetically propagating flame front subjected to different turbulent inflow conditions. Joint probability density functions of OH* concentration and heat release rate have been generated from the DNS results, showing a stronger scattering of the correlation curve compared to the corresponding laminar flame. As the chemiluminescence measurement gathers light only along one viewing direction, the line-of-sight integrated values of heat release and OH* concentration have been evaluated from the DNS, where the domain has been decomposed into a number of rays defined by a fixed viewing direction and a specific area. A quasi-linear relationship has been identified for these integral values, where the correlation becomes stronger for flames subjected to lower turbulence intensities or larger cross-section areas of the rays. A computational grid with 16 million finite volumes has been used for the DNS of the turbulent flames and the simulations have been performed in parallel with 3,600 processor cores from the Hazel Hen cluster of HLRS. Scale-up performance of the DNS code, which is based on the open-source program OpenFOAM, has been evaluated.

F. Zhang (✉) • T. Zirwes • P. Habisreuther • H. Bockhorn
Engler-Bunte-Institute, Division of Combustion Technology, Karlsruhe Institute of Technology, Engler-Bunte-Ring 1, 76131 Karlsruhe, Germany
e-mail: feichi.zhang@kit.edu

W.E. Nagel et al. (eds.), *High Performance Computing in Science and Engineering '16*, DOI 10.1007/978-3-319-47066-5_16

1 Introduction

Heat release is the major purpose of combustion processes, which can be used for heating, e.g. in heat exchangers, or converted to mechanical or electrical energy, e.g. in internal combustion engines or power plants. The rate of heat release is used to assess efficiency of the combustion process and to identify the location of the reaction zone of the flame, which is influenced by the interaction of fluid flow and chemical reactions. It is a fundamental property and of great importance for the theoretical and experimental investigation of combustion processes. Traditionally, high-speed imaging of the chemiluminescence of excited hydroxyl radicals OH* or methylidyne radicals (CH*) with intensified cameras is used to characterize the unsteady heat release in turbulent flames [1, 2]. This suffers from being a line-of-sight technique with limited capability for spatial resolution. Hence, only the integral or total heat release rate can be determined from this technique. The correlation between heat release and chemiluminescence is determined empirically in previous work [2, 3] and proportionality is commonly assumed which is not based on an understanding of the underlying transport and chemical process but rather sanctified by the obtained results. Therefore, there is a need to justify this general linear correlation of heat release and chemiluminescence emission in a more detailed way. In particular, the influence of turbulence or unsteady effect on this correlation is analyzed in this work, which has not been investigated in the literature before.

To accomplish this, direct numerical simulations (DNS) using complex reaction kinetics with the full reaction paths of the electronically excited OH* radical have been applied in the present work to simulate a synthetically propagating flame front of three-dimensions, which is perturbed artificially by different turbulent inflow conditions. The DNS relies on the numerical solutions of the governing balance equations without any simplifications. The full range of time and length scales of the turbulent flow as well as the chemical reaction system is resolved to a large extent. The fine-grained rendering of the interaction between the turbulent flow, molecular transport and complex chemistry in DNS provides greater insight and quantitative predictability, complementing measurements and less fine-grained turbulence and combustion models like Reynolds averaged Navier-Stokes (RANS) or large eddy simulation (LES) methodologies. The DNS is used in the present work to provide a quantitative statement of the correlation of heat release and chemiluminescence emission from turbulent combustion, whereas this is only qualitatively accessible in experiments.

2 Computational Methods

2.1 Governing Equations

The conservation equations for the total mass, the species masses, the momentum and the energy, together with the equation of state, constitute the basics for the detailed description of chemically reacting flows [4, 5]:

$$\frac{\partial \rho}{\partial t} = -\nabla \cdot (\rho \mathbf{v}) \tag{1}$$

$$\frac{\partial}{\partial t}(\rho \mathbf{v}) = -\nabla \cdot (\rho \mathbf{v}\mathbf{v}) - \nabla p + \nabla \cdot \boldsymbol{\tau} + \rho \mathbf{g} \tag{2}$$

$$\frac{\partial}{\partial t}(\rho Y_k) = -\nabla \cdot (\rho Y_k \mathbf{v}) - \nabla \cdot \mathbf{j}_k + \dot{r}_k, \quad k = 1 \ldots N-1 \tag{3}$$

$$\frac{\partial}{\partial t}(\rho h_s) = -\nabla \cdot (\rho h_s \mathbf{v}) - \nabla \cdot \dot{\mathbf{q}} + \frac{\mathrm{D}p}{\mathrm{D}t} + \dot{q}_r \tag{4}$$

$$p = \rho R T \tag{5}$$

where ρ and $\mathbf{v}$ are the density and velocity vector, p and T denote the static pressure and temperature and h_s the sensible enthalpy. Y_k and $\dot{r}_k$ indicate mass fraction and reaction rate of the species k. R is the specific gas constant. The gravitational force $\rho \mathbf{g}$ acts as an external force on the cell volume. The heat source from viscous dissipation and radiation are neglected.

The mixture-averaged diffusion flux $\mathbf{j}_k$, the viscous stress flux $\boldsymbol{\tau}$ for a Newtonian fluid and the diffusive heat flux $\dot{\mathbf{q}}$ are given by:

$$\mathbf{j}_k = -\rho D_{km} \nabla Y_k \tag{6}$$

$$\boldsymbol{\tau} = \mu[\nabla \mathbf{v} + (\nabla \mathbf{v})^T] - \frac{2}{3}\mu(\nabla \cdot \mathbf{v})\mathbf{I} \tag{7}$$

$$\dot{\mathbf{q}} = -\lambda \nabla T + \sum_{k=1}^{N} \mathbf{j}_k h_k \tag{8}$$

Here D_{km} is the mixture-averaged diffusion coefficient between the k-th species and the mixture, μ is the dynamic viscosity, λ is the thermal conductivity and h_k is the specific enthalpy of the k-th species. The reaction rate $\dot{r}_k$ in Eq. (3) is given by the rate law from reaction kinetics with the rate coefficient described by the extended Arrhenius law. Due to the usage of sensible enthalpies, the heat release caused by chemical reactions leads to a source term in Eq. (4):

$$\dot{q}_r = -\sum \dot{r}_k h_k^0 \tag{9}$$

where h_k^0 is the chemical enthalpy of species k.

2.2 Numerical Setups

The open-source code OpenFOAM [6] has been used to perform the three-dimensional DNS of turbulent combustion, where the detailed calculation of chemistry and molecular transport has been implemented in addition to its general capabilities for CFD modeling of non-reactive flows [7]. The code is capable of solving the compressible reactive flow Eqs. (1), (2), (3), (4), and (5) employing the finite volume method on unstructured grids. The detailed description of the chemistry, i.e. the reaction rates, and transport, i.e. the diffusion coefficients, has been accomplished by coupling with the open-source chemical kinetics library Cantera [8]. The mixture-averaged model is used in the current work for the diffusive mass flux, the viscous stress flux of a Newtonian fluid and the diffusive heat flux. A detailed reaction mechanism with 18 species and 69 fundamental reactions has been applied for premixed methane/air combustion. It consists of the reaction mechanism for methane/air combustion by Kee et al. [9] (17 species and 58 reactions) and adds the full reaction chain of the short-lived OH* radical [10] (1 species and 11 reactions). A general operator splitting technique has been used for the evaluation of chemical source terms, calculating the system of chemical reactions decoupled from the solution of the flow equations. In this case, a zero-dimensional batch reactor has been created for each discrete cell volume and the resulting kinetics equations are numerically integrated over the time step of the flow, thereby resolving the smallest time scales of the chemical reaction. The solver employs a fully implicit scheme of second order for the time derivative and a fourth order interpolation scheme for the discretization of the convective term. All diffusive terms are discretized with an unbounded scheme of fourth-order, too. The pressure-implicit split-operator (PISO) algorithm has been used for pressure correction. The reader is kindly referred to [4, 5, 11] for a detailed description of the governing equations and the numerical procedures. Informations about code validation can be found in [12, 13] and the references therein.

3 Correlation of Heat Release with Chemiluminescent Species

3.1 Local Correlation in Laminar Planar Unstrained Flames

A first numerical experiment has been performed in a one-dimensional freely propagating flame configuration, in order to assess the general local correlation between heat release and concentration of OH*. Premixed methane/air mixtures with equivalence ratios varying from lean to rich are considered. The fresh gases are burnt at an initial temperature of 300 K and 1 bar pressure. The one-dimensional flame calculations have been performed with the open-source thermo-chemical library Cantera [8]. Figure 1 compares profiles of heat release rate $\dot{q}$ and

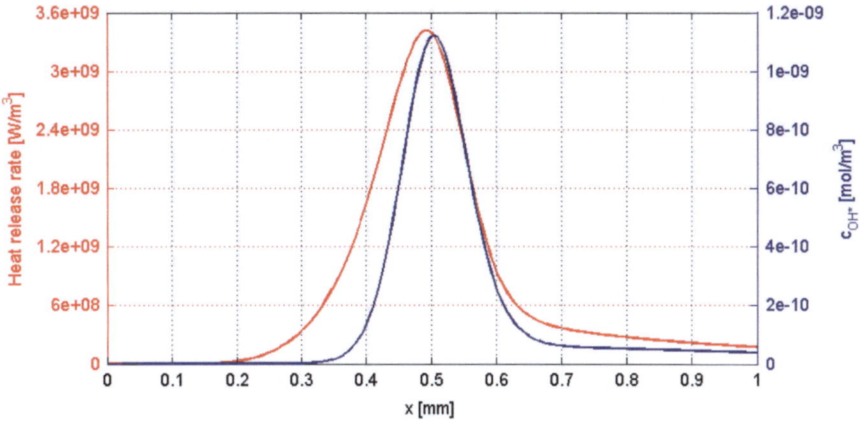

Fig. 1 Heat release rate and concentration of OH* along the flame coordinate for methane/air mixture at $\phi = 0.9$, temperature of 300 K and pressure of 1 bar

concentration of OH* c_{OH*} along the flame axis for a fixed equivalence ratio $\phi = 0.9$. It is clear that c_{OH*} starts to increase only after a considerable amount of heat has been released. c_{OH*}, however, rises more rapidly so that positions of peak values of $\dot{q}$ and c_{OH*} are very close together, with a displacement of approximately $10 - 20$ μm. Thereafter, both parameters decline sharply at a similar rate to zero. Similar results have been reported in [10] by simulations of one-dimensional methane/air flames employing the GRI-3.0 mechanism [14], where the appearance of OH* is found to be very close to the heat release location at different equivalence ratios.

Despite the fact that the evolution of c_{OH*} follows the generation of $\dot{q}$ quite well, a unique correlation between both parameters is not available. This becomes more evident when looking at Fig. 2, where local values of c_{OH*} are plotted against those of $\dot{q}$ leading to an enclosed envelope curve. The arrows in the figure show the reaction path from unburnt to burnt state. Obviously, there are generally two values of c_{OH*} assigned to a fixed $\dot{q}$ value and vice versa. On the right hand side of Fig. 2, $\dot{q}$ and c_{OH*} are scaled by their peak values and plotted against each other. The envelope curves coincide for lean flames with $\phi < 1$, indicating a similar correlation of $\dot{q}$ and c_{OH*} in this range. The correlation is attenuated for higher ϕ values, which can be identified by the increased distance between the lower and upper parts of the envelope curve, for example, by comparing the curves for $\phi = 1.1$ and $\phi = 1.2$ in Fig. 2 on the right. Although a direct proportionality cannot be observed for c_{OH*} and $\dot{q}$, the generation of OH* is strongly coupled with heat release, as shown in Figs. 1 and 2. Even a quasi-linear relationship can be identified for the upper part of the envelope curve, where c_{OH*} and $\dot{q}$ decrease from its maximum values.

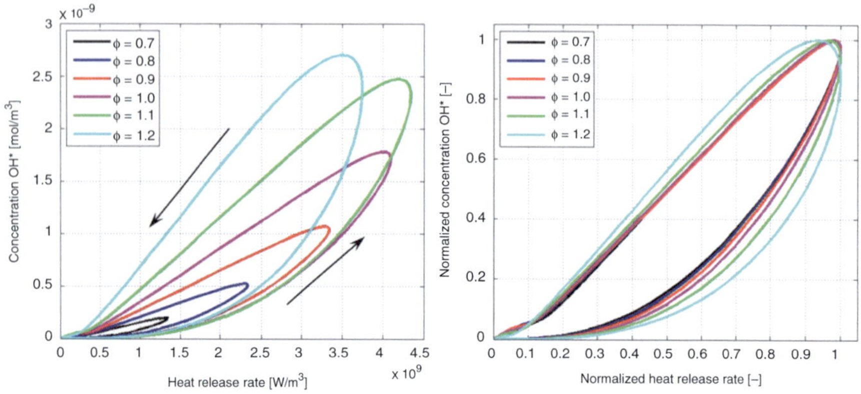

Fig. 2 Envelope curves of heat release rate and concentration of OH* obtained from one-dimensional flame calculations at different stoichiometries

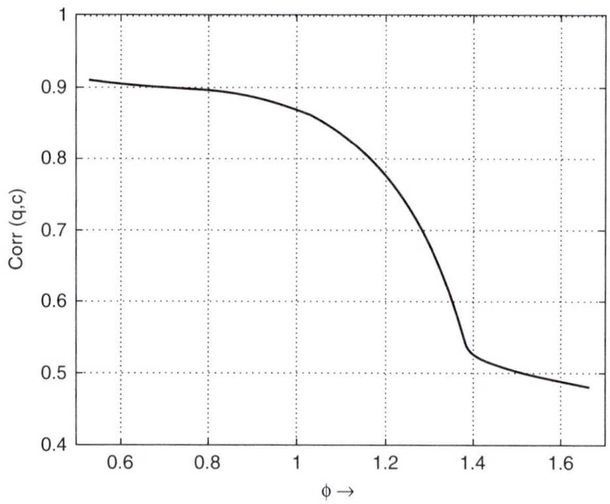

Fig. 3 Local correlation coefficient of heat release and OH* concentration

In Fig. 3, the correlation coefficients R are evaluated from data pairs of c_{OH^*} and $\dot{q}$ for different equivalence ratios, which are particularly high ($\approx$0.9) in regions with $\phi < 1$. This behavior can also be detected in the envelope curves in Fig. 2 on the left, where the upper and lower trajectories are closer to each other in case of lean flames. The correlation coefficient decreases in fuel-rich flames because intermediate species with higher hydrocarbons are formed lowering the level of released heat substantially. Hence, a quasi-proportionality relation between c_{OH^*} and $\dot{q}$ can be stated generally only for lean premixed flames. Because the chemiluminescence measurement in experiments gathers light only along one viewing direction, the integral or line-of-sight summed correlation of c_{OH^*} and $\dot{q}$ is studied in the following by three-dimensional DNS.

3.2 *Integral Correlation in 3D Turbulent Flames*

3.2.1 Simulation Setups

A synthetic flame front is considered which propagates freely in a cubic domain with the size of $5\times5\times5\,\text{mm}^3$. Along the stream-wise direction, fresh gas with methane/air mixture enters the domain at the inlet with $\phi = 0.9$, $T = 300\,\text{K}$ and $p = 1$ bar. The product gas leaves the domain on the other side (outlet). The lateral faces are defined as symmetry planes to avoid a loss of mass. The turbulence is prescribed by means of an inflow generator, which provides a spatially and temporally correlated velocity field at the inlet boundary for each time step [15, 16]. The bulk flow velocity and the turbulence properties at the inlet are adjusted so that the flame front cannot propagate out of the computational domain. This setup may be considered as a small segment of a real flame with a continuous counterflow of fresh gas to the flame front, as shown in Fig. 4.

Two turbulent Reynolds numbers with $Re_t = 15$ and $Re_t = 69$ are considered for the inflow condition, which are based on the integral length in the stream-wise direction ($l_x = 0.5$ and $2.5\,\text{mm}$) and the turbulence intensity ($u' = 0.5$ and $0.75\,\text{m/s}$). The length scales in the lateral directions l_r are set to 1 mm. The turbulence parameters are used as input for the inflow generator. A partially non-reflecting boundary condition (NRBC) proposed by Poinsot and Lele [17] has been applied to the inlet and outlet boundaries to avoid spurious reflection of pressure waves at those boundaries. The computational domain is discretized into 16 million finite cell volumes with an equidistant resolution of 20 μm in each direction, which is smaller than the Kolmogorov micro-scale estimated by $\eta = l_{x,r}/Re_t^{3/4}$ [4] and is able to resolve the planar unstrained reaction zone with approx. 20 cells. A uniform flow field and chemical scalars obtained from calculation of the corresponding one-dimensional laminar flame have been used to initialize the simulation. The DNS

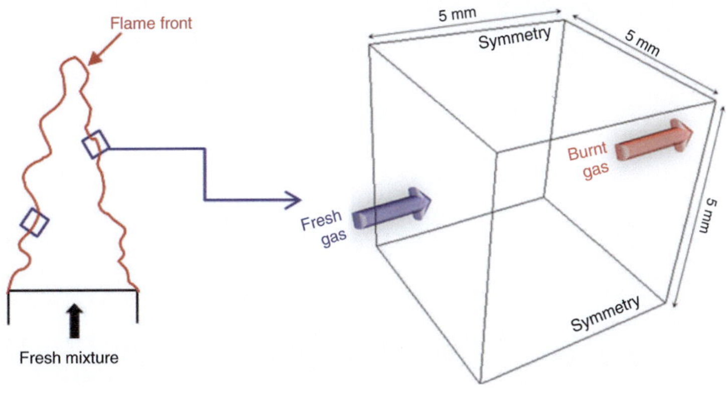

Fig. 4 Schematic illustration of the computational domain and boundary conditions used for DNS of a synthetically turbulent flame front

have been run for 40 ms with a time step of 0.5 μs, which allows a maximum CFL number of approx. 0.1.

3.2.2 Performance

In previous works [12, 13, 18], the implemented DNS solver in OpenFOAM has proven to exhibit an excellent parallel scalability on different supercomputers, e.g. the Cray XE6 (HERMIT) cluster maintained by the high performance computing center Stuttgart (HLRS) and the JUQUEEN cluster with the IBM Blue Gene/Q architecture from the Jülich Supercomputing Centre (JSC) [19]. Figure 5 shows a scalability anaysis of the DNS solver performed on the secondarily installed Cray XC40 machine (HORNET) from HLRS, where the test case is given by a three-dimensional hydrogen/air flame at laboratory scale with a computational grid consisting of 144 million cells [12]. A very good parallel performance is confirmed by running the code for this case with up to 14,400 processor cores. Even a super-linear behavior can be detected, indicating that the code is able to exploit the full capacity of the HPC machine. Therefore, the DNS solver is able to speed-up efficiently while running in parallel with a large number of processors. The DNS in the present work have been conducted on the Cray XC40 (HAZEL HEN) system [20]. For each case with $Re_t = 15$ and $Re_t = 69$, the DNS have been run with 3,600 processor cores for 3 computing days, therewith, consuming approx. 520,000 core hours in total.

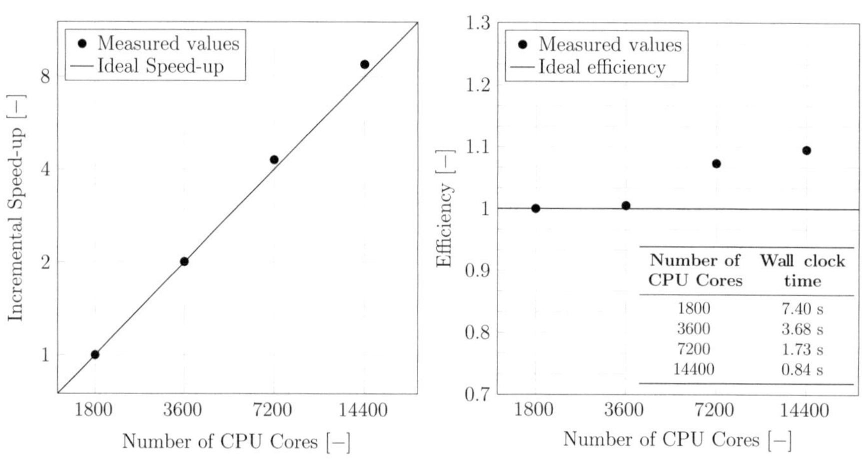

Fig. 5 Incremental speed-up (*left*) and efficiency (*right*) obtained from running the OpenFOAM based DNS code on the HPC platform Cray XC40 (HORNET) from HLRS [20] (normalized to 1800 processor cores)

3.2.3 Results

Figure 6 (left and middle) shows instantaneous contours of the heat release rate $\dot{q}$ and the OH* concentration c_{OH^*} at a slice passing through the centerline axis of the domain for $Re_t = 15$ (top) and $Re_t = 69$ (bottom). The flame front is corrugated due to the non-uniform inflow condition. The flame is more wrinkled in case of higher Re_t due to the more intensive turbulent fluctuations. Similar to the results obtained from the one-dimensional simulations in Sect. 3.1, $\dot{q}$ and c_{OH^*} have found to correlate strongly with each other for both Re_t numbers, which can be detected by the very similar contours of $\dot{q}$ and c_{OH^*}. Figure 6 on the right depicts the joint probability density function (PDF) of $\dot{q}$ and c_{OH^*} by using data pairs extracted from the entire flame volume ($\dot{q} > 0$). The solid lines indicate evolution of $c_{OH^*}(\dot{q})$ obtained from the corresponding one-dimensional simulation, as shown in Fig. 2, which is representative for the correlation of $\dot{q}$ and c_{OH^*} under turbulent conditions too. A scattering around the envelope curve from the laminar flame is however detected for the turbulent flame case. This is mainly attributed to the fact that the intrinsic flame structure, i.e., profiles of the chemical scalars along the flame's normal coordinate, is altered locally by the turbulent flow via stretching. Moreover, the flame undergoes a relaxation time to respond to the unsteady flow, leading to an effect of time history. The scattering is broader in case of the higher turbulence level

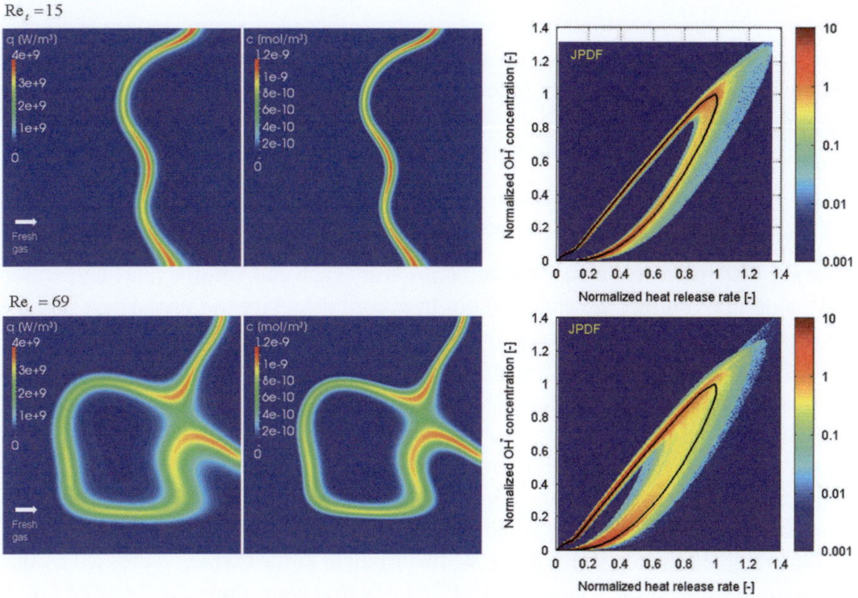

Fig. 6 Instantaneous contours of heat release rate (*left*) and OH* concentration (*middle*) as well as joint PDF of these parameters (*right*) for two different turbulent Reynolds numbers: $Re_t = 15$ at the *top* and $Re_t = 69$ at the *bottom*

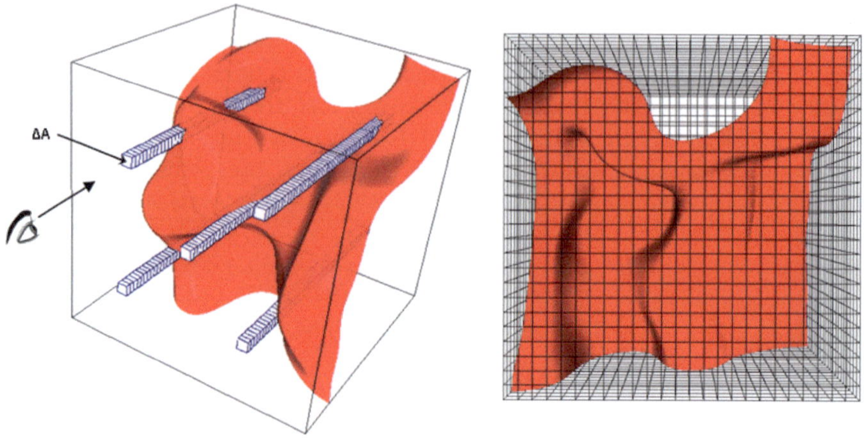

Fig. 7 Iso-surface of temperature and decomposition of the domain into finite rays along one line-of-sight direction

with $Re_t = 69$, because the turbulence intensity and turbulent time scale is larger in this case, leading to an increased stretching and response time of the flame.

Figure 7 presents a three-dimensional flame front defined by the $T = 1500\,\text{K}$ isotherm for the case with $Re_t = 69$. In order to analyse the line-of-sight correlation of $\dot{q}$ and c_{OH^*}, the computational domain is decomposed in a number of rays defined by a fixed viewing direction and an area ΔA, as illustrated in Fig. 7. The heat release and concentration are then integrated along these rays leading to their area-specific integral values:

$$\tilde{\phi} = \int \phi \cdot ds = \frac{1}{\Delta A} \sum \phi_i \cdot V_i, \quad \phi = \dot{q}, c_{OH^*} \tag{10}$$

In Eq. (10), discrete values of $\dot{q}$ and c_{OH^*} from each cell volume i, located within one single ray volume, are summed up. In accordance with the view angle and the instantaneous flame front shown in Fig. 7, the line-of-sight summed $\dot{q}$ and c_{OH^*} calculated from Eq. (10) are shown in Fig. 8 for two different averaging areas. A strong correlation of these integral values can be identified by comparing contours of $\tilde{\dot{q}}$ and $\tilde{c}_{OH^*}$ in Fig. 8. As expected, a sharp image can be obtained with thinner rays or smaller ΔA, respectively. The wrinkling of the flame front caused by flame-turbulence interaction leads to larger integral values of $\dot{q}$ and c_{OH^*}, because the flame surface may be passed through more frequently (more than once) by the rays. As illustrated on the top right of Fig. 9, the turbulent flame surface is crossed by one single ray for three times, leading to a triple reaction zone along this specific ray.

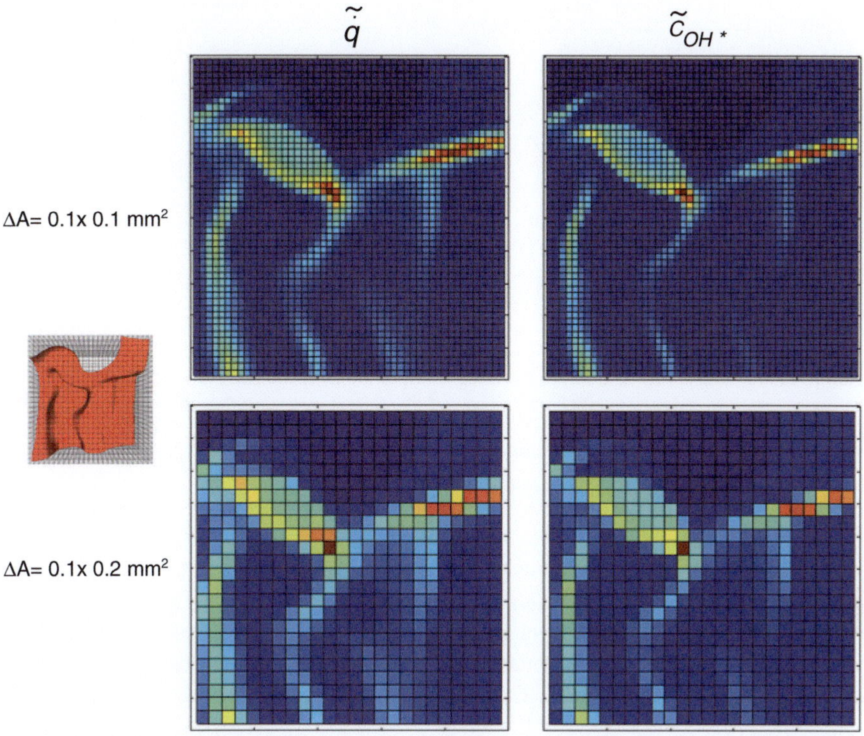

Fig. 8 Line-of-sight summed heat release rate (*left*) and OH* concentration (*right*) for two different specific areas $\Delta A = 0.1 \times 0.1\,\mathrm{mm}^2$ (*top*) and $\Delta A = 0.2 \times 0.2\,\mathrm{mm}^2$ (*bottom*)

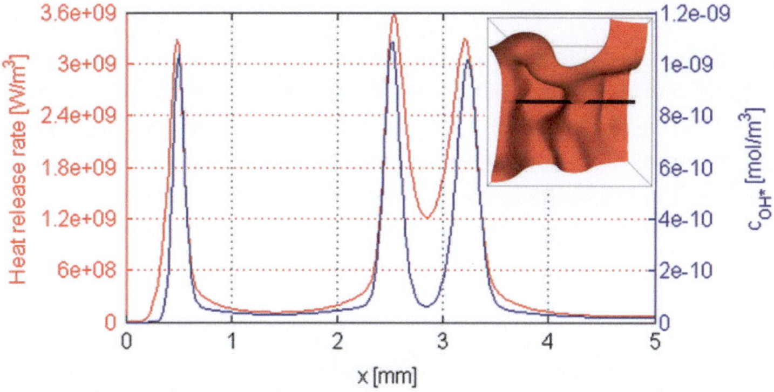

Fig. 9 Profiles of heat release rate and OH* concentration along one single ray passing through the flame surface

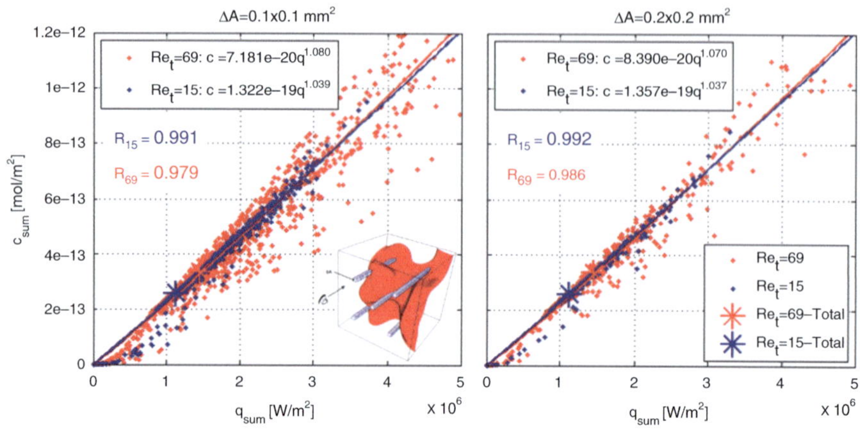

Fig. 10 Correlation of integral heat release rate and OH* concentration for different cross-section areas of rays and turbulent Reynolds numbers

The line-of-sight summed $\dot{q}$ and c_{OH^*} from different averaging areas are plotted against each other in Fig. 10. Due to the reduction of data from three-dimensions to two-dimensions by summing up along the viewing direction, the integrated values show a higher correlation than the local values, as demonstrated in Fig. 10 by the correlation coefficients. Consequently, fitting of the ray value pairs via a power function of the shape $\tilde{c}_{OH^*} = a \cdot \widetilde{\dot{q}}^{\,n}$ leads to exponents which are very close to unity (the fitting coefficients are displayed in the legend), indicating a quasi-linear relationship between $\widetilde{\dot{q}}$ and $\tilde{c}_{OH^*}$. The total volume integration of $\dot{q}$ and c_{OH^*} by considering a single ray spanning the whole domain is depicted by "*" symbols in Fig. 10, which lie fairly on the fitted curves too. The quasi-linear correlation is even stronger for lower turbulence levels and larger cross section areas ΔA, where the fitted exponent as well as the correlation coefficient is closer to unity. Although not shown here, similar results have been found when looking from other viewing angles. Therefore, the application of a proportional correlation between the line-of-sight summed concentration of chemiluminescent species and heat release is reasonable for turbulent combustion. This result is applicable to other lean equivalence ratios too, as long as a strong correlation exists locally in this case (see Fig.3).

3.2.4 Evaluation of Heat Release from Chemiluminescence Measurement

Based on the quasi-linear correlation between $\widetilde{\dot{q}}$ and $\tilde{c}_{OH^*}$ obtained in the previous section, the heat release may be evaluated from high-speed imaging of OH* chemiluminescence. As the intensity of light emitted by OH* is proportional to its concentration, i.e., $\widetilde{I} \propto \tilde{c} \propto \widetilde{\dot{q}}$, the heat release can be related to the intensity by $\widetilde{\dot{q}} = F \cdot \widetilde{I}$ with a proportionality factor F. The total heat release rate can then be formulated by summing up the integral intensities from each ray or pixel of the chemiluminescence imaging

$$\dot{Q}_t = \Delta A \sum_{k=1}^{N} \widetilde{\dot{q}_k} = \Delta A \sum_{k=1}^{N} F \cdot \widetilde{I}_k \tag{11}$$

with the total number of pixels N. The thermal load $\dot{Q}_{th}$ represents the time-mean value of the overall heat release rate, which is known from the operating conditions or the set mass flow of the fuel stream, respectively. For a time series of chemiluminescence snapshots, $\dot{Q}_{th}$ can be expressed as

$$\dot{Q}_{th} = \overline{\dot{Q}_t} = F \cdot \Delta A \sum_{k=1}^{N} \overline{\widetilde{I}_k} \tag{12}$$

leading to the proportionality factor F calculated by

$$F = \frac{\dot{Q}_{th}}{\Delta A \sum \overline{\widetilde{I}_i}} \tag{13}$$

In Eqs. (11), (12), and (13), "$\widetilde{\cdot}$" and "$\bar{\cdot}$" indicate line-of-sight summed and time-averaged values. ΔA is the pixel size of the camera in the experiment which represents the cross-section area of the rays discussed in Sect. 3.2.3. The line-of-sight summed intensity of light $\widetilde{I}$ is directly measured in the experiment. In this way, Eq. (11) predicts that the heat release rate is proportional to the chemiluminescent emission by a constant factor given by the ratio of their overall time-mean values.

4 Conclusions

Direct numerical simulations have been performed in connection with detailed treatment of chemical reactions and molecular transport, in order to find the correlation between heat release rate and the luminescent species in turbulent flames. The exact correlation was unknown before and has only been assumed to be linear until now. One-dimensional calculations of laminar unstrained flames have first been performed for different equivalence ratios of methane/air mixtures and it has been shown that the local generation of chemiluminescent species OH* and heat release are basically not uniquely assigned to each other. A correlation coefficient of approx. 0.9 between their local values has been confirmed for lean-premixed flames, which decreases with higher equivalence ratio. As the chemiluminescence measurement gathers light only along one viewing direction, the line-of-sight integrated values of heat release and OH* concentration have been evaluated from three-dimensional

DNS of a synthetically turbulent flame front for a lean methane/air flame. A quasi-linear correlation has been identified for these integral values, which has found to be stronger for flames subjected to lower turbulence intensities and larger cross-section area of the rays. Consequently, a proportionality relation has been used for the prediction of the heat release rate from the intensity measurement of luminescent emissions. The present work focused on lean-premixed flames with relatively low turbulence level, where a higher correlation for local generation of heat release and chemiluminescent species has been observed. The correlation decreases for rich equivalence ratios and high turbulence intensities. Therefore, further work is needed to validate the obtained results for fuel-rich flames and in more intense turbulent conditions.

Acknowledgements The authors wish to acknowledge the financial support by the German Research Council (DFG) through the Research Unit DFG-BO693/27 "Combustion Noise". This study has used computing resources from the High Performance Computing Center Stuttgart (HLRS) at the University of Stuttgart, Germany. The authors gratefully acknowledge assistance from these Communities.

References

1. Weyermann, F., Hirsch, C., Sattelmayer, T.: Influence of boundary conditions on the noise emission of turbulent premixed swirl flames. In: Schwarz, A., Janicka, J. (eds.) Combustion Noise, pp. 151–178. Springer, Berlin/Heidelberg (2009)
2. Copeland, C., Friedman, J., Renksizbulut, M.: Planar temperature imaging using thermally assisted laser induced fluorescence of OH in a methane-air flame. Exp. Therm. Fluid Sci. **31**, 221–236 (2007)
3. Lauer, M.R.W.: Determination of the heat release distribution in turbulent flames by chemiluminescence imaging. Ph.D. thesis, Technical University Munich (2011)
4. Poinsot, T., Veynante, D.: Theoretical and Numerical Combustion, 2nd edn. Edwards Inc., Philadelphia (2005)
5. Kee, R.J., Coltrin, M.E., Glarborg, P.: Chemically Reacting Flow: Theory and Practice. John Wiley & Sons Inc., Hoboken (2003)
6. OpenCFD Ltd.: OpenFOAM User Guide, Version 2.3.0 (2014)
7. Komen, E., Shams, A., Camilo, L., Koren, B.: Quasi-DNS capabilities of OpenFOAM for different mesh types. Comput. Fluids **96**, 87–104 (2014)
8. Goodwin, D.G.: Cantera C++ User's Guide. California Institute of Technology, California (2002)
9. Kee, R.J., Grcar, J.F., Smooke, M.D., Miller, J.A.: A Fortran Program for Modeling Steady Laminar One-Dimensional Premixed Flames. Report No. SAND85–8240. Sandia National Laboratories, Albuquerque (1985)
10. Kathrotia, T., Riedel, U., Seipel, A., Moshammer, K., Brockhinke, A.: Experimental and numerical study of chemiluminescent species in low-pressure flames. Appl. Phys. B **107**, 571–584 (2012)
11. Ferziger, J., Perić M.: Computational Methods for Fluid Dynamics. Springer, Berlin/New York (2002)

12. Zhang, F., Bonart, H., Zirwes, T., Habisreuther, P., Bockhorn, H., Zarzalis, N.: Direct numerical simulation of chemically reacting flows with the public domain code openfoam. In: Nagel, W.E., Kröner, D., Resch, M. (eds.) High Performance Computing in Science and Engineering'14, pp. 221–236. Springer, Berlin/Heidelberg (2015)
13. Zirwes, T.: Weiterentwicklung und Optimierung eines auf OpenFOAM basierten DNS Lösers zur Verbesserung der Effizienz und Handhabung. Bachelors thesis, Karlsruhe Institute of Technology, Karlsruhe (2013). http://digbib.ubka.uni-karlsruhe.de/volltexte/1000037538
14. Smith, G.P., Golden, D.M., Frenklach, M., Moriarty, N.W., Eiteneer, B., Goldenberg, M., Bowman, C.T., Hanson, R.K., Song, S., Gardiner, W.C., Lissianski, V.V., Qin, Z.: (1999). http://www.me.berkeley.edu/gri_mech/
15. Klein, M., Sadiki, A., Janicka, J.: A digital filter based generation of inflow data for spatially developing direct numerical or large eddy simulations. J. Comput. Phys. **286**, 652–665 (2003)
16. Zhang, F., Habisreuther, P., Bockhorn, H.: Application of the unified turbulent flame-speed closure (UTFC) combustion model to numerical computation of turbulent gas flames. In: Nagel, W.E., Kröner, D., Resch, M. (eds.) High Performance Computing in Science and Engineering'12, pp. 187–206. Springer, Berlin/Heidelberg (2012)
17. Poinsot, T., Lele, S.: Boundary conditions for direct simulation of compressible viscous flows. J. Comput. Phys. **101**, 104–129 (1992)
18. Zhang, F., Bonart, H., Habisreuther, P., Bockhorn, H.: Impact of grid refinement on turbulent combustion and combustion noise modeling with large eddy simulation. In: Nagel, W.E., Kröner, D., Resch, M. (eds.) High Performance Computing in Science and Engineering 13, pp. 259–274. Springer, Berlin/Heidelberg (2013)
19. IBM: IBM Blue Gene/Q – JUQUEEN. http://www.fz-juelich.de/ias/jsc/EN/Expertise/Supercomputers/JUQUEEN/
20. Cray Inc.: Cray XC40 – HAZEL HEN. http://www.hlrs.de/systems/cray-xc40-hazel-hen/

Direct Numerical Simulation of Non-premixed Syngas Combustion Using OpenFOAM

Son Vo, Andreas Kronenburg, Oliver T. Stein, and Evatt R. Hawkes

Abstract A direct numerical simulation (DNS) solver for turbulent reacting flows is developed using libraries and functions from the open-source computational fluid dynamics package OpenFOAM. The solver serves as a reference for developing sub-grid scale models for the large eddy simulation (LES) of turbulent flames. DNS typically requires spatial and temporal discretisation schemes of high order, which are not readily available in OpenFOAM. We validate our OpenFOAM solver by performing direct numerical simulations of a well-defined DNS case featuring non-premixed syngas combustion in a double shear layer. This configuration has previously been studied by Hawkes et al. (Proc Combust Inst 31:1633–1640, 2007) using a purpose-built, high-order DNS solver. Despite the lower discretisation schemes of OpenFOAM, simulation results agree very well with the reference DNS data. Local extinction and re-ignition of the syngas flame are captured and effects of differential diffusion are highlighted. Parallel scaling results using the *HazelHen* architecture of HLRS Stuttgart are reported.

1 Introduction

Reynolds-averaged Navier-Stokes (RANS) and large eddy simulation (LES) approaches are popular methods for the numerical modelling of turbulent flows. While RANS solves for the temporal averages of the variables of interest, LES uses a spatial filter to resolve the largest turbulent eddies and reverts to modelling of the small scales [1]. In reacting flows both methods encounter problems with the closure of the chemical source terms in the species and enthalpy transport equations due to their non-linear dependence on the local instantaneous species concentrations and temperature, which are not available in RANS or LES. As a result, turbulent combustion modelling for RANS and LES mainly focuses on the closure of the

S. Vo • A. Kronenburg (✉) • O.T. Stein
Institut für Technische Verbrennung, Universität Stuttgart, Herdweg 51, 70174 Stuttgart, Germany
e-mail: kronenburg@itv.uni-stuttgart.de; kronenburg@itv.uni-stuttgart.de

E.R. Hawkes
School of Mechanical and Manufacturing Engineering, University of New South Wales, 2052 Sydney, NSW, Australia

W.E. Nagel et al. (eds.), *High Performance Computing in Science and Engineering '16*, DOI 10.1007/978-3-319-47066-5_17

averaged/filtered reaction rates [2]. In principle, direct numerical simulation (DNS) can overcome these closure problems by solving the transport equations directly on very fine grids [3]. This is particularly important where experimental data is not available, or difficult to obtain [4]. DNS serves as a base for the development of RANS/LES models by providing DNS data as an input for modelling [5], or by offering an accurate reference for model validation [6]. Since reactive DNS is required to resolve the smallest scales of turbulence and the flame structure, its computational cost is very high, especially for high-Re number problems. A DNS of reacting flow with hundreds of million cells and 60 chemical reactions can take up a week on thousands of CPUs, producing terabytes of data. Therefore in turbulent combustion DNS is mostly applied for simulations of simple geometries using structured meshes and small chemical mechanisms. However, with the advent of large scale high-performance computing, more practical DNS is coming within reach [7].

This paper investigates the DNS capabilities of the widely-used open source CFD library OpenFOAM. OpenFOAM is becoming increasingly popular for the modelling of turbulent reacting flows, both for industrial and research applications. However, as the general design of the software accommodates the industrial requirement of geometrical flexibility, OpenFOAM is based on the finite volume method, which – in combination with unstructured meshes – limits the order of spatial discretisation. Moreover, parallel scaling and data handling of OpenFOAM computations is a topic of high relevance, in particular on large-scale HPC machines with a great number of users. Particularly the low order discretisation schemes have drawn repeated criticism in the DNS community, and the usefulness of using OpenFOAM as a DNS code has frequently been questioned. We study both accuracy and efficiency of OpenFOAM by conducting DNS of a well-characterised double shear layer configuration burning syngas ($H_2/CO/N_2$) in a preheated oxidizer, and the work shall serve as a reference for OpenFOAM's DNS capabilities for reacting flows. The set-up has previously been simulated by Hawkes et al. [10], where it was shown that significant local extinction and re-ignition occurs, which should be accurately captured by other modelling approaches *and* accurate DNS solvers. The original DNS has been performed with the well-established *S3D* software of Sandia National Laboratories, a dedicated DNS solver for turbulent reacting flows that uses high-order numerical schemes and has been demonstrated to provide accurate DNS data over more than a decade of combustion research [8]. The reference DNS used a uniform mesh with 150 million grid points to resolve all turbulent scales and the flame front. The same resolution is used in our OpenFOAM computations, alongside simulations using half the number of grid points in every direction (18M) to investigate grid resolution effects. The S3D solver is capable of accounting for differential diffusion and two OpenFOAM solver variants (with and without differential diffusion) are used to produce DNS results, which are validated by comparison to the earlier S3D predictions. Finally, strong and weak scaling tests using the OpenFOAM solver are conducted and results are reported.

2 Governing Equations

The governing equations for DNS of incompressible turbulent reacting flow are

$$\frac{\partial \rho}{\partial t} + \frac{\partial \rho u_j}{\partial x_j} = 0, \tag{1}$$

$$\frac{\partial \rho u_i}{\partial t} + \frac{\partial \rho u_i u_j}{\partial x_j} = -\frac{\partial p}{\partial x_i} + \frac{\partial \tau_{ij}}{\partial x_j} + \rho g, \tag{2}$$

$$\frac{\partial \rho Y_k}{\partial t} + \frac{\partial \rho u_j Y_k}{\partial x_j} = \frac{\partial}{\partial x_j}\left(\rho D_k \frac{\partial Y_k}{\partial x_j}\right) + \dot{\omega}_k, \tag{3}$$

$$\frac{\partial \rho h_s}{\partial t} + \frac{\partial \rho u_j h_s}{\partial x_j} = \frac{\partial}{\partial x_j}\left(\lambda \frac{\partial T}{\partial x_j} - \sum_{k=1}^{N} \rho D_k \frac{\partial Y_k}{\partial x_j} h_k\right) + \dot{\omega}_{h_s}, \tag{4}$$

where t is time, x_j is the spatial coordinate in the j-direction, ρ denotes density, u is velocity, p pressure, τ_{ij} the viscous stress tensor for a Newtonian fluid and g the gravity vector. Y_k, D_k and $\dot{\omega}_k$ are the mass fraction, mass diffusivity and chemical reaction rate of species k within the mixture of ideal gases. The remaining variables h_s, λ, T and $\dot{\omega}_{h_s}$ are sensible enthalpy, thermal conductivity, temperature and sensible enthalpy reaction source term, respectively, and N is the number of chemical species. For unity Lewis number calculations, D_k is considered to be identical for all species and calculated from the viscosity. Alternatively, differential diffusion can be considered by calculating individual mixture-averaged diffusion coefficients between the k-th species and the rest of the mixture [9], also considering all terms in Eq. (4).

3 Computational Configuration and DNS Solvers

The investigated set-up is a temporally evolving double shear layer burning syngas within two counterflowing streams of hot oxidizer, as shown in Fig. 1. This configuration is identical to case L described in [10], with a jet Reynolds number of 2510. Fuel and oxidizer move in opposite directions across the domain with a characteristic velocity $U = U_{fuel} - U_{oxidizer} = 145\,\text{m/s}$. To trigger the onset of turbulence from the initially laminar conditions, velocity perturbations with an amplitude of 0.05U and an integral length scale of $H/3$ are superimposed within the fuel stream, with H ($= 0.72\,\text{mm}$) being the width of the jet at $t = 0$. The dimensions of the computational domain are $L_x \times L_y \times L_z = 8.64 \times 10.065 \times 5.76\,\text{mm}^3$. The flame is initialized by setting a laminar mixture fraction profile and retrieving the initial species distributions from a pre-computed flamelet table. A reduced, non-stiff

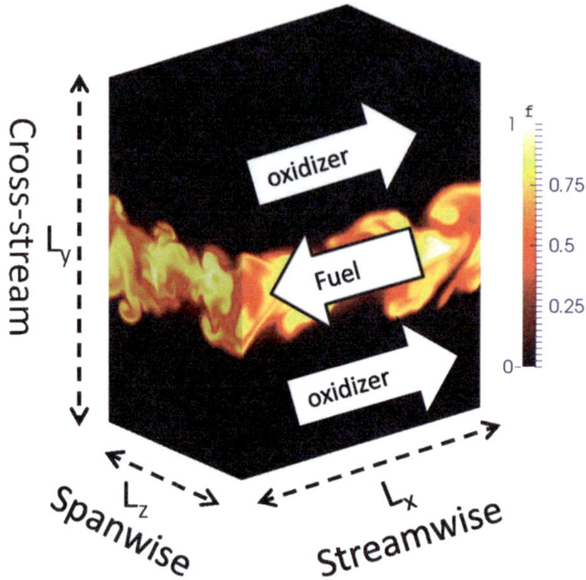

Fig. 1 Computational domain, illustrated by the instantaneous mixture fraction field at $t/t_j = 20$

11 species, 21 reactions chemical mechanism is used to describe syngas chemistry. The fuel mixture consists of 50 % CO, 10 % H_2 and 40 % N_2 by volume, whereas the oxidizer is 25 % O_2 and 75 % N_2, resulting in a stoichiometric mixture fraction of 0.42. The initial temperature of both streams is 500 K and pressure is atmospheric. The stream- and spanwise boundary conditions are periodic, while in the cross-stream direction the boundary condition is zero-gradient. The reference DNS of [10] has been performed with the well-established S3D solver of Sandia National Laboratories, which offers 8th order spatial and 4th order temporal accuracy, respectively. The S3D solver can also handle non-unity Lewis number cases. The DNS solver developed within the present work is based on the OpenFOAM C++ library, v2.4.x. OpenFOAM uses the finite volume method with a spatial accuracy of second order. Standard solver applications within the OpenFOAM library are based on the assumption of unity Lewis number and can therefore not account for differential diffusion effects. However, a coupled Cantera-OpenFOAM library has been developed by Zhang et al. [9], which allows for the calculation of individual mixture-averaged diffusion coefficients for each species with respect to the gas mixture using Cantera function calls. These routines are available in the present work and used to evaluate differential diffusion effects versus the standard unity Lewis number assumption. The main differences between the employed OpenFOAM DNS solver(s) and S3D are summarized in Table 1.

The computational domain of the original DNS was discretised on a uniform mesh with $576 \times 672 \times 384 \approx 150$M control volumes, which resulted in a grid spacing Δx of 15 microns. It was estimated that at the time of maximum local

Table 1 Comparison of the OpenFOAM DNS solver with the reference solver S3D

	S3D	OpenFOAM
Spatial discretisation	8th-order	2nd-order
Temporal discretisation	4th-order	2nd-order
Lewis number	Non-unity	Unity/non-unity

extinction ($t/t_j = 20$, where $t_j = H/U$) the Kolmogorov scale was resolved by a minimum of 1.2 cells. The flame structure was resolved by at least 10 grid points, considering the half-width of the OH reaction rate profile of a steady diffusion flame at half the extinction strain rate. It was also reported that cases run at half the resolution gave first and second moments of the solution variables in good agreement with the full resolution case [10]. For our OpenFOAM simulations we consider the identical $576 \times 672 \times 384 \approx 150$M uniform grid resolution, to allow for a direct comparison of S3D and OpenFOAM on the same grid. In addition, we run OpenFOAM simulations at half the original resolution in every coordinate direction, resulting in $288 \times 336 \times 192 \approx 18$M cells.

4 Results and Discussion

For our solver evaluation we compare the results from a set of four different DNS calculations. The datasets "ITV-OF-Le1 150M" (black lines, see Fig. 2) and "ITV-OF-Le1 18M" (red) are OpenFOAM calculations assuming unity Lewis number and using the 150M and 18M grid, respectively. The dataset "ITV-OF-DD 18M" (green) also uses OpenFOAM, but accounts for differential diffusion and is calculated on the 18M grid. The label "SAN-S3D-DD 150M" (blue) refers to the reference DNS from [10] using S3D, 150M and including realistic thermodynamic properties. In the following we evaluate the DNS resolution requirements first, followed by a discussion of the major flame characteristics and the level to which they are captured by the different DNS calculations.

4.1 DNS Resolution Requirements

The resolution requirements for our DNS are evaluated by comparing statistics of the scalar dissipation rate χ. The scalar dissipation rate is proportional to the square of the mixture fraction gradient and therefore a sensitive indicator of grid resolution effects. In addition, χ plays an important role for potential extinction and re-ignition of turbulent non-premixed flames. Figure 2 shows cross-stream profiles of the mean scalar dissipation rate at normalized jet times $10 \leq t/t_j \leq 40$. In this temporally evolving double shear layer configuration statistics are calculated by averaging across the homogeneous x-z-plane to obtain mean and RMS values

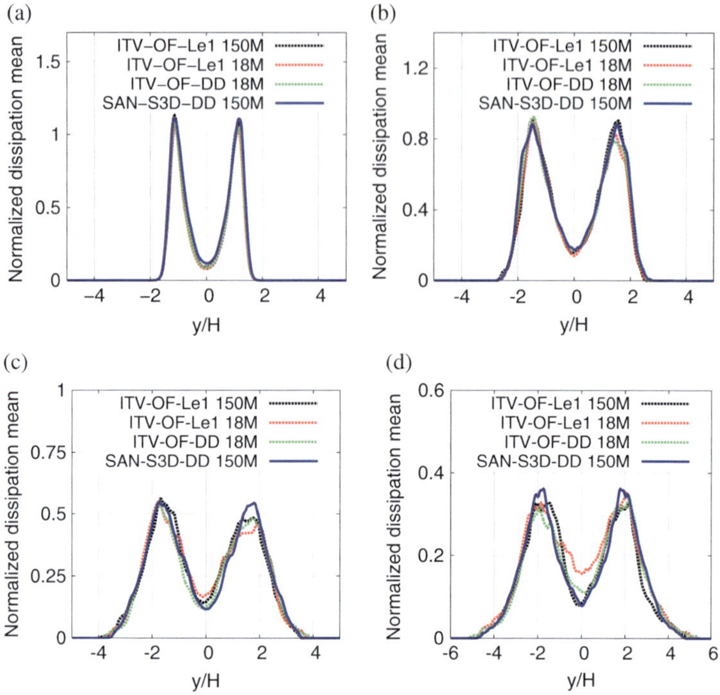

Fig. 2 Cross-stream profiles of the normalized mean scalar dissipation rate at (**a**) $t/t_j = 10$, (**b**) $t/t_j = 20$, (**c**) $t/t_j = 30$, (**d**) $t/t_j = 40$. The dissipation rate is normalized by its value at extinction of the corresponding laminar flamelet ($\chi_q = 2194$ 1/s)

at each fixed location y/H. It can be seen that all OpenFOAM calculations are in reasonable agreement with the S3D reference data, with only minor deviations becoming apparent at the late times $t/t_j = 30, 40$, where the 18M calculation assuming unity Lewis number shows the strongest (yet acceptable) discrepancies. A similar trend can be observed for the scalar dissipation rate RMS shown in Fig. 3. Here, small deviations from the reference dissipation RMS can already be observed at $t/t_j = 10$ (when turbulence develops). They increase, within acceptable bounds, until the latest time, $t/t_j = 40$. Even at this stage, after all four simulations have evolved independently from each other for 40 jet times, the scalar dissipation rate RMS profiles are in close agreement with each other, with the most pronounced deviations again for the 18M unity Lewis number run. Note that only axisymmetric cross-stream profiles from S3D are available, whereas the full y/H coordinate is plotted from the OpenFOAM simulations, explaining the perfect symmetry of the S3D results. Figure 4 shows PDFs of the scalar dissipation rate conditional on mixture fraction being near stoichiometric at $t/t_j = 20$, when the resolution requirements are most critical for capturing local extinction.

It can be seen that the scalar dissipation rate PDFs agree very well for a wide range of χ, with the high-order S3D simulation capturing extreme dissipation rate

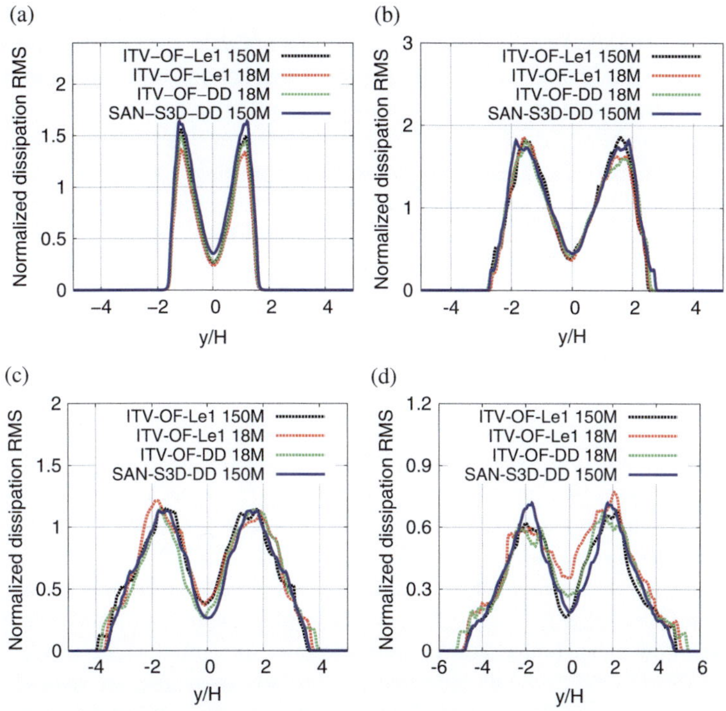

Fig. 3 Cross-stream profiles of the normalized scalar dissipation rate RMS at (**a**) $t/t_j = 10$, (**b**) $t/t_j = 20$, (**c**) $t/t_j = 30$, (**d**) $t/t_j = 40$. The dissipation rate is normalized by its value at extinction of the corresponding laminar flamelet ($\chi_q = 2194$ 1/s)

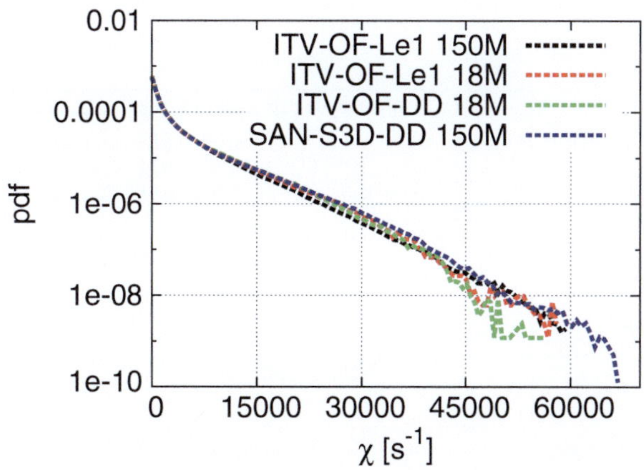

Fig. 4 PDF of the scalar dissipation rate conditional on mixture fraction being in the interval $f_{st} \pm 0.2$ (main reaction zone) at $t/t_j = 20$

events of the order of 70,000 1/s, followed by the 150M OpenFOAM simulation with a peak at 60,000 1/s, and the two 18M OpenFOAM runs recovering slightly smaller scalar dissipation rate peaks. A closer inspection shows that the discrepancies for extreme scalar dissipation events only affect considerably less than 1 % of the total number of dissipation rate samples. Overall, despite the significantly lower order of spatial and temporal discretisation available in OpenFOAM, scalar dissipation rate profiles are well resolved, and even simulations using half the reference resolution should provide adequate flame predictions.

4.2 Flame Characteristics

Figure 5 presents the maximum of the mean temperature as a function of normalized jet time from the four DNS calculations. The maximum mean temperature is obtained by taking the maximum value along y/H of each x-z-plane-averaged temperature. This quantity is calculated at each time t/t_j of the simulation and its temporal evolution can be taken as a global measure for capturing extinction and re-ignition [5]. Figure 5 shows that the maximum extinction (lowest maximum temperature) occurs at $t/t_j = 20$, followed by subsequent re-ignition, which leads to a maximum mean temperature of the order of the initial value at $t/t_j = 40$. All three OpenFOAM calculations faithfully follow the reference dataset, where both unity Lewis number calculations predict slightly stronger extinction and lower temperatures during the re-ignition phase. The prediction by the differential diffusion OpenFOAM solver is closest to the S3D dataset. In Fig. 6 cross-stream profiles of the first two moments of mixture fraction are compared among the simulations. At the time of maximum local extinction, $t/t_j = 20$, no significant difference between the predictions can be observed. At the end of the simulation,

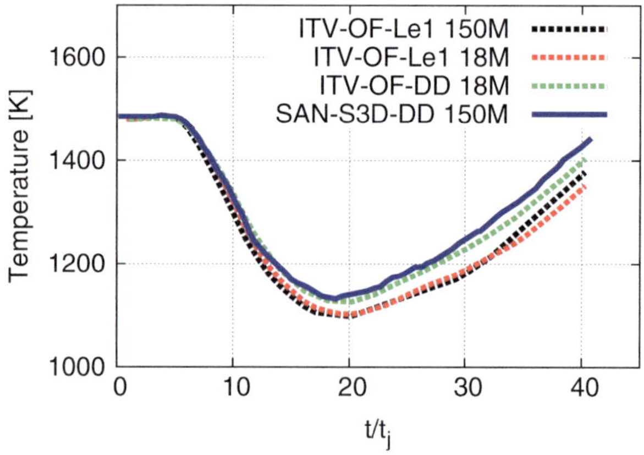

Fig. 5 Maximum of the mean temperature versus normalized time

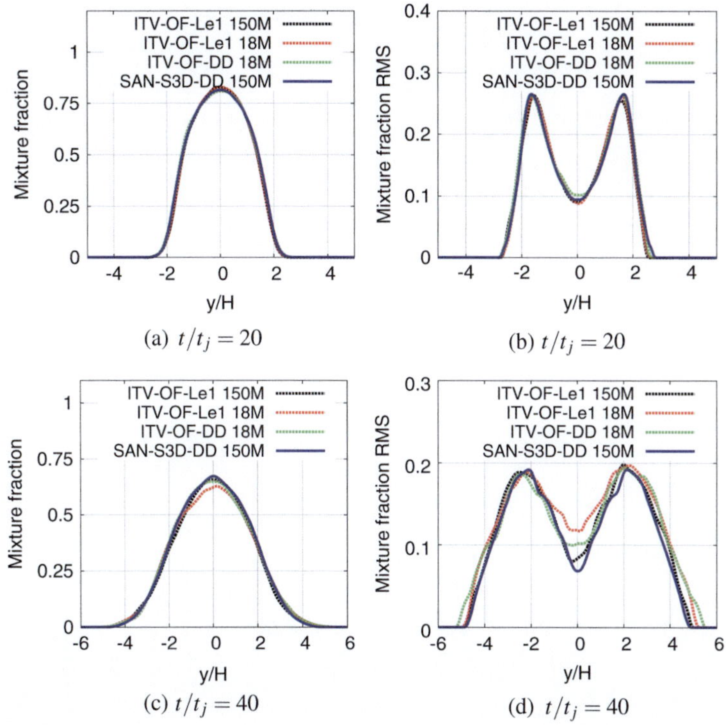

Fig. 6 Cross-stream profiles of the (**a**), (**c**) mean and (**b**), (**d**) RMS mixture fraction at $t/t_j = 20$ and $t/t_j = 40$

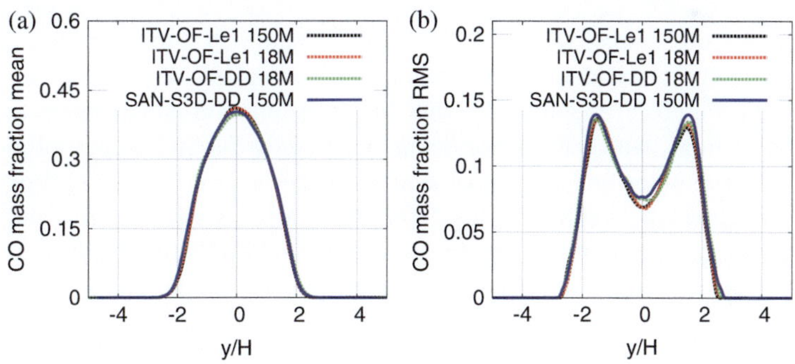

Fig. 7 Cross-stream profiles of the CO mass fraction at $t/t_j = 20$, (**a**) mean, (**b**) RMS

at $t/t_j = 40$, the mean profile of the coarse unity Lewis number simulation shows a slightly decreased peak value and a mild over-prediction of the mixture fraction RMS near the centre of the domain, while the RMS deviations from the reference data near $y/H = 0$ decrease by using more cells or accounting for differential diffusion in OpenFOAM. Figure 7 shows cross-stream profiles of the

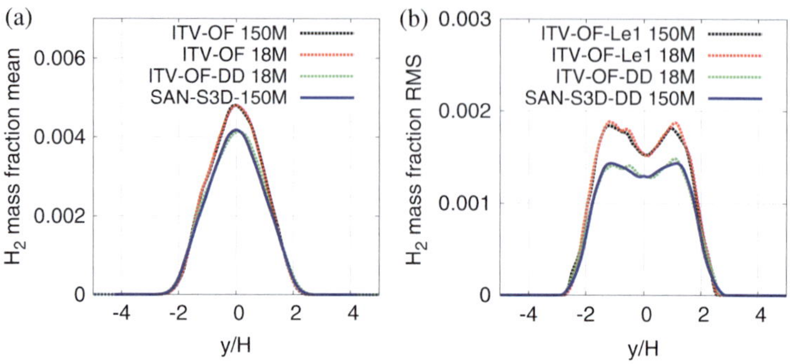

Fig. 8 Cross-stream profiles of the H_2 mass fraction at $t/t_j = 20$, (**a**) mean, (**b**) RMS

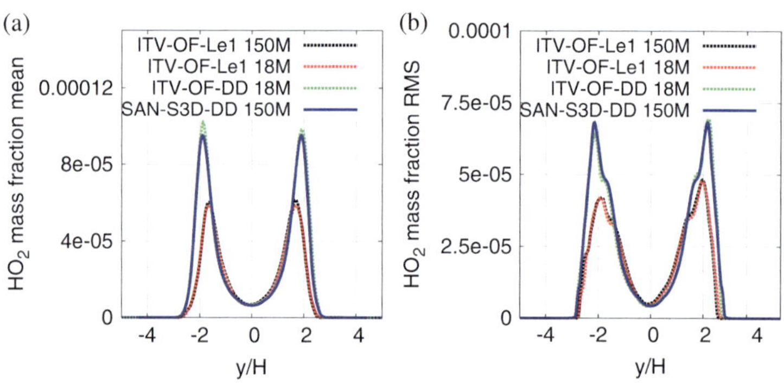

Fig. 9 Cross-stream profiles of the HO_2 mass fraction at $t/t_j = 20$, (**a**) mean, (**b**) RMS

CO mass fraction statistics at $t/t_j = 20$. While being an intermediate species of typical hydrocarbon oxidation, CO becomes an (abundant) fuel species in syngas combustion. Figure 7 demonstrates that fuel consumption is accurately captured by all simulations, with only minor deviations mainly due to lower resolution and for Le = 1. The other fuel species of syngas oxidation is molecular hydrogen, the statistics of which are plotted in Fig. 8. It can clearly be observed that the (light) hydrogen species is subject to significant differential diffusion, which leads to an almost perfect agreement of the two simulations accounting for this effect, whereas grid resolution seems to be less important, as both unity Lewis number simulations equally over-predict the mean and RMS of the CO mass fraction. Figure 9 shows the mean and RMS of the HO_2 mass fraction, which is an intermediate species of the oxidation process. Similar to the trend for H_2 in Fig. 8 accounting for differential diffusion yields an accurate prediction of HO_2 even at a lower grid resolution, while assuming unity Lewis number gives significant deviations from the reference DNS. Finally, Fig. 10 presents plots of the conditional mean temperature across mixture fraction at $t/t_j = 20$ and $t/t_j = 40$. At $t/t_j = 20$ it can be observed that all OpenFOAM simulations give overall reasonable results, but lead to slight under-predictions of the conditional mean temperature in mixture fraction space. Again,

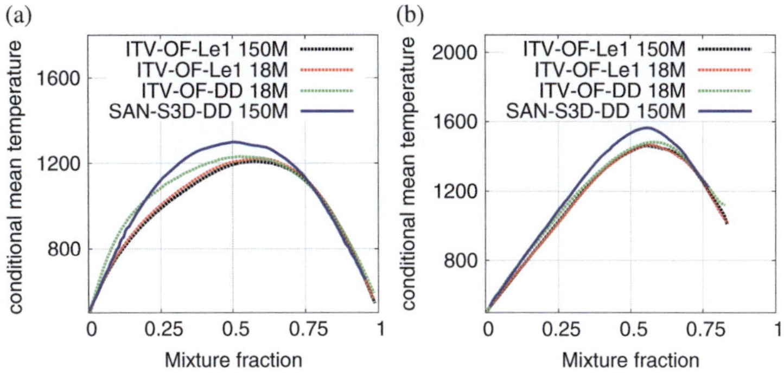

Fig. 10 Conditional mean temperature at (**a**) $t/t_j = 20$, (**b**) $t/t_j = 40$

considering differential diffusion improves the results, albeit not uniformly across mixture fraction space, but mainly on the lean side of stoichiometric. This is likely because including differential diffusion allows H_2 to diffuse faster from the centre of the domain towards the top and bottom boundary, i.e. into the oxidizer streams. At the late time $t/t_j = 40$ this effect is less pronounced, as the scalar fields are generally more homogenous and differential diffusion plays a less dominant role. Hence, all OpenFOAM simulations yield similar predictions, mildly under-predicting the conditional mean temperature of the reference DNS.

5 Parallel Performance

Strong and weak scaling tests are performed in order to assess the parallel performance of the unity Lewis number OpenFOAM solver on the *HazelHen* architecture of HLRS. For the strong scaling analysis the total number of CFD cells is kept constant at 150M and the number of requested computer cores is increased by constant factors of two from 128 up to 2048. Figure 11a plots the strong scaling efficiency (based on the 128 core run) versus the number of computer cores. It can be observed that the strong scaling efficiency drops significantly when moving from 128 to 256 and 512 cores, but it remains at a constant level of approximately 50 % when the number of cores is further increased to 1024 or 2048. Weak scaling was assessed by keeping a constant number of CFD cells per core (86 K) and performing DNS with 64, 216 and 1728 computer cores, which resulted in total problem sizes of 5M, 18M and 150M CFD cells, respectively. A similar analysis was carried out with increased numbers of CFD cells per core (172 K and 344 K), but the results did not change significantly. Figure 11b presents the results of the weak scaling study, based on the 64 core 5M cell run. It can be observed that the weak scaling efficiency remains high, only decreasing to 93 % for 216 and 90 % for 1728 cores. Standard computations of 150M cells have been carried out on 1024 codes at a

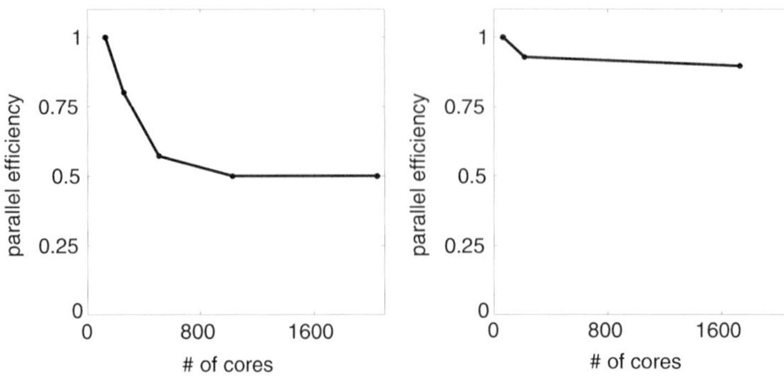

(a) strong scaling (b) weak scaling

Fig. 11 Parallel performance of the OpenFOAM DNS solver (for unity Lewis number): Parallel efficiency for (**a**) strong and (**b**) weak scaling

cost of approximately 20,000 CPU hours. Efficiencies are improved for current computations of two-phase flows due to the need to include more complex chemical kinetics for a realistic description of particle synthesis.

6 Conclusions

Direct numerical simulations of turbulent non-premixed syngas combustion in a double shear layer have been conducted using the OpenFOAM library and compared to a reference DNS database previously established by using a dedicated DNS solver for turbulent reacting flows. The effects of grid resolution and differential diffusion on flame physics were assessed and all three DNS datasets generated with OpenFOAM gave results in favourable agreement with the reference DNS. Despite the reduced order of spatial and temporal discretisation in OpenFOAM extinction and re-ignition events were accurately captured, even when using a reduced grid resolution, given that differential diffusion effects were considered. We can state that the discretisation schemes available in OpenFOAM do not unduly modify the statistics of the present simulation, the numerics do not seem to be excessively dissipative, and the paper may serve as a reference to demonstrate OpenFOAM's capability as a DNS code. It needs to be added, however, that OpenFOAM's low order discretisation schemes are likely to affect simulations with different set-ups, especially those where turbulence is not continuously generated at the largest scales.

Acknowledgements This work is supported by DFG (grant no. KR3684/4-1). We gratefully acknowledge the help of the research group headed by H. Bockhorn and P. Habisreuther at KIT for providing the Cantera-OpenFOAM library for our simulations including non-unity Lewis number effects.

References

1. Pope, S.B.: Turbulent Flows. Cambridge University Press, Cambridge (2000)
2. Maas, U., Warnatz, J., Dibble, R.W.: Combustion, 3rd edn. Springer, Berlin (2006)
3. Cant, R.S., Mastorakos, E.: An Introduction to Turbulent Reacting Flows. Imperial College Press, London (2008)
4. Attili, A., Bisetti, F., Mueller, M., Pitsch, H.: Damkoehler number effects on soot formation and growth in turbulent nonpremixed flames. Proc. Combust. Inst. **35**, 1215–1223 (2015)
5. Krisman, A., Tang, J., Hawkes, E.R., Lignell, D., Chen, J.H.: A DNS evaluation of mixing models for transported PDF modelling of turbulent nonpremixed flames. Combust. Flame **161**, 2085–2106 (2014)
6. Yang, Y., Wang, H., Pope, S., Chen, J.H.: Large-eddy simulation/probability density function modeling of a non-premixed CO/H_2 temporally evolving jet flame. Proc. Combust. Inst. **34**, 1241–1249 (2013)
7. Chen, J.H., Choudhary, A., de Supinski, B., DeVries, M., Hawkes, E.R., Klasky, S., Liao, W.K., Ma, K.L., Mellor-Crummey, J., Podhorszki, N., Sankaran, R., Shende, S., Yoo, C.S.: Terascale direct numerical simulations of turbulent combustion using S3D. Comput. Sci. Discov. **2**, 015001 (2009)
8. Chen, J.H.: Petascale direct numerical simulations of turbulent combustion – fundamental insights towards predictive models. Proc. Combust. Inst. **33**, 99–123 (2011)
9. Zhang, F., Bonart, H., Zirwes, T., Habisreuther, P., Bockhorn, H., Zarzalis, N.: Direct numerical simulation of chemically reacting flows with the public domain code OpenFOAM. In: High Performance Computing in Science and Engineering 2014, pp. 221–236. Springer, Heidelberg (2014)
10. Hawkes, E.R., Sankaran, R., Sutherland, J.C., Chen, J.H.: Scalar mixing in direct numerical simulations of temporally evolving plane jet flames with skeletal CO/H_2 kinetics. Proc. Combust. Inst. **31**, 1633–1640 (2007)

Numerical Simulations of Rocket Combustion Chambers with Supercritical Injection

Martin Seidl, Roman Keller, Peter Gerlinger, and Manfred Aigner

Abstract A thermodynamically consistent model has been implemented into the compressible, implicit combustion code TASCOM3D for the simulation of rocket combustion chambers with supercritical injection. The Soave-Redlich-Kwong equation of state is used, since it offers a good compromise between accuracy and numerical efficiency. Nonreactive and reactive high pressure test cases were simulated for the validation of the implemented model. Generally, a good agreement could be obtained for all test cases.

1 Introduction

The in-house CFD code TASCOM3D is used for numerical simulations of rocket combustion chambers. The main aspects to be considered for CFD simulations of rocket combustors are:

1. turbulence phenomena,
2. combustion processes,
3. thermodynamics and molecular transport properties,
4. grid resolution and discretization.

For an accurate simulation of rocket combustors it is essential to predict fluid properties and flow phenomena with sufficient accuracy. The fluid properties may differ significantly from an ideal gas behavior due to the extreme conditions in rocket engines. Pressures up to 100 bar and more and temperatures from below 100 K for the injected propellants up to about 4000 K within the reaction zone in the combustion chamber make these simulations very challenging. The fluids in the combustion chamber can be in different states of matter (gas-like or liquid-like) depending on the pressure and temperature. If a propellant is injected at cryogenic

M. Seidl (✉) • R. Keller • P. Gerlinger • M. Aigner
Institut für Verbrennungstechnik der Luft- und Raumfahrt, Pfaffenwaldring 38-40, 70569 Stuttgart, Germany
e-mail: martin.seidl@dlr.de

W.E. Nagel et al. (eds.), *High Performance Computing in Science and Engineering '16*, DOI 10.1007/978-3-319-47066-5_18

temperature and pressure below the thermodynamic critical pressure of the fluid, a discontinuous phase transition from liquid to gas will occur during heat up. The liquid and gaseous phase must be handled separately.

The focus of this work is on injection at cryogenic temperatures and pressures above the critical pressure of the fluid, where a continuous transition from a liquid-like state to a gaseous state is observed. The liquid-like and gaseous phase can no longer be distinguished and a combined treatment is required. This is achieved within a single-fluid model based on real gas thermodynamics.

The summary of this report is as follows: First, a brief introduction of the applied CFD code TASCOM3D is given in Sect. 2, followed by a description of the implemented real gas thermodynamics model in Sect. 3. Then, results of two test cases are presented in Sects. 4 and 5: a nonreactive liquid nitrogen jet injected into a warm nitrogen environment and a liquid oxygen/gaseous hydrogen model rocket combustor. Code performance issues are addressed in Sect. 6.

2 Numerical Method

The scientific in-house code TASCOM3D (Turbulent All Speed Combustion Multigrid Solver 3D) has been applied successfully during the last two decades to simulate reacting and non-reacting super- and subsonic flows. Reacting flows are described by solving the fully compressible Navier-Stokes, turbulence and species transport equations. Additionally, an assumed PDF (probability density function) approach is available to take turbulence-chemistry-interaction into account, though for ideal gas simulations only. The two-dimensional conservative form of the Reynolds-averaged Navier-Stokes equations in this work is given by

$$\frac{\partial \mathbf{Q}}{\partial t} + \frac{\partial (\mathbf{F} - \mathbf{F}_v)}{\partial x} + \frac{\partial (\mathbf{G} - \mathbf{G}_v)}{\partial y} = \mathbf{S}, \tag{1}$$

where

$$\mathbf{Q} = [\rho, \rho u, \rho v, \rho E, \rho K, \rho\omega, \rho Y_i]^T, \; i = 1, 2, \ldots, N_k - 1. \tag{2}$$

The conservative variable vector $\mathbf{Q}$ consists of the density ρ, the velocity components u and v, the total specific energy E, the turbulence variables K and ω and the species mass fractions Y_i. Depending on the chosen turbulence model, the variable K is either the turbulent kinetic energy k or its square root q. N_k is the total number of species. $\mathbf{F}$ and $\mathbf{G}$ are the vectors specifying the inviscid fluxes in the x- and y-direction, respectively. $\mathbf{F}_v$ and $\mathbf{G}_v$ are their viscous counterparts. The source vector $\mathbf{S}$ includes terms from turbulence and chemistry. It is given by

$$\mathbf{S} = [0, 0, 0, 0, S_K, S_\omega, S_{Y_i}]^T, \tag{3}$$

S_K and S_ω are the source terms of the turbulence variables and S_{Y_i} the source terms of the species mass fractions due to combustion. For turbulence closure, two-equation models are used, namely the $q - \omega$ model of Coakley [3], the $k - \omega$ model of Wilcox [18], and Menter's SST $k - \omega$ model [12].

The spatial discretization is performed on block structured grids based on a finite volume scheme. For the reconstruction of the cell interface values, MLP^{ld}(Multi-dimensional Limiting Process – low diffusion) [7] with up to fifth order is used to prevent oscillations at sharp gradients and discontinuities. MLP uses diagonal values to improve the TVD (Total Variation Diminishing) limiting behavior [20]. Using these interface values, the AUSM$^+$-up flux vector splitting [11] is employed to calculate the inviscid fluxes. The unsteady set of Eq. (1) is solved with an implicit Lower-Upper Symmetric Gauss-Seidel (LU-SGS) [8] algorithm. Furthermore, finite-rate chemistry is treated in a fully coupled manner. The code is parallelized with Message Passing Interface (MPI). More details concerning TASCOM3D may be found in Refs. [6, 8, 16].

3 Non-ideal Fluids in Rocket Combustors

3.1 Non-ideal Thermodynamics

For sufficiently low pressures and high temperatures, the thermodynamic relation between pressure, temperature, and density can accurately be described by the well-known ideal gas (IG) equation of state (EOS)

$$p = \rho \frac{R_u}{M_w} T \tag{4}$$

where p, T, M_w and R_u are the pressure, temperature, molecular weight of the mixture and the universal gas constant. However, with increasing pressure and decreasing temperature, deviations from this law are observed and cannot be neglected anymore for conditions like they occur in rocket combustors. This is due to the fact that the ideal gas law neglects the volume of the molecules and inter-molecular attractive forces. Both effects are important for high-density conditions. A large number of so-called 'real gas', better called 'real fluid', equations of state have been developed to account for these effects. One of the most famous and simplest is the Soave-Redlich-Kwong (SRK) EOS

$$p = \frac{\rho R_u T}{M_w - b\rho} - \frac{a}{M_w} \frac{\rho^2}{M_w + b\rho}. \tag{5}$$

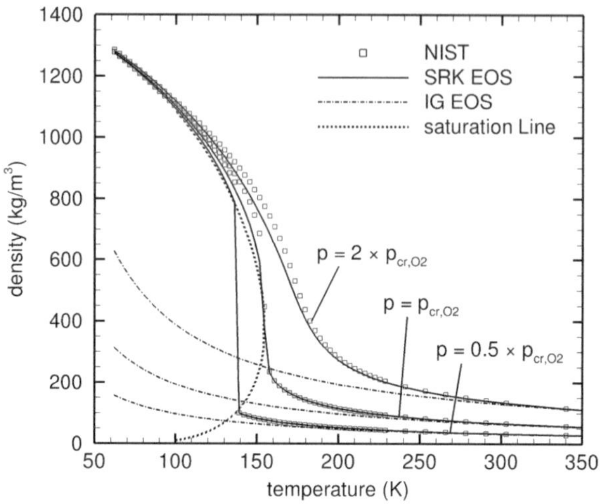

Fig. 1 Density of oxygen versus temperature for three different pressure levels

The parameters a (temperature dependent) and b for a mixture are obtained via mixing and combining rules from their pure species counterparts. The SRK is generally applicable to any pure fluid or mixture and continuously describes the p-ρ-T-relation for gases, liquids, and multi-phase regimes with remarkable accuracy over a wide range of thermodynamic states. For a more detailed description and a general introduction to real fluid properties, the interested reader is referred to textbooks on thermodynamics, e.g. [14]. Figure 1 shows the density-temperature relation for three different pressures for pure oxygen which is used as oxidizer in many rocket engines. Values from NIST database are plotted together with values predicted by the SRK EOS. A reasonable accuracy is achieved with this model and similar ones. Depending on the pressure level in the combustor and the injection temperatures, oxygen may be injected in a gas-like state (low density) or a liquid-like state (high density).

For chamber pressures below the thermodynamic critical pressure of oxygen at $p_{\mathrm{cr,O_2}} = 50.43$ bar and cryogenic injection temperatures below the saturation temperature, the liquid oxygen (LOX) will undergo a discontinuous phase transition during heat up in the chamber. Surface tension between the liquid and gaseous phase leads to an abrupt and distinct separation of both phases. Associated flow phenomena are primary and secondary atomization of the LOX jet into small ligaments and droplets and their final evaporation into the gas phase.

In contrast, for pressures above the critical pressure or sufficiently high temperatures of the injected oxygen, only a single phase will occur and a continuous transition from the cool injection conditions to the hot reaction zone is observed.

For a consistent implementation of a real fluid EOS into a CFD code, it is important to use Eq. (5) in combination with fundamental thermodynamic relations for the calculation of other thermodynamic properties like enthalpy or speed of sound. For more details see for example [14, 19].

3.2 Non-ideal Molecular Transport Properties

In addition to deviations of thermodynamic properties from ideal gases, also deviations of molecular transport properties like viscosity, thermal conductivity, and diffusion coefficients from ideal gas values have to be considered. These are usually calculated from empirical models. For example, for non-polar fluids or fluid mixtures, the model of Ely and Hanley [4, 5] may be used for the prediction of real fluid viscosities and thermal conductivities. Figure 2 shows a comparison of both properties for oxygen between values from NIST database and values obtained from the model of Ely and Hanley (E & H) in combination with the use of the SRK EOS.

3.3 Non-ideal Flow Phenomena

Apart from thermodynamic and transport properties, certain flow phenomena may become non-negligible for high-pressure and low-temperature conditions present in rocket engines. For example, the Soret effect (mass diffusion due to a temperature gradient) or the reciprocal Dufour effect (energy flux due to species concentration gradients) may become important in locally confined regions. Oefelein [13], however, observed that for injection of propellants with shear-coaxial injectors (typical for many rocket engines) their contribution may be neglected.

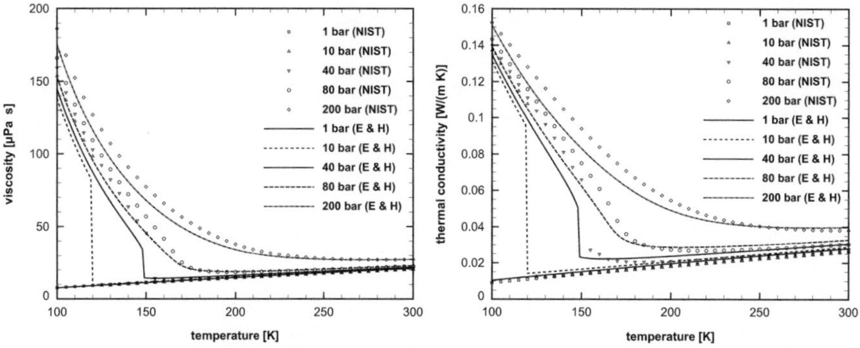

Fig. 2 Viscosity (*left*) and thermal conductivity (*right*) of oxygen versus temperature for various pressures

4 Simulation of Supercritical Nitrogen Jet

The non-reactive RCM-1-A test case presented at the 2nd International Workshop on Rocket Combustion Modeling [17] was chosen as a validation test case for the implemented real gas model. Cryogenic nitrogen at a temperature of about 120 K is injected through a circular duct of $d = 2.2$ mm diameter into a pressurized chamber at 40 bar. The chamber is filled with gaseous nitrogen at room temperature and has a diameter of 122 mm and a length of 1000 mm. There is some uncertainty concerning the actual injection temperature, which is supposed to lie within a range of 120.9 and 126.9 K. At the implied injection conditions close to the pseudo-boiling point [1], the density is very sensitive w.r.t. small changes in temperature.

Axial density distributions were measured in the experiment by Raman images (case 5 in [2]). Steady-state RANS simulations were performed in this study. The following setup is chosen for the presented simulation:

1. hexahedral grid with 95,000 elements and $y^+ \approx 1$ resolution at walls;
2. 2nd order spatial discretization of inviscid fluxes with low diffusion multi-dimensional limiting process (MLP^{ld}) [7];
3. Menter's $k - \omega$ SST turbulence model [12];
4. turbulent Prandtl number $Pr_t = 0.9$;
5. adiabatic walls for injector and faceplate, isothermal chamber wall (T = 297 K);
6. injection temperature T = 126.9 K.

Figure 3 displays the density and temperature distribution close to the injector. The high sensitivity of density w.r.t. temperature, as discussed before, is reflected in these contours. An increase of 10 K roughly halves the density right after the injection. A comparison of axial density profiles at the centerline is plotted in Fig. 4. The RANS simulation resembles the experimental data very well. The density at the centerline remains constant until $x/d \approx 10$. Further downstream, the liquid-like cold nitrogen core dissolves into the warm surrounding nitrogen and density decreases.

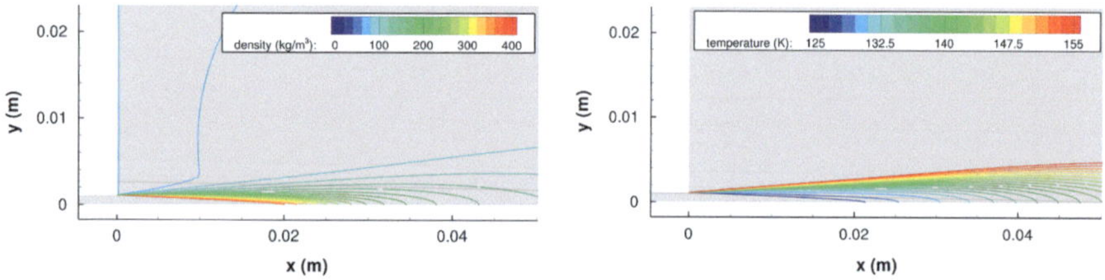

Fig. 3 Simulated density and temperature distribution close to the injector

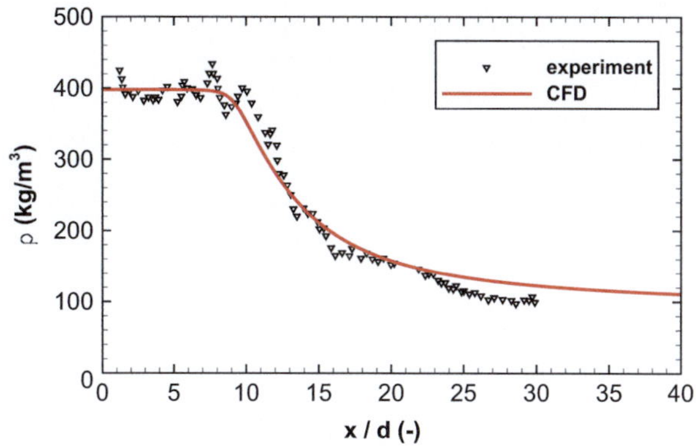

Fig. 4 Comparison of axial density profiles at the centerline between simulation and experiment

5 Simulation of Model Rocket Combustor

The DLR model rocket combustor investigated by Smith et al. [15] was studied numerically. Liquid oxygen and gaseous hydrogen are injected at cryogenic temperatures of 96 and 67 K and a pressure of 63 bar into a circular chamber with 50 mm diameter. Hydrogen is also used to cool the chamber walls. A 2D-axisymmetric simulation with a very fine grid (about 325,000 cells) is performed. Figure 5 displays the grid resolution close to the injector superimposed with contours of the oxygen radical. Tough the flame zone is thin, it is well resolved. The following setup is chosen for the presented simulation:

1. 5th order spatial discretization of inviscid fluxes with low diffusion multi-dimensional limiting process (MLP^{ld}) [7];
2. $k - \omega$ turbulence model of Wilcox [18];
3. turbulent Prandtl and Schmidt number $Pr_t = Sc_t = 0.7$;
4. adiabatic walls.

In the experiment, a highly turbulent and unsteady flame was observed. This is confirmed in the present simulation. Figure 6 presents contours of water mass fraction in the entire chamber. Instabilities in the mixing layer between the liquid oxygen and the hydrogen jet induce vortex roll-up, which improves mixing and thus combustion efficiency. In contrast to the experiment, no flame lift or even blow off close to the injector is observed in the simulation. Instead, it is stably anchored at the post-tip.

The interaction of large scale turbulent fluctuations with the flame leads to pulsations in the heat release, which in turn induces pressure oscillations and leads to unsteady injection conditions. This feedback mechanism can cause serious mechanical failure of the chamber structure when the frequencies of this physical phenomenon coincide with eigenfrequencies of the combustor geometry. In the

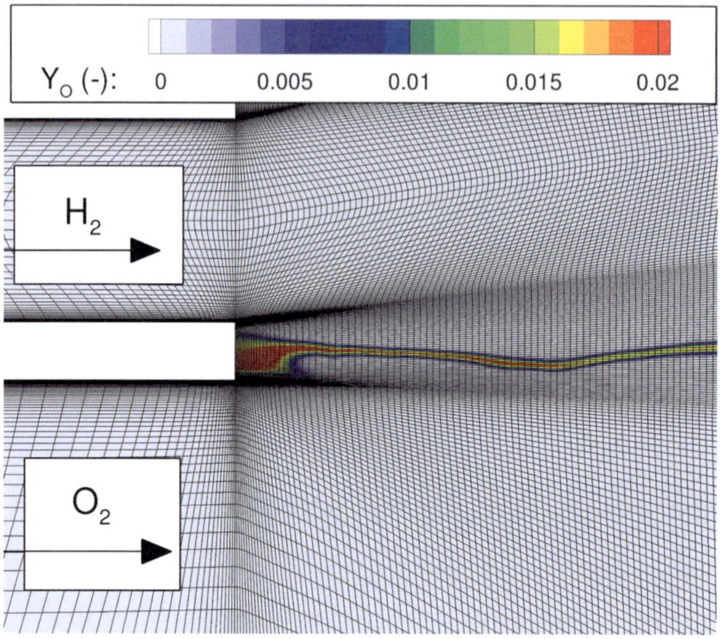

Fig. 5 Computational grid and contours of oxygen radical close to the injector in the DLR model rocket combustor

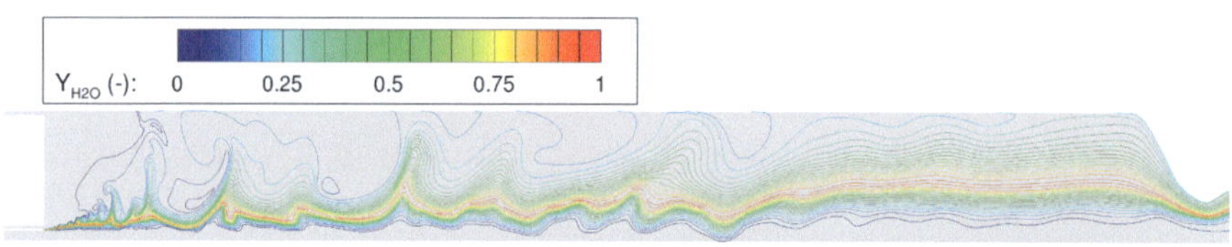

Fig. 6 Water mass fraction contours in the DLR model rocket combustor (compressed by factor of 2 in axial direction)

experiment, pressure oscillations with an amplitude of about 0.5 bar to 1.0 bar have been observed. This could also be confirmed in the simulation.

Temperature contours up to 500 K are plotted in Fig. 7. The highly turbulent and unsteady nature of the flame especially close to the injector is obvious. Low temperatures within the nozzle at the centerline indicate that some unburnt oxygen exits the combustor in the simulation. At the chamber wall, temperatures remain rather cool (below 300 K). This confirms the effective cooling with the cryogenic hydrogen film in the experiment.

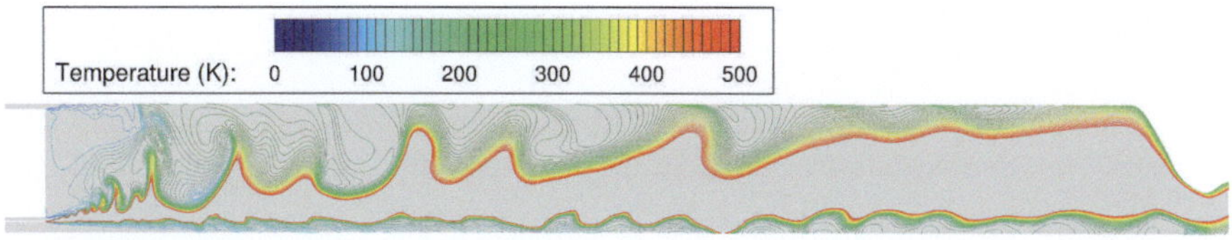

Fig. 7 Temperature contours in the DLR model rocket combustor (compressed by factor of 2 in axial direction)

6 Performance Comparison of HERMIT and HORNET

Supercritical injection conditions are present in most main stage rocket engines and therefore are very interesting for future research. Due to the complexity of the employed models and the resulting high computing times, the utilization of high performance computing systems is inevitable. Consequently, it is crucial to examine the performance of the code on the used platforms.

In the last two HLRS reports, comprehensive performance analysis for TASCOM3D on CRAY XE6 (HERMIT) and CRAY XC40 (HORNET) were performed [9, 10]. A very good scaling performance (strong and weak scaling) was observed.

During the last period, the focus for performance improvements in TASCOM3D was on parallel I/O using MPI. All data are now stored in a single binary file and handled by MPI I/O library routines.

7 Conclusion

A consistent thermodynamic model was implemented into the DLR in-house CFD code TASCOM3D for the simulation of rocket combustors with supercritical injection. A brief introduction to real gas thermodynamics and transport property modeling was given. Two high-pressure validation test cases were presented afterwards: a nonreactive cryogenic nitrogen jet dissolving into a warm nitrogen surrounding and a reactive simulation of liquid oxygen/gaseous hydrogen combustion in a model rocket combustor. The simulation results matched experimental observations very well in a qualitative and quantitative manner. The file handling of the code was improved by implementing parallel I/O routines, which utilize MPI I/O library routines.

Acknowledgements The presented work was performed within the framework of the SFBTR 40 funded by the Deutsche Forschungsgemeinschaft (DFG). This support is greatly appreciated. All simulations were performed on the Cray XE6 (HERMIT) and XC40 (HORNET/HAZEL HEN) cluster at the High Performance Computing Center Stuttgart (HLRS) under the grant number scrcomb. The authors wish to thank for the computing time and the technical support.

References

1. Banuti, D.T., Hannemann, K.: Thermodynamic interpretation of cryogenic injection experiments. In: 47th AIAA/ASME/SAE/ASEE Joint Propulsion Conference & Exhibit, San Diego (2011)
2. Branam, R., Mayer, W.: Characterization of cryogenic injection at supercritical pressure. J. Propuls. Power **19**(3), 342–355 (2003)
3. Coakley, T.J.: Turbulence modeling for high speed flows. AIAA Paper, No. 97-0436 (1992)
4. Ely, J.F., Hanley, J.M.: Prediction of transport properties. 1. Viscosity of fluids and mixtures. Ind. Eng. Chem. Fundam. **20**, 323–332 (1981)
5. Ely, J.F., Hanley, J.M.: Prediction of transport properties. 2. Thermal conductivity of pure fluids and mixtures. Ind. Eng. Chem. Fundam. **22**, 90–97 (1983)
6. Gerlinger, P.: Investigation of an assumed pdf approach for finite-rate chemistry. Combust. Sci. Technol. **175**(5), 841–872 (2003)
7. Gerlinger, P.: Multi-dimensional limiting for high-order schemes including turbulence and combustion. J. Comput. Phys. **231**, 2199–2228 (2012)
8. Gerlinger, P., Möbus, H., Brüggemann, D.: An implicit multigrid method for turbulent combustion. J. Comput. Phys. **167**(2), 247–276 (2001)
9. Keller, R., Lempke, M., Simsont, Y.H., Gerlinger, P., Aigner, M.: Parallelization and performance analysis of an implicit compressible combustion code for aerospace applications. In: High Performance Computing in Science and Engineering'14, Stuttgart, pp. 251–266. Springer (2015)
10. Keller, R., Seidl, M., Lempke, M., Gerlinger, P., Aigner, M.: Numerical simulations of rocket combustion chambers on massively parallel systems. In: High Performance Computing in Science and Engineering'15, Solán, pp. 251–266. Springer (2015)
11. Liou, M.S.: A sequel to AUSM, part II : AUSM$^+$-up for all speeds. J. Comput. Phys. **214**(1), 137–170 (2006)
12. Menter, F.R.: Zonal two equation k-ω turbulence models for aerodynamic flows. AIAA paper 93–2906 (1993)
13. Oefelein, J.C.: Large eddy simulation of a shear-coaxial $LOX-H_2$ jet at supercritical pressure. AIAA Paper 2002–4030 (2002)
14. Poling, B.E., Prausnitz, J.M., O'Connell, J.P.: The Properties of Gases and Liquids, 5th edn. McGraw-Hill, New York (2001)
15. Smith, J., Klimenko, D., Clauß, W., Mayer, W.: Supercritical Lox/hydrogen rocket combustion investigations using optical diagnostics. AIAA Paper 2002–4033 (2002)
16. Stoll, P., Gerlinger, P., Brüggemann, D.: Domain decomposition for an implicit LU-SGS scheme using overlapping grids. AIAA paper, pp. 97–1869 (1997)
17. Telaar, J., Schneider, G., Hussong, J., Mayer, W.: Cryogenic jet injection: decription of test case RCM-1. Technical report, 2nd International Workshop on Rocket Combustion Modeling, Lampoldshausen (2001)
18. Wilcox, D.C.: Formulation of the $k-\omega$ turbulence model revisited. AIAA J. **46**(11), 2823–2838 (2008)
19. Yang, V.: Liquid-propellant rocket engine injector dynamics and combustion processes at supercritical conditions. Technical report, Department of Mechanical Engineering, The Pennsylvania State University (2004)
20. Yoon, S.H., Kim, C., Kim, K.H.: Multi-dimensional limiting process for three-dimensional flow physics analyses. J. Comput. Phys. **227**(12), 6001–6043 (2008)

Two-Zone Fluidized Bed Reactors for Butadiene Production: A Multiphysical Approach with Solver Coupling for Supercomputing Application

Matthias Hettel, Jordan A. Denev, and Olaf Deutschmann

Abstract The application of multiphysical modelling is steadily increasing in the last decade, which also leads to a corresponding increase of the complexity and of the diversity of software packages used. To deal with this complexity, users of supercomputing clusters are often challenged to couple two or more software systems of different software vendors together. However, the combined use of complex software systems usually raises additional limitations, thus reducing considerably the efficiency of the parallel simulations. In the present work, an example of such complex software utilization has been shown and the particular limitations are identified. The most severe limitation for the current supercomputing simulations has been the relatively high RAM requirement per computing core. At this stage of the numerical investigation, in order to overcome the limitations, the software packages have been ported to a different, more suitable hardware architecture with increased RAM per node. This way, the efficient use of the parallel computational resources has been guaranteed which was confirmed by means of strong scaling tests.

Keywords CFD-DEM • Eulerian-Lagrangian approach • Strong scaling • Two-zone fluidized bed reactor • TZFBR

M. Hettel (✉) • O. Deutschmann
Institute for Chemical Technology and Polymer Chemistry (ITCP), Karlsruhe Institute of Technology, Engesserstr. 18/20, 76131 Karlsruhe, Germany
e-mail: matthias.hettel@kit.edu; olaf.deutschmann@kit.edu

J.A. Denev
Steinbuch Centre for Computing (SCC), Karlsruhe Institute of Technology, Hermann-von-Helmholtz-Platz 1, 76344 Eggenstein-Leopoldshafen, Germany
e-mail: jordan.denev@kit.edu

W.E. Nagel et al. (eds.), *High Performance Computing in Science and Engineering '16*, DOI 10.1007/978-3-319-47066-5_19

1 Introduction

The complexity of modern chemical processes and the corresponding technological equipment increases constantly. In accordance with this, and with the raising power of today's supercomputers, the modelling approaches become more comprehensive and more detailed. To achieve the desired complexity of modelling, very often two or more complex computer codes are combined together for a multiphysics simulation. The user then needs to ensure proper coupling and data exchange between the different software parts. While each of the codes usually is well-tuned for a special application area, their coherent work in a combined mode is not always a trivial task. The challenge becomes especially large, when the combined software should run efficiently also on supercomputers with a great number of CPUs/cores: in such a case non-expected limitations may suddenly arise and consequently new problems emerge with them.

The present work shows an example of such multiphysical modelling approach together with the challenges the authors were facing at the beginning of a new scientific project. The aim of the project is the numerical modelling of a two-zone fluidized bed reactor (TZFBR) in a laboratory-scale for production of the important basic chemical 1,3-butadiene from n-butane [5]. This process is a promising alternative to existing processes and sets an example for the possible production of numerous other chemicals in fluidized beds (e.g. olefin and synthesis gas production). The final target of the project is to model the complex interaction of all relevant processes including the gas-phase flow field, the movement of solid particles, the heterogeneously catalyzed reactions on the inner particle surface and the intra-particle transport phenomena.

The open-source platform CFDEM®coupling [1] is used in the project. The use of open public software for all levels of the simulations has two benefits. On the one hand, the whole source code is available and the implementation of additional features is straightforward. On the other hand, the findings will be helpful for the large community in science and technology which applies the aforementioned software tools.

As the work is at an early stage, the paper focuses on the calculation of the two-phase flow field. The limitations, standing in the way of the efficient use of a large number of cores, emerging from the increased complexity of the physical modelling as well as the combined software, are presented and discussed together with the solutions found so far. The computer resources required to run the codes on two different supercomputer architectures are compared.

2 The Engineering Problem and the Two-Zone Fluidized Bed Reactor

The importance of butadiene synthesis lies in the fact that it is one of the basic petrochemical products. Today's processes based on converting n-butane are exclusively implemented in two-stage procedures. The direct dehydrogenation of n-butane to 1,3-butadiene delivers very small yields, which is why the main products have to undergo a second dehydrogenation to attain 1,3-butadiene. Due to financial reasons, a single-stage process to produce 1,3-butadiene from n-butane is pursued. So far, no single-stage process is known for the synthetization of 1,3-butadiene where the yield and selectivity is large enough for an economic application.

Currently, two-zone fluidized bed reactors (see Fig. 1 left) show the highest yields as well as selectivity. The separation of the feeds oxygen and butane allows the creation of separated oxidation and reduction zones in the same reaction vessel, between which the catalyst is circulated, thus circumventing problems associated with the transfer of the catalyst between reactor and regenerator. The conversion takes place in the reaction zone located above the n-butane inlet using the lattice oxygen of the catalyst particles. In the regeneration zone at the lower part of the fluidized bed, coke depositions on the particles are burned and the lattice oxygen of the catalyst is filled up. After regeneration the catalyst particles penetrate again the reaction zone due to particle mingling inside the fluidized bed.

In the last years a two-zone fluidized bed reactor (TZFBR) was designed, built and experimentally investigated at the author's institute (ITCP) [5]. Figure 1 (left) shows a sketch of the reactor. The reactor consists of a 40 cm long quartz tube with an inner diameter of 28 mm. At the bottom of the reactor a frit (pore size 160–250 μm) holds the particles and homogenizes the incoming gas flow. Inside the reactor, a quartz tube with two holes at the end of the T-junction, serves as n-butane inlet. The product stream leaves the reactor at a side outlet.

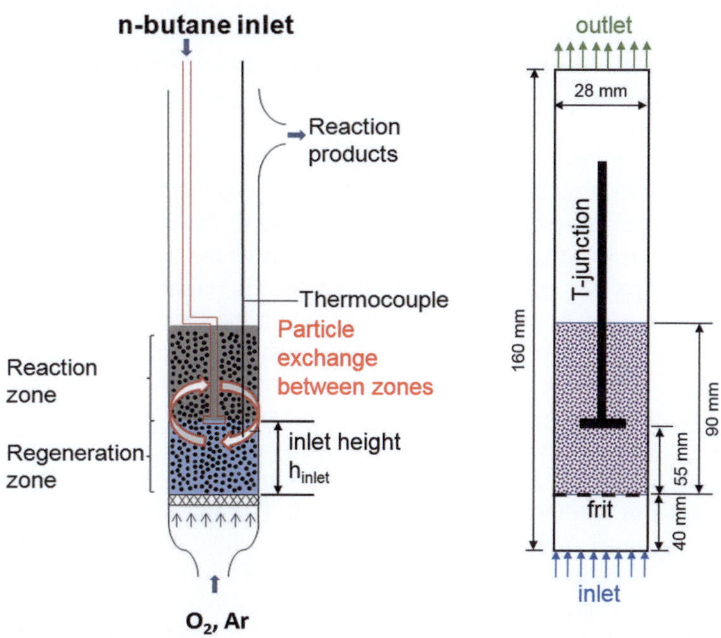

Fig. 1 *Left*: sketch of the two-zone fluidized bed reactor, measured experimentally [5]. *Right*: calculation domain and boundaries

The oxidative dehydrogenation of n-butane has been studied experimentally using various catalysts, among them two different Mo-V-MgO catalysts. At suitable conditions, the two-zone fluidized bed reactor can be operated at steady state performing chemical conversion and catalyst regeneration in a single vessel. The operating conditions temperature, flow velocity and oxygen/n-butane molar ratio were varied to maximize the 1,3-butadiene yield. Among other parameters, the height of the n-butane inlet is important for the process. Significant variations in conversion and even more in selectivity were observed dependent on these parameters. The experimental results indicate the strong sensitivity of the conversion from the subdivision of the bed into two zones, and the resulting flow and mixing conditions. The maximal yield of 1,3-butadiene was 32,7 %. So far, there has been no publication of similar large yields and selectivity, as the ones measured here on a TZFBR with the Mo-V-MgO catalysts.

3 Models and Computer Codes

The CFD-DEM method applied for the calculation of the two phase flow is a synthesis of CFD (Computational Fluid Dynamics) and DEM (Direct Element Method) to model coupled fluid-granular systems and is based on the Eulerian-Lagrange approach.

Fluid (gas or liquid) flows are governed by partial differential equations which represent conservation laws for the mass and momentum (Navier-Stokes Equations) and for additional scalar quantities, e.g. energy. Computational Fluid Dynamics (CFD) is the art of replacing such systems of partial differential equations by a set of algebraic equations which can be solved applying numerical methods using digital computers. Within the Eulerian approach, the gas phase is modeled as a continuum. The conservation laws (transport equations) are formally integrated over a finite volume and discretized on a numerical grid. Each node of the grid represents the volume averaged representation of a small section of the flow field. The solution procedure is always iterative and defective. The smaller the finite volumes (the larger the number of cells or grid points), the smaller the discretisation error.

The modeling of the particle phase is based on the Lagrangian approach. The motion of each particle of the system is calculated by integrating Newton‘s equation of motion. Various forces act on a single particle in a gas flow. The dominant forces in the present application are drag, contact and gravity. To describe the collision dynamics in the particulate flow the soft-sphere approach is used. The contact forces between particles are incorporated with mechanical models consisting of combinations of springs, dash-pots and sliders. The actual forces are calculated based on the small overlap between particles and allow direct integration of the particle displacement based on the contact forces. For a DEM simulation no numerical grid is necessary. The calculation domain is restricted by geometrical surfaces, with which the particles can interact (walls) or where they can leave the domain (openings).

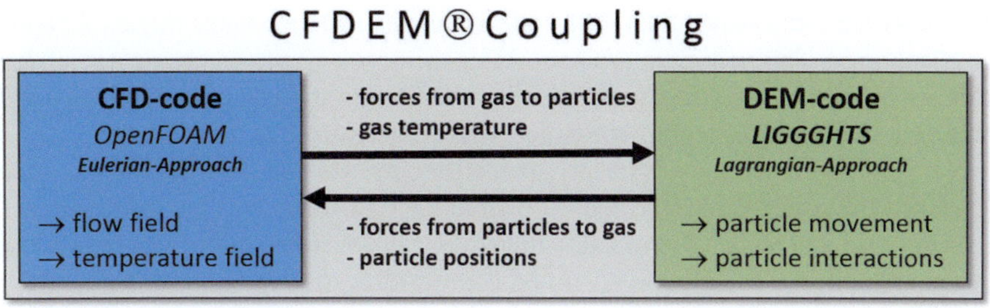

Fig. 2 Data flow between the software tools

We used the CFDEM®coupling software [1] (version 2.3.1) for calculation. This open-source platform couples the DEM engine LIGGGHTS®(version 3.3.0) [4] to the open source CFD code OpenFoam®(version 2.3.1) [7], see also Fig. 2. The name of the OpenFOAM solver is *cfdemSolverPiso*.

4 Calculation Procedure

The calculation domain (Fig. 1 right) comprises a cylinder with a length of 160 mm and a diameter of 28 mm. The frit acts as a wall for the particles and is positioned 40 mm above of the gas inlet. The T-junction is positioned at $h_{inlet} = 55$ mm above the frit.

The height of the bed is about 90 mm. The size of the particles in the real system is in the range of 160–250 μm. We used an average diameter of 205 μm, leading to a particle number of ca. 4.8E6.

The calculations were done for isothermal conditions (T = 500 K) without reaction (air only). The n-butane inlet was closed. The velocity at the lower inlet was 0.23 m/s. Under these conditions the flow is laminar and no turbulence model is needed. The gas leaves the calculation domain at the upper outlet.

Firstly, the reactor has to be filled with the particles which requires solely a DEM-calculation. After settling of the particles, the coupled CFD-DEM calculation can be started. On the one hand, the time-step of the particle movement has to follow the high frequency of the particle collision dynamics. On the other hand, the time-step has to be small enough to resolve the characteristic time in which the particles respond to a variation in the velocity of the surrounding flow. These restrictions lead typically to small time-steps for DEM calculations, 2.5E-6 s in our application.

The CFD-solver applies the PISO (Pressure-Implicit Split-Operator) approach for the pressure velocity coupling. Therefore, for the time-step of the CFD holds, that the Courant number (Co = time-step × velocity/cell-size) has to be smaller than one. We used a time-step of 2.5E-5 s.

The two codes calculate sequentially in cycles with a user defined coupling time of 2.5E-4 s. Within one cycle the CDF-code solves ten time-steps (in sum 2.5E-4 s physical time), afterwards the DEM-code calculates 100 time-steps (iagain 2.5E-4 s physical time). This is done alternately. After each cycle (2.5E-4 s physical time) the data which is necessary to capture the forces between the gas phase and the solid phase are interchanged among the codes. The information flux between the codes is depicted in Fig. 2. An anlysis of the calculation time yielded, that 53 % of the time is needed from the DEM-code and 47 % from the CFD-code.

After a typical physical simulation time of five physical seconds to put the system into operation, the calculation has to be continued to get time averaged quantities which can be analyzed and compared with experimental data. Therefore, a physical time of ca. 50 s is envisaged.

Figure 3 shows a snapshot of a calculation result. On the left side the fluidized bed including the T-junction is shown. The bed is divided in half vertically to illustrate the processes inside the bed. The colour represents the velocity magnitude of the particles. The regions with higher velocity (green/yellow/red color) indicate bubbly structures where the density of the particles per volume is smaller than in regions with lower velocity (blue and cyan color). In these structures gas is transported vertically through the bed. This process contributes to the mixture of the particles between the two regions of the bed. If a bubble reaches the surface of the bed, an eruption of particle clusters can be identified (right picture). To get an insight about the grid resolution, the right picture shows some surface cells of the reactor wall. As the geometrical data is in STL-format, each rectangular surface cell is divided with a diagonal line into two triangular subsections for the graphical representation.

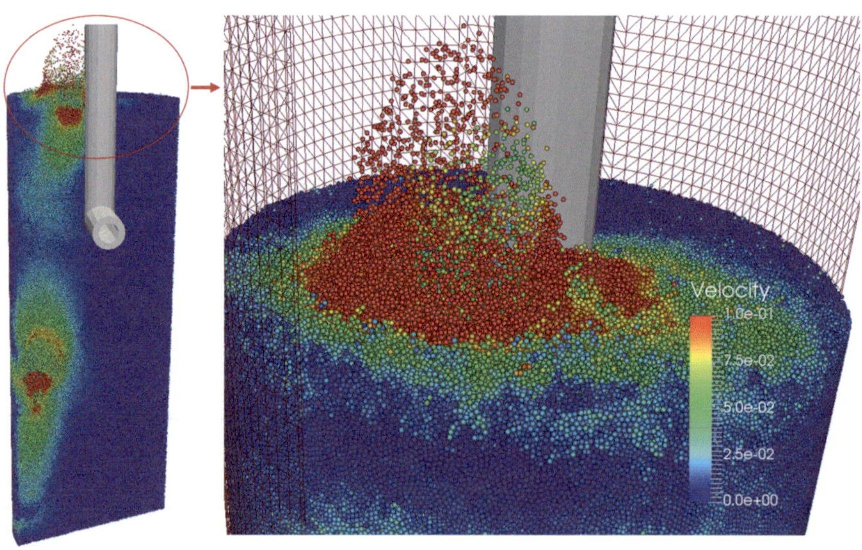

Fig. 3 Snapshot of results: fluidized bed as a whole (*left*) and detail near surface (*right*)

5 Computer Specifications

Two different supercomputers are used for the simulations. Their main features are briefly described below.

The research cluster of the state of Baden-Württemberg JUSTUS is located at the Communication and Information Center of the University of Ulm and is specialized for computational and theoretical chemistry. It is a high-performance massive parallel compute resource. Its intended use is mainly for chemistry-related jobs with high memory requirements (RAM and/or HDD). The supercomputer JUSTUS is suitable for user-jobs which have medium to low requirements to the node-interconnecting InfiniBand network.

In the present study computing nodes of JUSTUS with 128 GB DDR4-RAM have been used. Each node consists of two Intel Xeon E5-2630v3 (Haswell) processors (with 8 cores per processor, or, 16 cores per node) having a 2.4 GHz frequency and 20 MB cache per chip. The operating system is Red Hat Enterprise Linux 7. The interested reader can get further details from [3, 6].

OpenFoam 2.3.1 on this cluster was compiled with the Intel®compiler 15.0 and the corresponding MPI-library, version 5.0.3.

The massive parallel supercomputer ForHLR-I (recently being expanded by its second-stage complement ForHLR-II) has 512 nodes with 64 GB RAM and for the present study up to 32 of them have been used. Each node consists of 2 Deca-Core Intel Xeon E5-2670 v2 processors (Ivy Bridge) (with 10 cores per processor, or, 20 cores per node). The processors have a 2,5 GHz frequency (max. Turbo-frequency is 3,3 GHz).

Each one Deca-Core processor (Ivy Bridge) has 25 MB L3-Cache and operates the system bus with a frequency of 1866 MHz. Each Core has 64 KB L1-Cache and 256 KB L2-Cache memory, see also [2]. The network has one InfiniBand 4X FDR Interconnect. The operating system is Red Hat Enterprise Linux 6.x. For further details, please refer to [2].

OpenFoam 2.3.1 on ForHLR-I was compiled with the GNU compiler. Currently (april 2016), the default version of this compiler on the ForHLR-I supercomputer is version 4.9 and the default version of the Open MPI software is version 1.8.4.

6 Discussion of Current Limitations with Respect to the Efficient Use of the Targeted Parallel Computers

The limitations to be discussed in the following originate from different sources: the physical modelling, the software implementation, the supercomputers' architecture or combinations of them.

The CFD-DEM model requires, that the size of the particles is smaller than a portion of the fluid cell size. Optimally, the volume of the particles should not be larger than 30 % of the volume of a fluid cell. This is because the physical assumptions of the CFD-DEM method are only satisfied, if each cell contains a certain portion of fluid. If the cells are too small, it could happen, that a whole calculation cell is filled with solid material. For a particle size of 205 μm the minimal cell size has to be ca. 600 μm. For the modeling of the reactor (Fig. 1) a block-structured hexahedral grid with ca. 460.000 cells was generated. The smallest cell size in the grid was about 0.5 mm.

The number of fluid cells per computing core governs the relation between the amount of computational work on that core and the amount of the MPI-exchange of information: a small number of cells per core would lead to a small amount of work and a relatively large demand for information exchange. On the other hand, a large amount of cells per node will increase the total duration of the computations. A rule of thumb says that for a CFD calculation about 10.000–50.000 cells per core usually lead to a satisfactory ratio between computing time and communication, work thus ensuring an efficient use of the parallel resources with a good scalability. However, if the total number of fluid cells is limited, the number of total cores that can be utilized for the simulations necessarily becomes also limited. So, with 460.000 cells, the above rule returns a number between 9 and 46 cores. However, this number of cores is quite low, so that also larger number of cores have been utilized in the following tests. Thus, for the largest possible number of computing cores (256), the number of cells per core decreased to $\approx$1800.

Another restriction that needs to be considered here, is the restriction on the physical time step leading to an increase of the overall wall-clock time for the simulations. In the present simulations, there are two time-steps coupled together: for the fluid flow and for the particle tracking algorithm. The fluid flow time-step is restricted by the Courant condition, the corresponding time-step for the particle tracking is restricted by the model approach for the particle collision (see also Sect. 4).

Another, probably more severe restriction is coming from the handling of the Lagrangian particle tracking algorithm into the OpenFoam®. Currently, each domain for the Eulerian fluid flow contains the complete information about all particles. Because of the relatively high number of particles, the memory required per core remains nearly constant and decreases only very slightly with the number of cores. For the current calculations, each core needed $\approx$7 GB of RAM. This increases considerably the overall RAM requirements per node while quickly reaching the limits of the available RAM per node. For example, if a node of a supercomputer has 64 GB RAM (ForHLR-I, see Chap. 5), then only a number of 64 GB/7 GB $\approx$8 cores can be used. The rest of the cores (that means 8 out of 16 for ForHLR-I) are reserved, but not used. For this reason, the calculations shown later were performed on maximal 8 cores per node on ForHLR-I. On the other supercomputer – JUSTUS (see Chap. 5) there is no such limitation (128 GB RAM per node), but for reasons of compatibility of the results, tests with the same number of cores per node (8) have been made.

The CFDEM®Coupling software ensures a good distribution of the work load between the cores. The statistics given at the end of computations show that the cores are almost equally loaded: e.g. for the run on JUSTUS with 128 cores, the largest load ratio between any two cores is 1.005. Therefore, the load balance is not regarded as a factor which can decrease the efficiency of computations in the present study.

7 Results from the Strong Scaling

In the following, the results from the strong scaling tests on ForHLR-I and on JUSTUS are presented. The number of grid cells (Control Volumes) computed was kept constant, but the number of cores for the computations has been varied. The wall-clock time for each simulation had a duration of 12 h. The results from the simulations are measured and presented as the physical time in [s], advanced during the 12-h simulations, see Fig. 4.

The first performance test consists of using different number of nodes and different number of cores per node, while keeping the total number of cores constant. There are two opposite tendencies when using less cores per node: from one side, it leads to an increased MPI-communication through the node-interconnecting network while from the other side, the demand for accessing the RAM within each node decreases, which might become beneficial on the global level. This performance test was made only on JUSTUS: on the ForHLR-I there is not enough memory per node in order to carry out that test. Figure 4 reveals that for the present tests, using 8 cores per node (instead of 16) increases the physical time advanced on JUSTUS. Thus, the increased efficiency on the node level overpowers the increased communication need on the network level.

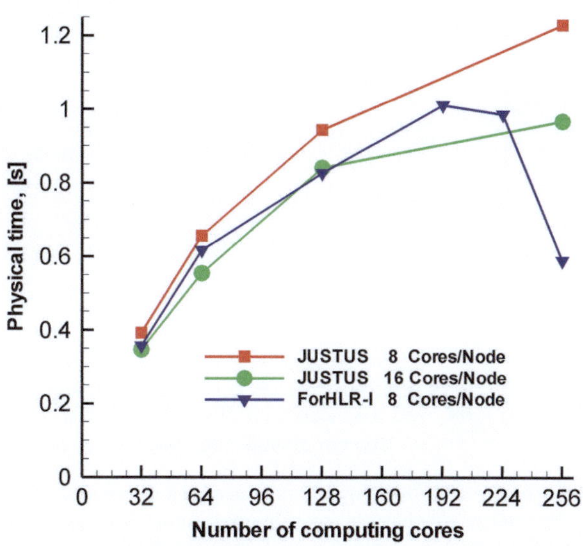

Fig. 4 Results from the strong scaling: physical time advanced vs. the number of cores which really took part in the computations

On the ForHLR-I supercomputer the performance with 32 and 64 computing cores (Fig. 4) is quite close to that of JUSTUS. The performance of ForHLR-I decreases noticeably for 128 computing cores and the physical time advanced even reduces for 256 cores. This was the reason to make two more additional tests on ForHLR-I with an intermediate number of computing cores (192 and 224). These additional tests allowed the more precise allocation of the performance reduction, which, according to Fig. 4 occurs around 200 computing cores. A possible explanation for this performance reduction might be the very small number of fluid cells per core leading to an increased need in MPI-communication through the InfiniBand network which is organized as a non-blocking two-level topology separated in groups of 18 nodes in one level. However, the proper investigation of this problem is a relatively time-demanding task which is planned to start in the near future.

For the above tests on ForHLR-I as well as the tests on JUSTUS with 8 cores per node, a great number of cores is reserved, but actually only a part of them is used for the real computational work. This certainly leads to a non-efficient use of the available resources. Thus, on JUSTUS, for the test with 8 cores per node, only 50 % of the cores are used (8 of totally 16 per node) and on the ForHLR-I – for all tests – only 40 % of the cores (8 of totally 20 per node) are used. Taking the total (reserved) number of cores into account, leads to the picture presented in Fig. 5. It can be seen in this Figure, that the computations with 16 cores per node on JUSTUS are definitely the most efficient when only a small number of total cores

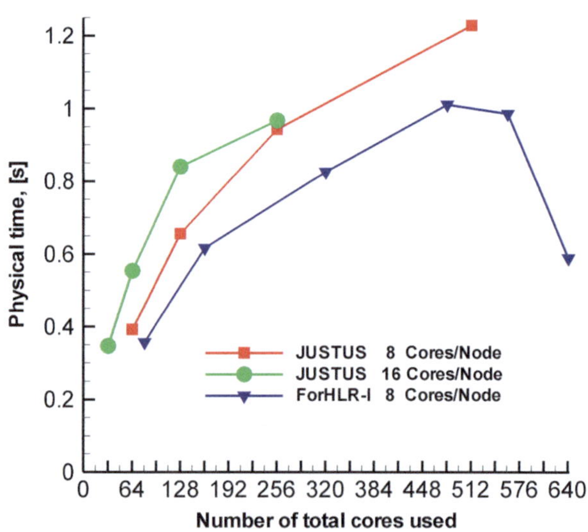

Fig. 5 Results from the strong scaling: physical time advanced vs. the total number of cores for a given simulation. This statistics includes all cores reserved for the particular simulation, although only a part of them took part in the computations. During the simulation, the all of the cores are not available to other users

is used (64 or 128). The two modes on JUSTUS (8 cores per node and 16 cores per node) become almost identical – in terms of advanced physical time – when 256 cores (total cores used) are taken (reserved) for the simulations. Unfortunately, the limitations on the maximum computing cores (see Sect. 6) did not allow following further the scaling trend on JUSTUS. The RAM limitations per node hindered a similar test (8 vs. 16 cores per node) to be carried out also on the ForHLR-I cluster.

As a whole, the scaling tests performed allowed gaining a first insight into the parallel efficiency of the simulations for the two-zone fluidized bed reactor. The limitations on each of the two massive parallel supercomputers have been identified and with this knowledge the actual production runs can be continued effectively. Using up to $\approx$200 computing nodes on ForHLR-I leads to a reasonable scaling of the simulations, however, there is a large overhead in terms of cores which are reserved, but not taking part in the computations. Therefore, moving the core of the simulations from ForHLR-I to JUSTUS is the best solution which allows the efficient use of up to 256 computing cores without any additional overhead.

8 Conclusions

The application of multiphysical modelling leads to a corresponding increase of the complexity and of the number of software packages used. In the present work, an example of such complex software combination has been shown and the particular limitations were identified. The most severe limitation for the current supercomputing simulations has been the RAM requirement per computing core. At this stage of the numerical modelling, in order to release the limitations, porting the software to a different, more suitable hardware platform with increased RAM per node turned out to be sufficient. This way, the efficient use of the parallel computational resources has been guaranteed.

One aim of the investigation has been to identify the limitations which hinder the efficient use of the parallel supercomputers. The limitations stem from combined features of the physical modelling, of the software implementation and of the supercomputers' hardware.

Although the CFD-approach demands, that the size of the fluids cells should be as small as possible, the size of the particles restricts the minimal dimension of the cells. This leads to a cell number of about 460.000 for the given size of the calculation domain. Consequently, the number of computing cores for the simulations is also limited. The usage of $\approx$256 cores seems to be a good compromise for the current application.

The size of the physical time step is dependent on the flow velocity, the particle dynamics and the numerical algorithm. These conditions limit the time-step for both, CFD and DEM, to a maximal value which should not be exceeded.

In the current version of the CFDEM®coupling software all particle data has to be available from each domain of the CFD-calculation. Because of the large number of particles (4.8E6), each core requires a RAM of $\approx$8 GB, independent from the number of cores used. The bottleneck for the RAM-usage lies in the coupling of the particle code to the CFD code. Here, much effect could be achieved, if every CFD-subdomain would only need the data for the particles which are inside this domain.

For the present investigation the JUSTUS supercomputer turns out to be more suitable than the ForHLR-I: the larger RAM per node on JUSTUS allows the efficient use of all cores in the nodes and a good scaling up to 256 computing cores. However, it is planned to integrate another, third software package to complement the existing two packages. This third software package allows considering chemical reactions and intra-particle transport phenomena, but also requires additional tests regarding the combined software performance.

Acknowledgements The simulations for the present work were partly supported by the bwHPC initiative and the bwHPC-C5 project [*A*1] provided through associated compute services of the JUSTUS HPC facility at the University of Ulm. The grant of supercomputing resources on the ForHLR-I supercomputer at the Steinbuch Centre for Computing of the Karlsruhe Institute of Technology for the project with acronym "butadiene" is highly appreciated.

The authors would like to thank Jürgen Salk from the Communication and Information Center of the University of Ulm (Competence Center for Computational chemistry), Alexandru Saramet from the University of Applied Sciences Esslingen (Competence Center for Engineering sciences) and Dr. Stefan Radl from the Graz University of Technology for their valuable help and advices during the software installation and the software adjustment processes. The authors would like to thank also to their colleagues Dr. Holger Obermaier and Richard Walter from SCC/SCS for the fruitful discussions.

The support of the Helmholtz programme "Supercomputing and Big Data" [*A*2] is also highly appreciated.

[A1] bwHPC and bwHPC-C5 (http://www.bwhpc-c5.de) funded by the Ministry of Science, Research and the Arts Baden-Württemberg (MWK) and the German Research Foundation (DFG).

[A2] The Programme "'Supercomputing & Big Data" https://www.helmholtz.de/en/research/key_technologies/supercomputing_big_data/

References

1. CFDEM®coupling Open Source CFD-DEM Framework: https://www.cfdem.com/ (2016)
2. Forschungshochleistungsrechner ForHLR I: https://www.scc.kit.edu/dienste/forhlr.php/ (2016)
3. Knowledge Base Wiki of Baden-Württemberg's HPC services: https://www.bwhpc-c5.de/wiki/index.php/Category:BwForCluster_Chemistry/ (2016)
4. Open Source Discrete Element Method Particle Simulation Code LIGGGHTS®: https://www.cfdem.com/ (2016)
5. Rischard, J., Antinori, C., Maier, L., Deutschmann, O.: Oxidative dehydrogenation of n-butane to butadiene with mo-v-mgo catalysts in a two-zone fluidized bed reactor. Appl. Catal. A: Gen. **511**, 23–30 (2016)
6. The bwForCluster for computational and theoretical Chemistry JUSTUS: https://www.uni-ulm.de/einrichtungen/kiz/service-katalog/wissenschaftliches-rechnen/justus.html/ (2016)
7. The Open Source CFD Toolbox OpenFOAM: https://www.openfoam.org/ (2016)

Part IV
Computational Fluid Dynamics

Ewald Krämer

A great number of research projects related to CFD with excellent scientific quality were run on the supercomputers of the HLRS in Stuttgart and of the SCC in Karlsruhe during the reporting period. Valuable fundamental as well as application-oriented knowledge could be attained from the simulation results, which became possible only through the extensive use of High Performance Computing. It is without saying that the access to supercomputers is crucial for successful research in Fluid Dynamics – today and even more in the future. This year, 37 annual reports had been submitted and underwent a peer review process. Due to limited space, only 17 contributions could be selected for publication in this book, which means that a number of high-qualified reports had to be rejected. Even though the presented collection cannot entirely represent an area this vast, the selected papers demonstrate the state-of-the-art use of high-performance computing in Germany. The spectrum of the projects is wide in several respects. Fundamental as well as application-oriented problems of industrial relevance were addressed using in-house, commercial, and open source codes (the latter two of which made up grounds with respect to massive parallel performance). Various established numerical methods as Finite Volume and Lattice Boltzmann methods, but also relatively new methods (at least in the context of CFD), as Smoothed Particle Hydrodynamics or Discontinuous Galerkin methods were employed. All CFD simulations presented in this book were either run on the Cray XC40 Hornet/Hazel Hen in Stuttgart (Europe's fastest supercomputer according to the HPCG benchmark) or on the ForHLR I in Karlsruhe.

E. Krämer (✉)
Institut für Aerodynamik und Gasdynamik, Universität Stuttgart, Pfaffenwaldring 21, 70550 Stuttgart, Germany
e-mail: kraemer@iag.uni-stuttgart.de

For many years, the working group of Munz at the Institute of Aerodynamics and Gas Dynamics (IAG), University of Stuttgart, has been developing a Discontinuous Galerkin based high-order simulation framework. DG methods can be considered as a combination of a finite-element scheme (with a continuous higher order polynomial in each grid cell) and a finite-volume scheme (allowing for discontinuities at the cell faces, which are handled by a Riemann solver) and provide a superior parallel performance if implemented appropriately. The latest fluid dynamics code from this framework, FLEXI, which uses a spectral element method (DGSEM), has increasingly been employed for real industrial application in recent years. One example is given by Hempert, Boblest, Hoffmann, Offenhäuser, Sadlo, Glass, Munz, Ertl, and Iben. They simulated a high-pressure throttle and jet flow, which serves as a simplified model for a gas injector in automotive combustion engines. Their studies assess the transient development and penetration of the gaseous jet. As shocks appear in such cases, a shock-capturing technique was applied based on a Finite Volume subcell method to avoid near shock oscillations and under-resolved scales. An efficient load-balancing strategy was implemented to remove the imbalances caused by the shock-capturing and to maintain the high parallel efficiency of the code. The ongoing work has been a cooperation between the IAG, the Robert Bosch GmbH, the Visualisation Research Center of the University of Stuttgart, the HLRS, and the Interdisciplinary Center for Scientific Computing of the Heidelberg University. The simulations were performed on the Hazel Hen.

The next two contributions are from the Institute of Thermal Turbomachines (ITS) of the Karlsruhe Institute of Technology (KIT). There, an inhouse code based on a Lagrangian, mesh-free Smoothed Particle Hydrodynamics (SPH) method has been developed during the last few years. In such methods, which are relatively new in the context of computational fluid dynamics, the spatial discretization of a computational domain is done via so-called particles, which represent a certain volume of the fluid. These Lagrangian particles move within the domain with the local flow velocity. The simulations were run on the ForHLR I cluster at the SCC displaying a very good parallel performance of the code. In the first report, Wieth, Braun, Chassonnet, Dauch, Keller, Höfler, Koch, and Bauer simulated the temporal evolution of droplet deformation at low aerodynamic loads, which plays a significant role in liquid fuel atomization processes. The deformation dynamics of single-fluid droplets as well as of fuel droplets with water added to the inside of the droplet was investigated. To validate the SPH-code for this type of application, a comparison of the results to well-known empirical findings was done for the pure liquid droplets, showing an excellent matching. The authors conclude that the SPH-code is capable of predicting droplet deformation dynamics physically correct.

The second SPH application, presented by Braun, Koch, and Bauer, deals with the numerical prediction of primary atomization taking place e.g. in air-assisted atomizer nozzles of jet engines. The focus is on the liquid disintegration processes, i.e. on the breakup behavior and the spray characteristics. The test case is derived from an experimentally investigated set-up and consists of up to 1.2 billion particles with a spatial resolution of roughly 5 μm. 2560 cores were used in parallel for these simulations. Comparisons to the experimentally observed features as well as to the results of established CFD tools using Volume-of-Fluid (VoF) solvers show good agreement, demonstrating that the SPH method is an adequate tool for predicting multi-phase flows.

The aim of the work described by Förster, Mink, and Krause from the Institute of Mechanical Process Engineering and Mechanics of the KIT is to achieve a more accurate characterization of the flow domain and flow dynamics especially in complex geometries. This shall be done by coupling existing (lower resolution) experimental data, e.g. obtained from Phase Contrast Magnetic Resonance Imaging (PC-MRI) in medical applications, with numerical simulations. The idea is to formulate this fluid flow domain identification problem as an optimization problem, which minimizes the differences between a given and a simulated flow field. The proposed gradient-based solution strategy makes use of an adjoint lattice Boltzmann method (ALBM). The authors' novel sensitivity based so-called first-optimize-then-discretize approach relies on first deriving an adjoint equation on a continuous basis and then discretizing it, which allows maintaining the excellent parallel efficiency known from LB methods in general. Using the open source software OpenLB, developed by the working group Computational Process Engineering at the KIT, a very good efficiency on massive parallel HPC has been achieved, and also the single core performance could be improved significantly. In the article, preliminary results are shown for a generic domain identification test case.

Ye and Tiedje from the Institute of Industrial Manufacturing and Management, University of Stuttgart, analyze in their contribution the dynamics of paint drops impacting onto dry surfaces. The special focus is on the air entrapment at the droplet-solid interface. Both, Newtonian and non-Newtonian droplets are simulated showing different results with respect to the creation of air discs and air bubbles during drop spreading. The VoF method implemented in the commercial CFD code ANSYS-FLUENT was used for a comprehensive parametrical study performed on the CRAY XC40 of the HLRS. The results of the investigations provide a new insight into the mechanism of air entrapment during drop impact onto solid surfaces.

Also Reitzle, Roth, and Weigand from the Institute of Aerospace Thermodynamics, University of Stuttgart, investigated the impact of droplets on dry solid walls. In their case, liquids are were used that show a non-Newtonian shear thinning behavior. Due to their different viscosities, their spreading behavior is slightly different. The simulations were performed on the CRAY XC40 using the in-house code Free Surface 3D (FS3D), which predicts incompressible multiphase flows based on the Volume-of-Fluid method. Scaling tests revealed that the speed-up of the code is not ideal due to the multigrid solver used to solve the pressure Poisson equation. However, a new multigrid solver library is being implemented, which is expected to significantly improve both the serial and the parallel performance of the code.

The reduction of viscous drag, especially turbulent skin-friction drag, is desirable for many fluid mechanical applications. During the last decades, various flow control strategies based on near-wall forcing have emerged, which show promising potential, at least for relatively low Reynolds numbers. However, due to missing experimental and numerical data, the efficiency of such control technologies at higher Reynolds numbers relevant for most industrial applications is still an open question. Davide Gatti from the Institute of Fluid Mechanics at the KIT therefore has addressed the effect of increasing Reynolds number on the achievable skin-friction drag reduction for a channel flow with enforced streamwise travelling waves of the spanwise wall velocity as control strategy. By means of Direct Numerical Simulations (DNS), he performed a comprehensive parameter study in two steps. First, 4020 cases were simulated in a small domain for different parameters of the spanwise forcing at two Reynolds numbers. These computations were performed as contemporaneous serial runs partly on the Blue Gene/Q system at the CINECA computing center in Bologna and partly on the For HLR I at the SCC in Karlsruhe. Additionally, a second set of computations for a few representative cases were conducted within a large domain. The results of both datasets are discussed in detail and maximum net saving rates are given. The author also derives an equation for the extrapolation of the drag reduction to higher Reynolds numbers. Based on this equation, he states that the decrease in drag reduction efficiency for higher Reynolds numbers is notably lower than the available pure low Reynolds number data bases suggest.

Control strategies for turbulent boundary layers have also been in the focus of Alexander Stroh of the same institute, who performed DNS computations for a turbulent channel flow. In contrast to Gatti, who applied spanwise forcing in a fully developed boundary layer along the whole wall area, Stroh has focused on localized control, which can easier be realized in industrial applications. In the work presented, he investigates two different drag reduction control methods for a spatially developing turbulent boundary layer and analyses in particular the flow behavior downstream of the control region. He also compares the efficiency of the flow control in a fully developed turbulent channel flow and in a developing boundary layer, finding that there are significant differences in the mechanisms behind the drag reduction. Up to 240 Mio grid nodes were used for his main configuration setup, and the simulations were performed on 256 parallel cores each, with different simulations running concurrently on the ForHLR I.

The next three contributions describe the results of different projects running under the biannual "Call for Large-Scale Projects" of the Gauss Centre for Supercomputing (GCS). Projects considered in these calls require more than 35 million core hours per year. The first paper by Axtmann and Rist from the IAG in Stuttgart presents a study of the scalability and MPI characteristics of OpenFOAM on the CRAY XC40 Hazel Hen at the HLRS. Direct Numerical Simulations for a three-dimensional laminar cavity flow and Large Eddy Simulations for a backward facing step were performed. Strong and weak scaling speedups as well as imbalance rates are displayed for two different compilers. In addition, the performance of the MPI routines ISend, Recv, and Waitall is compared. The tool CrayPAT was employed for doing profiling and tracing during the runs. The study gives insight in the parallel behavior of OpenFOAM version 2.3.0 on massively parallel computers.

Wilke and Sesterhenn, TU Berlin, Fachgebiet Numerische Fluiddynamik, summarize the work performed within two separate projects, both dealing with subsonic and supersonic jets impinging on a flat plate. The first part is dedicated to heat transfer enhancement, whereas the second part refers to sound source mechanisms.

In both cases, an in-house code was used that directly solves the governing Navier-Stokes equations in a characteristic pressure-velocity-entropy-formulation. To avoid Gibbs oscillations in the vicinity of shocks, an adaptive sock-capturing filter was used. An excellent scaling behavior of the code on the Hazel Hen is demonstrated up to 16,384 cores. For the highest Reynolds number investigated, a mesh with more than one billion grid points was used. Impinging jets are known to be an effective cooling means, and the amount of heat transfer can even be increased with pulsating inlets. The aim of the first project was to get some insight into the underlying physics behind this increase. Earlier investigations of a non-pulsating jet had revealed that periodically occurring vortex rings are responsible for an additional heat transfer. The authors show that the pulsation strongly amplifies these vortices. In the second part, the open question is addressed, how the sound waves in the feedback loop that is responsible for the generation of impinging tones, are produced. The authors could observe the feedback loop in their direct numerical simulations, and they can show that the interaction between vortices and stand-off shocks produce the sound waves by two different mechanisms, either by shock-vortex- or by shock-vortex-shock-interactions.

At the Institute of Aerodynamics of the RWTH Aachen, over many years a high-fidelity, massively parallelized flow solver using the MILES (monotone integrated LES) approach has been applied very successfully to various aerodynamic and aero-acoustic problems. The code runs on locally refined Cartesian hierarchical meshes. In their present contribution, Pogorelov, Cetin, Moghadam, Meinke, and Schröder describe latest results of their simulations of the flow fields and the acoustic fields of a ducted axial fan and a helicopter engine jet. For this purpose, a hybrid method was chosen, where the flow fields including the aero-acoustic sources were predicted by a highly resolved LES computation and, subsequently, the acoustic near and far fields were determined by solving the acoustic perturbation equations. The focus of the rotating fan simulations lay on the evaluation of the effect of the tip-gap size. It is shown that, in accordance to measurements, a larger tip-gap size produces stronger tip vortices and a higher broadband noise level. In the second part, jets from helicopter nozzles with different built-in components are compared to each other. The components have a strong impact on the acoustic near field, which is explained by its effects on the turbulent wake structures.

Not least owing to a long-lasting, very successful research work performed in the helicopter group of the IAG, the structured Finite-Volume code FLOWer, originally developed by the German Aerospace Center (DLR), has established as a very reliable, high-fidelity CFD-code for helicopter flow simulations. Many useful features have been implemented during the last years, among others a high-order reconstruction scheme, necessary for vortex dominated flow conservation. In their present contribution, Kowarsch, Hofmann, Keßler, and Krämer report on the latest enhancement, the implementation of unstructured grid handling into the code. This hybrid mesh approach allows for easier grid generation in the near body regions, whereas off-body regions can still be resolved with structured, preferably Cartesian meshes, in combination with computationally efficient higher-order numerical schemes. Validation was performed with a forward facing step, and results are shown for a complete helicopter in forward flight. Additional effort has been spent to further optimize the code with regard to its application on HPC systems. Multi-blocking and an efficient load balancing taking into account the respective mesh type and numerical scheme of the individual blocks are used. Furthermore, thanks to valuable support from the teams of HLRS and CRAY, the parallel performance on the CRAY XC40 Hazel Hen could be improved, facilitating the efficient use of more than 1000 nodes.

Chu and Laurien from the Institute of Nuclear Technology and Energy Systems, University of Stuttgart, investigated the heat transfer problem arising in the cooling system of nuclear power plants or heavy-duty coolers. Direct numerical simulations of supercritical carbon dioxide flow in a heated vertical pipe including buoyancy effects were performed for low Mach number flows with varying density using the open-source code OpenFOAM. Bulk properties, average flow field and secondary flow, and turbulence statistics are analyzed in detail. Scaling tests reveal a good speedup up to 1400 cores on the Hazel Hen. The findings of this work can help develop new turbulence models for this kind of practical applications.

OpenFOAM was also used by Stens and Riedelbauch from the Institute of Fluid Mechanics and Hydraulic Machinery, University of Stuttgart. They simulated a fast transition from pump mode to generating mode in a model scale reversible pump turbine. Such machines are used in pumped storage power plants, which are an efficient way to store energy at a large scale. However, the current procedures for changing from one operating mode to the other is still time consuming. The aim of the project is to understand the flow mechanisms during a change of operating modes, in order to develop faster maneuvers that do not damage the machine. Results for two different mesh sizes are presented for different monitor points. Furthermore, the flow field in the runner is analyzed at different points of time. The simulations were run on the ForHLR I at the SCC in Karlsruhe. Adequate speedups were achieved for 40 cores for the coarse and 120 cores for the fine mesh.

The next contribution is from the same institute. Here, Krappel and Riedelbauch present the results of their transient flow simulations in a Francis turbine at part load conditions. The flow field in the draft tube of the turbine at these conditions is dominated by the vortex rope phenomenon, which requires a very high resolution in space and time and an appropriate turbulence model. The authors applied the commercial code ANSYS CFX (in different versions) with two different turbulence models (the RANS-SST model and the scale resolving SST-SAS model). The meshes used in the study were in the range between 16 and 300 Mio nodes. The differences in the resolved flow structures are displayed for the various meshes and/or turbulence models. Additionally, different numerical schemes were used for the spatial discretization, which also have an effect on the predictions. The strong scaling behavior is shown for the different versions of ANSYS CFX, clearly indicating a significant parallel performance improvement from V16.0 to V17.0.

Mansour, Kaltenbach, and Laurien from the Institute of Nuclear Technology and Energy Systems, University of Stuttgart, present an application oriented CFD model for predicting the heat and mass transfer between large droplets and gas during the spray cooling process in an nuclear reactor containment with an Euler-Euler two-fluid approach. The resistance to droplet heating is taken into account, as this affects the phase change, too. In the context of reactor safety analyses, the application of CFD methods is quite new. Hence, experience has still to be gained in respect to the required mesh resolution for an appropriate prediction of the complex containment flow. In order to estimate the discretization error, a grid convergence study for a three dimensional natural convection flow in a model containment using the commercial ANSYS CFX code was performed, the results of which are displayed. A good scalability of the code on the CRAY XC40 is demonstrated.

At the IAG in Stuttgart, a working group is engaged in CFD simulation of wind turbines. As in the helicopter group, they routinely have used the FLOWer code, which is well validated for wind turbine applications and has a high parallel performance on the CRAY XC40 Hazel Hen. In the last contribution of the present section, Fischer, Klein, Lutz, and Krämer investigate the unsteady aerodynamics of a novel two-bladed and a three-bladed wind turbine. The latter is equipped with an innovative load reduction device, which consists of coupled leading and trailing edge flaps. To resolve the unsteady aerodynamic effects induced by the flap deflections properly, the focus of the investigation was on the influence of the temporal discretization. The two-bladed turbine was exposed to a 30° yawed inflow, and the induced unsteady loads were evaluated. It was found that the decrease in power output due to the yawed inflow is smaller than known from literature.

Not only the number, but also the large thematic variety of ambitious projects performed during the reporting period at the HLRS in Stuttgart as well as at the SCC in Karlsruhe in the field of Computational Fluid Dynamics is still impressive. None of them could have been realized without access to leading edge HPC facilities. This demonstrates the high value and the indispensability of supercomputing in this area. The upgrade of the CRAY XC40 at the HLRS from Hornet to Hazel Hen, which started in August 2015, has provided new opportunities to the CFD community, as simulation times have decreased and even more sophisticated fluid dynamic problems can be tackled now. It is without saying that the researchers have to permanently optimize their codes in order to achieve the optimal performance on the respective system architecture. Thanks are due to the staff of the HLRS and of CRAY for their valuable support to the individual projects in this respect.

High-Pressure Real-Gas Jet and Throttle Flow as a Simplified Gas Injector Model Using a Discontinuous Galerkin Method

Fabian Hempert, Sebastian Boblest, Malte Hoffmann, Philipp Offenhäuser, Filip Sadlo, Colin W. Glass, Claus-Dieter Munz, Thomas Ertl, and Uwe Iben

Abstract Industrial devices such as gas injectors for automotive combustion engines operate at ever-increasing pressures and already today reach regimes beyond the ideal-gas approximation. Numerical simulations are an important part of the design process for such components. In this paper, we present a case study with a computational fluid dynamics code based on the discontinuous Galerkin spectral element method with a real-gas equation of state. We assess a high-pressure throttle and jet flow as a basic model of a gas injector. We apply a shock-capturing method to achieve a robust simulation, and a newly developed method to maintain high efficiency despite load imbalances introduced by the shock capturing. The results indicate a dynamic mass flow rate at different pressure ratios between the inlet and outlet.

M. Hoffmann (✉) • C.-D. Munz
Institute for Aerodynamics and Gas dynamics, University of Stuttgart, Pfaffenwaldring 21, 70569 Stuttgart, Germany
e-mail: hoffmann@iag.uni-stuttgart.de; munz@iag.uni-stuttgart.de

S. Boblest • T. Ertl
Visualization Research Center, University of Stuttgart, Allmandring 19, 70569 Stuttgart, Germany
e-mail: sebastian.boblest@visus.uni-stuttgart.de; thomas.ertl@vis.uni-stuttgart.de

F. Hempert • U. Iben
Robert Bosch GmbH, Robert-Bosch-Campus 1, 71272 Renningen, Germany
e-mail: Fabian.Hempert@de.bosch.com; Uwe.Iben@de.bosch.com

P. Offenhäuser • C.W. Glass
High Performance Computing Center, University of Stuttgart, Nobelstrasse 19, 70569 Stuttgart, Germany
e-mail: offenhaeuser@hlrs.de; glass@hlrs.de

F. Sadlo
Interdisciplinary Center for Scientific Computing, Heidelberg University, Im Neuenheimer Feld 205, 69120 Heidelberg, Germany
e-mail: sadlo@uni-heidelberg.de

W.E. Nagel et al. (eds.), *High Performance Computing in Science and Engineering '16*, DOI 10.1007/978-3-319-47066-5_20

1 Introduction

Product development in all branches of engineering, and especially in the automotive industry, faces the challenge to further improve technical devices that already have reached a high level of sophistication, or to develop new technologies that outperform existing ones. To make such an ongoing improvement possible, it is vital to have a deep understanding of root-cause relationships in the involved physical processes. High-quality numerical simulations are a very promising choice to gain insights into many physical mechanisms that are hardly accessible by means of experiments. Hence, their influence on today's design processes continues to rise and makes virtual product development more and more feasible.

In the case of fuel injection, numerical simulations must be able to describe the physical behavior of the involved flow adequately. Further, the entire simulation framework needs to be able to compete with the high precision that is achievable in modern experimental setups or even outmatch them in some cases. It also has to be similarly cost-efficient as, for example, rapid prototyping techniques and must yield quantitative statements about device properties in at least a comparably small time frame. To achieve that, modern simulation codes must be adapted to the design of today's supercomputers with huge numbers of cores, with respect to their parallelizability and scalability, to reach maximum performance.

In the present paper, we study the capabilities of our computational fluid dynamics code FLEXI on the example of gas injection under high pressure, which is very relevant for present-day and future automotive combustion engines [1, 2, 22]. Several recent studies are concerned with the temporal development of gas jets [21, 23]. These studies asses the transient development and penetration of a gaseous jet.

One key property of injector components is the mass flow, because it, together with the injector opening, directly determines the amount of fuel available for combustion. Consequently, it has major influence on the power generated by the combustion engine and on the overall system behavior. For ideal gases, there exist a number of relations, which allow for the estimation of mass flow through throttles [4], but for real gases, the situation is more difficult. However, the injection pressures of modern gas combustion engines makes it indispensable to take real-gas effects into account [14]. Therefore, we evaluate the mass flow rate for a basic throttle geometry, which represents a simplified model of an injection system, at high pressures with a tabulated real-gas equation of state for methane.

The simulation of real-gas jet flow is complicated by the occurrence of shocks that need to be handled numerically, usually by employing shock-capturing techniques. These methods need to be flexible enough to cope with changing flow patterns. We use such a method in our simulations, together with a sophisticated load balancing strategy to remove the imbalances that it causes.

Altogether, the present paper aims at demonstrating a simulation framework based on the discontinuous Galerkin spectral element method, which is capable of representing real-gas flow with shocks, while maintaining an excellent parallelization efficiency.

2 Modeling, Discretization, and Visualization

We use the computational fluid dynamics (CFD) code FLEXI that we develop with the application in industrial environments in mind [6]. It is based on the discontinuous Galerkin spectral element method (DG SEM), which is a high-order accuracy method and yields great potential for parallel scaling [3, 8]. For a detailed description of DG SEM, together with a discussion of its parallelization efficiency, the reader is referred to Hindenlang et al. [11].

The basis for our calculations are the compressible Navier-Stokes equations (NSE). To close the NSE, an equation of state (EOS) is needed. This is achieved by either using an analytical formulation, or a tabulated approach, as we do it in the present paper. Our ansatz is based on the idea of Dumbser et al. [9]. We generate the data for our tabulated EOS with the CoolProp library [5]. This allows us to represent the EOS over a wide range of all thermodynamic variables with excellent accuracy. We consider methane, as it is the main component of natural gas [12].

To avoid Gibbs type oscillations near shocks or under-resolved scales, we use a detector proposed by Persson and Peraire [18] to detect regions, where such oscillations might occur, and then apply the finite volume (FV) subcell method [19, 20] in a slightly modified form to prevent their occurrence. The original FV method uses Gaussian-distributed subcells, while in our case they are distributed equidistantly for increased accuracy.

We developed a reader plugin for ParaView to visualize our simulations [6] that runs on the Hazel Hen. In recent years, several methods have been developed to directly visualize data from high-order CFD solvers without resampling [7, 13, 15–17].

Within that plugin, however, we still use a resampling method with user-defined resolution for DG elements and a fixed resolution for FV subcells. By also integrating our EOS tables into the plugin, we can visualize all simulation variables in ParaView, without the need to store all of them in our state files. This strategy keeps our storage requirements low.

3 Simulation Strategy

We use a throttle geometry with diameter $D = 0.5 \times 10^{-3}$ m and length $L = 4D$. This geometry is a simplified representation of a gas injector. The simulation domain is represented using an unstructured hexahedral mesh. An overview of the simulation domain, together with a section view of the mesh, is given in Fig. 1. The mesh has the highest resolution at the boundaries of the throttle wall, around the throttle exit, and downstream of the throttle. Downstream of the throttle, the gas is injected into the open and forms a jet. The computation mesh consists of 83,732 elements. For the current assessment, we apply a 4th-order spatial discretization, and use a 4th-order low-storage Runge-Kutta scheme for the temporal discretization.

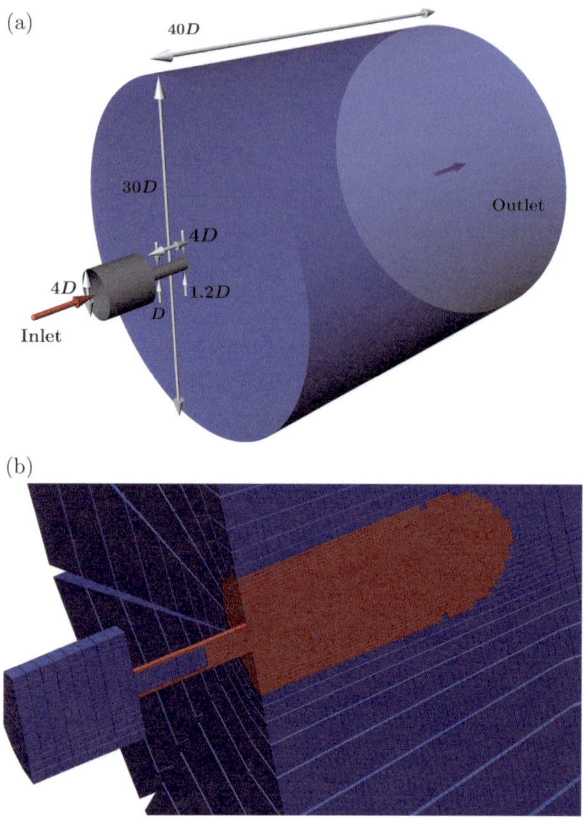

Fig. 1 (**a**) Overview of simulation domain. (**b**) Sectional view of the simulation mesh, with the high-resolution region in *red*

This corresponds to $4^3 = 64$ degrees of freedom (DOF) per element and 5,358,848 DOF in total. For the production runs, we used 1200 computation cores leading to 4465 DOF per core.

4 Results and Discussion

We study the described configuration for different pressure ratios $\mathcal{R}_p = p_i/p_o$ of inlet pressure p_i to outlet pressure p_o. The investigated cases are $\mathcal{R}_p = 1.25$, 1.67, 2.50, 2.86, 3.33, and 5.00, and include both subsonic and supersonic flow conditions. For approximately $\mathcal{R}_p \lesssim 2.50$ the flow is subsonic, for $\mathcal{R}_p \gtrsim 2.50$ the flow starts to become supersonic, and is eventually chocked at $\mathcal{R}_p \gtrsim 3.30$.

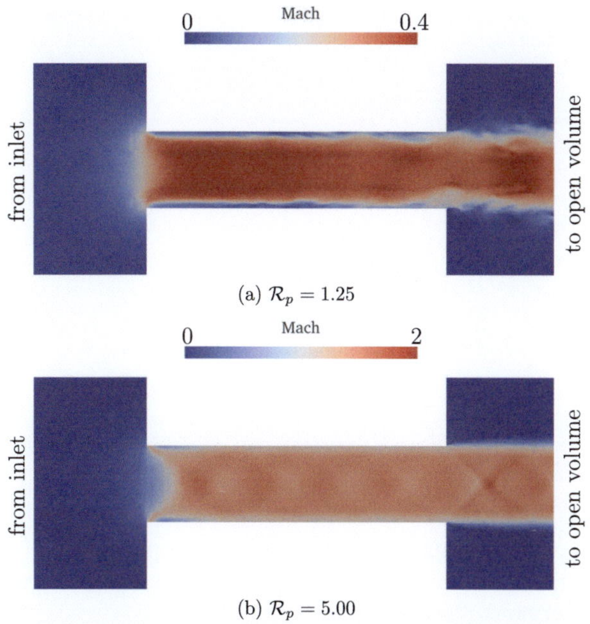

Fig. 2 Mach number on a section through the center of the nozzle at different pressure ratios $\mathcal{R}_p$ and corresponding subsonic (**a**) and supersonic (**b**) conditions. Please note the different Mach scales

Figure 2 shows the Mach number contours for a fully subsonic jet at $\mathcal{R}_p = 1.25$ (Fig. 2a) and a supersonic jet at $\mathcal{R}_p = 5.00$ (Fig. 2b). The subsonic jet exhibits a turbulent boundary layer, even though it is not fully developed because the throttle is too short for that. The throttle flow and the resulting jet are fully turbulent at this pressure ratio. For the supersonic jet, the shock systems are visible within the throttle, and a slightly under-expanded jet occurs downstream of the throttle. The flow is chocked and the critical cross section is at the inlet of the throttle. By increasing the pressure ratio $\mathcal{R}_p$ between inlet and outlet, the Reynolds number of the throttle flow and the jet increases. In the current investigation, we used the same computation mesh for all Reynolds numbers. For higher Reynolds numbers, this makes the simulation underresolved, especially in the jet region. However, this is not a real problem in our case, as we focus on the mass flow behavior through a throttle with a jet; for the lower pressure ratios, i.e., supersonic flow, the flow becomes choked and therefore the throttle inlet limits the mass flow. Consequently, at higher pressure ratios, the representation of the jet is less important for the determination of the mass flow and the resolution used is a reasonable trade-off between accuracy and computational cost.

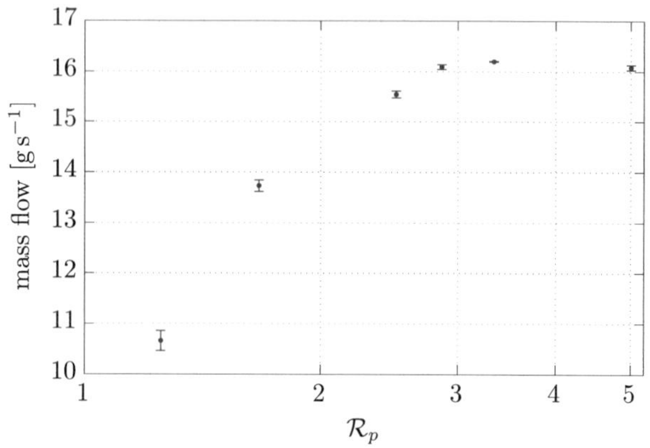

Fig. 3 Mass flow rate for different pressure ratios $\mathcal{R}_p$. All values are average values for the time interval $t \in [150–200\,\mu s]$, computed from 50 individual samples. *Error bars* denote two standard deviations with respect to all individual values

4.1 Mass Flow

For the design process of gas injectors, the accurate prediction of the mass flow is essential. While there are some analytical relations at lower pressures [4], the behavior at higher pressures is much less clear.

In the following, we focus on the mass flow in the quasi-stationary flow and on the transient behavior of the mass flow. The quasi-stationary mass flow of the investigated throttle is shown in Fig. 3. For lower pressure ratios, the flow becomes chocked and the mass flow is independent of $\mathcal{R}_p$. For an ideal gas, the critical pressure ratio of a restriction within a pipe is $\mathcal{R}_p^* < 2.44$ [4], however, in our case we find a larger value $\mathcal{R}_p^* \approx 2.86$. This is due to the sharp edges of the geometry and the real-gas effects, which have a non-negligible effect at these conditions.

The transient behavior of the mass flow can also be important, since an injection of gas commonly occurs at high frequencies. The temporal development of the mass flow rate for the different $\mathcal{R}_p$ is shown in Fig. 4. For $\mathcal{R}_p = 1.25$, the flow is fully subsonic and the mass flow reaches a quasi-constant value already at around $t > 23\,\mu s$. At $\mathcal{R}_p = 1.67$, we observe a similar overall temporal behavior to the $\mathcal{R}_p = 1.25$ case, apart from the significantly higher final value for the mass flow rate. With an even higher pressure ratio $\mathcal{R}_p = 2.50$, we find an initially similar rise of the mass flow rate as in the fully subsonic cases, however, it continues to rise until about $t \approx 120\,\mu s$. The flow is not fully chocked, but the mass flow is no longer limited by the conditions at the throttle exit but instead by those at the throttle inlet. Finally, the flow becomes chocked at the inlet at the even higher pressure ratios $\mathcal{R}_p = 2.86$, 3.33, and 5.00. Until $t \approx 40\,\mu s$, the mass flow rate here is lower than for $\mathcal{R}_p = 2.50$, and it takes significantly longer until the maximum value is reached, which is virtually independent of $\mathcal{R}_p$ in these three cases. All cases show a very dynamic initial mass flow behavior that strongly depends on whether the flow is sub- or supersonic. At later times, $t > 100\,\mu s$, the mass flow rate is nearly constant for all $\mathcal{R}_p$.

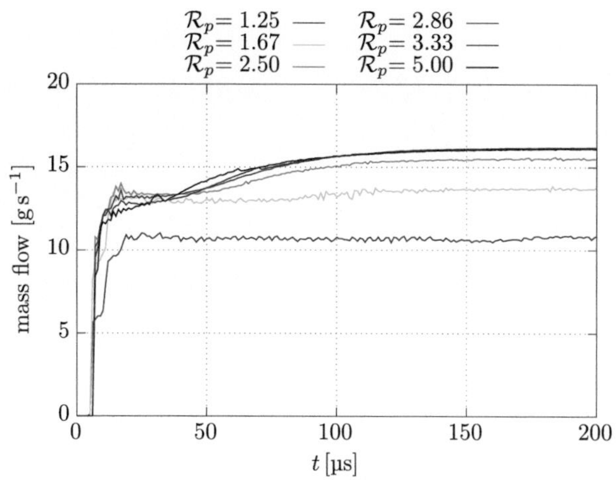

Fig. 4 Mass flow rate for different pressure ratios over time

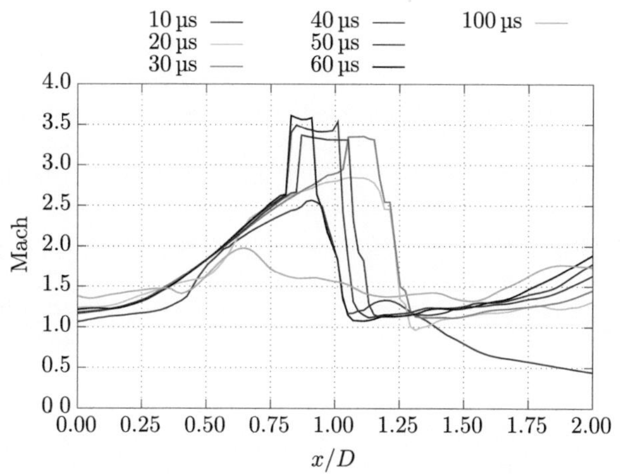

Fig. 5 Mach number along the centerline downstream of the throttle exit

4.2 Shock Representation

The position of the shocks, especially during the early stages, is very dynamic. In the following, we focus on the transient behavior of the first shock. The Mach number along the centerline downstream of the throttle exit is depicted in Fig. 5 at different times. At $t = 10\,\mu s$, the initial jet tip with a strong gradient is present at $x/D = 1$. At $t = 20\,\mu s$, a shock starts to form, which grows in strength over time while it moves upstream. Noticeably, for t = 30–60 μs, the maximum Mach number reaches a plateau. This plateau indicates that the flow enters the two-phase region. The velocity increase in the two-phase region is reduced and the speed of sound

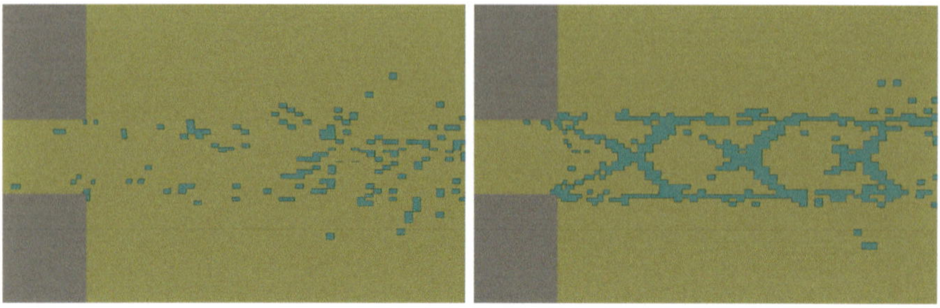

(a) $\mathcal{R}_p = 1.25$ (b) $\mathcal{R}_p = 5.00$

Fig. 6 Finite-volume subcell locations at one time instance for a subsonic jet (**a**) and a supersonic jet (**b**). In the supersonic jet, the locations of the FV cells reflect the typical criss-cross pattern of an under-expanded jet

remains nearly constant. Therefore, the Mach number only increases marginally in the two-phase region very close to the shock jump. Once the flow is developed, no normal shock is present anymore, since the jet is no longer under-expanded. Only weak oblique shocks still exist under these flow conditions.

The DG SEM needs stabilization for under-resolved scales and shocks, which we achieve by employing the aforementioned combination of the detector by Persson and Peraire [18] and the FV subcell method [19, 20]. Figure 6a shows the FV subcells for the subsonic jet at a given point in time. Even though no shocks are present, these subcells are used to stabilize the simulation at under-resolved scales, i.e., here mainly in the shear layer of the jet. For the supersonic jet with shocks, the FV subcells accomplish shock capturing, see Fig. 6b. Here, the distribution of the subcells shows the typical criss-cross pattern of an under-expanded jet.

4.3 HPC Assessment

The DG SEM is by construction a method that is very well parallelizable [11]. The supplementation of DG SEM with the FV subcell method enables the numerical simulation of complex flows with, e.g., occurring shocks [10]. However, without further measures, it also causes significant load imbalances, because a FV subcell is computationally more expensive than a DG element. Hence, if the mesh elements are distributed equally on all cores, those cores with many FV cells will take longer for their computation and hence decrease the performance of the entire simulation. A first step to reduce load imbalances is to take into account the higher computational cost of FV subcells in the initial distribution of mesh elements at the beginning of the simulation. However, this is not sufficient, because both the number of FV subcells as well as their locations within the simulation domain depend heavily on the flow conditions, and thus may change rapidly during the simulation, for example, during the emergence of shocks, see Fig. 6.

Hence, it is important to use a more sophisticated load balancing strategy to maintain DG SEM's excellent parallelizability properties in such complex flow simulations. We have developed such a new technique for dynamic load-balancing, and implemented it in FLEXI. For the initial element-to-core distribution, this technique takes into account the difference in computation cost between FV subcells and normal DG elements, by assigning a weight $w > 1$ to FV subcells, and $w = 1$ for DG elements. Then, the elements are distributed in such a way that the weight sums on all cores are as close to the average value as possible. During the simulation, when new FV subcells emerge or old ones become DG elements again, the distribution of elements on the cores is adapted. To do that, we shift elements from cores with high weight sum to cores with low weight sum until all cores have a weight sum as close to the average value as possible. We employ the shared memory window that has been introduced in MPI 3.0 on each node to make this element shifting as efficient as possible. The communication between nodes is performed with standard MPI routines.

In our current implementation, the adaptation of the element distribution is performed after fixed time-step intervals. Currently, we are investigating techniques to measure the load-imbalance and start adaptation if the load-imbalance reaches a certain threshold.

For a case study of the efficiency of our load-balancing strategy in its present form, we performed test calculations with 216 DOF per DG element. We scaled the number of cores from 96 to 1536, and executed the load-balancing every 1000 time steps. Table 1 shows the reduction of wall time compared to simulations without load balancing. Clearly, incorporating a higher number of cores increases load imbalance because the probability of one core receiving a large number of FV subcells rises. With our dynamic load balancing, we gain a significant reduction in wall time. As further illustration, Fig. 7a, b show the effect of our load balancing technique for one given timestep on 96 cores. In the example, 4.1 % of the elements are FV-subcells. In Fig. 7a, the mesh elements are evenly distributed on all cores (256 elements per core), ignoring the difference in numerical cost between DG and FV-subcell elements. This causes huge load imbalances and therefore a serious performance drop that can be removed by shifting elements between cores so that the load is evenly distributed (Fig. 7b), i.e., the numbers of elements can now differ significantly between cores (between 203 and 262, Fig. 7c).

Table 1 Reduction of wall time achieved with our load balancing strategy for different numbers of cores. In all cases, we performed a simulation with 24,576 elements and $6^3 = 216$ DOF per element

# Cores	DOF per core	Wall time reduction [%]
96	55,296	−3.4
192	27,648	−4.9
384	13,824	−6.4
768	6912	−7.3
1536	3456	−12.1

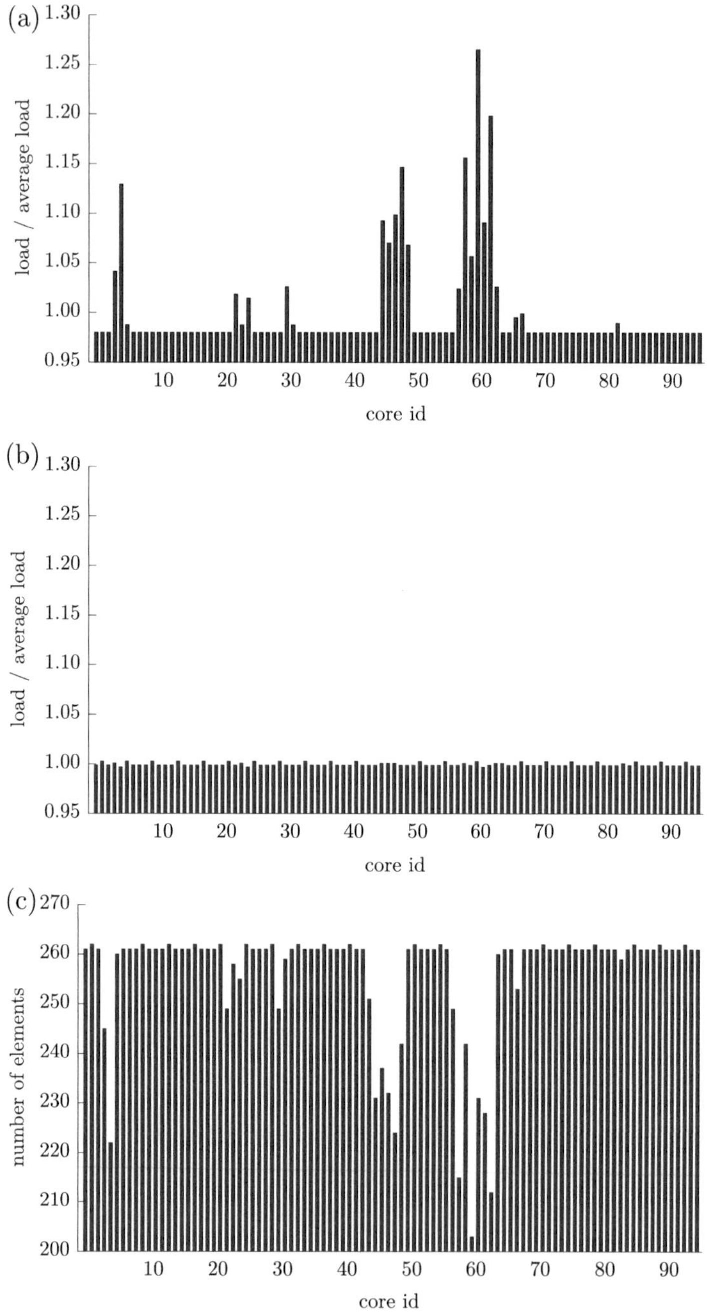

Fig. 7 Load distribution for one given timestep on 96 cores. (**a**) Load distribution without load balancing. (**b**) Load distribution with load balancing. (**c**) Number of elements per core with load balancing. The number of elements varies between 203 and 262, instead of being constant at 256 on each core if no load balancing is employed

5 Conclusion

In the present paper, we presented a framework which can efficiently simulate industrially relevant flow conditions.

We employed a highly accurate tabulated EOS for methane to be able to correctly and efficiently simulate real-gas effects. We used sub- and supersonic throttle flows to demonstrate the transient behavior of the mass flow rate at a high pressure. Additionally, the applied shock capturing was capable of enabling a stable and flexible simulation. Further, we developed a method, which allows the simulation to maintain its high efficiency on massive parallel systems despite the imbalances introduced by the shock capturing.

Acknowledgements This work is supported by the Federal Ministry of Education and Research (BMBF) within the HPC III project HONK "Industrialization of high-resolution numerical analysis of complex flow phenomena in hydraulic systems". We also thank the Gauss Centre for Supercomputing (GCS) which provided us with the necessary computing resources on the Hazel Hen.

References

1. Adolf, M., Bargende, M., Becker, M., Bender, T.B., Budde, M., Ebner, A., Feix, F., Figer, G., Heine, P., Jauss, A., Kehler, T., Keskin, M.T., Köhler, E., Kufferath, A., Langer, W., Lejsek, D., Petersen, C., Philipp, U., Sarikaya, A., Sauerstein, R., Schaarschmidt, M., Schenk, A., Volz, P., Weiske, S., Winke, F., Winkelmann, H., Wollenhaupt, H., Wunderlich, K.: Natural gas and renewable methane for powertrains: future strategies for a climate-neutral mobility. In: Vehicle Development for Natural Gas and Renewable Methane, pp. 229–458. Springer, Cham (2016)
2. Allgeier, T., Haug, M., Frehoff, R., Weikert, M., Kröger, K., Langer, W., Förster, J., Thurso, J., Wörsinger, J.: Gasoline engine management: systems and components. In: Operation of Gasoline Engines on Natural Gas, pp. 122–135. Springer, Wiesbaden (2015)
3. Altmann, C., Beck, A.D., Hindenlang, F., Staudenmaier, M., Gassner, G.J., Munz, C.-D.: An efficient high performance parallelization of a discontinuous galerkin spectral element method. Lect. Notes Comput. Sci. **7686**, 37–47 (2013)
4. Beater, P.: Pneumatic Drives System Design, Modeling and Control. Springer, Berlin/London (2007)
5. Bell, I.H., Wronski, J., Quoilin, S., Lemort, V.: Pure and pseudo-pure fluid thermophysical property evaluation and the open-source thermophysical property library coolprop. Ind. Eng. Chem. Res. **53**(6), 2498–2508 (2014)
6. Boblest, S., Hempert, F., Hoffmann, M., Offenhäuser, P., Sonntag, M., Sadlo, F., Glass, C.W., Munz, C.-D., Ertl, T., Iben, U.: Toward a discontinuous galerkin fluid dynamics framework for industrial applications. In: High Performance Computing in Science and Engineering'15, pp. 531–545. Springer, Berlin/New York (2016)
7. Bolemann, T., Üffinger, M., Sadlo, F., Ertl, T., Munz, C.-D.: Direct visualization of piecewise polynomial data. In: IDIHOM: Industrialization of High-Order Methods – A Top-Down Approach, pp. 535–550. Springer, Cham (2015)

8. de Wiart, C., Hillewaert, K.: Development and validation of a massively parallel high-order solver for DNS and LES of industrial flows. In: Kroll, N., Hirsch, C., Bassi, F., Johnston, C., Hillewaert, K. (eds.) IDIHOM: Industrialization of High-Order Methods – A Top-Down Approach. Volume 128 of Notes on Numerical Fluid Mechanics and Multidisciplinary Design, pp. 251–292. Springer, Cham (2015)
9. Dumbser, M., Iben, U., Munz, C.-D.: Efficient implementation of high order unstructured {WENO} schemes for cavitating flows. Comput. Fluids **86**, 141–168 (2013)
10. Hempert, F., Hoffmann, M., Iben, U., Munz, C.-D.: On the simulation of industrial gas dynamic applications with the discontinuous Galerkin spectral element method. J. Therm. Sci. **25**(3), 1–8 (2016)
11. Hindenlang, F., Gassner, G., Altmann, C., Beck, A., Staudenmaier, M., Munz, C.-D.: Explicit discontinuous Galerkin methods for unsteady problems. Comput. Fluids **61**, 86–93 (2012)
12. Huang, J., Crookes, R.: Assessment of simulated biogas as a fuel for the spark ignition engine. Fuel **77**(15), 1793–1801 (1998)
13. Martin, T., Cohen, E., Kirby, R.M.: Direct isosurface visualization of hex-based high-order geometry and attribute representations. IEEE Trans. Vis. Comput. Graph. **18**(5), 753–766 (2012)
14. McTaggart-Cowan, G., Mann, K., Huang, J., Singh, A., Patychuk, B., Zheng, Z.X., Munshi, S.: Direct injection of natural gas at up to 600 bar in a pilot-ignited heavy-duty engine. SAE Int. J. Engines **8**(3), 981–996 (2015)
15. Nelson, B., Kirby, R.M., Haimes, R.: Gpu-based interactive cut-surface extraction from high-order finite element fields. IEEE Trans. Vis. Comput. Graph. **17**(12), 1803–1811 (2011)
16. Nelson, B., Liu, E., Kirby, R.M., Haimes, R.: Elvis: a system for the accurate and interactive visualization of high-order finite element solutions. IEEE Trans. Vis. Comput. Graph. **18**(12), 2325–2334 (2012)
17. Pagot, C., Osmari, D., Sadlo, F., Weiskopf, D., Ertl, T., Comba, J.: Efficient parallel vectors feature extraction from higher-order data. Comput. Graph. Forum **30**(3), 751–760 (2011)
18. Persson, P.-O., Peraire, J.: Sub-cell shock capturing for discontinuous Galerkin methods. In: Proceedings of the American Institute of Aeronautics and Astronautics, Keystone, vol. 112 (2006)
19. Sonntag, M., Munz, C.-D.: Shock capturing for discontinuous Galerkin methods using finite volume subcells. In: Finite Volumes for Complex Applications VII-Elliptic, Parabolic and Hyperbolic Problems. Volume 78 of Springer Proceedings in Mathematics & Statistics, pp. 945–953. Springer, Cham (2014)
20. M. Sonntag and C.-D. Munz. Efficient parallelization of a shock capturing for discontinuous galerkin methods using finite volume sub-cells. J. Sci. Comput. 1–28 (2016)
21. Vuorinen, V., Yu, J., Tirunagari, S., Kaario, O., Larmi, M., Duwig, C., Boersma, B.: Large-eddy simulation of highly underexpanded transient gas jets. Phys. Fluids (1994-present) **25**(1), 016101 (2013)
22. Westerhoff, M., Holtmeier, G.: Erdgas Die greifbare Chance. MTZ – Motortechnische Zeitschrift **77**(2), 8–13 (2016)
23. Yu, J., Vuorinen, V., Kaario, O., Sarjovaara, T., Larmi, M.: Visualization and analysis of the characteristics of transitional underexpanded jets. Int. J. Heat Fluid Flow **44**, 140–154 (2013)

Modeling of the Deformation Dynamics of Single and Twin Fluid Droplets Exposed to Aerodynamic Loads

Lars Wieth, Samuel Braun, Geoffroy Chaussonnet, Thilo F. Dauch, Marc Keller, Corina Höfler, Rainer Koch, and Hans-Jörg Bauer

Abstract Droplet deformation and breakup plays a significant role in liquid fuel atomization processes. The droplet behavior needs to be understood in detail, in order to derive simplified models for predicting the different processes in combustion chambers. Therefore, the behavior of single droplets at low aerodynamic loads was investigated using the Lagrangian, mesh-free Smoothed Particle Hydrodynamics (SPH) method. The simulations to be presented in this paper are focused on the deformation dynamics of pure liquid droplets and fuel droplets with water added to the inside of the droplet. The simulations have been run at two different relative velocities.

As SPH is relatively new to Computational Fluid Dynamics (CFD), the pure liquid droplet simulations are used to verify the SPH code by empirical correlations available in literature. Furthermore, an enhanced characteristic deformation time is proposed, leading to a good description of the temporal initial deformation behavior for all investigated test cases. In the further course, the deformation behavior of two fluid droplets are compared to the corresponding single fluid droplet simulations. The results show an influence of the added water on the deformation history. However, it is found that, the droplet behavior can be characterized by the pure fuel Weber number.

1 Introduction

For an optimization of modern gas turbines, the atomization process needs to be understood in detail. The present investigation focuses on the behavior of single droplets at low aerodynamic loads. These conditions occur right after the jet breakup

L. Wieth (✉) • S. Braun • G. Chaussonnet • T.F. Dauch • M. Keller • C. Höfler • R. Koch • H.-J. Bauer
Institut für Thermische Strömungsmaschinen, Karlsruhe Institut für Technologie, Kaiserstrae 12, 76131 Karlsruhe, Germany
e-mail: lars.wieth@kit.edu

W.E. Nagel et al. (eds.), *High Performance Computing in Science and Engineering '16*, DOI 10.1007/978-3-319-47066-5_21

in a jet-in-crossflow configuration for example. Hence, they are crucial for the following evaporation and combustion process. Various experimental investigations of the behavior of droplets at aerodynamic loads have been conducted in the past, e.g. [11, 12, 14]. However, the experimental setup, which either relays on a shock tube experiment or a free falling droplet in a crossflow, does not allow for a detailed insight of the phenomena involved in the deformation and breakup process. Therefore, numerical investigations have been conducted to gain insight into the underlying physics, e.g. [17, 25, 30].

In order to predict all processes occurring in combustion chambers, from the liquid fuel injection to the combustion, commonly Euler-Lagrange methods are used. These methods predict the air flow on an Eulerian mesh, while the liquid fuel is inserted as Lagrangian parcels. To describe the behavior of the liquid fuel droplets, simplified models were derived using experimental and detailed numerical investigations. The most common models to describe the initial deformation phase are the Normal-Mode (NM) model and the Non-linear Taylor Analogy Breakup (NLTAB) model [27, 28], which is a nonlinear extension to the well known TAB model proposed by O'Rouke [24]. In all models it is assumed that after reaching a critical deformation, the Lagrangian parcel will undergo secondary breakup, which is described by empirical models as well (e.g. [2]).

The assumption of such empirical models is, that the droplet is exposed to a quasi-steady aerodynamic load. Therefore, the history and the temporal evolution of the droplet deformation is considered. This may lead to unphysical droplet drag predictions. In the present paper the temporal evolution of droplet deformation is investigated at low aerodynamic loads using the Lagrangian, mesh-free Smoothed Particle Hydrodynamics (SPH) method. The weakly compressible SPH code in use was developed and validated in order to predict the atomization process in gas turbine engines [13]. The main advantage of SPH over mesh-based methods is the inherent interface advection without the need of an interface capturing algorithm. Furthermore, the effect of water added to the inside of the liquid fuel droplet is investigated. Preliminary tests performed in heavy duty gas turbines showed that the addition of water has a positive effect on the thermal NOx emissions[18]. The addition of water to the fuel oil not only decreases the combustion temperature due to the heat of evaporation, but has a positive effect on the atomization process as well [8].

Therefore, the deformation of single, emulsified fuel droplets with an initial diameter of $d_0 \approx 60\,\mu\text{m}$ with different water volume fractions $\phi = V_W/(V_W + V_{Oil})(\phi = 0, \phi = 0.23, \phi = 1)$ exposed to different air velocities ($|\mathbf{v}_{Air}| = 22.5\,\text{m/s}$ and $|\mathbf{v}_{Air}| = 24.34\,\text{m/s}$) are investigated. Furthermore, the placement of the water inside the droplet is varied to determine its influence on the droplet deformation.

2 Methodology

The full Lagrangian, mesh-free Smoothed Particle Hydrodynamics (SPH) method has been developed in the late 1970s for the simulation of non-axisymmetric phenomena in astrophysics [10, 20].

In this approach integral equations or partial differential equations with boundary conditions are solved over an arbitrarily scattered set of movable discretization points. Those so called particles represent a finite volume of the fluid domain on a continuum scale. The physical quantities, e.g. position and velocity, as well as the fluid properties are assigned to the particles. The interaction of the particles is taken into account by a weighting function. Therefore, in contrast to common grid based techniques no complicated spatial discretization schemes are required. This is advantageous for the simulation of problems with free surfaces, high deformations and moving surfaces.

In the following the basics of the SPH method, which are needed to solve the conservation equations, will be presented.

2.1 *SPH Formulation*

The fact, that every spatial function $f(\mathbf{x})$ can be exactly reproduced by the convolution of the function itself with the Dirac delta function $\delta(\mathbf{x}-\mathbf{x}')$ is the basis of the SPH-interpolation:

$$f(\mathbf{x}) = \int_V f(\mathbf{x}')\delta(\mathbf{x}-\mathbf{x}')dx'. \tag{1}$$

The determination of a quantity at position $\mathbf{x}$ requires the quantities at the surrounding positions $\mathbf{x}'$ to be taken into account. Since the Dirac function is only valid at one point, it is replaced by a smooth weight function with similar properties, the so called kernel $W(\mathbf{x}-\mathbf{x}', h)$. The kernel assigns a weight to the neighboring particles depending on their distance $(\mathbf{x}-\mathbf{x}')$ from the center particle. To ensure stability and consistency of the method, the kernel has to fulfill certain requirements [19]. The kernel is compact. The maximum radius of influence is defined by the smoothing length h. In Fig. 1 the interpolation of a function at the position of particle i through the known functions at the positions of neighboring particles j is illustrated.

For the numerical determination of a quantity the integral approximation is replace by the summation over all neighbor particles j, the so called quadrature [21]:

$$f(\mathbf{x}) = \sum_j f(\mathbf{x}_j)V_jW(\mathbf{x}_i-\mathbf{x}_j, h) = \sum_j f(\mathbf{x}_j)\frac{m_j}{\rho_j}W(\mathbf{x}_i-\mathbf{x}_j, h), \tag{2}$$

where V is the volume, m is the mass and ρ the density of the particles.

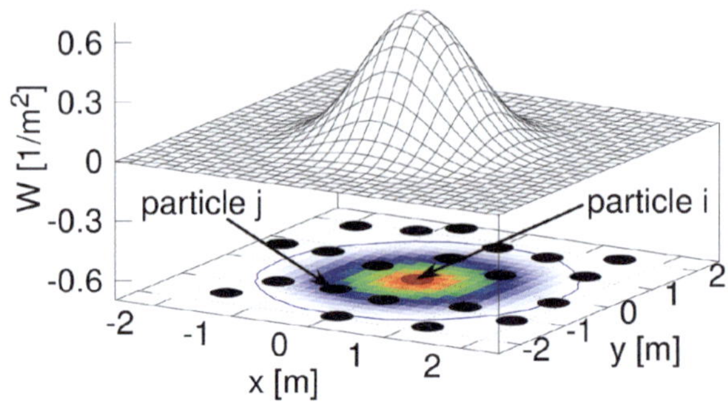

Fig. 1 Interpolation for the center particle (*red*) with a kernel function W

When using a differentiable kernel, the partial derivative of a function $\nabla f(\mathbf{x})$ is given by:

$$\nabla f(\mathbf{x}) = \sum_j f(\mathbf{x}_j)\frac{m_j}{\rho_j}\nabla W(\mathbf{x}_i - \mathbf{x}_j, h). \tag{3}$$

This is only valid if the domain of interpolation of a particle is not truncated by the boundary of the computational domain.

2.2 SPH Formulation of the Navier-Stokes Equations

The mathematical characterization of macroscopic flows are described by the Navier-Stokes-equations. They include three conservation equations: the continuity, momentum and energy equation, whereof the later will be neglected, because the flow is considered as isothermal.

Different SPH formulations of the Navier-Stokes equations can be found in literature based on the SPH-formulations (2) and (3) and further mathematical transformations [21]. The approximation of the density and the pressure gradient as well as the viscous term of the momentum equation are introduced briefly in the following. The SPH approximations will be indicated by brackets <>. The properties of the centered particle is denoted by the index i and the properties of the neighboring particles by the index j. The density of a particle is directly calculated by the summation over the weights of the neighboring particles:

$$\langle \rho \rangle_i = m_i \sum_j W(\mathbf{x}_i - \mathbf{x}_j, h). \tag{4}$$

This formulation conserves mass exactly and prevents a non-physical density gradient over the interface of multi-phase flows, in contrast to other formulations [16].

Various approaches for the approximation of gradients, like the pressure gradient term in the momentum equation $[-\nabla p/\rho]$ are available in literature. An approximation successfully applied to multi-phase problems was proposed by Colagrossi et al. [7]:

$$\left\langle -\frac{\nabla p}{\rho} \right\rangle_i = -\frac{1}{\rho_i} \sum_j \frac{m_j}{\rho_j} (p_i + p_j) \nabla W(\mathbf{x}_i - \mathbf{x}_j, h). \tag{5}$$

The viscous term of the momentum equation $[\nabla \tau / \rho]$ contains second-order derivatives. The approximation of this term was introduced to SPH by Morris et al. [23]. It is derived from the inter particle average shear stress using a combined viscosity:

$$\left\langle \frac{\nabla \tau}{\rho} \right\rangle_i = \sum_j m_j \frac{\mu_i + \mu_j}{\rho_i \rho_j} \frac{\mathbf{r}_{ij} \nabla W(\mathbf{x}_i - \mathbf{x}_j, h)}{\mathbf{r}_{ij}^2 + \eta^2} \mathbf{v}_{ij}. \tag{6}$$

Here μ denotes the dynamic viscosity, $\mathbf{r}_{ij}$ is the distance vector between the particles, V denotes the particle volume, $\mathbf{v}_{ij}$ denotes the velocity difference and η is a small parameter, which serves to avoid singularities.

In our SPH approach the weakly compressible SPH scheme is used, meaning that non-compressible liquids are modeled as weakly compressible. Therefore, the pressure p and the density ρ are linked through an equation of state. In this approach the density fluctuations are limited to $\Delta\rho/\rho = 1\,\%$ by imposing an artificial sound speed c which is approximately ten times higher than the maximum velocity $|\mathbf{v}_{max}|$. In general, this leads to an artificial sound speed, which is much lower than the physical one. By this approach small time steps due to the Courant-Friedrich-Levy (CFL) criterion are mitigated.

For the present investigation the equation of state in-use is a modified Tait equation, which was originally derived for water [3]:

$$p = \frac{\rho_{\text{nom}} c^2}{\gamma} \left[\left(\frac{V_{\text{nom}}}{V} \right)^{\gamma} - 1 \right]. \tag{7}$$

Here ρ_{nom} is the reference density, V_{nom} is the reference volume and γ is the polytropic exponent.

2.3 *Treatment of Interfacial Tension*

For the prediction of the complex physics leading to liquid atomization the modeling of surface tension effects plays a crucial role. This is due to the fact, that droplet deformation and disintegration are mainly determined by the force balance between the microscopic surface force acting in tangential and normal direction at the liquid-air interface and the shear forces acting on the droplet which are induced by the air flow. This balance is described by the dimensionless Weber number *We*, which can be used to estimate whether a droplet is exposed to an super- or subcritical aerodynamic load.

In our SPH code the surface tension is represented by the Continuum Surface Force (CSF) model, originally introduced by Brackbill et al. [4] in the framework of the VoF method. The CSF model adopted in our approach was proposed by Adami et al. [1]. The surface tension force is represented as a continuous force acting over a volume adjacent to the interface instead of a force acting directly on the surface of the droplet. Therefore, the surface tension is converted to a volumetric force $\mathbf{F}_{\mathrm{SF}}$ using a normalized delta-function δ_{S}, which has its peak at the interface:

$$\mathbf{F}_{\mathrm{SF}} = \sigma \kappa \hat{\mathbf{n}} \delta_{\mathrm{S}}. \tag{8}$$

In this formulation interfacial gradients are neglected assuming a constant surface tension coefficient. In Eq. (8) σ is the surface tension coefficient, $\hat{\mathbf{n}} = \mathbf{n}/|\mathbf{n}|$ is the normalized normal vector of the interface and $\kappa = -\nabla\hat{\mathbf{n}}$ the curvature. The interface normal vector $\mathbf{n}$ is determined from the gradient of a color index $\zeta_{\mathrm{j}}^{\mathrm{i}}$, which is assigned to each SPH particle according to the phase it belongs to. This results in the following SPH approximation scheme:

$$\mathbf{n} = \frac{1}{V_i} \sum_j \left(V_i^2 + V_j^2\right) \frac{\rho_i}{\rho_i + \rho_j} \zeta_j^i \nabla W(\mathbf{x}_i - \mathbf{x}_j, h). \tag{9}$$

The curvature κ is determined using another approximation, described in detail by Adami et al. [1]. The additional force to be added to the momentum equation has the following form:

$$\langle \mathbf{f} \rangle_{i,SF} = \frac{\mathbf{F}_{i,SF}}{\rho_i} = \frac{\sigma \kappa_i \hat{\mathbf{n}}_i |\mathbf{n}_i|}{\rho_i} = -\frac{\sigma}{\rho_i} \left(\nabla \hat{\mathbf{n}}_i\right) \mathbf{n}_i. \tag{10}$$

Wetting effects, which for example highly influence the primary atomization, are accounted for by using the model presented by Wieth et al. [29].

2.4 Boundary Conditions

The numerical representation of technical relevant systems require proper boundary conditions. All boundary conditions, which can be walls, inlets or outlets as well as periodic boundaries have to be treated specifically. The boundary conditions used for the present investigations, namely walls and permeable boundaries, will be introduced briefly in the following.

Since the SPH-method is commonly applied to unbounded or confined fluid problems, numerous treatments of wall boundaries can be found in literature. The approach in-use utilizes fixed pseudo particles placed outside of the boundary surface. These wall particles take part in the approximation to minimize to truncation error of the kernel. If a particle, representing the fluid, approaches the wall and undercuts a certain cutoff distance, a repulsive force is applied [22]. The additional force resembles a Lennard-Jones potential and acts on the center line between the fluid and wall particle under consideration.

Due to the Lagrangian nature of SPH, permeable boundary conditions cannot be handled straightforward like in Eulerian methods. Particles have to be generated at the inlet and removed at the outlet in a rate, which is equivalent to the physical flow rate. This is achieved by extending the numerical domain by so called buffer zones. The buffer zones are filled with particles, which take place in the approximation. The desired boundary conditions for the velocity $\mathbf{u}$, the pressure p and the temperature T are imposed onto these particles. The particles in the buffer zones are controlled by markers, which do not take place in the approximation and which are positioned right at the boundary surface. This procedure is suitable for arbitrarily shaped boundaries and enables the generation of particles at the inlet and the removal of particles at the outlet. A detailed description of the permeable boundaries method is given by Braun et al. [6].

3 Modeling of the Three Fluid Contact Line

The present investigation contains predictions of pure liquid droplets as well as of droplets containing a second phase. In the later case it is possible that three fluids, i.e. fuel, water and air, meet at one contact line. In the vicinity of this contact line, the fluids will form specific contact angles depending on the different interfacial tensions of the fluid pairings. To account for this effect, the formation of the contact angles has to be modeled. The modeling approach is presented briefly in the following.

Generally, the mechanical equilibrium state of three liquids can be expressed as a force balance of the interfacial tension forces at the contact line, where all three phases intersect. This force balance is illustrated in Fig. 2.

In Fig. 2 one phase (light grey, denoted as phase 1) is surrounded by two other immiscible phases (denoted as phase 2 and 3). The interfacial tension coefficients between the phases are denoted by σ_{ij}, where i and j represent the index of the phases considered. The interfacial tension forces ensure the formation of an equilibrium

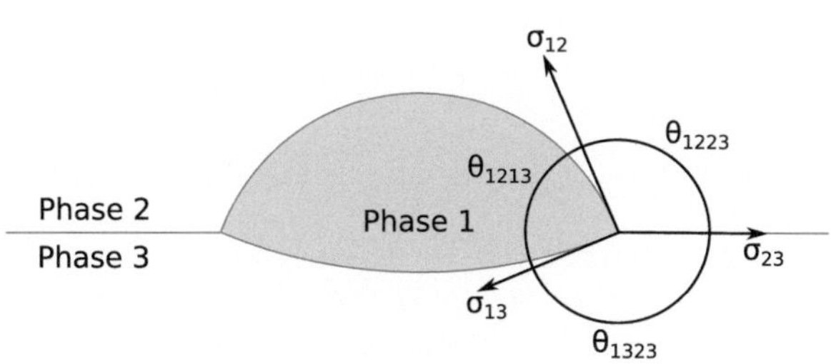

Fig. 2 Schematic of the force balance at the triple line for a general three-phase interaction

state leading to characteristic static contact angles inside the different phases. These angles θ are indexed by the interfacial forces, which span the angles. The force balance at the triple line results in a set of three equations, which relate the interfacial tension coefficients to the static contact angles. The geometric interpretation of this set of equations is known as the Neumann triangle. For dynamic contact angle simulations the static contact angles is set as initial condition.

In our approach the surface tension is represented by the CSF model, requires an additional acceleration in the momentum equation, which primarily depends on the interface normals $\mathbf{n}$ and their divergence (curvature), cf. Eq. (10). In the vicinity of the triple line the normal vectors are adjusted to introduce the desired static contact angles. Up to now the modeling of fluid interactions of three liquids and/or gases has not been realized on basis of the CSF model in the SPH framework. Hu and Adams [15] showed the applicability of a different approach, the Continuum Surface Stress (CSS) model to three phase interaction problems.

A schematic representation of the normal vector correction approach for a general three phase interaction is shown in Fig. 3.

The correction of the normal vectors is only applied to particles which are close to the triple line and if particles of the other phases are located within the radius of influence, like it is the case for the black particle indicated in Fig. 3. For each of those particles, two interface normal vectors $\mathbf{n}_1$ and $\mathbf{n}_2$ are calculated using (9). These span an angle α, which does not necessarily represent the contact angle θ. In order to impose a correct static contact angle, α has to be corrected as depicted in Fig. 3. Specifically, one of the two normal vectors, in this case $\mathbf{n}_2$, is rotated by an angle β to the corrected normal vector $\mathbf{n}_{2\mathrm{corr}}$. The normal vector used for the rotation is chosen by the strength of the kernel support. A higher kernel support is assumed to be more trustworthy. Following the rotation of the normal vector, the general approximations are used to calculate the curvature and then the acceleration.

The model presented yields excellent results (relative errors <5 %) in 2D and 3D for the formation of static contact angles in a water-air-alkane system. Details of the model and its validation shall be presented in a future publication.

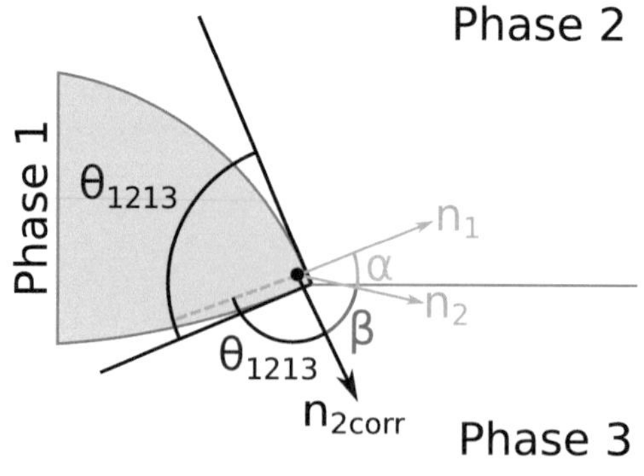

Fig. 3 Illustration of the normal vector correction for the three liquid interaction case

4 Numerical Setup

The simulations conducted for this study are focused on the deformation behavior of single droplets in an air flow. A confined channel domain with permeable boundary conditions (inlet, outlet) is used. A sketch of the investigated geometry is depicted in Fig. 4.

The channel has a square cross-section with the height of $h_c \approx 8.3d_0 = 0.5\,\text{mm}$ and a length of $l_c \approx 33.3d_0 = 2\,\text{mm}$. The initial droplet diameter is denoted by d_0. The channel is confined by moving walls in y- and z-direction, which have the same velocity as prescribed at the inlet. At the inlet a fixed velocity in x-direction u_{in} and at the outlet a fixed reference pressure p_{out} is prescribed. The velocities imposed in this study are $u_{in} = 22.5\,\text{m/s}$ and $u_{in} = 24.34\,\text{m/s}$. A summary of the fluid properties can be found in Table 1. The properties of air resemble combustion chamber conditions at $T = 700\,\text{K}$ and $p = 2\,\text{MPa}$.

The single droplet with a diameter of $d_0 \approx 60\,\mu\text{m}$ is initialized, after the air flow has reached a quasi-steady state. Hence, the droplet is initialized by a "stamp" once the air flow is settled. Within the stamp the fluid properties are changed from air to the desired liquid. The air is put to rest for a short time. During this time span, the spurious oscillations caused by the initialization of the droplet will relax and a steady state sphere is formed. Thereafter, the air and wall particles are set to the imposed inlet velocity, so that the droplet is exposed to a sudden aerodynamic load. This method provides a fast and simple way to create different types of droplets. In this paper fuel droplets with different water volume fractions of $\phi = 0$, $\phi = 0.23$ and $\phi = 1$ are investigated. This results in single fluid droplets for $\phi = 0$ (fuel)

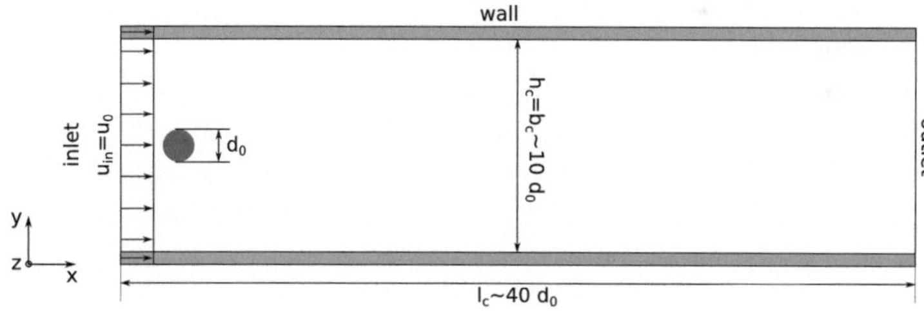

Fig. 4 Illustration of the geometry and boundary conditions

Table 1 Summary of fluid properties

	Air	Fuel	Water
Density ρ [kg/m^3]	9.95	825	1000
Dynamic viscosity μ [Pa s]	3.4×10^{-5}	2.93×10^{-3}	1.0×10^{-3}
Interfacial tension σ [N/m]	0.028	0.0263	
Surface tension air-water σ [N/m]		0.07	

and $\phi = 1$ (water), yielding Weber numbers for the pure fuel of $We_{\text{fuel}} \approx 10$ and $We_{\text{fuel}} \approx 12$ and for the water of $We_{\text{water}} \approx 4.1$ and $We_{\text{water}} \approx 4.8$. In the case of $\phi = 0.23$ a single water droplet is added to the interior of the fuel droplet. The influence of the placement of the water is investigated for different scenarios: centered, off center downstream (in flow direction, centered in the yz-plane), off center upstream (against the flow direction, centered in the yz-plane), off center perpendicular to the flow in both y-directions (centered in the xz-plane). Overall this results in 14 simulations with each having approximately 34 million particles using an initial particle distance of $dx = 2.5\,\mu\text{m}$. This spatial resolution is the absolute minimum required to correctly capturing the physics leading to droplet deformation or breakup. A finer spatial resolution would give even more reliable results. However, the computational effort would increase significantly. For the verification of the SPH code against common correlation other single fluid droplet simulations were conducted, which will not be specified in this context.

5 Computational Performance

The simulations were conducted using the SPH code developed by the Institut für Thermische Strömungsmaschinen (ITS) [13]. The parallel performance of this code was evaluated by strong scalability tests performed on the ForHLR I cluster at the Steinbuch Center for Computing in Karlsruhe [9]. The cluster is equipped with 512 thin nodes, each having 2 Deca-Core Intel® Xeon® E5-2670 v2 processors. The nodes are connected by an InfiniBand 4x FDR interconnect. For the scalability test, the SPH code was compared to the grid based VoF solvers of OpenFOAM® 2.3.0 (interFoam) and another commercial CFD package. The domain investigated was 2D and turbulence modeling was neglected. Details of this study can be found in [5]. All simulations were run for 1 h walltime, which easily enables a performance test over three orders of magnitude. Therefore, the simulation domain was divided into 1 up to 1000 subdomains. The results for the speedup and parallel efficiency are depicted in Fig. 5. The results are normalized by the performance of one node or 20 cores.

The results for OpenFOAM (indicated by squares) and the commercial software (indicated by diamonds) show a good scaling for up to 2 nodes (40 processors). The speedup for SPH is almost ideal for up to 200 cores and has not jet reached saturation at 1000 cores (cf. Fig. 5a). Consequently, the parallel efficiency of the SPH code stays above 0.9 till 200 cores, but is still over 0.6 at 1000 cores (cf. Fig. 5b). The efficiency of OpenFOAM and the commercial software indicates, that OpenFOAM has an excellent serial performance, while this is questionable for the commercial software. Beyond 100 cores the efficiency of both codes is severely reduced and the speedup reaches saturation. This yields a stagnation (commercial software) or even increase (OpenFOAM) of time needed for the simulations when increasing the computational effort.

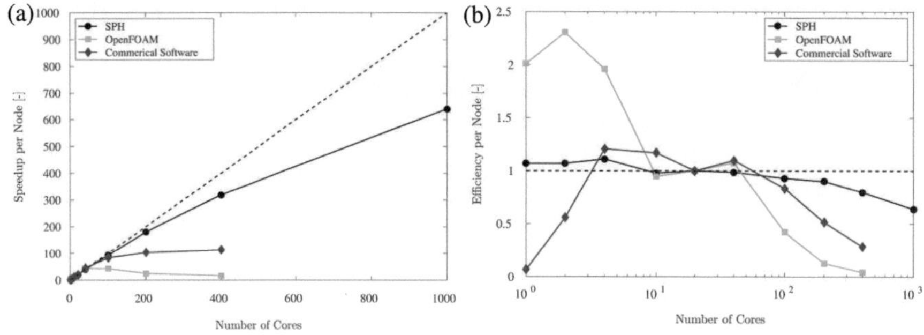

Fig. 5 Comparison of parallel performance for SPH, OpenFOAM and a commercial software. The data is normalized by the performance for one node with 20 cores. (**a**) Comparison of speedup per node. (**b**) Comparison of parallel efficiency per node

As it is evident, the SPH code in use shows a strong parallel performance even at high CPU numbers. Therefore, each simulation was conducted on 520 cores or 26 nodes, respectively. Depending on the simulation time, which is needed to advect the droplet through the whole domain, the average time required for the computation of each prediction is $t_{\mathrm{Comp}} \approx 17{,}000$ CPUh for the cases leading to $We_{\mathrm{fuel}} = 10$ and $t_{\mathrm{Comp}} \approx 14{,}500$ CPUh for the cases leading to $We_{\mathrm{fuel}} = 12$. Additionally, the computation of the quasi-stationary air solution as well as the decomposition and reconstruction of the simulation domain consumed computational resources. However, these are small compared to the resources required for the actual simulations.

6 Results and Discussion

At first the deformation dynamics of single fluid droplets is investigated to in order to verify the SPH method, since the SPH method is relatively new to CFD. The numerical findings are compared to the well-known correlations of Hsiang and Faeth [14]. Furthermore, an enhanced characteristic deformation time is extracted from the simulations. It is found to collapse the initial deformation dynamics for all single fluid test cases to one curve. Thereafter, the dynamics of two fluid droplets at the same aerodynamic loading are compared to those of single fluid droplets with the objective to determine the influence of adding a second fluid.

6.1 Single Fluid Droplet Simulations

The aerodynamic loads chosen for this investigation are characterized by Weber numbers below 12 for the pure liquids. In this We number range, droplets are

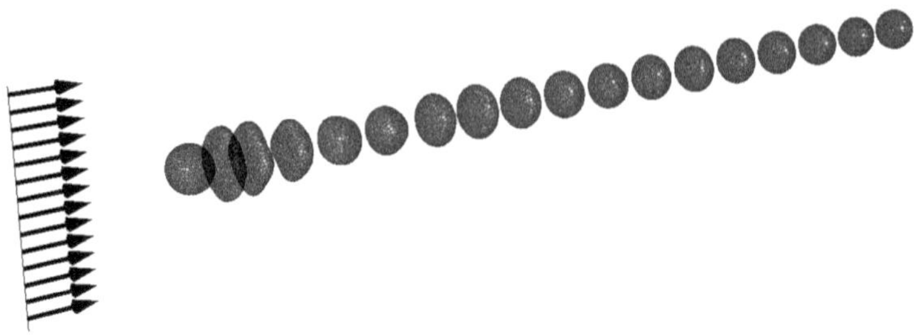

Fig. 6 Pure fuel droplet evolution for $We_{\text{fuel}} = 10$

expected to just show an oscillatory deformation. Breakup can only occur due to natural oscillations of the droplet [11]. The temporal evolution of the deformation as predicted by SPH for a pure fuel droplet at $We_{\text{fuel}} = 10$ is depicted in Fig. 6. The air as well as wall particles are omitted for the sake of clarity.

The aerodynamic load, indicated by black arrows, is impinging on the initial spherical drop in a shock-like manner. This leads to the deformation of the droplet to a flat disc. Thereafter, the surface tension causes a contraction of the droplet. The high dynamic viscosity of the fluid prevents the droplet from elongating in flow direction, yielding a spherical shape at the turning point of the deformation. Then, the drop starts to flatten again due to the forces imposed by the pressure of the air flow around the droplet and the surface tension force. Overall this leads to an oscillatory deformation, which is dampened severely by the viscosity of the fuel. A similar oscillatory deformation was also found for the other pure liquids. Thereby, the oscillation of the water droplets are of a higher frequency and dampened less as expected due to the higher surface tension and lower dynamic viscosity.

Overall the qualitatively observed deformation dynamics of the single fluid droplets perfectly reproduce the expected behavior found by experimental investigations. For the verification, that SPH is able to correctly predict the quantitative behavior of the drop deformation, the predictions are compared to empirical correlations in the following.

6.1.1 Comparison of SPH Results to Empirical Findings

Hsiang and Faeth [14] determined the maximum extent of the droplet $d_{\text{cross,max}}$ perpendicular to the flow direction as well as the minimum extent $d_{\text{str,min}}$ in flow direction experimentally. They found that the droplet extent perpendicular to the flow almost linearly increases with time until the maximum is reached. On a basis of a phenomenological analysis considering the surface tension and pressure forces, together with the experimental results they proposed the following correlation:

$$\frac{d_{\text{cross,max}}}{d_0} = \left(\frac{d_0}{d_{\text{str,min}}}\right)^{0.5} = 1 + 0.19We^{0.5}. \tag{11}$$

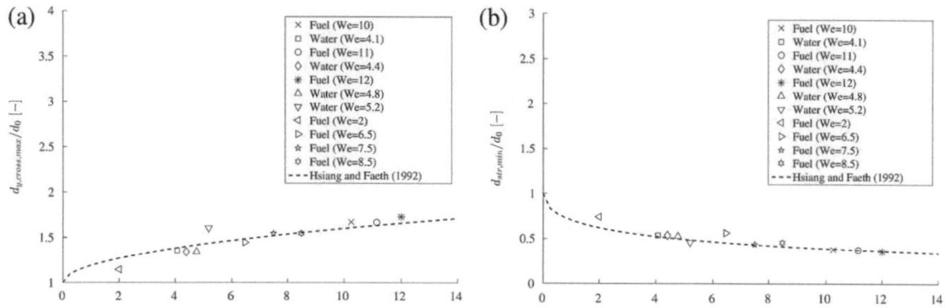

Fig. 7 Comparison of the single fluid simulation results to empirical findings of Hsiang and Faeth. (**a**) Maximum droplet extent perpendicular to the flow direction $d_{\text{y,cross,max}}$ over *We*. (**b**) Minimum droplet extent in flow direction $d_{\text{str,min}}$ over *We* Hsiang and Faeth [14]

The validity of this correlation is claimed to be for $We < 100$ and $Oh < 0.1$, where *Oh* represents the Ohnesorge number relating the viscous forces to the surface tension and inertial forces. The correlation claims, that the maximum cross-stream as well as the minimum stream-wise diameter are just dependent on *We*. Since the four cases addressing single fluid droplets are representing just four different *We* and two different *Oh*, more numerical predictions of single fluid droplets were conducted to cover a broader range of *We* as well as *Oh*. The simulations cover $2 < We < 12$ and $0.005 \leq Oh \leq 0.1$. The comparison of the numerical results to the correlation given by (11) is depicted in Fig. 7. The numerical results were obtained in a similar fashion as it is commonly done in experiments. Fixing the view to one observation plane (x-y in this case), the drop dimensions in x- and y-direction are measured. The findings for the maximum cross-stream extent in y-direction $d_{\text{y,cross,max}}$ are shown in Fig. 7a while the findings for $d_{\text{str,min}}$ are shown in Fig. 7b.

It is evident, that the numerical results fit the correlation extremely well. Minor deviations are observed for two points. One of them represents an extreme investigated. The droplet simulated at $We = 2$ features an Ohnesorge number $Oh = 0.1$, which is the limit of the correlation. This could explain the observed deviation. The second case is well in the range of validity of the correlation, having a $We \approx 5$ and $Oh \approx 0.016$. Therefore, this deviation only can be explained by numerical inaccuracies resulting from different numerical setups.

Altogether, it can be stated that SPH is able to capture quantitatively the correct physical behavior of drop deformation at low aerodynamic loads. The initial temporal dynamics of the single fluid drop deformation is discussed in the following.

6.1.2 Temporal Evolution of the Initial Drop Deformation

The temporal evolution of the drop deformation and thus the drag coefficient plays a major role when developing simplified models, which can be used in an Euler-Lagrange context. Hsiang and Faeth [14] observed a linear increase in cross-stream diameter with time until the maximum deformation is reached. They claim, that for a wide range of *We* the maximum deformation is always reached at approximately $t/t* \approx 1.6$, where $t* = d_0(\rho_{\text{liq}}/\rho_{\text{gas}})^{0.5}/u_0$ is the characteristic breakup time proposed by Ranger and Nicholls [26]. The temporal evolution of the deformation perpendicular to the flow in y-direction $d_{\text{y,cross}}$ for all single fluid cases is depicted in Fig. 8.

The numerical results for the cross-stream deformation over the dimensionless time $t/t*$ are shown in Fig. 8a. As evident, the dynamics of the deformation predicted by SPH is not linear and cannot be correlated using the commonly used characteristic time $t*$. Furthermore, overall the deformation as predicted seems to be faster than the experimental findings, exhibiting $t/t* \approx 1.6$ for the maximum deformation [14]. The maximum dimensionless time needed for the maximum deformation in SPH is $t/t* \approx 1.3$. The deviations observed may be due to the experimental setup used to acquire the data. Commonly shock tube experiments are used for the aerodynamic loading of droplets. These kind of experiments cannot guarantee correct boundary conditions in contrast to the numerical analysis. Furthermore, the droplets are introduced into the shock tube by a droplet chain using a vibrating capillary tube. This might lead to an interference due to previous droplets or the vibration might influence the initial drop shape.

A closer look to the dynamics in Fig. 8a indicates a dependence on *We* for the initial deformation dynamics. The smaller *We* the faster the maximum cross-stream deformation is reached. Correlating the numerical results with $t*$ and *We* using a 4th order polynomial leads to the following enhanced characteristic time:

$$T = We^{0.4} d_0(\rho_{\text{liq}}/\rho_{\text{gas}})^{0.5}/u_0 = We^{0.4} t*. \tag{12}$$

The plot of the predicted deformation dynamics over the new dimensionless time t/T are depicted in Fig. 8b. Now the deformation dynamics of all cases coincide on one curve. The only one case showing a deviation is the same as before ($We \approx 5$; $Oh \approx 0.016$). Here, too the probable cause is numerical inaccuracies.

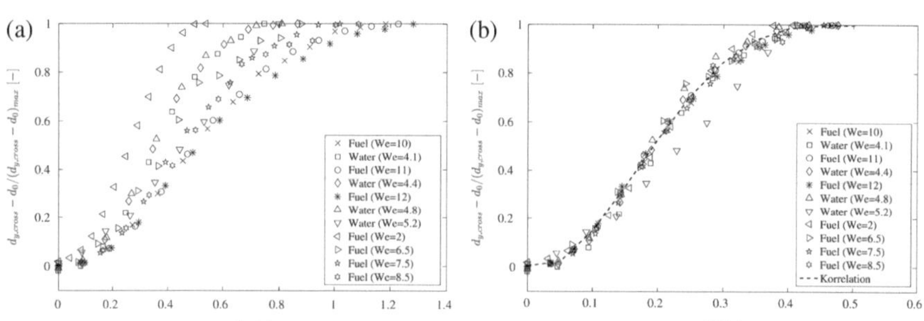

Fig. 8 Temporal evolution of the droplet deformation perpendicular to the flow direction. (**a**) Time evolution of the initial droplet deformation perpendicular to the flow in y-direction. (**b**) Correlated time evolution of the initial droplet deformation perpendicular to the flow in y-direction

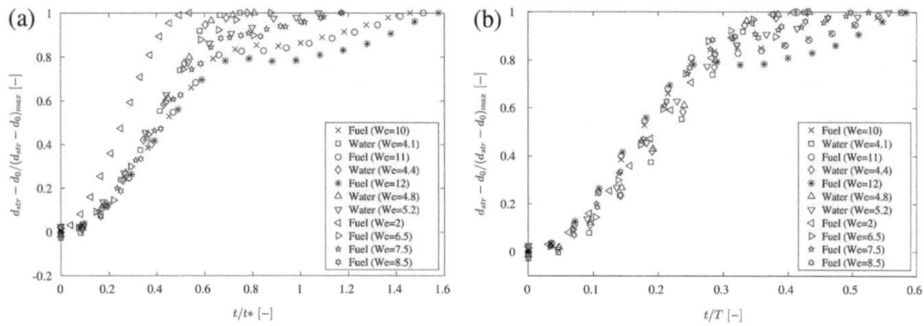

Fig. 9 Temporal evolution of the droplet deformation in flow direction. (**a**) Time evolution of the initial droplet deformation in flow direction. (**b**) Correlated time evolution of the droplet deformation dynamics in flow direction

The new characteristic time was derived using the data for the cross stream deformation of the droplets. In Fig. 9 the temporal evolution of the droplet dynamics in flow direction is plotted over the dimensionless time. In Fig. 9a $t/t*$ is used as dimensionless time while in Fig. 9b t/T is used.

Apparently, here too the new characteristic time serves to describe the dynamics of the deformation better, collapsing the temporal evolution to one curve. Even the formation of a small bag in upstream direction, which is observed in most cases, is reproduced quite well. The bag formation is indicated by the decrease of deformation at about $t/T \approx 0.3$ in Fig. 9b. The bag is formed by a faster acceleration of the outer part of the droplet compared to the core part. Although the small bag is formed, a droplet flattening is observed. Due to the ongoing droplet deformation perpendicular to the flow direction, the curvature on the upstream side of the droplet is decreased. Such a bag formation was not observed experimentally, whereas this phenomenon occurred in other numerical investigations as well [17, 30].

In summary it can be stated, that the dynamics of the initial deformation of single fluid droplets is dependent on *We* and is nonlinear with time. By introducing the new non dimensional time t/T the different cases considered can be correlated quite well.

6.2 *Two Fluid Droplet Simulations*

The influence of adding water to the droplet is investigated for the two different aerodynamic loads and five different placements of a single water droplet inside the fuel droplet. Exemplary the SPH prediction of temporal evolution of a fuel droplet with water added at the center at $u_0 = 22.5\,\mathrm{m/s}$ ($We_{\mathrm{fuel}} = 10$) is depicted in Fig. 10.

The addition of water causes a difference in deformation behavior. Due to the higher density as well as viscosity the dispersed water is deformed slower. This leads almost to a separation of the two phases (cf. second instance in Fig. 10), which is counteracted by interfacial forces acting on the different fluid pairings.

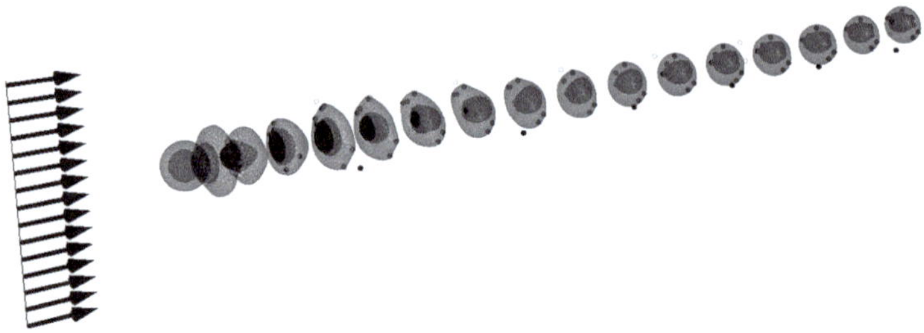

Fig. 10 Temporal droplet evolution of a fuel droplet with water centered at $u_0 = 22.5\,\mathrm{m/s}$

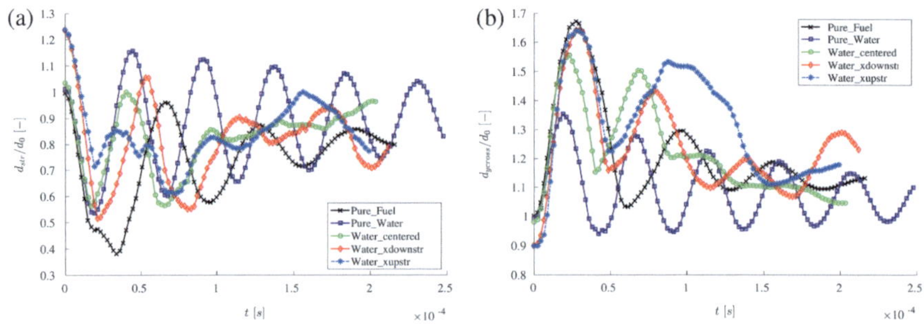

Fig. 11 Droplet deformation over time for all cases investigated at $u_0 = 22.5\,\mathrm{m/s}$. (**a**) Temporal evolution of the droplet deformation in flow direction d_{str}. (**b**) Temporal evolution of the droplet deformation perpendicular to the flow in y-direction $d_{\mathrm{y,cross}}$

In the further course of the simulation fuel and water are unified again and the droplet shows oscillatory deformations, like it was observed for the pure liquid cases. The deformation behavior for all the other two fluid droplet simulations is basically similar with minor differences due to the water placement. The water droplets placed off center perpendicular to the flow direction, additionally feature a rotation of the whole droplet. Due to the spatial resolution, which is not sufficient to properly resolve the three-liquid contact line, the behavior of these droplets might not be physical. Therefore, these results will be left out in the following.

In Fig. 11 the deformation of the droplets in flow direction d_{str} and perpendicular to the flow $d_{\mathrm{y,cross}}$ over time for an air velocity of $u_0 = 22.5\,\mathrm{m/s}$ is depicted. In Fig. 11a the deformation in flow direction while in Fig. 11b the deformation perpendicular to the flow in y-direction is plotted.

The pure liquid cases are indicated by black crosses (fuel) and blue squares (water). It is evident, that the water oscillates at a higher frequency and the oscillation is less dampened due to the lower Oh. In case of the two fluid droplets, the damping of the oscillation is similar to that of the pure fuel droplets. The frequency of the deformation seems to be in-between the frequencies of water and fuel for the first two oscillations. Afterwards, the oscillations almost vanish or show a behavior which cannot be classified by the behavior of neither fuel nor water. Possibly, the behavior is triggered by single water particles located at the surface of the fuel. These single particles do not represent a physical water droplet. They are rather the result of the insufficient spatial resolution of the three fluid contact line. Therefore, a non-physical acceleration due to the CSF model is imposed (cf. Fig. 10).

The deviations observed for the initial droplet extent perpendicular to the flow and in flow direction in Fig. 11 are due to the droplet initialization described previously. During the formation of the steady state droplet, some particles of the water droplet placed eccentric in the fuel droplet have air particles within their radius of influence. Therefore, due to the surface tension modeling an ellipsoid rather than a spherical droplet is formed.

The dynamics of the initial deformation shows an ambivalent behavior. In flow direction the deformation dynamics as well as the minimum extent of the two fluid droplets are similar to the results of the water. Perpendicular to the flow the droplet deformation and maximum droplet extent are similar to the observations made for the fuel droplet. At a first glance, this indicates similar drag coefficients for the two fluid and fuel droplets in the initial deformation stage. Therefore, a description of two fluid droplets might be possible by currently used simplified models.

Whether the same observations are true for the whole range of low $We \leq 12$, is revealed by the results with higher aerodynamic loading investigated in this study. In Fig. 12 the droplet deformation in flow direction d_{str} and perpendicular to the flow $d_{\text{y,cross}}$ over time is plotted for an air velocity of $u_0 = 24.34\,\text{m/s}$. Here too, the deformation dynamics in flow direction is shown in Fig. 12a and the deformation in cross-stream direction in Fig. 12b.

The deformation behavior of the pure liquids (water: blue squares and fuel black crosses) show a similar behavior as for the lower aerodynamic load. Similarly, the water droplet exhibits a distinct higher oscillation frequency and lower damping of the oscillation than the fuel droplet. Only the initial amplitude of the deformation is increased, as evident from correlation (12).

Regarding the deformation of the two fluid droplets, leaving out the cases with water placed off center perpendicular to the flow again, the initial deviations of the

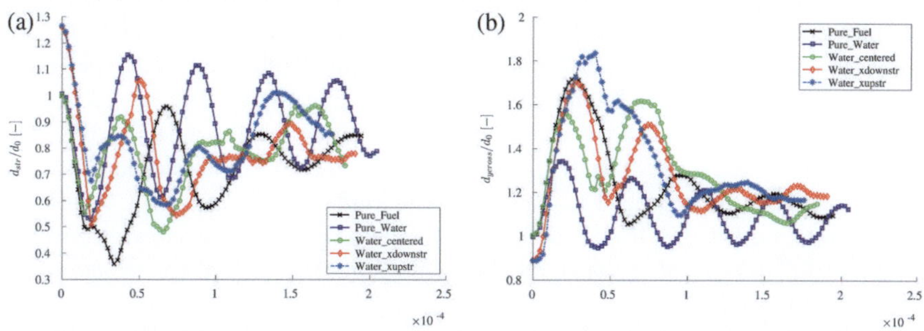

Fig. 12 Droplet deformation over time for all cases investigated at $u_0 = 24.34\,\text{m/s}$. (**a**) Temporal evolution of the droplet deformation in flow direction d_{str}. (**b**) Temporal evolution of the droplet deformation perpendicular to the flow in y-direction $d_{\text{y,cross}}$

droplet extents as well as the dynamics observed are similar to the cases with the lower aerodynamic load. Here too, the differences of the initial extents is also a result of the initialization of the droplet at the start of the simulation. The frequency of the first two oscillations is again somewhere in between the dynamics of pure fuel and pure water. Furthermore, the damping of the amplitude is as strong as in the case of the pure fuel, resembling the behavior observed before. Furthermore, the amplitude of the deformation exhibits the same ambivalent behavior as in the previous cases. The minimum extent in flow direction is similar to that of the pure water case, whereas the maximum cross stream extent is similar to the deformation of the pure fuel droplet.

In all cases investigated ($u_0 = 22.5\,\mathrm{m/s}$ and $u_0 = 24.34\,\mathrm{m/s}$) the droplets just experience deformation and no droplet breakup is observed. Since by SPH a droplet breakup is predicted at $We = 13$ it is assumed that the addition of 23 % volume fraction of water does not change the characteristics of the droplet dynamics described by the pure fuel Weber number We_{fuel}, at least in the deformation regime. Furthermore, it may be concluded from the deformation dynamics of the two fluid droplets, that they experience a similar drag coefficient as pure fuel droplets in the early stages of the deformation. This fact would allow to use the common correlations with minor adaptations in Euler-Lagrange investigations for two fluid droplets as well.

7 Conclusion

In this paper the dynamics of one and two fluid droplets at low aerodynamic loads is investigated numerically using the Smoothed Particle Hydrodynamics (SPH) method. The presented predictions are focused on the deformation dynamics of pure liquid and water-in-fuel droplets with a diameter of $d_0 \approx 60\,\mu\mathrm{m}$ exposed to two different air flow velocities: $u_0 = 22.5\,\mathrm{m/s}$ and $u_0 = 24.34\,\mathrm{m/s}$. As the SPH method is relatively new to CFD applications, first a verification of the code in-use is done comparing numerical results for pure liquid droplets to well known empirical findings. With few exceptions the predicted minimum initial deformation in flow direction d_{str} as well as the maximum cross-stream deformation d_{cross} perfectly matches the correlation of Hsiang and Faeth [14]. Deviations are observed mainly to occur at extreme conditions. The results demonstrate the capability of SPH for capturing the droplet deformation dynamics. Second, the dynamic deformation of the single fluid droplets was analyzed and found to be dependent on We. A new correlation for the droplet deformation was proposed. This correlation uses a modified definition of the characteristic time T compared to the correlations of Hsiang and Faeth [14].

For the prediction of two fluid droplets, a single water droplet with a volume fraction of 23 % was added to a fuel droplet and the placement was varied. For both aerodynamic loads the two fluid droplets show a behavior, which can be classified by the pure fuel Weber number We_{fuel}. The dynamic behavior of the two fluid droplets qualitatively feature a frequency in between the fuel and water droplet for the first oscillations, while the damping of the oscillations is similar to the fuel droplet. Furthermore, an ambivalent behavior is observed for the minimum extent in flow direction and the maximum cross-stream deformation. In the first case the two fluid droplets behave as the water droplet, while in the later case their behavior is similar to the fuel droplet.

In general, it can be stated that the SPH code is capable of predicting multiphase flows, like the droplet deformation, physically correct. Therefore, SPH will be used as tool to further investigate the dynamics of droplet deformation of mono- and two fluid droplets in order to derive simpler models, which can be used in typical CFD predictions of sprays.

Acknowledgements The financial support of the German Federal Ministry of Economics and Technology and Siemens AG within the cooperative research project 'Entwicklung von Verbrennungstechnologien im CEC für klimaschonende Energieerzeugung (03ET7011E)' is gratefully acknowledged.

This work was performed on the computational resource ForHLR Phase I, funded by the Ministry of Science, Research and the Arts Baden-Württemberg and DFG ("Deutsche Forschungsgemeinschaft").

References

1. Adami, S., Hu, X.Y., Adams, N.A.: A new surface-tension formulation for multi-phase SPH using a reproducing divergence approximation. J. Comput. Phys. **229**, 5011–5021 (2010)
2. Bartz, F.-O., Schmehl, R., Koch, R., Bauer, H.-J.: An extension of dynamic droplet deformation model to secondary atomization. In: 23rd Annual Conference on Liquid Atomization and Spray Systems, Brno (2010)
3. Batchelor, G.K.: An Introduction to Fluid Dynamics. Cambridge University Press, Cambridge (2000)
4. Brackbill, J.U., Kothe, D.B., Zemach, C.: A continuum method for modeling surface tension. J. Comput. Phys. **100**, 335–354 (1992)
5. Braun, S., Krug, M., Wieth, L., Höfler, C., Koch, R., Bauer, H.-J.: Simulation of primary atomization: assessment of the smoothed particle hydrodynamics (SPH) method. In: 13th Triennial International Conference on Liquid Atomization and Spray Systems, Tainan (2015)
6. Braun, S., Wieth, L., Koch, R., Bauer, H.-J.: A framework for permeable boundary conditions in SPH: inlet, outlet, periodicity. In: 10th International SPHERIC Workshop, Parma (2015)
7. Colagrossi, A., Landrini, M.: Numerical simulation of interfacial flows by smoothed particle hydrodynamics. J. Comput. Phys. **191**, 448–475 (2003)
8. Dryer, F.L.: Water addition to practical combustion systems – concepts and applications. Symp. Int. Combust. **16**(1), 279–295 (1977)
9. Forschungshochleistungsrechner ForHLR Phase I http://www.bwhpc-c5.de/wiki/index.php/ForHLR_Phase_I_Hardware_and_Architecture. Cited 04 Apr 2016
10. Gingold, R.A., Monaghan, J.J.: Smoothed particle hydrodynamics theory and application to non-spherical stars. Mon. Not. R. Aston. Soc. **181**, 375–389 (1977)
11. Guildenbecher, D.R., López-Rivera, C., Sojka, P.E.: Secondary atomization. Exp. Fluids **46**, 371–402 (2009)
12. Hinze, J.O.: Fundamentals of the hydrodynamic mechanism of splitting in dispersion processes. AIChE J. **1**, 289–295 (1955)

13. Höfler, C., Braun, S., Koch, R., Bauer, H.-J.: Modeling spray formation in gas turbines – a new meshless approach. J. Eng. Gas. Turb. Power **135**, 011503-1–011503-8 (2013)
14. Hsiang, L.-P., Faeth, G.M.: Near-limit drop deformation and secondary breakup. Int. J. Multiph. Flow. **18**(5), 635–652 (1992)
15. Hu, X.Y., Adams, N.A.: Angular-momentum conservative smoothed particle dynamics for incompressible viscous flows. Phys. Fluids **18**, 101702 (2006)
16. Hu, X.Y., Adams, N.A.: An incompressible multi-phase SPH method. J. Comput. Phys. **227**, 264–278 (2007)
17. Khare, P., Ma, D., Chen, X., Yang, D.: Breakup of liquid droplets. In: 12th Triennial International Conference on Liquid Atomization and Spray Systems, Heidelberg (2012)
18. Lechner, C., Seume, J.: Stationäre Gasturbinen. Springer, Heidelberg (2010)
19. Liu, M.B., Liu, G.R.: Smoothed particle hydrodynamics (SPH) an overview and recent developments. Arch. Comput. Method E **17**, 25–76 (2010)
20. Lucy, L.B.: A numerical approach to the testing of the fission hypothesis. Astron. J. **82**, 1013–1024 (1977)
21. Monaghan, J.J.: Smoothed particle hydrodynamics. Annu. Rev. Astron. Astrophys. **30**, 543–574 (1992)
22. Monaghan, J.J.: Simulating free surface flows with SPH. J. Comput. Phys. **110**, 399–406 (1994)
23. Morris, J.P., Fox, P.J., Zhu, Y.: Modeling low Reynolds number incompressible flows using SPH. J. Comput. Phys. **136**, 214–226 (1997)
24. O'Rourke, P.J., Amsden, A.A.: The TAB method for numerical calculation of spray droplet breakup. In: International Fuels and Lubricants Meeting and Exposition, Toronto (1987)
25. Quan, S., Schmidt, D.P.: Direct numerical study of a liquid droplet impulsively accelerated by gaseous flow. Phys. Fluids **18**, 102103 (2006)
26. Ranger, A.A., Nicholls, J.A.: Aerodynamic shattering of liquid drops. AIAA J. **7**(2), 285–289 (1969)
27. Schmehl, R.: Advanced modeling of droplet deformation and breakup for CFD analysis of mixture preparation. In: 18th Annual Conference on Liquid Atomization and Spray Systems, Zaragoza (2002)
28. Schmehl, R., Maier, G., Wittig, S.: CFD analysis of fuel atomization, secondary droplet breakup and spray dispersion in the premix duct of a LPP combustor. In: 8th International Conference on Liquid Atomization and Spray Systems, Pasadena (2000)
29. Wieth, L., Braun, S., Koch, R., Bauer, H.-J.: Modeling of liquid-wall interaction using the meshless Smoothed Particle Hydrodynamics (SPH) method. In: 26th European Conference on Liquid Atomization and Spray Systems, Bremen (2014)
30. Zaleski, S., Li, J., Succi, S.: Two-dimensional Navier-Stokes simulation of deformation and breakup of liquid patches. Phys. Rev. Lett. **75**(2), 244–247 (1995)

Smoothed Particle Hydrodynamics for Numerical Predictions of Primary Atomization

Samuel Braun, Rainer Koch, and Hans-Jörg Bauer

Abstract A code framework based on the Smoothed Particle Hydrodynamics (SPH) method has been used to investigate the liquid disintegration processes of an air-assisted atomizer. As the flow physics includes spatial and temporal scales which cover at least 4 orders of magnitude, the use of HPC resources is indispensable. The application of the SPH method is rather new to computational fluid dynamics (CFD). We therefore compare our in-house code to established CFD tools in order to assess the computational performance as well as the quality the physical results. It can be shown, that SPH is able to outperform commonly used grid based methods concerning the scalability behavior as well as the absolute computing speed.

The three dimensional test case to be presented consists of 1.2 billion particles. The simulation has been run on the ForHLR I cluster, where 2560 cores have been used for 60 days. The simulation is the most detailed numerical investigation of a prefilmer based atomizer and one of the largest SPH multi-phase flow simulations ever. It did capture the experimentally observed bag breakup regime with good agreement of the spatial liquid disintegration and the breakup time scales.

1 Introduction

Jet engines for civil aircrafts have to provide the highest possible efficiency over a broad range of operating conditions and, at the same time, a low emission footprint. This is thermodynamically contradictory. Particularly, the reduction of thermally induced nitric oxides (NOx) needs special attention. One major key to

S. Braun (✉) • R. Koch • H.-J. Bauer
Institut für Thermische Strömungsmaschinen, Karlsruher Institut für Technologie, Kaiserstr. 12, 76131 Karlsruhe, Germany
e-mail: samuel.braun@kit.edu; rainer.koch@kit.edu; hans-joerg.bauer@kit.edu

W.E. Nagel et al. (eds.), *High Performance Computing in Science and Engineering '16*, DOI 10.1007/978-3-319-47066-5_22

comply with those requirements is a proper distribution and positioning of the fuel inside the combustion chamber in order to control the combustion process. Therefore, the performance of the fuel nozzles has to be known and optimized. Up to now, the atomization process of air assisted atomizers is not understood in detail. Furthermore, numerical simulations were prohibitive due to the enormous computational costs which result from the multi-scale characteristic of primary atomization. The length scales to be considered cover the range between 1 μm, which is probably the diameter of the smallest droplets and several centimeters, which corresponds to the atomizer dimensions.

With the steadily increasing availability of HPC resources such simulations have come into reach. However, the algorithms, methods and codes must be able to fully exploit the hardware capabilities. Current state of the art flow solvers are mostly grid based, often use implicit time stepping schemes and require huge sets of linear equations to be solved. Concerning the treatment of multi-phase flows, the Eulerian frame of reference requires an appropriate tracking, caption and reconstruction of the phase interfaces between the liquid and the ambient air.

Explicit Lagrangian particle methods feature a great potential to efficiently use the massively parallel architecture of HPC systems. In general, these methods have a comparatively low memory footprint and are easy to parallelize. Due to the Lagrangian frame of reference, there is no need to separately track or capture the phase interfaces. One of the most popular particle methods for computational fluid dynamics has become the Smoothed Particle Hydrodynamics method [12, 15]. It originates from an astrophysical context, where its main benefit is to avoid the discretization of void (interstellar) spaces. In the context of terrestrial CFD, the main areas of application are coastal protection, naval engineering or avalanche predictions. Also in computer games or in movies, SPH-like techniques are used to animate liquids or smoke.

The present study is one of the first attempts to numerically predict the atomization behavior of an experimentally investigated planar prefilming air-blast atomizer [9–11]. Whereas 2D studies were already able to reproduce some of the experimentally observed breakup features [4, 5], only 3D simulations will capture all relevant physical effects. This inevitably leads to the necessity of HPC resources. The test case to be presented is derived from an experimentally investigated setup. It consists of 1.2 billion particles with a spatial resolution of roughly 5 μm.

This report is structured as follows. In Sect. 2, the numerical method SPH and the relevant physical models are briefly introduced. In Sect. 3, the experimental setup and the derived numerical model are presented. Section 4 will give an impression of the computational performance of our in-house SPH code, which has been initialized at the Institut für Thermische Strömungsmaschinen 4 years ago. In Sect. 5, the simulation of an planar prefilming airblast atomizer, the computational requirements and selected results are presented. Section 6 will give an outlook to simulations planned for the near future.

2 Numerical Method

The SPH method has been developed in the late 1970s in the context of astrophysics [12, 15]. The spatial discretization of a computational domain is done via so called particles, which represent a certain volume of the fluid. These Lagrangian particles move within the computational domain with the actual fluid velocity. The governing equations describing the flow physics are the Navier Stokes equations. The main idea behind the SPH formalism is to evaluate the physical property of a particle or its derivative by interpolating over neighbor particles within a certain cut-off radius. Equation (1) represents the basic interpolation formalism for a particle with index i.

$$\langle \Phi \rangle_i = \sum_j V_j \Phi_j W(\mathbf{x}_i - \mathbf{x}_j, h) \tag{1}$$

In Eq. (1) Φ is a physical quantity, V_j is the volume of an adjacent particle j and $W(\mathbf{x}_i - \mathbf{x}_j, h)$ is a weighting function (kernel), depending on the positions $\mathbf{x}_{i/j}$ and the smoothing length h, which defines the cut-off radius. Using Eq. (1) and its spatial derivative, the continuity equations can be reformulated. Regarding mass conservation, SPH offers several possibilities to determine the density. For multiphase flows with density ratios in the range of 1000, Eq. (2) gives a much better stability than integrating the substantive derivative [14]. It is therefore used in the context of this project.

$$\rho_i = m_i \sum_j W(\mathbf{x}_{ij}, h) \tag{2}$$

With Eq. (2), the particle's density ρ_i can be directly calculated using its mass m_i and the available volume, which is defined by the positions of the surrounding particles j. The conservation of momentum results in the following contributions to the total particle acceleration. Equation (3) accounts for pressure gradients,

$$\mathbf{a}_{i,\nabla p} = -\sum_j \frac{m_j (p_i + p_j)}{\rho_i \cdot \rho_j} \nabla W(\mathbf{x}_{ij}, h) \tag{3}$$

where p_i and p_j are the pressures of the regarded particle i and its neighbor particles j respectively. $\nabla W(\mathbf{x}_{ij}, h)$ denotes the spatial derivative of the kernel. Shear forces result in accelerations given by Eq. (4) [18], d denotes the dimensionality.

$$\mathbf{a}_{i,\mu \Delta v} = \sum_j 2 m_j (d+2) \left(\frac{\nu_i + \nu_j}{\rho_i + \rho_j} \frac{\mathbf{v}_{ij} \cdot \mathbf{r}_{ij}}{r_{ij}^2 + \eta^2} \right) \nabla W(\mathbf{x}_{ij}, h) \tag{4}$$

Note that instead of using the second derivative or a nested sum, an alternative expression is used, where a standard SPH first derivative is combined with a finite difference approximation of a first derivative. This reduces the effects of kernel interpolation errors and also reduces the computational costs.

One crucial aspect for simulating atomization effects is the modeling of surface tension. In the present study, the Continuum Surface Force (CSF) method [3] is used, which has been adapted to SPH by Morris [16] and Adami [1]. The basic idea is to convert a surface force into a volumetric force $\mathbf{f}_{i,SF}$, which is then applied to particles in vicinity of the phase interface.

$$\mathbf{f}_{i,SF} = \frac{\mathbf{F}_{i,SF}}{\rho_i} = -\frac{\sigma}{\rho_i} (\nabla \hat{\mathbf{n}}_i)\, \mathbf{n}_i \tag{5}$$

In Eq. (5), $\mathbf{n}_i$ denotes the surface normal of the interface and $\nabla \hat{\mathbf{n}}_i$ its normalized curvature. The surface normal is obtained using a color function ζ_j^i, by which the different fluids are identified.

$$\mathbf{n}_i = \frac{1}{V_i} \sum_j \left(V_i^2 + V_j^2\right) \frac{\rho_i}{\rho_i + \rho_j} \zeta_j^i \nabla W(\mathbf{x}_{ij}, h) \tag{6}$$

The evaluation of Eqs. (5) and (6) can be limited to particles in vicinity of the interface, the overall computational impact of surface tension modeling can therefore be reduced to a minimum.

In the context of this work, a weakly compressible SPH formulation is considered. Thus, solving the pressure Poisson equation is not necessary, but a suitable equation of state must be provided in order to close the Navier Stokes equations. The pressure is calculated using a modified Tait equation, which directly links the density to the pressure.

$$p = \frac{c^2 \rho_0}{\gamma} \left[\left(\frac{\rho}{\rho_0} \right)^{\gamma} - 1 \right] \tag{7}$$

In Eq. (7), c is a numerical speed of sound, ρ_0 is the reference density and γ influences the degree of compressibility. The flows to be considered in this project are subsonic with a Mach number of $0 < \mathrm{Ma} < 0.3$. Time integration is done via a predictor corrector scheme. No turbulence models are used in the context of this investigations, as the spatial discretization allows to consider a DNS.

Until now, the application of SPH to technical problems was limited by the lack of suitable boundary conditions. Only recently, flexible and robust inflow and outflow boundaries have been introduced to the method [7]. The 3D simulation to be presented is the first of its kind, incorporating multiple phases, physically realistic fluid properties as well as proper inflow and outflow boundary conditions.

3 Reference Setup

The test case to be regarded is based on an experimentally investigated generic planar atomizer, which is a two-dimensional abstraction of an annular airblast atomizer [9–11]. Figure 1 gives an impression of the geometric features of the experimental setup and the numerical abstraction. An airfoil shaped prefilmer is exposed to an air stream. At the upper side of the prefilmer surface, a liquid film is fed through small drill holes, which are located approximately at the first third of the prefilmer length. High aerodynamic forces push the liquid film to the trailing edge of the prefilmer. Here, the liquid accumulates and forms flapping ligaments, which finally detach from the prefilmer lip. This detachment and disintegration of the ligaments is called primary atomization.

The numerical domain covers the region in vicinity of the trailing edge. At the inlet, the air velocity is prescribed by a piecewise defined profile with a maximum velocity of $50\,\mathrm{m\,s^{-1}}$. The liquid phase enters with a constant velocity of $0.617\,\mathrm{m\,s^{-1}}$. On top and on bottom the computational domain is confined by static walls. Table 1 summarizes the geometric features and the fluid properties. With an inter-particle spacing of $dx = 5\,\mu\mathrm{m}$, the two-dimensional domain consists of approximately 1.5×10^6 particles. In three dimensions, the span-wise extent of $\mathrm{d_{lateral}} = 4\,\mathrm{mm}$ yields 1.2×10^9 particles. The lateral faces are handled by periodic boundary conditions.

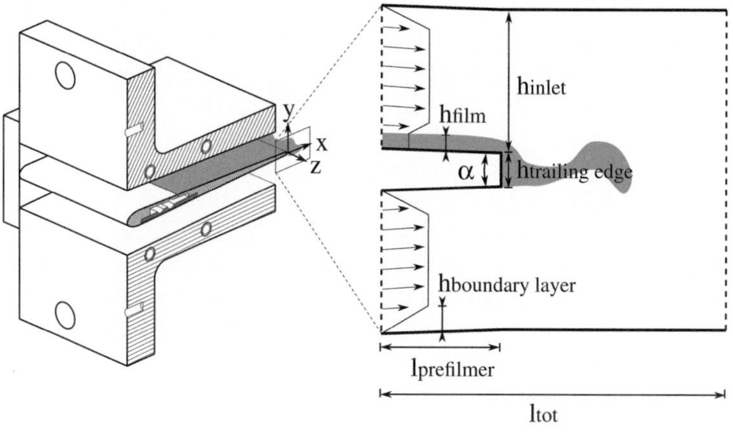

Fig. 1 Experimental setup (*left*) and two-dimensional numerical abstraction of the region in vicinity of the trailing edge (*right*)

Table 1 Experimental setup, corresponding computational domain and fluid properties

Geometry		Fluid properties	Air		Liquid
$h_{\text{trailing edge}}$ [µm]	230	$\rho\,[\text{kg}\,\text{m}^{-3}]$	1		770
h_{inlet} [mm]	3	μ [Pa s]	1.8e − 5		1.56e − 3
h_{film} [µm]	80	Velocity@inlet $[\text{m}\,\text{s}^{-1}]$	0–50		0.617
$h_{\text{boundary layer}}$ [mm]	0.5	$\sigma_{\text{air–liquid}}\,[\text{N}\,\text{m}^{-1}]$		0.0275	
l_{tot} [mm]	6	Contact angle$_{\text{wall–liquid}}$ [°]		60	
$l_{\text{prefilmer}}$ [mm]	2				
d_{lateral} [mm]	4				
α [°]	4.29				
dx [µm]	5				

4 Code Framework and Performance

The SPH simulations have been conducted using an in-house code, which has been initialized at the Institut für Thermische Strömungsmaschinen 4 years ago. It is written in C++. Parallelization is done via domain decompositioning and MPI. Due to a lack of creativity, it is named `super_sph`. Before running large 3D simulations, the performance of the code has been investigated. Within the reporting period, the serial performance has been optimized by improving the cache efficiency. Most importantly, switching from an array of structures data layout to a structure of arrays layout and a spatial particle sorting improved the serial performance by a factor of 4. In order to analyze the serial performance, different instrumentation and sampling tools like e.g. gprof [13] and MAQAO [2] have been applied.

The parallel code performance has been investigated by scalability tests (strong scalability). In order to classify the SPH results of the multi-phase flow simulations, comparative runs have been performed using the VoF solvers of a commercial code and OpenFOAM® 2.3.0, which are both grid based codes. A structured mesh has been used, where the number of cells for both codes was identical to the number of particles in the SPH simulations. The boundary conditions were identical to the SPH simulation, turbulence modeling has been disabled. Using the two-dimensional computational domain as described in Fig. 1 and Table 1 together with a high number of cores, a maximum communication to computing ratio can be achieved, which served to provoke and to identify communication bottlenecks. The tests have been run on the thin nodes of the ForHLR I cluster, which are equipped with 2 Deca-Core Intel® Xeon® E5-2670 v2 processors. For jobs with less than 20 cores, a single node has been used exclusively. Two types of scalability tests have been performed, using different termination criteria. For the comparison of the grid based tools and SPH, 1 h wall clock time was set to terminate the simulations. The speedup and the numerical efficiency have been calculated using the number of time steps, which could be performed during this period of time. For the second scalability test, the number of time steps has been fixed. The calculation of the speedup is based on the resulting runtimes.

It is to be emphasized that the following results might not be representative for the maximum achievable performance of the commercial code and OpenFOAM. Both codes have been used to the best of the author's knowledge. The solver settings correspond to production run settings, which are typically used at our institute. Furthermore, small domain sizes of only 1.5×10^6 cells are usually not considered to be run on more than 2 nodes.

In Figs. 2 and 3 the speedup and the parallel efficiency are depicted respectively. The reference performance is given by the number of time steps achieved with 20 cores (1 node) within 1 h wall clock time. OpenFOAM and the commercial

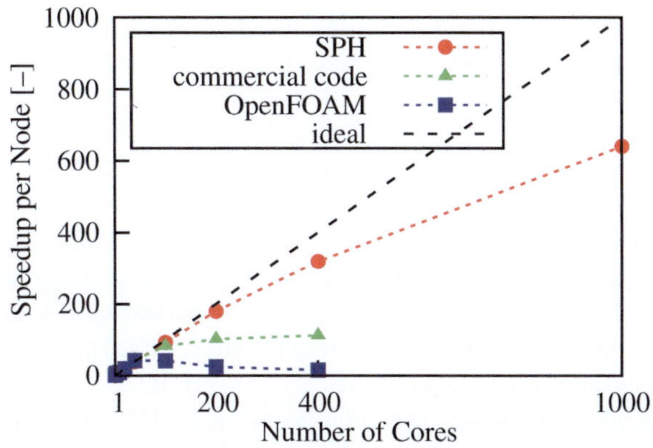

Fig. 2 Speedup behavior of three different codes. The termination criterion of the simulations is 1 h wall clock time. The graphs are normalized by the performance of a single node with 20 cores

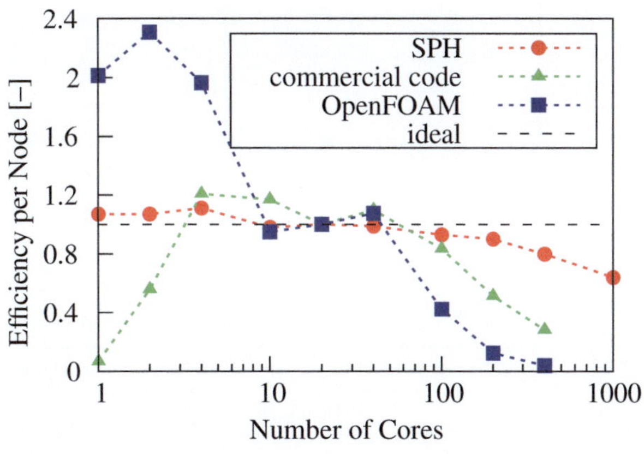

Fig. 3 Parallel efficiency behavior of three different codes. The termination criterion of the simulations is 1 h wall clock time. The graphs are normalized by the performance of a single node with 20 cores

code show a reasonable scaling till 2 nodes, where OpenFOAM profits especially from the fast intra-socket communication. The achieved serial performance of the commercial code might be questionable. The parallel efficiency of SPH remains above 0.9 till 200 cores. At 1000 cores, the parallel efficiency is still above 0.6 and the speedup did not yet reach saturation. Please note, that even at a sub-domain size of less than 1500 particles, a further acceleration of the simulation can be achieved. The speedup of the investigated grid based methods saturated at approximately 100 cores, a further increase of computational resources would not reduce the computing time of the simulations any more.

Typically, scalability tests are run using a fixed number of time steps as termination criterion and not using a fixed wall clock time. This ensures that temporal variations of the computational effort do not affect the scalability results. An example of a temporal increased computational effort would be the occurrence of a breakup event within the atomization process, which temporally leads to an enlarged gas-liquid interface and, therefore, to higher computational costs.

In Fig. 4 the parallel efficiency of SPH is displayed using both termination criteria. The red line indicates the efficiency results obtained after 1 h wall clock time. The green line and the blue line are obtained after 20,000 time steps. The error bars indicate a temporal variation of ± 10 s, which seems to be a reasonable deviation when reading hundreds of sub-domain initialization files from a non-exclusive file system. The wall clock time for 20,000 time steps varies between 12 h at 1 core and 97 s at 1000 cores. The reference performance for the green line is given by 1 core, for the blue and the red line by 1 node, i.e. 20 cores. Due to the high communication to computing ratio, the effect of the fast intra-socket communication is clearly perceptible. However, the InfiniBand® 4X FDR interconnect ensures a nearly constant efficiency up to 200 cores. In summary it can be stated, that the SPH

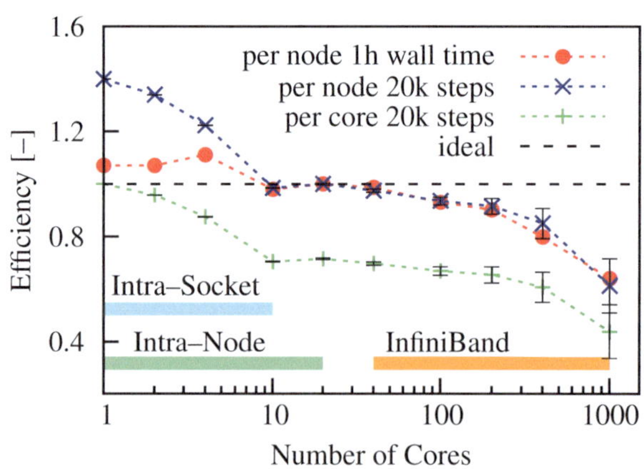

Fig. 4 Parallel efficiency of our SPH code. Comparison of a test run with fixed wall clock time (*red curve*) and 20,000 time steps (*blue* and *green curves*) as termination criterion. Reference performance is given by one core (*green*) or one node (*blue* and *red*)

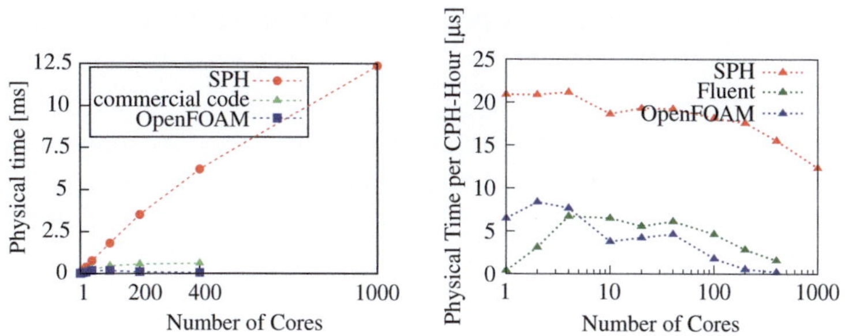

Fig. 5 Simulated physical time within 1 h wall clock time (*left*) and within one CPU-hour (*right*) for three different codes

code shows a decent strong scalability over 3 orders of magnitude, with sub-domain sizes ranging from 1.5×10^6 down to 1500 particles.

When doing simulations, the values of speedup and efficiency are only of minor interest. The benefit of a simulation is rather the achievable physical time, which can be simulated within a certain time span or spending a certain amount of CPU-hours. Therefore, in Fig. 5 the computed physical time is depicted, which can be achieved within 1 h wall clock time or by spending one CPU-hour. The simulations with the commercial code used a fixed time step size of 2.5×10^{-8} s, SPH and OpenFOAM used an adaptive time stepping with mean time increments of 1.6×10^{-8} s (SPH) and 4.3×10^{-8} s (OpenFOAM), respectively. The maximum achievable physical time which can be computed within 1 h is limited to 0.2 ms using OpenFOAM and 0.65 ms using the commercial code. With SPH 12.3 ms have been computed using 1000 cores and saturation is not yet reached. The physical time per CPU-hour, which can be interpreted as costs per simulated physical time, reaches up to 21 µs/CPUh in the case of SPH. Even at 1000 cores, 12.3 µs can be achieved per CPU-hour. The optimum values for OpenFOAM and the commercial code are 8.5 µs and 6.7 µs per CPU-hour, respectively.

Concerning the physical results of the simulation like velocity fields, breakup frequencies, ligament lengths, droplet sizes and droplet numbers, the commercial code and SPH show very similar results. OpenFOAM, however, predicts very stable liquid ligaments which resist disintegration. This results in fewer droplets of larger mean diameters compared to the other two methods [6].

Three-dimensional test cases have not been subject to comparisons with the grid based tools. However, 3 different 3D test simulations have been performed with SPH. The test cases consisted of (a) 75 million, (b) 150 million and (c) 1.2 billion particles respectively. The two small simulations were run on 400 cores, the large one on 2560 cores. The obtained particle iteration frequencies are 79,896 (a), 77,542 (b) and 82,310 (c) particle-steps per CPU-second, which corresponds to 12.5, 12.9, 12.1 CPUµs per step and per particle.

5 Simulation of a Planar Prefilming Airblast Atomizer

Up to now, aircraft industry relies on experience and correlations when it comes to designing of new atomizer nozzles. Often, these correlations for spray characteristics were derived from completely different nozzle designs. Numerical predictions of the primary atomization were not yet feasible. However, the subsequent simulations of combustion and heat release rely on precise predictions of the fuel positioning. This means that droplet sizes, trajectories and breakup frequencies have to be provided. These features are thus the target of atomization simulations.

The simulation to be presented is an abstraction of a generic planar atomizer, which has been designed for good optical accessibility. Except the planar geometry, boundary conditions and fluid properties are comparable to real aircraft applications.

5.1 Computation Details

The two-dimensional geometry as depicted in Fig. 1 has been extruded in spanwise direction by 4 mm. An average inter-particle spacing of 5 µm yields 1.2×10^9 particles. The computational domain has been split into 2560 sub-domains, as 128 nodes is the maximum for regular ForHLR I users. During the first 490 h a quintic interpolation function with a smoothing length of 5 µm was applied. Within 305,522 time steps 3.17 ms physical time could be simulated. Afterwards, a so called Wendland kernel with a smoothing length of 6.5 µm has been implemented and used, due to its lower computational costs. Within 960 h, 790,160 time steps have been simulated, which corresponds to 11.43 ms of physical time. On average 43,000 time steps have been achieved during a 2-day run. However, there were 3 exceptions, where only 16,000–21,000 time steps have been achieved. This may be attributed to some file system problems during that time as the simulation results did not show any computationally intensive breakup events. Using the Wendland kernel, the computational effort for one particle iteration is 8.5×10^{-6} CPUs/step/particle.

Altogether, 1,095,682 time steps have been calculated within 1450 h. This corresponds to 3.71×10^6 CPUh. With a data output interval of 1000 time steps, 1113 time steps have been dumped to the file system (during the first run, the output interval was 250). The size of one time step is approximately 62 GB, the entire simulation data has a size of 69 TB.

5.2 Breakup Behavior and Spray Characteristics

The actual target quantities of the simulation are droplet sizes and trajectories, breakup frequencies and other characteristic length- and time-scales. As the liquid volume only represents roughly 1 % of the entire computational domain, the data

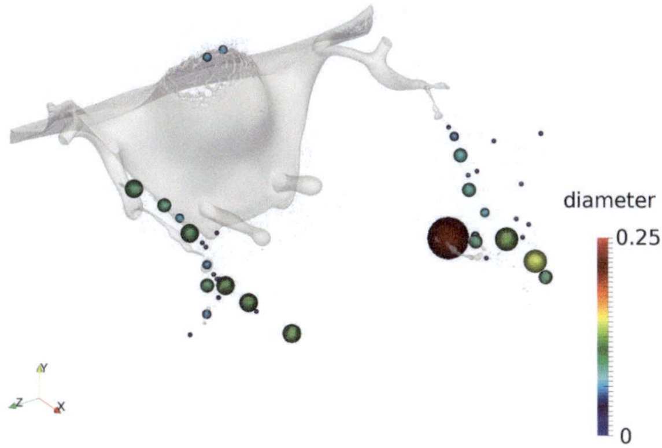

Fig. 6 Replacement of connected liquid structures by representative single particle droplets of different sizes. The colors correspond to the droplet diameters

size to be evaluated reduces dramatically. Typically 10 million particles per time step are subject to spray analysis post-processing. In order to derive droplet parameters, the droplets first have to be identified. Therefore, a Connected Component Labeling [17] technique is applied to the particles representing the liquid phase. This procedure can be performed on a standard desktop computer, where the analysis of one time step requires approximately 1 min per core. As a result, all particle agglomerations representing a droplet or a ligament can be replaced by a single particle with corresponding position, velocity, mass and deformation index. Consequently, the spray data size is further reduced to only a few megabytes. In Fig. 6, the replacement of particle agglomerations by corresponding single particle droplets is illustrated. The color represents the droplet diameter. The identified droplets can be analyzed statistically e.g. by generating droplet size probability graphs or by calculating representative mean droplet sizes. These post-processing steps are currently ongoing.

Concerning the understanding of the breakup mechanism, an appropriate visualization of the liquid phase is required. However, most post-processing tools available rely on computational grids or tessellated data. The direct depiction of particles generally obstructs the visual sensation of depth or creates a false impression of surface texture. For visualization purposes, the surfaces of the liquid particle agglomerations are therefore tessellated. This can be done e.g. by applying the α-shape algorithm [8]. In Fig. 7 a detail of the simulation is depicted using particles and tessellated surfaces. The advantage of the latter visualization technique is obvious.

The visualization of the gaseous phase is currently limited to the depiction of slices or segments, as the generation of e.g. iso-contours would require to build connectivity lists of the particles. The interpolation onto a Cartesian grid can be done efficiently and with very low memory requirements, however, this introduces some interpolation artifacts. In Fig. 8, the tessellated liquid surface is combined with

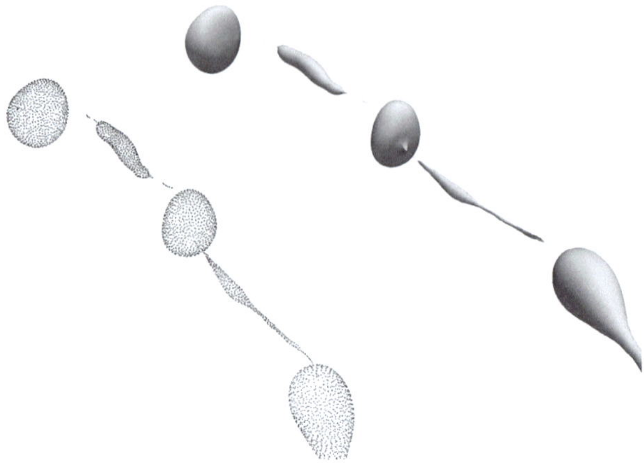

Fig. 7 Detail of a Rayleigh disintegration process and the formation of satellite droplets. Rendering using particles (*left*) or tessellated surfaces

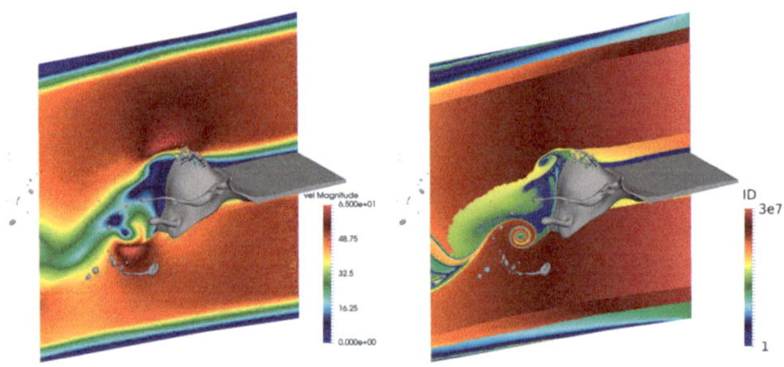

Fig. 8 Bag breakup event. The liquid phase and the wall of the atomizer lip are depicted by tessellated surfaces. Confining *upper* and *lower walls* are not depicted. The gaseous phase is visualized using a slice. The coloring of the *left* figure denotes the velocity magnitude. The coloring of the *right* figure represents the particle IDs

a slice of the gaseous phase. The confining upper and lower walls are not depicted. The coloring of the left figure indicates the velocity magnitude, the slice of the right figure is colored by the particle ID. In the example given, the depicted ID denotes the order of particle creation for every sub-domain located at the inflow region. This means for the core flow, that in every sub-domain more than 30 million particles have been released into the computational domain.Using time dependent IDs or float values allows a descriptive depiction of vortices, recirculation zones, dead wakes or residence times.

In Fig. 9, a time series of a bag breakup event is depicted. The time distance between two consecutive images is 74.5 µs. Identical atomization characteristics are observed in the experimental investigations and have been identified as main breakup mechanism for prefilmer based atomizers [9–11]. Numerical predictions of the generation and blowing up of the bag shaped structures are very sensitive to the spatial discretization. If the inter-particle distance or the mesh size is too coarse, the

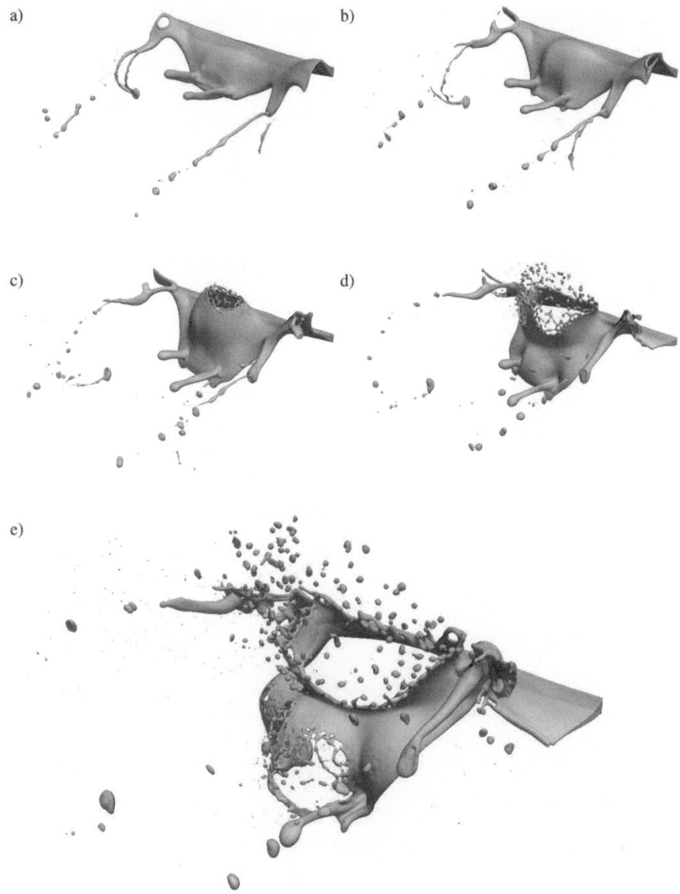

Fig. 9 Sequence of a bag breakup event. Time increment between two consecutive images is 74.5 μs

bag will not be formed or it will burst too early. When compared to experimental high speed videos, the sizes of the bag shaped structures in Fig. 9 seem to be too small. This indicates, that the inter-particle spacing of 5 μm is still not sufficient to properly represent the liquid skin of the bubbles. However, it is questionable, if the very small droplets resulting from the breakdown of this liquid skin can be experimentally captured at all. The quantitative comparison of the droplet spectra from experiments and simulations will therefore be limited to droplets bigger than 14 μm in diameter, which is the spatial resolution of the high speed camera used for image recording.

In Fig. 10, three consecutive top view snapshots of the test case are depicted. The left half of the snapshots are experimentally obtained, using a high speed camera. The right half shows the simulative results, where the data has been duplicated in

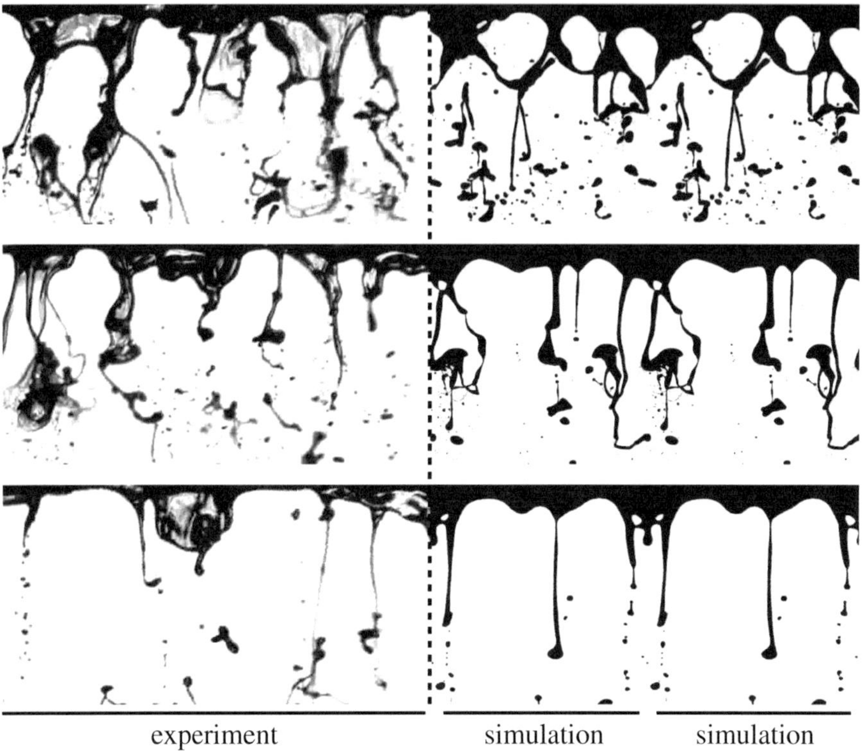

Fig. 10 Visual comparison of experiment (*left half*) and simulation (*right half*). Three temporal *top view* snapshots are depicted. The size of the displayed experimental section is 8 ×2 mm. The simulation data with a spanwise extent of 4 mm is duplicated in span-wise direction (Images of the experimental investigations: courtesy of Sebastian Gepperth)

span-wise direction. In general, the predicted frequencies, length scales and breakup mechanisms match very well. However, the images clearly show that the very thin skin of the liquid bags can not be taken into account using an inter-particle distance of 5 μm. In future numerical setups we will therefore use an inter-particle distance of 2.5 μm, which will allow to better investigate these structures.

6 Conclusion and Outlook

The numerical prediction of air assisted atomizers has come into reach due to steadily growing HPC resources. The results and insights which have been gained by the simulations presented in this report will help to reduce the environmental impact of civil aviation.

The SPH method has proven to be an adequate tool, when it comes to predicting multi-phase flow phenomena. Despite being a relatively young method, it can compete with established methods. Particularly, the excellent use of the hardware resources leads to both, fast and efficient numerical simulations. Furthermore, the method seems to be well suited for the upcoming heterogeneous computer systems. Although the current HPC facilities in Baden-Württemberg do mainly comprise multi-core systems, many-core and GPU-accelerated clusters will be the predominant facilities in the near future. Particle based methods seem to be tailored for such computer architectures.

Regarding the scientific investigation of air assisted atomizers, further simulations are planned with a higher spatial resolution. These simulations will clarify, whether the proper representation of the experimentally observed liquid bubbles and their skin does affect the overall spray properties substantially.

Acknowledgements This work was performed on the computational resource ForHLR Phase I funded by the Ministry of Science, Research and the Arts Baden-Württemberg and DFG ("Deutsche Forschungsgemeinschaft"). We greatly acknowledge the excellent technical support provided by the Steinbuch Centre for Computing (SCC) at the Karlsruhe Institute of Technology.

References

1. Adami, S., Hu, X.Y., Adams, N.A.: A new surface-tension formulation for multi-phase SPH using a reproducing divergence approximation. J. Comput. Phys. **229**(13), 5011–5021 (2010)
2. Bendifallah, Z., Jalby, W., Noudohouenou, J., Oseret, E., Palomares, V., Rubial, A.C.: PAMDA: performance assessment using MAQAO toolset and differential analysis. In: Tools for High Performance Computing 2013, pp. 107–127. Springer, Berlin/New York (2014)
3. Brackbill, J.U., Kothe, D.B.: Dynamical Modeling of Surface Tension. NASA Conference Publication, pp. 693–700 (1996)
4. Braun, S., Höfler, C., Koch, R., Bauer, H.-J.: Modeling fuel injection in gas turbines using the meshless smoothed particle hydrodynamics method. In: ASME Turbo Expo 2013: Turbine Technical Conference and Exposition, pp. V01AT04A001-V01AT04A001. American Society of Mechanical Engineers, New York (2015)
5. Braun, S., Wieth, L., Koch, R., Bauer, H.-J.: Influence of trailing edge height on primary atomization: numerical studies applying the smoothed particle hydrodynamics (SPH) method. In: 13th International Conference on Liquid Atomization and Spray Systems, Taiwan (2015)
6. Braun, S., Krug, M., Wieth, L., Höfler, C. Koch, R., Bauer, H.-J.: Simulation of primary atomization: assessment of the smoothed particle hydrodynamics (SPH) method. In: 13th International Conference on Liquid Atomization and Spray Systems, Taiwan (2015)
7. Braun, S., Wieth, L., Koch, R., Bauer, H.-J.: A framework for permeable boundary conditions in SPH: inlet, outlet, periodicity. In: 10th International SPHERIC Workshop, Parma (2015)
8. Edelsbrunner, H., Kirkpatrick, D.G., Seidel, R.: On the shape of a set of points in the plane. IEEE Trans. Inf. Theory **29**(4), 551–559 (1983)
9. Gepperth, S., Guildenbecher, D., Koch, R., Bauer, H.J.: Pre-filming primary atomization: experiments and modeling. In: 23rd European Conference on Liquid Atomization and Spray Systems (ILASS-Europe 2010), Brno, Sept 2010, pp. 6–8
10. Gepperth, S., Müller, A., Koch, R., Bauer, H.-J.: Ligament and droplet characteristics in prefilming airblast atomization. In: International Conference on Liquid Atomization and Spray Systems (ICLASS), Heidelberg, Sept 2012, pp. 2–6

11. Gepperth, S., Koch, R., Bauer, H.-J.: Analysis and comparison of primary droplet characteristics in the near field of a prefilming airblast atomizer. In: ASME Turbo Expo 2013: Turbine Technical Conference and Exposition, pp. V01AT04A002-V01AT04A002. American Society of Mechanical Engineers, New York (2013)
12. Gingold, R.A., Monaghan, J.J.: Smoothed particle hydrodynamics: theory and application to non-spherical stars. Mon. Not. R. Astron. Soc. **181**(3), 375–389 (1977)
13. Graham, S.L., Kessler, P.B., Mckusick, M.K.: Gprof: a call graph execution profiler. ACM Sigplan Not. **17**(6), 120–126. ACM (1982)
14. Hu, X.Y., Adams, N.A.: An incompressible multi-phase SPH method. J. Comput. Phys. **227**(1), 264–278 (2007)
15. Lucy, L.B.: A numerical approach to the testing of the fission hypothesis. Astron. J. **82**, 1013–1024 (1977)
16. Morris, J.P.: Simulating surface tension with smoothed particle hydrodynamics. Int. J. Numer. Methods Fluids **33**, 333–353 (2000)
17. Rosenfeld, A., Pfaltz, J.L.: Sequential operations in digital picture processing. J. ACM (JACM) **13**(4), 471–494 (1966)
18. Szewc, K., Pozorski, J., Minier, J.P.: Analysis of the incompressibility constraint in the smoothed particle hydrodynamics method. Int. J. Numer. Methods Eng. **92**(4), 343–369 (2012)

Towards Solving Fluid Flow Domain Identification Problems with Adjoint Lattice Boltzmann Methods

Mathias J. Krause, Benjamin Förster, Albert Mink, and Hermann Nirschl

Abstract A novel strategy towards solving fluid flow domain identification problems for incompressible Newtonian fluids is proposed and investigated in this paper. The resulting numerical approach is of great importance for academic studies as well as for medical and industrial applications. For example, it can be used in combination with *Phase Contrast MRI* measurements to characterise flow dynamics as well as flow domains highly accurately. The problem is formulated as a optimisation problem which minimised the distance between a given and a simulated flow field, whereby the latter one is the solution of a parameterised porous media BGK-Boltzmann model. The parameter represents the porosity distributed in the domain and its distribution is obtained as the final result of the optimisation problem. The proposed gradient-based solution strategy makes use of an adjoint lattice Boltzmann method (ALBM). Due to their similar structure to lattice Boltzmann methods (LBM), they also show excellent parallelisation behaviour. In this preliminary work, first validation results are presented as well as performance results and improvements for both single core and parallel implementation. In particular, with a simple domain identification test case a cube is being identified by position and shape inside a wind tunnel in only few optimisation steps, even with only partial flow data being available.

1 Introduction

Solving fluid flow domain identification problems numerically is of great importance for academic studies as well as for medical and industrial applications. With it, e.g. the resolution of *Phase Contrast MRI* measurements can be improved dramatically. Coupling measurement and simulation enables significant progress

M.J. Krause (✉) • B. Förster
Institute for Mechanical Process Engineering and Mechanics, Institute for Applied and Numerical Mathematics, Karlsruhe Institute of Technology, Karlsruhe, Germany
e-mail: mathias.krause@kit.edu

A. Mink • H. Nirschl
Institute for Mechanical Process Engineering and Mechanics, Karlsruhe Institute of Technology, Karlsruhe, Germany

W.E. Nagel et al. (eds.), *High Performance Computing in Science and Engineering '16*, DOI 10.1007/978-3-319-47066-5_23

concerning the accurate characterisation of flow domains and flow dynamics especially in complex, e.g. patient-specific or filter, geometries, even in situations of low image contrast or fragmentary data. This promises to expand the area of its application and further enables saving costly and limited resources or improving the effectiveness of their use. Thus, its economic and social impact is potentially enormous. However, solving such problems numerically is still a challenging task in many respects. A fluid flow domain identification problem can be formulated as a fluid flow control problem which is a restricted optimisation problem. Whereby, the flow domain is to be determined such that the difference of the simulated and the measured fluid flow field is minimised. Due to the complexity of the underlying fluid flow problem, which is usually governed by non-linear partial differential equations, they demand very large computer resources. Therefore, an integrative solution strategy which incorporates a highly efficient and scalable parallelisation is vital.

The main aim of this paper is to propose and investigate a generic strategy to solve fluid flow domain identification problems for incompressible Newtonian fluids. Based on that, its performance is to be improved before it is to be combined and tested with MRI measurement data.

During the last decade, LBM have become a widely accepted numerical tool in fluid dynamics, for instance to solve incompressible Navier–Stokes equations [1]. The simplicity of the core algorithm as well as the local computations result equally in an outstanding method for HPC and a well suited tool for various relevant physical problems [6, 7]. However, their application to solve fluid flow control and optimisation problems has not been discussed in a general framework, although dedicated strategies for certain problems have been considered before. Pingen et al. propose approaches for $2D$ topology and design optimisation problems [13–15] which are extended by Kirk et al. for transient flows [9]. Tekitek et al. deal with parameter identification problems [19]. These works have in common that an adjoint equation is derived on a discrete basis. This results in a linear system, which can be solved e.g. by applying a Schur-complement method [14], or in a backwards in time evaluation for transient problems, which can be computed by matrix-vector products [9]. In contrast, the strategy presented in this article follows a sensitivity-based *first-optimise-then-discretise* approach, which relies on deriving an adjoint equation on a continuous basis and discretising it afterwards. The expected advantage is that a novel discretisation strategy similar to LBM will lead to algorithms as efficient as LB algorithms due to similar locality properties. Solving a linear system and computing matrix vector products will not be required.

In Sect. 2 we present a parameterised BGK-Boltzmann model for fluid flow through porous media and how to solve and discretise it. In Sect. 3 we propose a sensitivity-based strategy to solve and discretise fluid flow domain identification problems with the help of an adjoint lattice Boltzmann method and its parallel implementation. In Sect. 4 we present the numerical results of a domain identification test case as well as single core performance improvements and parallel efficiency and scaling analysis.

2 Parameterised Fluid Flow Simulations with a Porous Media BGK-Boltzmann Model

In this section, a porous media BGK-Boltzmann model is introduced in order to simulate fluid flow on a mesoscopic scale. The resulting parameterised family of porous media BGK-Boltzmann equations is discretised by a lattice Boltzmann method consisting of collision and streaming steps. Due to the local character of these equations, this method can be parallelised extremly well.

2.1 Mesoscopic Modeling

LBM are strategies for discretising a family of Boltzmann equations or of other mesoscopic Boltzmann-like equations which is related to a target equation in a certain limit which is usually macroscopic in nature. The basic and common idea of all LBM is the coupling of discretisation parameters with those parameters characterising the limit process. The modality of the connection depends on the regime in which the macroscopic target equation is reached by the elements of the family of mesoscopic equations.

An important subclass of LBM, which is frequently used and investigated, enables simulating the dynamics of incompressible Newtonian fluids which is usually described macroscopically by an initial value problem governed by a Navier-Stokes equation. Methods of this subclass can be regarded as discretisation strategies for families of Porous-Media-BGK-Boltzmann (*PM-BGK-Boltzmann*) equations $\mathcal{F}$ (cf. [10]) which in the diffusive limit case are related to an incompressible Navier-Stokes equation as shown by Saint-Raymond in [17].

The particular *diffusive limit family of PM-BGK-Boltzmann equations*

$$\mathcal{F} := \left(h^2 \frac{d}{dt} f + \frac{1}{3\nu} \left(f - M^{eq}_{f,d_h} \right) = 0\,, \quad f \in V(I \times \Omega \times \mathbb{R}^d) \right)_{h>0} \tag{1}$$

is obtained by setting the *mean free path*

$$l_f = \sqrt{\frac{5}{3} RT} := \sqrt{\frac{24}{\pi}} \nu h$$

with the *kinematic viscosity* being $\nu := \frac{\pi}{8} \overline{c} l_f$ where $\overline{c} = \sqrt{\frac{8}{\pi} RT}$ is the *mean absolute thermal velocity* one obtains the *speed of sound* according to

$$c_s = \sqrt{3RT} = \frac{1}{h}\,.$$

Here, for any model parameter $h \in \mathbb{R}_{>0}, f = f(t, \boldsymbol{r}, \boldsymbol{c})$ is the *ρ_f-particle distribution function* in a transient phase space of dimension $2d$ with time $t \in I = [t_0, t_1) \subseteq \mathbb{R}_{\geq 0}$, position $\boldsymbol{r} \in \Omega \subseteq \mathbb{R}^d$ and velocity $\boldsymbol{c} \in \mathbb{R}^d$. The total derivative of f is denoted by $\frac{d}{dt}f = \left(\frac{\partial}{\partial t} + \boldsymbol{c} \cdot \nabla_{\boldsymbol{r}} + \frac{\boldsymbol{F}}{m} \cdot \nabla_{\boldsymbol{c}}\right) f$. The *particle density* ρ_f and *macroscopic velocity* $\boldsymbol{u_f}$ of the Newtonian fluid are obtained as moments of f as follows:

$$\rho_f := \int_{\mathbb{R}^d} f(v)dv \quad \text{and} \quad \boldsymbol{u_f} := \frac{1}{\rho_f} \int_{\Omega} v f(v)dv \ .$$

Furthermore,

$$M^{eq}_{f,d_h} = \frac{\rho_f \, h^d}{\left(\frac{2}{3}\pi\right)^{d/2}} \exp\left(-\frac{3}{2}\left(\boldsymbol{c}\, h - d_h \boldsymbol{u_f}\, h\right)^2\right) \qquad \text{in } I \times \Omega \times \mathbb{R}^d$$

denotes the *Porous Media Maxwellian distribution*, where the *porosity* $d_h : \Omega \to [0, 1]$ is defined by

$$d_h(r) = 1 - h^{d-1} \frac{\frac{1}{h}\nu(3\frac{\nu}{h} + \frac{1}{2})}{K} \tag{2}$$

with *permeability* K. Porosity values of $d_h := d_h(r)$ are to be interpreted as solid ($d_h = 0$), fluid ($d_h = 1$) and porous ($d_h \in (0, 1)$) at point $r \in \Omega$.

2.2 Lattice Boltzmann Algorithm: Collide and Stream

The coupling of the model parameter h to the discretisation parameter leads to LBM. The continuous space $I \times \Omega \times \mathbb{R}^d$ is replaced by a discrete space $I_h \times \Omega_h \times Q$ where h is identified with the model parameter and is now called the discretisation parameter. The position space Ω_h is chosen as a uniform grid with spacing $\delta r_1 = \delta r_2 = \ldots = \delta r_d = h$ and the discrete time interval is set to $I_h := \left\{t \in I : t = t_0 + kh^2, k \in \mathbb{N}\right\}$. The velocity space Q consists of $q \in \mathbb{N}$ directions $\boldsymbol{c_i}$ $(i = 0, 1, \ldots, q-1)$ which link dedicated neighbouring positions in such a way that for $\boldsymbol{r} \in \text{int}\ \Omega_h$ it holds $\boldsymbol{r} + \boldsymbol{c_i}h^2 \in \Omega_h$, i.e. $\boldsymbol{c_i} \sim h^{-1}$. The resulting discrete phase space is called the *lattice* and denoted by *DdQq*. To reflect the discretisation of the velocity space, the continuous distribution function f is replaced by a set $\boldsymbol{f^h}$ of q distribution functions f_i $(i = 0, 1, \ldots, q-1)$, representing an average value of f in the vicinity of the velocity $\boldsymbol{c_i}$.

The iterative process in an *LB algorithm* can be written in two steps as follows, the *collision step* (3) and the *streaming step* (4):

$$\tilde{f}_i(t, \boldsymbol{r}) = f_i(t, \boldsymbol{r}) - \frac{1}{3\nu + 1/2}\left(f_i(t, \boldsymbol{r}) - M^{eq}_{f_i,d_h}(t, \boldsymbol{r})\right) \ , \tag{3}$$

$$f_i(t + h^2, \boldsymbol{r} + \boldsymbol{c_i}h^2) = \tilde{f}_i(t, \boldsymbol{r}) \tag{4}$$

for $i = 0, 1, \ldots, q-1$, where

$$M^{eq}_{f_i,d_h}(t,r) := \frac{w_i}{w}\rho_{f^h}\left(1 + 3h^2\, c_i \cdot d_h u_{f^h} - \frac{3}{2}h^2(d_h u_{f^h})^2 + \frac{9}{2}h^4\left(c_i \cdot d_h u_{f^h}\right)^2\right)$$

is a *discretised Porous-Media Maxwell distribution* with moments ρ_{f^h} and u_{f^h} which are defined as

$$\rho_{f^h} := \sum_{i=0}^{q-1} f_i \quad \text{and} \quad u_{f^h} := \frac{1}{\rho_{f^h}} \sum_{i=0}^{q-1} c_i f_i \,.$$

The variable u_{f^h} corresponds to the *macroscopic fluid velocity* and ρ_{f^h} to the *mass density*. The kinematic fluid viscosity ν is assumed to be given, and the terms w_i/w, $c_i h$ $(i = 0, 1, \ldots, q-1)$ are model dependent constants. An exhaustive derivation of various LB equations can be found e.g. in [1, 5, 18].

In [10] Krause shows for the *D2Q9* and *D3Q27* that the truncation error comparing an element of the diffusive limit family of BGK-Boltzmann equations with its corresponding discrete LB term is of second order. In most previously published derivations of LBM, macroscopically motivated assumptions are made which is in contrast to Krause's approach. This is important to note since the derivation of the ALBM, presented later on in this article, will follow the approach in [10].

2.3 *Parallel Implementation*

Most of the computation time in LB simulations is spent performing the collision step (3) and the streaming step (4). Since the collision step is purely local and the streaming step only requires data of $q-1$ neighbouring nodes, parallelising by domain partitioning leads to low communication costs and is therefore efficient. This is widely discussed, e.g. in [11, 12, 16, 21]. Krause et al. propose in [8] a general and highly efficient hybrid parallelisation strategy for LBM which is dedicated for modern hardware technologies that blur the line of separation between architectures with shared and distributed memory. The concept is also based on domain partitioning and realised in the framework of the open source library *OpenLB*, taking advantage of object-oriented and template-based programming techniques.

3 A Sensitivity-Based Strategy to Solve Fluid Flow Domain Identification Problems

In this section, a strategy for solving fluid flow domain identification problems is presented using a method similar to LBM. Therefore, a general fluid flow optimisation problem is formulated, which is then discretised step-by-step by a first-optimise-then-discretise approach. A continuous solution strategy for the optimisation problem is given by formulating a primal and dual problem. The specific domain identification problem equations are then formulated and discretised with an adjoint lattice Boltzmann method (ALBM). Implementation details are provided regarding the ALBM and its parallelisation.

3.1 *Formulation of a General Fluid Flow Control Problem*

In the following, a strategy to solve optimal flow control and flow optimisation problems of incompressible Newtonian fluids numerically is presented. The class considered consists of *constrained optimisation problems* which can be formulated in an abstract manner according to

$$\left.\begin{array}{l}\text{find control } \boldsymbol{\alpha} \text{ and state } f \text{ which}\\ \text{minimise } J(f,\boldsymbol{\alpha}) \text{ and fulfill } \boldsymbol{G}(f,\boldsymbol{\alpha}) = \boldsymbol{0}\,.\end{array}\right\} \tag{5}$$

The particle distribution function f is said to be the *state*, the vector $\boldsymbol{\alpha}$ the *control*, the functional J the *objective* or *cost functional* and $\boldsymbol{G}(f,\boldsymbol{\alpha}) = \boldsymbol{0}$ the *constraint* or *side condition*. Here, the side condition couples the control $\boldsymbol{\alpha}$ with the state f in terms of a BGK-Boltzmann equation which is chosen as an element of the corresponding diffusive limit family of BGK-Boltzmann equations. This is in contrast to the classical macroscopic approach where the constraint is typically governed by a Navier-Stokes equation.

Problems of this class can be solved numerically in two steps by a procedure often referred to as the *first-optimise-then-discretise* strategy [4]. For the first step, it is proposed to solve the optimisation problem iteratively by applying a *line search algorithm* as presented in Algorithm 1. In particular a gradient-based method like *steepest descent* or *BFGS* in combination with e.g. the *Armijo* or the *Wolfe-Powell rule* can be chosen (e.g. [3]). Methods of this type have in common that solely the evaluation of the goal functional J and its total derivative $\frac{d}{d\alpha}J$ are required to determine the descent direction $\boldsymbol{d}^k$ and the step length δ^k in every optimisation step $k = 1, 2, \ldots$.

The evaluation of the goal functional J requires solving the side condition $\boldsymbol{G}(f,\boldsymbol{\alpha}^k) = \boldsymbol{0}$ to obtain $f(\boldsymbol{\alpha}^k)$ which corresponds to solving a fluid flow problem in every optimisation step $k = 1, 2, \ldots$. This can be done numerically after discretisation as illustrated in Sect. 2.

Algorithm 1 Line search

Set $\boldsymbol{\alpha}^0 = \boldsymbol{\alpha}_0$ as initial guess for control variable
Set $k = 0$
while Termination condition not fulfilled **do**
  1. Compute the descent direction $\boldsymbol{d}^k$
  2. Compute step length δ^k
  3. Set $\boldsymbol{\alpha}^{k+1} = \boldsymbol{\alpha}^k + \delta^k \boldsymbol{d}^k$
  4. Set $k = k + 1$

The descent direction $d^k = \frac{d}{d\boldsymbol{\alpha}} J(f(\boldsymbol{\alpha}^k), \boldsymbol{\alpha}^k)$ is obtained by the *optimality condition*

$$\frac{d}{d\boldsymbol{\alpha}} J(f(\boldsymbol{\alpha}), \boldsymbol{\alpha}) = \boldsymbol{\varphi} \cdot (\nabla_{\alpha} \otimes \boldsymbol{G}(f(\boldsymbol{\alpha}), \boldsymbol{\alpha})) + \nabla_{\alpha} J(f(\boldsymbol{\alpha}), \boldsymbol{\alpha}) \,, \tag{6}$$

where φ is the solution of the *adjoint equation*

$$\frac{\partial}{\partial f(\boldsymbol{\alpha})} J(f(\boldsymbol{\alpha}), \boldsymbol{\alpha}) = -\boldsymbol{\varphi} \cdot \frac{\partial}{\partial f(\boldsymbol{\alpha})} \boldsymbol{G}(f(\boldsymbol{\alpha}), \boldsymbol{\alpha})) \,. \tag{7}$$

In the following subsection, we consider the special problem class of domain identification. We therefor provide particular definitions of the goal functional, the side condition and the optimisation parameter as well as (6) and (7).

3.2 *Objective and Dual Problem Formulation for Domain Identification Problems*

Using the porous media model requires the control parameter $\alpha(r) \in \mathbb{R}$ to be projected onto the porosity parameter $d_h(r) \in [0, 1]$ for all points $r \in \Omega$ through an operator $B\alpha = d_h$. Finding an appropriate operator B with sensitive optimisation behaviour is subject to current research within this project.

With *domain identification problems* the goal functional J is defined by

$$J(f, \alpha) = \frac{1}{2} \int_{\Omega} (u_f - u^*)^2 dr \tag{8}$$

with its derivative

$$\frac{\partial}{\partial f} J(f, \alpha) = -\frac{(u^* - u_f)(v - u^*)}{\rho_f} \,,$$

where u^* is the measured flow field (e.g. of an MRI scan). Note, subdomains of Ω may also be used.

With the side condition

$$G(f(\alpha),\alpha) = h^2 \frac{d}{dt} f + \frac{1}{3\nu}\left(f - M^{eq}_{f,B\alpha}\right) , \tag{9}$$

and the goal functional (8), the optimality condition (6) formulates as follows:

$$\frac{d}{d\boldsymbol{\alpha}} J(f(\boldsymbol{\alpha}),\boldsymbol{\alpha}) = u\tau \int_{\mathbb{R}^d} \varphi \; 3h^2(v - B\alpha\, \boldsymbol{u}_f) M^{eq}_{f,B\alpha}\, dv + \frac{\partial}{\partial f} J(f,\alpha) . \tag{10}$$

φ is determined by solving the *adjoint PM-BGK-Boltzmann equation* (cf. (7))

$$-(\frac{\partial}{\partial t} + v \cdot \nabla_r)\varphi = dQ(\varphi) - \frac{\partial}{\partial f} J \tag{11}$$

$$dQ(\varphi) = -\frac{1}{3\nu}(\varphi - dM^{eq}_{f,B\alpha}) \tag{12}$$

$$dM^{eq}_{f,B\alpha} = \int_{\mathbb{R}^d} \varphi(\hat{\boldsymbol{v}}) \frac{\left(3h^2\left(\boldsymbol{u}_f - \boldsymbol{v}\right)\left(B\alpha\boldsymbol{u}_f - \hat{\boldsymbol{v}}\right) B\alpha + 1\right)}{\rho} M^{eq}_{f,B\alpha}(\hat{\boldsymbol{v}})\, d\hat{\boldsymbol{v}} . \tag{13}$$

Through these definitions and Algorithm 1 we have obtained a sequence of continuous equations (primal and dual) which still need to be discretised and solved. A discretisation strategy for the primal problem is discussed in Sect. 2. In the following subsection, a method is introduced to solve and discretise the dual problem.

3.3 Adjoint Lattice Boltzmann Method: Collide and Stream

The *adjoint lattice Boltzmann equation* (ALB equation) in discrete time and phase space reads

$$\begin{aligned}\varphi_j(t) - \varphi_j(t-h^2) &= \frac{1}{3\nu + 1/2}\left(\varphi_j(t) - dM^{eq}_{f^h,B\alpha}(t)\right) \\ &\quad - \frac{6\nu}{6\nu+1} h^2 dJ_{f^h,B\alpha}(t) \quad \text{for } t \in I_h,\ j = 0, 1, \ldots, q-1 ,\end{aligned} \tag{14}$$

where $\varphi_j(t,\boldsymbol{r}) := \varphi(t,\boldsymbol{r},\boldsymbol{c}_j)$ and $c_j \in Q$.

The transient phase space $I \times \Omega \times \mathbb{R}^d$ is discretised by $I_h \times \Omega_h \times Q$ exactly as described in Sect. 2. Here, $h \in \mathbb{R}_{>0}$ denotes the discretisation parameter which is coupled to a particular adjoint BGK-Boltzmann equation (14). As for the LBM (cf. [1, 18, 20]), the particular choice of $I_h \times \Omega_h \times Q$ sets up an ALBM model which is denoted by *DdQq* with d representing the dimension and q the cardinal number of $Q \subseteq \mathbb{R}^d$. Commonly applied models are *D2Q9*, *D3Q19* and *D3Q27*.

The *velocity discrete adjoint Maxwellian distribution* $dM^{eq}_{f^h,B\alpha}$ which belongs to the adjoint Maxwellian distribution $dM^{eq}_{f,B\alpha}$ is defined in $I \times \Omega \times Q$. By setting $\varphi_j(t, \boldsymbol{r}) := \varphi(t, \boldsymbol{r}, \boldsymbol{c}_j)$ for all $t \in I, \boldsymbol{r} \in \Omega$ and $\boldsymbol{c}_j \in Q$ it reads

$$dM^{eq}_{f^h,B\alpha}(\boldsymbol{c}_j) := \sum_{i=0}^{q-1} \varphi_i(\boldsymbol{c}_i) \frac{\left(3\left(h\boldsymbol{u}_{f^h} - \tilde{\boldsymbol{c}}_j\right)\left(h\boldsymbol{u}_{f^h} - \tilde{\boldsymbol{c}}_i\right) + 1\right)}{\rho_{f^h}} M^{eq}_{f^h,B\alpha} \tag{15}$$

for all $\boldsymbol{c}_j \in Q$ in $I \times \Omega$.

$$dJ_{f^h,B\alpha}(c_j) = -\frac{(u^* - u_{f^h})(c_j - u^*)}{\rho_{f^h}}.$$

3.4 *Adjoint Lattice Boltzmann Algorithm and Its Parallel Realisation*

The structure of an ALB equation like (14) is very similar to that of a standard LB equation. The main differences are its time reverse character and the additional term $\frac{6\nu}{6\nu+1}h^2 dJ_{f^h}$. However, its locality properties basically remain the same. This encourages the implementation of ALBM with a similar algorithm to that for LBM presented in Sect. 2.

An iterative algorithm can be derived from (14). It is executed step by step for decreasing $t \in I_h$. In each single time step two operations are to be performed for all $r \in \Omega_h$ and every $j = 0, 1, \ldots, q-1$, namely the *adjoint collision step*

$$\tilde{\varphi}_j(t, \boldsymbol{r}) = \varphi_j(t, \boldsymbol{r}) - \frac{1}{3\nu + 1/2}\left(\varphi_j(t, \boldsymbol{r}) - dM^{eq}_{f^h}(t, \boldsymbol{r})\right) + \frac{6\nu}{6\nu + 1} h^2 dJ_{f^h}(t) \tag{16}$$

and the *adjoint streaming step*

$$\varphi_j(t - h^2, \boldsymbol{r} - h^2\boldsymbol{c}_j) = \tilde{\varphi}_j(t, \boldsymbol{r}) \tag{17}$$

which is alternatively referred to as *adjoint propagation step*.

As the collision step (3) in an LB algorithm, the adjoint collision step (16) has a local character with respect to the position space. For this step the solution of the corresponding primal problem f_i $(i = 0, 1, \ldots, q-1)$ needs to be provided. It is of key importance that for the computation of $\tilde{\varphi}_j(t, \boldsymbol{r})$ $(j = 0, 1, \ldots, q-1)$ at a particular $t \in I_h$ and $r \in \Omega_h$ only the $f_i(t, r)$ $(i = 0, 1, \ldots, q-1)$ for the same values of t and r are required. In order to obtain an efficient implementation, it is therefore recommendable to take advantage of this property. This can be realised by, for example, a local storage of the solution f_i with respect to φ_j $(i, j = 0, 1, \ldots, q-1)$ respecting the memory hierarchy. When it comes to realising a parallel approach, it is expected that this leads to a scalable implementation with respect to the memory consumption.

While an adjoint streaming step (17) at a certain $(t, \boldsymbol{r}) \in I_h \times \Omega_h$ is performed, φ_j is manipulated only at directly neighbouring nodes. This also holds true for a LB streaming step (4). However, the propagation takes place reversely.

Due to the structure of both steps, (16) and (17), combined with its mentioned locality properties an ALB algorithm can be implemented similarly to an LB algorithm. In particular, the data structure design and the hybrid parallelisation strategy proposed early for standard LBM in [8] can be applied. Executing an ALB scheme, it is expected that it qualitatively performs as efficiently as an LB scheme.

4 Numerical Experiments

First, the results of reconstruction of a partially given flow field are presented, in order to validate the proposed method. Afterwards, the single core performance improvements of the open-source LBM implementation OpenLB[1] are presented. The results of the most recent version 1.0 are compared to the version 0.9 of OpenLB. Finally, a comprehensive scaling study on the HPC cluster FH1 is shown.

4.1 Domain Identification Test Case

In a simple test scenario, the validity of the method is demonstrated. For this purpose, artificial "experimental" flow data u^* is being generated by computing the flow field around a solid cube in a virtual wind tunnel with fixed velocity values at all boundaries. The tunnel is constructed to be 125 times as big as the cube.

The cube is then to be identified by the adjoint lattice Boltzmann algorithm (Algorithm 1). The cube is to be identified in terms of position and shape within the design domain, an area around the cube 9 times as big as the cube. Thereafter, the algorithm is provided with only partial data of the simulated flow field, meaning that the goal functional (8) integrates only over a subdomain $\bar{\Omega} \subset \Omega$ instead of Ω (see blue colored regions on the left hand side of Fig. 1). The object is still expected to be identified.

Figure 1 shows the reconstruction of the cube with different amounts of "experimental" data u^* being provided to the domain identification algorithm. Even in the case of only quarter of the overall data being provided at distance of the cube, the object can be reconstructed within 20 optimisation steps. These results show the high potential of a porous-media-based adjoint lattice Boltzmann method to improve noisy MRI data.

[1] www.openlb.net

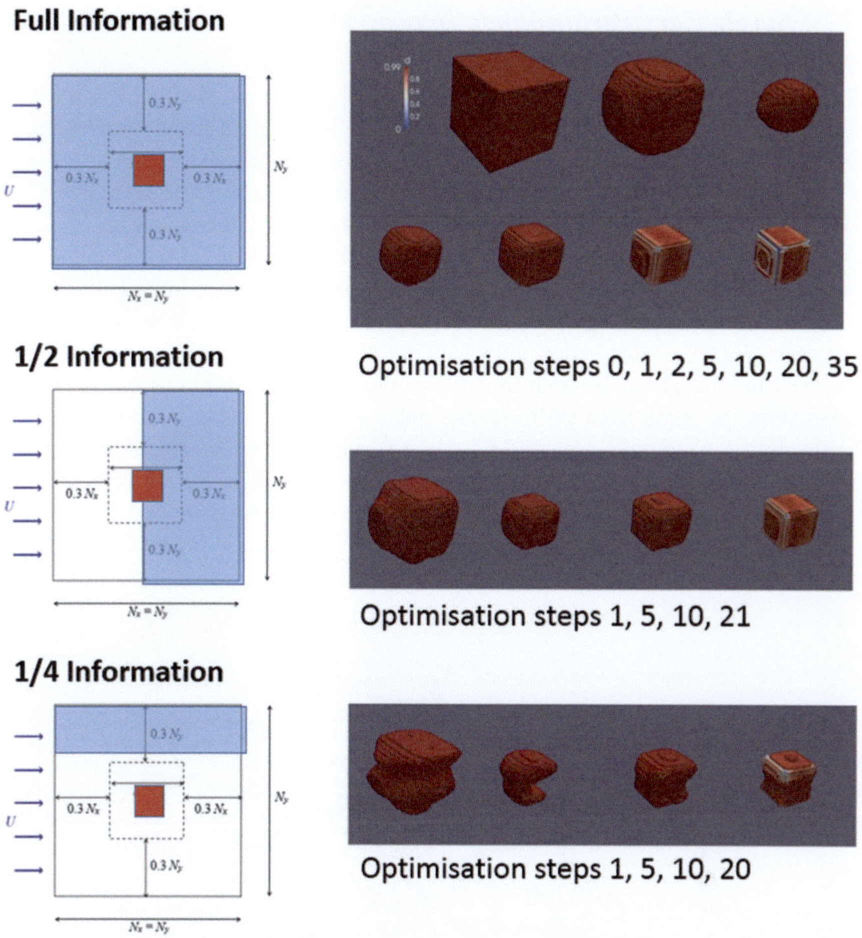

Fig. 1 Identification of a cube through the porous media adjoint lattice Boltzmann method. On the *left hand side*, the (sub-)domain Ω of the goal functional (8) is marked *blue*, with the cube (*red*) being surrounded by the design domain (*dotted line*). The goal functional (8) computes the error between the measured input flow data and the simulated optimisation flow data inside the *blue*-colored area. The control parameter α^k determines the lattice porosity d_h^k in the design domain at the k-th optimisation step, where low porosity values are to be interpreted as solid. α^k is determined by Algorithm 1 for every optimisation step k. The fluid flow is said to enter the virtual wind tunnel from left at a fixed velocity u for both artificial flow data generation and optimisation. The *right hand side* shows the resulting reconstruction of the cube after various optimisation steps k. Even in the case of only quarter of the overall data being provided at distance of the cube, the object can be reconstructed within 20 optimisation steps

4.2 Single Core Performance Improvements

Many HPC applications use sophisticated parallelisation strategies and communication layers. OpenLB is a high performance implementation of LBM, implementing MPI and OpenMP for its massive parallelisation. Fluid cells are bundled into `Blocks` for shared memory applications. At a more abstract level, these `Blocks` are orchestrated by `SuperBlocks` for the use of distributed memory [7].

OpenLB developers have spent large amounts of work into improving single core performance in the most recent version 1.0 of OpenLB. By unrolling and hardcoding loops of the collision step (3), the LBM algorithm runs about 30 % faster for the common $D3Q19$ discretisation model. In addition, cache-friendly memory access and well-thought reduction of arithmetic operations also contribute to the presented improvements. Significant performance enhancement has also been obtained by hardcoding the computation of $M^{eq}_{f_i,d_h}$ in (4) and simultaneously deploying the particular structure of the $D3Q19$ directions. As a result, a lot of multiplicative operations containing a zero term are omitted, e.g. for $c_i \cdot u_{fh}$ (Fig. 2).

Figure 3 shows a performance increase of 30 % for the most recent $D3Q19$ LBM implementation of version 1.0 of OpenLB.

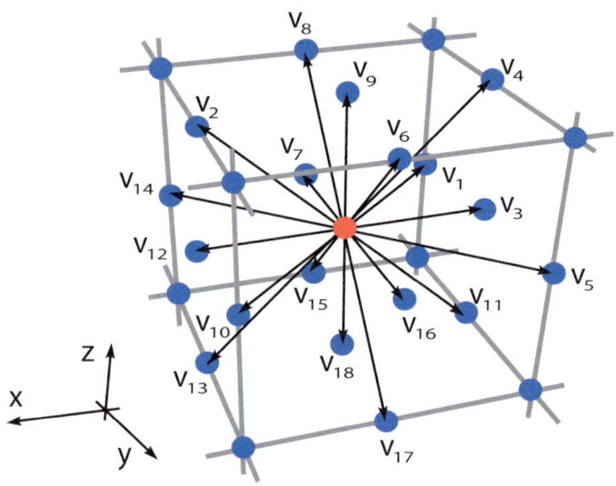

Fig. 2 Discrete velocities ξ_i for lattice arrangement $D3Q19$

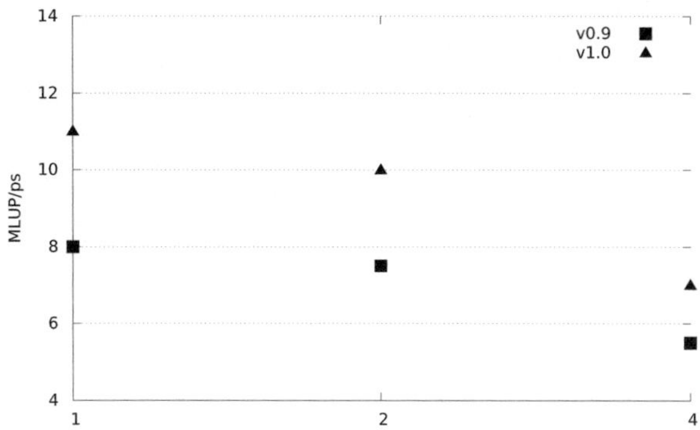

Fig. 3 Million Lattice Updates (*MLUP*) per process and second as a function of the allocated cores. Graph shows that the recent version 1.0 of OpenLB performs about 30 % faster than version 0.9 The problem size is fixed to 125,000 lattice nodes, see cylinder3d example of OpenLB. Computations are performed on Intel i7-4790 compiled with gcc 5.3

4.3 Parallel Efficiency and Scaling

Scaling studies on the HPC cluster FH1[2] are promising with the open-source software OpenLB, which is developed by the working group Computational Process Engineering (CPE) at the Karlsruhe Institute of Technology (KIT). As shown in Sect. 3.4, ALBM based optimisation requires remarkable computation time. For every optimisation step, a 3D fluid flow problem is to be solved numerically. As a consequence HPC infrastructure is an elementary component for the proposed method to tackle relevant problems.

Key index of performance is $MLUP/ps$ (Mega Lattice Updates per process and second, which is proportional to "mega FLOP per second and per core"), denoting the number of fluid cells computed in one second by a single core. Two conclusions can be drawn from the following discussion:

(a) Running the algorithm on an increasing number of cores with a fixed *overall* problem size increases the amount of necessary communication and results in lower $MLUP/ps$ (*strong scaling*).
(b) Running the algorithm on an increasing number of cores with a fixed problem size *per core* leads to constant $MLUP/ps$ (*weak scaling*).

[2]https://www.scc.kit.edu/dienste/forhlr.php/

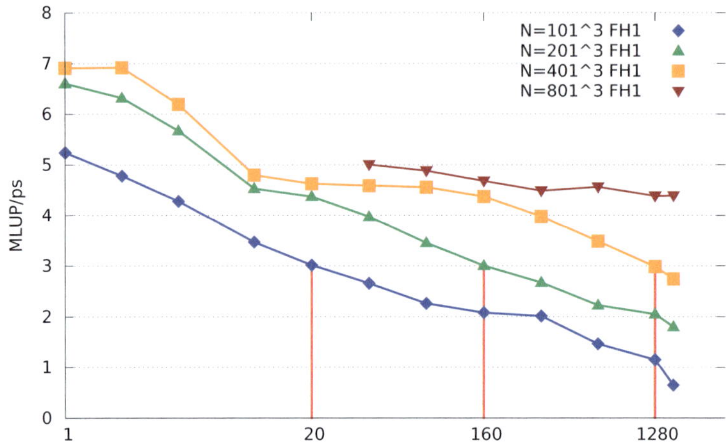

Fig. 4 Simulation of lid driven cavity on HPC cluster FH1 using open-source software OpenLB. The graph shows for varying fluid cell number N, the performance index $MLUPS/ps$ (computed cells per second as a function of allocated cores). For fixed overall problem size N (*strong scaling*), a decrease of $MLUPS/ps$ is observed (*horizontal lines*). However, for *weak scaling* (constant problem size per core) a constant $MLUPS/ps$ is seen (*vertical lines*)

4.3.1 Strong scaling

For a problem size of 101^3 computed on a single compute node (2 deca-core CPUs), a $MLUP/ps$ of 3 is obtained (see Fig. 4), meaning the algorithm simulates 60 million fluid cells per second on a single compute node. In comparison, for a problem of size 401^3, which is a typical size for LBM applications, 4.8 $MLUP/ps$ have been observed using one compute node (20 cores) and 4.5 with four compute nodes (80 cores). Due to the favourable relation between communication and computation it holds: The bigger the problem, the higher the performance in terms of $MLUP/ps$. In fact, OpenLB provides a very good incremental speed-up of about 46 or an efficiency of 0.57. The biggest problem currently being considered is of size 801^3 and shows an excellent efficiency of 0.88 from 40 to 1600 cores.

4.3.2 Weak Scaling

Focusing on communication analysis, the configuration of interest is limited to a constant problem size per core. By fixing the number of fluid cells per core, the computation load remains the same with increasing node size. The communication effort increases slightly, since one global commutation step is required in each time step. The domain decomposition is done in three dimension optimising the communication cost while tolerating an only small imbalance in the computation load [2, 10]. Remarkably, OpenLB provides constant $MLUP/ps$ in the weak

Table 1 Nearly constant performance index $MLUP/ps$ for varying core numbers with fixed problem size per core. This indicates that the OpenMPI implementation of OpenLB provides outstanding scaling properties and benefits particularly from HPC infrastructure

Number of cores	Fluid cells per core	$MLUP/ps$
20	$5 * 10^4$	3.01
160	$5 * 10^4$	3.0
1280	$5 * 10^4$	2.98
1	10^6	5.2
8	10^6	5.0
64	10^6	4.9
512	10^6	4.9

scaling (see Table 1). Therefore, OpenLB is very well suited for massively parallel infrastructure.

5 Conclusion

In this preliminary work, a novel solution strategy for domain identification problems is presented. In combination with modern Phase Contrast MRI technology, the holistic approach promises crucial improvements of accuracy for the characterisation of fluid flow dynamics and domains. The paper provides a comprehensive study using OpenLB for innovative application of adjoint lattice Boltzmann methods (ALBM), with particular focus on HPC and validation. Based on a porous media model formulated on a mesoscopic scale, it is shown that the corresponding ALBM-based optimisation approach is able to reconstruct a flow field of a partially given data set. After 20 optimisation steps, the method is able to reconstruct the object. For practical use, an efficient MPI implementation and HPC infrastructure is crucial, since every step requires to solve a $3D$ flow field problem as well as its adjoint problem. The presented scaling results show that the realised approach is a cutting edge implementation of the LBM algorithm concerning elaborated MPI implementation and efficient single core performance. Evidence is provided, that big problems benefit significantly from massive parallel computer infrastructure, which is seen by almost perfect *weak scaling*. Also it is shown that a sophisticated implementation of the standard LBM algorithm provides a single core speed up of 30 %. These promising results encourage further research and development of the ALBM towards a combined measurement and simulation tool.

References

1. Chopard, B., Droz, M.: Cellular Automata Modeling of Physical Systems. Cambridge University Press, Cambridge/New York (1998)
2. Fietz, J., et al.: Optimized hybrid parallel lattice Boltzmann fluid flow simulations on complex geometries. In: Kaklamanis, C., Papatheodorou, T., Spirakis, P.G. (eds.) Euro-Par 2012 Parallel Processing. Lecture Notes in Computer Science, vol. 7484, pp. 818–829. Springer, Berlin/Heidelberg (2012). ISBN:9783642328190, doi:10.1007/9783642328206_81, http://dx.doi.org/10.1007/9783642328206_81
3. Geiger, C., Kanzow, C.: Numerische Verfahren zur Lösung un-restringierter Optimierungsaufgaben. Springer-Lehrbuch. Springer, Berlin (1999). ISBN:3540662200, http://swbplus.bszbw.de/bsz080178243inh.htm%20;%20http://swbplus.bszbw.de/bsz080178243cov.htm
4. Gunzburger, M.D.: Perspectives in flow control and optimization. Advances in design and control. Society for Industrial and Applied Mathematics, Philadelphia (2002). http://www.ulb.tu-darmstadt.de/tocs/129935174.pdf
5. Hänel, D.: Molekulare Gasdynamik. Springer (2004)
6. Henn, T., et al.: Parallel dilute particulate flow simulations in the human nasal cavity. Comput. Fluids **124**, 197–207 (2016). ISSN:0045-7930, doi:http://dxdoiorg/10.1016/jcompfluid2015.08002, http://www.sciencedirect.com/science/article/pii/S0045793015002728
7. Heuveline, V., Strauss, F.: Shape optimization towards stability in constrained hydrodynamic systems. J. Comput. Phys. **228**, 938–951 (2009)
8. Heuveline, V., Krause, M.J., Latt, J.: Towards a hybrid parallelization of lattice Boltzmann methods. Comput. Math. Appl. **58**, 1071–1080 (2009). doi:10.1016/j.camwa2009.04001, URL:http://dx.doi.org/10.1016/j.camwa.2009.04.001
9. Kirk, A., et al.: Lattice Boltzmann topology optimization for transient flow. In: MAESC 2011 Conference May 3, 2011. Christian Brothers University Memphis, Tennessee (2011). http://wwwmaescorg/maesc11/Papers/Kirk_Kreissl_Pingen_Maute_LatticeBoltzmannTopologyOptimizationForTransientpaper.pdf
10. Krause, M.J.: Fluid flow simulation and optimisation with lattice Boltzmann methods on high performance computers: application to the human respiratory system. Eng. http://digbib.ubka.uni-karlsruhe.de/volltexte/1000019768. PhD thesis, Karlsruhe Institute of Technology (KIT), Universität Karlsruhe (TH), Karlsruhe, July 2010. http://digbib.ubka.uni-karlsruhe.de/volltexte/1000019768
11. Massaioli, F., Amati, G.: Achieving high performance in a LBM code using OpenMP. In: EWOMP 2002, Rome (2002)
12. Ni, J., et al.: Parallelism of lattice Boltzmann method (LBM) for Lid- driven cavity flows. In: High Performance Computing and Applications (HPCA2004), Shanghai, 8–10 Aug 2004. Accepted and being published in lecture note in computer science (LNCS). Springer, Heidelberg (2004)
13. Pingen, G., Evgrafov, A., Maute, K.: Topology optimization of flow domains using the lattice Boltzmann method. Struct. Multidiscip. Optim. **34**(6), 507–524 (2007)
14. Pingen, G., Evgrafov, A., Maute, K.: A parallel Schur complement solver for the solution of the adjoint steady-state lattice Boltzmann equations: application to design optimisation. Int. J. Comput. Fluid Dyn. **22**(7), 457–464 (2008)
15. Pingen, G., Evgrafov, A., Maute, K.: Adjoint parameter sensitivity analysis for the hydrodynamic lattice Boltzmann method with applications to design optimization. Comput. Fluids **38**(4), 910–923 (2009). ISSN:0045-7930, doi:10.1016/jcompfluid200810.002, http://www.sciencedirect.com/science/article/B6V264-TTMJN3-1/2/16383afe088243863f7bc5f569da1279
16. Pohl, T., et al.: Performance evaluation of parallel large-scale lattice Boltzmann applications on three supercomputing architectures. In: Proceedings of the ACM/IEEE SC2004 Conference on Supercomputing, Washington, DC, p. 21 (2004)

17. Saint-Raymond, L.: From the BGK model to the Navier-Stokes equations. Annales Scientifiques de l'École Normale Supérieure **36**(2), 271–317 (2003). ISSN:0012-9593, doi:10.1016/S0012-9593(03) 00010-7, http://www.sciencedirect.com/science/article/B6VKH48HS9DK5/2/4b7102c9ed9f501112dc9b08b7c9ae3d
18. Sukop, M.C., Thorne, D.T.: Lattice Boltzmann Modeling. Springer, Berlin/New York (2006)
19. Tekitek, M.M., et al.: Adjoint lattice Boltzmann equation for parameter identification. Comput. Fluids **35**, 805–813 (2006)
20. Wolf-Gladrow, D.A.: Lattice-Gas, Cellular Automata and Lattice Boltzmann Models, An Introduction. Lecture Notes in Mathematics. Springer, Heidelberg/Berlin (2000). ISBN:3540669736
21. Zeiser, T., Götz, J., Stürmer, M.: On performance and accuracy of lattice Boltzmann approaches for single phase flow in porous media: a toy became an accepted tool – how to maintain its features despite more and mor complex (physical) models and changing trends in high performance computing!? In: Krause, E., Shokin, Y.I., Shokina, N. (eds.) Computational Science and High Performance Computing III, Proceedings of 3rd Russian-German Workshop on High Performance Computing, Novosibirsk, 23–27 July 2007. Notes on Numerical Fluid Mechanics and Multidisciplinary Design, vol. 101. Springer (2008)

Investigation on Air Entrapment in Paint Drops Under Impact onto Dry Solid Surfaces

Qiaoyan Ye and Oliver Tiedje

Abstract The present annual report summarises the purpose of the project and the ongoing investigations performed at the Institut für Industrielle Fertigung und Fabrikbetrieb Universität Stuttgart (IFF) on the numerical study of paint drop impacting onto dry solid surfaces. Both Newtonian and yield-stress viscous droplets were applied. Detailed numerical observations of the drop impact dynamics with the focus of air entrapment were obtained. It has been found that at the early stage of the droplet spreading there is no contact line movement, but only direct contact of the droplet outline with the substrate, which results in the formation of an air disc under the impact point. The maximum air disc is reached, when the drop spreading is driven by the movement of the fully wetted contact line. Numerical results showed much more bubble entrapment at the interface between liquid and solid for Newtonian droplets. For shear thinning non-Newtonian fluids the created air disc and air bubbles during drop spreading are reduced tremendously because of the quite low liquid viscosity. The effects of the drop properties, impact velocity and static contact angles on the maximum air disc and on the air bubble release from the droplet film were analysed.

1 Introduction

Droplet impingement and spreading on a solid surface are phenomena that occur frequently in many industrial applications, such as coating processes using liquid sprays. The paint film quality of such coating processes is affected by the entrapment of air bubbles in the liquid film which release later in the drying process, resulting in pinholes in the dry paint film. One of the presumptions where air bubbles come from is the air entrapment resulting from the impact of the liquid drops, which has been experimentally observed by many researchers using different liquid materials and impact velocities.

Q. Ye (✉) • O. Tiedje
Institut für Industrielle Fertigung und Fabrikbetrieb, Universität Stuttgart, Nobelstr. 12, D-70569 Stuttgart, Germany
e-mail: qiaoyan.ye@ipa.fraunhofer.de

W.E. Nagel et al. (eds.), *High Performance Computing in Science and Engineering '16*, DOI 10.1007/978-3-319-47066-5_24

Experimental observations of the impact of liquid drops onto dry solid surfaces at room temperature with the analysis of air entrapment have been reported extensively [1, 4, 10, 12–14]. By using flash photographic methods and high speed cameras, as well as different light settings, such as back, or oblique lighting, with and without light diffuser, the authors observed bubble formation at the stagnation point and assumed bubble formation because of a dimple created at the drop surface at impact point [1, 12]. In investigations using viscous drops, Thoroddsen et al.[14] found much more bubble entrapment during the drop spreading process, resulting from the localised contacts of the levitated lamella with the solid substrate, especially for intermediate values of the Reynolds number (Re $\sim$ 250–350). Similar researches have also been carried out by Palacios et al. [10]. Besides myriad of air bubbles at the interface between liquid and solid, they also observed two rings of micro-bubbles under the drop of glycerol/water, impacting onto a dry glass surface at Reynolds and Weber numbers around the splashing/deposition threshold and analysed the behaviour of these rings, depending on the drop impact velocity and on the ranges of relevant dimensionless numbers. However, the quality of the time-resolved imaging depends strongly on the used facilities, namely the flash photographic methods and high speed cameras, as well as the different light settings. In general, large drops, e.g. $d > 500\,\mu m$, have to be used in the experiment. For small drops (50–300 μm), especially for opaque liquids, like in spray painting processes, it is very difficult to experimentally get high quality time-resolved imaging of the entrapment of air bubbles by drop impingement.

There are not so many numerical studies that focus on the air entrapment under the drop impact. Mehdi-Nejad, Mostaghimi and Chandra [8] simulated the impact of water, n-heptane, and molten nickel droplets on a solid surface using two-dimensional computational domains. They included the effect of the gas around the droplets and predicted the formation of air bubbles at the solid-liquid interface. The impact dynamics of non-Newtonian drops, namely, yield-stress fluid droplets, have been studied experimentally [5, 9] and numerically [6]. The latter study evaluated the influence of the rheological parameters on the droplet spreading and recoiling processes.

Although the air entrapment phenomenon under drop impact onto a solid surface is well known experimentally, the knowledge about the detailed processes and mechanisms underlying air bubble entrapment and release from the liquid film is still limited, especially for high-viscous and non-Newtonian liquids. In this project we have carried out numerical studies on Newtonian viscous drops (0.04–1 Pa·s) and yield-stress drops impacting onto dry smooth solid surface. Parameters of impact velocity and droplet diameter were selected (Fig. 1) by taking into account the spray painting applications. Comparison of air entrapment and bubble release from liquid films between Newtonian and non-Newtonian liquids was carried out.

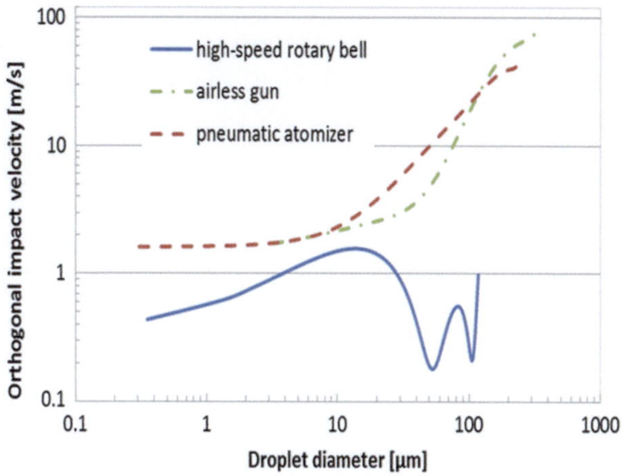

Fig. 1 Droplet impact velocity vs. droplet diameter of the different atomizers in coating industry [16]

2 Numerical Method

2.1 Governing Equations

The droplet impact and spreading on a surface is an example of an interfacial flow problem that can be calculated using the Volume of Fluid (VoF) method, a surface tracing technique, with which two or more immiscible fluids can be modelled by solving a single set of momentum equations and tracking the volume fraction of each of the fluids throughout the domain. The numerical simulations in this work were carried out with the commercial CFD code ANSYS-FLUENT based on the finite-volume approach. The flow field and the liquid-gas interfaces during the droplet impact process are solved by the volume fraction equation and single momentum equations:

$$\frac{1}{\rho_q}\left[\frac{\partial}{\partial t}(\alpha_q \rho_q) + \nabla \cdot (\alpha_q \rho_q \mathbf{v}_q) = 0\right] \quad (1)$$

$$\frac{\partial}{\partial t}(\rho \mathbf{v}) + \nabla \cdot (\rho \mathbf{v}\mathbf{v}) = -\nabla p + \nabla \cdot \left[\mu\left(\nabla \mathbf{v} + \nabla \mathbf{v}^T\right)\right] + \rho \mathbf{g} + \mathbf{F} \quad (2)$$

where $\mathbf{v}$ denotes the velocity vector, t the time, ρ the density, μ the dynamic viscosity, p the pressure and α_q the qth volume fraction of the fluid in the cell. The resulting velocity field is shared among the phases. Time-dependent VoF calculations with variable step sizes from 0.01 to 10 µs were carried out using an explicit scheme. A geometric reconstruction scheme for the volume fraction discretization [3] was used, ensuring a sharp interface between liquid and gas phase. PRESTO scheme was applied for the pressure discretization. For the momentum equation we used QUICK scheme that is based on a weighted average of second order upwind and central interpolation of the variable. This discretization scheme will be typically more accurate on quadrilateral and hexahedral meshes aligned with the flow direction.

2.2 Computational Domain, Mesh and Boundary Conditions

In contrast to the mentioned experiments which used large droplets (droplet diameters of 1–4 mm), the present numerical study uses 10, 50, 100 and 300 μm drops, corresponding to the droplet sizes in coating processes. The smaller droplet diameter will also save computational capacity, if micro-sized bubbles produced by a drop impacting on a solid surface should be observed. Thereby, a computational domain of $1400 \times 1400 \times 380\,\mu m^3$ with Cartesian grid (cut cell) was created. A structured grid with cell resolution $D/\Delta x = 80$–150 in the region around the droplet was found to be necessary to avoid grid effects and get accurate results, where D is the diameter of drop, Δx the grid size. Far away from the liquid-air interface coarse hexahedral meshes were used to reduce the total number of the cells. Corresponding to the parameters used, 20–120 million cells were used in the present numerical study.

Based on common spray painting conditions, droplet impact velocities 0.1–80 m/s were applied. The numerical initial drop injection position was $D/10\,\mu m$ away from the wall surface, so that the surrounding gas field can be calculated, which is absolutely necessary in order to study the mechanism of air bubble entrapment under droplet impact. Initial pressure inside the drop because of surface tension of the liquid was calculated. Atmospheric pressure was set on the boundaries of the computational domain. A dry smooth wall with no-slip boundary condition was used.

At first, model viscous liquids with variant Newtonian viscosity were used in the calculation. Static contact angles (SCA) have to be specified on the wall, although experimental investigations (Šikalo et al. 2005 [11]) show that dynamic contact angles (DCA) differ appreciably from both the static advancing and receding value. The comparison of the evolution of drop shapes obtained using SCA and DCA in the numerical study carried out by Lunkad et al. [7] showed that the difference in drop spreading is not so significant, especially for the large SCA and in the early impact stage. The objective of the present study is mainly to report the mechanism of air entrapment under drop impact, which does not, as shown later, depend on wettability. However, the speed of bubble formation and bubble release from the liquid will be influenced by the wettability. Therefore the SCA as parameter was included in our numerical study. SCA-values of 60° are quite close to practical applications in painting processes, whereas 30° corresponds to well wettable systems, such as a liquid on some glass target.

For drop impact of non-Newtonian fluids two paint liquids were used in the simulations. The corresponding rheological properties were experimentally obtained using a rotation viscometer.

Calculations were carried out using Cray XC40 (Hazelhen) at High Performance Computing Center Stuttgart. Figure 2 shows the evaluation of the code performance that was made using a grid size of 80 million cells and a time steps of 1e-6 s. The wall-clock time with 1200 and 2400 cores was tremendously decreased. However, the performance using more than 2400 cores was worse. Simulations were mainly carried out with reasonable parallel processors of 1200, which has already speeded up the parameter study tremendously. The CPU-times for calculating one second droplet impact process are summarized in Table 1. In most cases we reach the equilibrium state in the calculation after 0.1 s of the process, which results in 20 CPU-hours per case using 1200 cores.

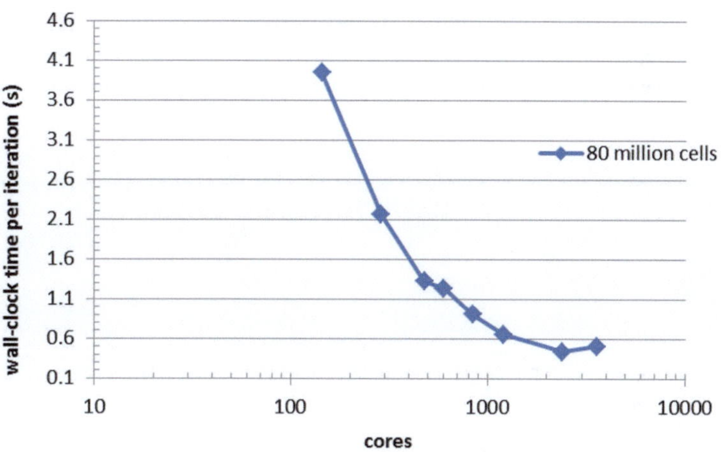

Fig. 2 Performance of parallel processors for a test case of droplet impact calculation

Table 1 CPU time information

Grid size	Time steps (s)	Cray XC40, 24 cores/node, 50 nodes CPU time (hours/s calc.)	Cray XC40, 24 cores/node, 100 nodes CPU time
80 million cell elements	1e-6	183	123

3 Simulation Results

3.1 Some Validation of Simulation Results

Slow deposition of drops onto a near-complete wetting solid substrate was experimentally and numerically investigated by Ding et al. [2]. They observed the typical droplet shape evolution of pinch-off process and the occurrence of droplet ejections from the mother drop in rapid droplet spreading, as shown in Fig. 3. Pinch-off criteria was analysed. Under certain conditions, six stages of pinch-off with droplet ejections could be observed.

In the present investigation, simulation of water droplet spread with the zero impact velocity has been carried out. Droplet diameter of 300 μm and static contact angle of 30° were applied. Figure 4 shows the calculated drop shape evolution. Comparing to the experimental results [2], a qualitative identical behaviour of the pinch-off process can be observed. Since the parameters used in the simulation are not identical to the experiment from Ding et al., the production of daughter droplets with only two stages of droplet ejections was observed. Figure 4 shows also the air entrapment by the coalescence of the daughter and mother drops.

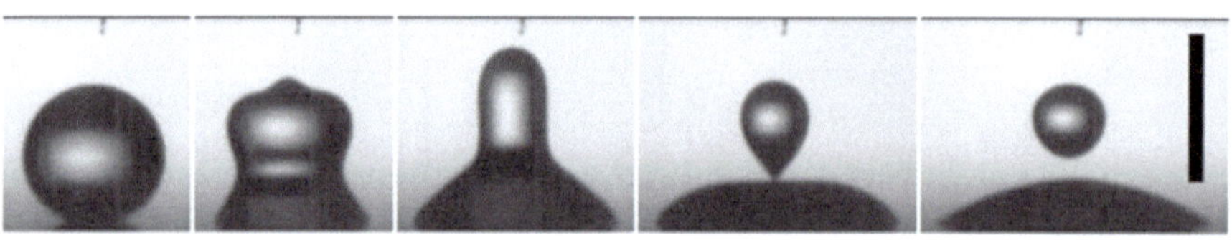

Fig. 3 First-stage pinch-off for a water drop of 486 μm in diameter, u = 0 m/s, Oh = $\mu/\sqrt{\sigma d\rho} = 1.68e-4, \theta s = 12°$ [2]

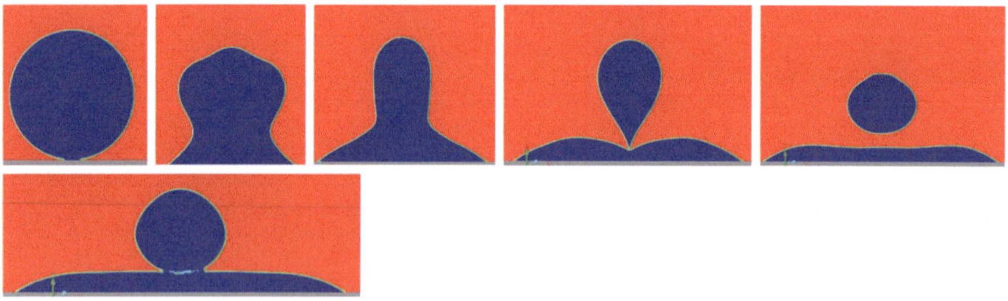

Fig. 4 Simulated first pinch-off for a water drop of 300 μm in diameter, u = 0 m/s, Oh = 6.78 e-3, θs = 30° (Contours of volume fractions: *red*: air, *blue*: water)

Table 2 Parameters used in the simulation for Newtonian drops

case name	liquid	$\rho(kg/m^3)$	D (μm)	μ (Pa·s)	σ (N/m)	U (m/s)	SCA (°)	Re	We	Oh
lh2o	Water	1000	300	0.001	0.0725	1	60	300	4.14	0.0068
l1a	A	1000	300	0.04	0.0725	1	60	7.5	4.14	0.271
l1b	B	1000	300	0.04	0.025	1	60	7.5	12	0.462
l1b_2	B	1000	50	0.04	0.025	1	60	1.25	2	1.13
l1b_3	B	1000	300	0.04	0.025	0.5	60	3.75	3	0.462
l1c	C	1000	300	1.0	0.025	1	60	0.3	12	11.55
l1d	D	1000	300	0.005	0.065	1	60	60	4.6	0.036
l1e	E	1000	300	0.01	0.065	1	60	30	4.6	0.072
l2b	B	1000	300	0.04	0.025	10	60	75	1200	0.462
l3b	B	1000	300	0.04	0.025	1	30	7.5	12	0.462

3.2 *Newtonian Drop Impact on a Dry Solid Surface*

Table 2 summarizes the parameters used for Newtonian liquids in this study. The corresponding dimensionless numbers are Reynolds number Re $= \rho\mathrm{D}u/\mu$, Weber number We $= \rho u^2 D/\sigma$, Ohnesorge number Oh $= \mu/\sqrt{\sigma d \rho}$, where D, ρ, μ and σ are the diameter, density, viscosity and surface tension of the drop, respectively; u is the drop impact velocity. For viscous liquids we have, Re $<$ 75, We $<$ 1200 and Oh $=$ 0.271 $\sim$ 11.55. A high and a relative low viscosity, e.g. 1 and 0.04 Pa·s, and impact velocity of 1 and 10 m/s, were applied. For comparison we also carried out drop impact simulations of water drops with Re $=$ 300. Clearly, the regime of drop impact presented in this section is mainly the droplet spreading on the wall without breakup, especially for viscous liquids.

3.2.1 Mechanism of Air Entrapment Under Droplet Impacting on a Solid Surface

In order to understand the mechanism of air entrapment by droplet impacting on a solid surface, simulation results using water liquid and viscous drop are analysed in detailed especially around the impact point. Detailed analysis of velocity and pressure distribution close to the impact point was reported by Ye and Tiedje [15]. The shear thinning viscosity behaviour will be discussed in the Sect. 3.3.

Figure 5 shows the early stage of a water drop impact on the wall. The clear interface contour lines were obtained by showing contours of the air volume fractions scaled from 0.01 to 0.8 in a centre cross-section. Just before impacting, a slight flattening at the bottom of the drop (Fig. 5a) can be observed. In none of our simulations did we observe a dimple created at the drop surface at impact point assumed by Chandra and Avedisian [1]. The experimental observations made by Thoroddsen et al. [13] have shown different flatness of bottom curvature of water drops before the impact, which results in different sizes of the entrapped air

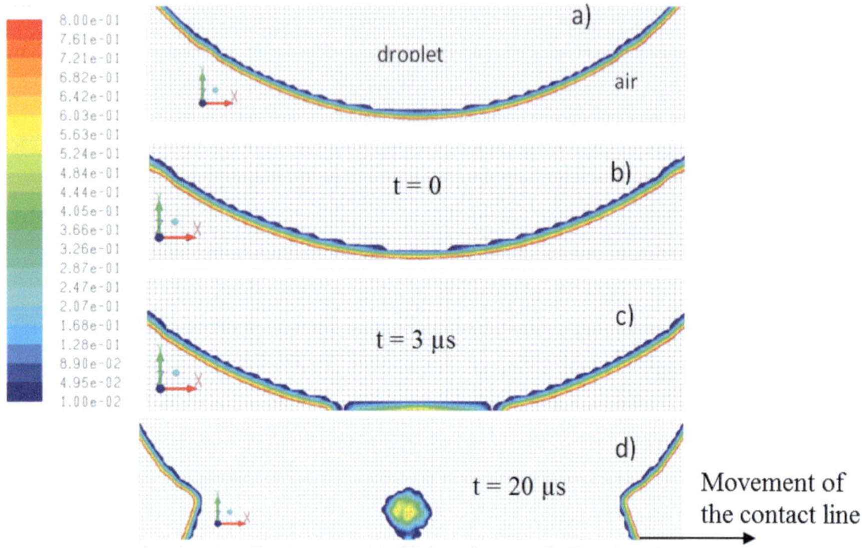

Fig. 5 Detailed view of contours of air volume fractions scaled from 0.01 to 0.8 for the impact of a water drop (case: lh2o from Table 2: D = 300 μm, impact velocity = 1 m/s, corresponding to Re = 300, We = 4). (**a**): 1 μs before impact, (**b**): droplet contacts just with the wall, (**c**): maximal air disc on the wall, (**d**): air bubble under the bottom centre of the drop

disc under water drops at the initial contact. The droplet shape, however, could be unstable because of the surrounding experimental conditions in a droplet free-fall. The large droplet usually used in experimental research makes its shape change easily. Since the free-fall distance in the present simulation is quite small, only one tenth of the drop diameter, the spherical droplet is always ensured before the droplet impact in the calculation.

An initial air disc (Fig. 5b) with a radius of 11 μm was obtained. The air disc is enlarged continuously during the drop spreading until a fully wetted contact line is created. The subsequent spreading is driven by the contact line movement. Such wetting process can be seen more clearly later for the case of viscous drop. The maximal radius of the captured air disc, as shown in Fig. 5c, is about 32 μm with the thickness < 1 μm. This air disc contracts into a bubble whose equivalent diameter is about 13 μm under the bottom centre of the drop. The time interval of the contraction is ca. 20 μs.

The simulation results using a viscous fluid listed in the case l1b in Table 2 are shown in Fig. 6 with the focus on the phase contours close to the solid wall. Compared to the water drop, a slightly weaker flatness and smoothness of the curvature around the impact point can be observed in at t = 0. The initial air disc radius is about 9 μm and is enlarged continuously, as shown in Fig. 6 at t = 0.022 ms. The air contracts into bubbles during the spread process, resulting in a partly wetted region. The maximal air disc on the wall (strictly speaking, the region with bubbles

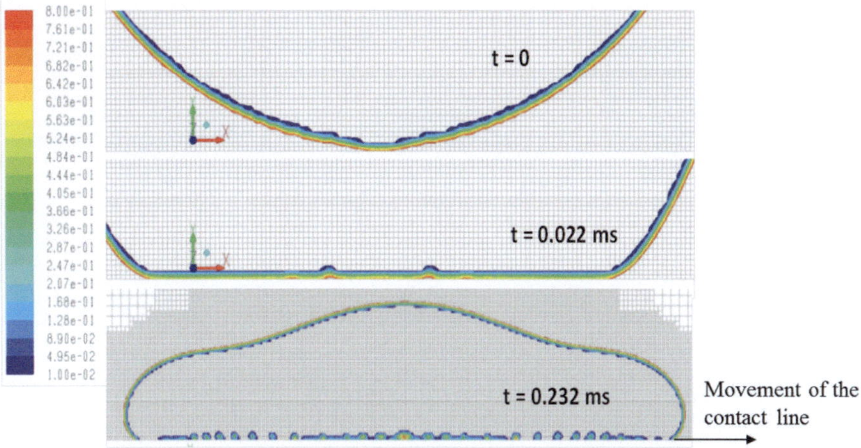

Fig. 6 Viscous droplet: detailed view of contours of air volume fractions scaled from 0.01 to 0.8 (case: 11b: D = 300 μm, impact velocity = 1 m/s, corresponding to Re = 7.5, We = 12

and partly wetted area) was observed at t = 0.23 ms to have a radial size of 253 μm (Fig. 6). Clearly, there is always a thin air layer or air disc under droplet impact onto solid surface. This thin air layer results from the direct contact between the droplet outline and the substrate, even for a near-completely wetting solid substrate. The maximum air disc is reached if the wetted contact line moves. Of course, the size of air disc and the release of air bubbles depend on material properties and application parameters, which will be discussed in the following sections

3.2.2 Air Bubble Formation and Release Under Droplet Impacting on a Solid Surface

As shown in Figs. 5 and 6, the air layer contracts into bubbles. Figure 7 shows the evolution of the water drop impact with the focus particularly on the droplet shape, air bubble formation and release. The bubble created by the air disc could not drift up at once and is located under the centre of the drop because of the symmetrical down flow inside the droplet. With the decreasing of the apex height of the drop, the bubble leaves the liquid film (Fig. 7c). In this case, inertia force is lower than the large surface tension, and high SCA, namely worse wettability, yielding a strong contraction of the liquid film, which in turn results in droplet breakup (Fig. 7d). Formation of new small bubbles during the coalescence of drops on the solid surface was observed (Fig. 7e). During the advance and recoil of the droplet, the bubbles drift up. An air-bubble-free condition was observed after approx. 2 ms by examining the 3d-region of the liquid phase.

In contrast to the water droplet, the release of bubbles from the viscous droplet is quite difficult, which can be observed in Figs. 8 and 9. Much more air bubbles are

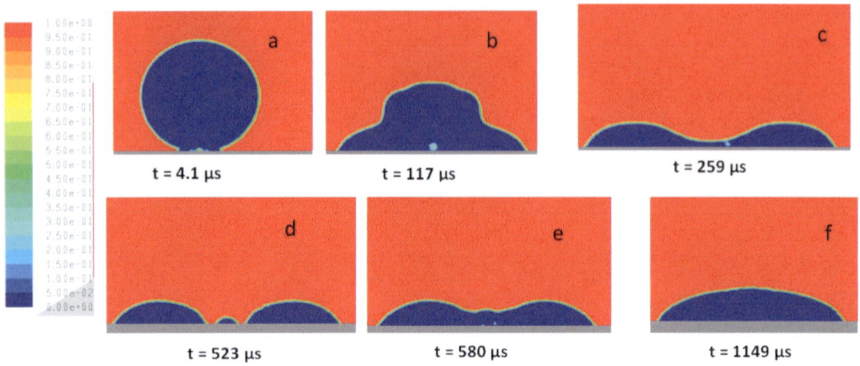

Fig. 7 Contours of volume fractions (1: air, 0: water liquid), impact of a water drop (D = 300 μm, impact velocity = 1 m/s, corresponding to Re = 300, We = 4), t is the real impact time

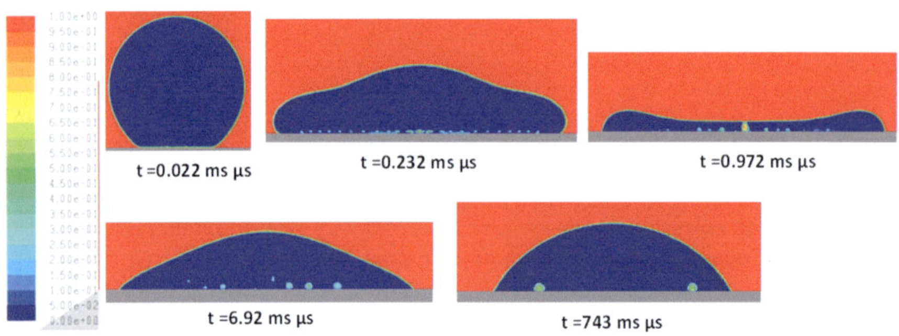

Fig. 8 Contours of air volume fraction (cross section view, *red*: air, *blue*: liquid), impact of a viscous drop (D = 300 μm, impact velocity = 1 m/s, corresponding to Re = 7.5, We = 12), t is the real impact time

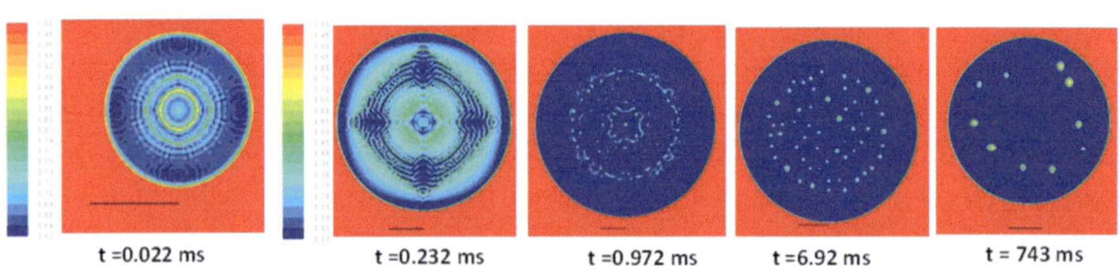

Fig. 9 Contour lines of air volume fraction (bottom view, 1: air, 0: liquid), impact of a viscous drop (D = 300 μm, impact velocity = 1 m/s, corresponding to Re = 7.5, We = 12). Scale line is 100 μm

entrapped in the liquid-solid interface for the viscous drop, which is in accordance with experimental observations [10, 14]. During the advance and recoil processes micro-bubbles combine and some of them are able to leave the liquid film if the height of the film decreases sufficiently. During the first advance scenario, the lowest apex height of drop in the centre is reached, resulting in the escape of large bubbles located in the centre at first (Fig. 8 at t = 0.972 ms). The remaining bubbles move radially outward by the oscillation of the drop spreading and drift up as soon as the drift forces are strong enough to overcome the adhesion force. Figure 9 shows detailed air bubble formation on the solid surface. Automatic scaling of air volume fraction was applied, 0.62–1 for the sequence at t = 0.022 ms, 0–1 for the rest of sequences. Some air bubble patterns, e.g. centre bubble rings and cartwheel patterns in Fig. 9 at t = 0.022 ms, can be observed, which is similar to the experimental observations reported by Palacios et al. [10]. At t = 0.743 s, a quasi-equilibrium phase, there are still fairly large air bubbles on the substrate, the release of such bubbles becomes more slowly and more difficult.

3.2.3 Effect of Liquid Property and Impact Velocity on the Air Entrapment

Based on the present simulation results, it is found that the initial radius of the air disc at the impact point for the 300 μm droplet is 10 ± 2 μm for all the test cases, due to the nearly spherical shape of the original drop in the simulation. However, the maximal air disc caught under the droplet depends on the droplet properties, the impact velocity, as well as the wettability, namely the substrate properties. This maximal air disc results finally in a myriad of micro-bubbles at the interface between the liquid and the substrate. The size of such air discs, or air regions of viscous drops, is plotted against Ohnesorge number and Reynolds number with a relationship of $Oh \times Re^{0.8}$ in Fig. 10. In general, the size of the air region is inversely proportional

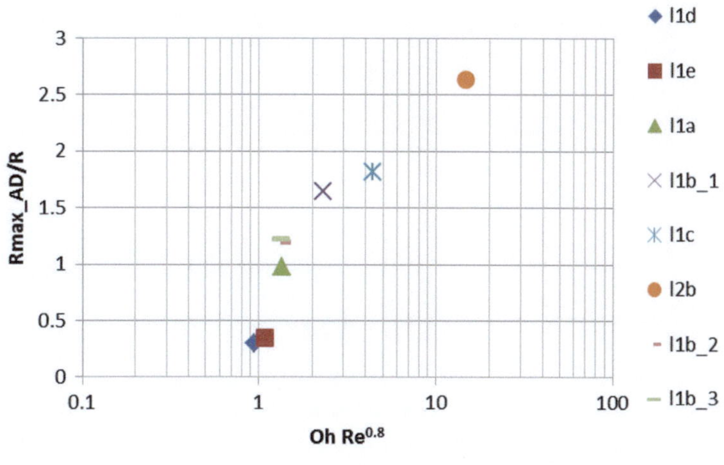

Fig. 10 The maximal radius of the air disc of Newtonian viscous drops vs. $Oh \times Re^{0.8}$

to the surface tension of the fluid and increases with impact velocity and liquid viscosity.

The effect of static contact angle SCA is also investigated in the present study. With decreasing SCA, i.e. improving wettability, the maximal air disc reduces from 253 μm in case 11b to 238 μm in case 13b. Figures 11 and 12 show the entrapped air bubbles under the drop at t = 0.18 s in detail. On the substrate there are only two visible small bubbles for the case with SCA = 30°, whereas many more large bubbles can be observed for the case with SCA = 60°. The small SCA helps the bubbles to break the adhesion force and leave the solid surface much easier. The decreasing height of the droplet film (small SCA) also makes bubbles drift up quickly.

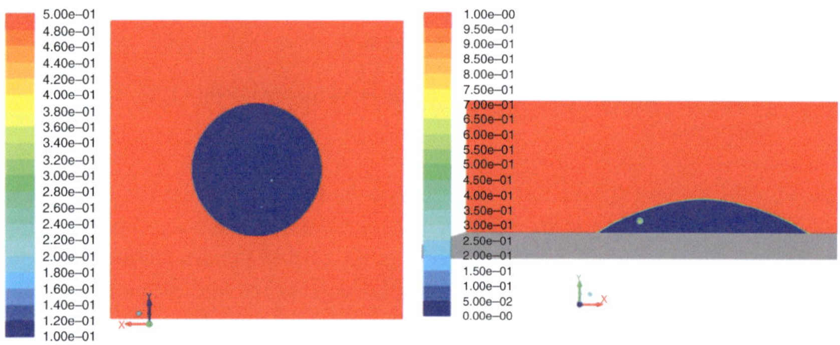

Fig. 11 Contours of air volume fractions for the viscous drop with SCA = 30° at t = 0.18 s (case 13b)

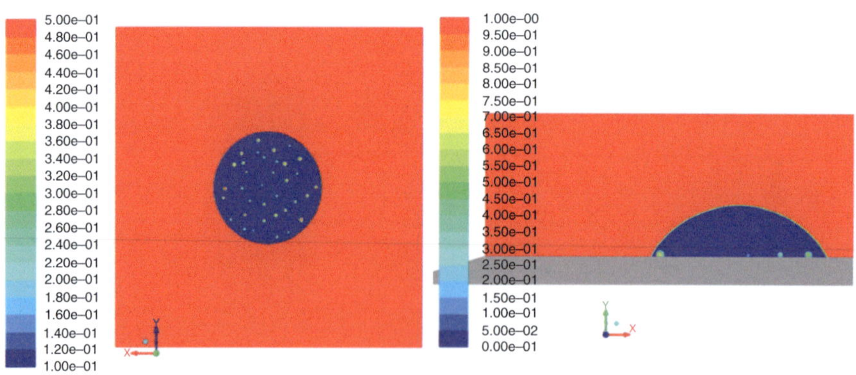

Fig. 12 Contours of air volume fractions for the viscous drop with SCA = 60° at t = 0.18 s (case 11b)

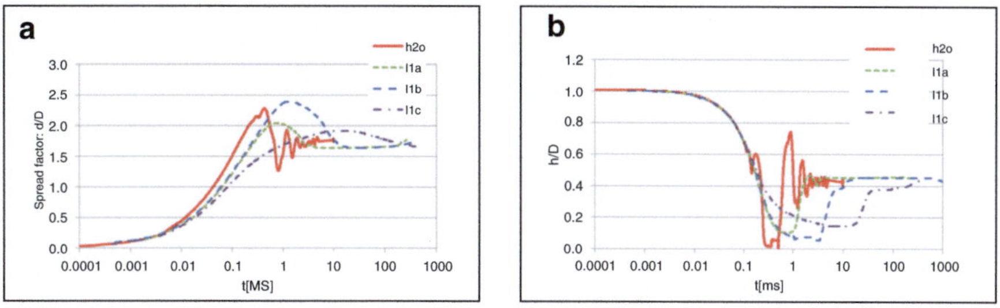

Fig. 13 Spreading factors (d/D and h/D) of different liquids (impact velocity = 1 m/s, D = 300 μm and SCA = 60°)

3.2.4 Effect of Liquid Property and Impact Velocity on the Spreading Factors

The droplet impact dynamics, namely the evolution of droplet shapes during advancing and recoiling scenarios, are evaluated (Fig. 13) using spreading factors defined by d/D and h/D, where d is spreading diameter and h the apex height. There are no large differences in spreading factor distributions at the beginning, e.g. $t < 0.01$ ms, since the inertia force dominates the droplet spreading at this time. Significant oscillations of spreading factors for water drop can be observed, which can promote the release of air bubbles from the water film. In contrast, for viscous liquids there is only one period of advancing and recoiling scenario. Significant differences of d/D and h/D distributions in dependence on the viscosity and the surface tension can be observed for $t > 0.1$ ms.

3.3 *Yield-Stress Drop Impact onto a Dry Solid Surface*

Paint liquids used in industrial applications exhibit usually non-Newtonian behaviour. For example most of them have shear-thinning behaviour, in which the viscosity decreases with increasing shear rate. In this section we focus on shear-thinning non-Newtonian fluids. Impact dynamics, air entrapment and bubble release from the liquid film with two rheological parameters and various impact velocities are investigated numerically. The air entrapment behaviour of shear-thinning fluid drops is compared with that of Newtonian droplets.

3.3.1 A Rheological Model of Yield-Stress Fluid

The Herschel-Bulkley model that is described as follows was used in the present numerical study for paint liquids.

$$\dot{\gamma} = 0 \quad \text{for} \quad \tau < \tau_0 \tag{3}$$

$$\tau = \tau_0 + k\dot{\gamma}^n \quad \text{for} \quad \tau > \tau_0 \tag{4}$$

The corresponding viscosity model with the limit value for $\dot{\gamma} \to 0$ is given by

$$\mu(\dot{\gamma}) = (\frac{\tau_0}{\dot{\gamma}} + k\dot{\gamma}^{n-1}) \cdot (1 - e^{-\epsilon\dot{\gamma}}) \tag{5}$$

$$\mu(\dot{\gamma} \to 0) = \epsilon\tau_0 \tag{6}$$

In above equations, $\dot{\gamma}$ is the shear rate (s^{-1}), and τ the shear stress (Pa). τ_0, k and n are rheological parameters that represent the yield-stress magnitude (Pa), the consistency factor (Pa·s^n) and the power law index, respectively. The function of the limit value, i.e. the second bracket in equation (5), is necessary, since droplet impact dynamics is calculated until quasi-equilibrium state. ϵ is used for building the function of the limit value. An increase in τ_0 induces an increase in additional plastic-like dissipation, and an increase in k represents an increase in the apparent viscosity. The power law index n is related to the shear-thinning behaviour (fluid viscosity becomes lower as n decreases).

Two paint liquids were used. The rheological parameters are shown in Table 3. Clearly, paint_f has higher apparent viscosity than paint_t. The slight difference of value n indicates a similar shear-thinning behaviour of both paint liquids. Unless otherwise specified, the droplet density, surface tension and static contact angle are always 1000 kg/m^3, 0.025 N/m and 60°, respectively. Parameter study was carried out mainly with different drop diameter and impact velocity. A non-Newtonian Reynolds number $Re_n = \rho\, D^n U^{(2-n)}/k$ and Weber number We defined in Sect. 3.2 are used for the discussion, where D, ρ, U, n and k are the diameter, density, drop impact velocity and rheological parameters, respectively.

3.3.2 Simulation Results

Figures 14 and 15 show the velocity field of a paint droplet impact onto solid surface and the corresponding shear rate around the impact point. The maximum pinch-off air velocity from the impact region is 3.97 m/s and the maximum shear rate of gas-liquid mixture reaches 2.4e6 (1/s), the corresponding viscosity is about 4 mPa·s. Therefore, the liquid viscosity around the impact region reduces tremendously. The evolution of droplet shape and viscosity, the formation of the air disc as well as the bubble release from the liquid film are show in Fig. 16. At the early stage of droplet impact, the viscosity is quite low around the impact region because of the high shear rate. Higher viscosity is located in the gas-liquid interface on the droplet top. The diameter of the maximum air disc is about 72 μm and contracts subsequently

Table 3 Parameters used in the simulation for yield-stress drops

Name	τ_0(Pa)	k(Pa s^n)	n	ϵ(s)	D(μm)	U (m/s)	Re_n	We
paint_t	0.214	0.1046	0.742	20	50,100,300	0.1–50	1-3e3	0.12-3e4
paint_f	0.455	0.604	0.658	20	10,50,100,300	0.1–80	0.1-3e3	0.02-7e4

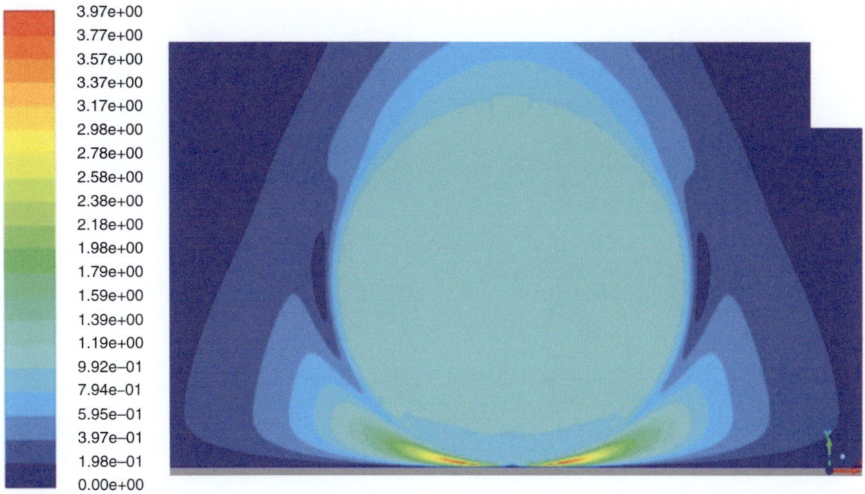

Fig. 14 Contours of velocity magnitude (m/s), contacting just with the wall (paint_f, D = 300 μm, U = 1 m/s)

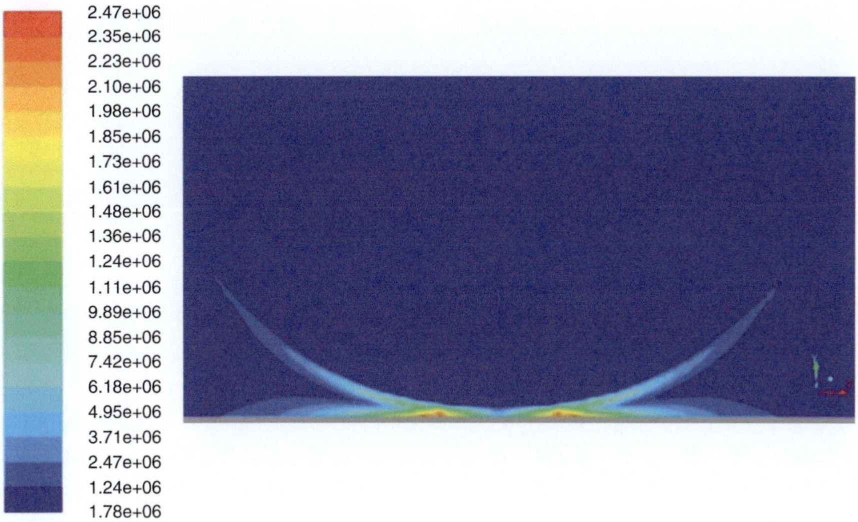

Fig. 15 Contours of shear rate (1/s), contacting just with the wall (paint_f, D = 300 μm, U = 1 m/s)

into a bubble and releases totally from the liquid film at the quasi-equilibrium state (t = 40 ms). In the previous case of the viscous droplet impact with Newtonian fluid (Fig. 8), the maximum air disc with diameter of 506 μm was obtained, and at the quasi-equilibrium state (t = 743 ms) there were still some large bubbles on the solid surface. In addition, for the non-Newtonian case, the maximum air disc

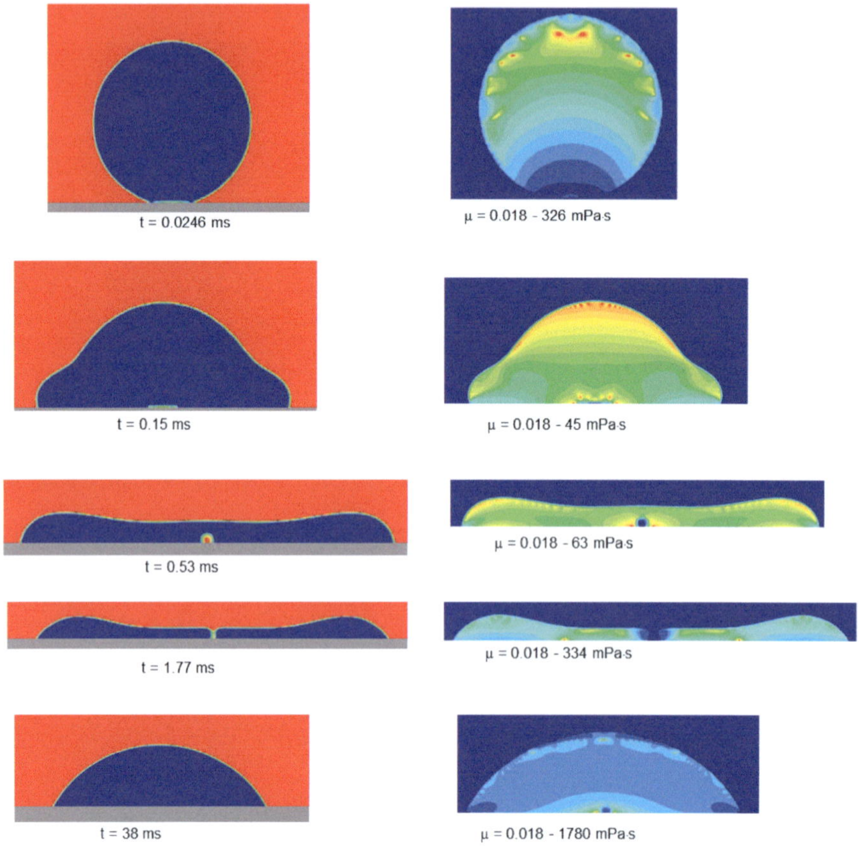

Fig. 16 Paint drop impact (paint_t: D $=$ 300 μm, impact velocity = 1 m/s), *Left*: Contours of volume fraction (*red*: air, blue: liquid), *right*: Contours of molecular viscosity (mixture, *blue*: air, $\mu = 0.018\,\text{mPa}\cdot\text{s}$, *red*: maximum liquid viscosity at the time)

formed at the early stage of droplet impact for the shear thinning liquids is almost independent on the dimensionless numbers, such as Re_n and We. The dimensionless air disc defined as a ratio of the air disc to droplet radius, Rmax_AD/R, is about 0.2 ± 0.1.

At high impact velocities droplet splashing occurs. Figure 17 shows the phase view on the target wall. The air disc breaks up into many small bubbles that can still easily release from the liquid, since the quite low apex height of drop in the centre is reached in this case. Because of splashing the created lamellas contact with the solid surface, which entraps again many small bubbles near the outside of the liquid film, as shown in Fig. 17. However, these small bubbles can escape from the liquid during the droplet recoiling process. At the quasi-equilibrium state the wall is bubble free.

At quite low impact velocities the air disc contracts into a bubble that sticks still hard on the wall at the equilibrium state, as shown in Fig. 18a. By decreasing the

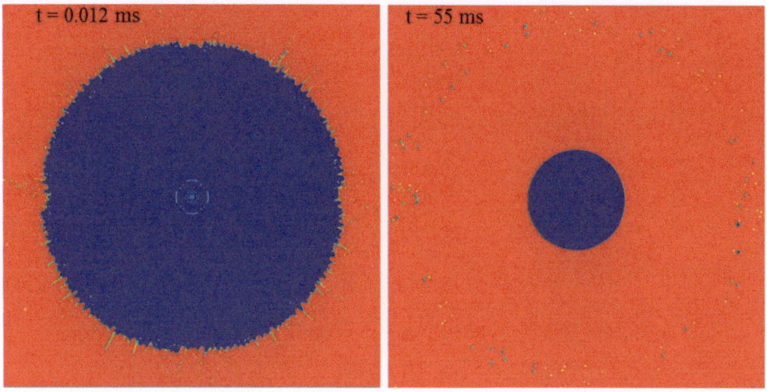

Fig. 17 Contours of volume fraction on the wall (*red*: air, *blue*: paint_f, D = 300 μm, impact velocity = 80 m/s, We = 7.68e4)

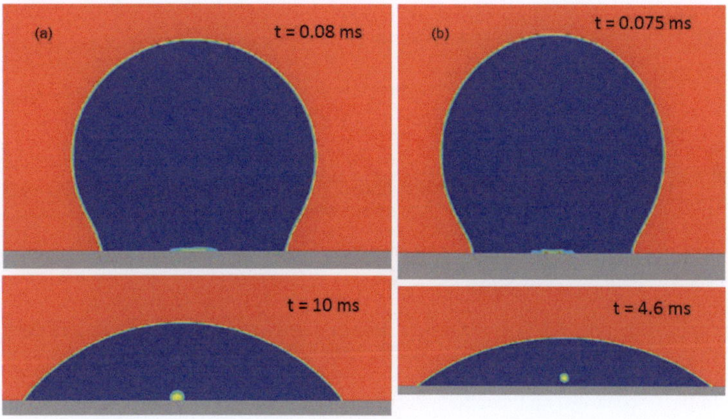

Fig. 18 Contours of volume fraction in a cross-section (*red*: air, *blue*: paint_f, D = 300 μm, impact velocity = 0.5 m/s, We = 3), *Left*: SCA = 60°, *right*: SCA = 30°

static contact angle, such as SCA = 30°, the initial air disc is similar to that with SCA = 60°, since the early stage of droplet impact, especially the formation of air disc, depends mainly on the droplet inertia and the viscosity. During the droplet recoiling process with SCA = 30°, the bubble, as shown clearly in Fig. 18b, escapes already from the wall at t = 4.6 ms. The bubble therefore becomes easier free from the liquid for the case in Fig. 18b than 18a.

Figure 19 shows the effects of rheological properties on spread factor d/D during the drop impact with a diameter of 300 μm and an impact velocity of 1 m/s. There is no difference of the spread factor at $t < 0.3$ ms, since the inertia force dominates the spreading process in the stage of $t = 0 - 0.3$ ms. After $t = 0.3$ ms, effects of viscous and surface tension forces increase. The liquid film contracts again with the help of surface tension. The lower apparent viscosity of paint_t makes the contraction easier, which results in an early recoiling process. The entrapped dimensionless air

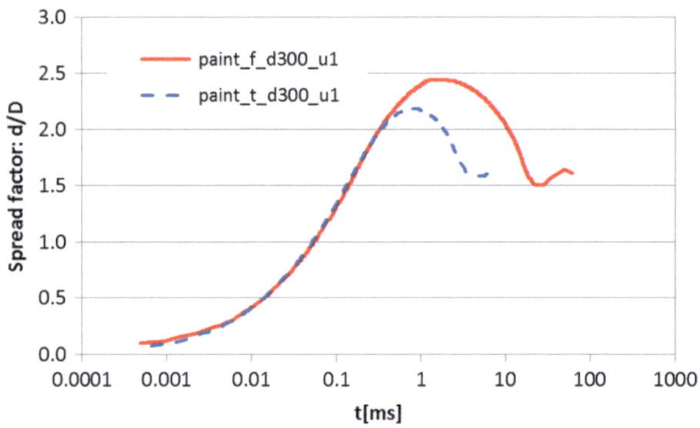

Fig. 19 Effect of liquid rheological properties on the spread factor (D = 300 μm, impact velocity U = 1 m/s, SCA = 60°)

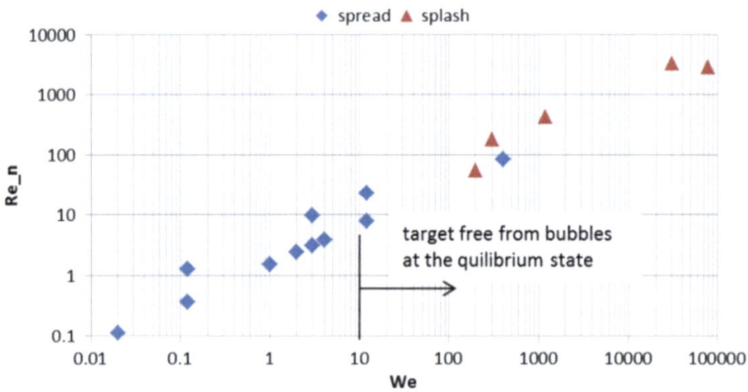

Fig. 20 Impact scenarios and bubbles free condition for yield-stress droplets in relation to dimensionless numbers

disc d/D is the same, about 0.2 for both liquids. The corresponding time is located within the kinematic phase of the drop impact (t < 0.3 ms).

A summary of drop impact dynamics concerning different impact scenarios in relation to dimensionless numbers of Re_n und We is shown in Fig. 20. The bubble free condition is also indicated in the figure. It was found that the entrapped air bubbles can escape from the wall in the quasi-equilibrium state, if the Weber-number satisfies We > 10.

4 Conclusions

For the first time, a numerical simulation on the time-resolved imaging of the air entrapment and bubble movement under drop impacting onto dry solid surfaces was carried out. Both Newtonian and yield-stress viscous droplets were applied in the study

Based on the simulation results, the mechanism of air entrapment during drop impact onto solid surfaces can be figured out. Basically, there is always air entrapment between the droplet-solid interface, which does not depend strongly on the surface wettability. Thin air layers result from the direct contact between the droplet outline and the substrate. The maximum air disc is reached if the wetted contact line moves. The size of air disc, the contraction of the air disc to bubbles and the release of air bubbles, however, depend on material and target properties and application parameters. The size of the entrapped air disc is inversely proportional to the surface tension of the fluid and severely increases with liquid viscosity and impact velocity. The air disc contracts and breaks into bubbles during the advancing phase, which can escape from the liquid film at the equilibrium state under certain conditions. Decreasing static contact angle of the liquid will enhance the bubble release from the target wall as well as from the liquid film. For Newtonian viscous droplets the maximum dimensionless air discs in dependence on the dimensionless numbers were made. At the equilibrium state there were still fairly bubbles on the wall, especially for high viscous drops.

For yield-stress liquids the wetting of solid wall was tremendously improved because of the high shear rate and subsequently quite low viscosity at the early stage of droplet impact. The dimensionless maximum air disc was quite small and almost constant 0.2 ± 0.1. The impact scenarios and effects of rheological properties on the time-dependent spread factor were analyzed. Bubble free conditions were also discussed. It was found that the target wall could be bubbles free if the Weber number is larger than 10. According to the simulation results, assumptions could be made, for instance, the trend of air entrapment at the solid-liquid interface by using pneumatic atomizer and airless gun that create impact droplets with large We-number is lower than by using high-speed rotary bell. Bubbles that are created by the droplet impacting onto solid surfaces and still adhere on the surface at the quasi-static state provoke pinhole formation after baking process. Of course, air entrapment occurs also by drop impact onto wet solid surface, which will be further investigated in future.

Acknowledgements The author would like to thank the steering committee for the supercomputing facilities at the Höchstleistungsrechenzentrum (HLRS) Stuttgart, Germany.

References

1. Chandra, S., Avedisian, C.T.: On the collision of a droplet with a solid surface. Proc. R. Soc. Lond. Ser. A **432**, 13 (1991)
2. Ding, H., Li, E.Q., Zhang, F.H., Sui, Y., Spelt, P.D.M., Thoroddsen, S.T.: Propagation of capillary waves and ejection of small droplets in rapid droplet spreading. J. Fluid Mech. **697**, 92–114 (2012)
3. Ansys-Fluent 17.0 User Manual
4. Fujimoto, H., Shiraishi, H., Hatta, N.: Evolution of liquid/solid contact area of drop impinging on a solid surface. Int. J. Heat Mass Transf. **43**, 1673–1677 (2000)

5. German, G., Bertola, V.: Impact of shear-thinning and yield-stress drops on solid substrates. J. Phys.: Condens. Matter **21**, 375111 (2009)
6. Kim, E., Baek, J.: Numerical study of the parameters governing the impact dynamics of yield-stress fluid droplets on a solid surface. J. Non-Newton. Fluid Mech. **173–174**, 62–71 (2012)
7. Lunkad, S.F., Buwa, V.V., Nigam, K.D.P.: Numerical simulations of drop impact and spreading on horizontal and inclined surfaces. Chem. Eng. Sci. **62**, 7214–7224 (2007)
8. Mehdi-Nejad, V., Mostaghimi, J., Chandra, S.: Air bubble entrapment under an impacting droplet. Phys. Fluids **15**(1), 173–183 (2003)
9. Nigen, S.: Experimental investigation of the impact of an (apparent) yield-stress material. Atomization Sprays **15**, 103–117 (2005)
10. Palacios, J., Hernandez, J., Gómez, P., Zanzi, C., Lopez, J.: Experimental study on the splash/deposition limit in drop impact onto solid surfaces. Exp. Fluids **52**, 1449–1463 (2012)
11. Šikalo, Š., Tropea, C., Ganic, E.N.: Dynamic wetting angle of a spreading droplet. Exp. Therm. Fluid Sci. **29**, 795–802 (2005)
12. Thoroddsen, S.T., Sakakibara, J.: Evolution of the fingering pattern of an impacting drop. Phys. Fluids **10**(6), 1359–1374 (1998)
13. Thoroddsen, S.T., Etoh, T.G., Takehara, K., Ootsuka, N., Hatsuki, Y.: The air bubble entrapped under a drop impacting on a solid surface. J. Fluid Mech. **545**, 203–212 (2005)
14. Thoroddsen, S.T., Takehara, K., Etoh, T.G.: Bubble entrapment through topological change. Phys. Fluids **22**(051701), 1–4 (2010)
15. Ye, Q., Tiedje, O.: Numerical study on air entrapment in droplets under impact onto a solid surface. In: Proceeding of ILASS – Europe 2013, 25th European Conference on Liquid Atomization and Spray Systems, Chania, 1–4 Sept 2013
16. Ye, Q., Burk, S., Domnick, J.: Analysis of droplet impingement of different atomizers used in spray coating processes. In: 13th Triennial International Conference on Liquid Atomization and Spray Systems, Tainan, 23–27 Aug 2015

Numerical Study of the Impact of Praestol® Droplets on Solid Walls

Martin Reitzle, Norbert Roth, and Bernhard Weigand

Abstract The behaviour of droplets consisting of two different Praestol® solutions impacting on a dry solid wall was studied. They show a similar but not identical spreading behaviour. This difference is due to the shear-thinning characteristics of Praetol® 2540 where lower viscosities are reached for smaller shear rates compared to Praetol® 2500. The results may serve as a basis for future comparisons with Newtonian liquids. The in-house code FS3D was used which is based on a Volume of Fluid method. The parallel performance of the code was analysed by studying the strong and weak scaling behaviour on the Cray CX40 system at the HLRS Stuttgart. Non-ideal speed-up was found which is mostly due to the high communication load in the multigrid solver. A new solver will overcome these limitations in the future.

1 Introduction

As many technical liquids, as for instance paints, show non-Newtonian behaviour it is of great interest to be able to handle such substances numerically. In this study some aspects of two liquids with different shear thinning behaviour are compared during wall impact. For one liquid the decrease in viscosity is observed for lower shear rates than for the other. The spreading of the liquid on the solid wall is studied in detail, which is important for coating processes.

Solutions of non-ionic flexible polyacrylamides in water were used for all calculations presented in this report. The solutions were Praestol® 2540 0.05 % and Praestol® 2500 0.8 % (Stockhausen Inc., Krefeld, Germany). Typically, these substances are used in waste-water treatments and show a shear thinning behaviour. These liquids show a non-Newtonian shear thinning behaviour, which was modelled in the numerical simulations.

For the numerical simulations in this study the in-house DNS Code FS3D was used. This code was developed at the Institute of Aerospace Thermodynamics (ITLR), University of Stuttgart for about twenty years. The implementation of the model for non-Newtonian fluids in FS3D is shown in detail in [2]. First numerical

M. Reitzle (✉) • N. Roth • B. Weigand
Institut für Thermodynamik der Luft- und Raumfahrt, Universität Stuttgart, Pfaffenwaldring 31, 70569 Stuttgart, Germany

W.E. Nagel et al. (eds.), *High Performance Computing in Science and Engineering '16*, DOI 10.1007/978-3-319-47066-5_25

calculations of droplet impact with shear thinning liquids are presented in [13]. Here in this work the focus lies on the difference between the two Praestol® solution, which have different shear thinning behaviour. We intent to present the validation of the numerical code with experiments on the European Conference on Liquid Atomization and Spray Systems (ILASS) 2016 conference in Brighton.

2 Numerical Method

Free Surface 3D (FS3D) is a code to numerically predict incompressible multiphase flows based on the volume of fluid (VOF) method. A wide variety of physical problems can be investigated among which are droplet deformations [10], droplet wall interactions [11], droplet film interactions [5], droplet collisions [12], bubbles [16] and also more recently rigid particle interactions [9]. Furthermore, heat and mass transfer can be included what allows, e.g., to simulate evaporating droplets [6, 14] or liquid solid phase change problems [8]. An overview over the capabilities of FS3D can be found in [1].

In FS3D the dimensional, incompressible Navier-Stokes equations for the mass and the momentum conservation are solved:

$$\frac{\partial \rho}{\partial t} + \nabla \cdot (\rho \mathbf{u}) = 0, \tag{1}$$

$$\frac{\partial (\rho \mathbf{u})}{\partial t} + \nabla \cdot [(\rho \mathbf{u}) \otimes \mathbf{u}] = -\nabla p + \nabla \cdot \mathbf{S} + \rho \mathbf{g} + \mathbf{f}_\gamma . \tag{2}$$

where $\mathbf{u}$ denotes the velocity vector, t the time, ρ the density, p the pressure, $\mathbf{g}$ the gravitational acceleration and $\mathbf{f}_\gamma$ is a body force which is used to model surface tension according to the CSF model [3]. Furthermore, $\mathbf{S}$ is the viscosity stress tensor, which has the form

$$\mathbf{S} = \mu \left(\nabla \mathbf{u} + (\nabla \mathbf{u})^T \right) . \tag{3}$$

2.1 *Simulation of Multiphase Flows with VOF and PLIC*

In the VOF method [7] a scalar field variable f is introduced which represents the volume fraction of a fluid in each computational cell. This variable is unity inside the liquid and zero in the gaseous phase. It can therefore directly be used to identify the location of the interface in the computational domain and also allows to compute geometrical properties such as normal vectors or curvature.The variable f is therefore defined as:

$$f(\mathbf{x}, t) = \begin{cases} 0 & \text{outside the liquid phase,} \\ (0, 1) & \text{at the interface,} \\ 1 & \text{inside the liquid phase.} \end{cases}$$

Using this definition, physical properties of the liquid and gaseous phase can be expressed in a continuous way across the interface, e.g. the density reads

$$\rho(\mathbf{x}, t) = \rho_g + (\rho_l - \rho_g) f(\mathbf{x}, t), \tag{4}$$

where the subscript l denotes the liquid phase and g the gaseous phase.

The VOF variable f is transported by an additional transport equation:

$$\frac{\partial f}{\partial t} + \nabla \cdot (\mathbf{u} f) = 0. \tag{5}$$

Note, that the right hand side is zero only if no phase change takes place.

Combining the VOF method with a finite volume discretization, a volume conserving numerical method can be guaranteed. Additionally, the advection Eq. (5) is solved using fluxes calculated by using the method of piecewise linear interface reconstruction (PLIC). This is necessary in order to suppress numerical diffusion of the interface since without an accurate information about the spatial distribution of the liquid phase, it is impossible to precisely determine how much liquid and how much gas is transported across a cell boundary in each time step.

As mentioned above, the normal vectors at the interface $\mathbf{n}_\gamma = \nabla f / |\nabla f|$ can be easily calculated. As of now, the 26 surrounding cells are used to evaluate the gradient operator. The PLIC surfaces are then constructed using this normal vector to define a plane in each cell which sharply separates the liquid and the gaseous phase. The position of this plane in a local coordinate system in each cell can be analytically determined. A 2D example of a PLIC reconstructed interface is shown in Fig. 1b.

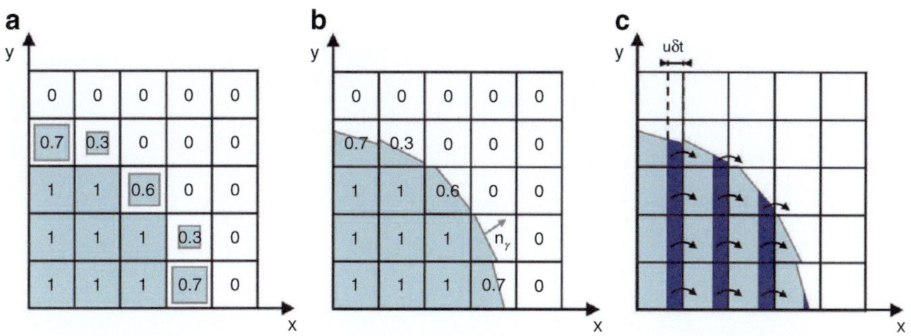

Fig. 1 (**a**) f-field without interface information; (**b**) interface reconstruction with the PLIC-method; (**c**) calculation of the f-flux $u\delta t$ with δt being the timestep from a PLIC reconstructed interface

2.2 Treatment of Non-Newtonian Shear-Thinning Liquids

In the case of shear thinning or shear thickening liquids, the viscosity is no longer constant, but a function depending on the shear rate. In FS3D the Carreau-Yasuda model (see e.g. [15] and [2]) is used:

$$\frac{\mu(\dot{\gamma}) - \mu_\infty}{\mu_0 - \mu_\infty} = \left[1 + (\tau\dot{\gamma})^a\right]^{(n-1)/a}. \tag{6}$$

Here, the subscripts 0 and ∞ denote the viscosities at zero and very large shear rates, respectively; τ, a and n are parameters depending on the rheometric characteristics of the liquid.

The liquid properties are listed in Table 1 where σ denotes the surface tension and the corresponding plot of the viscosities as a function of the shear rate is shown in Fig. 2. Note, that the lower limit of the shear rate for Praestol® 2500 is in close proximity to the viscosity of pure water.

Table 1 Liquid properties

	ρ	μ_0	σ	n	a	τ	μ_∞
	[kg/m^3]	[Pa s]	[mN/m]	[−]	[−]	[−]	[Pa s]
Praestol® 2500 0.8 %	1000.9	0.7588	75.55	0.515	1.0	1.4	0.001
Praestol® 2540 0.05 %	998.8	1.5208	76.51	0.267	1.0	9.5	0.052

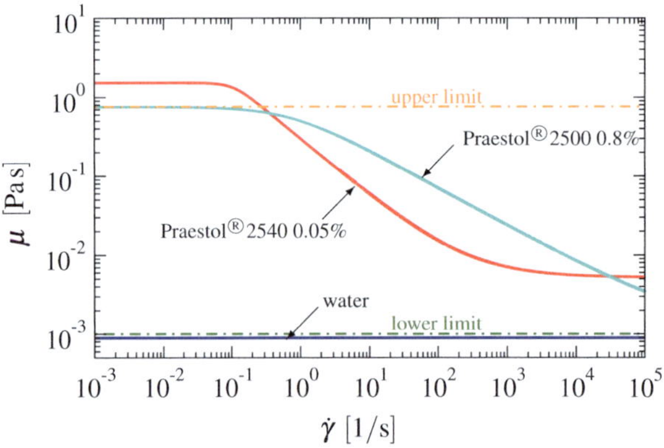

Fig. 2 Evaluation of the viscosity due to shear rates according to the Carreau-Yasuda model for 0.8 % Praestol® 2500 (turquoise) and 0.05 % Praestol® 2540 (*red*) in water. The *dash-dotted lines* represent the lower and upper limits for the former. The upper and lower limit for the latter liquid are 1.5208 Pa s and 0.052 Pa s, respectively

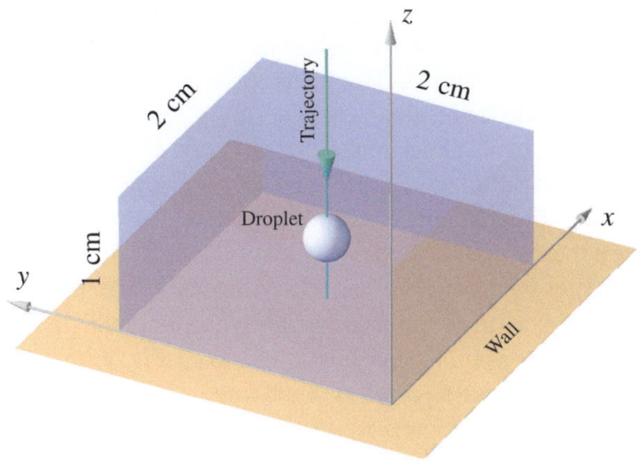

Fig. 3 Schematic view of the computational domain and the initial position of the impacting droplet

3 Numerical Setup

The setup for the numerical simulations consists of a computational area of 2 cm in both the x- and y-direction, and 1 cm in the z-direction. The solid wall is located in the x, y plane. The initial position of the droplet centre is 0.45 cm above the wall. A schematic view of the setup is depicted in Fig. 3.

The grid in the x- and y-direction is equidistant and consists of 512 grid cells. In the z-direction 256 grid cells are used. However, towards the wall the grid is refined linearly. The grid cell at the wall has a height of 10 μm. This allows a good resolution of the velocity gradient in the boundary layer and the shear rate, which determines the viscosity of the droplet liquid due to their shear thinning behaviour.

4 Results

In the numerical simulations the initial droplet diameter was in the range of $D_0 \approx$ 3–3.5 mm. The initial droplet velocity in the z-direction was varied from $v_0 \approx$ 0.4–3.6 m/s. This resulted in a variation of the impact Weber number $\mathrm{We} = \rho_l D_0 v_0^2/\sigma \approx$ 6 to 555, with ρ_l the liquid density and σ the surface tension. Depending on the impact Weber number and on the Praestol® solution different spreading behaviours of the droplet liquid on the solid surface could be observed. The impact process can be separated into different phases according to [13]. Phase A is the part of the impact process where the droplet approaches the wall. In phase B the droplet touches the surface and a disk begins to form. This phase ends at the non-dimensional time $t^* = tv_0/D_0 = 2$. The disk then spreads over the surface in phase C until the maximum disk diameter $d = d_{max}$ is reached. Afterwards, the recoiling process is charaterises in phase D. The recoiling phase D is mainly determined by the contact angle between the droplet liquid and the solid surface. In the numerical simulations of this study the contact angle was fixed to $\alpha = 90°$, which is only a very rough estimation. In future calculations this restriction will be removed and a more physical contact angle model will be implemented. On the experimental side the wall has to be cleaned accurately before each droplet impact, what will be performed in future experiments. Therefore, the results shown below concentrate mainly on the well validated phases. The final phase E describes the behaviour of the droplet, when the disk has collapsed.

Figure 4 elucidates the development of the viscosity inside the droplet during the first phases of the impact process. It further shows the essential differences between the different Praestol® solutions. It is apparent from Fig. 2, that the viscosity of the Praestol® 2540 solution decreases at much lower shear rates than the Praestol® 2500 solution. Therefore, the viscosity of the Praestol® 2540 solution reaches its lower limit shortly after the droplet impact as can be seen from Fig. 4. However,

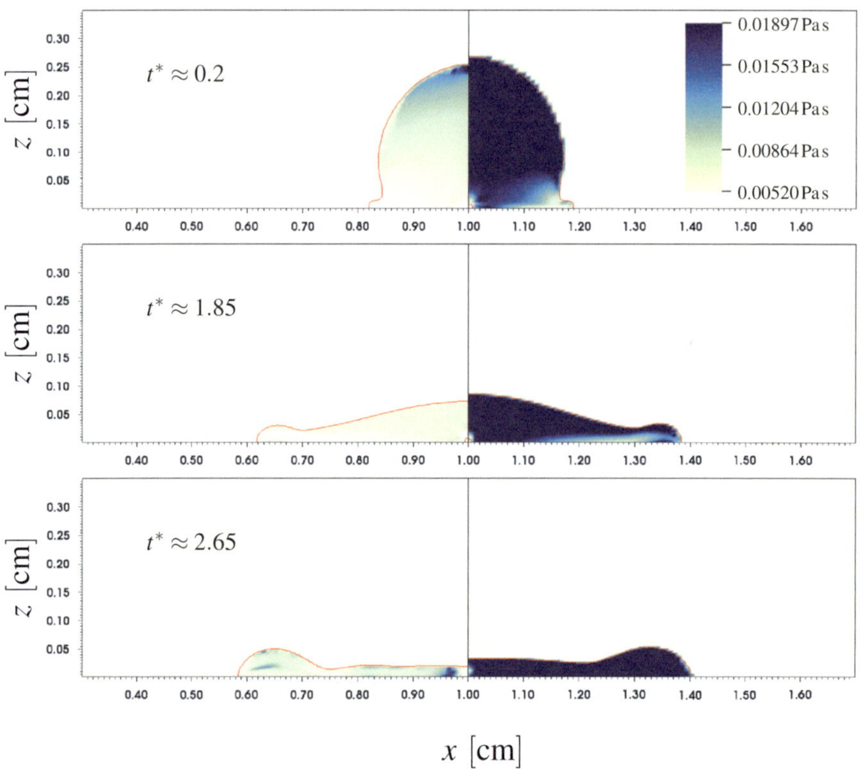

Fig. 4 Intersections through the centre of the droplets at different times t^*. The *orange solid line* marks the border of the droplet liquid for impact Weber numbers $We \approx 67$. On the *left hand side* results for the 0.8 % solution of Praestol 2540 and on the *right hand side* results for the 0.005 % solution of Praestol 2500 are shown. The colours indicate the viscosity of the droplet liquids. In regions of the *darkest blue* the viscosity is larger than 0.01897 Pa s, which is 2.5 % of the upper limit of the Praestol 2500 solution. In regions of the *lightest yellow* the viscosity is lower than 0.0052 Pa s, which is the lower limit of the Praestol 2540 solution

at $t^* \approx 2.65$, when the recoiling process has started, the shear rate decreases and the viscosity begins to increase. The viscosity of the Praestol® 2500 solution is at a much higher level, in many regions above 0.01897 Pa s. However, in close proximity to the wall low viscosity values are obtained.

Figure 5 gives an impression of the velocity distribution in a cut through the centre of the droplet for the 0.005 % solution of Praestol® 2500 for an impact Weber

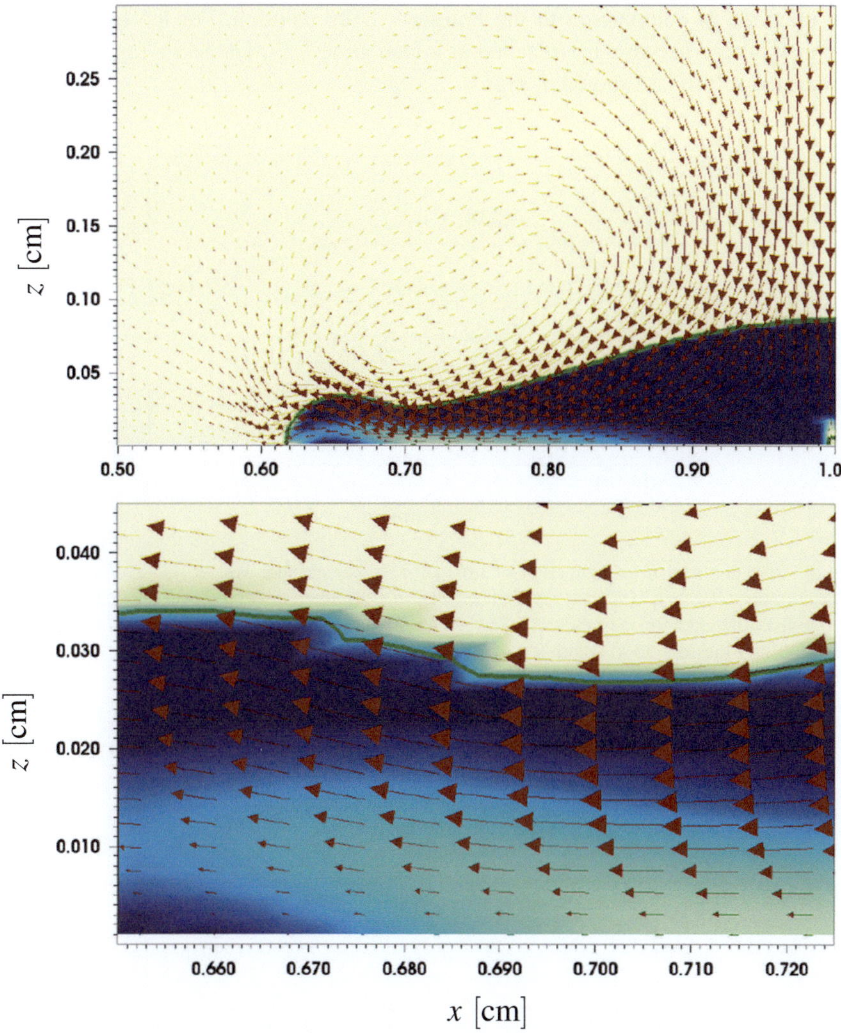

Fig. 5 Intersections through the centre of the droplets at $t^* = 1.85$. The *green solid line* marks the border of the droplet liquid. Shown is the result for an impact Weber number $We \approx 67$ for the 0.005 % solution of Praestol® 2500. The colours indicate the viscosity of the droplet liquids. The legend can be found in Fig. 4. The velocity field is indicated by *red arrows* where the *longest arrows* corresponds to approx. 1 m/s. The lower picture shows a zoom into the boundary layer close to the rim

number $We \approx 67$. Inside of the droplet a hydrodynamic boundary layer close to the wall is formed where the viscosity decreases due to the larger velocity gradients.

The development of the disk diameter d with respect to time t is shown in Fig. 6 for two different Weber numbers and both Praestol® solutions.

For higher Weber numbers the disk grows faster and higher maximum disk diameters d_{max} are obtained. Due to the lower viscosity of the Praestol® 2540 solution during the impact process (compare Fig. 4) higher maximum disk diameters are obtained in comparison to the Praestol® 2500 solution. For the lower Weber numbers, these maxima are reached at a later time t_{max}. The same results in non-dimensional form are shown in Fig. 7.

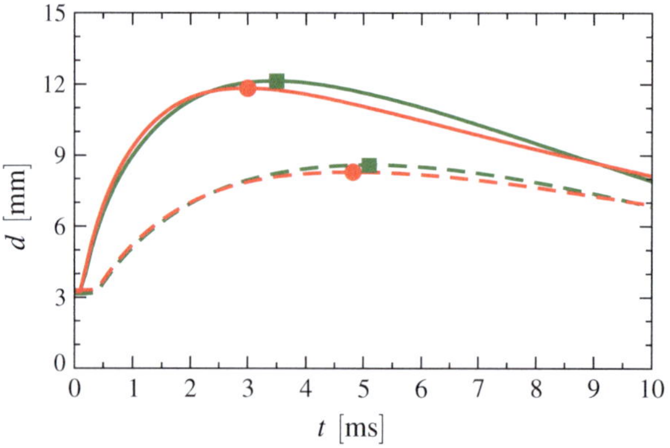

Fig. 6 Disc diameter d as a function of time t for Weber number We $\approx$ 430 (*solid lines*) and Weber number We $\approx$ 67 (*dashed lines*). The *squared symbols* in *green* indicate the maximum d_{max} of the disk diameter for the Praestol® 2540 solution and the *circled symbols* in *red* indicate the same for the Praestol® 2500 solution

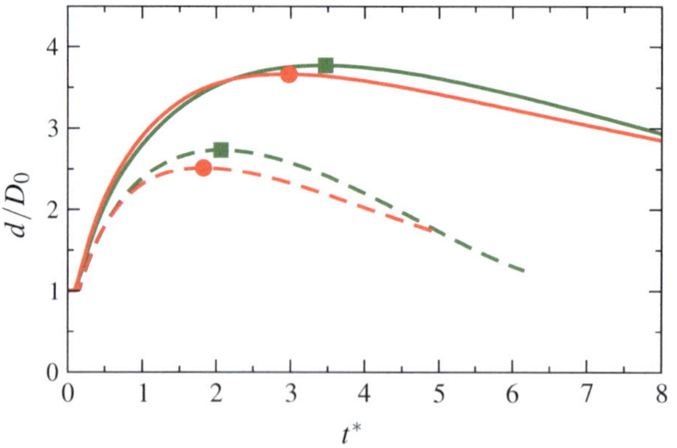

Fig. 7 Non-dimensional disk diameter d/D_0 as a function of non-dimensional time t^* for Weber number We $\approx$ 430 (*solid lines*) and Weber number We $\approx$ 67 (*dashed lines*). The *squared symbols* in *green* indicate the maximum d_{max} of the disk diameter for the Praestol® 2540 solution and the *circled symbols* in *red* indicate the same for the Praestol® 2500 solution

The non-dimensional disk diameters d/D_0 are reached at lower non-dimensional times t^*_{max} for lower Weber numbers in contrast to Fig. 6.

The non-dimensional maximum disk diameter is presented in Fig. 8 as a function of Weber number. A non-linear increase of d_{max}/D_0 with Weber number can be observed with slightly higher values for the Praestol® 2540 solution. The non-dimensional time t^*_{max}, at which the maximum disk diameter is reached, is depicted in Fig. 9. Here, again a non-linear increase of t^*_{max} with Weber number can be observed. For the same Weber number the maxima of the disk diameters are reached at later non-dimensional times t^*_{max}.

When the non-dimensional maximum disk diameter d_{max}/D_0 is plotted against the non-dimensional time t^*_{max}, as done in Fig. 10, an approximately linear behaviour is found for both Praestol® solutions. The results for both Praestol® solutions are only slightly different.

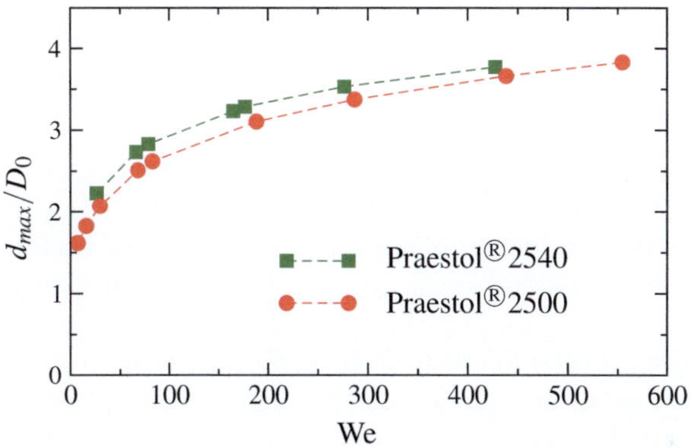

Fig. 8 Non-dimensional maximum disk diameter d_{max}/D_0 as a function of Weber number We

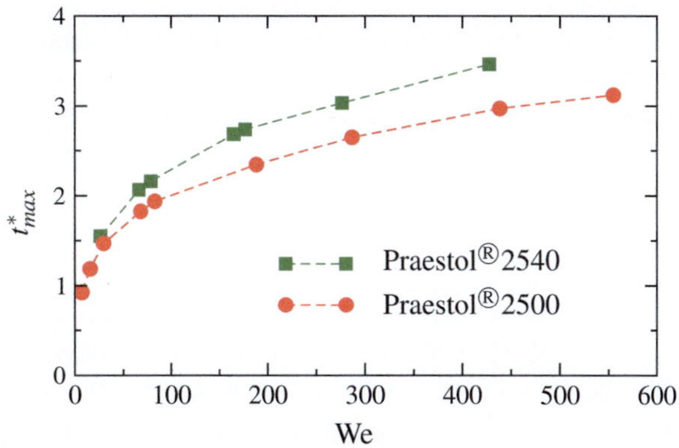

Fig. 9 Non-dimensional time t^*_{max}, at which the maximum disk diameter has been reached as a function of Weber number We

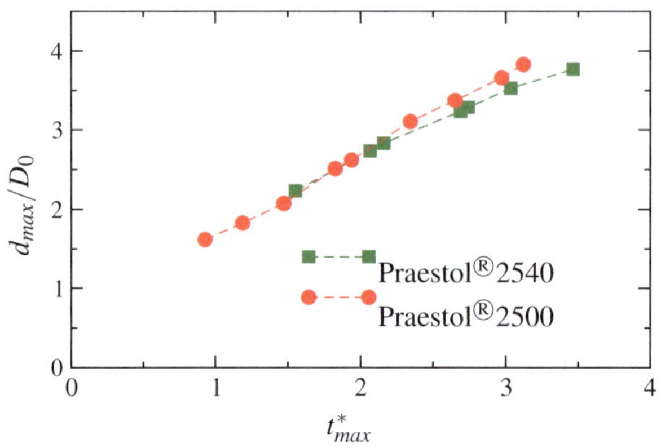

Fig. 10 Non-dimensional maximum disk diameter d_{max}/D_0 as a function of non-dimensional time t^*_{max}, at which this maximum has been reached

5 Computational Performance

For the performance analysis, an arbitrary setup is created consisting of two spheres of radius $r_0 = 0.5$ cm which are placed into a domain of size $[3 \times 3 \times 3]$ cm. They are initially located at the centre in the y- and z-direction but slightly shifted in the x-direction, so that there is a small overlap of $1/8r_0$. Both spheres are given an initial velocity of 1.2 m/s towards the centre of the domain. The resulting impact and lamella formulation is the basis of the performance analysis. In accordance to the results shown in the previous section, the non-Newtonian Praestol® 2500 0.8 % solution is used for both droplets. A schematic setup is shown in Fig. 11.

The baseline case is computed on a Cartesian grid of 512^3 grid cells. Even though hybrid parallelisation is possible (spatial domain composition with MPI data exchange and OpenMP on a loop level), we used only spatial domain composition here since previous calculations showed a decrease in parallel efficiency if OpenMP is used additionally [4].

Strong scaling was investigated by increasing the number of cores while keeping the problem size and resolution constant. The speed-up S was evaluated, defined as

$$S = \frac{N_i}{N_{32}}. \tag{7}$$

Here, N_i is the number of completed computation cycles (i.e. timesteps) in 2 h on i cores. The relatively long runtime allows to neglect the initialization time at the beginning of the calculation. The reference case was calculated on 32 cores since memory limitations did not allow to go lower than that. Parallel efficiency E was additionally evaluated based on the definition

$$E = \frac{32S}{i} \tag{8}$$

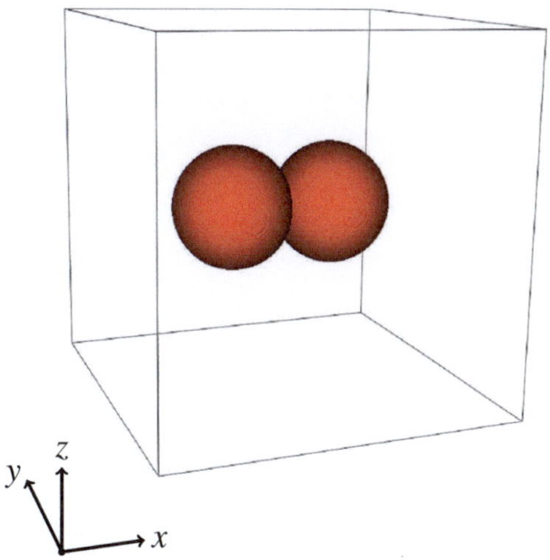

Fig. 11 Schematic representation of the setup for the performance analysis. The domain size is [3 × 3 × 3] cm

Table 2 Speed-up and parallel efficiency of FS3D. Shown are the results for different numbers of cores relative to the baseline case

Rel. number of cores	1	2	4	8	16	32	64
Speed-up S	1.0	1.95	3.59	6.28	11.26	13.91	24.91
Parallel efficiency E	100 %	97.2 %	89.6 %	78.4 %	70.4 %	43.5 %	38.9 %

Note, that instead of the classical definition of both the speed-up and the parallel efficiency in terms of computational times, the number of completed cycles was taken which is, however, equivalent.

From Table 2 and Fig. 12 it is apparent that the speed-up is not ideal. In fact, the parallel efficiency deviates from the ideal speed-up more and more with an increasing number of cores. This is mostly due to the multigrid solver used in FS3D to solve the pressure Poisson equation. Here, about 80 % of the total computational time is spent. Within this solver a high communication load arises due to multiple point-to-point and global MPI data exchanges. Furthermore, as the grid coarsens in the multigrid cycles, the ratio of calculation to communication overhead gets worse. Additionally, a solver for multiphase flows has inevitably different numerical schemes that are applied consecutively. These schemes have different serial fractions which corrupt the parallel efficiency. Currently, a new multigrid-solver library is implemented in cooperation with the University of Frankfurt, G. Wittum, with the hope of greatly improving both the serial performance and the parallel efficiency.

For the weak scaling shown on the left of Fig. 12 and in Table 3 the number of cells per core was kept constant, while the size of the problem was varied. The baseline case consisted of [256 × 256 × 512] grid cells calculated on 128 cores. Each case was run for 2 h, analogous to the strong scaling, and the total number

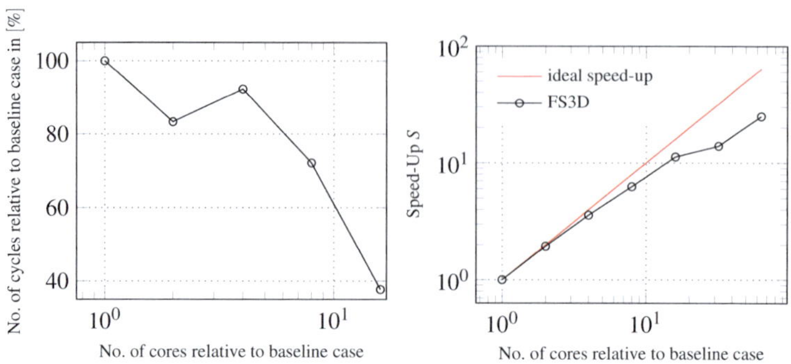

Fig. 12 Weak (*left*) and strong scaling (*right*) behaviour for FS3D relative to the baseline case of non-Newtonian droplet impact. On the *left* the number of cycles relative to the baseline case on [512 × 512 × 512] cells is shown. The *right side* shows the speed-up as defined in Eq. (7) as a function of the number of cores relative to the baseline case

Table 3 Weak scaling performance in terms of total number of cycles in 2 h runtime. Shown are the results relative to the case with [256 × 256 × 512] grid cells, herein called the baseline case

Rel. number of cores	Cells in x	Cells in y	Cells in z	E_{weak} (%)
1	256	256	512	100
2	256	512	512	83.30
4	512	512	512	92.25
8	512	512	1024	72.16
16	512	1024	1024	37.71

of completed timesteps were compared. The parallel efficiency for weak scaling is defined as

$$E_{weak} = \frac{N}{N_{baseline}} \cdot 100\,\%. \tag{9}$$

Note, that E_{weak} is dropping for a rising number of cores with the exception of the case with 512 processors. This is due to the rising number of coarsening levels and consequently the varying number of cycles in the multigrid solver. We hope to overcome this problem in the near future with the new solver for the Poisson equation.

6 Conclusions

A comparison of the shear thinning behaviour of droplets impacting on a dry wall was performed for two different Praestol® solutions. All simulations were done using the in-house code FS3D. The Praestol® solutions show a similar but not identical behaviour. In future studies, the behaviour of Newtonian fluids shall be compared with the results shown in this report. Furthermore, a performance analysis was done for two non-Newtonian droplets colliding without the influence of gravity. The strong scaling analysis shows good parallel efficiency up to 512 cores for the given case. However, FS3D still shows speed-up for more cores even though the parallel efficiency drops. In the near-future, a new multigrid solver for the pressure equation will be implemented with the hopes of greatly increasing parallel-efficiency.

Acknowledgements The authors kindly acknowledge the *High Performance Computing Center Stuttgart* (HLRS) for support and supply of computational time on the Cray XC40 platform under the Grant No. FS3D/11142 and the financial support by the Deutsche Forschungsgemeinschaft (DFG) for the Collaborative Research Center SFB-TRR75.

References

1. Eisenschmidt, K., Ertl, M., Gomaa, H., Kieffer-Roth, C., Meister, C., Rauschenberger, P., Reitzle, M., Schlottke, K., Weigand, B.: Direct numerical simulations for multiphase flows: an overview of the multiphase code fs3d. J. Appl. Math. Comput. **272**(2), 508–517 (2016). doi:10.1016/j.amc.2015.05.095
2. Ertl, M., Roth, N., Brenn, G., Gomaa, H., Weigand, B.: Simulations and experiments on shape oscillations of newtonian and non-newtonian liquid droplets. In: ILASS 2013, Chania, p. 7 (2013)
3. Francois, M.M., Cummins, S.J., Dendy, E.D., Kothe, D.B., Sicilian, J.M., Williams, M.W.: A balanced-force algorithm for continuous and sharp interfacial surface tension models within a volume tracking framework. J. Comput. Phys. **213**(1), 141–173 (2006)
4. Galbiati, C.M.E., Tonini, S., Cossali, G.E., Weigand, B.: DNS investigation of the primary breakup in a Conical Swirled Jet. In: High Performance Computing in Science and Engineering '15 Transactions of the High Performance Computing Center, Stuttgart (HLRS), pp. 333–347 (2016)
5. Gomaa, H., Stotz, I., Sievers, M., Lamanna, G., Weigand, B.: Preliminary Investigation on diesel droplet impact on oil wallfilms in diesel engines. In: ILASS – Europe 2011, 24th European Conference on Liquid Atomization and Spray Systems, Estoril, Sept 2011
6. Hase, M., Weigand, B.: A numerical model for 3D transient evaporation processes based on the volume-of- fluid method. In: ICHMT International Symposium on Advances in Computational Heat Transfer, Istambul, pp. 1–23 (2004)
7. Hirt, C.W., Nichols, B.D.: Volume of fluid (VOF) method for the dynamics of free boundaries. J. Comput. Phys. **39**(1), 201–225 (1981). doi:10.1016/0021–9991(81)90145–5
8. Rauschenberger, P., Weigand, B.: A volume-of-fluid method with interface reconstruction for ice growth in supercooled water. J. Comput. Phys. **282**, 98–112 (2015). doi:10.1016/j.jcp.2014.10.037
9. Rauschenberger, P., Weigand, B.: Direct numerical simulation of rigid bodies in multiphase flow within an Eulerian framework. J. Comput. Phys. **291**, 238–253 (2015). doi:10.1016/j.jcp.2015.03.023
10. Rieber, M., Graf, F., Hase, M., Roth, N., Weigand, B.: Numerical simulation of moving spherical and strongly deformed droplets. In: Proceedings ILASS-Europe, Darmstadt, pp. 1–6 (2000)
11. Roth, N., Schlottke, J., Urban, J., Weigand, B.: Simulations of droplet impact on cold wall without wetting. In: ILASS, Como Lake, pp. 1–7 (2008)

12. Roth, N., Gomaa, H., Weigand, B.: Droplet collisions at high weber numbers: experiments and numerical simulations. In: Proceedings DIPSI Workshop 2010 on Droplet Impact Phenomena & Spray Investigation. Bergamo (2010)
13. Roth, N., Meister, C., Gomaa, H., Ertl, M., Weigand, B.: Numerical simulation of shear thinning liquids impacting on dry solid walls. In: Proceedings 26th Europe Conference on Liquid Atomization and Spray Systems. ILASS, Bremen (2014)
14. Schlottke, J., Rauschenberger, P., Weigand, B., Ma, C., Bothe, D.: Volume of fluid direct numerical simulation of heat and mass transfer using sharp temperature and concentration fields. In: ILASS – Europe 2011, 24th European Conference on Liquid Atomization and Spray Systems, Estoril (2011). http://www.ilass.uci.edu/
15. Tanner, R.I.: Engineering Rheology, 2nd edn. Oxford Engineering Science Series. Oxford University Press, New York (2002)
16. Weking, H., Schlottke, J., Boger, M., Munz, C.D., Weigand, B.: DNS of rising bubbles using VOF and balanced force surface tension. In: High Performance Computing on Vector Systems (2010). Springer, Berlin/Heidelberg/New York

Turbulent Skin-Friction Drag Reduction at High Reynolds Numbers

Davide Gatti

Abstract Direct Numerical Simulation (DNS) of turbulent channel flows at moderately high values of the Reynolds number (*Re*) are performed to examine how *Re* affects the capabilities of wall-based spanwise-forcing techniques to achieve turbulent skin-friction drag reduction. With the present new data, a relationship could be derived and validated, which predicts the amount of drag reduction at several values of *Re*. The present study shows that a drag reduction of nearly 30 % would still be possible for an airplane at flight Reynolds numbers thanks to the spanwise forcing.

1 Introduction

The present manuscript briefly summarizes the research performed utilizing the computational resources of the ForHLR I computer cluster within the project "reef-fect". A more detailed description of the present research has been already published in Gatti and Quadrio [11] and in Stroh, Gatti, Hasegawa and Frohnapfel [22].

In the last few decades, fundamental research efforts in turbulent skin-friction drag reduction met with considerable success, and several viable strategies to reduce drag have been introduced. Due to the shrinkage of space- and time- scale of wall turbulence in laboratory implementations, or to the fast growth of the computational costs with increasing *Re* in numerical experiments, such studies are typically limited to low-Reynolds number flows. Therefore, the question naturally arises how to extrapolate the observed performance to the higher values of the Reynolds number *Re* typical of most industrial applications.

In simple and well-controlled laboratory flows like a channel flow the friction drag reduction is typically characterized in terms of the drag reduction rate $\mathscr{R}$, defined as the relative change of skin-friction coefficient C_f between the controlled

D. Gatti (✉)
Institute of Fluid Mechanics (ISTM), Karlsruhe Institute of Technology (KIT), Karlsruhe, Germany
e-mail: davide.gatti@kit.edu

W.E. Nagel et al. (eds.), *High Performance Computing in Science and Engineering '16*, DOI 10.1007/978-3-319-47066-5_26

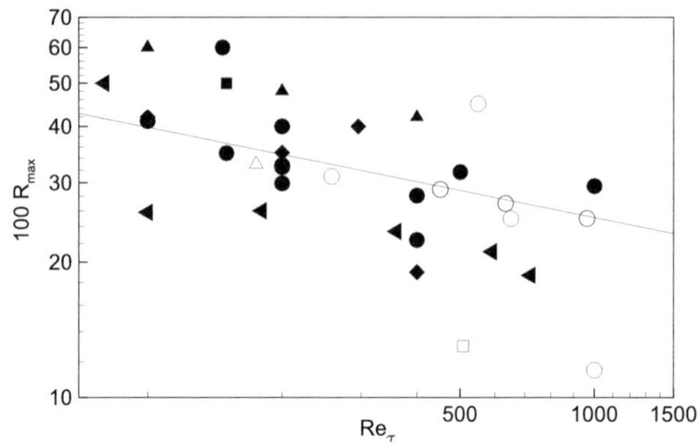

Fig. 1 Literature data for maximum drag reduction rate $\mathscr{R}_{\text{max}}$ versus Re_τ for spanwise-forcing techniques. *Black* (*white*) symbols indicate results from DNS (experimental) studies. We explicitly note that the forcing amplitude is not always identical among different datasets. ○: oscillating wall [4–6, 12, 15, 17, 18, 20, 21, 23–25]; △: streamwise-traveling waves [1, 19]; □: spanwise-traveling waves [7, 8]; ◇: Lorentz force [2, 16]; ◁: reactive opposition control [3]. The *solid line* is $\mathscr{R}_{\text{max}} \sim Re_\tau^{-0.2}$ (Figure taken from Gatti and Quadrio [10]. Reprinted with permission of AIP Publishing)

and the reference flow:

$$\mathscr{R} = 1 - \frac{C_f}{C_{f,0}}. \tag{1}$$

In this definition, the subscript "0" indicates a quantity measured in the reference flow, and the skin-friction coefficient is defined as

$$C_f = 2\frac{\tau_w}{\rho U_b^2}, \tag{2}$$

where τ_w is the wall-shear stress, ρ is the fluid density and U_b the bulk velocity.

The low-*Re* laboratory and numerical evidence available shows (Fig. 1) that the maximum drag reduction $\mathscr{R}_{\text{max}}$ obtained with various control techniques based on near-wall forcing decreases for increasing Reynolds numbers, which poses the question whether sizeable drag reduction is still achievable at high values of *Re* and hence worth pursuing.

A lively debate is taking place in the scientific community regarding the high-*Re* behaviour of the last generation of control techniques for turbulent skin-friction drag reduction, in particular the active open-loop ones (i.e. those requiring additional power to be transferred to the flow). Such techniques operate by enforcing suitable temporal and spatial distributions of velocity perturbations at the wall. Assessing their potential for achieving sizeable benefits at high *Re* is of paramount importance to motivate further research in this field.

2 Goals and Methods

The goal of the resent research is to address the effect of increasing the Reynolds number on the achievable turbulent skin-friction drag reduction. We take a particularly promising control strategy as model for the present investigation: the streamwise-travelling waves of spanwise wall velocity [19], which consists on the following spanwise wall velocity distribution:

$$W_w(x,t) = A\sin(\kappa x - \omega t). \tag{3}$$

In the above expression for the wall forcing, W_w is the spanwise velocity enforced at the wall, A is the amplitude of the forcing, κ is the streamwise wavenumber and ω is the angular frequency. x and t are the streamwise coordinate and time respectively. The forcing, sketched in Fig. 2, consists in a wall distribution of streamwise-modulated waves of the spanwise (z) velocity component with wavelength $\lambda = 2\pi/\kappa$ and period $T = 2\pi/\omega$, which travel at speed $c = \omega/\kappa$ forward ($c > 0$) or backward ($c < 0$) with respect to the direction x of the mean flow. The three independent parameters (for example A, κ, ω) of the control law (3) combined with the Reynolds number Re define a 4-dimensional parameter space, whose complete investigation represents a computational challenge. A large number of Direct Numerical Simulations (DNS) of the turbulent flow in a doubly periodic channel modified by streamwise-travelling waves of spanwise wall velocity are performed either a constant flow rate (CFR) or a constant pressure gradient (CPG).

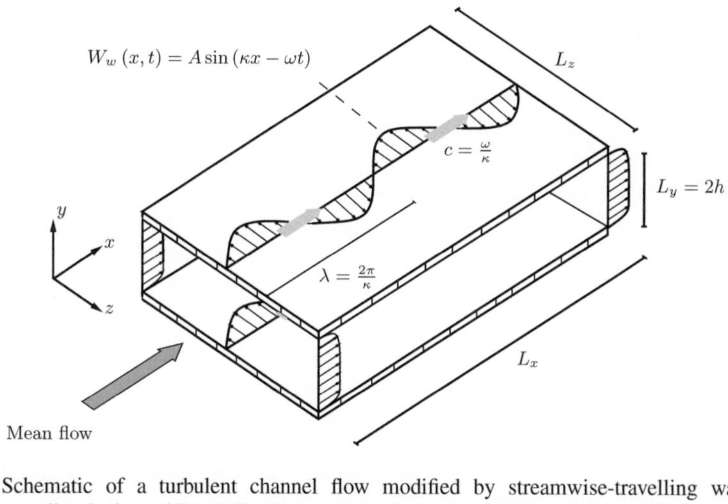

Fig. 2 Schematic of a turbulent channel flow modified by streamwise-travelling waves of spanwise wall velocity, with amplitude A, streamwise wavenumber κ and angular frequency ω. λ is the streamwise wavelength and c is the phase speed of the waves. L_x, $L_y = 2h$ and L_z are the dimensions of the computational domain in the streamwise, wall-normal and spanwise directions, respectively

Table 1 Details of the small-box (upper half) and large-box (lower half) simulations. Every caseset is detailed in terms of simulation type (CFR or CPG), number of cases N_{cases}, values of bulk Reynolds number Re_b and friction Reynolds number Re_τ, length and width of the computational domain in inner and outer units, number of Fourier modes in the homogeneous directions (additional modes are used for dealiasing, according to the 3/2 rule) and collocation points in the wall-normal direction

Type	N_{cases}	Re_b	Re_τ	L_x/h	L_z/h	L_x^+	L_z^+	$N_x \times N_y \times N_z$
CFR	1530	6627	203.0	1.59π	0.80π	1015	507	$96 \times 100 \times 96$
CFR	480	6627	203.4	2.05π	1.02π	1308	654	$128 \times 100 \times 128$
CFR	1530	39,333	905.6	0.32π	0.16π	906	453	$96 \times 500 \times 96$
CFR	480	39,333	948.3	0.43π	0.22π	1290	645	$128 \times 500 \times 128$
CFR	5	6360	199.9	4π	2π	2512	1256	$256 \times 128 \times 256$
CPG	5	6358	200.0	4π	2π	2513	1257	$256 \times 128 \times 256$
CFR	5	39,980	1000.0	4π	2π	12,566	6283	$1024 \times 500 \times 1024$
CPG	5	39,900	998.6	4π	2π	12,549	6274	$1024 \times 500 \times 1024$

The aim is to obtain and compare two comprehensive sets of cases at $Re_\tau = 200$ and $Re_\tau = 1000$, where $Re_\tau = u_\tau h/\nu$ is the Reynolds number based on the channel half-height h, the friction velocity u_τ of the uncontrolled flow and the kinematic viscosity ν of the fluid. The initial condition is that of an uncontrolled turbulent flow. The spatial resolution in wall units is always better than $\Delta x^+ = 12.3$ and $\Delta z^+ = 6.1$ (or $\Delta x^+ = 8.2$ and $\Delta z^+ = 4.1$ if the additional modes used to completely remove the aliasing error are considered). Δy^+ smoothly varies from $\Delta y^+ \approx 1$ near the wall to $\Delta y^+ \approx 7$ at the centerline. Time integration is carried out with a partially implicit approach, with a Crank-Nicolson scheme for the viscous terms and a third-order Runge–Kutta scheme for the convective terms. The CFL number is set at unity; the consequent average size of the timestep is always below $\Delta t^+ = 0.17$ for the low-*Re* cases, and below $\Delta t^+ = 0.1$ for the high-*Re* cases. The integration time is at least 24,000 viscous time units, and in certain cases it increases up to 80,000 viscous units. For each value of *Re*, the computational study considers two distinct sets of simulations, described below, details of which are reported in Table 1.

The first set (upper half of Table 1) is a parameter study designed to produce a massive database of drag reduction data (4020 cases overall); the parameter space includes the forcing wavenumber κ, the forcing angular frequency ω and, for the first time, the forcing amplitude A too. For this set of calculations, carried out under the CFR condition, a relatively small computational domain is employed: the consequent savings in computing time are key to make this huge parameter study possible.

The second set of simulations (lower half of Table 1) employs a larger domain size. For both *Re* we consider the reference uncontrolled case, and four other cases at the amplitude $A^+ = 7$. One case is for the oscillating wall at nearly optimal period $T^+ = 75$, one case with oscillating wall at the larger period $T^+ = 250$, one case with travelling waves with large drag reduction ($\omega^+ = 0.0239$ and $\kappa^+ = 0.01$)

and one case for travelling waves with drag increase ($\omega^+ = 0.12$ and $\kappa^+ = 0.01$). Each case is run under both CFR and CPG (and for the latter the forcing parameters listed above are to be intended in actual wall units), for a total of 20 simulations featuring the larger computational domain.

3 Computational Details

The computation of the small-box large database, which involved a total of 4020 simulations, were partially performed on a Blue Gene/Q system at the CINECA computing centre in Bologna and partially on the ForHLR supercomputer of the Steinbuch Centre for Computing (SCC) in Karlsruhe. The simulations have been run with the solver for the incompressible Navier–Stokes equations developed by [13]. The large number of relatively inexpensive simulation, due to the small domain size, allows to run them contemporaneously as 4020 serial simulations. Automated procedures in bash and Python have been developed to control the whole workflow, collect and postprocess the results easily. The simulation run for 720 wall clock hours on 4020 cores, totalling 2.9 Mio CPU hours.

The large-box small database requires more resources and ad-hoc parallelization to be generated. The smaller number (20) of very large simulation, each one consisting of 524.3 Mio grid points, of course impedes to adopt the same strategy used for the small-box simulations, i.e. by simultaneously running a large number of serial computations. An hybrid shared-memory and distributed-memory parallelization has been employed, which relies on MPI and OpenMP, in order to perform the computation efficiently. The performance of the distributed matrix transpose required in the pseudo-spectral convolutions has been improved by adopting a copy-free algorithm which relies on MPI derived datatypes. Data are sent and received in the appropriate order, which automatically results into a transposition, without the requirement for manual packing and unpacking of send and receive buffers. These computations have been entirely performed on the ForHLR supercomputer of the SCC in Karlsruhe. Typically, each simulation was run on 140 CPUs, organized in 35 proceeses à 4 treads, for about 3.5 months, totalling about 7 Mio CPU hours. The total volume of generated data is 4 TB.

4 Results

Figure 3 globally represents the whole DNS dataset as isosurfaces of drag reduction rate in the control parameter space. The cloud of black dots represents the 2024 datapoints used for interpolation at each *Re*. This overview already confirms that the drag reduction rate decreases throughout the whole dataset when the *Re* is increased from $Re_\tau = 200$ to $Re_\tau = 1000$. For instance, The connected region

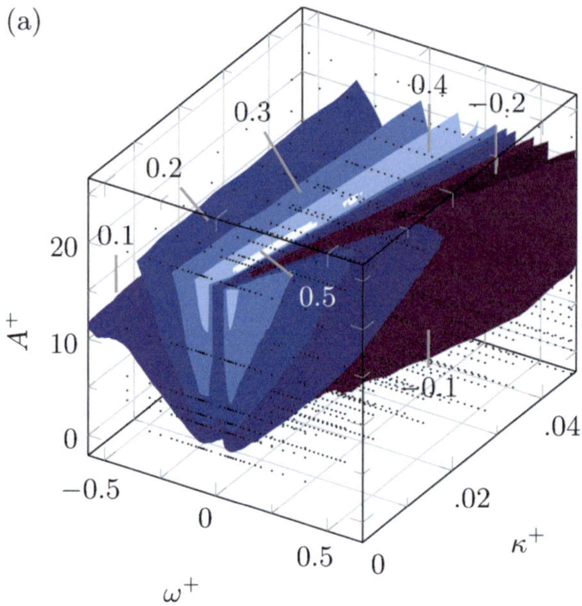

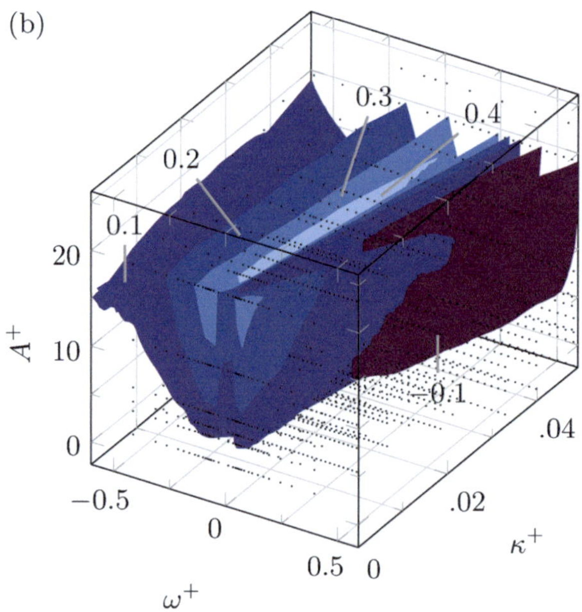

Fig. 3 Isosurfaces of drag reduction $\mathscr{R}$ in the three-dimensional parameter space $(\omega^+, \kappa^+, A^+)$ for $Re_\tau = 200$ (**a**) and $Re_\tau = 1000$ (**b**). Isosurface from dark to light range from $\mathscr{R} = -0.2$ to 0.5 in steps of 0.1. The cloud of dots represents the 2010 data points where, at each Re, a DNS has been carried out (Figure taken from Gatti and Quadrio [11]. Reprinted with permission from Cambridge University Press)

where $\mathscr{R} > 0.5$ at $Re_\tau = 200$ is not visible at $Re_\tau = 1000$. Interestingly, the region of drag increase is most affected by the change in Reynolds number. The isosurface at $\mathscr{R} = -0.2$ disappears at $Re_\tau = 1000$ and the one at $\mathscr{R} = -0.1$ shrinks significantly. The rate at which the drag reduction decays with Re is traditionally described after [6], i.e. by assuming a power-law decay as $\mathscr{R} = Re_\tau^\gamma$, with the exponent $\gamma = -0.2$. However, the present results show that the exponent γ is not constant but is itself a function of all control parameters and the Reynolds number Re. As a result, the power law $\mathscr{R} = Re_\tau^\gamma$ can not be used to predict the high-Re behaviour of drag reduction and other approaches are required. Thanks to the results of the large-box small database and the computations run on ForHLR, a different approach to describe the effect of Re on $\mathscr{R}$ has been suggested.

In analogy with surface roughness, the effect of drag-reducing control can be quantified via the so-called roughness function and considered as a positive (upward) shift ΔB^+ of the velocity profile in the logarithmic layer. This known result is systematically checked in Fig. 4a–d. The mean velocity profiles are obtained at both Re from the large-box simulations.

Following a procedure already done for riblets e.g. by [14] and by [9], it is possible to derive a relationship which links the vertical shift ΔB^+ of the mean velocity profile in its logarithmic region, the drag reduction rate $\mathscr{R}$ and the Reynolds number (via the skin-friction coefficient of the uncontrolled flow, which is a unique function thereof). Further details of the derivation can be found in [11].

If the uncontrolled and controlled flows are compared under the CFR constraint, this relationship reads:

$$\Delta B^+ = \sqrt{\frac{2}{C_{f,0}}}\left[(1-\mathscr{R})^{-1/2} - 1\right] - \frac{1}{2k}\ln(1-\mathscr{R})\,. \tag{4}$$

If on the other hand the pressure gradient is kept constant across the comparison (CPG), then by definition $Re_\tau = Re_{\tau,0}$, and the above equation further simplifies to:

$$\Delta B^* = \sqrt{\frac{2}{C_{f,0}}}\left[(1-\mathscr{R})^{-1/2} - 1\right]. \tag{5}$$

The data presented in Gatti and Quadrio [11] show that ΔB is a function of the control parameters only and does not depend upon Re, if the Reynolds number is large enough for the Prandt-von Kármán friction relation to reasonably hold. Therefore, the relationship (4) and (5) can be utilized, once ΔB is known, to predict the behaviour of $\mathscr{R}$ at large Re.

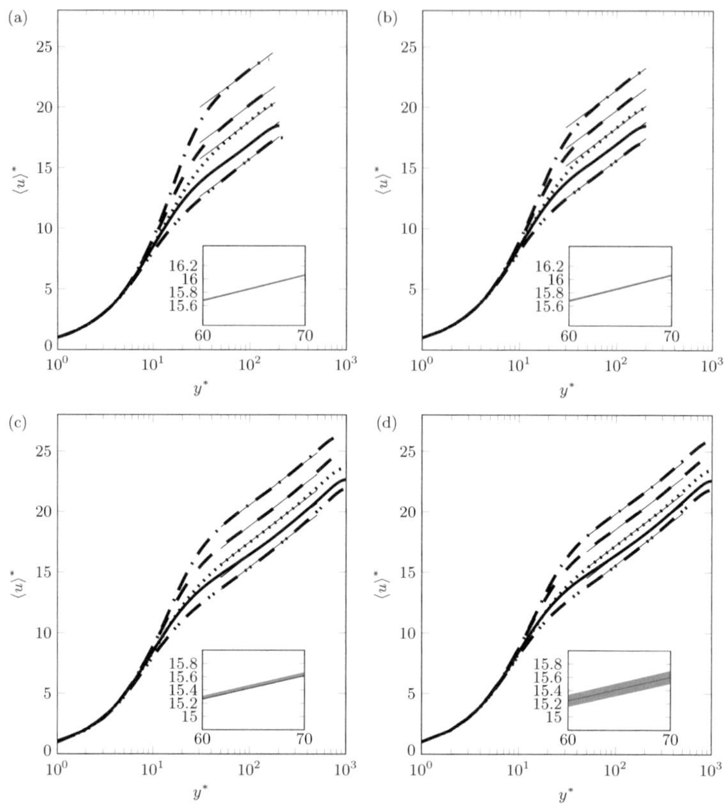

Fig. 4 Mean velocity profiles obtained from the large-domain simulations reported in the lower half of Table 1. *Top*: $Re_\tau = 200$; bottom: $Re_\tau = 1000$. *Left*: CFR cases; *right*: CPG cases. The *solid line* is the reference case and the other lines correspond to control yielding both drag reduction and drag increase (see text). The insets enlarge a portion of the logarithmic layer to show the (very small) statistical uncertainty at 95 % confidence, denoted by the *shaded area* (Figure taken from Gatti and Quadrio [11]. Reprinted with permission from Cambridge University Press)

5 Conclusion

In this study a large drag reduction DNS database has been produced for a turbulent plane channel flow subject to a spanwise forcing. Four-thousand and twenty simulations have been used to describe how increasing the value of the Reynolds number from $Re_\tau = 200$ to $Re_\tau = 1000$ affects drag reduction, and to propose a rationale behind the observed performance deterioration. To the authors' knowledge, this is the first study on spanwise forcing that includes a wide range of forcing amplitudes, as well as Constant Pressure Gradient (CPG) data at different values of *Re*.

The existing information regarding spanwise forcing has been significantly extended. The classic argument linking the skin-friction drag changes of a rough wall to the vertical shift ΔB of the logarithmic portion of the mean velocity profile has been shown to apply to the case of spanwise forcing. A non-linear expression has been derived that can be specialized to the CFR or CPG cases.

Under the assumption that ΔB^* measured in the present work at $Re_\tau = 1000$ is already Re-independent, Eq. (5) can be used to extrapolate drag reduction at higher Re_τ. It can be shown that a drag reduction of $\mathscr{R} = 0.5$ at $Re_\tau = 1000$ translates into $\mathscr{R} = 0.34$ at $Re_\tau = 10^5$. The decrease is still significant but not as dramatic as the low-Re evidence suggests.

References

1. Auteri, F., Baron, A., Belan, M., Campanardi, G., Quadrio, M.: Experimental assessment of drag reduction by traveling waves in a turbulent pipe flow. Phys. Fluids **22**(11), 115103/14 (2010)
2. Berger, T.W., Kim, J., Lee, C., Lim, J.: Turbulent boundary layer control utilizing the Lorentz force. Phys. Fluids **12**(3), 631–649 (2000)
3. Chang, Y., Collis, S.S., Ramakrishnan, S.: Viscous effect in control near-wall turbulence. Phys. Fluids **14**, 4069–4080 (2002)
4. Choi, K.S., Graham, M.: Drag reduction of turbulent pipe flows by circular-wall oscillation. Phys. Fluids **10**(1), 7–9 (1998)
5. Choi, K.S., DeBisschop, J., Clayton, B.: Turbulent boundary-layer control by means of spanwise-wall oscillation. AIAA J. **36**(7), 1157–1162 (1998)
6. Choi, J.I., Xu, C.X., Sung, H.J.: Drag reduction by spanwise wall oscillation in wall-bounded turbulent flows. AIAA J. **40**(5), 842–850 (2002)
7. Du, Y., Karniadakis, G.E.: Suppressing wall turbulence by means of a transverse traveling wave. Science **288**, 1230–1234 (2000)
8. Du, Y., Symeonidis, V., Karniadakis, G.E.: Drag reduction in wall-bounded turbulence via a transverse travelling wave. J. Fluid Mech. **457**, 1–34 (2002)
9. García-Mayoral, R., Jiménez, J.: Drag reduction by riblets. Phil. Trans. R. Soc. A **369**(1940), 1412–1427 (2011)
10. Gatti, D., Quadrio, M.: Performance losses of drag-reducing spanwise forcing at moderate values of the Reynolds number. Phys. Fluids **25**, 125109(17) (2013)
11. Gatti, D., Quadrio, M.: Reynolds-number dependence of turbulent skin-friction drag reduction induced by spanwise forcing. J. Fluid Mech. **802**, 553–582 (2016)
12. Jung, W., Mangiavacchi, N., Akhavan, R.: Suppression of turbulence in wall-bounded flows by high-frequency spanwise oscillations. Phys. Fluids A **4**(8), 1605–1607 (1992)
13. Luchini, P., Quadrio, M.: A low-cost parallel implementation of direct numerical simulation of wall turbulence. J. Comput. Phys. **211**(2), 551–571 (2006)
14. Luchini, P., Manzo, F., Pozzi, A.: Resistance of a grooved surface to parallel flow and cross-flow. J. Fluid Mech. **228**, 87–109 (1991)
15. Nikitin, N.V.: On the mechanism of turbulence suppression by spanwise surface oscillations. Fluid Dyn. **35**(2), 185–190 (2000)
16. Pang, J., Choi, K.S.: Turbulent drag reduction by Lorentz force oscillation. Phys. Fluids **16**(5), L35–L38 (2004)
17. Quadrio, M., Ricco, P.: Critical assessment of turbulent drag reduction through spanwise wall oscillation. J. Fluid Mech. **521**, 251–271 (2004)
18. Quadrio, M., Sibilla, S.: Numerical simulation of turbulent flow in a pipe oscillating around its axis. J. Fluid Mech. **424**, 217–241 (2000)

19. Quadrio, M., Ricco, P., Viotti, C.: Streamwise-traveling waves of spanwise wall velocity for turbulent drag reduction. J. Fluid Mech. **627**, 161–178 (2009)
20. Ricco, P., Quadrio, M.: Wall-oscillation conditions for drag reduction in turbulent channel flow. Int. J. Heat Fluid Flow **29**, 601–612 (2008)
21. Ricco, P., Wu, S.: On the effects of lateral wall oscillations on a turbulent boundary layer. Exp. Therm. Fluid Sci. **29**(1), 41–52 (2004)
22. Stroh, A., Gatti, D., Hasegawa, Y., Frohnapfel, B.: Influence of drag-reducing near-wall turbulence control on spectral properties of Reynolds shear stress. In: Proceedings of the 11th ETMM, Palermo (2016)
23. Tamano, S., Itoh, M.: Drag reduction in turbulent boundary layers by spanwise traveling waves with wall deformation. J. Turbul. **13**, N9 (2012)
24. Touber, E., Leschziner, M.: Near-wall streak modification by spanwise oscillatory wall motion and drag-reduction mechanisms. J. Fluid Mech. **693**, 150–200 (2012)
25. Trujillo, S., Bogard, D., Ball, K.: Turbulent boundary layer drag reduction using an oscillating wall. AIAA Paper 97–1870 (1997)

Control of Spatially Developing Turbulent Boundary Layers for Skin Friction Drag Reduction

Alexander Stroh

Abstract This project comprises direct numerical simulations (DNS) of turbulent boundary layer flows. The wall along which the flow develops is modified in some parts in order to introduce control techniques that aim at a reduction of skin friction drag. The obtained results are used in two ways. They are first compared with turbulent channel flows, in which the applied control schemes were originally developed. Second, the flow development after a controlled section in a turbulent boundary layer is analyzed. The detailed scientific results that were obtained based on the generated data are published in Stroh et al. (Phys Fluids 27(7):075101, 2015; J. Fluid Mech 805:303–321, 2016).

1 Introduction

A broad variety of control methods aimed at the reduction of skin friction drag in turbulent boundary layers was introduced over the past few decades [1–4]. Since the majority of these control methods are proposed for a configuration of a periodic fully developed turbulent channel flow (TCF) controlling the entire wall area, the knowledge about local control application is still limited. However, localized control is more realistic from the engineering point of view. In this case the flow alteration outside of the control region also has to be taken into account for the overall control performance estimation. In the present work two locally applied drag reducing control methods with entirely different control mechanisms are investigated in the framework of spatially developing turbulent boundary layers (TBL) in order to analyse the flow behaviour downstream of the control region. In addition, the global performance of these flow control techniques is evaluated.

A. Stroh (✉)
Institute of Fluid Mechanics (ISTM), Karlsruhe Institute of Technology (KIT), Karlsruhe, Germany
e-mail: alexander.stroh@kit.edu

W.E. Nagel et al. (eds.), *High Performance Computing in Science and Engineering '16*, DOI 10.1007/978-3-319-47066-5_27

2 Description and Goals

The HPC-project investigates the effects of drag reducing turbulence control applications in spatially developing turbulent boundary layers. The investigation is carried out using direct numerical simulation (DNS) with the main aim to clarify the local drag reduction effect and the effect on the flow field far downstream induced by an application of several (re-) active control techniques.

Following goals have been identified for the present project:

- comparison of drag reducing control application in TBL to the results of drag reducing control application in TCF;
- clarification of the flow behaviour downstream of the control region and estimation of the control influence on this far flow field;
- investigation of Reynolds-number dependency of achievable global and local drag reduction and its mechanisms.

3 Numerical Procedure

The investigation is performed using DNS of a turbulent boundary layer with zero pressure gradient (ZPG). The coordinate system of the numerical domain and its geometry are illustrated in Fig. 1, where x, y and z correspond to the streamwise, wall-normal and spanwise directions respectively. For an incompressible fluid, the Navier-Stokes equations for a constant property Newtonian fluid and the continuity equation are required:

$$\frac{\mathrm{D}u_i}{\mathrm{D}t} = \mu \frac{\partial^2 u_i}{\partial {x_j}^2} - \frac{\partial p}{\partial x_i} \quad \text{and} \quad \frac{\partial u_i}{\partial x_i} = 0, \tag{1}$$

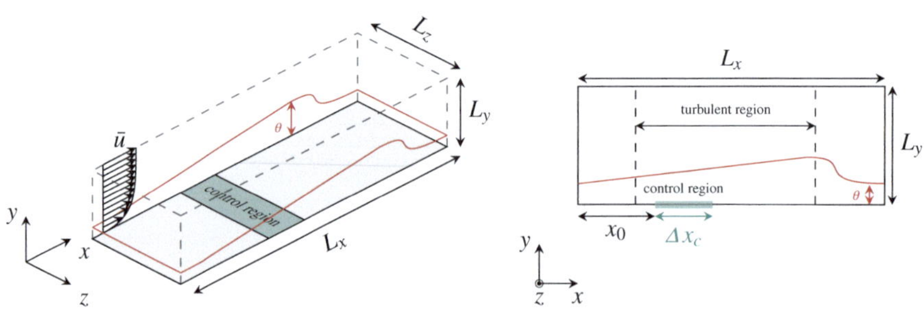

Fig. 1 Schematic of simulation domain and control placement

where p is the static pressure and μ is the dynamic viscosity. The Reynolds numbers for a boundary layer flow are defined as

$$\mathrm{Re}_{\delta,0} = \frac{U_\infty \delta_0^*}{\nu}, \qquad \mathrm{Re}_\theta = \frac{U_\infty \theta}{\nu} \qquad \text{and} \qquad \mathrm{Re}_\tau = \frac{u_\tau \delta_{99}}{\nu}, \tag{2}$$

where U_∞, δ_0, θ and ν are the streamwise free-stream velocity at $x = 0$, the undisturbed displacement thickness at $x = 0$, the local momentum thickness and the kinematic viscosity, respectively. u_τ is the friction velocity and δ_{99} represents the boundary layer thickness based on $0.99U_\infty$. A non-dimensionalization based on the viscous scales of the flow field (u_τ and ν/u_τ) is denoted with superscripted plus sign ($^+$) throughout the paper.

The implementation is based on a pseudo-spectral solver for incompressible boundary layer flows [5, 6]. The Navier-Stokes equations are numerically integrated using the velocity-vorticity formulation by a spectral method with Fourier decomposition in the horizontal directions and Chebyshev discretization in the wall-normal direction. For temporal advancement, the convection and viscous terms are discretized using the 3rd order Runge-Kutta and Crank-Nicolson methods, respectively. The fringe region technique is used in the present DNS to generate turbulence [7]. The non-physical phenomena occurring in the fringe region do not invalidate the solution in the physically useful part of the computational domain. The flow is bounded by the wall ($y = 0$), while the spanwise and streamwise boundary conditions are periodic. At the wall, no-slip conditions are applied except for the velocity component to which the control input is imposed. A Neumann condition for the wall-normal derivative based on the Falkner–Skan–Cooke solution is utilised at the free-stream boundary of the numerical domain. An adaptive adjustment of computational time step is utilized during the simulation. The detailed properties of the grid resolution in the area of interest and simulation domain are summarised in Table 1.

A modification of the flow field in terms of skin friction drag due to the application of flow control has to be quantified through the definition of several control performance indices. For TCF the following definitions are applied [3]. With

Table 1 Properties of considered simulation configurations for TBL. Viscous lengthscale is based on the average u_τ in the turbulent region of the TBL simulation. Grey shade highlights the main configuration setup

	Grid size	Domain size	Resolution			Height	Grid nodes
#	$N_x \times N_y \times N_z$	$L_x \times L_y \times L_z$	Δx^+	Δy^+	Δz^+	$\frac{L_y}{\delta_{99}^{\max}}$	$N \times 10^6$
1	$512 \times 129 \times 128$	$600 \times 30 \times 34$	23.8	$0.1 - 8.2$	5.9	2.25	8.6
2	$1024 \times 257 \times 128$	$1200 \times 60 \times 34$	23.8	$0.1 - 8.2$	5.9	2.88	33.6
3	$3072 \times 301 \times 256$	$3000 \times 100 \times 120$	17.8	$0.1 - 13.3$	8.9	2.32	236.7

respect to the uncontrolled case, the reduction rate of skin friction drag is given by

$$R = 1 - \frac{c_f}{c_{f,0}} \qquad \text{with} \qquad c_f = \frac{\tau_w}{0.5\rho U_b^2}, \tag{3}$$

where c_f denotes the skin friction coefficient, U_b is the bulk mean velocity and the subscript "0" denotes the uncontrolled value. If the flow rate in a channel flow is kept constant (CFR), the modification of the skin friction coefficient is reflected in a change in $\tau_w = \sqrt{u_\tau/\rho}$ or u_τ:

$$R = 1 - \frac{\tau_w}{\tau_{w,0}} = 1 - \left(\frac{u_\tau}{u_{\tau,0}}\right)^2. \tag{4}$$

Similarly, the control performance indices are introduced in TBL using U_∞ instead of U_b, so the local driving power is given as

$$P(x) = U_\infty(x)\,\tau_w(x). \tag{5}$$

and the drag reduction rate is alternatively given by

$$R = 1 - \frac{P}{P_0}. \tag{6}$$

Control is applied locally in the streamwise direction, while the spanwise extent of the control area covers the total domain width (Fig. 1). All control types are placed at the same position, x_0, with the same control area extension, Δx_c. The location is defined by the control input profile:

$$f(x) = \begin{cases} 1, & \text{for } x_0 \le x \le x_0 + \Delta x_c \\ 0, & \text{otherwise.} \end{cases} \tag{7}$$

The control amplitude is smoothly increased and decreased at the edges of the control area using a hyperbolic tangent function. Three control techniques are considered for the present investigation: opposition control, body force damping and uniform blowing.

Opposition control [1] is one of the most prominent classical reactive control schemes. Control activation is performed by local suction and blowing in the wall-normal direction at the wall surface, so as to suppress the sweep and ejection events in the near-wall region and reduce the skin friction drag. In TCF the control is commonly applied to the entire area of the wall, imposing wall-normal or spanwise velocity opposite to the velocity captured at a prescribed sensing plane y_s. The wall-normal control input at the wall is given by

$$v(x, 0, z, t) = -\alpha \cdot f(x) \cdot v(x, y_s, z, t), \tag{8}$$

where α is a positive amplification factor and the control placement is defined by $f(x)$. Application of opposition control provides $R = 20\,\%–30\,\%$. The scheme is well studied and can be used as a reference for comparison due to the presence of a broad literature database [1, 8–10].

The scheme of body force damping utilizes volume forces for the modification of the flow. The reactive scheme is introduced by Satake and Kasagi [11] for the damping of the spanwise velocity fluctuations. Similarly to opposition control, the control law aims at the suppression of turbulent fluctuations in the near-wall region and uses velocity as the sensor information. The control input is given in the form of a body force in i.e. wall-normal direction for a damping layer with thickness $y < \Delta y_c$:

$$b_y(x, y, z, t) = -\frac{f(x)}{\Phi} \cdot v(x, y, z, t), \tag{9}$$

with the forcing time constant Φ. The scheme is very efficient in terms of drag reduction (up to $R = 75\,\%$ for $\Delta y_c^+ = 60$) [12–14]. The technique reproduces effects of various near-wall reactive control schemes such as opposition control or suboptimal control [2] and provides more flexibility in terms of tuning and easier implementation than velocity-based control schemes.

The most prominent example of drag reducing flow control is the uniform blowing at the wall of a flat plate boundary layer [4, 15, 16]. The control scheme can be also considered the most realistic one, since it does not utilize any information about the instantaneous flow field and thus can be classified as a predetermined active control technique. The control can be imagined to be implemented in reality by transpiration through a porous wall or by direct suction or blowing through a slot on the wall surface. The wall-normal velocity profile at the wall is given by

$$v\,(x, y = 0, z, t) = V_w \cdot f\,(x)\,, \tag{10}$$

where V_w is the velocity amplitude. Depending on the velocity amplitude and Reynolds number, uniform blowing delivers up to $R = 70\,\%–80\,\%$ [4].

4 Computational Details

The solver is implemented using Fortran and utilizes OpenMP, MPI or hybrid (MPI with OpenMP) parallelization paradigms. The code introduces one-dimensional and two-dimensional domain decomposition for MPI parallelization model.

Smaller simulation configurations with 8.6 and 33.6 Mio. grid nodes (see Table 1) have been used for tests, development and preliminary investigation of the parameter set utilizing 16–32 and 64–128 CPU-cores per job, correspondingly. Main simulations with 236.6 Mio. grid nodes has been carried out with 256 CPU-cores per job, which has been found to be an optimal trade-off between queuing time and simulation run time. One-dimensional and two-dimensional domain decomposition with MPI-parallelization has been utilized in the study. Due to the presence of higher CFL numbers close to the wall for wall-based control application, simulations with opposition control & uniform blowing utilize smaller time steps and hence have to be executed for a longer time period to achieve the same statistical integrational time in comparison to simulations with body force damping. However, since at least three control configuration cases for each control technique had to be tested, it was possible to run several (up to ten) 256-CPU-cores cases simultaneously. Table 2 presents the summary of the computational details for the carried out simulations.

Figures 2 and 3 demonstrate the strong scaling for the main simulation configuration with 236.7 Mio. grid nodes. Due to the dimension of the computational domain and the specifics of utilized parallelization the amount of used CPUs

Table 2 Computational details of the performed simulation. Grey shade highlights the main configuration setup

#	Grid nodes $N \times 10^6$	CPU-cores `procs`	Process memory `pmem`, Mb	Initial field size	Mean time-to-solution, days per case
1	8.6	16,20,32	768	194 Mb	14
2	33.6	64,128	1024	773 Mb	40
3	236.7	256	1536	5.3 Gb	60

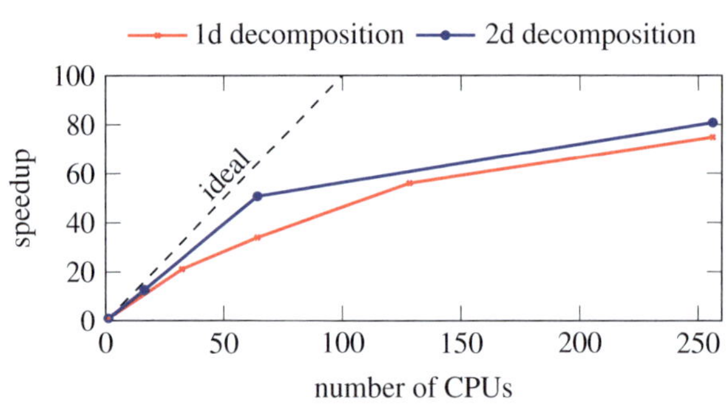

Fig. 2 Speedup of the utilized numerical code for the main simulation configuration on ForHLR I

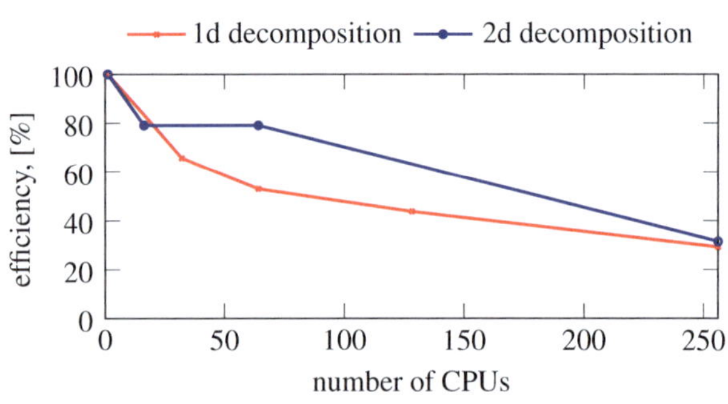

Fig. 3 Efficiency of the utilized numerical code for the main simulation configuration on ForHLR I

cannot exceed 256 for 1d decomposition. It is evident that two-dimensional domain decomposition provides better speedup and efficiency, especially for the mid-range of the considered amount of CPUs. The largest tested amount of CPUs ($n = 256$) yields a similar speedup ($\approx$80) and efficiency ($\approx$30 %) for both decomposition paradigms.

5 Results

The scientific results of the study have been published in [17] and [18]. Therefore, the current section provides only a condensed summary about the simulation results and contains text segments and figures from these publications. For further details please refer to the journal publications.

5.1 Opposition Control in Spatially Developing Turbulent Boundary Layers

Although turbulent boundary layers and turbulent channel flows reveal many similarities in the corresponding flow statistics of near-wall turbulence, some principal differences for these two flows are known to exist even in the uncontrolled state [19]. The present project aims at understanding how opposition flow control designed to reduce skin friction drag acts in both flows and whether fundamental differences of the control mechanism can be identified.

In order to perform a direct comparison between TCF and TBL at a number of different friction Reynolds numbers, five DNS of TCF (each driven by a prescribed flow rate) are carried out. In TBL control is applied partially in the streamwise direction, while the spanwise extension of the control area covers the total domain width. All control areas begin at $x_0 = 186$ corresponding to $\mathrm{Re}_\tau = 188$ as shown in Fig. 1. Three different control areas with a streamwise extension of $\Delta x_c = 100, 150$ and 200 are introduced in TBL. The Reynolds numbers of the TCF are chosen in such a way that the friction based Reynolds numbers for the uncontrolled TCF are within the range found for the uncontrolled TBL. Statistical averaging for TCF and TBL simulations is performed during 100–150 eddy turnover times after the controlled flow reaches an equilibrium state.

Figure 4 shows the distribution of the local drag reduction rate for the three control area lengths along the streamwise coordinate within the turbulent region of the TBL in comparison to TCF results. It can be seen that very similar results in terms of R are obtained for TBL and TCF.

Further insight into the mechanism how this drag reduction rate is generated in this flow is provided through a decomposition of the skin friction coefficient into its contributing parts as originally suggested by Fukagata et al. [21]. Their original

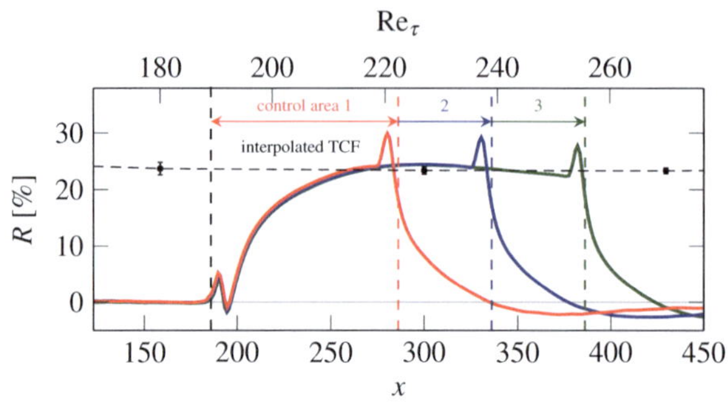

Fig. 4 Comparison of skin friction drag reduction distribution in TBL with interpolated controlled TCF results at $\mathrm{Re}_\tau = 150, 180, 227, 270, 300$. Error bars represent a 3σ-confidence interval for TCF data [20] (The figure is taken from [17]. Reprinted with permission from AIP Publishing LLC)

formulation is modified in such a way that the centerline velocity, $U_{\mathrm{cl}} = \bar{u}(\delta)$, (instead of the bulk velocity) is used as a normalisation factor in TCF, which corresponds to the free-stream velocity in TBL. Accordingly, the skin friction coefficient in TCF is defined by $c_f = \tau_w/0.5\rho U_{\mathrm{cl}}{}^2$. Consequently, the following form of the FIK-identity in TCF for the newly defined c_f can be derived [17]:

$$c_f = \underbrace{\frac{2}{3}\left(-\frac{\partial \bar{p}}{\partial x}\right)}_{\substack{c_f^P \\ \text{pressure development} \\ \text{contribution}}} + \underbrace{\frac{4\,(1-\delta_d)}{\mathrm{Re}_c}}_{\substack{c_f^L \\ \text{laminar} \\ \text{contribution}}} + \underbrace{4\int_0^1 (1-y)\left(-\overline{u'v'}\right) \mathrm{d}y}_{\substack{c_f^T \\ \text{Reynolds shear stress} \\ \text{contribution}}}, \tag{11}$$

where y is normalised with the channel half-height δ and $\mathrm{Re}_c = U_{cl}\delta/\nu$. This division shows that c_f in the TCF consists of the laminar (c_f^L) and turbulent (c_f^T) contributions. In contrast to TCF, the FIK-identity for TBL is given by [17]:

$$c_f(x) = \underbrace{\frac{4\,(1-\delta_d)}{\mathrm{Re}_{\delta_{99}}}}_{\substack{c_f^\delta \\ \text{boundary layer} \\ \text{contribution}}} + \underbrace{4\int_0^1 (1-y)\left(-\overline{u'v'}\right) \mathrm{d}y}_{\substack{c_f^T \\ \text{Reynolds shear stress} \\ \text{contribution}}} + \underbrace{4\int_0^1 (1-y)\left(-\bar{u}\bar{v}\right) \mathrm{d}y}_{\substack{c_f^C \\ \text{mean convection} \\ \text{contribution}}} \tag{12}$$

$$\underbrace{-2\int_0^1 (1-y)^2 \left(\frac{\partial \bar{u}\bar{u}}{\partial x} + \frac{\partial \overline{u'u'}}{\partial x} - \frac{1}{\mathrm{Re}_{\delta_{99}}}\frac{\partial^2 \bar{u}}{\partial x^2} + \frac{\partial \bar{p}}{\partial x}\right) \mathrm{d}y}_{\substack{c_f^D \\ \text{spatial development} \\ \text{contribution}}},$$

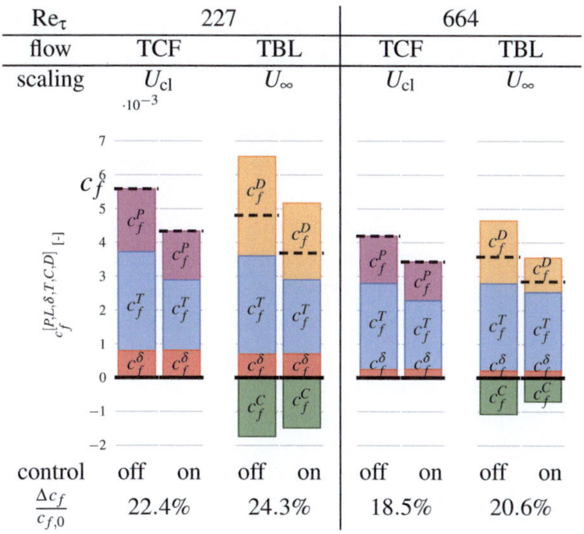

Fig. 5 Comparison of dynamical contributions to c_f in uncontrolled and controlled TCF and TBL at $\mathrm{Re}_\tau = 227$ and $\mathrm{Re}_\tau = 664$. The figure is part of a figure in [17] (Reprinted with permission from AIP Publishing LLC)

where δ_d represents the displacement thickness. In this equation all variables are non-dimensionalised by U_∞ and δ_{99}. The turbulent contribution, c_f^T, is obviously present for the TCF and TBL cases, while the boundary layer contribution, c_f^δ, from TBL can be compared with the laminar contribution, c_f^L, in TCF. For TBL two additional terms, namely c_f^C and c_f^D, are present.

A comparison for opposition control in TCF and TBL is shown in Fig. 5 where the skin friction decomposition for the uncontrolled and controlled flow states are shown at a fixed Reynolds number. For TCF, the reduction of c_f^T is the main control effect. In contrast, for TBL the suppression of the turbulent contribution c_f^T is weaker while changes in the boundary layer specific terms, namely c_f^C and c_f^D, also contribute to changes in skin friction drag. This difference between TCF and TBL becomes more pronounced at higher Reynolds number.

Based on the obtained result, it is expected that the present scenario for drag reduction does not change significantly for a further increase of Reynolds numbers. Meanwhile, the fact that drag reduction in TBL is achieved through the interaction of different dynamic contributions might eventually lead to different drag reduction rates for TCF and TBL.

5.2 *Downstream Behaviour of Locally Controlled Spatially Developing Turbulent Boundary Layers*

Two skin friction drag reducing control schemes with essentially different control mechanisms are investigated in turbulent boundary layers (TBL). While the first control type, uniform blowing, affects the convective contribution to the skin friction coefficient by introduction of additional mass flux, the second type, body force damping, aims at direct reduction of c_f^T. Since all control will end at some point on a surface, we investigate how the boundary layers develop after they have passed the controlled sections and how this flow development influences the global control performance [18].

The control placement corresponds to the previous study control area 3 ($x_0 = 186$, $\Delta x_c = 200$). Equation (10) defines the control input for the uniform blowing with blowing intensity, V_w, set to 0.5 % of U_∞. The reactive scheme of body force damping is based on the definition from Eq. (9) with the forcing time constant Φ fixed to $5/3$ in order to yield a drag reduction similar to the uniform blowing case. The body force is applied up to $y^+ \approx 40$. For both control schemes the control amplitude is increased and decreased smoothly within a spatial extent of $10\delta_0^*$ at the edges within the control area using a hyperbolic tangent function.

Figure 6 shows the influence of the applied control on the turbulent structures of the flow. Due to cancellation of the wall-normal fluctuations in the near-wall region, a strongly pronounced attenuation of turbulent activity can be observed for body force damping. The effect is also visible over a certain area downstream of the control region, where a retransition of the flow occurs. In contrast, the application of uniform blowing rather leads to visible thickening of the TBL due to additional

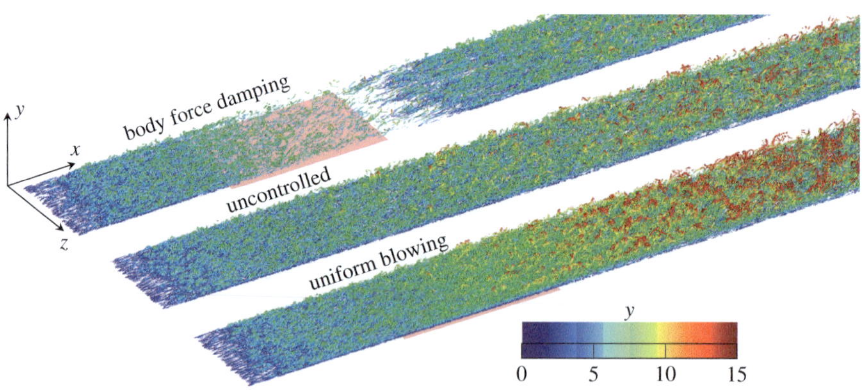

Fig. 6 Flow structure in uncontrolled and controlled cases represented by the isosurfaces of λ_2-criterion ($\lambda_2 = -0.005$) coloured by the wall-normal coordinate. *Red shaded area* at the wall marks the location of the applied control (Figure taken from [18]. Reprinted with permission from Cambridge University Press)

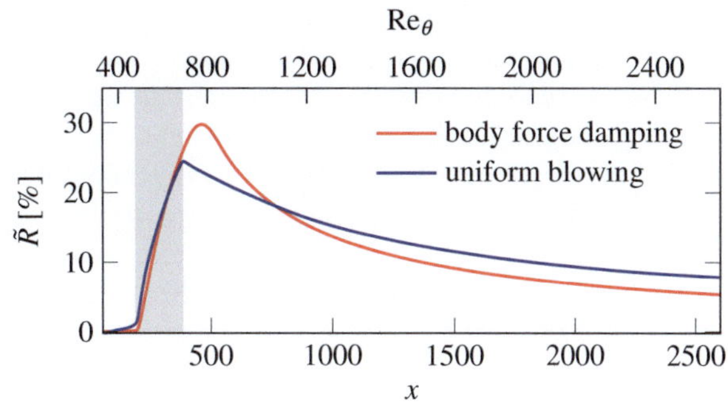

Fig. 7 Streamwise development of integral drag reduction rate. *Shaded area* marks the location of control region (Figure taken from [18]. Reprinted with permission from Cambridge University Press)

wall-normal mass and momentum, which is accompanied by an enhancement of turbulent activity.

Since the aim of the present investigation is to examine the global effect of the introduced control on TBL, and integral drag reduction rate is proposed [18]:

$$\tilde{c}_f(x) = \int_{x_s}^{x} c_f(x)\,\mathrm{d}x, \qquad \tilde{R}(x) = 1 - \frac{\tilde{c}_f(x)}{\tilde{c}_{f,0}(x)}. \tag{13}$$

The two control types are adjusted to yield very similar R in the control region. However, as seen in Fig. 7 they show significant differences downstream of the control section. It can be shown that the resulting R far downstream of the control section can actually be predicted when one quantity of the control is evaluated. This essential quantity is a virtual shift Δx_v that is introduced by the control. One can imagine that the controlled flow eventually returns to a canonical state when the control is no longer present. This state is the same as the one found for an uncontrolled flow at a different location along the plate. Uniform blowing inserts a positive shift, while body force damping leads to a negative shift. Due the this difference the global performance strongly depends on the length of the uncontrolled section after the control. Once Δx_v is identified from the simulation results it can be used to predict R on any longer plate. For details on the estimation methodology please refer to the journal publication [18].

6 Conclusions and Outlook

Application of localized drag reducing control is analyzed in the framework of a spatially developing turbulent boundary layer using direct numerical simulation. It is found that the opposition control scheme yields similar drag reduction rates if compared at the same friction Reynolds numbers to a fully developed turbulent channel flow. However, a detailed analysis of the dynamical contributions to the skin friction coefficient reveals significant differences in the mechanism behind the drag reduction. While drag reduction in turbulent channel flow is entirely based on the attenuation of the Reynolds shear stress, the modification of the spatial flow development is essential for the turbulent boundary layer in terms of achievable drag reduction.

Comparison of a global drag reduction rate between the control designed to damp near-wall turbulence and the control inducing constant mass flux in the wall-normal direction reveal significantly different flow development downstream of the control section. It is shown that the far downstream development of the TBL after the control region can be described by a single quantity, namely a streamwise shift of the uncontrolled boundary layer, i.e. a changed virtual origin. Based on this result, local and global drag reduction rate can be estimated without the need of conducting expensive simulations or measurements far downstream of the control region.

An analysis of spectral properties (spectra of $\overline{u'u'}$ and co-spectra $\overline{u'v'}$) of the flow field and their relation to the reduction of the skin friction coefficient is planned to be performed to further clarify the global control effect.

References

1. Choi, H., Moin, P., Kim, J.: Active turbulence control for drag reduction in wall-bounded flows. J. Fluid Mech. **262**, 75–110, 10 (1994)
2. Lee, C., Kim, J., Choi, H.: Suboptimal control of turbulent channel flow for drag reduction. J. Fluid Mech. **358**, 245–258, 3 (1998)
3. Kasagi, N., Suzuki, Y., Fukagata, K.: Microelectromechanical systems-based feedback control of turbulence for skin friction reduction. Annu. Rev. Fluid Mech. **41**, 231–251 (2009)
4. Kametani, Y., Fukagata, K.: Direct numerical simulation of spatially developing turbulent boundary layers with uniform blowing or suction. J. Fluid Mech. **681**, 154–172 (2011)
5. Lundbladh, A., Berlin, S., Skote, M., Hildings, C., Choi, J., Kim, J., Henningson, D.: An efficient spectral method for simulation of incompressible flow over a flat plate. Technical report (1999)
6. Skote, M.: Studies of turbulent boundary layer flow through direct numerical simulation. PhD thesis, Royal Institute of Technology, Stockholm (2001)
7. Nordström, J., Nordin, N., Henningson, D.: The fringe region technique and the Fourier method used in the direct numerical simulation of spatially evolving viscous flows. SIAM J. Sci. Comput. **20**, 1365–1393 (1999)
8. Chang, Y., Collis, S., Ramakrishnan, S.: Viscous effects in control of near-wall turbulence. Phys. Fluids **14**(11), 4069–4080 (2002)
9. Iwamoto, K., Suzuki, Y., Kasagi, N.: Reynolds number effect on wall turbulence: toward effective feedback control. Int. J. Heat Fluid Flow **23**(5), 678–689 (2002)
10. Pamiès, M., Garnier, E., Merlen, A., Sagaut, P.: Response of a spatially developing turbulent boundary layer to active control strategies in the framework of opposition control. Phys. Fluids **19**(10), 108102 (2007)
11. Satake, S., Kasagi, N.: Turbulence control with wall-adjacent thin layer damping spanwise velocity fluctuations. Int. J. Heat Fluid Flow **17**(3), 343–352 (1996)

12. Lee, C., Kim, J.: Control of the viscous sublayer for drag reduction. Phys. Fluids **14**(7), 2523–2529 (2002)
13. Iwamoto, K., Fukagata, K., Kasagi, N., Suzuki, Y.: Friction drag reduction achievable by near-wall turbulence manipulation at high Reynolds numbers. Phys. Fluids **17**(1), 011702–011702 (2005)
14. Frohnapfel, B., Hasegawa, Y., Kasagi, N.: Friction drag reduction through damping of the near-wall spanwise velocity fluctuation. Int. J. Heat Fluid Flow **31**(3), 434–441 (2010)
15. Park, J., Choi, H.: Effects of uniform blowing or suction from a spanwise slot on a turbulent boundary layer flow. Phys. Fluids **11**(10), 3095–3105 (1999)
16. Kim, K., Sung, H.J., Chung, M.K.: Assessment of local blowing and suction in a turbulent boundary layer. AIAA J. **40**(1), 175–177 (2002)
17. Stroh, A., Frohnapfel, B., Schlatter, P., Hasegawa, Y.: A comparison of opposition control in turbulent boundary layer and turbulent channel flow. Phys. Fluids **27**(7), 075101 (2015)
18. Stroh, A., Hasegawa, Y., Schlatter, P., Frohnapfel, B.: Global effect of local skin friction drag reduction in spatially developing turbulent boundary layer. J. Fluid Mech. **805**, 303–321 (2016)
19. Jiménez, J., Hoyas, S., Simens, M.P., Mizuno, Y.: Turbulent boundary layers and channels at moderate Reynolds numbers. J. Fluid Mech. **657**, 335–360 (2010)
20. Oliver, T.A., Malaya, N., Ulerich, R., Moser, R.D.: Estimating uncertainties in statistics computed from direct numerical simulation. Phys. Fluids **26**(3), 035101 (2014)
21. Fukagata, K., Iwamoto, K., Kasagi, N.: Contribution of Reynolds stress distribution to the skin friction in wall-bounded flows. Phys. Fluids **14**, L73–L76 (2002)

Scalability of OpenFOAM with Large Eddy Simulations and DNS on High-Performance Systems

Gabriel Axtmann and Ulrich Rist

Abstract OpenFOAM (Open Field Operation and Manipulation) is a complete open-source framework for the solution of Partial Differential Equations (PDE) using the Finite Volume Method. It is one of the most popular open source tools used in Continuum Mechanics and Computational Fluid Dynamics (CFD). In this study, we used DirectNumerical Simulation and Large Eddy Simulation to investigate the scalability and MPI characteristics of OpenFOAM. Semi-implicit methods were applied to two representative benchmark problems. Three-dimensional laminar cavity flow, solved by direct numerical simulation, and turbulent backward facing step, solved by LES. The latter problem represents a configuration with common features found in many engineering applications. Strong and weak scaling behaviour using GNU and Intel compiler are compared and MPI routines are traced by CRAY's profiling tools in detail.

1 Introduction

OpenFOAM is a complete open-source framework for numerical simulation in several areas of CFD and engineering. Modern programming techniques in the sense of OOP (Object Oriented Programming) are used to increase flexibility and performance. High modularity is achieved by mimicking the mathematical notation of tensor algebra and PDE Solutions [7]. It consists of many libraries grouped by functionality. Some of them are common to all solvers like mesh manipulation and parallelization. This non-monolithic software approach makes it easy to implement and parallelize own solvers. The parallelization of OpenFOAM is performed by MPI (Message Passing Interface) and is generally linked to OpenMPI. A master-slave configuration is used for this, which leads to non-blocking and blocking send/receive functions on any core. Advantageously, all these bindings are encapsulated in one functional library, which makes optimization easy. However, such flexibility increases the complexity compared to other common MPI software codes. OpenFOAM's parallel behaviour is not well understood when run

G. Axtmann (✉) • U. Rist
Institute of Aerodynamics and Gas Dynamics, Pfaffenwaldring 21, 70569 Stuttgart, Germany
e-mail: axtmann@iag.uni-stuttgart.de

W.E. Nagel et al. (eds.), *High Performance Computing in Science and Engineering '16*, DOI 10.1007/978-3-319-47066-5_28

on massively parallel systems. Therefore several studies were performed in the last years. The CSC IT Center of Science in Finland run benchmark of the cavity test case up to 22 million of cells. They reached nearly super linear scalability up to 1024 CPUs [2]. Duran et al. investigated the scalability for bio-medical flows. Using icoFoam as laminar, incompressible flow solver (DNS), they achieved in their study even super linear behaviour up to 2048 cores [3]. Pringle [5] investigated in his study the scalability of the cavity benachmark from 4 to 4096 cores. The mesh size was increased from 100^3 to 200^3 cells. Super linear behaviour was achieved up to 1024 cores.

In this study, the parallel performance of OpenFOAM has been investigated on the HPC system Cray XC40 Hazel Hen (Stuttgart). Hazel Hen is a massivly parallel computer with 7712 nodes, each with two 12-core Intel Xeon E5-2680 b3 CPU's and 128 GB of memory per node. The interconnect consists of a Cray Aries network with Dragonfly topology. One node consists of two sockets each with 12 cores. Filling these nodes completely is beneficial with respect to both communication and fragmentation of the job queue. Unless otherwise explicitly written, all cases are run with OpenFOAM version 2.3.0 compiled with GNU/Intel compiler.

2 Turbulent Backward Facing Step

2.1 LES Principles and Modelling

The basic equations for LES were first formulated by Smagorinsky [6] in the early 1960s. Since computational resources were severely limited by that time an alternative to resolving all the scales of motion had to be conceived. Based on the theory of Kolmogorov [4], that the smallest scales of motion were uniform and the assumption that these small scales serve mainly to drain energy from the larger scales through a cascade process, it was felt that the small scales could be successfully approximated. The large scales of motion, which contain most of the energy, perform of the transport and are affected the strongest by the boundary conditions, such that they should therefore be calculated directly, while the small scales are represented by a model. This is the basis of LES.

In order to separate the large scales of motion from the small ones, some kind of averaging must be done. In LES, this is locally derived by a weighted average of flow properties over a volume fluid. The filtering process is performed with a filter width Δ. This represents a characteristic length scale. Thus scales, larger than Δ are retained in the filtered flow field, while scales smaller than Δ must be modeled by a Sub-Grid Scale (SGS) model. Formally in LES, any flow variable f is decomposed in larger and small scales via:

$$f = \bar{f} + f', \tag{1}$$

where the prime denotes the small scales and the overbar the larger ones. In order to extract the large-scale components a filter operation is applied:

$$\bar{f}(x) = \oint G(x, x'; \Delta) f(x') dx', \tag{2}$$

where Δ is the filter width proportional to the wavenlength of the smalles scales, retained by the filtering operation $G(x, x'; \Delta)$. The most common filters that have been applied to LES are the Gaussian filter and the top-hat filter. The latter one is the common choice for finite volume methods, because the average is over a grid volume where the flow variables are a piecewise function of x. This implies that the filter width Δ is equal to the grid-spacing.

Next, this filtering process is applied to the Navier-Stokes equations. For incompressible flow they are:

$$\nabla \cdot \overline{u} = 0 \tag{3}$$

$$\frac{\partial \overline{u}}{\partial t} + \nabla \cdot (\overline{uu}) = -\frac{1}{\rho} \nabla \overline{p} + \nabla \cdot \nu (\nabla \overline{u} + \nabla \overline{u}^T) \tag{4}$$

Since $\overline{uu} \neq \overline{u}\,\overline{u}$ a modelling approximation must be introduced, accounting for the difference for the two sides of inequality:

$$\tau = \overline{uu} - \overline{u}\,\overline{u}. \tag{5}$$

In LES, τ is known as the sub-grid scale stress. In the limit for small mesh spacings, where $|\ \tau\ | \rightarrow 0$ as $\Delta \rightarrow 0$, a DNS solution is retained. This is similar to Reynolds stress modelling in RANS. However, the SGS stresses here represent a much smaller part of turbulent energy spectrum than in RANS turbulent energy. Of course, this modelling leads a higher buildup of energy in resolved scales and can produce instabilities. Decomposing the SGS stress results in three separate terms:

$$\tau = \overline{(\overline{u} + u')(\overline{u} + u')} - \overline{u}\,\overline{u} = (\overline{\overline{u}\,\overline{u}} - \overline{u}\,\overline{u}) + (\overline{\overline{u}u'} + \overline{u'\overline{u}}) + \overline{u'u'}, \tag{6}$$

where the first term represents the interaction of resolved eddies (Leonard term), the second term the energy transfer between the resolved and unresolved scales (cross term) and the last term the effect of small eddy interaction (SGS Reynolds stress).

The main role of the SGS model is to extract energy from the resolved scales and model the drain associated with the energy cascade. This can be done with an eddy-viscosity model (similar to the RANS turbulence modeling approach). The normal stresses are taken as isotropic and can be expressed in terms of SGS kinetic energy:

$$\tau - \frac{1}{3} tr(\tau) I = \tau - \frac{2}{3} K I = -\nu_{SGS} (\nabla \overline{u} + \nabla \overline{u}^T) = -2 \nu_{SGS} \overline{S}, \tag{7}$$

where $\overline{S}$ is the strain rate tensor defined as:

$$\overline{S} = \frac{1}{2}(\nabla\overline{u} + \nabla\overline{u}^T). \tag{8}$$

Smagorinsky proposed a first relation for the sub-grid scale eddy-viscosity. He assumed that small scales are in equilibrium and dissipate entirely and instantaneously the energy received from the resolved scales. The formulation of Smagorinsky leads to the following general model:

$$\nu_{SGS} = (C_S\Delta)^2 \mid \overline{S} \mid \tag{9}$$

$$\mid \overline{S} \mid = (\overline{S} : \overline{S})^{0.5} \tag{10}$$

where C_S is the Smagorinsky constant typically chosen between 0.1 and 0.2. For further information we refer to Smagorinsky [8].

2.2 Numerical Setup

The backward facing step testcase is supplied with OpenFOAM (pitzDailys) and is an example of a LES simulation. The solver used is pisoFoam and solves the Poisson equation by using a Pressure Implicit stepping method (PISO). For turbulence modeling, the k-equation eddy viscosity model with a cube root of the cell volume is used as LES filter width Δ. The schematic view of the domain is shown in Fig. 1. Top and bottom walls are set to non-slip conditions, while periodic boundary conditions on the sides are used. The inlet boundary condition contains an artificial noise of 2 % of the velocity. For the outlet condition a pressure-driven type is used. The Reynolds number Re_h is 13333 with respect to the step height h. The Pressure equation is solved by the Pre-conditioned Conjugate Gradient solver (PCG). All other fields with the preconditioned Bi-Conjuage Gradient (PBiCG) solver. In all cases the CFL number is less than 1.0. A brief summary of geometrical dimensions and parameters is given in Table 1. For benchmarking, five meshes with different resolutions were examined. Beginning from one million of hexahedral cells, the size was doubled up to 16 million cells. Within the meshes, the x, y and z-discretization is adapted for higher resolution near the step region. Runs on 1, 2, 4, 9, 18, 36, 27 and 144 nodes were performed, summarized in Table 2. Here, one node consists of 24 cores, thus the number of MPI tasks range up from 24 to 3456. With increasing mesh

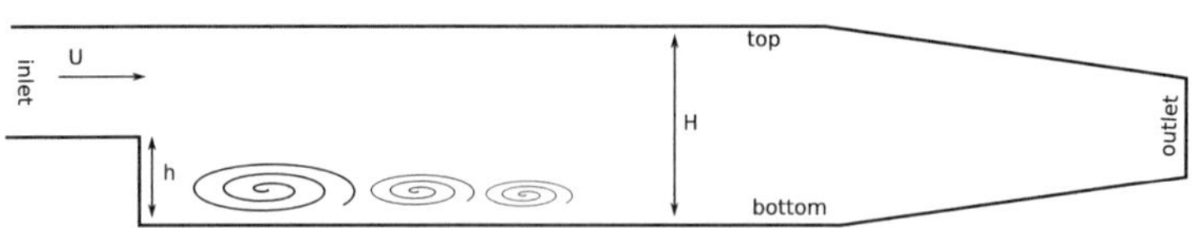

Fig. 1 Schematic Setup backward facing step benchmark

Table 1 Backward facing step dimensions and parameters

Reynolds number Re_h	13333
Kinematic viscosity	1.5e-05 m^2/s
Cube dimensions	$0.3 \times 0.04 \times 0.04$ m
Step height h	0.02 m
Velocity U	10 m/s
Timestep	1e-06 s
Solver for pressure eqn.	PCG/PBiCG
Decomp. method	Scotch

Table 2 Investigated test cases, M: = mesh size in million cells, N = Number of nodes (2 $*$ 12 = 24 CPU)

	1 M	2 M	4 M	8 M	16 M
1N	x	x	x	x	x
2N	x	x	x	x	x
4N	x	x	x	x	x
9N	x	x	x	x	x
18N	x	x	x	x	x
36N	x	x	x	x	x
72N	x	x	x	x	x
144N	x	x	x	x	x

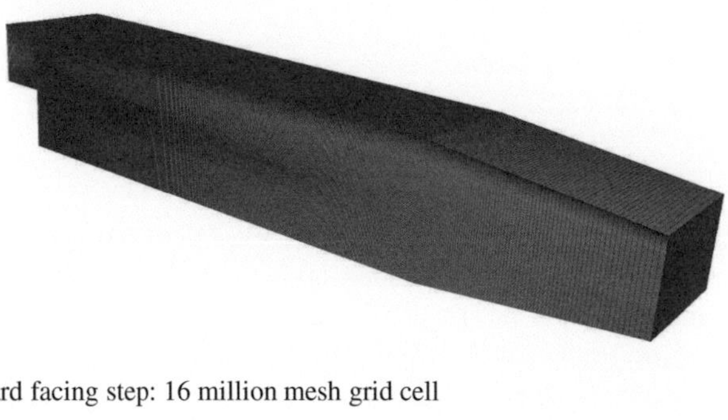

Fig. 2 Backward facing step: 16 million mesh grid cell

resolution, the number of cells and aspect ratios were adjusted to be conform with the smallest mesh. An example for the discretization of the 16 million cell mesh is given in Fig. 2. Strong- and weak scaling studies were performed with these meshes. In addition MPI routines were traced by using CRAY's performance measurement and analysis tool CrayPAT. This is a suite of optional utilities that enable tracing and analyzing performance data [1]. To enable this utility, the user has to compile the code of interest with additional flags. On HazelHen precompiled versions of OpenFOAM v2.3.0 and v2.4.0 are already compiled for profiling via CAE modules. CrayPAT identifies bottlenecks, collects statistics and helps to optimize parallel efficiency. Since the focus of this report is on the scalability of OpenFOAM just a brief visualization of the flow results is given in Fig. 3a in terms of ν_{SGS} and (b) Line Integral Convolution (LIC) visualization of the velocity field U. This shows that OpenFOAM is capable resolving such highly complex flows and features.

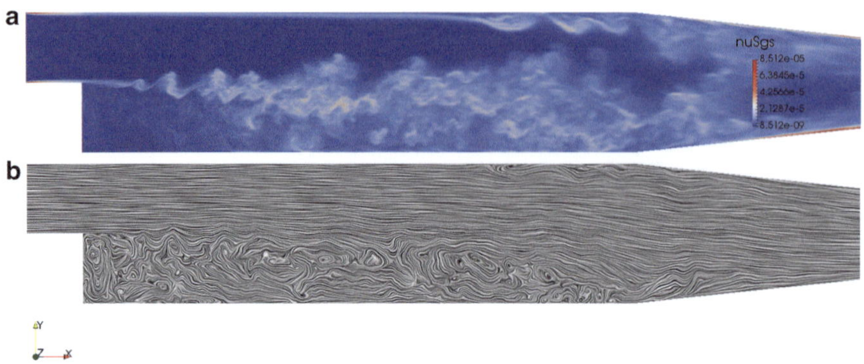

Fig. 3 (**a**) Turbulente kinetic viscosity ν_{SGS} and (**b**) LIC visualization of velocity U at center plane at $t = 0.11$ s

2.3 Performance Results

First, an investigation of strong scaling is performed. Here, the solution time varies with the number of processors for a fixed total problem size. For comparison of data, the speedup is defined as:

$$S_p = \frac{T_1}{(f + \frac{1-f}{p})T_1} \tag{11}$$

where S_p is theoretical speedup, p number of cores, T_1 computing time on a single node, T_p computing time on multiple nodes, and f fraction of serial processes. Ideal speedup is given by $S_p = p$.

A task-parallel program is more efficient than a data-parallel program due to cash effects. Parallel codes can sometimes achieve super-linear behaviour due to efficient cache usage per worker. This behaviour is described by the parallel efficiency:

$$E = \frac{S_p}{p}, \tag{12}$$

where a program that scales linearly has an efficiency of $E = 1$.

From Fig. 4a we can see that the pisoFoam solver compiled with GNU scales up to 864 CPUs super ideal. At this peak we get a maximal cell number distribution for each core of 18.500 cells. With 1728 and 3456 MPI tasks it slows down. This is due to a too small testcase, where the cell numbers of each core is less than 10,000 and the overhead intercommunication increases. Surprisingly, the Intel compiler even performs worse than the one with the GNU compiler. Here, the maximum scaling

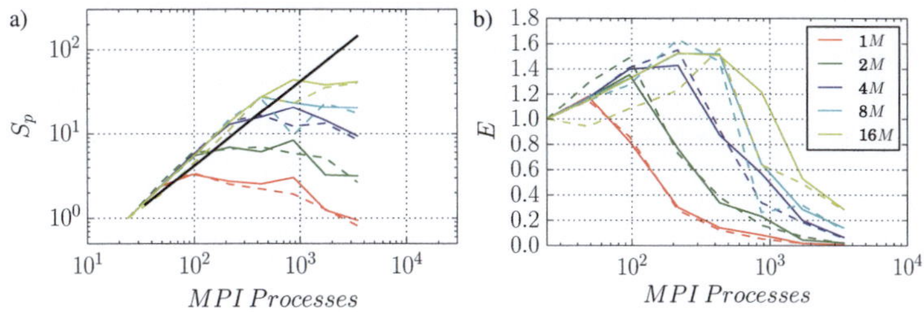

Fig. 4 (**a**) Strong scaling speedup S_p and (**b**) parallel efficiency E for backward facing step benchmark: GNU compiler, - - Intel compiler

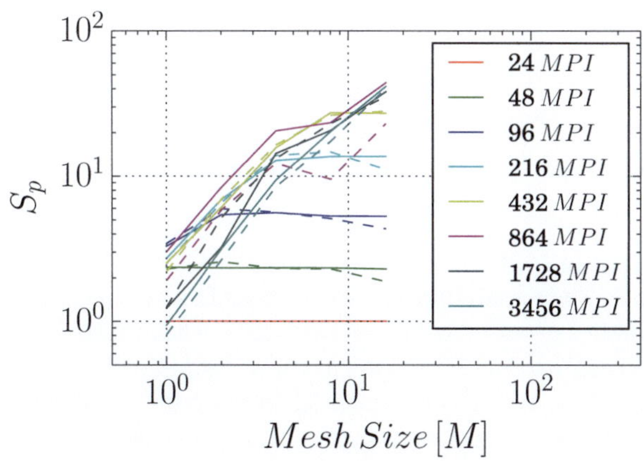

Fig. 5 Weak scaling speedup S_p for backward facing step benchmark: GNU compiler, - - Intel compiler

performance is achieved at 432 MPI tasks. Only between 24 and 432 MPI tasks, a small performance increase compared to the code compiled with GNU compiler is observed. Cache effects are for both, GNU and Intel compiler present, as shown in Fig. 4b. Between 24 and 432 MPI tasks, they ramp up until 1.5 for GNU and even 1.6 for Intel compiler. With increasing MPI tasks, these effects get insignificant. Next, we show the results of scaling study in terms of weak scaling. The advantage of weak scaling is that this reveals problems which are not related with load imbalance due to small domains. It shows how the solution time varies with the number of MPI tasks for a fixed problem size per processor. The comparison between GNU and Intel compiler is shown in Fig. 5. From 24 to 216 MPI tasks a decrease in scalability with respect to increasing mesh size is observed, while from 432 to 3456 MPI tasks GNU and Intel compiler scale reasonably well. Comparing GNU and Intel with each other, a small benefit of Intel compiler in the range lower than 432 MPI tasks is observed. For higher MPI tasks, the GNU compiler scales better.

Table 3 Time spent in MPI ALL, ETC, USER and IO routines in relation to total time for backward facing step benchmark GNU compiler, M: = mesh size in million cells, N = Number of nodes (2 × 12 = 24 CPU)

Case mesh/nodes	MPI ALL[%]	ETC [%]	USER [%]	IO [%]
1M 2N	16.3	2.5	80.4	0.8
1M 36N	60.7	28.5	7.6	3.2
1M 72N	91.8	1.3	4.0	2.9
1M 144N	94.5	1.2	1.9	2.4
8M 2N	6.9	0.0	92.5	0.6
8M 36N	47.8	1.8	48.8	1.6
8M 72N	77.9	1.0	19.6	1.5
8M 144N	90.7	0.0	7.1	2.2
16M 2N	6.3	93.2	0.0	0.5
16M 36N	19.0	19.6	61.2	0.2
16M 72N	61.4	0.0	37.4	1.2
16M 144N	70.7	0.0	27.3	2

The time spent in MPI ALL, ETC, USER and IO in [%] for GNU compiler is given in Table 3. The parallel processing MPI ALL is maximal for the 144 nodes benchmarks and lies in between 70.7 % and 94.5 %. The averaged IO produced over all test cases is calculated at 1.6 % and quite low. Additionally, imbalance sampling rates of several MPI routines were measured. In Fig. 6 the imbalance sampling rates for three different mesh sizes over 2, 36, 72 and 144 nodes for Intel and GNU compiler are shown. For the 1M mesh, most overhead is produced by MPI Isend and MPI Recv with 53 % and 49 % for both compilers. Furthermore, with higher node numbers, the calls of MPI Waitall increases rapidly and generates overhead up to 45 % for Intel and 48 % for GNU compiler.

By increasing the mesh size up to 16 million cells, the imbalance of all MPI routines is decreasing to 42 % maximal. This is caused by better intercommunication between the different subdomains. Most significant overheads are again observed for MPI Isend, MPI Recv and MPI Waitall. Comparing these results to the different compilers for the 16 million cells mesh higher imbalance rates are observed by Intel compiler. Again most overhead is produced in the MPI Isend and MPI Recv and MPI Waitall routines. Here, higher performance rates of Intel compiler do not show up for increasing node numbers. For example for the mesh size 16M calculated with 144 nodes, the imbalance of the MPI Waitall routine by GNU compiler is 30 %, while the imbalance using the Intel compiler results in 42 %. This implies a big improvement in scalability of the open-source GNU compiler in the last years.

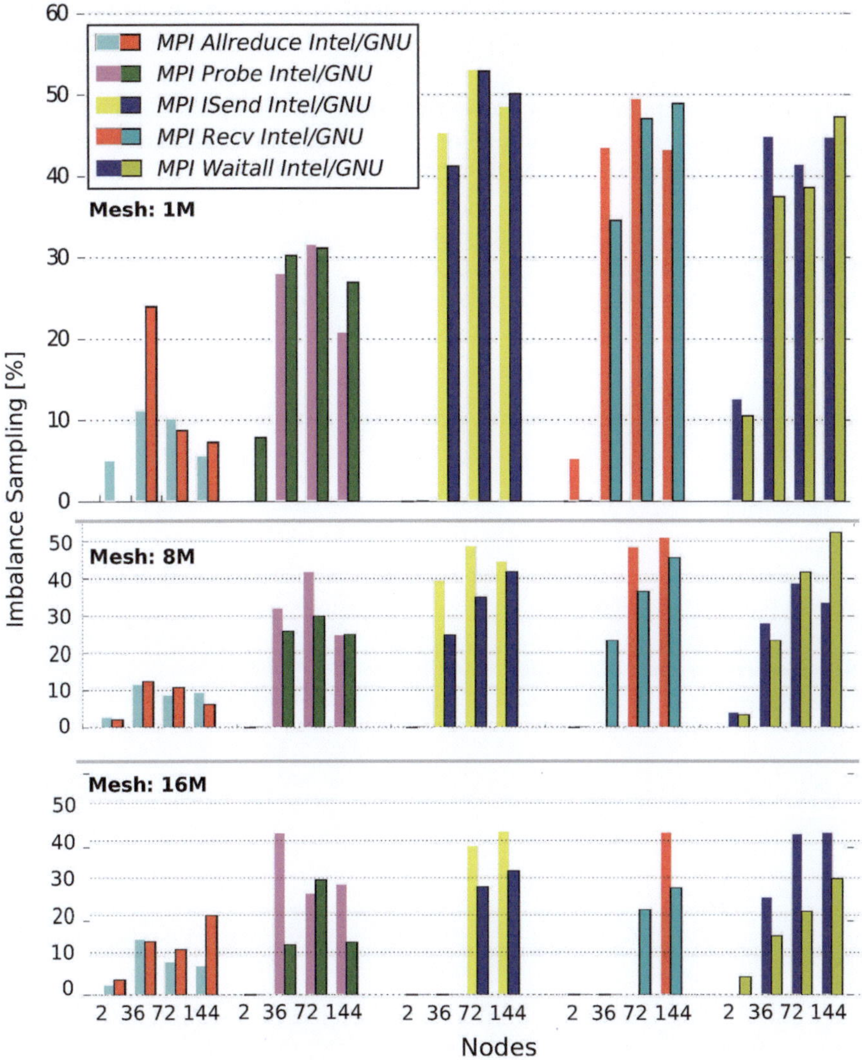

Fig. 6 MPI routines measured by CrayPAT for Intel and GNU compiler for 1M, 8M and 16M mesh size over 2, 36, 72 and 144 nodes

3 Laminar Lid-Driven Cavity Flow

In addition to the study above, a second benchmark using a solver, which solves the Navier-Stokes-Equations directly, is investigated. Therefore, the classical cavity tutorial supplied with OpenFOAM is extended from two to three dimensions and used as a benchmark. The front and back patches are converted to walls, such that the domain is a cube with five steady and one moving wall. For further information the reader is referred to the OpenFOAM Documentation [8]. The Reynolds number has been increased from 10 to 1000. Since the flow is laminar and incompressible, icoFoam is used as solver. Here, the performance study was only done with GNU compiler. Some important parameters of the simulation are given in Table 4. The investigated cases (mesh sizes in million is indicated by suffix M and the number of nodes is indicated by suffix N, the number of MPI Processes is N times 24 are presented in Table 4.

From Fig. 7a we can see that the icoFoam solver compiled with GNU scales well up to 3456 MPI tasks. Super ideal scaling is observed up to 1728 tasks. Since the solver does not include any turbulence modeling and solves the Navier-Stokes-Equation directly, less overhead in comparison to the LES benchmark is produced.

Table 4 Lid-driven cavity setup parameters and test case matrix

Reynolds number	1000
Kinematic viscosity	1e-04 m^2/s
Cube dimensions	0.1 × 0.1 × 0.1 m
Velocity U	1 m/s
Timestep	1e-04 s
Solver for pressure eqn.	PCG w/DIC
Decomp. method	Simple

Case	1M	3.4M	8M	15.6M	27M
1N	x	x	x	x	x
2N	x	x	x	x	x
4N	x	x	x	x	x
9N	x	x	x	x	x
18N	x	x	x	x	x
27N	x	x	x	x	x
36N	x	x	x	x	x
72N	x	x	x	x	x
144N	x	x	x	x	x

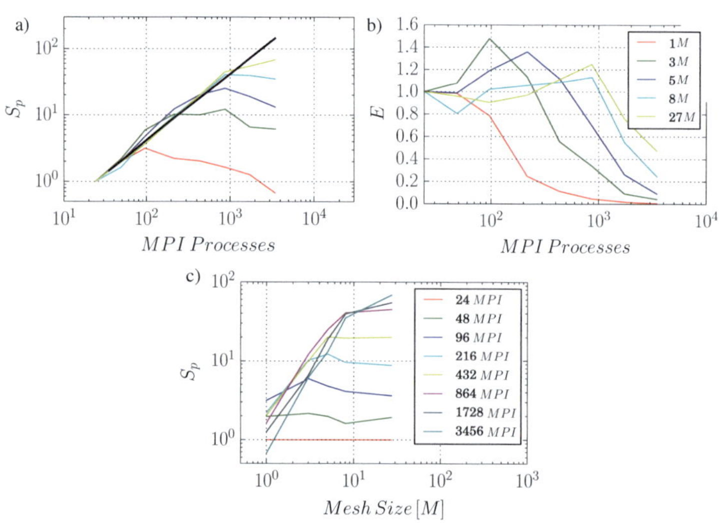

Fig. 7 (**a**) Strong scaling speedup S_p, (**b**) parallel efficiency E and (**c**) weak scaling speedup for lid-driven cavity benchmark: GNU compiler

At the performance peak for the 27 million cells mesh with 1728 MPI tasks, we get a maximal cell number distribution for each core around 15,000 cells. With more than 1728 tasks, the performance decreases. Again, this is due to a too small testcase, where the cell numbers on each core drops below than 10,000 and the overhead intercommunication increases. Cache effects are present, as shown in Fig. 7b. Between 24 and 100 MPI tasks, they reach until 1.5. With increasing MPI tasks, these effects decrease. The weak scaling is shown in Fig. 7c. Almost linear scaling for the one million cells mesh is observed for 96 MPI tasks. By increasing mesh size, there is a strong dependency between cell number distribution for each core and MPI tasks. Best results are given for 864, 1728 and even 3456 MPI tasks for higher mesh sizes.

4 Conclusion

Understanding the scalability of OpenFOAM is still a challenging task. The internal structure and flexibility offered by OpenFOAM, comes with high complexity of parallel behaviour. We have to point out that the main issues are due to the MPI ISend MPI Recv and MPI Waitall routines. Further research is needed on this. Regarding IO the performance is not a real issue during execution time. The average IO amounts to 1.6 %. Comparing GNU and Intel compiler to each other, the GNU compiler performs for higher mesh sizes better than the Intel compiler. Regarding scalability we observed super ideal scaling up to 1000 MPI tasks. This is comparable to other commercial finite volume solvers. For the lid-driven cavity benchmark, an overall scaling up to 3456 MPI tasks can be seen. It is important to note that the performance strongly depends on the cell number on each core. As a rule of thumb, OpenFOAM performs best if the number of cells on each core is between 15,000 and 20,000 cells. Lastly, we have to emphasize that all these results are only possible by doing profiling and tracing. CrayPAT is an amazing tool, which produces here the required information.

Acknowledgements We greatly acknowledge the provision of supercomputing time and technical support by the High Performance Computing Center Stuttgart (HLRS).

References

1. Cray Research Inc.: Optimizing applications on the Cray X1 system. URL http://docs.cray.com/books/S-2315-52/html-S-2315-52/index.html (2002)
2. CSC IT Center: OpenFOAM -CSC (2010). https://research.csc.fi/-/openfoam
3. Duran, A., Celebi, M.S., Piskin, S., Tuncel, M.: Scalability of OpenFOAM for bio-medical flow simulations. J. Supercomput. **71**(3), 938–951 (2015). doi:10.1007/s11227-014-1344-1, http://dx.doi.org/10.1007/s11227-014-1344-1

4. Pope, S.B.: Turbulent flows (2000). doi:10.1088/0957-0233/12/11/705, https://books.google.com/books?hl=fr&lr=&id=HZsTw9SMx-0C&pgis=1$\delimiter"026E30F$nhttp://www.mendeley.com/catalog/turbulent-flows-19/, arXiv:1011.1669v3
5. Pringle, G.J.: Porting OpenFOAM to HECToR. A dCSE project (2010). http://www.hector.ac.uk/cse/distributedcse/reports/openfoam/openfoam.pdf
6. Smagorinsky, J.: General circulation experiments with the primitive equations. Mon. Weather Rev. **91**(3), 99–164 (1963). doi:10.1175/1520-0493(1963)091<0099:GCEWTP>2.3.CO;2, http://journals.ametsoc.org/doi/abs/10.1175/1520-0493%281963%29091%3C0099%3AGCEWTP%3E2.3.CO%3B2
7. Tvergaard, V., Hutchinson, J.W.: Two mechanisms of ductile fracture: void by void growth versus multiple void interaction. PhD thesis (2002). doi:10.1016/S0020-7683(02)00168-3
8. Weller, H.G., Tabor, G., Jasak, H., Fureby, C.: A tensorial approach to computational continuum mechanics using object-oriented techniques. Comput. Phys. **12**(6), 620–631 (1998). doi:10.1063/1.168744

Numerical Simulation of Subsonic and Supersonic Impinging Jets II

Robert Wilke and Jörn Sesterhenn

Abstract This report covers two aspects of impinging jets: heat transfer enhancement and sound source mechanisms. Recent experimental investigations indicate a possible increase of up to 40 % of heat transfer efficiency due to a pulsation of the inlet. However, the underlying physical effects are still unclear. Performing direct numerical simulations, we were able to compute the eigenfrequencies of the impinging jet. Our hypothesis is that pulsating with that frequency leads to a maximal increase of ring vortices and consequently of the heat transfer at the impinging plate. First results of a pulsed impinging jet are shown. In addition, impinging compressible jets may cause deafness and material fatigue due to immensely loud tonal noise. It is generally accepted that a feedback mechanism is responsible for impinging tones. However, it is being discussed which mechanism creates those strong pressure waves. Using direct numerical simulations we were able to identify the source mechanism for under-expanded impinging jets with a nozzle pressure ratio of 2.15 and a plate distance of 5 diameters. We found two different types of interactions between vortices and shocks to be responsible for the generation of the impinging tones.

Keywords Direct numerical simulation • Impinging jet • Heat transfer • Pulsed • Computational aeroacoustics • Impinging tones

1 Introduction

Within this report, our in-house DNS code is used for the investigation of two topics concerning impinging jets. For this reason, the code is described first (Sect. 2). Afterwards, there is one section in which we look at a possible increase of the efficiency due to pulsation (Sect. 3) and another one addressing the sound source mechanism of impinging tones (Sect. 4). Each section has its own introduction and conclusion.

R. Wilke (✉) • J. Sesterhenn
Technische Universität Berlin, Fachgebiet Numerische Fluiddynamik, Müller-Breslau-Str. 12, 10623 Berlin, Germany
e-mail: robert.wilke@tnt.tu-berlin.de; joern.sesterhenn@tu-berlin.de

W.E. Nagel et al. (eds.), *High Performance Computing in Science and Engineering '16*, DOI 10.1007/978-3-319-47066-5_29

2 Numerical Method and Code Performance

The governing Navier-Stokes equations are formulated in a characteristic pressure-velocity-entropy-formulation, as described by Sesterhenn [14] and solved directly numerically. This formulation has advantages in the fields of boundary conditions, parallelization and space discretisation. No turbulence modelling is required since the smallest scales of turbulent motion are resolved. The spatial discretisation uses 6th order compact central schemes for the diffusive terms and compact 5th order upwind finite differences for the convective terms. To advance in time a 4th order Runge-Kutta scheme is applied. In order to avoid Gibbs oscillations in the vicinity of the standoff shock an adaptive shock-capturing filter developed by Bogey et al. [1] that automatically detects shocks is used.

The impinging jet with a Reynolds number of 8000 has to be resolved with more than one billion grid points in order to achieve an adequate spacial resolution of the Kolmogorov scales. Storing one time step with the necessary five variables (pressure, velocity (x, y, z) and entropy) requires 41 GB of storage. It is not possible today to store thousands of time steps so as to do statistical analysis as post-processing. Therefore we compute statistical variables e.g. mean values, variances or complicated budget terms on-the-fly. Applying this strategy, the required storage is reduced to a fraction.

Investigating physics by means of direct numerical simulation requires huge computing capacity, which can only be provided by the most powerful high performance computers that are available nowadays. The Kolomgorov scales that need to be resolved lead at high Reynolds numbers to capacities of multiple million core hours per computation. The load is partitioned between a huge number of processes, e.g. 8192 or 16,384. Each process solves the Navier-Stokes equations for a fractional part of the computational domain (block). This approach is referred to as domain decomposition, see [6]. In order to calculate derivatives, information from the adjacent blocks are needed. Therefore the decomposed domain is rearranged so that each process receives grid lines that span the entire domain in the particular direction. The total number of grid points per process remains constant and is typically between 32^3 and 64^3. Figure 1 exemplary shows the transformation

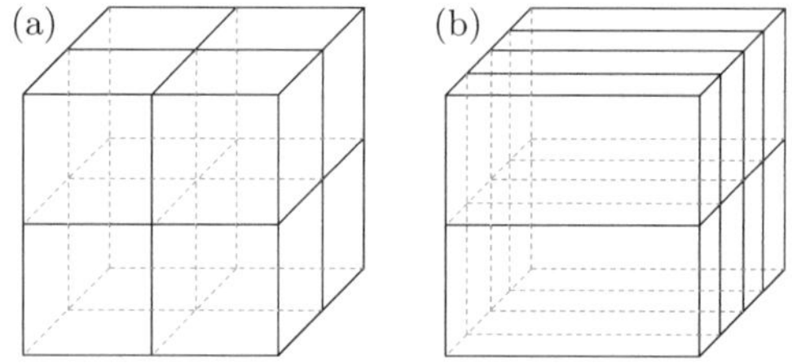

Fig. 1 Domain decomposition of a three-dimensional domain. (**a**) Original decomposition. (**b**) Transformated decomposition for the calculation of derivatives in x-direction

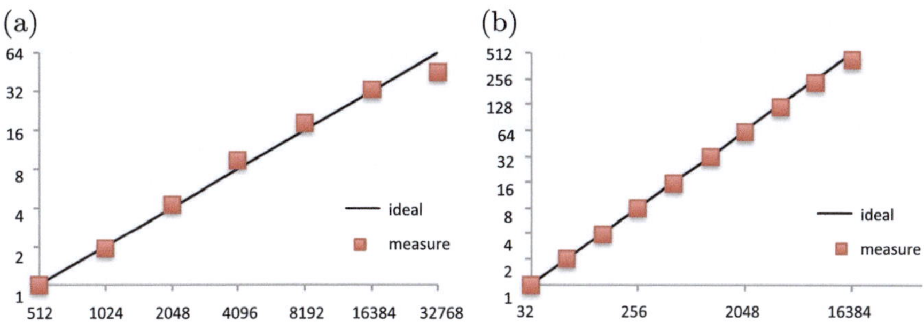

Fig. 2 Strong and weak scaling of the code on CRAY XC40 (Hazelhen). (**a**) Strong scaling; simulations run with 1024^3 grid points. (**b**) Weak scaling; simulations run with 64^3 grid points per core

Table 1 Scaling of the code on CRAY XC40 (Hazelhen). Upper part: strong scaling, simulations run with $n = 1024^3$. Lower part: weak scaling, simulations run with $n/i = 64^3$. n and i denote the total number of grid points respectively the number of used cores

i	$(n/i)^{1/3}$	Wall time per time step [s]	Speedup	Ideal speedup	Efficiency
512	128	166	1.00	1	1.00
1024	102	85.6	1.94	2	0.97
2048	81	39.2	4.23	4	1.06
4096	64	17.9	9.28	8	1.16
8192	51	9.1	18.3	16	1.14
16,384	40	5.1	32.6	32	1.02
32,768	32	3.7	44.9	64	0.70
i	n	Wall time per time step [s]	Speedup	Ideal speedup	Efficiency
32	8.4×10^6	16.7	1.00	1	1.00
64	1.7×10^7	16.6	2.02	2	1.01
128	3.4×10^7	17.0	3.93	4	0.98
256	6.7×10^7	17.1	7.84	8	0.98
512	1.3×10^8	17.1	15.7	16	0.98
1024	2.7×10^8	17.0	31.4	32	0.98
2048	5.4×10^8	17.2	62.1	64	0.97
4096	1.1×10^9	17.9	120	128	0.94
8192	2.1×10^9	18.5	231	264	0.90
16,384	4.3×10^9	20.1	425	512	0.83

from the original decomposition to the decomposition used for the calculation of derivatives in x-direction. The required inter-process communication is managed via MPI libaries.

The code is successfully used on CRAY XC40 (Hazelhen). Figure 2 shows nearly perfect linear scaling up to 16384 cores on that machine. Detailed run times can be found in Table 1. The scaling was made for the case of an impinging jet. Using auto-vectorisation, the efficiency with 16,384 cores is 102 % (strong) respectively 83 %

(weak). Grids with 512^3 (1024^3) points are typically parallelized on $16^3 = 4096$ ($32 \times 16 \times 16 = 8192$ or $32 \times 32 \times 16 = 16{,}384$) cores. The preferred wall time interval is 24 h.

The computational domain has the size of $12 \times 5 \times 12$ diameters. The cuboid is delimited by four non-reflecting boundary conditions: one isothermal wall which is the impinging plate and one boundary consisting of an isothermal wall and the inlet. The walls are fully acoustically reflective. The location of the nozzle is defined using a hyperbolic tangent profile with a disturbed thin laminar annular shear layer as described in [16].

A sponge region is applied for the outlet area $r/D > 5$ that smoothly forces the values of pressure, velocity and entropy to reference values. This destroys vortices before leaving the computational domain. The reference values at the outlet were obtained by a preliminary large eddy simulation of a greater domain.

The grid is refined in the wall-adjacent regions in order to ascertain a maximum value of the dimensionless wall distance y^+ of the closest grid point to the wall not larger than one for both plates. For the wall-parallel-directions a slight symmetrical grid stretching is applied, which refines the shear layer of the jet. The refinements use hyperbolic tangent respectively hyperbolic sin functions resulting in a change of the mesh spacing lower than 1 % for all directions and cases. The physical and geometrical parameters of the simulations are given in Tables 2 and 3.

Table 2 Geometrical and physical parameters of the simulations. $p_0, p_\infty, T_0, T_\infty, T_W, Re, Pr, \kappa, R$ denote total- and ambient pressure, total-, ambient and wall temperature, Reynolds number, Prandtl number, ratio of specific heats and the specific gas constant. Ma, $(\rho, v\rho \cdot v)_\infty$ are the theoretical values of the Mach number, density, axial velocity and specific mass flow computed from T_0, p_∞ and p_0. All values refer to the time span of the open valve

N°	*Re*	η_∞ [Kg m^{-1} s^{-1}]	p_0/p_∞	*Ma*	ρ_∞ [Kg]	v_∞ [m s^{-1}]	$\rho_\infty \cdot v_\infty$ [Kg m s^{-1}]
#1	8000	0.0423	1.5000	0.7837	1.3346	253.82	338.74
#2	3300	0.1026	1.5000	0.7837	1.3346	253.82	338.74
#3	3300	0.0513	1.1217	0.4084	1.2282	137.90	169.37
#4	6600	0.0513	1.5000	0.7837	1.3346	253.82	338.74
N°	Type	Pulsating	Grid points	Max. y^+ [*D*]	Grid width x,z [*D*]		Grid width y
#1	DNS	No	1024^3	0.58	0.0099 .. 0.0296		0.0009 .. 0.0078
#2	DNS	No	512^3	0.63	0.0165 .. 0.0388		0.0017 .. 0.0159
#3	DNS	No	512^3	0.62	0.0184 .. 0.0636		0.0017 .. 0.0159
#4	DNS	Yes	1024^3	*	0.0093 .. 0.0307		0.0009 .. 0.0078
N°	p_∞ [Pa]	T_o [K]	$T_\infty = T_W$ [K]	*Pr*	κ	*R* [J Kg^{-1} K^{-1}]	Domain size [D]
All	10^5	293.15	373.15	0.71	1.4	287	$12 \times 5 \times 12$

*No value available (computation is still running)

Table 3 Geometrical and physical parameters of the simulations. $p_0, p_\infty, T_0, T_\infty, T_W, Re, Pr, \kappa, R$ denote total- and ambient pressure, total-, ambient and wall temperature, Reynolds number, Prandtl number, ratio of specific heats and the specific gas constant. M_j is the fully expanded jet Mach number, computed from T_0, p_∞ and p_0

Re	$T_\infty = T_W$		Grid points [D]		Max. y^+ [D]	Grid width x,z	Grid width y
8000	293.15		1024^3		1.02	0.0099 .. 0.0296	0.0012 .. 0.0072
p_∞ [Pa]	T_o [K]	p_0/p_∞	M_j	Pr	κ	R [J Kg^{-1} K^{-1}]	Domain size [D]
10^5	293.15	2.15	1.1056	0.71	1.4	287	$12 \times 5 \times 12$

3 Heat Transfer

3.1 Introduction

An effective cooling of turbine components is necessary for the success of new engine and combustion concepts, e.g. pulsed combustion, which is studied within the Collaborative Research Centre 1029. Therefore efficient cooling mechanisms have to be developed and optimized. A promising approach is the use of pulsating impinging jets.

Impinging jets have been studied for decades. General information including schematic illustrations of the flow fields as well as distributions of local Nusselt numbers for plenty of different geometrical configurations and Reynolds numbers Re can be found in several reviews, such as [15] based on experimental and numerical results. Since experiments cannot provide all quantities of the entire flow domain spatially and temporally well resolved, the understanding of the turbulent flow field requires simulations. Most existing publications of numerical nature use either turbulence modelling for the closure of the Reynolds-averaged Navier-Stokes (RANS) equations, e.g. [21], or large eddy simulation (LES), e.g. [3]. Almost all available direct numerical simulations (DNS) are either two-dimensional, e.g. [2], or do not exhibit an appropriate spatial resolution in the three-dimensional case, e.g. [7]. Recent investigations come from Dairay et al. [4]. He conducted a DNS of a round impinging jet with a nozzle to plate distance of $h/D = 2$ and focused on the secondary maximum of the heat transfer distribution and the connection to elongated structures.

Janetzke [11] investigated impinging jets with pulsating inlets experimentally. He found that pulsating with a Strouhal number of around 0.9 at maximal amplitude (on/off) it is possible to increase the heat transfer compared to a non-pulsating jet of 40 %, see Fig. 3. The reason for this behaviour remained unclear.

The aim of this project is to clarify the underlying physics behind the increase of the heat transfer efficiency. Therefore a pulsating inlet was applied to the impinging jet. The frequency was chosen based on the results of the simulations using a stationary inlet. The mass flow was kept constant. In this report, we present the key results obtained from the non-pulsed jets and first results of the pulsed jet.

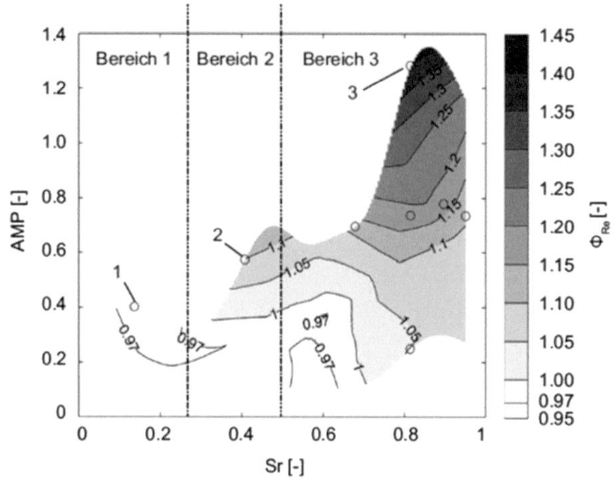

Fig. 3 Increase of heat transfer effectivity Φ_{Re} of an pulsating impinging jet related to a stationary one, depending on the Strouhal number *Sr* and amplitude *AMP* (Modified from [11])

3.2 *Results*

3.2.1 Non-pulsed Impinging Jet

In this project, three direct numerical simulations of subsonic non-pulsed impinging jets were performed. All concern a nozzle-to-plate distance of $h/D = 5$. Two different Reynolds numbers (3300, 8000) were investigated. For each Reynolds number a simulation with a Mach number of $Ma = 0.7837$ was carried out. Additional, $Ma = 0.4084$ was performed for the smaller Reynolds number. In this section, a brief summary of the results motivating the approach of the pulsating impinging jet is given. Parts of the results have already been published in: [16–18].

The flow characteristics described in the following apply to all three simulations. The heat transfer at the impinging plate is strongly related to the vortical structures of the turbulent flow field. Figure 4 shows the life cycle of the vortex rings on a x–y plane through the center of the jet. The temperature is shown in the background, where blue indicates cold and red hot fluid. The Nusselt number is shown on the x–z plane at the wall. Black represents high positive heat transfer (cooling of the wall) and white no heat transfer. In the shear layer of the jet (primary) ring vortices develop and grow until they collide with the wall and then stretch and move in radial direction. As soon as the primary toroidal vortex passes the deceleration area of the wall jet (a) the flow separates and forms a new secondary counter-rotating ring vortex that enhances the local heat transfer, directly followed by an annular area of poor heat transfer due to separation. Travelling downstream the vortex pair increases in strength and ability of heat transfer (b). Moving on the pair separates (c,d) and dissipates. The cycle restarts.

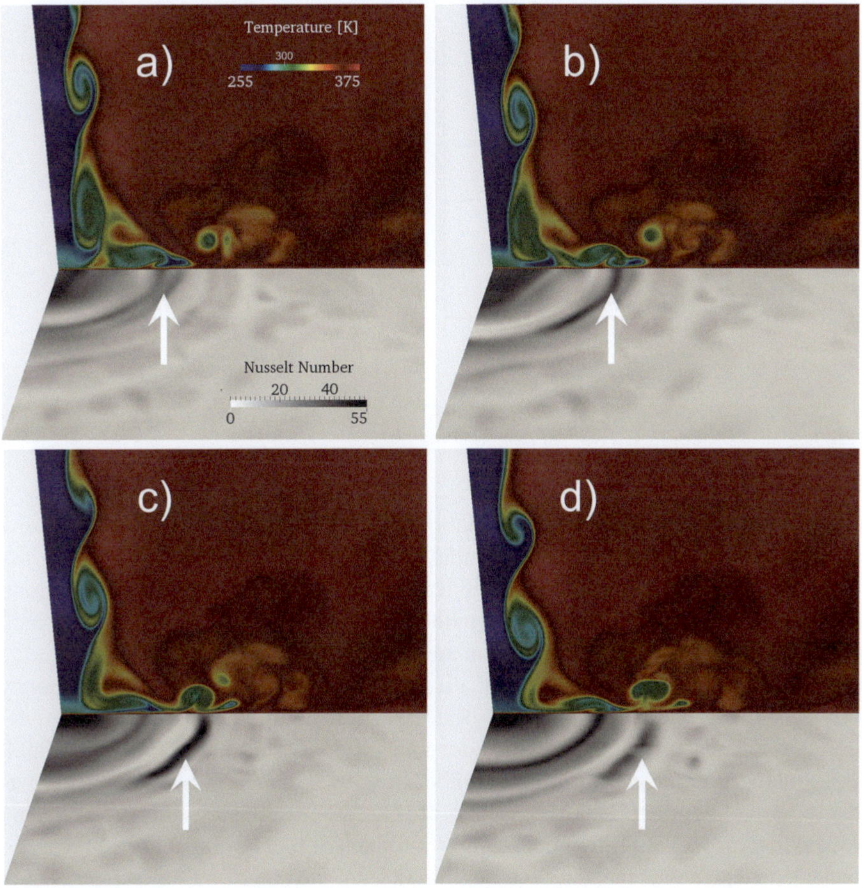

Fig. 4 Life cycle of the secondary vortex ring and connection to the local heat transfer. Simulation #2

The phenomenon leads to high fluctuations of the temperature and the axial velocity and consequently to high values of the turbulent heat flux at radii of $r/D = 1..1.6$, as shown in Fig. 5a. As a consequence, the decreasing trend of the Nusselt number with increasing radius is strongly weakened in this area (Fig. 5b). This means that the ring vortices participate positively to the heat transfer. As a consequence, we aim to maximally increase the ring vortices by applying a pulsation at the inlet. We expect a strong positive influence to heat transfer at the impinging plate.

Only DNS is able to correctly predict the effect of the vortex pairs on the Nusselt number. Dairay et al. [5] compared large eddy simulations using different subgrid scale models with DNS. They observed that non of the tested models was able to clearly predict the secondary peak in the Nusselt number distribution, as measured and computed with DNS. This investigation is the only one in literature, where

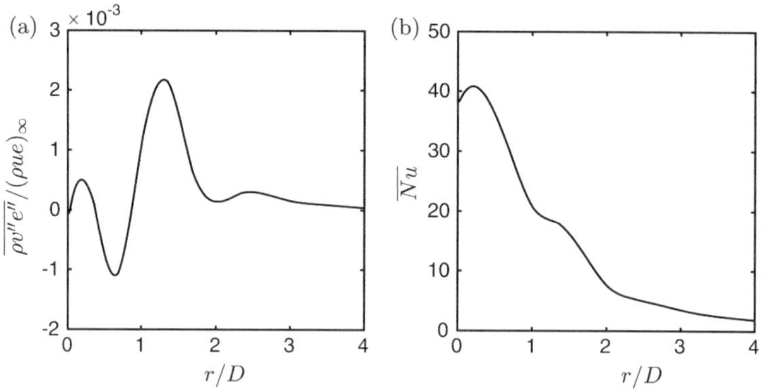

Fig. 5 Time averaged local heat transfer of the DNS #3, $Re = 3300$, $Ma = 0.408$. (**a**) Turbulent heat flux at $y/D = 0.05$. (**b**) Nusselt number

large eddy- and direct numerical simulations are directly compared for an impinging jet. Given that well-resolved LES computations fail for heat transfer prediction, the usage of RANS cannot be recommended for the given case as long as common models are not adapted. Our DNS provide a database for the improvement of such models. More quantities for validation are given in [20].

Conducting direct numerical simulations, we were able to identify the frequency of the vortical system. Therefore we performed a FFT of the Nusselt number at the impinging plate and a dynamic mode decomposition (DMD) of the entire flow field. Both methods revealed a Strouhal number of 0.46 and its first harmonic (0.92) as the important frequencies. This numbers are based on simulation #1 and #2. We can record that the Reynolds number has no significant influence on the phenomena and the eigenfrequency of the subsonic impinging jet in the range of $3300 \leq Re \leq 8000$. This allows us to proceed with the lower Reynolds number for further investigations.

Comparing simulations #2 and #3, we see an influence of the Mach number. In the case of $Ma = 0.78$ the dynamic mode decomposition reveals clearly only one dominant frequency: $Sr = 0.46$ and its first harmonic. At lower speed ($Ma = 0.41$) this frequency remains, but an additional dominant frequency appears: $Sr = 0.59$. According to the coefficients of the DMD, the mode with $Sr = 0.59$ is even more relevant and was therefore used for the pulsating impinging jet, described in Sect. 3.2.2.

3.2.2 Pulsed Impinging Jet

In order to reach a maximal amplification of the ring vortices, the pulsation amplitude used is 100 % (on/off). The profile in time is a smoothed (hyperbolic tangent) rectangular function, which approximates an opening and closing of a valve. The Reynolds number (defined at the nozzle inlet) periodically fluctuates between zero and 6600, so that the average remains constant at 3300. Compared

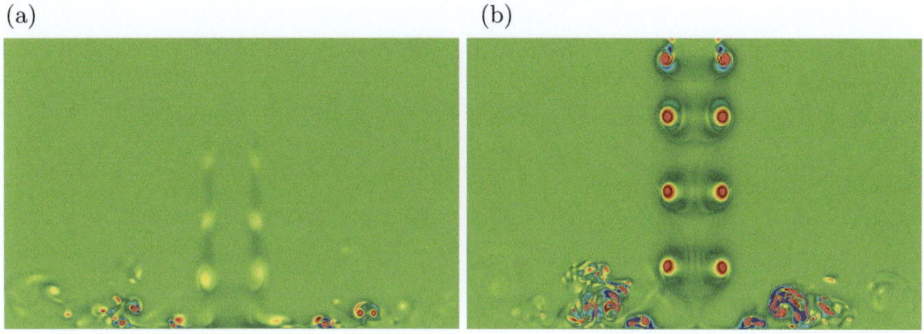

Fig. 6 Vortical structure represented by Q [s^{-2}] ranging from -10^6 (*blue*) to 10^6 (*red*) on a cut through the jet axis. Both simulations have the same mass flow. (**a**) #3 stationary inlet. (**b**) #4 pulsed inlet

to the non-pulsating case, the double resolution in each space direction is needed in order to ensure the resolution of the Kolmogorov length scale. The simulations to be compared are #3 (non-pulsed) and #4 (pulsed). Both have the same mass flow and dynamic viscosity. In order to avoid supersonic flow, the maximal nozzle pressure ratio (NPR) for the pulsed jet was chosen equal to the simulations #1 and #2: NPR$= p_0/p_\infty = 1.5$. As result of these conditions, the NPR for the low Mach number case is 1.1217. This avoids another simulation as reference for the pulsed case. The values given in Table 2 refer to the time span in which the valve is open. For the non-pulsed jets those values are, apart from fluctuations due to acoustic waves reaching the nozzle, constant and therefore also the average. In contrast, the mass flow of the pulsed case is half of the value given in the table, since the time spans of closed and open valve are equal.

In Fig. 6 the vortical structure of the non-pulsed (a) and the pulsed (b) case are confronted. A strong increase of Q of the pulsed case indicates that the eigenfrequency is a reasonable choice for the pulsation frequency. Statistical values are not available at this stage of work. However, we expect an increase of the integral Nusselt number for the pulsed case.

3.3 Conclusion

Vortex rings are responsible for an additional heat transfer at the wall due to a positive contribution of the turbulent heat flux. Those vortex rings occur periodically. The frequency is not dependent on the Reynolds number in the range $3300 \leq Re \leq 8000$. On the contrary, the Mach number plays a role. In the high subsonic regime one mode is dominant ($Sr = 0.46$), whereas at lower Mach number, a second one occurs additionally ($Sr = 0.59$) and exceeds the importance regarding heat transfer of the lower frequent mode. The frequency of this mode ($Sr = 0.59$)

was applied to a pulsed inlet. As a result of the pulsation, the ring vortices were strongly amplified. Quantitative results (e.g. Nusselt number profile) are due, since the simulation is presently running.

4 Impinging Tone

4.1 Introduction

This section is based on [20].

A jet impinging on a flat plate may emanate incredibly loud tonal noise if the Mach number is sufficiently high ($M \gtrsim 0.7$) and the plate is less than about 7.5 diameters away from the nozzle [10].

The loud tonal components in the sound spectrum (impinging tones) were early found to be due to a feedback loop involving a shear layer instability travelling downstream and some acoustic wave travelling upstream in some, necessarily subsonic part of the flow [13]. The same idea was convincingly applied by Ho and Nosseir 1981 [10] as well as Henderson and Powell [8, 9], but it remained unclear who are the culprits for the feedback loop at the wall. Ho and Nosseir identified primary vortices impinging on the wall as a possible link in the feedback chain. Powell and Henderson on the contrary identified standoff shock oscillations as the responsible mechanism within the loop.

Using direct numerical simulations, we are able to identify the sound source mechanism of the impinging jet for the configuration with NPR $= 2.15$ and $h/D = 5$. We expect this result to hold for low NPR and sufficiently high h/D. Two different sound source mechanisms exist. Sound waves are emitted either by shock-vortex- or shock-vortex-shock-interactions. The shock-vortex-interaction is similar to screech in free shear layers but differs significantly as the shock involved is the standoff shock ahead of the wall and not part of the shock cell structure. Shock-vortex-shock-interaction is entirely new and can in short be described as the quenching of the sonic line in between two standoff shocks by the passing vortex. In this report, we concentrate on the description of the two sound source mechanisms. Sections are taken from [19], where a more detailed description of the impinging tones is given.

4.2 Results

4.2.1 Shock-Vortex-Interaction

This kind of sound-emitting interaction requires two components: One shock and one vortex or an aggregation of vortices. The computational results show that multiple shocks can occur near by the stagnation point. Usually two or three shocks are simultaneously present. The system of the shocks is highly unsteady within a periodical cycle.

Shock-vortex-interactions occur also in free jets, as described by Fernandez and Sesterhenn [12]. However, the strength of the shock due to the impinging plate is much stronger than the one in the shock-cell-system due to the under-expansion of the jet. This results in much higher sound pressure levels in the case of a present impinging plate, on which we concentrate in this paper. Therefore the term *shock* refers here always to *standoff-shock*.

This sound source mechanism can involve either the main vortical structure of the impinging jet, which are the vortex rings or a vortex within a turbulent aggregation of vortices. The first case is typical for low Reynolds numbers, like $Re = 3300$ and was found by Wilke and Sesterhenn [18]. With increasing Reynolds numbers, the phenomenon shifts to the second case. In the following, the mechanism is explained using Fig. 7 which shows snapshots of the simulation with $Re = 8000$. All snapshots are a section of a slice through the jet axis. In the first column normalised values of Q and of the divergence of the velocity field $div(u)$ are shown. At the starting point (first row) three shocks are present. For this mechanism only the upper one ($y/D \approx 0.85$) plays a role. For simplicity only that one is shown in the sketch. Additionally a vortex ring (1a,1b) is present, which is slightly asymmetric. The center of the ring in the left shear layer (1a) is at the same height of shock, whereas the center of the ring in the right side (1b) is closer to the wall. A bunch of turbulent vortices (3) is above the shock. The vortex (2a) is a fragment that is left from the next vortex ring that lost its symmetric structure due to leap-frogging. At this point in time the shock keeps its position due to an equilibrium between the stagnation pressure pushing the shock up and the flow pushing the shock down to the wall. The vortices however are transported by the jet with high velocity and approach the impinging plate. The vortex ring (1a,1b) is transported in wall normal direction around the shock, without interaction. Vortex (3) on the contrary crashes into the right end of the shock. As a consequence, the shock looses its equilibrium, turns to the left and strongly accelerates. This can be seen in the second row of Fig. 7. At this point in time the vortex bunch (3) already cut the right end of the shock. The shock transformed into a pressure wave and is now (third row of Fig. 7) in between the two vortices (1a) and (3), moving in north-west direction. At this point there are two possibilities for the pressure wave. The first option is shown in the forth row of Fig. 7: no vortex is in the way and the pressure wave can expand without disturbance. Here, the wave can pass between vortices (1a) and (2a). In this case, the wave leaves the jet and does not trigger a feedback loop. More often is the case that there is no gap for the wave to escape and the wave interacts with another vortex, that changes the direction of the wave. In this case, the wave goes through the whole jet and triggers another instability at the nozzle lip.

Important for this mechanism is a flow field that is at least slightly asymmetric. At low Reynolds number ($Re = 3300$), we observe a flow field that switches between a mainly symmetric and a clear asymmetric state. Also the mainly symmetric state is slightly distorted, so that one side of the vortex ring touches the shock slightly before the other side and leads to the described sound wave.

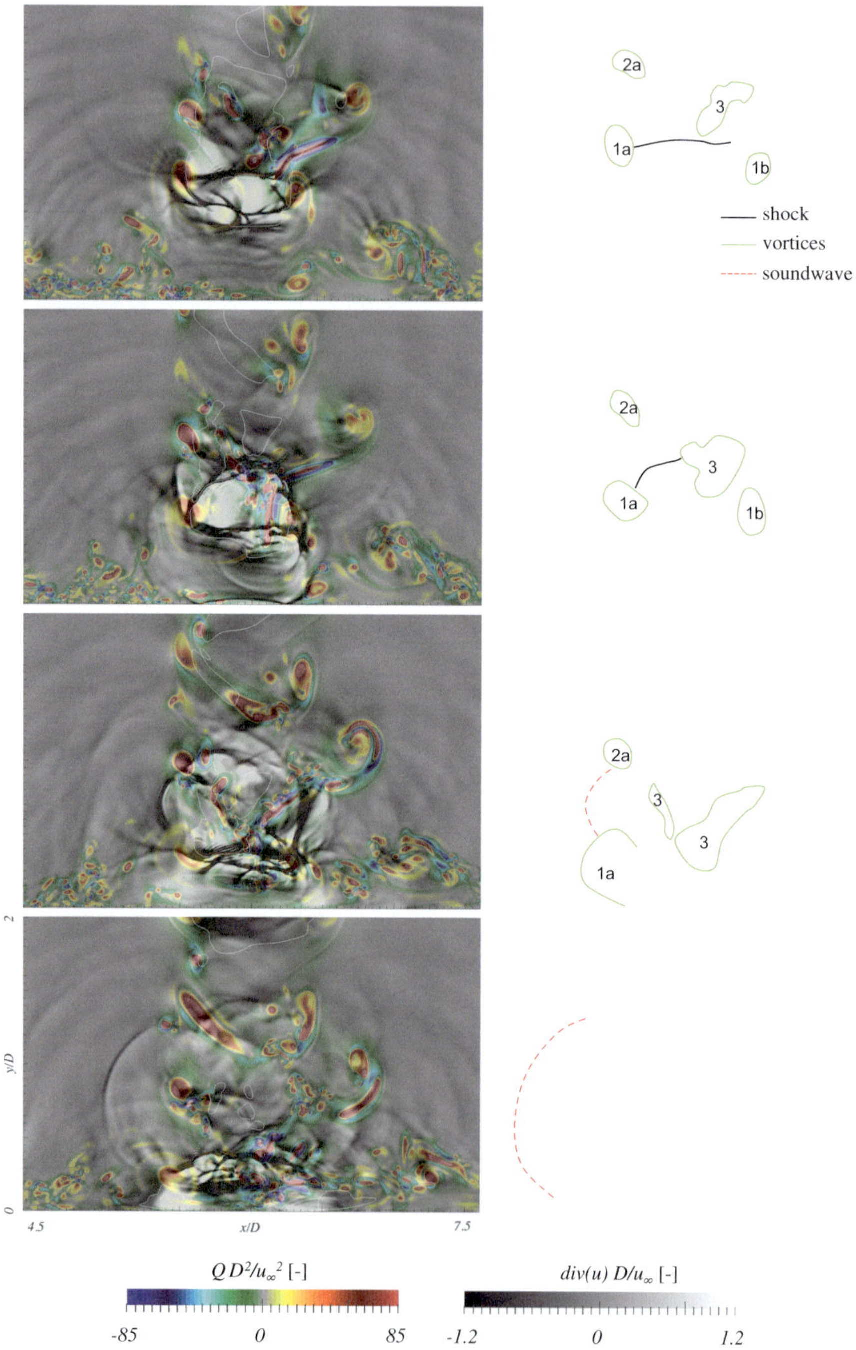

Fig. 7 Shock-vortex-interaction ($Re = 8000$). *First column*: normalised values of Q and of the divergence of the velocity field $div(u)$. *Second column*: sketch. The snapshots (rows) are in consecutive order

4.2.2 Shock-Vortex-Shock-Interaction

The second kind of interaction that produces strong acoustic waves involves two shocks, a vortex ring and a sonic line. Figure 8 shows snapshots of the simulation with $Re = 8000$. All snapshots are a section of a slice through the jet axis. In the first column normalised values of Q and of the divergence of the velocity field $div(u)$ are shown. This mechanism requires a periodical appearance and disappearance of the supersonic zone close to the stagnation point. We start from a point in time where the supersonic zone close to the stagnation point was destroyed and a new one is transported downstream by the jet. This zone is circumscribed by the sonic line ($M = 1$). As long as no obstacles are in the way, the sonic line travels together with vortex rings, but slightly ahead of them. Travelling further downstream the supersonic zone encounters zones of high pressure, which are fragments of the high pressure at the stagnation point. As mentioned, typically there are multiple of such zones. In our example, we have three of them. Each time the sonic line faces a zone of high pressure, it stops its downstream movement for a while until the jet pushes the sonic line over the shock by continuously delivering new fluid. The vortex rings travel in the shear layer, which is outside of the high pressure zone formed only in the core of the jet. Thus they are not affected by those high pressure zones. As a consequence, the vortex rings approach the sonic line and interact. This means they influence the shape of the sonic line due to its rotating velocity components. In the first row of Fig. 8 the sonic line is confined by the shear layer of the jet in radial direction. Streamwise it consists of three parts: on the left side, the sonic line coincides with the upper shock, whereas on the right side, it coincides with the lower shock. The crossover coincides with the inner border of the left side of the vortex ring. The sound wave is produced when this arrangement collapses: The vortex is not able anymore to separate the sub- and supersonic areas. This can be seen in the following two time steps (second and third row of Fig. 8). The sonic line looses its connection to the vortex ring and the upper shock and jumps to the lower shock so that the upper shock gets embedded in the supersonic zone. Thereby a subsonic area is initially embedded and then collapses. A strong spheric pressure wave expands from that point. This goes through the whole jet and reaches the nozzle. The phenomenon therefore triggers new instabilities of the shear layer and is part of a feedback mechanism.

4.2.3 Emanated Sound

In order to obtain the sound spectra, the pressure was recorded in the near-field on three different cylinders around the jet axis at distances of two, three and four diameters. For the presented results, the position $r/D = 4$ and $y/D = 5$ was chosen. The upper wall has the advantage, that the velocity is zero and no flow disturbs the acoustic measurements. The choice of the radius does not influence the investigated tones (frequencies), since the different distances only move the sound pressure level up and down. For each of the 256 circumferential positions,

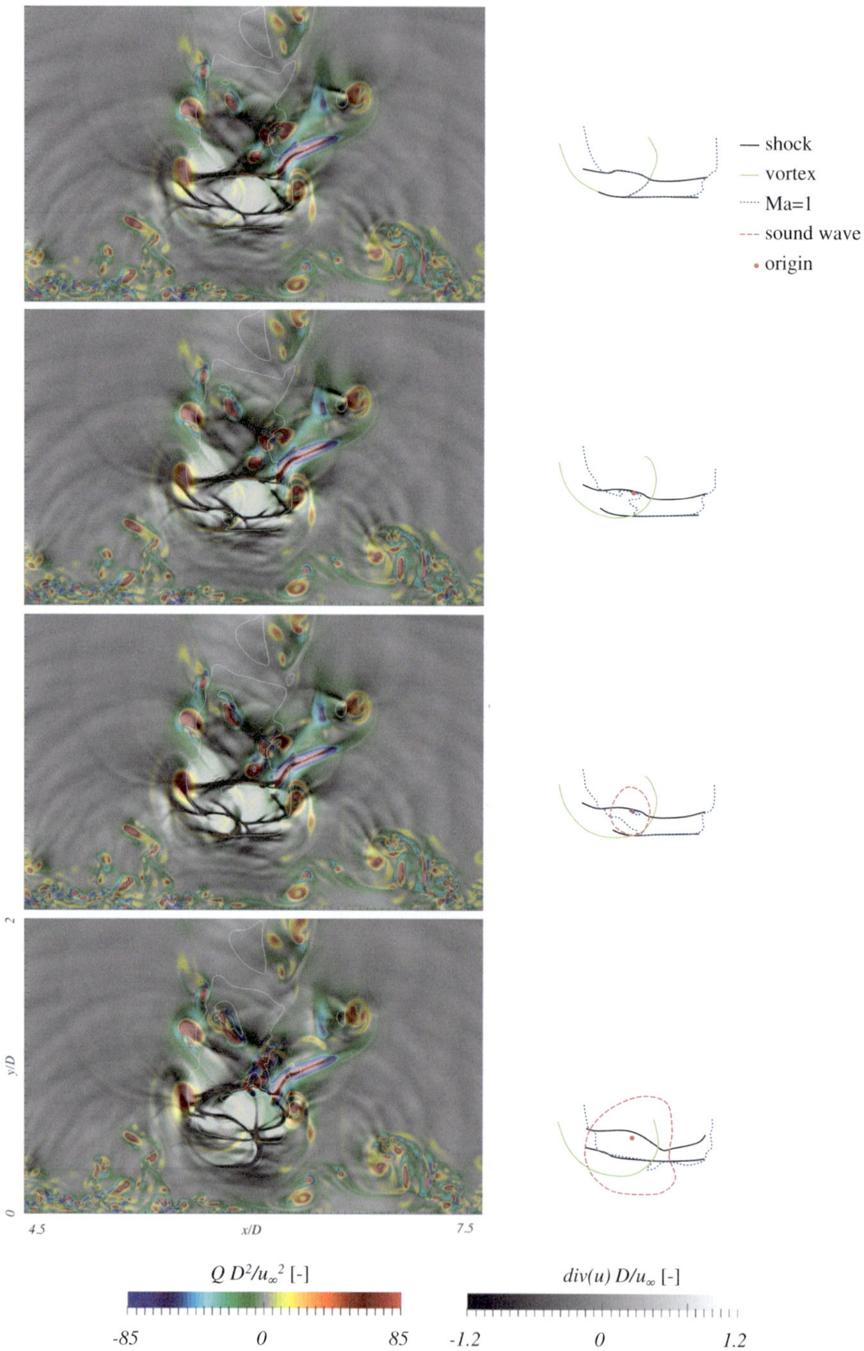

Fig. 8 Shock-vortex-shock-interaction ($Re = 8000$). *First column*: normalised values of Q and of the divergence of the velocity field $div(u)$. *Second column*: sketch. The snapshots (rows) are in consecutive order

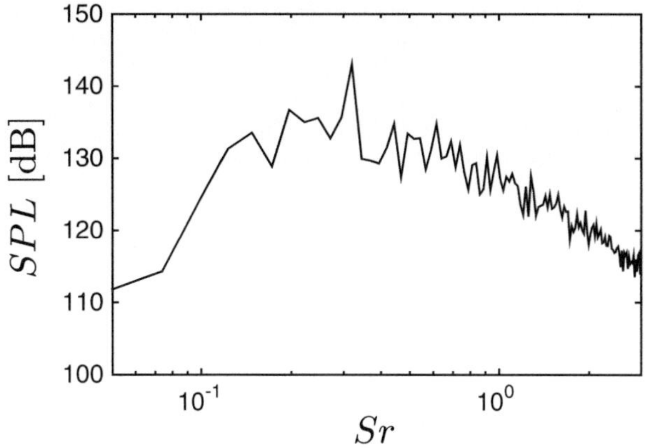

Fig. 9 Sound pressure level (SPL) of the supersonic impinging jet with $Re = 8000$. Reference pressure: $p_{ref} = 2 \cdot 10^{-5}$ Pa

the spectra was computed using a fast Fourier transform (FFT). The spectra were then averaged. Figure 9 shows the sound pressure level depending on the Strouhal number. The impinging tone can be clearly observed at $Sr = 0.32$. A prove that the two sound source mechanisms found correspond to this frequency is given in [20].

4.3 Conclusion

Despite the general accordance that impinging tones are produced due to a feedback loop, inconsistent statements about the production of the sound waves can be found in literature. In addition, no consensus could be found if standoff shocks are present in the pre-silence zone, a regime in NPR, where tones can be observed.

In order to clarify the open questions, we performed a direct numerical simulation with a nozzle pressure ratio of 2.15 and a nozzle-to-plate distance of five diameters at Reynolds number of 8000. Analysing the data, we find that standoff shocks periodically appear, disappear and move between the impinging plate and the shock cell system. Multiple standoff shocks can exist simultaneously, usually two or three are present for the chosen set of parameters. Concerning the generation of impinging tones, we clearly observe the feedback loop and prove that the interaction between vortices and standoff shocks produce the sound waves via two different mechanisms. One of the two mechanism can analogously be found in free jets and is responsible for screech. The difference however is that not the shock diamonds, but the standoff shock is involved in the interaction with the vortices. The impinging tone is not related to screech. The mode of the impinging jet is axisymmetrical.

Acknowledgements The simulations were performed on the national supercomputer Cray XC40 (Hornet, Hazelhen) at the High Performance Computing Center Stuttgart (HLRS) under the grant numbers GCS-NOIJ/12993 and GCS-ARSI/44027.

The authors gratefully acknowledge support by the Deutsche Forschungsgemeinschaft (DFG) as part of collaborative research center SFB 1029 "Substantial efficiency increase in gas turbines through direct use of coupled unsteady combustion and flow dynamics".

References

1. Bogey, C., de Cacqueray, N., Bailly, C.: A shock-capturing methodology based on adaptative spatial filtering for high-order non-linear computations. J. Comput. Phys. **228**(Nr. 5), 1447–1465 (2009). http://dx.doi.org/http://dx.doi.org/10.1016/j.jcp.2008.10.042, doi:http://dx.doi.org/10.1016/j.jcp.2008.10.042, ISSN 0021–9991
2. Chung, Y.M., Luo, K.H.: Unsteady heat transfer analysis of an impinging jet. J. Heat Transf. **124**, 12(Nr. 6), 1039–1048 (2002). ISBN 0022–1481
3. Cziesla, T., Biswas, G., Chattopadhyay, H., Mitra, N.: Large-eddy simulation of flow and heat transfer in an impinging slot jet. Int. J. Heat Fluid Flow **22**(Nr. 5), 500–508 (2001). http://dx.doi.org/http://dx.doi.org/10.1016/S0142-727X(01)00105-9, doi:http://dx.doi.org/10.1016/S0142--727X(01)00105--9, ISSN 0142–727X
5. Dairay, T., Fortuné, V., Lamballais, E., Brizzi, L.: LES of a turbulent jet impinging on a heated wall using high-order numerical schemes. Int. J. Heat Fluid Flow **50**(Nr. 0), 177–187 (2014). http://dx.doi.org/http://dx.doi.org/10.1016/j.ijheatfluidflow.2014.08.001, doi:http://dx.doi.org/10.1016/j.ijheatfluidflow.2014.08.001, ISSN 0142–727X
4. Dairay, T., Fortuné, V., Lamballais, E., Brizzi, L.-E.: Direct numerical simulation of a turbulent jet impinging on a heated wall. J. Fluid Mech. **764**(2), 362–394 (2015). http://dx.doi.org/10.1017/jfm.2014.715, doi:10.1017/jfm.2014.715, ISSN 1469–7645
6. Eidson, T.M., Erlebacher, G.: Implementation of a fully balanced periodic tridiagonal solver on a parallel distributed memory architecture. Concurr.: Pract. Exp. **7**(Nr. 4), 273–302 (1995)
7. Hattori, H., Nagano, Y.: Direct numerical simulation of turbulent heat transfer in plane impinging jet. Int. J. Heat Fluid Flow **25**(Nr. 5), 749–758 (2004). http://dx.doi.org/http://dx.doi.org/10.1016/j.ijheatfluidflow.2004.05.004, doi:http://dx.doi.org/10.1016/j.ijheatfluidflow.2004.05.004, ISSN 0142–727X. Selected papers from the 4th International Symposium on Turbulence Heat and Mass Transfer
9. Henderson, B.: The connection between sound production and jet structure of the supersonic impinging jet. J. Acoust. Soc. Am. **111**,(Nr. 2), 735–747 (2002). http://dx.doi.org/http://dx.doi.org/10.1121/1.1436069, doi:http://dx.doi.org/10.1121/1.1436069
8. Henderson, B., Powell, A.: Experiments concerning tones produced by an axisymmetric choked jet impinging on flat plates. J. Sound Vib. **168**(Nr. 2), 307–326 (1993). http://dx.doi.org/http://dx.doi.org/10.1006/jsvi.1993.1375, doi:http://dx.doi.org/10.1006/jsvi.1993.1375, ISSN 0022–460X
10. Ho, C.-M., Nosseir, N.S.: Dynamics of an impinging jet. Part 1. The feedback phenomenon. J. Fluid Mech. **105**(4), 119–142 (1981), http://dx.doi.org/10.1017/S0022112081003133, doi:10.1017/S0022112081003133, ISSN 1469–7645
11. Janetzke, T.: Experimentelle Untersuchungen zur Effizienzsteigerung von Prallkühlkonfigurationen durch dynamische Ringwirbel hoher Amplitude, TU Berlin, Diss. (2010)
12. Peña Fernández, J.J., Sesterhenn, J.: Interaction between the shear layer, shock-wave and vortex ring in a starting free jet injecting into a plenum. In: European Turbulence Conference, Delft (2015)
13. Rockwell, D., Naudascher, E.: Self-sustained oscillations of impinging free shear layers. Annu. Rev. Fluid Mech. **11**(Nr. 1), 67–94 (1979)

14. Sesterhenn, J.L.: A characteristic–type formulation of the Navier–Stokes equations for high order upwind schemes. Comput. Fluids **30**(Nr. 1), 37–67 (2001)
15. Weigand, B., Spring, S.: Multiple jet impingement – a review. Heat Transf. Res. **42**(Nr. 2), 101–142 (2011). ISSN 1064–2285
16. Wilke, R., Sesterhenn, J.: Direct numerical simulation of heat transfer of a round subsonic impinging jet. In: Active Flow and Combustion Control 2014, pp. 147–159. Springer, Cham (2015)
17. Wilke, R., Sesterhenn, J.: Numerical simulation of impinging jets. In: High Performance Computing in Science and Engineering '14, pp. 275–287. Springer, Cham (2015)
18. Wilke, R., Sesterhenn, J.: Numerical simulation of subsonic and supersonic impinging jets. In: High Performance Computing in Science and Engineering´ 15, pp. 349–369. Springer, Cham (2016)
19. Wilke, R., Sesterhenn, J.: On the origin of impinging tones at low supersonic flow (2016). arXiv preprint, arXiv:1604.05624
20. Wilke, R., Sesterhenn, J.: Statistics of fully turbulent impinging jets (2016). arXiv preprint, arXiv:1606.09167
21. Zuckerman, N., Lior, N.: Impingement heat transfer: correlations and numerical modeling. J. Heat Transf. **127**(Nr. 5), 544–552 (2005). ISBN 0022–1481

Aeroacoustic Simulations of Ducted Axial Fan and Helicopter Engine Nozzle Flows

Alexej Pogorelov, Mehmet Onur Cetin, Seyed Mohsen Alavi Moghadam, Matthias Meinke, and Wolfgang Schröder

Abstract The flow and the acoustic field of an axial fan and a helicopter engine jet are computed by a hybrid fluid dynamics – computational aeroacoustics method. For the predictions of the flow field a high-fidelity, parallelized solver for compressible flow is used in the first step. In the second step, the acoustic field is determined by solving the acoustic perturbation equations. The axial fan is investigated at a Reynolds number of $Re = 9.36 \times 10^5$ for two tip-gap sizes, i.e., $s/D_o = 0.001$ and $s/D_o = 0.01$ at a fixed flow rate coefficient $\Phi = 0.195$. A comparison of the numerical results of the pressure spectrum and its directivity with measurements show a good agreement which confirms the correct identification of the sound sources and the accurate prediction of the acoustic duct propagation. Furthermore, the results show in agreement with the experimental data a higher broadband noise level for the larger tip-gap size. In the second application, jets from three different helicopter engine nozzles at a Reynolds number of $Re = 7.5 \times 10^5$ are investigated, showing an important dependence of the jet acoustic near field on the presence of the nozzle built-in components. The presence of the centerbody increases the OASPL compared to the clean nozzle, where the inclusion of struts reduces the OASPL compared to the centerbody nozzle owing to the increased turbulent mixing caused by the struts which lesses the length and time scales of the turbulent structures shed from the centerbody.

1 Introduction

The prediction and reduction of noise generated by turbulent flows has become one of the major tasks of todays aircraft development and is also one of the key goals in European aircraft policy. Compared to the year 2000 the perceived noise level of flying aircraft should to be reduced by 65 % until the year 2050. To comply with new noise level regulations, reliable, efficient and accurate aeroacoustic predictions are required, i.e., for low noise design of technical devices such as axial fans or helicopter engine nozzles.

A. Pogorelov (✉) • M. Onur Cetin • S. Mohsen Alavi Moghadam • M. Meinke • W. Schröder
Institute of Aerodynamics, RWTH Aachen University, Wüllnerstr. 5a, 52062 Aachen, Germany
e-mail: a.pogorelov@aia.rwth-aachen.de

W.E. Nagel et al. (eds.), *High Performance Computing in Science and Engineering '16*, DOI 10.1007/978-3-319-47066-5_30

Fan industry increasingly demands for quieter and more efficient axial fans in a wide range of applications. A systematic quiet fan design, however, requires prediction methods for the acoustic field and sufficient details of the flow field to understand the intricate flow mechanisms, e.g. in the tip-gap region of the fan blade. Since measurements of the flow field in the rotating fan environment are difficult to perform, time-accurate numerical simulations such as highly resolved large-eddy simulations (LES) have shown to successfully predict the main flow phenomena [22–24], especially those in the tip-gap region since these can be a significant source of aerodynamic losses and noise emission.

Appreciable progress has been achieved over the last 20 years in the decrease of jet noise by using various noise reduction techniques such as high bypass ratio and design variations on the nozzle casing. These techniques have primarily focused on increasing the turbulent mixing by altering the nozzle design. In modern engines, the bypass ratio has already reached the limiting value and any further increase will aggravate the engine performance. Flow control inside the nozzle by additional built-in components such as wedges vanes etc. is an alternative approach and increasingly used to suppress the noise in the jet near field [14, 20].

The overall reliability of an acoustic prediction is prominently restricted with the quality of the flow field solution. To accurately capture the essential part of the turbulent spatial and temporal scales generated in the flow field highly resolved LES calculations are a must. That is such aeroacoustic analyses of high Reynolds number flow with complex geometries included in the computational domain require advanced computing resources.

In this paper the acoustic fields of a ducted axial fan and a helicopter engine nozzle are predicted by a hybrid fluid-dynamics-acoustics method. In a first step, large-eddy simulations are performed to determine the acoustic sources. In a second step, the acoustic field on the near and far-field is determined by solving the acoustic perturbation equations (APE) [6] on a mesh. The acoustic results of the axial fan are compared to experimental data [27].

This paper is organized as follows. First, the numerical methods are presented in Sect. 2. Subsequently, the LES and aeroacoustic results of the axial fan and nozzle-jet simulations are discussed in Sects. 3 and 4. Computational features and scalability analysis are given in Sect. 5. Finally, some conclusions are outlined in Sect. 6.

2 Numerical Method

An LES model based on a finite volume method is used to simulate the compressible unsteady turbulent flow by solving the Navier-Stokes equations. For the LES an implicit grid filter is assumed and the monotone integrated LES (MILES) approach [2] is adopted, i.e., the dissipative part of the truncation error of the numerical method is assumed to mimic the dissipation of the non-resolved subgrid-scale stresses. This solution method has been validated and successfully used, e.g., in [1, 16]. The governing equations are spatially discretized by using the modified advection upstream splitting method (AUSM) [19]. The cell center gradients are computed using a second-order accurate least-squares reconstruction scheme [10], i.e., the overall spatial approximation is second-order accurate. For stability reasons, small cut-cells are treated using an interpolation and flux-redistribution method [25]. A second order 5-stage Runge-Kutta method is used for the temporal integration. A parallel grid generator is used to create a computational hierarchical Cartesian mesh featuring local refinement [18]. The interested reader is referred to [19] for the details of the numerical methods, i.e., the discretization and computation of the viscid and inviscid fluxes. To determine the sound propagation and to identify the dominant noise sources the acoustic perturbation equations (APE) are applied. Since a compressible flow problem is considered, the APE-4 system is used [6].

To accurately resolve the acoustic wave propagation described by the acoustic perturbation equations in the APE-4 formulation [15] a sixth-order finite difference scheme with the summation by parts property [13] is used for the spatial discretization and an alternating 5–6 stage low-dispersion and low-dissipation Runge-Kutta method for the temporal integration [11]. On the embedded boundaries between the inhomogeneous and the homogeneous acoustic domain an artificial damping zone has been implemented to suppress spurious sound generated by the acoustic-flow-domain transition [26]. A detailed description of the two-step method and the discretization of the Navier-Stokes equations and the acoustic perturbation equations is given in [7].

3 Effect of Tip-Gap Size on Fan Aeroacoustics

In this subsection, a rotating low Mach number axial fan is investigated. In the first subsection, it is discussed how the gap size between blade tip and the outer casing wall affects the flow field at different operating conditions. All computations are performed at a fixed Reynolds number based on the rotational velocity and the diameter of the outer casing wall $Re = \frac{\rho \pi D_o^2 n}{\eta} = 9.36 \times 10^5$ and a fixed Mach number $M = \frac{\pi D_o n}{a} = 0.136$. Afterwards, the acoustic field is analyzed at the flow rate coefficient $\Phi = \frac{4\dot{V}}{\pi^2 D_o^3 n} = 0.195$ for two tip-gap widths $s/D_o = 0.001$ and $s/D_o = 0.01$.

3.1 Effect of Tip-Gap Size on the Overall Flow Field

The axial fan investigated in this section is shown in Fig. 1. The fan has five twisted blades out of which only one has been resolved in both LES and CAA computations to reduce the computational costs. The diameter of the outer casing wall is $D_o = 300$ mm and the inner diameter of the hub is $D_i = 135$ mm. The rotational

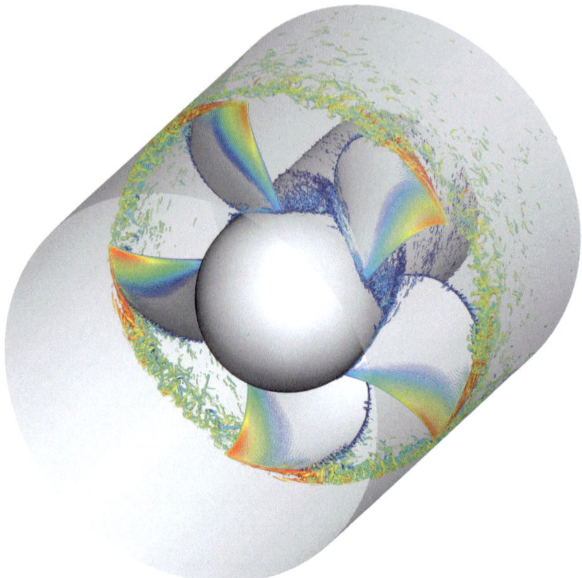

Fig. 1 Instantaneous contours of the Q-criterion inside the ducted axial fan configuration colored by the relative Mach number showing the vortical structures generated by the tip leakage flow at $\Phi = 0.195$ and $s/D_o = 0.005$

speed is $n = 3000$ rpm. As depicted in Fig. 1 for $\Phi = 0.195$ and $s/D_o = 0.005$, the existence of a gap between the blades tip and the outer casing wall and the pressure difference between the pressure and the suction side of the blades, lead to the development of a tip-gap vortex. Depending of the operating conditions the tip-gap vortex can be a major noise source in the axial fan, especially at low flow rate coefficients Φ, as demonstrated in [22] at $\Phi = 0.165$ and a tip-gap size of $s/D_o = 0.01$. At low flow rate coefficients the highly unsteady turbulent wake generated by the tip-gap vortex is shifted further upstream and impinges upon the leading edge of the neighboring blade. The intermittent interaction leads to a cyclic transition on the suction side of the blade. Acoustic measurements have shown broadband peaks in the specific sound power spectrum at frequencies corresponding to these phenomena. The decrease of the tip-gap width from $s/D_o = 0.01$ to $s/D_o = 0.005$ at $\Phi = 0.165$, stabilizes the tip-gap vortex and reduces the wandering motion of the turbulent wake such that the interaction with the leading edge of the neighboring blade and the cyclic transition triggered by this interaction vanish as discussed by Pogorelov et al. [23]. Instead, a permanent turbulent transition, which is triggered by a separation bubble at the leading edge was observed. The reduction of the tip-gap width leads to a strong decrease of the noise level. However, for the smaller tip-gap size, the turbulent wake still interacts with the pressure side of the blade. To separate the noise generated by the interaction and the phenomena triggered by this interaction from the self-generated noise of the tip-gap vortex

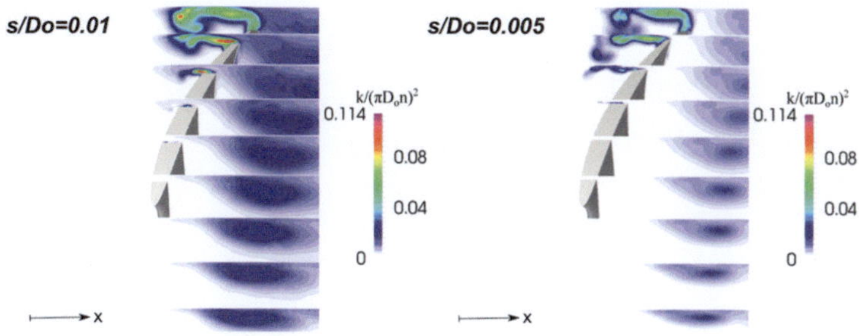

Fig. 2 Turbulent kinetic energy contours in several radial planes from $\theta = 30°$ to $\theta = 70°$, for $s/D_o = 0.01$ (*left*) and $s/D_o = 0.005$ (*right*)

it is important to analyze the acoustic field at higher flow rate coefficients and small tip-gap widths where no interaction with the neighboring blades is evident. Pogorelov et al. [24] analyzed the flow field at $\Phi = 0.195$ for the tip-gap widths $s/D_o = 0.005$ and $s/D_o = 0.01$. This study has demonstrated the strong impact of the tip-gap width on the size and shape of the tip-gap vortex. It has been shown, that due to the stronger curvature and the smaller diameter of the tip-gap vortex for $s/D_o = 0.005$, the entire turbulent wake passes the neighboring blade without any interaction, where for $s/D_o = 0.01$ several vortical structures of the turbulent wake reach the trailing edge of the blade at the pressure side, as depicted in Fig. 2. Therefore, for tip-gap sizes below $s/D_o = 0.005$ no interaction with the neighboring blades is expected. In the following subsection, the acoustic field of the flow field at $\Phi = 0.195$ for $s/D_o = 0.001$ and $s/D_o = 0.01$ is analyzed. For the source computation, required for the acoustic analysis LES have been conducted for both operating conditions. The computational mesh resolving one out of five blades has approx. 140 million grid points. Two full rotations have been required to obtain a fully developed flow field. Data from another two full rotations has been used for statistical analysis. In total, 1440 samples were recorded which required 8.6 TB of disc space. The CPU time was approx. 200 h and the computations were conducted on approx. 6000 CPUs.

3.2 *Effect of Tip-Gap Size on the Acoustic Field*

In the following, the acoustic field is numerically analyzed by a hybrid fluid-dynamics-aeroacoustics method. The acoustic field on the near field and far field is determined by solving the APE [6] in the rotating frame of reference on a mesh for a single blade consisting of approx. 1060×10^6 grid points which comprises a 72° segment of a rotating axial fan with periodic boundary conditions in the azimuthal direction. The computations are performed for two tip-gap sizes namely, $s/D_o = 0.001$ and $s/D_o = 0.01$ at the flow rate coefficient $\Phi = 0.195$. Based on the LES solution of the turbulent flow field, from which the acoustic sources are

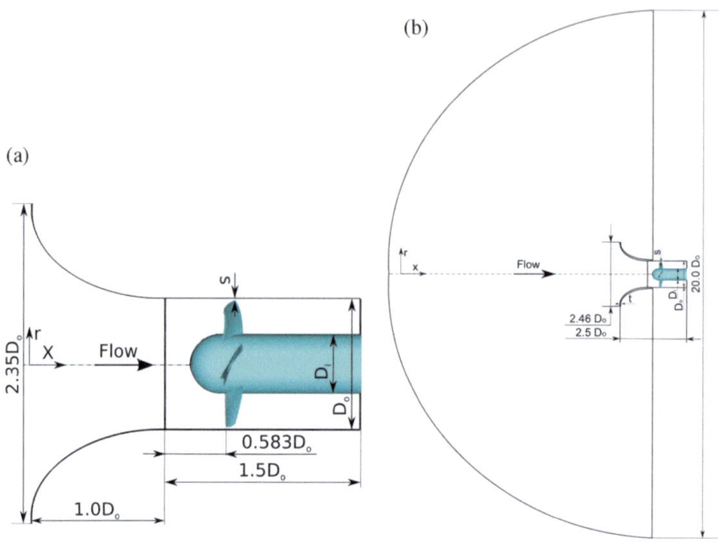

Fig. 3 (**a**) Schematic view of the LES and (**b**) the acoustic configuration of an axial fan

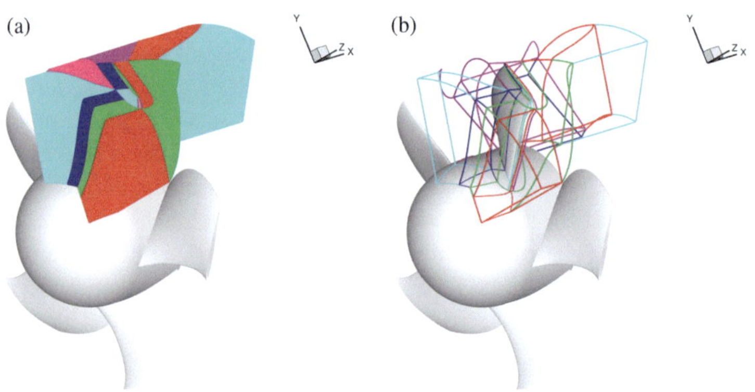

Fig. 4 The multi-block structured mesh in the acoustic source region resolving one out of five blades of the axial fan; (**a**) view of the overall mesh; (**b**) detailed topological view of the mesh

determined, the near far-field acoustics is computed by solving the APE-4 system. Since the contribution of entropy and non-linear terms can be neglected in this study, only the vortex sound sources are taken into account.

A schematic view of the present computational setup is shown in Fig. 3 In a first step, the turbulent flow fields are determined by LES for the two configurations for 24 full rotations. Subsequently, the source terms are computed in the source region which contains approximately 122 million grid points with the same mesh resolution as the corresponding LES mesh. Figure 4 shows the computational mesh used for computing the source terms. The instantaneous distribution of the dominant

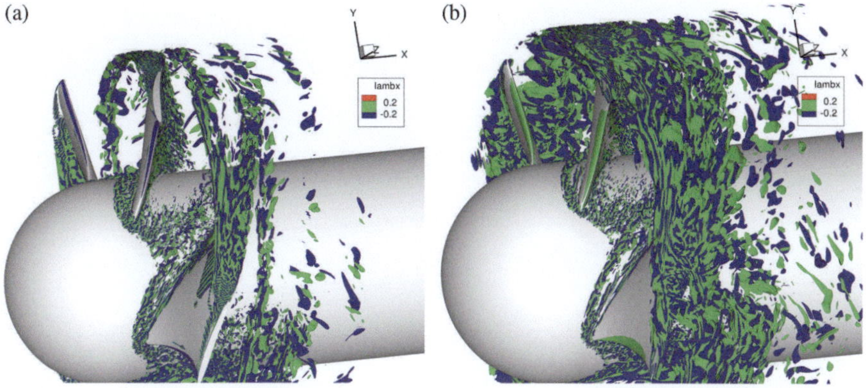

Fig. 5 Instantaneous contours of the Iso-surface of axial component of the fluctuating Lamb vector showing the major sound sources around the blade; (**a**) configuration $s/D_o = 0.001$; (**b**) configuration $s/D_o = 0.01$

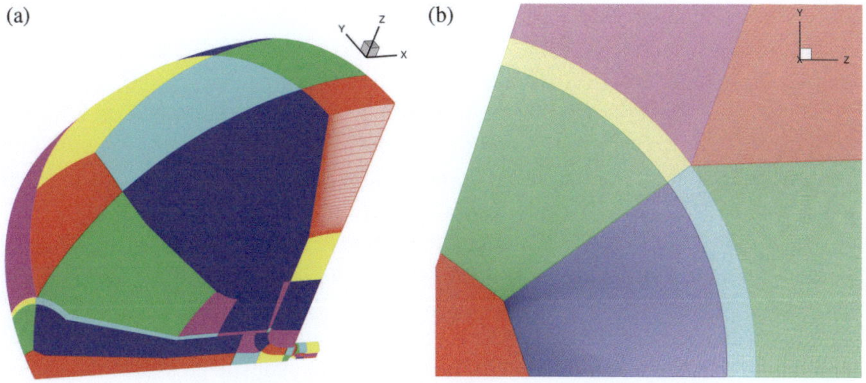

Fig. 6 The multi-block structured mesh for the acoustic domain resolving one out of five blades of the axial fan; (**a**) view of the overall mesh; (**b**) detailed view of the mesh at far-field

fan noise sources, which is the fluctuating Lamb vector $\mathbf{L}' = (\Omega \times \mathbf{u})'$, for the two configurations is shown in Fig. 5

It is clearly visible that the strongest sources occur in regions with the highest turbulent kinetic energy, i.e., in the tip vortex, blade wake and on the hub region. Moreover, the noise sources generated by the bigger tip-gap size $s/D_o = 0.01$ exhibits higher amplitudes compared to the smaller tip-gap size $s/D_o = 0.001$. In a second step, the acoustic field is predicted based on the corresponding LES results. The computational mesh used for the LES is extended in the axial and radial direction up to $20D_o$. The grid spacing around the microphones positions is $\Delta x_{mic}/D_o \approx 5\times10^{-3}$, so that for 10 points per wavelength, the maximum frequency resolvable by the grid is about 10 kHz. The acoustic mesh including some details of the mesh resolution in the far-field are shown in Fig. 6. The time step of the

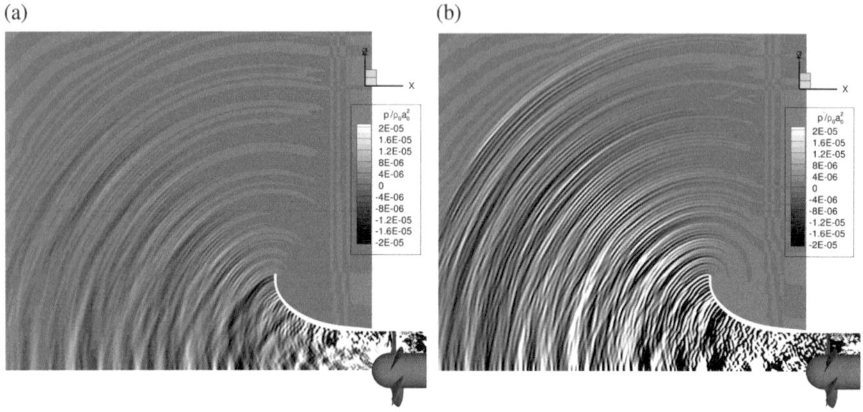

Fig. 7 Instantaneous contours of the fully developed acoustic field showing the acoustic duct propagation into the far-field; (**a**) configuration $s/D_o = 0.001$; (**b**) configuration $s/D_o = 0.01$

acoustic analysis is $\Delta t = 4.613 \times 10^{-3} D_o/a_\infty$ to ensure a fully stable numerical solution. Based on 1500 LES snapshots at a time interval of $\Delta t_{src} = 0.0224 D_o/a_\infty$, the source terms are computed and a least square optimized interpolation filter [9] using $N = 10$ source samples is used to provide source fields at every Runge-Kutta time-integration step. The acoustic computations are run for a non-dimensional time period of $39 D_o/a_\infty$. Explicit low-pass filtering at every 5th Runge-Kutta time-integration step is used to avoid numerical oscillations. Additionally, a sponge layer is used in order to damp acoustic wave reflections at far-field and downstream of the fan.

In Fig. 7 the acoustic fields generated by the turbulent structures of the rotating axial fan for the two configurations are illustrated. The acoustic pressure field shows noise generation at a higher frequency for the configuration $s/D_o = 0.01$ with the bigger tip-gap size and a noise generation at lower frequency for the configuration $s/D_o = 0.001$ with the smaller tip-gap size. In the following acoustic analysis, the computed sound pressure spectra at the circle C1 which is defined in Fig. 8, are compared with the experimental data [27]. For the comparison of the numerical results with experimental data, the acoustic signals are analyzed on circle C1 and circle C2, which are located 1.30 and 1.0 m from suction mouth of the fan. The acoustic measurements were carried out in the fixed frame of reference. In order to compare the computed sound spectra in rotating frame of reference with the experimental findings, 1001 probes are equally distributed on each circle of 72°. First, the position of the microphones are calculated in the fixed frame of reference and then sound pressure spectrum for all processed microphones are computed. Finally, the sound pressure spectrum of all microphones are averaged. The computed sound pressure spectra at the circle C1 and circle C2 are shown in Figs. 9 and 10 respectively.

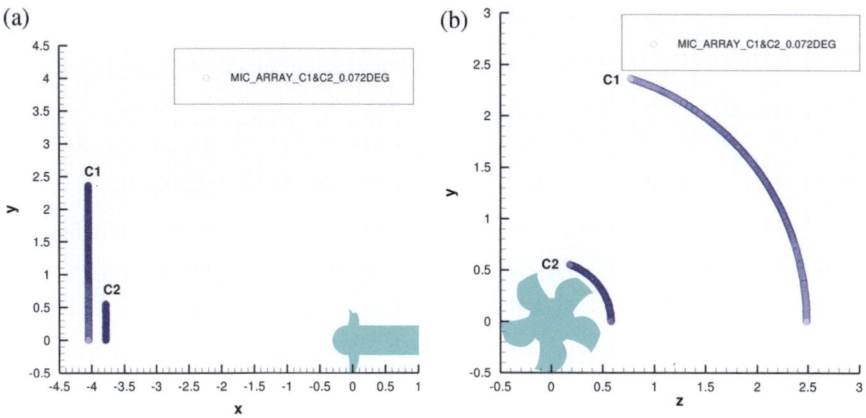

Fig. 8 Schematic of the virtual microphone positions for the two acoustic configurations; (**a**) side view; (**b**) front view

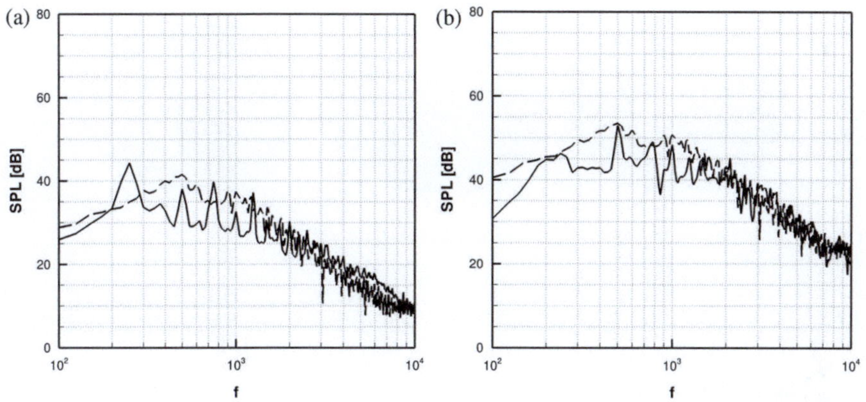

Fig. 9 Sound spectra at the far-field locations circle C1; (**a**) configuration $s/D_o = 0.001$; (**b**) configuration $s/D_o = 0.01$; comparison of the (––) numerical results with the (—) experimental results [27]

The evaluation of the sound pressure level at the circle C1 and the circle C2 show a convincing agreement especially at the broadband noise level. However, considering the circle C2 towards center line of the axial fan, the computed sound pressure level at the lower frequencies deviate from the experimental measurements which is due to the fact that one blade acoustic simulations using periodic boundary condition lacks certain low wave number ranges which is clearly observable in corresponding spectral analysis. In addition, a higher noise level of the case with the bigger compared to the smaller tip-gag size is clearly reproduced by the numerical simulation method.

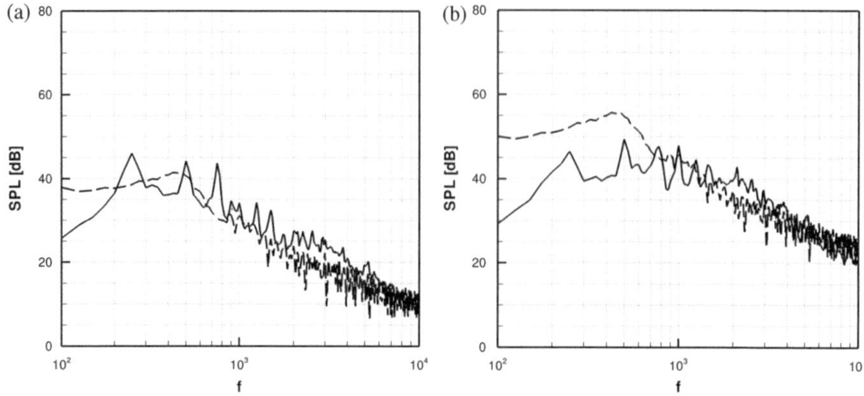

Fig. 10 Sound spectra at the far-field locations circle C2; (**a**) configuration $s/D_o = 0.001$; (**b**) configuration $s/D_o = 0.01$; comparison of the (––) numerical results with the (—) experimental results [27]

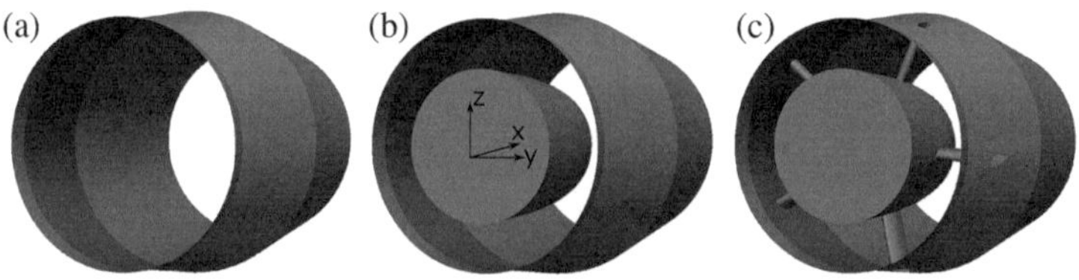

Fig. 11 Rear section of the nozzle geometry (**a**) clean nozzle hj_1, (**b**) centerbody nozzle hj_2, (**c**) centerbody-plus-strut nozzle hj_3

4 Effect of the Interior Nozzle Geometry on Jet Aeroacoustics

In this section, simulation results of round jets emanating from a three variants of non-generic nozzle are presented. First the flow field of the three nozzle configurations at a Reynolds number of $\mathrm{Re} = 7.5 \times 10^5$ and a Mach number of $\mathrm{M} = 0.341$ are conducted and thereafter, the acoustic field is computed whose acoustic source terms are determined by LES data.

4.1 Flow Field

The nozzle geometry corresponds to a divergent helicopter engine nozzle. Figure 11 shows the interior of three variants of the engine nozzle, the clean nozzle hj_1, the centerbody nozzle hj_2, and the centerbody-plus-strut nozzle hj_3 which are identical except for the centerbody and the struts which support the centerbody.

Table 1 Simulation features and mesh parameters of the flow and the acoustic field solutions

	Clean nozzle (hj_1)	Centerbody nozzle (hj_2)	Centerbody-plus-strut nozzle (hj_3)
	Flow field		
Mach number M_j	0.341	0.341	0.341
Reynolds number Re_{D_e}	750,000	750,000	750,000
Mesh points	335×10^6	329×10^6	328×10^6
Number of samples	2251	2251	2251
	Acoustic field		
Mesh points	108.5×10^6	108.5×10^6	108.5×10^6

Fig. 12 Contours of the Q-criterion color coded by density for three geometries (**a**) hj_1, (**b**) hj_2, (**c**) hj_3

The operating conditions of the last turbine stage are set at the inlet boundary which were taken from the measurements of a full-scale turbo-shaft engine [21]. Isotropic synthetic turbulence is injected at the inlet plane with approx. 10 % turbulence intensity [17]. For the outflow and lateral boundaries of the jet domain, static pressure is kept constant and other variables are extrapolated from the internal domain. To damp the numerical reflections at the boundaries, sponge layers are prescribed [8]. At the nozzle-wall a no-slip condition with a zero pressure and density gradient is applied. Hierarchically refined Cartesian meshes are used for the flow field computations and a grid convergence study of the centerbody nozzle hj_2 configuration is studied in [4, 5]. The essential mesh and simulation parameters of the analysis of the flow and the acoustic fields are summarized in Table 1.

The overall turbulent structures in the jet are visualized in Fig. 12 by the contours of the instantaneous Q-field [12] for the three configurations. Since the same threshold value for the Q-contours is used, the various widening of the free jets can be deduced from this illustration. In other words, Q-fields evidence the smaller spreading of the jet exhausting from the clean nozzle hj_1.

The modified turbulence field influences the jet characteristics downstream of the nozzle exit. This is illustrated by the contours of the mean axial velocity in the free jet region in Fig. 13. The mean velocity on the centerline decreases much more strongly for the hj_2 and hj_3 geometries than for the clean nozzle which possesses a standard jet plume shape. Furthermore, the asymmetric velocity distribution caused by the struts is visible in the jet field just downstream of the exit. However, further downstream hardly any asymmetric influence of the struts is observed.

The mean axial velocity distribution normalized with the average nozzle exit axial velocity $u_{ne} = \frac{1}{A} \int \mathbf{u} \cdot \mathbf{n} dA$ on the centerline starting at the rear face of the

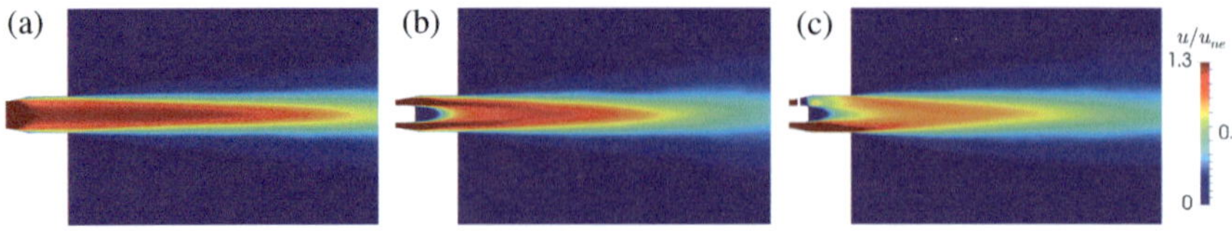

Fig. 13 Contours of the mean axial velocity in the free jet region for three geometries (**a**) hj_1, (**b**) hj_2, (**c**) hj_3

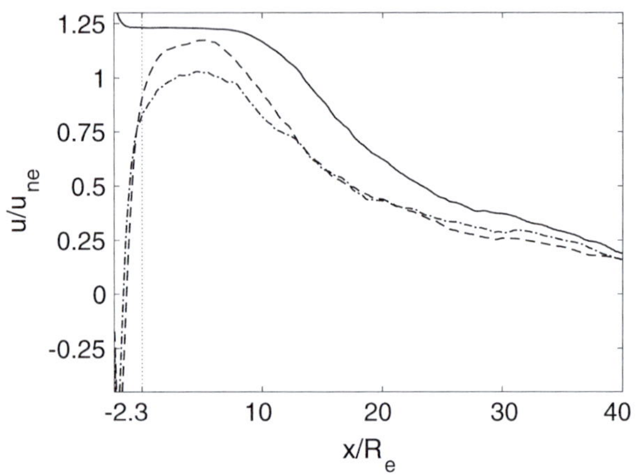

Fig. 14 Streamwise distribution of the axial velocity on the *centerline* $r/R_e = 0$ for (—) hj_1, (– –) hj_2, (-·-) hj_3

centerbody $x/R_e = -2.3$ where $R_e = D_e/2$ is the nozzle exit radius, is presented in Fig. 14. Note that the decreasing distribution between $-2.3 < x/R_e < -1.1$ of the clean nozzle hj_1 is due to the divergence of the nozzle casing. Besides the impact of the diverging part of the nozzle, the velocity distribution on the centerline of the clean nozzle hj_1 undergoes the standard decay. Downstream of the exit of the nozzle the centerline velocity remains constant till the free-shear layers start to merge causing the decay of the centerline velocity. For the centerbody and the centerbody-plus-strut configurations hj_2 and hj_3 the distribution of the streamwise velocity on the centerline is characterized by the pronounced recirculation in the base region of the centerbody. Downstream of this reversal flow neither the hj_2 nor the hj_3 centerline velocity reach the value of the clean nozzle. To be more precise, the peak value of the centerline velocity of the hj_2 solution is diminished by 11 % and that of the hj_3 solution by 22 % compared to the hj_1 value. When the velocity decay sets in the hj_2 and hj_3 solutions approach the hj_1 distribution such that at $x/R_e \geq 35$ the centerline velocities almost agree.

Figure 15 shows the streamwise distribution of the axial and radial turbulence intensity on the centerline. The intensity of the axial velocity fluctuations in Fig. 15a rises rapidly downstream of the nozzle exit. At the nozzle exit the centerbody nozzle hj_2 and the centerbody-plus-strut nozzle hj_3 solutions possess much higher

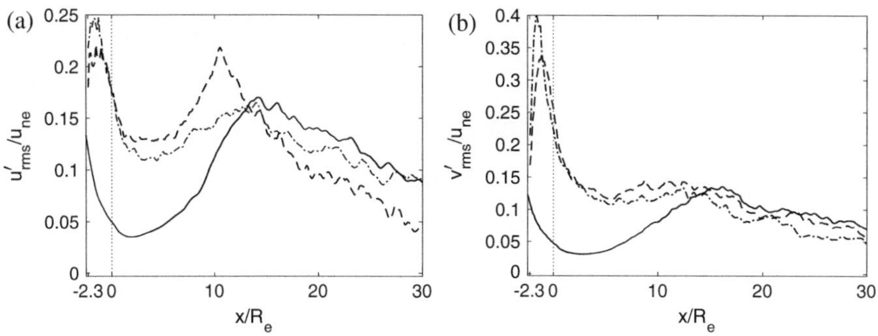

Fig. 15 Streamwise distribution at $r/R_e = 0$ of (**a**) the rms axial velocity and (**b**) the rms radial velocity for (—) hj_1, (——) hj_2, (-··-) hj_3

turbulence intensity than the clean nozzle hj_1 solution due to the enhanced turbulent mixing caused by the centerbody and the struts. Further downstream of $x/R_e \geq 15$ all profiles of the rms axial and radial velocities show a similar decaying trend.

4.2 Acoustic Field

The acoustic perturbation equations (APE) are applied to determine the sound propagation and to identify a dominant noise source excited by the hot jets. Since a compressible flow problem is tackled the APE-4 system is used [6].

For the computations a time step $\Delta t = 0.011R_e/a_\infty$ is chosen to obtain stable numerical solutions. The acoustic analyses include the sound waves whose maximum wavenumber $k_{\max} = 2\pi/\lambda_{\min}$ is approximately $0.36\pi/R_e$. The source fields are provided for all Runge-Kutta steps using a least squares optimized interpolation algorithm [9]. The time interval reconstructed by the 2251 LES snapshots is $T_{\text{total}} = 148.5R_e/u_e$.

The acoustic simulation setup and mesh details are discussed at length in [3].

In Fig. 16 the acoustic field determined by the aforementioned numerical schemes is illustrated. The contours of the acoustic pressure are ranged in $p' \leq 5 \times 10^{-6}\rho_0 a_0^2$ near the jet nozzle region. The acoustic pressure of the configuration hj_1 possesses smaller amplitudes than the other two configurations hj_2 and hj_3. At the nozzle exit in Fig. 13 the mean axial velocity in the radial direction ($r = \sqrt{y^2 + z^2}$) decreases for the clean nozzle configuration hj_1. The turbulent fluctuations in the shear layer are less pronounced for the single jet hj_1 as discussed in Fig. 15. These are the major reason of a low acoustic energy in the single jet hj_1.

The overall acoustic level in Fig. 17 evidences the low acoustic emission of the single jet hj_1. The profiles of three acoustic fields are obtained by the microphones aligned in the axial direction at the sideline location $8R_e$ away from the jet centerline. The dominant wave radiation occurs in the upstream position due to

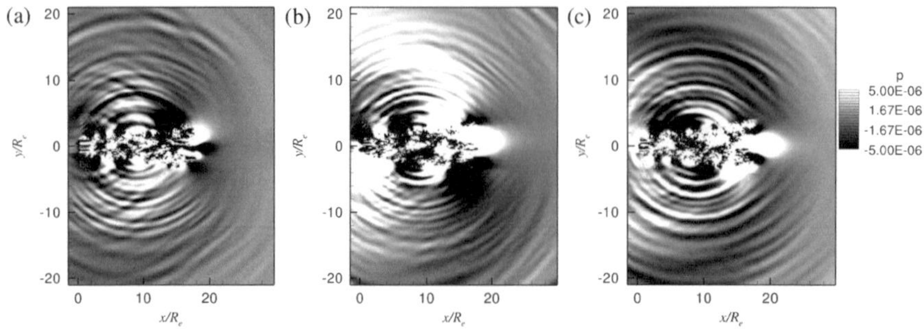

Fig. 16 Acoustic pressure contours in the range of $|p'/\rho_0 a_0^2| \leq 5 \times 10^{-6}$ on the $z = 0$ plane, (**a**) hj_1, (**b**) hj_2, and (**c**) hj_3

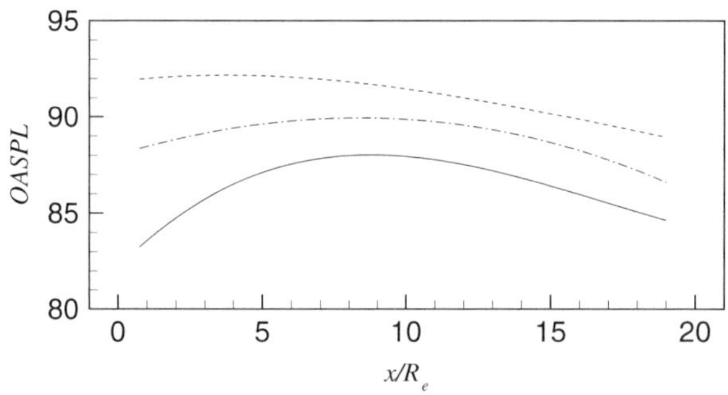

Fig. 17 Overall sound pressure level in dB at the radial distance of $8R_e$ from the jet centerline, (—) hj_1, (– –) hj_2, (-·-) hj_3

the unperturbed jet core in the nozzle exit area. The microphone in a downstream location captures the acoustic waves at a relatively farther distance from the end of the jet core. The centerbody nozzle configuration hj_2 generates the most powerful acoustics which shows 3 dB larger OASPL at a streamwise position $x = 10R_e$ compared to the single jet hj_1. The additional turbulence mixing by struts in the configuration hj_3 reduces the acoustic generation by approximately 2–4 dB over the streamwise position $R_e \leq x \leq 19R_e$. The acoustic directivity of the single jet hj_1 shows a silent zone in the upstream position $x \leq 5R_e$. Compared with the findings of the single jet (hj_1) the axial profiles of the other jets (hj_2 and hj_3) show an approximately 2–9 dB higher acoustic pressure.

In Fig. 18 the acoustic spectra of a single and two centerbody jets are compared. The sound pressure is determined at the coordinates ($x = 3R_e$, $r = 8R_e$) for the sideline acoustics in Fig. 18a and ($x = 18R_e$, $r = R_e$) for the downstream acoustics in Fig. 18b. The sideline acoustics in Fig. 18a display a large increase of power spectral density in the frequency range $fD_e/u_e = 0.3 \sim 0.8$, where f is the frequency and u_e nozzle exit average velocity. The peaks are located at $fD_e/u_e = 0.45$ for the single jet hj_1 and at $fD_e/u_e = 0.5 \sim 0.6$ for the jets with a centerbody hj_2,

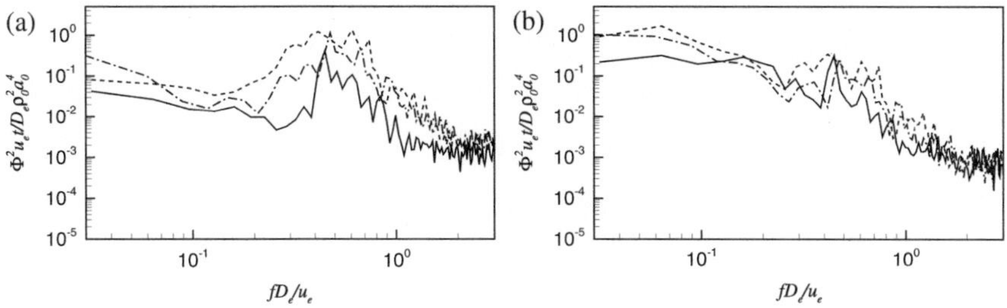

Fig. 18 Power spectra of the acoustic pressure signals determined at the coordinates (**a**) $x/R_e = 3,\ r/R_e = 8$ and (**b**) $x/R_e = 18,\ r/R_e = 8$: (—) hj_1, (– –) hj_2, (-··-) hj_3

hj_3. The downstream acoustics in Fig. 18b shows the pronounced low frequency radiation at $fD_e/u_e \leq 0.1$. The acoustic peaks occur at the same frequency range identified in the sideline acoustics. As indicated by the spectra of hj_2 and hj_3 the increase of the acoustic power becomes more prominent when the turbulent fluctuations increase. The sound generation of a hot jet includes two features. The first feature is the downstream acoustics due to the large scale turbulence in the shear layers and the second one is the sideline acoustics enhanced by the temperature gradient. Figure 18a illustrates the differences of the sideline acoustics. The acoustic radiation almost perpendicular to the jet axis is clearly intensified for the jets with a centerbody hj_2, hj_3 more than that of the single jet hj_1. Besides, in the frequency band $0.1 \leq fD_e/u_e \leq 0.5$ the acoustic level of the centerbody-plus-strut configuration hj_3 is reduced compared to that of the centerbody configuration hj_2.

5 Computational Specifications and Scalability Analysis

The simulations of the acoustic field were carried out on the CRAY XC40 at HLRS Stuttgart, containing two socket nodes with 12 cores at 2.5 GHz. Each node is equipped with 128 GB of RAM, i.e., each core has 5.33 GB of memory available for the computation. Strong scaling experiments were conducted to demonstrate the scalability of the APE-4 solver. Five core numbers were used, i.e., 512, 1024, 2048, 4096, and 8192. Furthermore, the results are based in 100 integrated time steps using a mono-block cubic grid with 256^3 grid points and periodic boundary conditions. The overall speedup as a function of the number of cores shown in Fig. 19 proves the good scalability of the code.

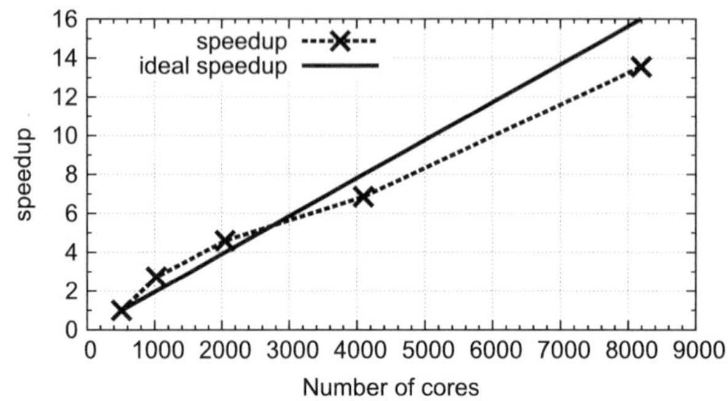

Fig. 19 Strong scaling experiment; Simulations were performed for 100 integrated time steps using five number of cores, i.e., 512, 1024, 2048, 4096 and 8192

6 Conclusion

The flow and the acoustic field of a ducted axial fan and a subsonic jet including the nozzle geometry were simulated by a hybrid CFD/CAA method. First, the flow field was computed by an LES and subsequently, the acoustic field was determined by solving the APE.

For the axial fan, two configurations with different tip-gap sizes, i.e. „$s/D_o = 0.001$ and $s/D_o = 0.01$ at the flow rate coefficient $\Phi = 0.195$ were performed and the results were compared to reference data. The findings showed that the diameter and strength of the tip vortex increase with the tip-gap size, while simultaneously the efficiency of the fan decreases. Increasingly the tip-gap size led to the strongest sound sources occur in the tip-gap regions as well as at wake of the fan blade. In the second step, acoustic field was determined by solving APE-4 system in rotating frame of reference. The overall agreement of the pressure spectrum and its directivity with measurements confirm the correct identification of the sound sources and accurate prediction of the acoustic duct propagation. The results show that the larger the tip-gap size the higher the broadband noise level.

Next, three turbulent jets emanating from of a clean divergent annular reference nozzle, a configuration with a centerbody and a geometry with a centerbody plus 5 equidistantly distributed struts were considered. The results showed an important dependence of the jet acoustic near field on the presence of the nozzle built-in components. For example, on the one hand, the presence of the centerbody increased the OASPL up to 6 dB compared to the clean nozzle, on the other hand, inclusion of the 5 struts reduced the OASPL up to 4 dB compared to the centerbody nozzle owing to the increased turbulent mixing caused by the struts which lessen the length and time scales of the turbulent structures shed from the centerbody.

Acknowledgements The research has received funding by the German Federal Ministry of Economics and Technology via the "Arbeitsgemainschaft industrieller Forschungsvereinigungen Otto von Guericke e.V." (AiF) and the "Forschungsvereinigung Luft- und Trocknungstechnik e.V."

(FLT) under the grant no. 17747N (L238) as well as from the European Community's Seventh Framework Programme (FP7, 2007–2013), PEOPLE program under the grant agreement No. FP7-290042 (COPAGT project). Computing resources were provided by the High Performance Computing Center Stuttgart (HLRS) and by the Jülich Supercomputing Center (JSC).

References

1. Alkishriwi, N., Meinke, M., Schröder, W.: Large-eddy simulation of streamwise-rotating turbulent channel flow. Comput. Fluids **37**, 786–792 (2008)
2. Boris, J.P., Grinstein, F.F., Oran, E.S., Kolbe, R.L.: New insights into large eddy simulation. Fluid Dyn. Res. **10**, 199–228 (1992)
3. Cetin, M.O., Koh, S.R., Meinke, M., Schröder, W.: Numerical analysis of the impact of the interior nozzle geometry on the jet flow and the acoustic field. Flow Turbul. Combust. (2016). doi:10.1007/s10494-016-9764-z
4. Cetin, M.O., Pauz, V., Meinke, M., Schröder, W.: Computational analysis of nozzle geometry variations for subsonic turbulent jets. Comput. Fluids **136**, 467–484 (2015)
5. Cetin, M.O., Pogorelov, A., Lintermann, A., Cheng, H.J., Meinke, M., Schröder, W.: Large-scale simulations of a non-generic helicopter engine nozzle and a ducted axial fan. In: High Performance Computing in Science and Engineering´ 15, pp. 389–405. Springer, Cham (2016)
6. Ewert, R., Schröder, W.: Acoustic perturbation equations based on flow decomposition via source filtering. J. Comput. Phys. **188**(2), 365–398 (2003)
7. Ewert, R., Schröder, W.: On the simulation of trailing edge noise with a hybrid LES/APE method. J. Sound Vibr. **270**(3), 509–524 (2004)
8. Freund, J.B.: Proposed inflow/outflow boundary condition for direct computation of aerodynamic sound. AIAA J. **35**(4), 740–742 (1997)
9. Geiser, G., Koh, S.R., Schröder, W.: Analysis of acoustic source terms of a coaxial helium/air jet. AIAA Paper 2011–2793 (2011)
10. Hartmann, D., Meinke, M., Schröder, W.: An adaptive multilevel multigrid formulation for Cartesian hierarchical grid methods. Comput. Fluids **37**(9), 1103–1125 (2008)
11. Hu, F.Q., Hussaini, M.Y., Manthey, J.L.: Low-dissipation and low-dispersion Runge-Kutta schemes for computational acoustics. J. Comput. Phys. **124**(1), 177–191 (1996)
12. Jeong, J., Hussain, F.: On the identification of a vortex. J. Fluid Mech. **285**, 69–94 (1995)
13. Johansson, S.: High order finite difference operators with the summation by parts property based on DRP schemes. Technical report, 2004–036 (2004)
14. Johnson, A.D., Xiong, J., Rostamimonjezi, S., Liu, F., Papamoschou, D.: Aerodynamic and acoustic optimization for fan flow deflection. AIAA paper, 2011–1156 (2011)
15. Koh, S.R., Geiser, G., Schröder, W.: Reformulation of acoustic entropy source terms. AIAA paper, 2011–2927 (2011)
16. Konopka, M., Meinke, M., Schröder, W.: Large-eddy simulation of shock-cooling-film interaction. AIAA J. **50**, 2102–2114 (2012)
17. Kunnen, R.P.J., Siewert, C., Meinke, M., Schröder, W., Beheng, K.D.: Numerically determined geometric collision kernels in spatially evolving isotropic turbulence relevant for droplets in clouds. Atmos. Res. **127**, 8–21 (2013)
18. Lintermann, A., Schlimpert, S., Grimmen, J.H., Günther, C., Meinke, M., Schröder, W.: Massively parallel grid generation on HPC systems. Comput. Methods Appl. Mech. Eng. **277**, 131–153 (2014)
19. Meinke, M., Schröder, W., Krause, E., Rister, T.: A comparison of second- and sixth-order methods for large-eddy simulation. Comput. Fluids **31**(4–7), 695–718 (2002)
20. Papamoschou, D., Shupe, R.S.: Effect of nozzle geometry on jet noise reduction using fan flow deflectors. AIAA paper 2006–2707 (2006)

21. Pardowitz, B., Tapken, U., Knobloch, K., Bake, F., Bouty, E., Davis, I., Bennett, G.: Core noise – identification of broadband noise sources of a turbo-shaft engine. AIAA paper 2014–3321 (2014)
22. Pogorelov, A., Meinke, M., Schröder, W.: Cut-cell method based large-eddy simulation of tip-leakage flow. Phys. Fluids **27**(7), 075106 (2015)
23. Pogorelov, A., Meinke, M., Schröder, W.: Effects of tip-gap width on the flow field in an axial fan. Int. J. Heat Fluid Flow (2016). doi:10.1016/j.ijheatfluidflow.2016.06.009
24. Pogorelov, A., Meinke, M., Schröder, W., Kessler, R.: Cut-cell method based large-eddy simulation of a tip-leakage vortex of an axial fan. AIAA paper 2015-1979 (2015)
25. Schneiders, L., Hartmann, D., Meinke, M., Schröder, W.: An accurate moving boundary formulation in cut-cell methods. J. Comput. Phys. **235**, 786–809 (2013)
26. Schröder, W., Ewert, R.: LES-CAA coupling. In: Large-Eddy Simulations for Acoustics. Cambridge University Press (2005)
27. Zhu, T., Carolus, T.H.: Experimental and numerical investigation of the tip clearance noise of an axial fan. GT2013-94100 (2014)

Adding Hybrid Mesh Capability to a CFD-Solver for Helicopter Flows

Ulrich Kowarsch, Timo Hofmann, Manuel Keßler, and Ewald Krämer

Abstract The enhancement of the so far structured Computational Fluid Dynamics solver FLOWer to enable the use of hybrid meshes and its advantage to numerical helicopter simulations is presented. The improvement is conducted by the implementation of unstructured grid handling into the existing code framework. The aim of the implementation is to reduce meshing effort in near body regions requiring the mapping of complex surfaces including boundary layer extrusion. Using the hybrid mesh approach, off-body regions can still be solved with structured meshes using computationally efficient higher order methods. This off-body region can be meshed automatically using Cartesian grids. The unstructured module features a second-order reconstruction scheme with an efficient GMRES implementation to solve linear systems of equations. Efficient high performance computation is ensured by multi-blocking and efficient load balancing considering the computational effort of the block according to the mesh type and numerical methods applied to. A forward facing step test case provides a reliable reproduction of different physical phenomena. An application-oriented complete helicopter simulation with particular use of unstructured body grids demonstrates the benefit of the hybrid mesh approach regarding our regular work flow.

1 Introduction

The helicopter aerodynamics are characterized by a highly unsteady flow field around a very complex geometry. Besides the high requirements of the code's numerics – such as ALE formulation to consider grid movements, the Chimera method to enable relative grid movements, and higher-order methods for vortex-afflicted flow conservation – high quality meshes mapping the complex geometry have to be provided. So far, the considered CFD-code supports only the computation of structured meshes which take a lot of time to be created manually for near body meshes.

U. Kowarsch (✉) • T. Hofmann • M. Keßler • E. Krämer
Institut für Aerodynamik und Gasdynamik, Universität Stuttgart, Pfaffenwaldring 21, D-70569 Stuttgart, Germany
e-mail: kowarsch@iag.uni-stuttgart.de

W.E. Nagel et al. (eds.), *High Performance Computing in Science and Engineering '16*, DOI 10.1007/978-3-319-47066-5_31

However, with an increasing use of high-fidelity CFD in the early helicopters design phase, more flexibility is required. Therefore, a work flow with an efficient numerical code enabling rapid response to arising issues is of significant concern. A reduction of the human workload can be achieved by the usage of unstructured meshes, often enabling an automatized mesh generation for even complex geometries. Yet, the use of unstructured meshes requires a higher computational effort, therefore it is not inevitably the better choice compared to structured meshes. In addition, higher-order methods are much more costly for unstructured meshes and therefore often out of scope for application-oriented simulations, whereas structured meshes enable efficient implementations of advanced numerical schemes. This comparison shows that both methods do have their unique advantage over each other and are both favourable in their own ways especially for helicopter simulations. Therefore, the aim is to provide a hybrid code enabling *easy to mesh* unstructured body grids for complex geometries and *computationally efficient* higher-order computed structured meshes in the remaining body sections and in the Cartesian off-body mesh.

The following chapters will provide a summary of the code extension, a validation case and an application-oriented test case with focus to achieve high-performance on the HLRS cluster platform CRAY XC40 Hazelhen.

2 Initial Numerical Code

The hybrid mesh treatment is implemented in the structured finite-volume flow solver FLOWer [8], originally developed by the German Aerospace Center (DLR) and enhanced with various functions by the Institute of Aerodynamics and Gas Dynamics (IAG) of the University of Stuttgart. The code discretizes the unsteady Reynolds-averaged Navier Stokes equations with different spatial orders for the flux computation. Besides the standard second-order central-difference JST scheme [5], a fifth order Weighted Essentially Non-Oscillatory (WENO) scheme [9] is available. The time discretization is achieved by merging the governing differential equation in space with the implicit dual time-stepping approach according to Jameson [4], which transforms each time step into a steady-state problem. The steady-state problems can then be solved with a conventional time stepping scheme. In case of FLOWer a Runge-Kutta scheme is used. To support an efficient computation, convergence accelerators like multigrid and residual smoothing are implemented in the code. Essential for helicopter flows, fluxes due to grid movements are taken into account using an Arbitrary Lagrangian Eulerian (ALE) approach. In addition, the Chimera technique for overset grids enables relative movements between grids, which allows for example the simulation of rotor-fuselage configurations. Mandatory for a reliable representation of the helicopter's physics is the consideration of the helicopter's flight state and aeroelasticity. Therefore, flight mechanics are taken into account to compute the helicopter's orientation in space due to the acting aerodynamic loads. Structural dynamics are considered by the deformation of the body meshes to include the aeroelasticity [3, 6]. An efficient computation is achieved by a multi-block structure of the grid to enable parallel computing with a satisfying scaling beyond 24,000 cores [7]. This comprehensive features of the code make it one of very few codes world-wide for high-fidelity helicopter simulations. With this unique characteristics the code is the optimal basis for the extension, although the code's architecture is designed to process structured meshes only.

3 Hybrid Mesh Implementation

The unstructured extension is implemented as an additional module extending the code's features and preserving the current properties for the structured mesh treatment. Therefore, using a hybrid mesh discretization does not affect structured meshed areas and all implemented features and extensions like higher-order methods are still applicable.

The unstructured module allows a mesh creation using different cell types, like tetrahedrons, prisms, pyramids, and hexahedrons, which can be mixed within a mesh.

The unstructured spatial discretization of the unsteady Reynolds-averaged Navier Stokes equations is based on the cell-centred scheme as for the structured code. The convective fluxes are determined by a Godunov-type HLLC Riemann solver according to Toro [11], which is also used by the WENO scheme for structured blocks. Second-order accuracy is achieved by piecewise linear reconstruction (PLR) according to Barth and Jespersen [2]. The gradients of the conservative variables, which are required by the reconstruction, are evaluated by the least-squares approach [1]. In principle, second- and higher-order upwind schemes tend to generate oscillations and spurious solutions in regions with large gradients. To avoid the creation of such extrema, a limiter function is required. However, this function reduces the order and therefore the accuracy of the spatial discretization. For the unstructured enhancement of FLOWer, two different limiter functions are implemented. The limiter according to Barth and Jesperson [2] is one of the simplest function and enforces a monotone solution. The limiter is rather dissipative and smears gradients and discontinuities. The function according to Venkatakrishnan [12, 13] allows the user to decrease the limitation by a parameter and achieve the theoretical order of the numerical scheme. Viscous fluxes of unstructured blocks are determined with flow quantities and their first derivatives. To compute the gradients, the least-squares method for the conservative variable's gradients is reused.

For the convective variables, the implicit dual time-stepping approach according to Jameson is applied for the unstructured blocks as it is done for the structured blocks. The steady-state problem is also evaluated by a explicit Runge-Kutta scheme. To improve the convergence of the pseudo time step, convergence acceleration techniques for unstructured blocks are implemented. The local time-stepping method allows every control volume to determine its ideal time step. The implicit residual smoothing shifts the characteristic of the explicit Runge-Kutta scheme towards an implicit method. Hence higher CFL numbers can be used. Multigrid methods would further accelerate the convergence of the unstructured block handling. This will be a topic for further development of the unstructured module.

The turbulence in unstructured blocks is modelled by the two-equation Wilkox $k-\omega$ turbulence model [14]. However, the turbulence model for unstructured blocks can be selected independently from the turbulence model of structured blocks. The convective and viscous fluxes are approximated by first-order methods. Time discretization for turbulence variables is achieved by a dual-time stepping scheme with an implicit treatment of the pseudo time step. The implicit method is more robust than using an explicit method for the turbulence variables, which is applied to the structured blocks. However, contrary to the implicit operator constituting a block-diagonal matrix for structured meshes for the equation, unstructured meshes lead to a sparse, non-symmetric block matrix, with a quasi-random distribution of non-zero elements. This requires much more focus on solving the linear system of equation than for the structured code, which can easily be solved by the performance-efficient Thomas algorithm. For the unstructured blocks, the equation is solved by an iterative GMRES(m) (Generalised Minimal Residual) algorithm suggested by Saad and Schulz [10]. The efficiency of the algorithm is further increased by an ILU(0) (Incomplete Lower Upper) pre-conditioner. Non-zero elements in the matrix represent the grid connectivity and therefore in a single row the corresponding cell neighbours. To reduce the memory requirements drastically, only (ncell)×(7) elements are stored instead of a full (ncell)×(ncell) matrix. Considering up to hexahedral cells, all non-zero entries in a row can thus be mapped containing the entry of the cell itself on the main diagonal and a maximum of six neighbour cells. However, the compressed storage scheme requires additional decompressing information stored in an additional array which can be stored memory-efficient with 1 byte integers. This approach results in less memory bandwidth and increased efficiency of the equation system solver. Furthermore, a restarted GMRES(m) method is used to limit the amount of Krylov subspaces and the ILU(0) pre-conditioner creates no additional fill-in.

The hybrid mesh capability requires an interfacing between the unstructured and structured meshed areas, which is solved using the already available Chimera interpolation method. The method enables the interpolation between arbitrary grid overlaps by the transformation of meshes into point clouds. Therefore, the currently implemented Chimera method is already capable to handle the overlap between a structured and unstructured meshed area.

3.1 High Performance Computing and Parallelization

Besides the numerical methods required for the processing of unstructured grids, the efficient parallelization of the work is a key feature of the code to be applicable for current and upcoming research and development. Therefore, the implementation is integrated in the existing parallelization process for structured grids. By splitting the grid into sub-grids, so called blocks, the grid can be distributed over several computation units executing the numerics of the sub-grids separately. With so called ghost-layers which consist of dummy cells at the block-boundaries, the information is exchanged between the different sub-grids enabling the exchange of numerical flux between blocks. In case of structured grids the block splitting process is performed in the grid generation tool. Since this functionality is not available for unstructured grids with the grid generation programs used, a pre-processing tool was created. The computational workload is defined by the number of cells of the sub-grid/block. With the input of the desired cell size per block and the grid in CGNS-format, a block splitting using the widely utilized METIS library is performed. The resulting mesh is prepared for the FLOWer simulation with specific output including the ghost-layer information used for data-exchange over the block boundaries. A workload factor considers the additional workload required, compared to structured methods, during the distribution of the grid block in case of a hybrid mesh. This approach ensures an equal workload for each process. The workload factor is measured by a single-core single-block computation applying structured numerics compared to a computation applying unstructured numerics to a hexahedral mesh. The computation time of an iteration is measured and the workload factor of an unstructured computation is determined by the ratio to the computation time using the structured numerics. Compared to the standard second-order JST scheme, the unstructured computation requires 2.5 times more computational effort. This is equal to the additional effort required for the higher-order WENO scheme, which is extensively used in current helicopter simulations. With a consideration of these factors during load balancing, there is no influence of the unstructured approach on the parallelization logic.

4 Validation Case

In this chapter a test case is presented showing the successful validation of the implementation. A representative test case for the numerical challenges faced by a CFD-code is the computation of the viscous flow over a forward facing step (FFS). The front side of the step leads to a stagnation of the flow including a recirculation area. The sharp edge challenges the numerics for its capability of representing viscous effects resulting in flow separation with a long recirculation vortex on the step's upper side. The reference of this validation is the flow field of the structured computation using the standard second-order JST method. The simulations are performed using 3-D meshes with equal mesh resolutions. However, the flow over a

FFS has a two-dimensional flow characteristic. The free-stream Mach number is set to 0.2, leading to a subsonic flow with very slight compressible effects, representing a usual on-flow towards a helicopter geometry. The same turbulence model (Wilkox $k-\omega$) for the unsteady RANS simulation is applied to concentrate on the differences due to the flux computation approaches.

Figure 1 shows the comparison of the resulting flow field using the different computation methods with comparable meshes. For the unstructured computation the PLR scheme is used in combination with the Venkatakrishnan limiter. The flow characteristics show a very good agreement between the structured and unstructured computation. The recirculation area in front of the step is comparable in shape and magnitude. Most important for the flow field characteristic of the FFS is the separation behind the step. Comparing this area shows a very good accordance in terms of the extension of the separation vortex in wall normal and downstream direction. The position of the reattaching point of the flow is similar. An additional important characteristic is the increase of momentum thickness after the step as a result of the viscous effects over the step. Comparing the velocity profile at the most downstream position, a good accordance between the wall normal distance at which the velocity drops significantly compared to the scaled free stream velocity of 1.0 is achieved.

(a)
(b)

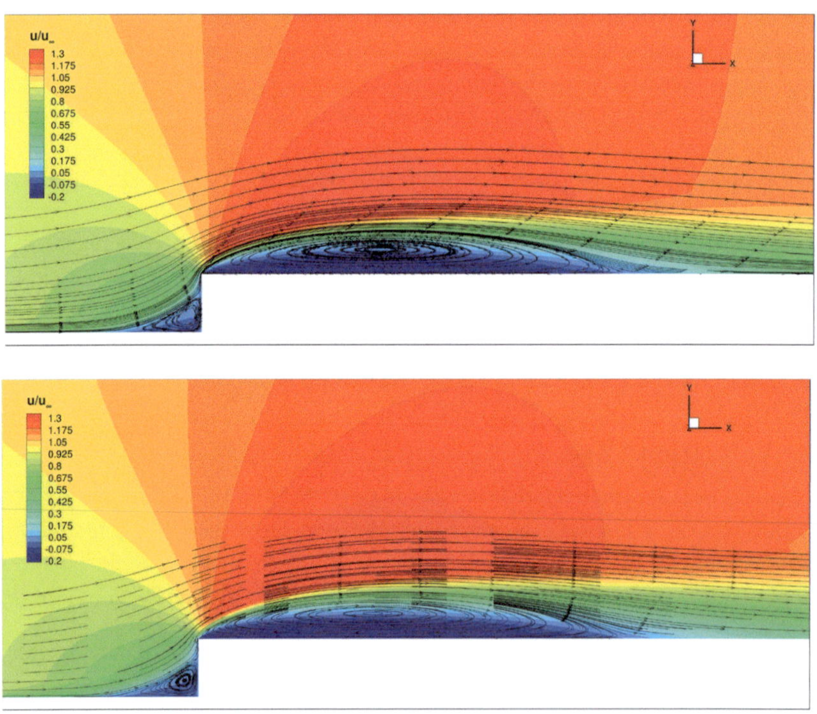

Fig. 1 Flow solution of the forward facing step validation case. (**a**) Unstructured computation using PLR and VK-Limiter of 200. (**b**) Structured computation using JST-scheme

Fig. 2 Surface of the geometry components of the helicopter considered for the simulation

5 HPC Simulation of a Hybrid-Meshed Helicopter

In the following section a simulation of a helicopter flow using the hybrid mesh approach is presented. The unstructured extension is aimed to be applied in near-body areas on geometries which are complex to be meshed. The considered simulation is a helicopter configuration including the main aerodynamic components: the main rotor, airframe, and tail rotor (cf. Fig. 2). This configuration is commonly used to get a first impression of the flow field as well as an estimation for the loads acting on the helicopter.

5.1 Mesh Generation

Figure 3 shows the area where the unstructured mesh ability is applied. At the area of the engine inlet, several geometric features would force a structured grid with a disproportionate amount of grid cells to reproduce the geometry. Therefore, an unstructured patch (red) is embedded into the structured meshed airframe in this region enabling a fast and efficient meshing of this area. As already mentioned, the interface between structured and unstructured meshes is performed via the Chimera method available in FLOWer. Therefore, overlapping regions of the grids are required where the data exchange takes place. The orange marked structured mesh region is considered in both, the unstructured and structured mesh leading to a congruent mesh area to ensure an accurate and conservative data exchange.The extrusion normal to the surface is performed in the same manner as for the structured grid. After the discretization of the boundary layer using prisms, the unstructured mesh topology switches to tetrahedrons for further extrusion. After several boundary layer heights, the Chimera interface into the structured Cartesian off-body mesh is applied. On this off-body mesh a higher-order scheme may be applied to ensure a low dissipation of the convecting flow.

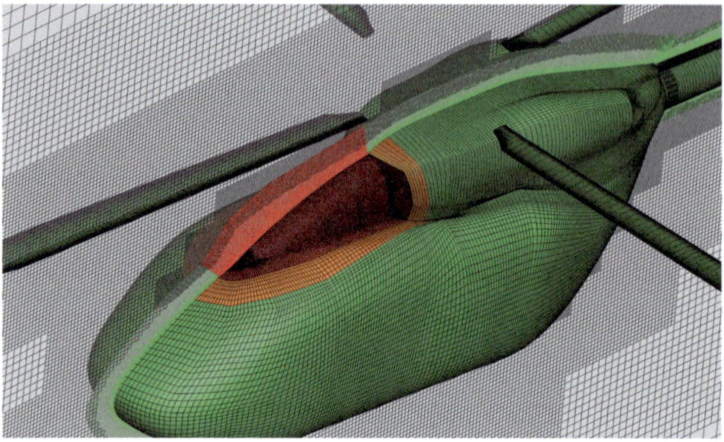

Fig. 3 Application of unstructured mesh in complex geometry regions. *Red* marks unstructured, *green* structured and *orange* interpolation areas. Slice made through the volume mesh

However, this application is performed using the second-order JST scheme in the structured meshes. The simulation is performed on the Cray XC40 Hazelhen system using 1200 cores. Both simulation strategies show a comparable computational effort. The higher computational workload of the unstructured mesh treatment is compensated by the slightly lower amount of grid cells required compared to the structured meshing. However, in summary the benefit is found in the human work load during mesh generation with is significantly lower for unstructured grids.

5.2 *Evaluation*

For evaluation purposes the computed flow field using the hybrid mesh approach is compared to a simulation with a structured only grid. The structured simulation can be seen as a reference which is extensively validated for its correctness against flight test and wind tunnel data.

Figure 4 gives an overview of the flow field in terms of a vortex visualization for the two simulation strategies. Both simulations show very similar results with minor influence of the unstructured mesh region on the vortex field around the helicopter.

In both cases, the area computed structured with the characterizing blade tip vortices shows no substantial differences. In the region of the engine inlet with its unstructured discretization in the hybrid mesh case, slight differences influence the vortex field around the helicopter. Minor differences in the flow separation region behind the edge downstream of the inlet are found. A more detailed flow field around the inlet is depicted in Fig. 5. A slice through the engine inlet plane shows the pressure levels for the two simulation methods. In both cases the region with higher pressure is found in front of the inlet, which is caused by the passing

Fig. 4 Comparison of the flow field in terms of vortex visualization using λ_2-criterion (*red*: hybrid, *green*: structured)

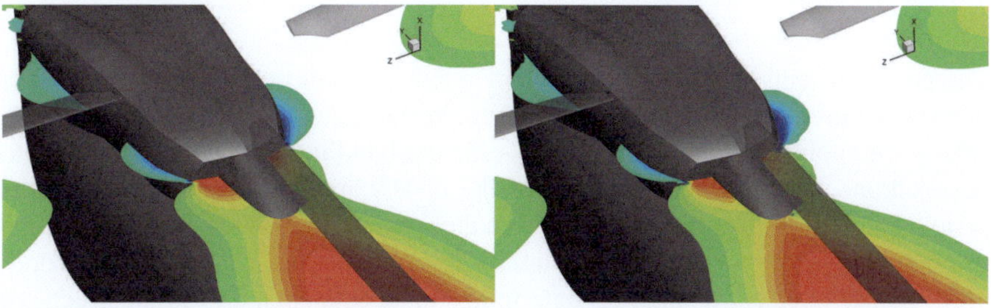

Fig. 5 Comparison of a slice through the engine inlet plane coloured with the pressure

blade at the considered time step. The expected character of the inlet is seen by a positive pressure in the region of the stagnation zone and subsequent flow separation with negative pressure after the edge to the engine cowling further downstream. In both cases a comparable magnitude in pressure is found leading to similar flow characteristics. The subsequent flow field downstream the fuselage shows the same properties, implying that no substantial deviating disturbances are introduced by the engine inlet using the different mesh approaches.

6 Advances in Code Optimization

Besides the implementation of further numerical methods, the current state of the code was investigated with regard to optimization potential running the code on HPC systems. In the course of a *"bring your own code"* optimization workshop organized by the HLRS and Cray in Stuttgart, a detailed profiling of the code was conducted to investigated bottlenecks and weak points. Main focus was set on high parallelization computations beyond 1000 nodes on the Cray XC40 Hazelhen system at the HLRS. By identifying optimization potential in the MPI communication, 5 % overall performance could be gained by using sub-world communicators for the parallelization which was presented at the last years HELISIM annual report [7]. The scaling characteristic in terms of strong- and weak-scaling shows the same behavior of the code as presented in [7]. Further, manual loop decompositions and restructuring allowed an improved cache reusing in runtime relevant routines leading to an additional performance increase of 20 %. This speed-up independent from code-parallelization enables a more efficient use of the resources available on each core.

Overall the workshop showed a high benefit in knowledge transfer from the HLRS and Cray staff to the users on the Cray XC40 Hazelhen system, giving them deeper insight into how to use the system's capability most efficiently.

7 Conclusions

The paper presents the implementation of a hybrid mesh treatment for the former block-structured only CFD-Code FLOWer. Various numerically optimized algorithms are applied to a computationally efficient handling of cell topologies up to hexahedrons. With the consideration of the code's application on highly parallel systems, the extensions are embedded into the communication structure of the code to enable massively parallel computations. The computational effort for the second-order unstructured computation is determined to be equal to a fifth-order structured computation, which can be applied at the same time to structured meshed regions.

Validation of the code in terms of a forward facing step shows very good results for the hybrid mesh approach. A full helicopter simulation with an unstructured meshed engine inlet shows the capability to represent the physical behaviour with good accuracy using the hybrid mesh approach, enabling the discretization of complex areas using an unstructured discretization.

Acknowledgements The investigations are based on the long-standing cooperation with the High Performance Computing Center (HLRS) in Stuttgart who provided us with support and service to perform the computations on their high performance computing system Cray XC40 Hazelhen. We greatly acknowledge the German Aerospace Center (DLR) making us their CFD-code FLOWer available for advancements and research purpose, which we would like to thank for.

References

1. Barth, T.J.: Aspects on unstructured grids and finite volume solvers for the Euler and Navier-Stokes equations, AGARD report 787, pp. 6.1–6.61. VKI special course on unstructured grid methods for advection dominated flows (1992)

2. Barth, T.J., Jespersen, D.C.: The design and application of upwind schemes on unstructured meshes. AIAA paper 89-0366 (1989)
3. Busch, R.E., Wurst, M.S., Keßler, M., Krämer, E.: Computational aeroacoustics with higher order methods. In: Nagel, W.E., Kröner, D.H., Resch, M. (eds.) High Performance Computing in Science and Engineering '12, pp. 239–253. Springer, Berlin/New York (2012)
4. Jameson, A.: Time dependent calculations using multigrid, with applications to unsteady flows past airfoils and wings. In: Proceedings of the 10th AIAA Computational Fluid Dynamics Conference, Honolulu (1991)
5. Jameson, A., Schmidt, W., Turkel, E.: Numerical solution of the Euler equations by finite volume methods using Runge-Kutta time-stepping schemes. In: 14th AIAA Fluid and Plasma Dynamic Conference, Palo Alto (1981)
6. Kranzinger, P.P., Keßler, M., Krämer, E.: Advanced CFD-CSD coupling – generalized, high performant, radiual basis function based volume mesh deformation algorithm for structured, unstructured and overlapping meshes. In: Proceedings of the 40th European Rotorcraft Conference, Southampton (2014)
7. Kranzinger, P.P., Kowarsch, U., Schuff, M., Keßler, M., Krämer, E.: Advances in parallelization and high-fidelity simulation of helicopter phenomena. In: Nagel, W.E., Kröner, D.H., Resch, M. (eds.) High Performance Computing in Science and Engineering '15, Stuttgart (2015)
8. Kroll, N., Eisfeld, B., Bleeke, H.M.: The Navier-Stokes Code FLOWer. Notes on Numerical Fluid Mechanics, vol. 71, pp. 58–68. Vieweg, Braunschweig/Wiesbaden (1999)
9. Liu, X.-D., Osher, S., Chan, T.: Weighted essentially non-oscillatory schemes. J. Comput. Phys. **115**, 200–212 (1994)
10. Saad, Y., Schulz, M.H.: GMRES: a generalized minimal residual algorithm for solving nonsymmetric linear systems. SIAM J. Sci. Stat. Comput. **7**, 856–869 (1986)
11. Toro, E.F.: Riemann Solvers and Numerical Methods for Fluid Dynamics. Springer, Berlin (1997)
12. Venkatakrishnan, V.: On the accuracy of limiters and convergence to steady state solutions. AIAA paper 93-0880 (1993)
13. Venkatakrishnan, V.: Convergence to steady-state solutions of the Euler equations on unstructured grids with limiter. J. Comput. Phys. **118**, 120–130 (1995)
14. Wilcox, D.C.: Re-assessment of the scale-determining equation for advanced turbulence models. AIAA J. **26**, 1299–1310 (1988)

Direct Numerical Simulation of Heated Pipe Flow with Strong Property Variation

Xu Chu, Eckart Laurien, and Sandeep Pandey

Abstract Using supercritical fluid as coolant in a power cycle is generally considered as an advanced solution for energy conversion. When the pressure is above the critical point (P_c), thermo-physical properties vary significantly with temperature, which leads to complicated heat transfer phenomena. In the current project, direct numerical simulation (DNS) in a horizontal heated pipe has been developed for supercritical CO_2 using the numerical solver based on OpenFOAM. DNS enables us to investigate the detailed turbulence modulation and heat transfer characteristics. The horizontal layout of the pipe leads to a flow stratification, which is not observed in the vertical pipes from the report in the last year. Furthermore, the obtained turbulence data are serving for the development of advanced turbulence models.

1 Introduction

Using supercritical fluids in a power cycle is widely considered as an advanced solution. High efficiency, compact size, and reduced complexity are the main advantages of these cycles [6]. State-of-the art fossil power plants use the supercritical water Rankine cycle to increase the thermal efficiency to about 45 % [7]. Compared with water (critical pressure $P_c = 22.06$ MPa, critical temperature $T_c = 647.1$ K), CO_2 ($P_c = 7.38$ MPa, $T_c = 304.1$ K) has a lower critical pressure and critical temperature [1]. Supercritical fluids have distinctive properties. At supercritical pressure, the fluid phase change from liquid to gas does not exist as in subcritical flows. When the temperature rises across the pseudo-critical point (T_{pc}), the density (ρ), the thermal conductivity (κ) and the dynamic viscosity (μ) decrease drastically, and the specific heat capacity (C_p) shows a peak in a very narrow temperature range. The variable properties of CO_2 as a function of the temperature (T) at a constant pressure ($P_0 = 8$ MPa) above the critical pressure has been introduced in [4].

X. Chu (✉) • E. Laurien • S. Pandey
Institute of Nuclear Technology and Energy Systems, University of Stuttgart, Pfaffenwaldring 31, 70569 Stuttgart, Germany
e-mail: xu.chu@ike.uni-stuttgart.de; sandeep.pandey@ike.uni-stuttgart.de; eckart.laurien@ike.uni-stuttgart.de

W.E. Nagel et al. (eds.), *High Performance Computing in Science and Engineering '16*, DOI 10.1007/978-3-319-47066-5_32

Based on the previous experience [3, 9, 18, 19], dealing with steep property variation and related complicated flow phenomenon is beyond the ability of Reynolds-averaged modeling (RANS). Even if a certain turbulence model has shown some satisfying results in a few cases, its superiority may not be achieved in other cases. On the other hand, only a few experimental studies delivered detailed hydraulic resistance, mean and turbulent velocity, and temperature fields. The technical difficulties and high cost required for developing such techniques have practically limited the progress of experimental works according to Yoo [19]. Jackson [10] suggested using high-fidelity DNS or LES to investigate the heat transfer to supercritical fluids and provided a reliable data base for modeling validation and improvement, which has been proved to feasible in He et al. [9]. According to the authors' knowledge, no DNS about the supercritical fluid flow in a horizontal pipe has been published, which offers an insight into the detailed flow mechanisms without any turbulence modeling. The current study is aimed at elucidating the flow pattern of a heated supercritical fluid in a horizontal pipe. Various simulation conditions will be reported. The pipe geometry is adjusted to $D = 1, 2\,\text{mm}$, which is in the range of printed circuit heat exchanger (PCHE) channels. The influence of buoyancy on the heat transfer and flow turbulence of supercritical fluid is going to be our major consideration.

2 Computational Details

2.1 Governing Equations

In the present DNS study, supercritical CO_2 in the pipe is intensively heated by the constant and uniform wall heat flux q_w, which leads to significantly variable properties. Considering this, the Navier-Stokes equations are formulated in low-Mach-number form (Eqs. 1, 2, and 3), in which the compressibility effect due to temperature change at constant pressure P_0 is included. Li et al. [11] use the full compressible Navier-Stokes equations in DNS of supercritical CO_2, and proved the validity of their assumption in low-Mach cases. This form of governing equations is also applied by other authors [2, 13] in this area:

$$\frac{\partial \rho}{\partial t} + \frac{\partial(\rho U_j)}{\partial x_j} = 0 \tag{1}$$

$$\frac{\partial \rho U_i}{\partial t} + \frac{\partial(\rho U_i U_j)}{\partial x_j} = -\frac{\partial P}{\partial x_i} + \frac{\partial}{\partial x_j}(\mu(\frac{\partial U_i}{\partial x_j} + \frac{\partial U_j}{\partial x_i})) \pm \rho g \delta_{i1} \tag{2}$$

$$\frac{\partial \rho h}{\partial t} + \frac{\partial(\rho U_j h)}{\partial x_j} = \frac{\partial}{\partial x_j}(\kappa \frac{\partial T}{\partial x_j}) \tag{3}$$

$$h = h(P_0, T), T = T(P_0, h), \rho = \rho(P_0, h), \mu = \mu(P_0, h), C_p = C_p(P_0, h), \kappa = \kappa(P_0, h). \tag{4}$$

2.2 Numerical Method

The governing equations, Eqs. (1, 2, and 3) are discretized with the open-source finite-volume code OpenFOAM V2.4 [16]. The Pressure-Implicit with Splitting of Operators (PISO) algorithm is applied for the pressure-velocity coupling. The temporal term is discretized with the second-order implicit differencing scheme. The spatial discretization is handled with a central differencing scheme and the third-order upwind scheme QUICK is adopted for the convective term in the energy equation.

Figure 1 shows the pipe geometry and the boundary conditions. At the inlet, an inflow generator of the length $L_1 = 5D$ with an isothermal wall is adopted to generate approximately fully developed inflow turbulence. A recycling/rescaling procedure [12] is applied in this domain, which does not require a priori knowledge of turbulent flow profiles. For accelerating the turbulence development, the velocity field is initialized with the perturbation method introduced by Schoppa and Hussain [15]. In the second section of the pipe $L_2 = 30D$, a constant wall heat flux q_w is applied. The boundary condition for the velocity field at the outlet is the convective boundary condition $\frac{\partial \phi}{\partial t} + U_c \frac{\partial(\phi)}{\partial x} = 0$, where ϕ can be any dependent variable, e.g. the velocity U.

The cylindrical pipe is discretized with a total of 80 Mio. structured hexahedral mesh. The mesh resolution is identical in all the simulation cases. The resolution is equivalent to approximately $168 \times 172 \times 400$ (radial r, circumferential θ and axial z direction) for the inflow domain and $168 \times 172 \times 2400$ for the heated domain, when converted from Cartesian to Cylindrical coordinates. The grid mesh is uniform spaced in the axial direction, and refined near the wall in the radial direction with a stretching ratio of 10, which corresponds to a dimensionless resolution of $0.11\,(\text{wall}) < \Delta y^+ < 1.1\,(\text{center})$, $(R\Delta\Theta)^+ \approx 6.5$, $\Delta z^+ = 4.6$ in wall units, i.e., $y^+ = yU_{\tau,0}/\nu_0$, where $U_\tau = \sqrt{\tau_w/\rho}$ based on the inlet Reynolds number $Re_0 = U_0 D/\nu_0 = 5400$. Compared with the DNS study of Bae et al. [2] at the same simulation conditions except the vertical placement of the pipe, the current DNS shows significant improvement of resolution in all three directions and time considering the same second order accuracy in both studies. Cumulatively, the total mesh number in the heated domain is about 10 times that of Bae et al. [2]. At the outlet of the pipe, a rise of Reynolds number should be considered in the mesh resolution. The dimensionless mesh resolution here is still higher

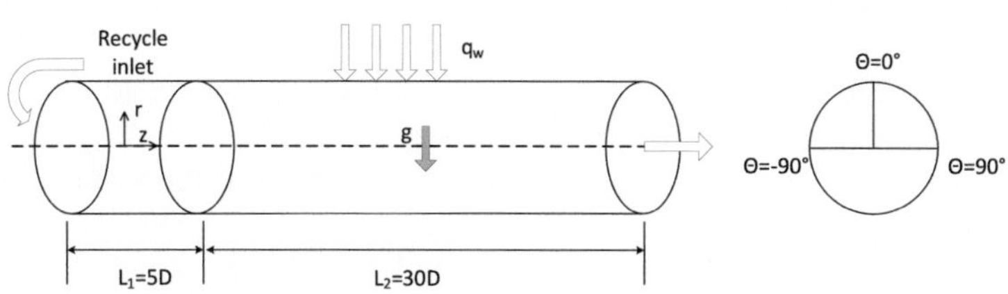

Fig. 1 Flow domain and boundary conditions

Table 1 Simulation conditions, identical inlet conditions $Re_0 = 5400$, $P_0 = 8$ MPa

case	Type	D (mm)	q_w (kW/m^2)	q^+ 10^4	T_0 (K)	$U_{z,0}$ (m/s)
SC160	Mixed	1	61.74	1.44	301.15	0.452
SC230	Mixed	2	30.87	1.44	301.15	0.225
SC230F	Forced (g = 0)	2	30.87	1.44	301.15	0.225
SC260	Mixed	2	61.74	2.88	301.15	0.225

than the reference work at the inlet, especially in radial and streamwise direction. Therefore, it is expected that the current mesh is fine enough for handling this simulation conditions. In the post processing, the mesh coordinate transformation from Cartesian coordinate to Cylindrical coordinate is necessary. The flow statistics are obtained through averaging in time.

This numerical procedure has been applied to the DNS of heated vertical pipe with air at $Re_0 = 4200, 6000$ [5], where the DNS is validated with experimental results. The variable properties of air are comparable with those of supercritical CO_2. Various flow statistics including heat transfer results and flow profiles match well with the experimental data. Besides, vertical pipe flow cases with supercritical CO_2 have been also investigated in our previous study [4] and validated with existing DNS work [2, 13]. Significant flow relaminarization and transition are observed in this study. Furthermore, the obtained turbulence data is serving for advanced turbulence modeling by Pandey and Laurien [14].

An introduction of simulation conditions is given in Table 1. Under the condition of the same inlet $Re_0 = 5400$, the pipe diameter D and the wall heat flux q_w are set to different values. The pipe diameter is considered to be an important parameter for the buoyancy effect. The fixed wall heat flux q_w results in a streamwise-distribution of wall temperature T_w. The dimensionless heat flux q^+ is defined as $q^+ = q_w/(\rho_0 U_0 C_{p,0} T_0)$. In the forced convection case SC230F, buoyancy is totally absent by omitting the gravity term ($g = 0$) in Eq. 2.

2.3 *Inflow Turbulence*

The resolution applied in the present DNS exceeds the previously used reference DNS of Eggels et al. [8]. Therefore, the quality of the inflow turbulence is validated with better resolved reference DNS data by Wu and Moin [17]. This DNS is obtained using a second-order finite difference method. Grid points of $256 \times 512 \times 512$ (r, θ and z direction) are spaced in the $L = 7.5D$ long pipe at $Re = 5300$. The root-mean-square velocity in dimensionless form $U^+ = U/U_\tau$ of three directions is shown in Fig. 2. The best agreement is observed in axial direction z, because current dimensionless resolution $z^+ = 4.5$ is similar and even slightly better than the reference work $z^+ = 5.3$. In circumferential direction θ, a small difference is observed because lack of resolution ($\theta^+ = 6.5$ compared with $\theta^+ = 2.2$ in Wu and Moin [17]).

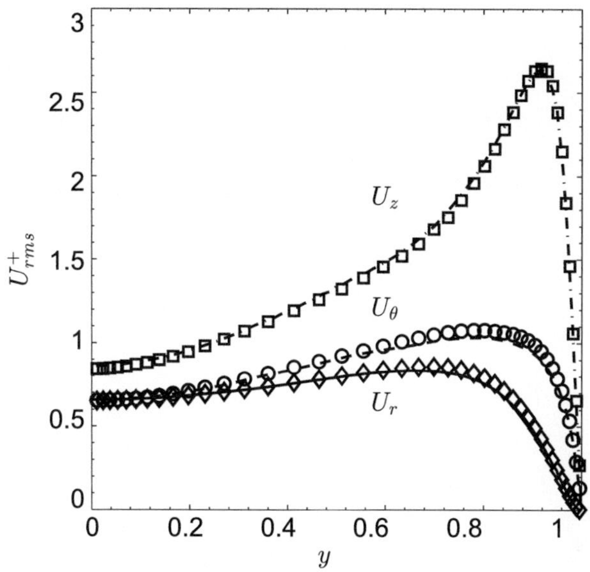

Fig. 2 Inflow turbulence validation, dimensionless velocity fluctuation U^+_{rms} in r, θ and z direction, lines: current DNS at $Re_0 = 5400$, symbols: DNS data from Wu and Moin [17] at $Re = 5300$

3 Results and Discussion

3.1 Bulk Properties

Figure 3a summarizes the development of wall temperature T_w on top- and bottom surface of the pipe. T_w is homogeneously distributed in circumferential direction (θ) in forced-convection case SC230F, but buoyancy leads to a non-uniform distribution of wall temperature in this direction. In SC160, SC230 and SC260, T_w is significantly higher on the top surface than the bottom surface. On the top surface, T_w shows a monotonically rising tendency in three cases, where the highest T_w distribution is found in SC260 due to high q_w. At the end of the pipe $z = 30D$, the temperature difference ΔT_w between top- and bottom surface is 365.2K (SC260), 234.2K (SC230) and 136.1K (SC160). The skin friction coefficient $C_f = 2\tau_w/(\rho_b U_b^2)$ distribution based upon local wall shear stress τ_w, local ρ_b and U_b is summarized in Fig. 3b. At the inlet, $C_{f,0} = 0.00896$ matches the Blasius estimation $C_f = 0.079\text{Re}^{-0.25} = 0.00897$ with 0.15 % difference. In the downstream direction, C_f on the bottom of pipe is higher than on the top surface in SC160 and SC230. On bottom surface, C_f in SC230 and SC260 shows similar development. But on the top surface, SC260 shows an obvious increasing tendency after about $z = 3D$, which is not clearly observed in SC230.

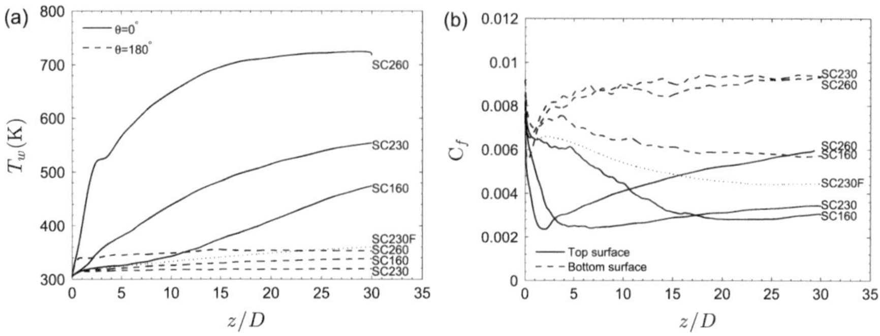

Fig. 3 Development of T_w (**a**) and C_f (**b**) in downstream direction, forced-convection case SC230F shows no differences in the circumferential direction

3.2 Average Flow Field and Secondary Flow

In the turbulence statistics below, we define the mean quantities with Reynolds- and Favre averaging, where $\bar{\phi}$ is the Reynolds average of any quantity and $\tilde{\phi} = \frac{\overline{\rho\phi}}{\bar{\rho}}$ is the mass-weighted (Favre) average. The corresponding fluctuations are denoted by $\phi' = \phi - \bar{\phi}$ and $\phi'' = \phi - \tilde{\phi}$. Figure 4 demonstrates the development of various average flow profiles in downstream direction of SC230. From top to bottom, velocity $\widetilde{U_z}/U_{z,0}$, temperature $\overline{T}$ (K), density $\overline{\rho}/\rho_0$, thermal capacity $\overline{C_p}/C_{p,0}$ are presented. In the following subsections, each case will be discussed separately.

Compared with SC160, a stronger buoyancy effect in SC230 leads to a deformation of the average velocity profile as can be seen in the first row of Fig. 4. At $z = 10D$, high-velocity flow with low density begins to concentrate in the bottom section and low-velocity flow with low density occupies the upper part of the pipe cross section. High-velocity flow takes a crescent shape at this position. At $z = 15D$ and $20D$, a small area of high velocity flow is developed close to the top wall surface and it connects with the major part of high-velocity flow at $z = 25D$. The high-velocity flow is found to be an anchor shape at this position. The quantitative analysis of the velocity field at $z = 25D$ is shown in Fig. 5a. At $\theta = 0°$, a velocity peak is observed at about $r/R = 0.75$, which corresponds to the high-velocity region near the top wall. Compared with that, the velocity profile at $\theta = 45°$ shows a low value from $r/R = 0.4$ to $r/R = 0.9$, which is also visualized in Fig. 4. This can be explained by the transport of secondary flow. Low-velocity flow close to the circumferential wall flows upwards due to low density and drops down at about $\theta = 45°$. Therefore, a low velocity region is developed here. The stratification of the temperature field is similar to that observed in SC160. The hot flow gathers near the top surface and it shows a significant temperature difference against the cold flow on the bottom. Compared with SC160, this hot layer becomes thicker. This change of the temperature field is also reflected in the density field in the third row. Due to buoyancy, high-temperature CO_2 with low density concentrates on the upper side of

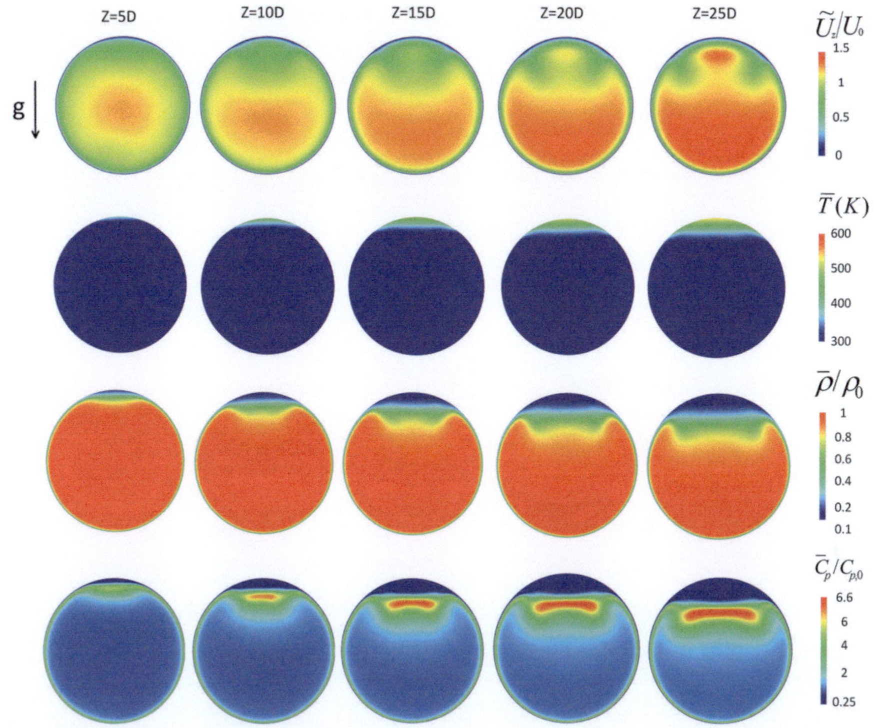

Fig. 4 Flow field of SC230 in downstream direction, velocity $\widetilde{U}_z/U_{z,0}$, temperature $\overline{T}$ (K), density $\overline{\rho}/\rho_0$, special thermal capacity $\overline{C}_p/C_{p,0}$

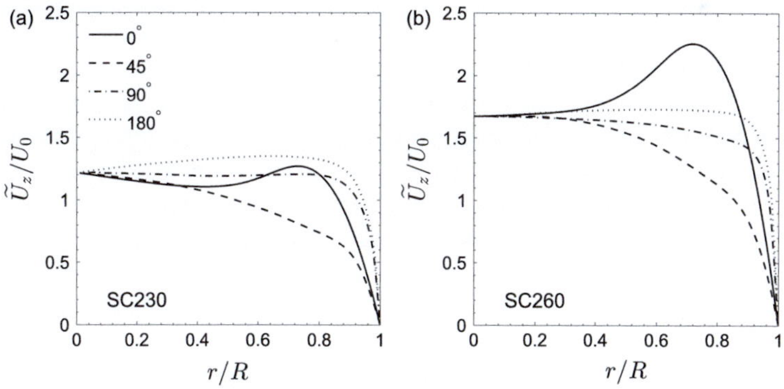

Fig. 5 Velocity profile $\widetilde{U}_z/U_0$ at $z = 25D$, (**a**): SC230, (**b**): SC260

cross section. With the input of wall heat flux, the low density layer is growing in downstream direction.

Vector plots of the 2-D average velocity field over the cross section are given in Fig. 6. The lines are colored with the normalized density field $\overline{\rho}/\rho_0$. The

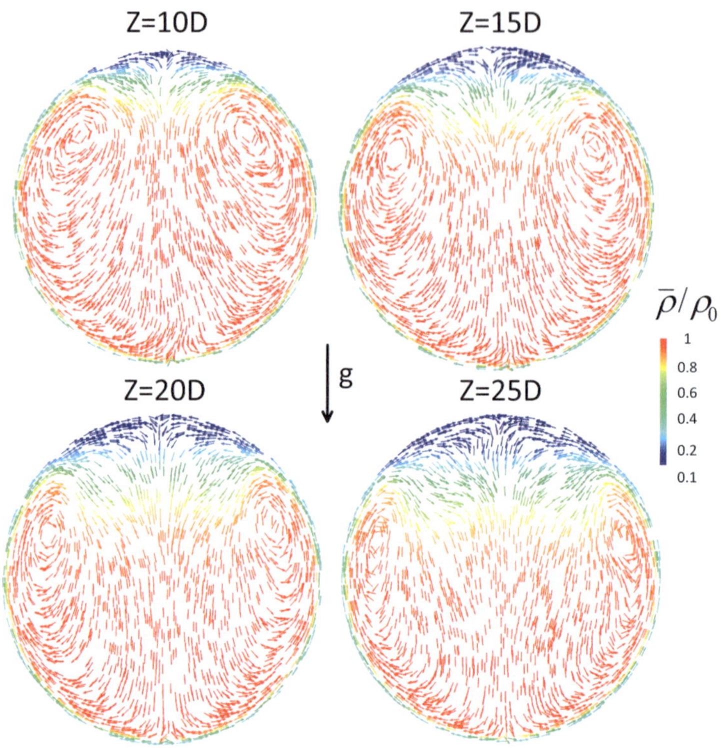

Fig. 6 Vector plot of the two-dimensional average velocity field of SC230 at various downstream positions

visualization shows that buoyancy brought by enormous density difference leads to the formation of a secondary flow. Following the path of velocity in all four figures in SC230 (Fig. 4), it is observed that the flow near the circumferential wall (marked in blue) is heated by the wall heat flux q_w firstly, which leads to a significant decrease of the density. As a result of buoyancy, this low-density flows near the wall flow upward along the respective wall surface, and meet near the top surface. Then it falls down in the gravitational direction along the centerline. The center of the vortex pair secondary flow is located nearly axis-symmetrically on the lateral sides. At these four streamwise positions, the positions of each vortex center are slightly different. Comparing the figures horizontally ($z = 10D$ to $z = 15D$, $z = 20D$ to $z = 25D$), the vortex center moves downwards. In downstream direction, the stratified layer with low-density flow is growing progressively, but the center of each vortex of the secondary flow is filled with high-density fluid (colored in red) while located just slightly below a layer between high- and low density, which is colored in yellow in the figure.

3.3 Turbulence Statistics

Figure 7 shows the evolution of the turbulent kinetic energy $TKE = \frac{1}{2}\overline{\rho U_i'' U_i''}$, which indicates the intensity of the velocity fluctuations in downstream direction. Generally, the TKE shows a decreasing tendency in downstream direction in all three cases. Because of the same inlet Reynolds number $Re_0 = 5400$, they are expected to give a similar distribution of TKE in the inlet section. After a length of five diameters in downstream direction, TKE shows the fastest decrease in SC260. Besides, the TKE is no more homogeneous in circumferential direction in SC260. Near the top surface, a region of low TKE appears, which is less obvious in SC230 at this position. In SC230, the ring of high TKE starts to deform at about $z = 10D$. It is broken by the low TKE region near the top surface and bended to the pipe center at the breakpoints. Similar distribution of the high TKE ring is observed in its downstream direction at $z = 15D$, $20D$ and $25D$. In SC160, the reduction of the TKE is also observed near the top surface starting at about $z = 15D$. The TKE distribution in SC260 is qualitatively similar to SC230, but it is noticeable that starting from $z = 20D$ near the top wall surface, a region of high TKE begins to build up, which cannot be clearly identified in SC230.

A quantitative analysis of the TKE at $z = 25D$ in various circumferential direction is shown in Fig. 8. The profile from isothermal flow at $z = 0D$ is given with the symboled line as a reference. At $z = 25D$, the TKE at all circumferential directions in these cases is reduced compared with that of isothermal flow. In the direction of $\theta = 0°$, the original peak value of TKE near the wall disappears in SC160 and SC230. In SC260, TKE shows a character of two peaks instead of a single peak in this direction. The peak near the wall ($0.8 < r/R < 0.9$) corresponds to the recovery of TKE in the last figure in the third row from Fig. 7.

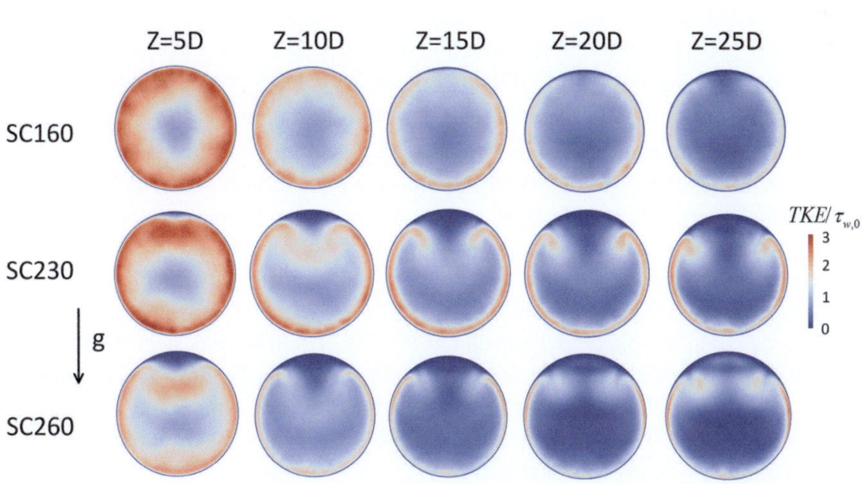

Fig. 7 Evaluation of normalized turbulent kinetic energy $TKE/\tau_{w,0}$ in downstream direction

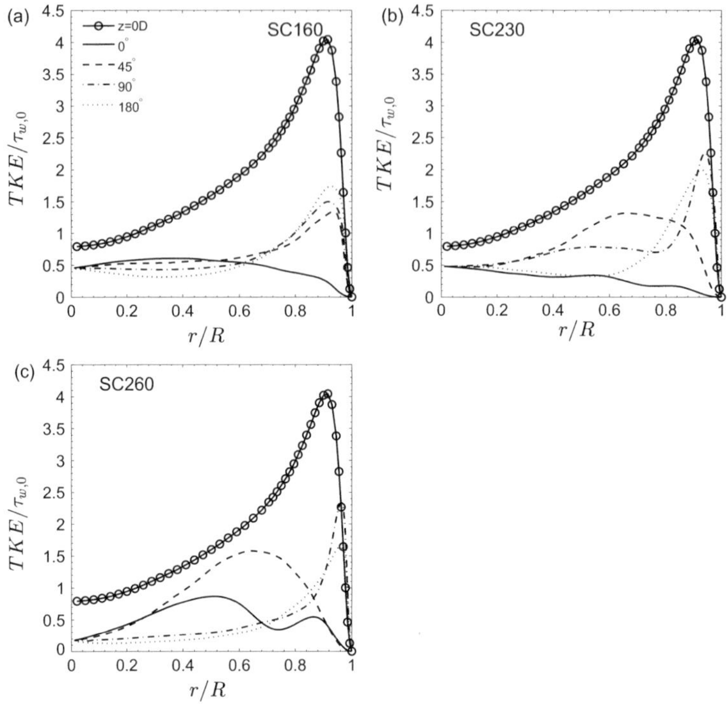

Fig. 8 $TKE/\tau_{w,0}$ at $z = 25D$ of SC160 (**a**), SC230 (**b**) and SC260 (**c**), legend is identical as shown in (**a**)

It is also the position, where a strong velocity gradient brought by flow acceleration is observed in Fig. 4. In SC230 and SC260, a broad peak value away from the wall ($0.6 < r/R < 0.8$) is found in the direction of $\theta = 45°$, which is absent in SC160.

The shear production rate of turbulent kinetic energy (Pk) at various circumferential positions at $z = 25D$ is shown in Fig. 9, where Pk is defined as $Pk = \overline{\rho U_i'' U_j''}\widetilde{\frac{U_i}{x_j}}$. The isothermal flow at $z = 0D$ is marked with a symbol as a reference. In SC230, Pk almost vanishes at $\theta = 0°$, which explains the significantly reduced TKE at this position in Fig. 8. The profile at $\theta = 45°$ shows a sign change near $r/R = 0.8$, which is relevant with the secondary flow at this position. Pk at $\theta = 90°$ is with a reduced peak value, while Pk at 180° shows a higher peak. For the pipe bulk area $0 < r/R < 0.9$, P_k is significantly reduced at $\theta = 0°$, 90°, and 180°. In SC260, P_k shows a slight double peak character at $\theta = 0°$. The first peak near the wall can be explained with the increased velocity gradient brought by flow acceleration as shown in Fig. 5. At $\theta = 45°$, Pk shifts its peak to $r/R = 0.7$ under the influence of secondary flow. At $\theta = 90°$ and 180°, narrow peak with a maximum close to the original value is observed in the figure.

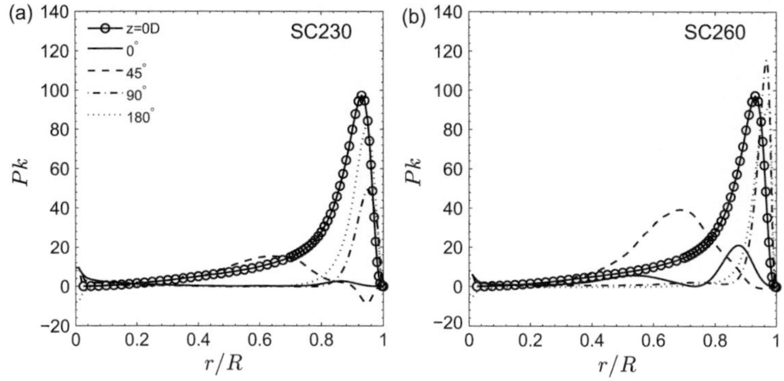

Fig. 9 Circumferential distribution of *Pk* at $z = 25D$ of SC230 (**a**), and SC260 (**b**), legend is identical as shown in (**a**)

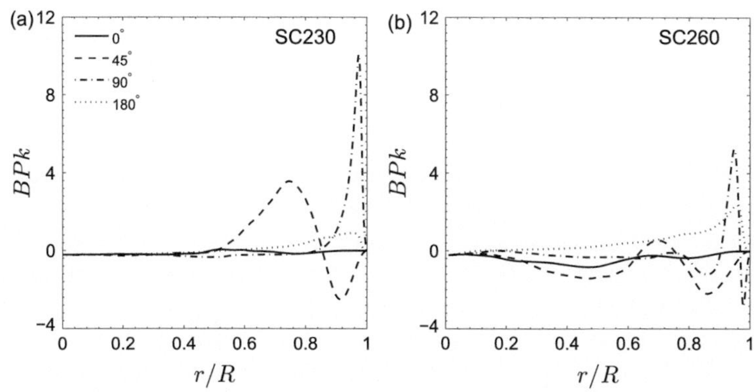

Fig. 10 Circumferential distribution of *BPk* at $z = 25D$ of SC230 (**a**), and SC260 (**b**), legend is identical as shown in (**a**)

Buoyancy production of turbulence ($BPk = g\overline{\rho U_i'}$) is depicted in Fig. 10. Compared with shear production for turbulence *Pk*, *BPk* is an order of magnitude lower. It points out that the direct contribution from buoyancy for turbulence is small. In $\theta = 0°$, *BPk* shows a flat distribution close to zero. The peak of *BPk* at $\theta = 90°$ corresponds to the secondary flow along the pipe wall. At $\theta = 45°$, the sign change of *BPk* indicates a damping of turbulence close to the wall and an enhancement next to it.

4 Computational Performance

The parallel computational performance will be discussed in this chapter. The hardware utilized for the computations is Hazel Hen located at the High-Performance Computer Center Stuttgart (HLRS, Stuttgart). Hazel Hen is a Cray XC40 system

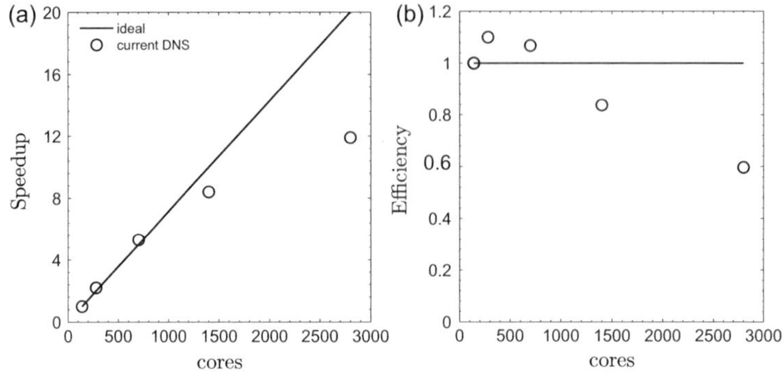

Fig. 11 HPC performance of current DNS case (strong scaling, 80 Mio. cells) on Hazel Hen, speedup (**a**), efficiency (**b**)

that consists of 7712 compute nodes. Each node has two Intel Haswell processors (E5-2680 v3, 12 cores) and 128 GB memory, and the nodes are interconnected by a Cray Aries each network with a Dragonfly topology. This amounts to a total of 185,088 cores and a theoretical peak performance of 7.4 PFlops.

Parallel scalability of the current numerical solver has been tested on the Hazel Hen platform, as shown in Fig. 11. Under the condition of the present mesh size (80 Mio. cells), the solver shows a linear, even super linear, scalability until 700 cores. A considerable speedup can be expected at 1400 cores (80 % efficiency) and 2800 cores (60 % efficiency). At 2800 cores, about 28000 cells are distributed on a single computational core. In a daily job, it costs about 4 days on 1400 cores for running 10 flow through time in the pipe. In the foreseeable future, the mesh resolution will increase to 300 Mio. aimed at a higher Reynolds number and an improved resolving of Kolmogorov scale and Batchelor scale.

5 Conclusions

In the current research, heat transfer to supercritical CO_2 in a horizontal pipe has been investigated using direct numerical simulation (DNS) for the first time. A well resolved DNS eliminates the uncertainty brought by turbulence modeling and gives the opportunity to discover the stratification in the turbulent flow field directly. The small pipe diameter ($D = 1, 2$ mm) with moderately low inlet Reynolds number ($Re_0 = 5400$) is similar as the channel flow in the compact heat exchanger (PCHE). Inlet flow temperature (T_0) is slightly lower than the pseudo-critical temperature T_{pc}. Some interesting results have been found and discussed. The open-source code OpenFOAM runs on the HPC platform Hazel Hen with an excellent scalability with up to 2800 cores. It shows also potential for efficiently dealing larger problem with more computational resource. Compared with vertical orientation, flow stratification was observed in horizontal layout. In addition to this, 'M' shaped velocity profile as a result of buoyancy in vertical layout, was missing in horizontal orientation. In the next step, the mesh resolution will increase to 300 Mio. aiming at a higher Reynolds number and an improved resolving of Kolmogorov scale and Batchelor scale.

Acknowledgements The research presented in this paper is supported by the Forschungsinstitut fuer Kerntechnik und Energiewandlung e.V., for project DNSTHTSC. The authors would like to thank to the HLRS and Cray team for their kind support.

References

1. NIST Chemistry Webbook: In: Lemmon, E., McLinden, M., Friend, D., Linstrom, P., Mallard, W. (eds.) NIST Standard Reference Database Number 69. National Institute of Standards and Technology, Gaithersburg (2011). http://webbook.nist.gov/chemistry/
2. Bae, J.H., Yoo, J.Y., Choi, H.: Direct numerical simulation of turbulent supercritical flows with heat transfer. Phys. Fluids **17**(10), 105104 (2005)
3. Cheng, X., Kuang, B., Yang, Y.: Numerical analysis of heat transfer in supercritical water cooled flow channels. Nucl. Eng. Des. **237**(3), 240–252 (2007)
4. Chu, X., Laurien, E.: Investigation of convective heat transfer to supercritical carbon dioxide with direct numerical simulation. In: High Performance Computing in Science and Engineering'15, pp. 315–331. Springer, Cham (2016)
5. Chu, X., Laurien, E., McEligot, D.M.: Direct numerical simulation of strongly heated air flow in a vertical pipe. Int. J. Heat Mass Transf. **101**, 1163–1176 (2016)
6. Dostal, V., Driscoll, M.J., Hejzlar, P.: A supercritical carbon dioxide cycle for next generation nuclear reactors. Ph.D. thesis, Massachusetts Institute of Technology (2004)
7. Duffey, R.B., Pioro, I.L.: Experimental heat transfer of supercritical carbon dioxide flowing inside channels (survey). Nucl. Eng. Des. **235**(8), 913–924 (2005)
8. Eggels, J.G., Unger, F., Weiss, M.H., Westerweel, J., Adrian, R.J., Friedrich, R., Nieuwstadt, F.: Fully developed turbulent pipe flow: a comparison between direct numerical simulation and experiment. J. Fluid Mech. **268**, 175–210 (1994)
9. He, S., Kim, W.S., Bae, J.H.: Assessment of performance of turbulence models in predicting supercritical pressure heat transfer in a vertical tube. Int. J. Heat Mass Transf. **51**(19–20), 4659–4675 (2008)
10. Jackson, J.D.: Fluid flow and convective heat transfer to fluids at supercritical pressure. Nucl. Eng. Des. **264**, 24–40 (2013)
11. Li, X., Hashimoto, K., Tominaga, Y., Tanahashi, M., Miyauchi, T.: Numerical study of heat transfer mechanism in turbulent supercritical CO2 channel flow. J. Thermal Sci. Technol. **3**(1), 112–123 (2008)
12. Lund, T.S., Wu, X., Squires, K.D.: Generation of turbulent inflow data for spatially-developing boundary layer simulations. J. Comput. Phys. **140**(2), 233–258 (1998). http://dx.doi.org/10.1006/jcph.1998.5882
13. Nemati, H., Patel, A., Boersma, B.J., Pecnik, R.: Mean statistics of a heated turbulent pipe flow at supercritical pressure. Int. J. Heat Mass Transf. **83**, 741–752 (2015)
14. Pandey, S., Laurien, E.: Heat transfer analysis at supercritical pressure using two layer theory. J. Supercrit. Fluids **109**, 80–86 (2016)
15. Schoppa, W., Hussain, F.: Coherent structure dynamics in near-wall turbulence. Fluid Dyn. Res. **26**(2), 119–139 (2000)

16. Weller, H.G., Tabor, G., Jasak, H., Fureby, C.: A tensorial approach to computational continuum mechanics using object-oriented techniques. Comput. Phys. **12**(6), 620–631 (1998)
17. Wu, X., Moin, P.: A direct numerical simulation study on the mean velocity characteristics in turbulent pipe flow. J. Fluid Mech. **608**, 81–112 (2008)
18. Yang, J., Oka, Y., Ishiwatari, Y., Liu, J., Yoo, J.: Numerical investigation of heat transfer in upward flows of supercritical water in circular tubes and tight fuel rod bundles. Nucl. Eng. Des. **237**(4), 420–430 (2007)
19. Yoo, J.Y.: The turbulent flows of supercritical fluids with heat transfer. Ann. Rev. Fluid Mech. **45**, 495–525 (2013)

CFD Analysis of Fast Transition from Pump Mode to Generating Mode in a Reversible Pump Turbine

Christine Stens and Stefan Riedelbauch

Abstract To improve the flexiblity in the operation of pumped storage power plants, it is necessary to understand the flow phenomena during a change of operating mode. In this work, a fast transition from pump mode to generating mode in a reversible pump turbine is investigated with the open source code OpenFOAM®. The analysis is run on two different meshes for a constant guide vane opening on the ForHLR 1 cluster. A speedup test is employed to test scalability and determine a suitable number of cores. Results are presented for different monitor points in the machine. Furthermore, the flow field in the runner is analysed for different points of time. The coarse mesh is generally able to give the same trends as the fine mesh, but with an offset in absolute value during parts of the transient.

1 Introduction

Pumped storage power plants are an efficient way to store energy at a large scale. Their importance increases with a growing share of renewables in the grid, as excessive energy can be stored in times of high production and released when the demand exceeds production. An optimal storage cycle in terms of profit is around 6 h [8]. However, the current procedure for changing from one operating mode to the other is still time consuming. It is therefore desirable to develop faster manoeuvres.

In order not to damage the machine, it is important to understand the flow mechanisms during a change of operating modes. An overview of possible flow phenomena in reversible pump turbines is given in [5]. CFD has proven to be a suitable tool for gaining such information and various authors have investigated hydraulic machines under time varying conditions such as runaway [3, 4, 6], start-up [7] and speed-no load [2] conditions.

This project investigates a fast transition from pump mode to generating mode in a model scale reversible pump turbine with a linear variation of rotational speed and a fixed guide vane opening. Due to the comparably large number of time steps

C. Stens (✉) • S. Riedelbauch
Institute of Fluid Mechanics and Hydraulic Machinery, Pfaffenwaldring 10, 70569 Stuttgart, Germany
e-mail: christine.stens@ihs.uni-stuttgart.de; riedelbauch@ihs.uni-stuttgart.de

W.E. Nagel et al. (eds.), *High Performance Computing in Science and Engineering '16*, DOI 10.1007/978-3-319-47066-5_33

Table 1 Number of mesh points in each domain

Domain	SC	SV/GV	RUN	DT	Total
2.5M	577,317	811,074	503,370	567,580	2,459,341
20M	2,479,508	4,295,628	5,107,571	8,374,898	20,257,605

and the model size, a suitable parallelization is required. The present work focuses on setup, a comparison of different meshes and flow field analysis. Results from a pre-study on a coarse mesh are found in [10]. A preliminary evaluation of related results from a model test is presented in [9] and a comparison between simulation and experiment is published in [11].

2 Computational Mesh

Two block structured meshes are generated for the analysis, a coarse mesh containing approximately 2.5 million points (2.5M) and a refined mesh with 20 million points (20M). The geometry is split into four domains, spiral case (SC), twin cascade (SV/GV), runner (RUN) and draft tube (DT). Domains are connected with each other via interfaces. The number of mesh points per domain is listed in Table 1.

During the meshing process, special attention was paid to keeping the cells close to the walls comparable between the meshes in the guide vanes and the runner. Average y^+ values are between 20 and 65 in the runner and between 20 and 50 in the guide vanes, depending on the time. In the draft tube, the coarse mesh showed low y^+ values with a maximum average over time of 30. Therefore, wall distance of the first point was increased in the fine mesh, leading to average values between 40 and 180.

3 Numerical Setup and Methodology

Simulations are carried out using OpenFOAM® 2.3. The single domains are connected via the arbitrary mesh interfaceİ (AMI). Flow rate is prescribed at the respective inlet, i.e. at the draft tube outlet in pump mode and at the spiral case in generating mode, together with a zero gradient condition for pressure. At the remaining outlet, a constant average pressure and a zero gradient condition for velocity are applied. Time varying values for flow rate and rotational speed during the transient are prescribed via table files, where a linear variation of rotational speed is chosen and flow rate is determined by the test rig conditions. It shows a large gradient during the first half of the transient, while the change in operating mode is moderate. The guide vane angle is fixed to 25°.

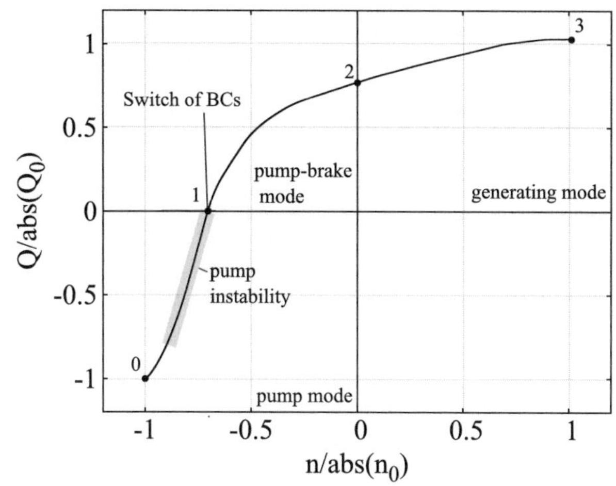

Fig. 1 Representation of the transient in a four quadrant plot

Figure 1 shows the transient in a four quadrant plot. The transient starts in the lower left corner with negative flow rate and negative rotational speed, i.e. normal pump mode. As the rotational speed decreases, the machine passes to the next quadrant. Flow direction is now from the spiral to the draft tube, while the runner continues its rotation in the same direction as before (pump brake or dissipation mode). Finally, the runner reverses its rotational direction and the upper right quadrant is reached. This represents normal generating mode.

The solution procedure follows the SIMPLE algorithm, a semi-implicit segregated approach. To account for the time dependent behaviour, the transient solver transientSimpleDyMFoam is employed, where DyM indicates the solver's capability to deal with moving meshes. The k-omega-SST model is chosen for turbulence.

Discretisation is first order accurate in time, first/second order accurate for the convection term and first order accurate for turbulent quantities. Higher order schemes could not produce converged solutions under the highly unsteady flow conditions.

Time step is constant throughout the complete transient to facilitate FFT of the result quantities. It is chosen to be $5 \cdot 10^{-4}$ s, equalling 2.8° at the maximum rotational speed at the beginning of the transient. This leads to a maximum CFL number of 110 at the beginning and 65 at the end of the transient. Peak values of 130 and high amplitude flucutuations are found at the beginning of generating mode between 5.8 and 6.5 s.

The influence of time step size is tested for the unstable operating regime in pump mode between 2 and 3 s. A time step of $2 \cdot 10^{-4}$ s equalling 0.9° per time step is used for comparison. Although the flucutuations for head, torque and pressure at the monitor points differ over time, mean values remain unchanged. As an example, head is presented in Fig. 2.

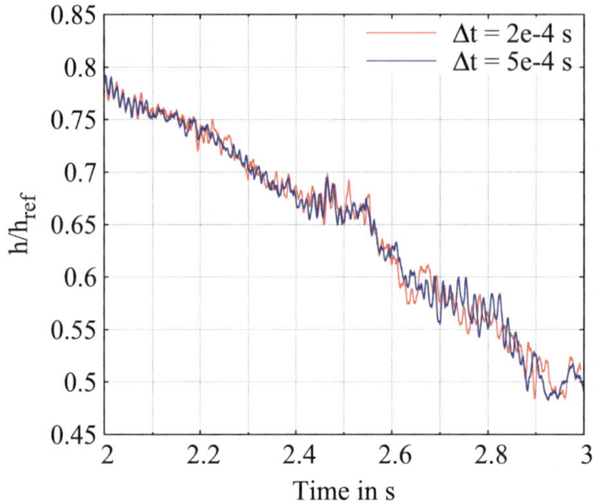

Fig. 2 Influence of time step size on simulated head for the fine mesh

The SIMPLE algorithm requires a number of so called outer corrections, i.e. the equations for pressure, velocity and turbulent quantities need to be solved multiple times during each time step until a converged solution is reached. The number of iterations is dependent on mesh size and mesh quality. A study of the development of head and torque over the number of outer corrections during an unstable phase of the transient leads to the conclusion that seven steps are sufficient for the coarse mesh and eleven are required for the fine mesh. The coarse mesh additionally requires two corrector steps due to mesh non-orthogonality. In both cases, relaxation factors of 0.3 for pressure and 0.7 for velocity and turbulent quantities are employed.

4 Computational Resources

Simulations were run on the ForHLR 1 cluster at SCC Karlsruhe. A speedup test was carried out for both meshes in order to investigate the scalability of OpenFOAM® 2.3 and determine a suitable number of cores for the subsequent simulations. The results are presented in Fig. 3 for both meshes. For the coarse mesh, scalability is nearly ideal up to 40 cores, equalling 62,500 mesh points per core. The same test for the fine mesh reveals that this number is not achieved for the higher number of overall mesh points. The behaviour is ideal up to only 80 cores and acceptable up to 120 cores, equalling 250,000 and 167,000 mesh points per core, respectively. This leads to computation times of approximately four days for the complete transient with the coarse mesh and 15 days with the fine mesh.

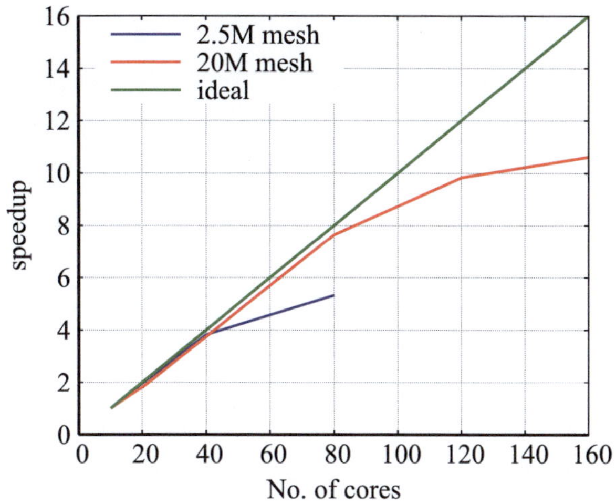

Fig. 3 Speedup results for the coarse and fine mesh

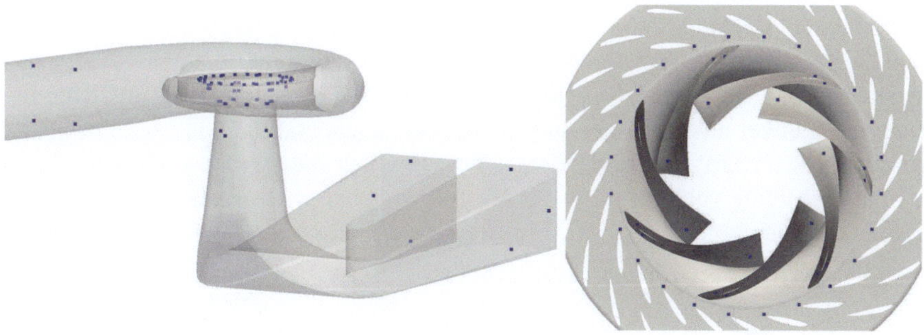

Fig. 4 Position of the pressure monitor points

5 Simulation Results for the Transient

For evaluation, a number of pressure monitor points are defined along the machine as shown in Fig. 4. There are two points on each side of each runner blade, one at the top of each guide vane channel and four in the draft tube below the runner. The points in the spiral case and at the draft tube outlets are used to calculate head. Additionally, simulated head and torque are analysed and compared between the meshes.

To ensure a converged solution at each time step, initial and final residuals are monitored during the solution. For velocity, highest residuals appear at zero flow rate, while pressure residuals show a minimum at zero rotational speed. The predefined final residual is reached for all variables in every time step. The number of iterations required to meet that target in the last iteration is approximately six for pressure during the transient and rises up to 20 during the pump instability. From

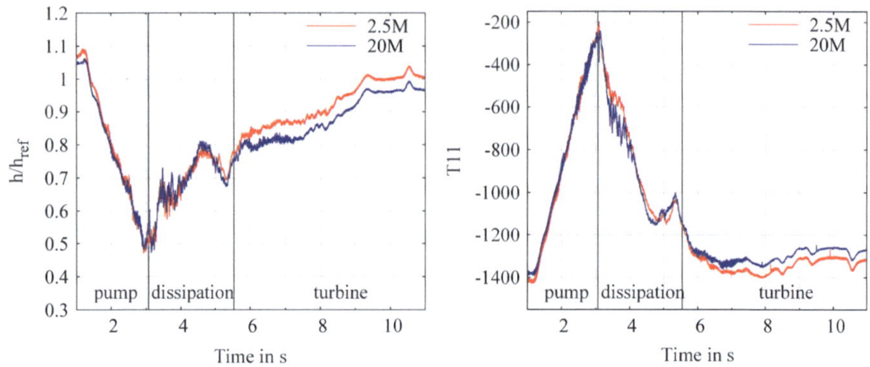

Fig. 5 Simulated head and dimensionless torque over time. A constant head is used for the normalization of torque

residuals and number of iterations, flow direction from spiral to draft tube seems to be numerically more stable than in opposite direction.

Simulated head and torque give a first indication of the behaviour of the machine and are presented in Fig. 5. A comparison of the results shows that the coarse mesh is already able to capture the general trends and agrees well with results from the finer mesh under unstable operating conditions. In generating mode, both head and torque show a nearly constant offset, where the finer mesh gives a lower head and a higher absolute value of torque. The offset in head is approximately 5 % of the reference head.

5.1 *Results in the Guide Vanes*

As volume flow rate decreases to small values in pump mode, large fluctuations occur in torque and pressure on the runner blades, which continue in the first half of the pump brake quadrant. This is a result of stall in the guide vanes. While flow is evenly distributed between the guide vane channels in pump mode, it concentrates on single channels during the pump instability while other passages are nearly blocked. Figure 6 shows the torque on the guide vanes during the instability. Irregularities can be tracked across various adjacent channels, starting from 1.9 s. This indicates rotating stall in the guide vanes, a phenomenon that has been found before in centrifugal pumps and pump-turbines [1, 12]. In pump mode, a passing of the disturbances from high to low channel numbers signifies that the phenomenon is moving in the rotational direction of the runner. The constant distance between the lines shows that the absolute values of torque are similar for all guide vanes, with the exception of guide vane number four, which contained an error in the setup. This behaviour is found independently of the mesh size, but the onset of the phenomenon

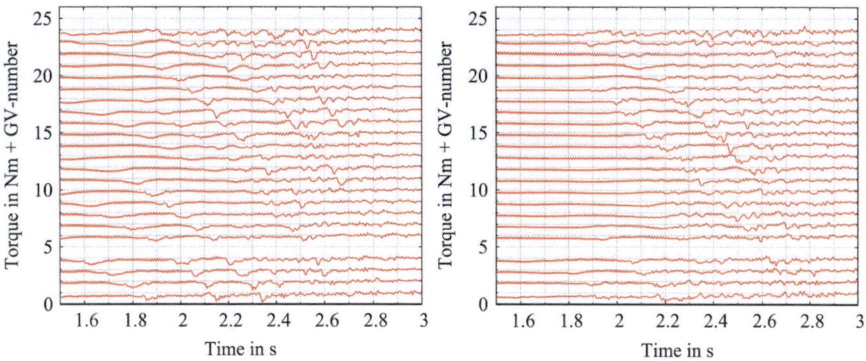

Fig. 6 Disturbances in the guide vane torque signal passing through the channels during the pump instability. The torque signal has been offset by the respective guide vane number. *Left picture*: coarse mesh, *right*: fine mesh

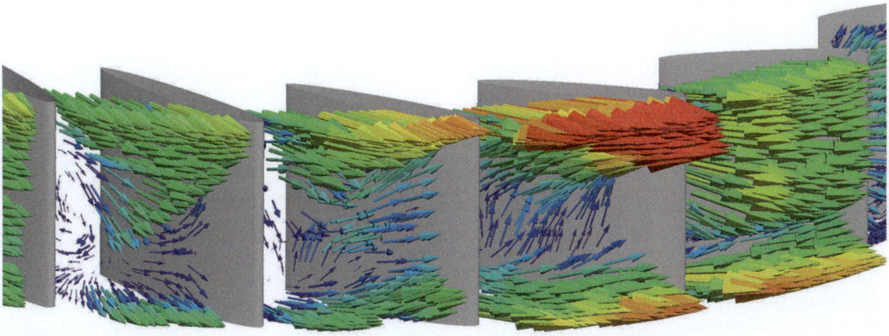

Fig. 7 Flow visualization in the guide vanes at t = 2.45 s. Arrows coloured by flow velocity from 1 to 7 m/s

is earlier in the coarse mesh. The speed of propagation decreases with decreasing rotational speed of the runner.

Another indicator of rotating stall is the flow rate through each of the guide vane channels. It gives similar results, but is less easily visualized as the differences between the channels interfere with the disturbances caused by stall. In single channels, zero flow or even backflow occurs while the global flow rate is still at 40 % of its initial value. Flow visualizations as in Fig. 7 show that at the beginning, stall occurs near the bottom ring and in the middle of the channels, while flow near the head cover side remains stable. At lower flow rates, outward flow concentrates near the head cover and bottom ring meridionals, with almost no flow or slight backflow in the middle of the guide vane channels. It is therefore interesting to evaluate the third possible variable to track rotating stall, namely pressure at the top of the guide vane channels. While torque and flow rate describe the integral result of the flow in the entire channel, pressure is evaluated locally. Although located in a region where flow stays stable for a longer time, the start of irregularities in the signal coincides in time with those in flow rate and torque.

An FFT of short periods of the signal reveals that during the rest of the transient, flow through the guide vanes is dominated by the passing of the runner blades, especially in pump brake mode. Here, flow is forced outward (pump direction) near the runner blades, but inward between the runner blades. This leads to large fluctuations in pressure and torque on the guide vanes.

5.2 *Results in the Runner*

In the runner, the fluctuations in the torque contribution for each blade change in the period where rotating stall in the guide vanes is detected. However, the contribution to overall torque is still evenly distributed between all seven blades during the relevant period from 1.9 to 2.7 s. Only at very low flow rates, curves start to deviate from each other. Differences are random rather than passing from one blade to the next. The pressure sensor on the pressure side near the guide vanes confirms this tendency.

As in head, a constant offset exists between the two meshes in generating mode at the pressure side sensor as shown in Fig. 8. On suction side, the offset disappears between 8.5 and 8.6 s, where the curve for the finer mesh jumps back to the one of the coarser mesh. The sudden change in pressure in simulation results from the fact that in the upper part of the suction side, flow is able to follow the blade contour, while in the lower part of the channel, it detaches from the suction side. The jump signals that the border between the two has moved further downward and the sensor has passed from the stall zone to the one with attached flow. In the coarse mesh, the general flow is comparable to the fine mesh, but the pressure gradient along the

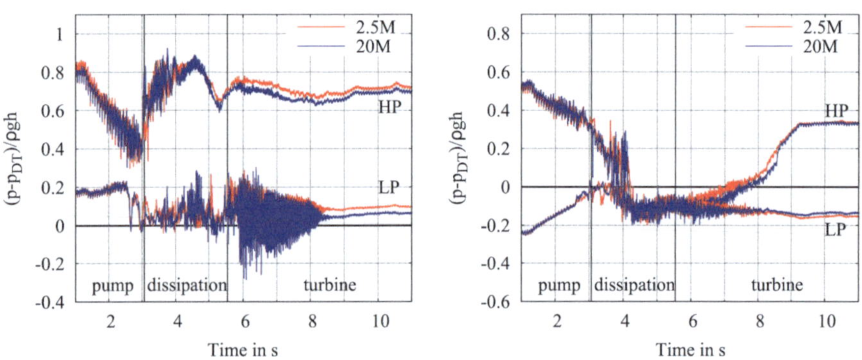

Fig. 8 Pressure on the pressure side of a runner blade (*left*) and the suction side (*right*). High pressure side (HP) close to the guide vanes, low pressure side (LP) near the draft tube

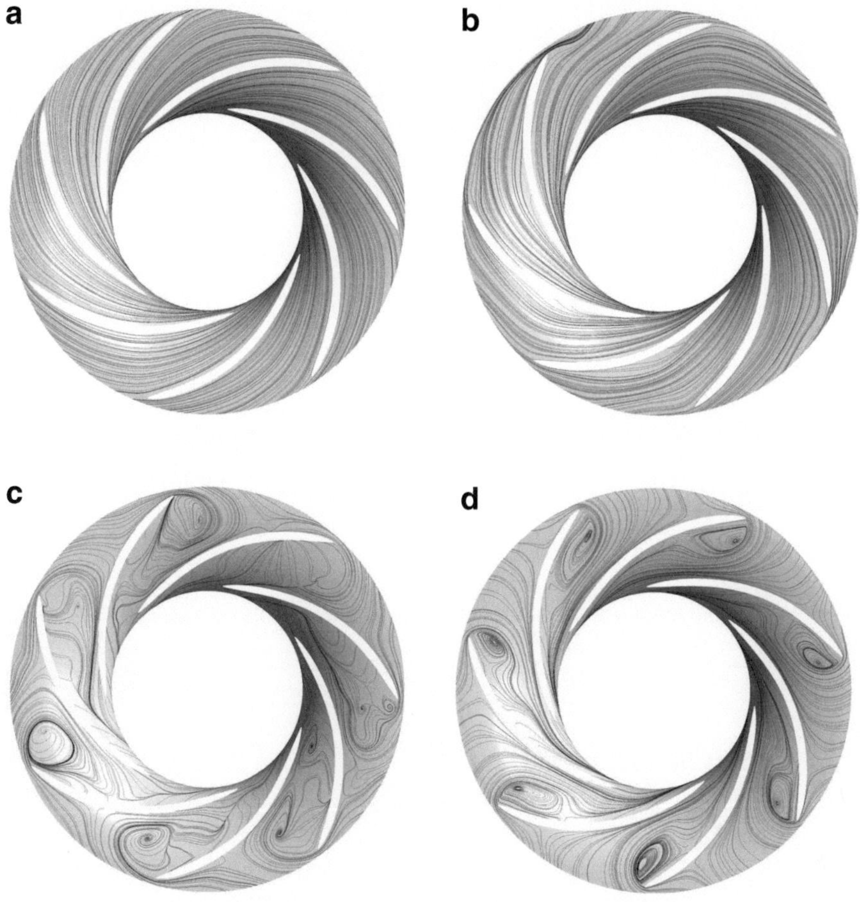

Fig. 9 (**a**) t = 1.0 s (pump mode). (**b**) t = 2.6 s (pump mode, instability). (**c**) t = 4.0 s (dissipation mode). (**d**) t = 6.0 s (generating mode, low flow rate)

channel height is less steep between the two zones, resulting in a more constant rise of mean pressure.

Figure 9 gives an impression of the flow field in different operating regimes. It shows the streamlines in the midplane of the runner. The first picture shows the starting point of the transient in pump mode, with flow evenly distributed between the channels. As detected by the pressure sensors on the runner blades, flow during the pump instability is slightly influenced near the guide vanes, but stable in the rest of the runner channel. In pump brake or dissipation mode at 4 s, flow hits the runner blades at approximately one third of chord length. Flow is strongly three dimensional, as vortices form as well around a vertical as around a horizontal axis. In generating mode, vortices still exist near the guide vanes, but are more stable

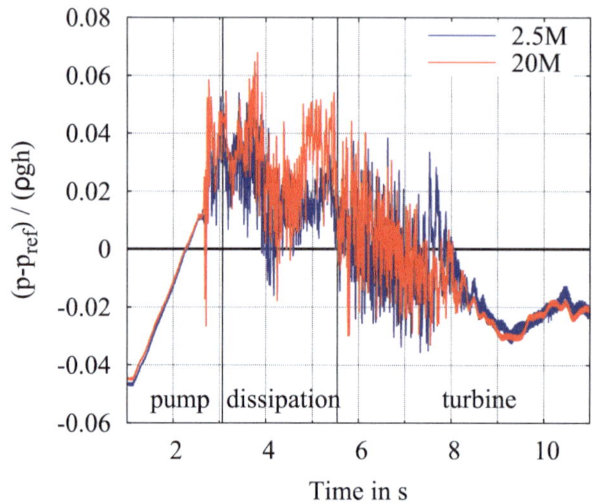

Fig. 10 Pressure fluctuations in the draft tube below the runner

in size, form and location, leading to smaller fluctuations at the respective pressure sensors.

5.3 *Results in the Draft Tube*

In the draft tube, four monitor points are located below the runner positioned at 90° from each other. Figure 10 provides the signal of the first pressure sensor for both meshes. As in the other domains, there is a good agreement between the results from the different cell sizes concerning the general behaviour.

The behaviour itself is characterized by high fluctuations that start in pump mode and disappear after a certain flow rate and rotational speed are reached in generating mode. An FFT of the signal shows high amplitudes at low frequencies in the middle of the transient, but without a singular identifiable frequency that could give evidence for a vortex rope rotating at a defined speed. Compared to the coarse mesh, the finer mesh gives higher amplitudes at a larger number of frequencies.

Figure 11 shows the axial velocity in the draft tube on a line parallel to the draft tube channels in the plane of the pressure sensors below the runner at different times. A positive velocity signifies an upward flow, i.e. towards the runner. At the beginning of pump mode, flow direction is upward in the complete draft tube. With decreasing flow rate, it starts to detach from the draft tube walls and a swirling flow away from the runner develops at the draft tube wall, while in the middle of the draft tube, the fluid is moving towards the runner. The detachment starts at the bottom of the draft tube and expands upwards until reaching the evaluation plane at approximately 2.7 s. This causes the large fluctuations observed in the pressure

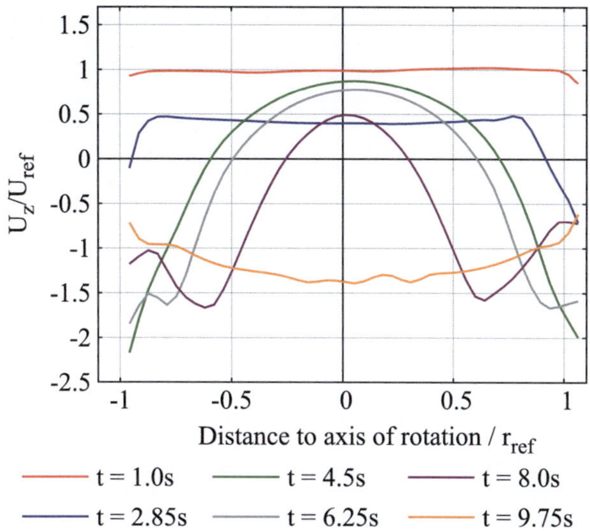

Fig. 11 Axial velocity in the draft tube normalized to mean velocity at t = 1.0 s. Radius is normalized to the runner outlet radius

monitor points. During the transient, the region with downward flow grows until finally the flow direction has reversed in the complete cross section.

6 Conclusion and Outlook

Investigations on the flow through a model scale reversible pump turbine during a change from pump mode to generating mode are carried out using a coarse and a fine mesh and the open source code OpenFOAM®. Input data for flow rate and rotational speed was taken from experiment.

Comparing pressure at several monitor points in the machine shows that the coarse mesh is generally able to predict tendencies in mean pressure and amplitudes of fluctuations. However, a refined mesh gives different values e.g. on the runner blades in generating mode. This is important for a correct prediction of the mechanical loads caused by the fluid. As shown in [11], the values obtained from the fine mesh are in better agreement with experimental data for head and pressure in the guide vane channels.

The simulations for the fine mesh were run on 120 cores with a simulation time of approximately two weeks. The number of cores was chosen based on a speedup test. In future work, the simulation is to be coupled with a 1D model of the test rig, so that no experimental data is necessary to predict the behaviour of the machine.

Acknowledgements The authors would like to thank the European Commission for funding wihin the HYPERBOLE project (ERC/FP7-ENERGY-2013-1-Grant 608532). Part of this work was performed on the computational resource ForHLR Phase I funded by the Ministry of Science, Research and the Arts Baden-Württemberg and DFG ("Deutsche Forschungsgemeinschaft").

References

1. Braun, O.: Part load flow in radial centrifugal pumps. Ph.D. thesis, STI, Lausanne (2009)
2. Casartelli, E., Mangani, L., Romanelli, G., Staubli, T.: Transient simulation of speed-no load conditions with an open-source based C++ code. In: Proceedings of 27th Symposium of Hydraulic Machinery and Systems, Montreal (2014)
3. Cherny, S., Chirkov, D., Bannikov, D., Lapin, V., Skorospelov, V., Eshkunova, I., Avdushenko, A.: 3D numerical simulation of transient processes in hydraulic turbines. IOP Conf. Ser. Earth Environ. Sci. **12**(1), 012071 (2010)
4. Fortin, M., Houde, S., Deschênes, C.: Validation of simulation strategies for the flow in a model propeller turbine during a runaway event. In: Proceedings of 27th Symposium of Hydraulic Machinery and Systems, Montreal (2014)
5. Kerschberger, P., Gehrer, A.: Hydraulic development of high specific-speed pump-turbines by means of an inverse design method, numerical flow-simulation (CFD) and model testing. IOP Conf. Ser. Earth Environ. Sci. **12**(1), 012039 (2010)
6. Li, J., Yu, J., Wu, Y.: 3D unsteady turbulent simulations of transients of the francis turbine. IOP Conf. Ser. Earth Environ. Sci. **12**(1), 012001 (2010)
7. Nicolle, J., Morissette, J.F., Giroux, A.M.: Transient CFD simulation of a francis turbine startup. In: 26th IAHR Symposium on Hydraulic Machinery and Systems, Beijing (2012)
8. Rapp, C., Zeiselmair, A., Halblaub, A.B.: Überlegungen zur Abschätzung der Wirtschaftlichkeit von Pumpspeicherkraftwerken. WasserWirtschaft **2**, 68–74 (2016)
9. Ruchonnet, N., Braun, O.: Reduced scale model test of pump-turbine transition. In: Lipej, A., Muhic, S. (eds.) Cavitation and Dynamic Problems: 6th IAHR Meeting of the Working Group, pp. 264–272. IAHR, Ljubljana (2015)
10. Stens, C., Riedelbauch, S.: CFD simulation of the flow through a pump turbine during a fast transition from pump to generating mode. In: Lipej, A., Muhic, S. (eds.) Cavitation and Dynamic Problems: 6th IAHR Meeting of the Working Group, pp. 264–272. IAHR, Ljubljana (2015)
11. Stens, C., Riedelbauch, S.: Investigation of a fast transition from pump mode to generating mode in a model scale reversible pump turbine. In: Proceedings of 28th IAHR Symposium of Hydraulic Machinery and Systems, Grenoble (2016)
12. Xia, L.S., Cheng, Y.G., Zhang, X.X., Yang, J.D.: Numerical analysis of rotating stall instabilities of a pump-turbine in pump mode. IOP Conf. Ser. Earth Environ. Sci. **22**(3), 032020 (2014)

Scale Resolving Flow Simulations of a Francis Turbine Using Highly Parallel CFD Simulations

Timo Krappel and Stefan Riedelbauch

Abstract In this paper, transient flow simulations of a Francis turbine in part load conditions are presented. The dominating flow phenomenon, the vortex rope, leads to a very complex flow field, especially in the draft tube of the turbine. As the resolution of turbulence is important, the Scale Adaptive Simulation (SAS) approach is used. The mesh size of the entire Francis turbine is up to 300 million mesh nodes. The commercial CFD code Ansys CFX version 17.0 is used, which performs up to a few thousands of cores for this kind of application.

1 Introduction

In the last years, the operation of Francis turbines is more and more in off-design conditions. Therefore, it is important to reach a better understanding of the flow behaviour at operating points, like part load conditions, which is focus of this paper. As computational resources have increased, a transient, turbulence resolving flow simulation using thousands of cores in parallel [9, 16] is conducted.

The flow field in the draft tube of a Francis turbine at part load conditions is dominated by the vortex rope phenomenon. This leads to a complex and three-dimensional flow field, which has to be resolved properly in space and time, as well as with turbulence models being able to resolve a large amount of turbulence. Good results could be achieved by using hybrid RANS-LES models, like the SAS turbulence model in the research field of hydraulic turbines [7], which is also chosen and investigated within this paper.

T. Krappel (✉) • S. Riedelbauch
Institute of Fluid Mechanics and Hydraulic Machinery, Pfaffenwaldring 10, 70550 Stuttgart, Germany
e-mail: timo.krappel@ihs.uni-stuttgart.de

W.E. Nagel et al. (eds.), *High Performance Computing in Science and Engineering '16*, DOI 10.1007/978-3-319-47066-5_34

2 Numerical Methods

2.1 Flow Solver

All flow simulations of the Francis turbine were performed using different versions (16.0, 17.0-pre-release and 17.0) of the commercial CFD code Ansys CFX [1]. The CFD code is able to handle the rotation of the turbine runner and to couple different meshes by an general-grid-interface. The finite-volume method is used for discretisation based on an implicit pressure-based formulation, while the volumes of discretisation are built around the cell nodes. A coupled algebraic multigrid (AMG) linear solver [15] is used with an ILU based solver.

2.2 Turbulence Modelling

Two turbulence models are applied in this work, namely the RANS-SST [10] (Reynolds-averaged Navier-Stokes) and the SAS-SST [3, 4, 11] (Scale Adaptive Simulation) turbulence model. Within the SAS framework, the unsteady SST RANS turbulence model is able to operate in SRS (Scale Resolving Simulation) mode [12], resolving small turbulent structures similar to a LES turbulence model. This is achieved by introducing the source term Q_{SAS} into the transport equation of turbulence eddy frequency ω of the SST model. The additional source term leads to a reduction of the turbulent eddy viscosity, which may be overestimated for fine meshes at smallest turbulent scales. Therefore, a high wave-number limit based on the WALE model is used [14] in such way that the effective eddy viscosity will not fall below the LES eddy viscosity. Further details are referred to above mentioned references.

2.3 Temporal and Spatial Discretisation

For temporal discretisation a second order backward Euler scheme is used. For spatial discretisation different schemes are applied and investigated. For simulations with RANS turbulence modelling a high-resolution scheme (HR) [2] is used. In the framework of SAS turbulence modelling, less dissipative schemes, which are formal second order, should be used to allow turbulent structures to evolve. The first one is the bounded second order central differencing scheme [5] (BCD). This scheme is based on the normalized variable diagram approach together with the convection boundedness criterion. The second one is a hybrid convection scheme [17] (hybCon), which is a combination of the HR-scheme and the central differencing scheme (CD). The blending between those schemes is mainly based on vortex detection parameters. For the turbulence quantities a bounded second order backward Euler scheme is applied for the temporal discretisation and a first order scheme for the spatial discretisation [13].

3 Francis Turbine Case

3.1 Computational Setup

The geometry of the Francis turbine being used in this study is depicted in Fig. 1. The different parts are the spiral casing, stay and guide vanes, runner and draft tube with expansion tank (in streamwise direction). According to these parts, the domain is divided into four domains of hexahedral meshes coupled with a general-grid-interface. At the inlet of the spiral casing typical steady-state boundary conditions are applied for the velocity profile and the turbulent quantities.

The meshes used in this study are in the range between 16 and 300 million mesh nodes (see Table 1). The 16M-mesh has a near-wall resolution of $y^+ = 9$–16 and all other meshes of $y^+ = 1$. The mesh refinement between the meshes strongly focuses on the draft tube domain, almost reaching LES-like resolution in the boundary layer. Further details of the mesh are referred to in [9]. The time step is chosen to keep the Courant number below one in the whole computational domain.

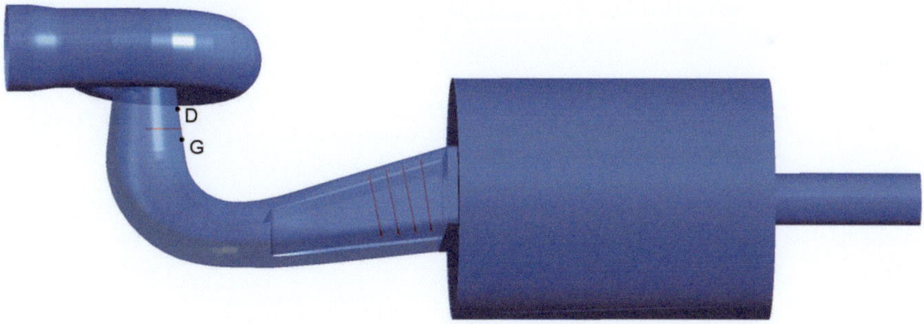

Fig. 1 Visualisation of the computational domain of the Francis turbine, the *red lines* indicate the evaluation lines, points D and G are used for wall-pressure evaluation

Table 1 Description of different grid sizes for different domains in million nodes and the corresponding time steps, see also [9]

Name	Spiral casing	Stay&guide vanes	Runner	Draft tube	Total	#Time steps /rev	Δt in °/time step
16M	1.02	3.70	3.78	8.09	16.20	180	2.0
50M	4.54	10.18	13.47	22.14	50.33	720	0.5
150M	7.29	17.95	29.90	98.81	153.95	840	0.43
300M	11.84	27.92	54.98	211.62	306.36	1000	0.36

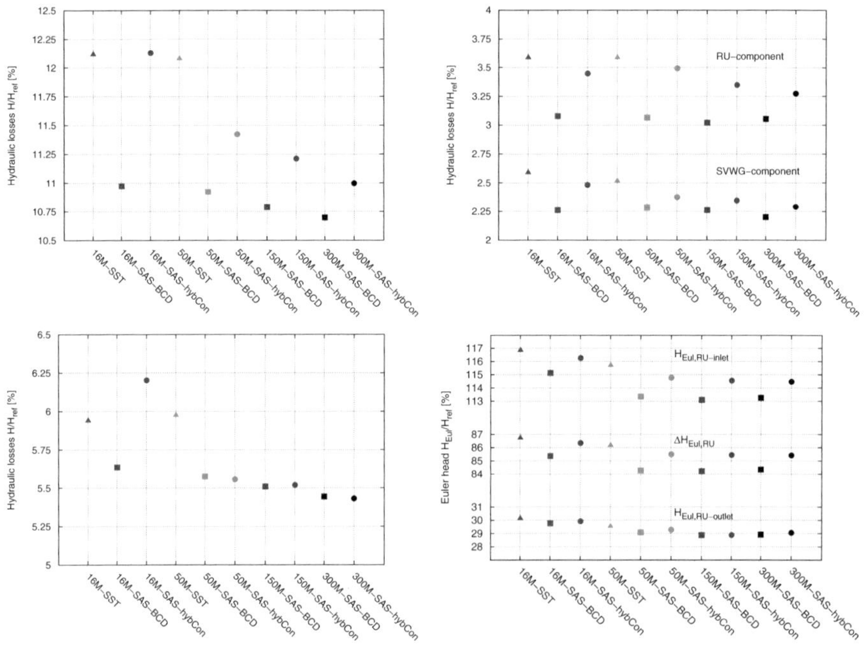

Fig. 2 Hydraulic losses for different simulation approaches: total machine (*top*, *left*), SVWG- and RU-component (*top*, *right*), DT-component (*bottom*, *left*) and Euler head of the runner (*bottom*, *right*)

3.2 Global Machine Data

The comparison of hydraulic losses and Euler head of the different simulations of the Francis turbine are depicted in Fig. 2. The Euler head is defined as:

$$H = 1/g(u_1 c_{u1} - u_2 c_{u2}) \tag{1}$$

whereas index 1 indicates the runner inlet and 2 the runner outlet.

The hydraulic losses of the total machine exhibit higher values for the simulations applying the SST turbulence model. The simulations with the SAS-turbulence model lead to lower hydraulic losses. The losses obtained with the BCD-scheme are lower than those of the hybrid convection scheme (the reason is discussed later). For both convection schemes using the SAS-turbulence model no strict grid convergence is reached, even for the 300M-mesh, with the lowest loss values.

The results with the SST-model, especially with the coarse mesh, predict the highest losses in all components. This is explained by the dissipative character of a RANS model and its inability to resolve the turbulent flow structures. The 16M-SAS-hybCon-simulation predicts higher draft tube losses. This might be caused by the deviant tangential velocity component in the draft tube cone (see Fig. 3).

The draft losses obtained by using the SAS-model decrease with larger meshes. The losses of the upstream components stay- and guide vanes (SVWG) and runner (RU) depend on the convection scheme. The BCD-scheme predicts quite similar results for all meshes. The losses using the hybCon-scheme decrease with larger meshes. This might be explained by the nature of the convection scheme, as it switches from a HR-scheme at the inlet to a CD-scheme (beside the boundary layer) where turbulent structures are resolved. The coarse mesh simulations are closer to the SST-results and the losses of the fine meshes are lower.

There is still quite an offset between the BCD- and hybCon-scheme for the runner losses. Whereas the losses obtained by the BCD-scheme are quite constant for all meshes, the losses for the hybCon-scheme decrease with larger mesh sizes. This might be explained by the Euler head at the runner inlet, which shows a similar trend. The Euler head at the runner outlet is quite the same for the larger meshes, for which reason the flow distribution into the draft tube should be quite similar. The Euler head difference between runner inlet and outlet indicates the resulting torque predicted by the simulations. This trend is similar to the trend of the Euler head at the inlet as the Euler head at the outlet is quite constant.

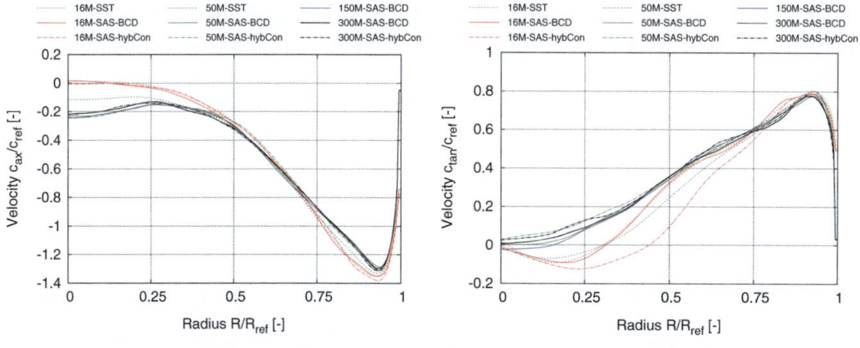

Fig. 3 Time-averaged normalised axial (*left*) and circumferential velocity component (*right*) in the draft tube cone

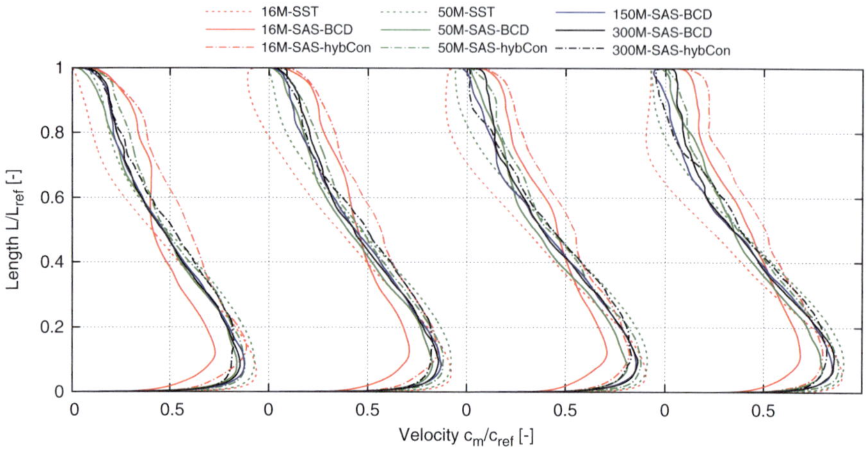

Fig. 4 Velocity distributions in the diffuser of the stream-wise velocity component for different stream-wise positions

3.3 Flow Analysis

In the draft tube cone and diffuser a flow analysis is done for time-averaged velocity components, which are depicted in Figs. 3 and 4. The positions of the evaluation lines are according to Fig. 1. The simulation time for all configurations equals 40 runner revolutions of time-averaging.

The axial velocity component in the cone is quite similar for all results of the 16M-mesh. The simulations on the finer meshes predict a higher axial component in the centre of the cone, except for the 50M-SST-simulation. For higher radii this trend is inverted.

The tangential velocity component is more or less similar for all finer meshes. The results of the 16M-SST and 16M-SAS-BCD simulation are almost the same with lower values in the centre. The 16M-SAS-hybCon predicts an even lower swirl in the centre of the cone.

At the end of the draft tube diffuser, the SST-simulations predict separation at the upper wall (L/L_{ref}=1). The 16M-SAS-BCD simulation predicts lower values at the lower part. The other simulation approaches show a quite similar flow distribution.

3.4 Vortex Rope Induced Pressure Pulsations

The vortex rope induced pressure pulsations are evaluated with the wall-pressure signal at two positions in the draft tube cone somewhat above the evaluation line in Fig. 1 named D and somewhat below named G. The results of the time signal and FFT can be seen in Fig. 5. As the results of the 150M-mesh are quite similar to the results of the 50M-mesh, they are not discussed in this section.

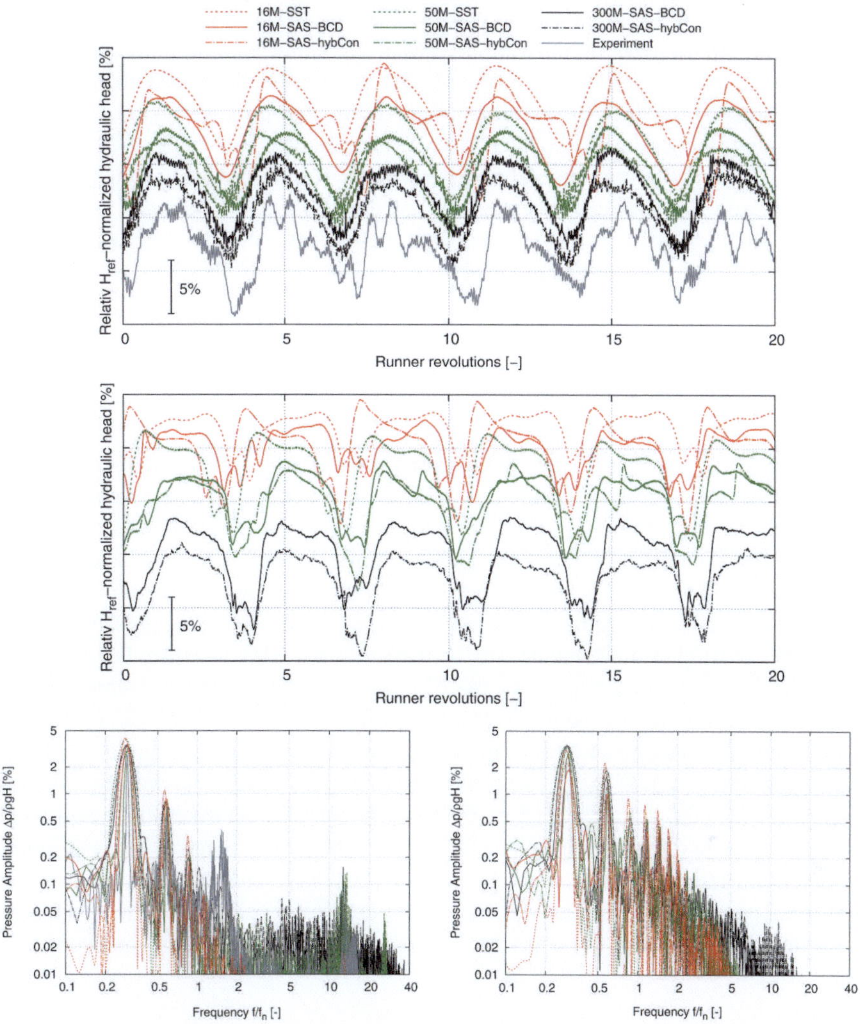

Fig. 5 Wall-pressure evaluation in the draft tube cone with comparison to experimental results in point D; *top*: time signal in point D, *middle*: time signal in point G, *bottom*: FFT-analysis in point D (*left*) and point G (*right*); legend is the same for all figures

At the upper part of the draft tube cone, the results are compared with measurements done at the closed loop test rig at the laboratory at the Institute of Fluid Mechanics and Hydraulic Machinery, University of Stuttgart. The wall-pressure pulsation in point D is measured with piezo-resistive pressure transducers. At this point the wall-pressure signal has almost sinusoidal shape, mainly consisting of the first and second mode of the vortex rope. The first mode is at around $f/f_n = 0.3$. The frequencies induced by the runner blade wakes at $f/f_n = 13$ and $f/f_n = 26$ can only be resolved by the SAS turbulence model with larger meshes. The simulations fit quite well with the experimental results. The low frequency pressure oscillation of the first modes and of the runner blades are quite similar predicted by the simulations, except for the 16M-SAS-hybCon and 50M-SST-simulation.

At the end of the draft tube cone at point G, the wall-pressure signal consists of several dominating modes, like the first six to nine modes. The shape of the pressure signal varies for different simulation approaches. The higher frequencies between $f/f_n = 5$ and $f/f_n = 10$ are only predicted by the 300M-simulation and even better by using the hybCon-scheme. The origin of these frequencies is a better resolution of the vortex rope rotation around itself.

3.5 *Turbulence Evaluation*

The RANS-SST-simulations predict higher values of turbulent eddy viscosity (see Fig. 6). The 16M-mesh in combination with the hybrid convection scheme also predicts quite high values of turbulent eddy viscosity in the cone. The reason for this is that in the upstream components of the draft tube the convection scheme uses the more dissipative HR-scheme. Therefore, the SAS-model is not able to switch into SRS-mode in this region. The results of the other meshes using the SAS-model show that the higher the mesh density is, the lower the eddy viscosity becomes. As the hybrid convection scheme uses the CD-scheme in the draft tube, which is less dissipative than BCD, the eddy viscosity is further reduced.

For the visualisation of the turbulent flow structures the Q-criterion is used [6]. Q is defined as $0.5\,(\Omega^2 - S^2)$, where S and Ω are the symmetric and asymmetric components of the velocity gradient tensor . The large structure of the vortex rope is dominating in the draft tube cone (see Fig. 7). The large structures decay to small turbulent structures in the draft tube elbow. Only large (turbulent) structures are predicted by using RANS. The influence of the convection scheme on the simulation with the 16M-mesh for the SAS-model is also visible. As the hybCon-scheme is

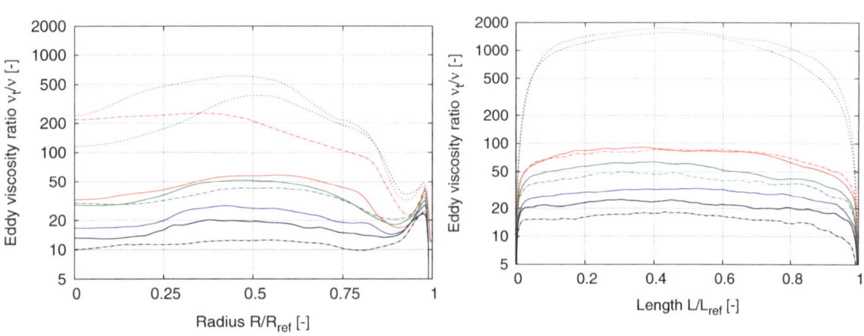

Fig. 6 Turbulent eddy viscosity ratio in the draft tube cone (*left*) and diffuser (*right*); legend is the same as in Fig. 3

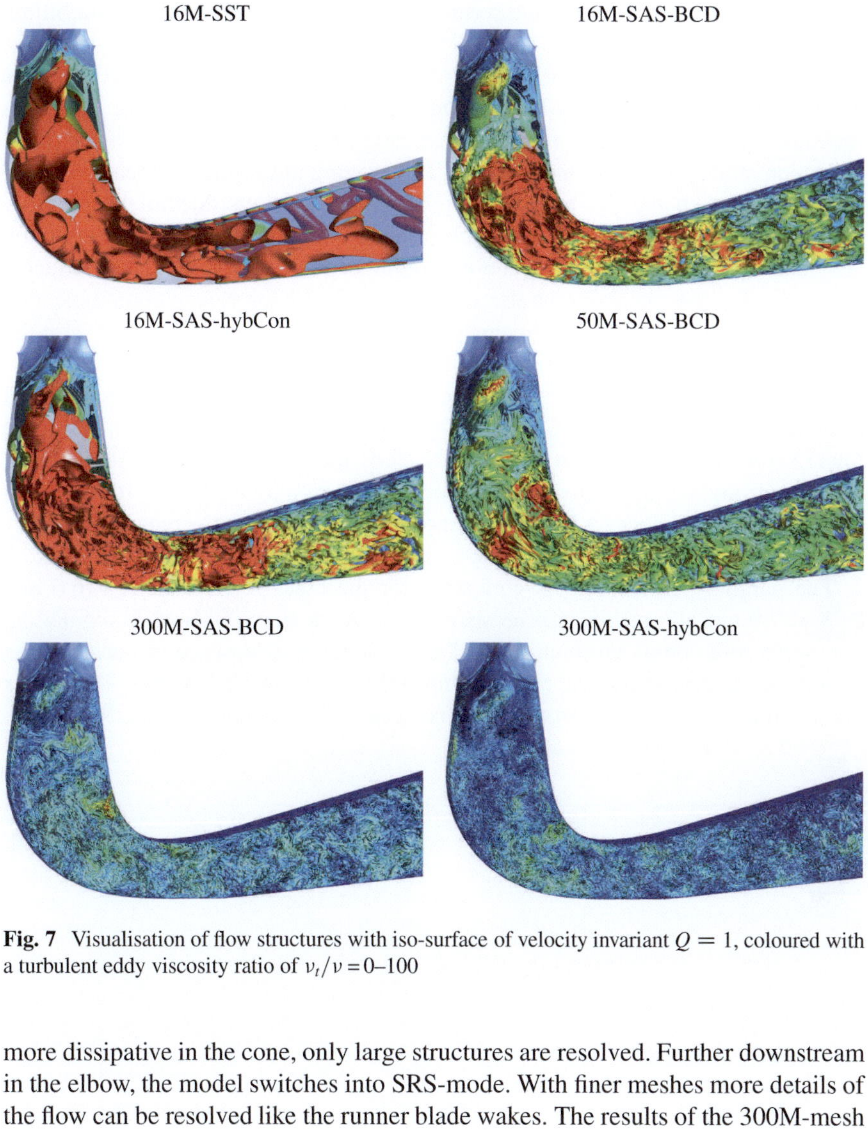

Fig. 7 Visualisation of flow structures with iso-surface of velocity invariant $Q = 1$, coloured with a turbulent eddy viscosity ratio of $\nu_t/\nu = 0$–100

more dissipative in the cone, only large structures are resolved. Further downstream in the elbow, the model switches into SRS-mode. With finer meshes more details of the flow can be resolved like the runner blade wakes. The results of the 300M-mesh show very fine flow structures, whereas with the hybCon-scheme even smaller flow structures can be resolved.

4 Parallelisation and Computational Resources

The CFD solver is highly optimized for large scale parallel systems using the SPMD (Single Program Multiple Data) parallelisation approach, combined with the common MeTiS [8] domain decomposition method. The partitioning topology is created with an upfront partitioning run. Partitioning is possible up to one billion mesh vertices.

For an efficient simulation of an entire Francis turbine some improvements of the code had to be done. This includes the efficient usage of MPI collective routines, IO improvements, new communication methods and hierarchical AMG collection strategies in the linear solver. The moving mesh interface leads to large overlapping mesh regions between neighbouring partitions due to the arbitrary relative position between stationary and rotating domains. This negative effect on the parallel performance in the equation assembly and the linear solution is reduced. The code also benefits from the use of Cray MPI for interprocess communication on CRAY XC40. All these improvements made this project possible, otherwise the wall clock time would not be manageable.

All flow simulations were performed on the CRAY XC40 Hornet installed at the HLRS Stuttgart. The CRAY XC40 Hornet has 3944 compute nodes with Intel®Xeon® processors with 12 cores, 128GB memory and Aries interconnect. For the pre-processing steps interpolation and partitioning as well as for the post-processing, special nodes with large memory are necessary, as the required memory for the 300M mesh is roughly 512GB.

The parallel performance is compared for the code versions CFX-v16.0 and CFX-v17.0 and the results can be seen in Fig. 8 for the 300M-mesh simulation. For lower core counts up to around 2000 cores, the performance of code version CFX-v17.0 is somewhat improved compared to version CFX-v16.0. The major improvement between the versions can be seen for larger core counts. In contrast

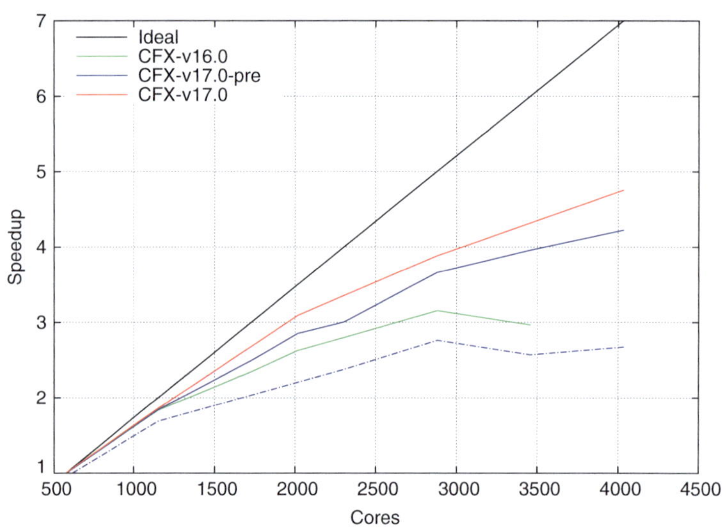

Fig. 8 Speed up tests for the transient flow simulations of the Francis turbine using the SAS turbulence model for the 300M-mesh with different code versions; the *dash-dotted line* indicates simulations with extensive data recording

to the performance of version CFX-v16.0, which decreases from around 3000 cores on, the newer version CFX-v17.0, the preliminary (pre) and final version, still has an increasing speedup up to at least 4000 cores. Speedup-tests with larger core counts were not possible due to licensing limitations. The parallel performance for simulations with extensive data recording impairs, especially for larger core counts. This means that for around 33,000 defined points, mostly in the draft tube domain, physical data, like velocity, pressure and turbulent quantities, is recorded for each time step.

5 Conclusion

Flow simulations of a Francis turbine at part load operating conditions were performed using the commercial CFD code Ansys CFX. For the resolution of the turbulent structures the SAS turbulence model was applied for meshes up to 300 million mesh nodes. The code scales up to a few thousand cores for the largest mesh.

It could be shown that numerical settings can have an influence on the predicted operating point. The SAS approach leads to a reduction of the hydraulic losses compared to RANS. The formal second order convection schemes in combination with the SAS model changes the Euler head at the runner inlet. The RANS simulations overestimate the separation in the draft tube diffuser leading to a worse pressure recovery. Only with the largest mesh (300M), the pressure pulsation in the draft tube cone could be resolved to significant higher frequencies. The evaluation of turbulent quantities revealed that it is necessary to have as few dissipation as possible from both the mesh and convection scheme to resolve turbulent structures. If the dissipation is too high, less or even no turbulent structures are resolved.

Acknowledgements The authors gratefully acknowledge the High Performance Computing Center Stuttgart (HLRS) for providing computational resources. The research leading to the results presented in this paper is part of a common research project of the Institute of Fluid Mechanics and Hydraulic Machinery, University of Stuttgart, Voith Hydro Holding GmbH & Co. KG and Ansys Germany GmbH.

References

1. ANSYS Inc.: ANSYS CFX Version 17.0 (2016)
2. Barth, T.J., Jesperson, D.C.: The design and application of upwind schemes on unstructured meshes. AIAA Paper 89-0366 (1989)
3. Egorov Y., Menter, F.R.: Development and application of SST-SAS turbulence model in the DESIDER project. In: Peng, S.-H., Haase, W. (eds.) Advances in Hybrid RANS-LES Modelling. Notes on Numerical Fluid Mechanics and Multidisciplinary Design: Papers contributed to the 2007 Symposium of Hybrid RANS-LES Methods, Corfu, vol. 97, pp. 261–270. Springer, Berlin/Heidelberg (2008)

4. Egorov Y., Menter, F.R., Cokljat, D.: The scale-adaptive simulation method for unsteady turbulent flow predictions. Part 2: application to aerodynamic flows. J. Flow Turbul. Combust. **85**(1), 139–165 (2010)
5. Jasak, H., Weller, H.G., Gosman, A.D.: High resolution NVD differencing scheme for arbitrarily unstructured meshes. Int. J. Numer. Methods Fluids **31**, 431–449 (1999)
6. Jeong, J., Hussain, F.: On the identification of a vortex. J. Fluid Mech. **285**, 69–94 (1995)
7. Jost, D., Skerlavaj, A., Lipej, A.: Numerical flow simulation and efficiency prediction for axial turbines by advanced turbulence models. In: 26th IAHR Symposium on Hydraulic Machinery and Systems, Beijing (2012)
8. Karypis, G., Kumar, V.: MeTiS: unstrucured graph partitioning and sparse matrix ordering system. University of Minnesota (1995)
9. Krappel, T., Ruprecht, A., Riedelbauch, S.: Turbulence resolving flow simulations of a francis turbine with a commercial CFD code. In: High Performance Computing in Science and Engineering'15. Springer, Berlin (2016)
10. Menter, F.R.: Two-equation eddy-viscosity turbulence models for engineering applications. AIAA J. **32**(8), 269–289 (1994)
11. Menter, F.R., Egorov Y.: The scale-adaptive simulation method for unsteady turbulent flow predictions. Part 1: theory and model Description. J. Flow Turbul. Combust. **85**(1), 113–138 (2010)
12. Menter, F.R., Schütze, J., Gritskevich M.: Global vs. zonal approaches in hybrid RANS-LES turbulence modelling. In: Fu, S., Haase, W., Peng, S.-H., Schwamborn, D. (eds.) Progress in Hybrid RANS-LES Modelling: Papers Contributed to the 4th Symposium on Hybrid RANS-LES Methods, Beijing. Notes on Numerical Fluid Mechanics and Multidisciplinary Design, vol. 117, pp. 15–28. Springer, Berlin/Heidelberg (2012)
13. Menter, F.R.: Best practice: scale-resolving simulations in ANSYS CFD version 1.0 ANSYS Germany GmbH, April 2012
14. Nicoud, F., Ducros F.: Subgrid-scale stress modelling based on the square of the velocity gradient tensor. Flow Turbul. Combust. **62**, 183–200 (1999)
15. Raw, M.J.: Robustness of coupled algebraic multigrid for the Navier-Stokes equations. In: AIAA 96-0297, 34th Aerospace and Sciences Meeting & Exhibit, Reno (1996)
16. Pacot, O., Kato, C., Avellan, F.: High-resolution LES of the rotating stall in a reduced scale model pump-turbine. In: 27th IAHR Symposium on Hydraulic Machinery and Systems, Montreal (2014)
17. Strelets, M.: Detached eddy simulation of massively separated flows. In: AIAA Paper 2001-0879, 39th Aerospace Sciences Meeting and Exhibit, Reno (2001)

CFD Simulations of Thermal-Hydraulic Flows in a Model Containment: Phase Change Model and Verification of Grid Convergence

Abdennaceur Mansour, Christian Kaltenbach, and Eckart Laurien

Abstract Two-phase flows with water droplets greatly affect the thermal-hydraulic behaviour in the containment of a Pressurized Water Reactor *PWR*. Such flows occur, inter alia, in French PWR in the form of spray cooling. Spray cooling ensures in case of a leak in the primary circuit the reduction of increased pressure and temperature in the containment due to the released steam. Purpose of the current paper is to present an application-oriented *CFD* model concerning heat and mass transfer between droplets and gas during the spray cooling process with an Euler-Euler two-fluid approach. In the current model, the resistance to droplet heating is taken into account. A grid convergence study *GCI* was also performed to quantify the spatial discretization error for a three dimensional natural convection flow simulation using the commercial *CFD* package Ansys *CFX* 16.1. Five numerical grids with up to $39.73 \cdot 10^6$ elements have been considered to perform this study. Low grid convergence indexes were reported for the fine-mesh comparisons of $7.11 \cdot 10^6$–$16.85 \cdot 10^6$ and $16.85 \cdot 10^6$–$39.73 \cdot 10^6$, resulting in averaged *GC*I values of less than 1 % for all considered flow variables. The parallel scalability of the simulations was also investigated in this work. Due to the large size and complexity of containment simulations as well as the physically complex flow phenomena in nuclear applications, numerical meshes with large cell numbers may have to be generated in order to minimize the numerical errors. Hence, efficient parallel computing is very important to get realistic computing time. Good scalability of *CFX* 16.1 is achieved up to 1800 computational cores on a mesh with $83 \cdot 10^6$ elements and $24 \cdot 10^6$ nodes.

A. Mansour (✉) • C. Kaltenbach • E. Laurien
Institute of Nuclear Technology and Energy Systems, University of Stuttgart, Pfaffenwaldring 31, 70569 Stuttgart, Germany
e-mail: abdennaceur.mansour@ike.uni-stuttgart.de; christian.kaltenbach@ike.uni-stuttgart.de; eckart.laurien@ike.uni-stuttgart.de

W.E. Nagel et al. (eds.), *High Performance Computing in Science and Engineering '16*, DOI 10.1007/978-3-319-47066-5_35

1 Introduction

One severe accident in a containment could be a leak in the primary circuit of a *PWR*. As a result, hot steam is injected into the plant room and mixes with the air, which increases the containment pressure and could affect its functionality. In addition, this pressure increase could cause the opening of the burst disc. Due to the density difference between the hot and cold humid air in the plant room and the operating room, a natural convection flow between those two rooms is initiated. To prevent this, spray cooling has been proven as an effictive method to reduce the containment pressure and temperature. For instance, spray cooling systems are installed in the upper section of the containment. Spray activation affects thermal-hydraulic processes in the containment. In some areas, condensation proceeds due to the supersaturation of the humid air gas atmosphere. In contrast, in areas where droplets reach saturation temperature, they evaporate and transfer mass to the gas phase. Merely bigger droplets retain in their disperse shape and heat up but without phase change. For the understanding and prediction of those containment flow phenomena, the methods of computational fluid dynamics *CFD* have been recently used [1, 2]. Those methods are expected to have a better accuracy and a larger applicability than the traditional methods (lumped parameter system codes) [3] used in the reactor safety analysis, which are based on one-dimensional models of transport processes and have limitations regarding the conservation of momentum [4]. *CFD* methods, however, are characterized by very large computing time due to the complexity of nuclear containment applications and the required high grid resolution. Hence, the estimation of the discretization error and the investigation of parallel computing are very important in containment applications, in order to get reliable results in realistic computing times.

In the past thermal-hydraulic investigations concerning heat and mass transfer between droplets and a humid air atmosphere were carried out by different authors. Babić et al. [5] used a single-phase approach to model condensation and evaporation of droplets in THAI and TOSQAN. In this simplified treatment, the involved gas and droplet phase share the same velocity field. This approach is only valid for very small droplets and a negligible small Stokes number and therefore not appropriate for spray modeling. To take into account different velocity fields for gas and droplets, Zhang and Laurien [6] used an Euler-Euler two phase approach for volume condensation modeling. They characterized the gas as continuous and droplets as disperse phase with monosized droplet diameters up to 150 μm. Mimouni et al. [7] used an Euler-Euler two phase approach to simulate a spray in the French TOSQAN facility. A spray with monosized droplets up to a diameter of 200 μm is assumed. The heat and mass transfer is described based on diffusion. This is valid for small droplets. For larger ones, the heating process up to saturation temperature must be considered. As a result of the SARNET-2 spray benchmark, Malet et al. [8] worked out that droplet diameter modeling has a large impact on spray flows. Monosized droplets have almost the same trajectories due to the same mass. This leads to a high concentration of droplets in the spray envelope. Polydisperse sprays with different droplet sizes in contrast provide droplets in the center of the spray due to less inertia forces.

In order to investigate the thermal-hydraulic behavior in a nuclear reactor, different *CFD* containment simulations have been performed. Using the commercial *CFD* code CFX 4.4, a model containment of a nuclear power plant was used in the Petten Research Center to calculate an accident scenario, which has been performed earlier with the lumped parameter code SPECTRA [4]. For the 3D geometry, a mesh with approx. 680,000 hexahedral cells was generated. The results of the *CFD* model and the lumped parameter code were qualitatively close to each other, although a quantitative discrepancy was observed due to the absence of an evaporation model. The spatial discretization errors due to the coarse numerical grid used for this investigation can be another reason for this discrepancy. Due to the large computational efforts and the absence of hardware with high-level computational capacity, many works could not study and quantify the discretization error in *CFD* simulations for nuclear applications [1, 4, 9]. For the quantification of the spatial discretization error, the Grid Convergence Index *GCI* has been proposed by Roache, [10]. This method has been recommended for the estimation of discretization, since it has been tested in many cases [11]. However, meshes with large element numbers have to be generated in order to carry out a GCI study in containment calculations, due to the complexity of both nuclear physics and geometries. Hence, efficient parallel computing is very important in order to reduce the resulting high computing time. To study the parallel performance of Ansys *CFX* 14.0, calculations using a mesh with approx. $10.2 \cdot 10^6$ of a *PWR* containment were performed on the Cray *XE*6 Hermit Cluster at the HLRS Stuttgart [6]. The speedup and efficiency of those parallel calculations were significantly away from the ideal values. For 80 cores it was approx. 35/80 and for 160 cores, 43/160. A comparison of the parallel efficiency of Ansys *CFX* 14.5 and $OpenFOAM16 - ext$ was carried out also on the Cray *XE*6 Hermit Cluster through transient *CFD* simulations in a Francis turbine [12]. A numerical Mesh with $40 \cdot 10^6$ cells was used for this investigation. The results showed a relatively poor parallel behavior of *CFX* compared with *OpenFOAM*. The optimum speedups were achieved at 192 cores for *CFX* and 768 Cores for *OpenFOAM*. However, many improvements in terms of parallel performance should have been added to the new versions of *CFX*, namely the version 16.1 used in this work.

The aim of this paper is to present an application-oriented Euler-Euler model for Ansys CFX 16.1, which describes the heat and mass transfer for containment spray cooling applications with larger droplets (up to 1250 μm). In the present model, the heating process of droplets is additionally taken into account which affects the phase change. This model will be used to simulate monosized as well as polysized sprays in the model containment THAI. Another aim of this work is to estimate the numerical discretization error in a natural convection flow simulation based on the theory of the grid convergence index. The applicability of this theory on the two-room geometry $THAI^+$ and the complicated convection flow is considered. In addition, the parallel efficiency of Ansys *CFX*16.1.will be investigated.

2 Computational Model

2.1 Mathematical Approach and Droplet Modeling

The basic mathematical approach for this work is the Euler-Euler two-fluid model, which was developed for multiphase flows by Ishii and Hibiki [13]. Each fluid is considered as a continuous phase and has a complete set of conservation equations for mass, momentum and energy. Due to the interpenetrating continuum, each phase is indicated with the so-called volumetric fraction α_k. The subscript k indicates the phase state gas G or liquid L. The continuous gas phase (humid air) is a mixture of dry air and water vapor. The liquid is handled as disperse with a fixed droplet diameter. The contact area between the phases is denoted by the interfacial area density A_{KK}. Through the interface area, interactions between the phases can be taken into account. The postulated phase change model for evaporation and condensation is implemented via source terms for mass (Γ_k) and energy (E_k) in Ansys CFX. In the following the basic equations for Ishii's two-fluid model are explained. The mass conservation is described by

$$\frac{\partial(\rho_k \alpha_k)}{\partial t} + \nabla(\rho_k \alpha_k \overrightarrow{u}^k) = \Gamma_k, \tag{1}$$

ρ_k represents the density of phase k, $\overrightarrow{u}^k$ stands for the averaged velocity for phase k and t is the physical time. The momentum conservation is described by the following equation:

$$\begin{aligned} \frac{\partial(\rho_k \alpha_k \overline{u}_m^k)}{\partial t} + \nabla(\rho_k \alpha_k \overrightarrow{u}^k_m \overline{u}_m^k) = \\ -\frac{\partial(\alpha_k p)}{\partial x_m} + \nabla[(\alpha_k \underline{\underline{\tau}}^k + \underline{\underline{\tau}}^{Re,k})]_m + \overline{u}_m^k \Gamma_k + \alpha_k \rho_k g_m + M_{k,m}, \end{aligned} \tag{2}$$

p is the pressure, $\underline{\underline{\tau}}^k$ and $\underline{\underline{\tau}}^{Re,k}$ represent the molecular and the turbulent Reynolds stresses of phase k and g is the acceleration of gravity. $M_{k,m}$ is the momentum source term and must be modeled. The energy equation is specified with the enthalpy $\overline{e}_k$

$$\frac{\partial(\rho_k \alpha_k \overline{e}^k)}{\partial t} + \nabla(\rho_k \alpha_k \overrightarrow{u}^k \overline{e}^k) = \nabla[\alpha_k(\overline{q}^k + \overline{q}^{Re,k})] + \overline{e}^k \Gamma_k + E_k. \tag{3}$$

Here $\overline{q}^k$ and $\overline{q}^{Re,k}$ are the molecular and the turbulent heat fluxes. E_k represents the source term for the energy.

Due to the application of the two-fluid model, there are several secondary conditions, which must also confirm conservation. All volume fractions α_k must sum to one and all mass source terms Γ_k have to yield zero.

The droplets are modelled as disperse phase with a fixed diameter d. The interfacial momentum transfer term $M_{k,m}$ contains the interfacial drag force $M^D_{k,m}$. The interfacial drag force $M^D_{k,m}$ can be described by the following equation:

$$M^D_{k,m} = M^D_{G,m} = -M^D_{L,m} = \alpha_L \frac{3\rho_G}{4d_L} c_D |\overrightarrow{u}^L - \overrightarrow{u}^G| (\overrightarrow{u}^L - \overrightarrow{u}^G) \quad , \tag{4}$$

where c_D is the drag coefficient. A correlation for c_D is denoted according Schiller and Naumann in [14]

$$c_D = \frac{24}{Re}(1 + 0.15 \cdot Re^{0.687}) \quad . \tag{5}$$

The correlation is valid for Reynolds numbers up to 800. Beyond, the Euler-Euler two-fluid model is based on the Unsteady Reynolds Averaged Navier Stokes (URANS) equations. Therefore the Reynolds stresses $\overline{\tau}^{Re,k}$ and the turbulent heat fluxes $\overline{q}^{Re,k}$ must be modeled. In the current work this is done with the shear stress turbulence (SST) model, which was developed by Menter [15] and is based on two equations.

2.2 Grid Convergence Index

The Method of the Grid Convergence Index was introduced by Roache [10] as a uniform criterion to estimate the spatial discretization error in *CFD* applications. The *GCI* is based on the theory of the Richardson Extrapolation

$$f_{exact} \approx f_1 + \frac{f_1 - f_2}{r^p - 1} \quad , \tag{6}$$

where f_1 and f_2 are solutions of the considered variables (in this investigation: temperature, velocity, pressure and relative humidity) on two different grids with discrete spacings h_1 (fine grid) and h_2 (coarse grid), respectively. $r = \frac{h_2}{h_1}$ represents the grid refinement ratio and p stands for the accuracy order of the numerical method. The objective of the Richardson extrapolation, according to Eq. (6), is to provide a more accurate estimation f_{exact} of the exact solution, using the two numerical solutions f_1 and f_2. The relative error between the Richardson Extrapolation estimation f_{exact} and the fine grid solution f_1 is defined as follows:

$$E_1 = \frac{f_{exact} - f_1}{f_1} = \frac{\varepsilon}{r^p - 1} \quad . \tag{7}$$

In Eq. (7), ε is the relative error between the fine and coarse grid solutions f_1 and f_2

$$\varepsilon = \frac{f_1 - f_2}{f_1} \quad . \tag{8}$$

The estimator ε would be only accepted by most *CFD* users as a good error estimation for a grid doubling/halving ($r = 2$) and a code with 2nd-order accuracy ($p = 2$). In this case, cumulative experience has demonstrated the reasonability of the indicator ε [10]. For other cases, i.e. $r \neq 2$ or $p \neq 2$, ε seems not to be an appropriate error estimator. On the one hand, it does not take into account r or p and on the other hand it is not always conservative with respect to E_1. The last issue, however, relates also to E_1 which can be conservative and optimistic with an equal probability of 50 %. For this reason, E_1 cannot be a well-founded criterion such as for example the 2σ indicator for statisticians [10]. The idea behind the Grid Convergence Index is to combine both error estimators ε and E_1. Suppose, we have performed a mesh study with any r and p and determined the error indicator E_1. The *GCI* will be equal to an equivalent ε_{eq} which would produce the same E_1 for the same problem and on the same mesh but for $r = 2$ and $p = 2$.

$$GCI = F_s \frac{|\varepsilon|}{r^p - 1} \tag{9}$$

The safety factor F_s is set to 1.25 since more than 2 meshes are used in the current study [10]. The *GCI* can be understood as a measure, which indicates how far a computed solution from the asymptotic numerical value is. To perform a *GCI* study, the following procedure has been adopted [11]. Suppose that for a specific *CFD* calculation we generated 3 meshes. N_1, N_2 and N_3 are the total cell numbers for mesh 1 (fine), mesh 2 (middle) and mesh 3 (coarse). First, one calculates the averaged grid spacing for each mesh

$$h_i = \left[\frac{1}{N}\sum_{i=1}^{N_i} \Delta V_i\right]^{1/3} , \tag{10}$$

where ΔV_i the volume of each mesh cell. After calculating the grid refinement ratios $r_{21} = \frac{h_2}{h_1}$ and $r_{32} = \frac{h_3}{h_2}$, one should determine the observed accuracy order p using the numerical solutions f_1, f_2 and f_3.

$$p = \frac{\left| ln\left|\frac{f_3 - f_2}{f_2 - f_1}\right| + ln\left(\frac{r_{21}^p - s}{r_{32}^p - s}\right)\right|}{ln(r_{21})} \tag{11}$$

$$s = sign\left(\frac{f_3 - f_2}{f_2 - f_1})\right) \tag{12}$$

Equation (11) can be solved using fixed-point iteration. The fact that $\left(\frac{f_3 - f_2}{f_2 - f_1}\right) < 0$ indicates an oscillatory convergence, which should also be reported. When the observed order of accuracy agrees with the theoretical order of the numerical method, then the grids are assumed to be within the asymptotic range. The next step is to calculate the extrapolated value f_{exact} for the fine mesh using Eq. 7 and the

GCI:

$$GCI_{21} = F_s \frac{|\varepsilon_{21}|}{r_{21}^p - 1} \quad , \tag{13}$$

$$GCI_{32} = F_s \frac{|\varepsilon_{32}|}{r_{32}^p - 1} \quad . \tag{14}$$

2.3 Parallelization

The flow simulations described below are performed on the CRAY XC40 *HazelHen* of the High Performance Computing Center Stuttgart *HLRS*. This is a supercomputer with 7712 compute nodes. Each node contains 24 processing cores and has 128 *GB* memory. In order to measure the parallel performance of *CFX* 16.1, one considers the speedup which can be defined as

$$Sp = \frac{T_{120}}{T_p} \quad . \tag{15}$$

T_{120} is the reference speedup, which denotes the wall clock time needed on 120 processing cores, while T_p is the wall clock time for p cores. For the partitioning operation, *CFX* uses a node-based partitioning method. The default partitioner is the Multilevel Graph Partitioning Software *MeTiS* [18]. To improve the parallel performance of CFX, the expert parameter *parallel optimization level* was set to its upper bound 3 and the *large Problem partitioner* was selected.

3 Phase Change Model

3.1 Consideration of Droplet Heating

The implemented phase change model is based on heat transfer between the droplet and gas phase. Droplets can't evaporate until they heat up to saturation temperature. A constant temperature distribution over the whole droplet is assumed. Based on the convective heat transfer between droplets and gas (radiation is neglected), we can perform a simple assessment of the droplet thermal response time according to Crowe et al. [14]. We assume the droplets have spherical shape

$$\frac{\partial \overline{T}_L}{\partial t} = \frac{6 \cdot Nu \cdot \lambda}{c_L \cdot d_L^2 \cdot \rho_L} (\overline{T}_L - \overline{T}_G) \quad . \tag{16}$$

Table 1 Fluid properties for thermal response time assessment

Property	Value	Unit
Re	800	[-]
Pr	0.7	[-]
λ	0.025	[W/m K]
ρ_L	1000	[kg/m^3]
c_L	4200	[J/kg K]

In Eq. (16), $\overline{T}_L$ is the temperature of the droplet, λ is the thermal conductivity, c_L represents the specific heat capacity of water and Nu stands for the Nusselt number, which is defined for a droplet of spherical shape by the Ranz-Marshall correlation [16]

$$Nu = 2 + 0.6 \cdot Re^{\frac{1}{2}} \cdot Pr^{\frac{1}{3}} \quad . \tag{17}$$

The temperature change of a droplet is dependend on the temperature-difference prefactor in Eq. (16). The prefactor is therefore a criterion of thermal inertia and is denoted as the thermal response time τ_T

$$\tau_T = \frac{c_L \cdot d_L^2 \cdot \rho_L}{6 \cdot Nu \cdot \alpha} \quad . \tag{18}$$

The spray nozzle in the used validation experiment is characterized with a Sauter droplet diameter of $d_{32} = 830\,\mu\text{m}$. For evaluation of τ_T we use additional fluid properties, see Table 1.

Based on the assumptions, we obtain a thermal response time in the order of one second. Due to the spray velocity of roughly 20 m/s at the spray nozzle outlet, each droplet covers a significant distance until it reaches saturation temperature. It is therefore mandatory to consider droplet heating.

3.2 Phase Change Model and Relevant Equations

The present phase change model is based on the Euler-Euler two-fluid model. In the case of spray cooling, droplet heating with subsequent evaporation and volume condensation due to supersaturation of the gas phase is possible. In Fig. 1 the temperature distribution around a single droplet for the two thermodynamic processes are shown schematically. Figure 1 (left) shows a cold droplet with temperature $\overline{T}^L$ entering in a hot gas atmosphere, which holds $\overline{T}^G$. When droplets reach T_{sat}, the heating process is finished and evaporation occurs due to saturation. In the case of condensation, Fig. 1 (right), droplets possess $\overline{T}_{sat}$. $\overline{T}^G$ around the droplet is lower than T_{sat} and therefore the latent heat due to the phase change can be released.

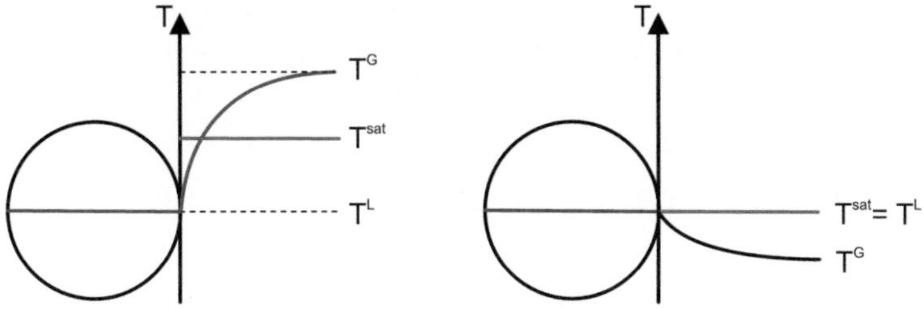

Fig. 1 Temperature distribution during droplet heating and evaporation *(left)* and during condensation *(right)*

Important for the process of heat up and evaporation is the temperature difference between the droplets and the surrounding humid air atmosphere (mixture of water vapor and air as non-condensable gas). The interfacial heat flux E_L due to the temperature difference in Eq. (19) drives the droplet heating and the following evaporation process:

$$E_L = -E_G = \alpha_{HT} \cdot \alpha_L \cdot \frac{6}{d_L} \cdot (\overline{T}^G - \overline{T}^L) \quad . \tag{19}$$

In Eq. (19), α_{HT} represents the heat transfer coefficient, which is based on the Nusselt number

$$\alpha_{HT} = \frac{\lambda}{d_L}(2 + 0.6 \cdot Re^{\frac{1}{2}} \cdot Pr^{\frac{1}{3}}) \quad . \tag{20}$$

The described interfacial heat flux E_L heats the droplet up to saturation temperature T_{sat}. It is determined with the gas-atmosphere conditions and can be defined by Antoine's equation [17]

$$T_{sat} = \frac{1687.537}{5.11564 - log(p_{sat})} - 230.17 + 273.15 \quad [K] \quad . \tag{21}$$

In Eq. (21) p_{sat} is the saturation pressure and is calculated with the assumption, that each partial pressure of water vapor in the humid gas atmosphere is equal to the saturation pressure

$$p_{Sat} = c_{vap} \cdot \overline{p} \quad , \tag{22}$$

where c_{vap} stands for the volume fraction of water vapor and $\overline{p}$ is the absolute pressure in the containment.

The following term for the mass transfer Γ_G due to evaporation or condensation is only valid, when the droplets achieve T_{sat}. Is $\overline{T}^L$ less than T_{sat}, phase change is not taken into account and droplets are only heated due to the temperature difference between gas and droplets. When droplets achieve saturation temperature T_{sat}, they evaporate in the containment atmosphere and the temperature difference drives the phase change. Having regard to the latent heat Δh_{LG}, one can obtain the evaporated mass Γ_G, which is incorporated into the gas phase:

$$\Gamma_G = -\Gamma_L = \frac{6 \cdot \alpha_{HT} \cdot \alpha_L \cdot (\overline{T}^G - T_{sat})}{\Delta h_{LG} \cdot d_L} \quad . \tag{23}$$

In case of condensation, the temperature difference is negative and mass is transferred to the droplet out of the humid-air atmosphere.

4 Numerical Method

4.1 Geometry and Boundary Conditions

THAI$^+$ is a two-room geometry composed of two cylindrical vessels: *THAI* (Thermal-hydraulics, H2, Iodine and Aerosols) and *PAD* (Parallel Attachable Drum). *THAI* is 9.2 m high and has a diameter of 3.2 m while height and diameter of *PAD* are 9.8 and 1.6 m, respectively. Both vessels are interconnected by large horizontal *DN* 500 pipes with diameters of 0.5 m, see Fig. 2, [19]. There are various installations inside *THAI*: an open inner cylinder with a diameter of 1.4 m and a height of 4 m. Four equally spaced condensate trays are located between the inner cylinder and the THAI walls and enclose an angle of 60° each. The open sectors of 30° present flow paths between the upper and lower annulus. There are also condensate gutters on the *THAI* walls in the upper annulus and at the

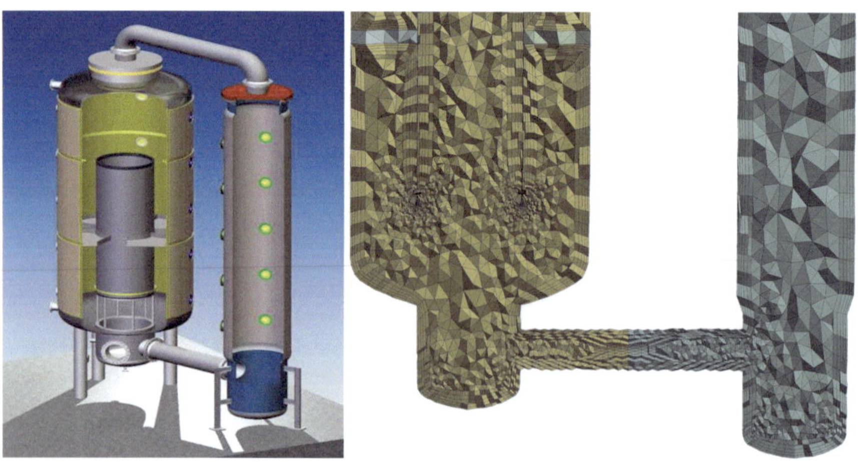

Fig. 2 *THAI*$^+$ facility with the two vessels *THAI* and *PAD* (*left*), [19]; Unstructured grid at the bottom of *THAI* vessel (*right*)

bottom on the inner and outer wall of the inner cylinder. The *THAI* vessel with those internal installations represents the plant room, while *PAD* vessel corresponds to the operating room. The geometry of *THAI*$^+$ was created with the *Ansys DesignModeler* originating from a *CAD* model of Becker Technologies GmbH. The 3D mesh was generated using the Ansys Meshing Tool. This is an unstructured grid composed of tetrahedral elements in the mesh volume and prism layers on the walls. The structure of this grid is shown on Fig. 2. The initial state is described by an overall temperature of 92.6 °C and a pressure of 2 *bars*. The air throughout the facility is considered to be saturated at the beginning, i.e. the relative humidity amounts to 100 %. The THAI walls are then heated up to 105 °C. Starting from this initial state, a transient simulation is run until thermal equilibrium was reached, i.e. the temperature is almost equal to 105 °C everywhere in *THAI*$^+$. Due to the high temperatures, no steam condensation occurred during this simulation.

4.2 Numerical Setup

The transient simulations were carried out using the commercial *CFD* package Ansys *CFX* 16.1. For turbulence and buoyancy, the *SST* turbulence model and full buoyancy model were used. An advection scheme with high resolution was employed for the spatial discretization of the *URANS*. This adaptive method uses a blend factor β, which varies between 0 and 1 and aims to maintain the solver accuracy as much second order as possible. In regions with low gradients, β is set to values near to 1, i.e. the method has a second order accuracy. However, in regions with high gradients, the first order upwind scheme will be used for more robustness, i.e. $\beta = 0$. The high resolution scheme was also used in the turbulence numerics for the spatial discretization of the turbulent terms. The second order backward euler method was employed for the temporal discretization of the transient terms, while a scheme of first order was set to solve the temporal turbulent terms. The total physical time for the simulations was set to 3.36 h.

5 Results

5.1 Results of the Grid Convergence Index

To perform the *GCI* study, three unstructured meshes with approx. $1.26 \cdot 10^6$, $3.02 \cdot 10^6$ and $7.11 \cdot 10^6$ tetrahedral and prism elements (see Fig. 2) were initially generated and a grid independence study was carried out, in which a comparison of temperature, velocity, pressure and relative humidity in different points was performed. In Fig. 3 temperature profiles over time in two measurement points *MP* are shown. One recognizes no significant difference between the two coarse meshes.

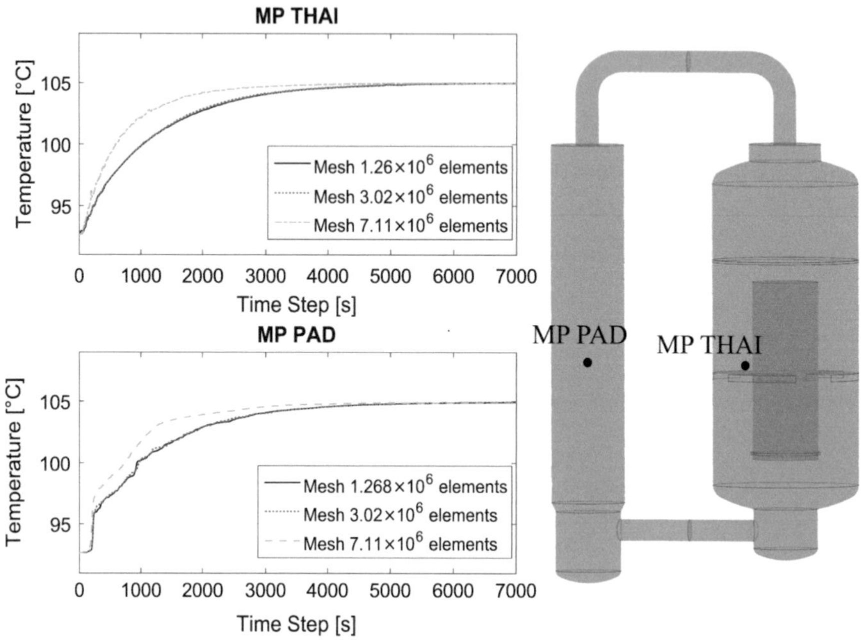

Fig. 3 Grid independence study using 3 meshes with $1.26 \cdot 10^6$, $3.02 \cdot 10^6$ and $7.11 \cdot 10^6$ elements: comparison of temporal temperature profiles in 2 Measurement points in *THAI* and *PAD* (*left*); *M*P locations in $THAI^+$ (*right*)

Table 2 Grid information

	$V_{mesh}[m^3]$	Element number	h_i
Mesh 5	78.610	1268908	0.040
Mesh 4		3020178	0.030
Mesh 3		7118431	0.022
Mesh 2		16856294	0.017
Mesh 1		39731271	0.013
r_{21}	1.331		
r_{32}	1.333		
r_{43}	1.331		
r_{54}	1.335		

In contrast, the temperature for the fine mesh with $7.11 \cdot 10^6$ elements shows a big deviation from the coarse meshes. Due to this behavior, which has also been detected in almost all evaluated points, a further mesh refinement was necessary in order to make sure that the asymptotic range was achieved using the mesh with $7.11 \cdot 10^6$ elements. For this reason, two other grids with approx. $16.85 \cdot 10^6$ and $39.73 \cdot 10^6$ elements were generated, see Table 2. The corresponding grid spacings according to Eq. 10 and the refinement ratios are also reported in Table 2. Celik [11] recommended refinement ratios greater than 1.3 for a better accuracy of the *GCI* results. The five grids in this study were generated based on this recommendation. In addition, the *CFL* stability condition is an important numerical issue related to the refinement ratio

$$CFL = \frac{u \cdot \Delta t}{\Delta h} \quad . \tag{24}$$

When refining the mesh, Δh will decrease and the *CFL* number may be very large, which could affect the stability of the numerical method. Especially for explicit methods, the *CFL* condition is a very important numerical issue. Even if *CFX* uses an implicit code, the refinement ratios were selected to be as small as possible ($\approx$1.33), so that the variations in the *CFL* number are kept in the small range of 1.33.

In this study, the five grids are compared at approx. 936 points for the variables temperature, relative humidity, pressure and vertical velocity. A major problem for the *GCI* evaluations was to determine the observed order of accuracy p_{obs} according to Eqs. (11) and (12). In fact, this results in high p values significantly greater than the maximum theoretical order of *CFX*, i.e. $p_{max} = 2$. These high p_{obs} values result in very low "non-realistic" *GCI* values. This problem has been encountered in several complex turbulent flows in complicated geometries [20], which is the case of this natural convection flow. The main contributors of this so called *noisy Grid Convergence* could be the lack of the geometrical similarity of the unstructured grids and the use of damping functions and switches in the turbulence models [20]. One approach to deal with this issue was suggested by Roache [20] and consists in estimating the *GCI* using the theoretical order of accuracy [21]. The evaluation of the blend factor β of the high resolution scheme (see Chap. 4.2) results in an averaged theoretical order of accuracy $p_{theo} = 1.926$ [22]. This value is used to assess the *GCI*'s in the further investigations. The results reported in Table 3, show that the *GCI* values are over all conservative compared to ε. Successive grid refinement leads to a minimization of both ε and *GCI*. An exception is the

Table 3 The averaged ε, *GCI* measures and percentage of oscillatory convergence *OC* over the 936 considered points for temperature T, relative humidity ϕ, pressure p and vertical velocity v_y

	ε_{54} [%]	ε_{43} [%]	ε_{32} [%]	ε_{21} [%]	GCI_{54} [%]	GCI_{43} [%]	GCI_{32} [%]	GCI_{21} [%]
T	0.104	1.119	0.178	0.079	0.175	1.906	0.229	0.134
ϕ	0.385	4.138	0.351	0.285	0.645	7.047	0.513	0.485
p	0.016	0.298	0.011	0.002	0.026	0.507	0.018	0.003
v_y	372	534	139	180	1521	896	236	308

	OC 5-4-3 [%]	*OC* 4-3-2 [%]	*OC* 3-2-1 [%]
T	8.25	14.25	9.25
ϕ	8.79	13.33	12.52
p	25	100	62.50
v_y	70.83	47.92	59.03

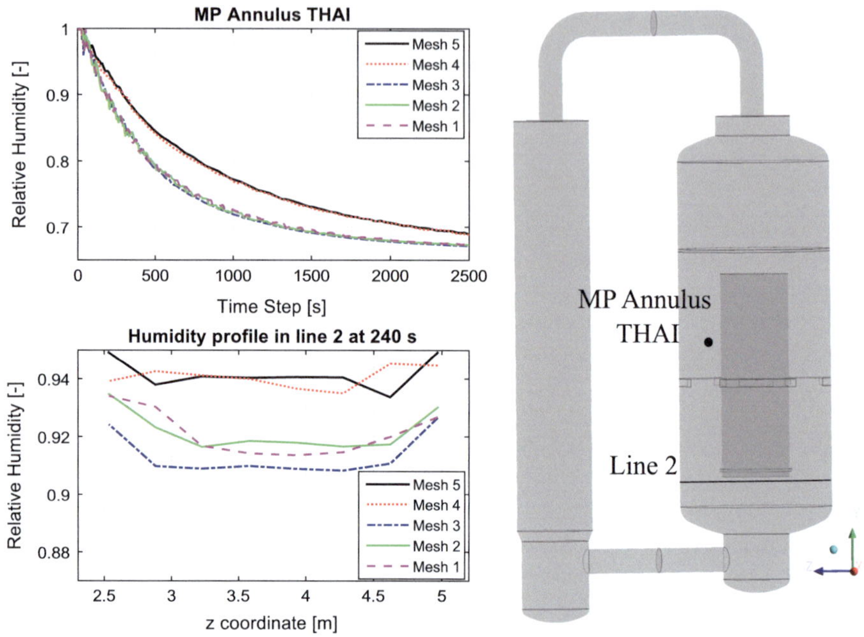

Fig. 4 Comparison of the relative humidity profiles over time in *MP* Annulus *THAI* and in line 2 at 240 s (*left*); Locations of the *MP* Annulus *THAI* and Line 2 (*right*)

transition from grid no. 4 to 3. This issue was already mentioned at the beginning of this section. A large deviation between those two grids results in large ε and *GCI* values for more than 7 % for the relative humidity, see Fig. 4. A possible reason for this issue may be the numerical resolution of vortices on the mesh no. 3 with $7.11 \cdot 10^6$, which could not have been resolved on the coarse meshes 4 and 5. For a high-quality grid independence study, the asymptotic range is assumed to be reached for a numerical error ε less than 1 % [21]. Even if the *GCI*'s are more conservative than ε, grid convergence index values less than 1 % have been achieved for the grid pairs 3-2 and 2-1. For instance, a comparison of the relative humidity profiles in Fig. 4 confirms this statement. Indeed, no significant changes between the fine meshes 3, 2 and 1 are detected. This behavior has been identified for almost all considered profiles. The only exception is the vertical velocity v_y, where very large ε and *GCI* values (more than 300 % for the fine grid set 1-2) are reported. In effect, the flow circulation in this natural convection problem is too slow resulting in very low velocities with an averaged value $v_{ave} \approx 0.028$ m/s. In addition, the velocities are characterized by high fluctuations due the complexity of the geometry and the several internal installations in THAI vessel, see Fig. 5. For these reasons, the velocity does not seem to be an appropriate field variable for the quantification of the grid convergence index in the current natural convection flow. However, the comparison of the transient velocity profiles on Fig. 5 shows that the vertical velocity decreases nearly to zero when the solution is close to the

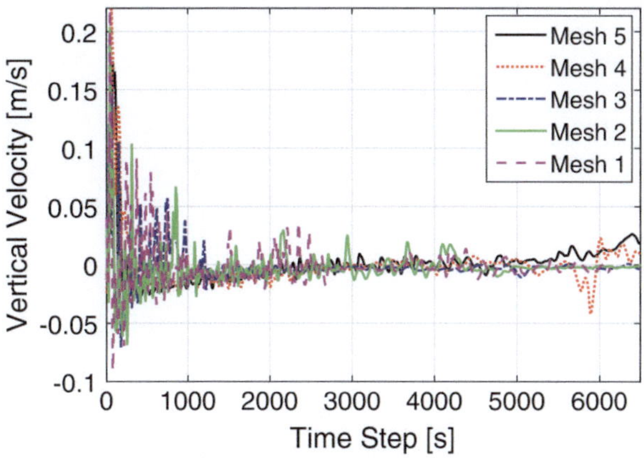

Fig. 5 Comparison of the simulated velocity time diagrams for meshes 1–5 in a measurement point in the *THAI* annulus

thermal equilibrium (after approx. 5000 s) in the fine grids 1-2-3, while it is always oscillating in the coarse grids 4-5. On the basis of the above results of the Grid Convergence Index and the mesh independence study, it can be concluded that the asymptotic range has been reached using mesh no. 3 with approximately $7.11 \cdot 10^6$ elements.

5.2 *Parallelization*

The grid used in this investigation is an unstructured grid composed of tetrahedral elements in the mesh volume and prism layers on the walls, see Fig. 2. This was generated using the *Ansys Meshing Too*l and has approximately $83 \cdot 10^6$ elements and $24 \cdot 10^6$ nodes. The scaling tests have been performed up to 2880 cores. The results on Fig. 6 show, that up to 710 cores, *CFX* 16.1 is characterized by a super-linear speedup. A possible reason behind this behavior could be the cache effect, which enables a quicker access to the stored data in the *CPU* cache memory without the need to recompute them again. With increasing core number, *CFX* speedup decreases, but remains always close to the ideal speedup. It reaches a maximum at 1800 cores, which is equal to the ideal speedup at a ratio of approximately 11.4/15. For higher core numbers, the speedup decreases, i.e. the communication time between the cores outweighs the computational time of each core. For the maximum speedup at 1800 cores, the computational time is equal to the physical time at a ratio of 44:1.

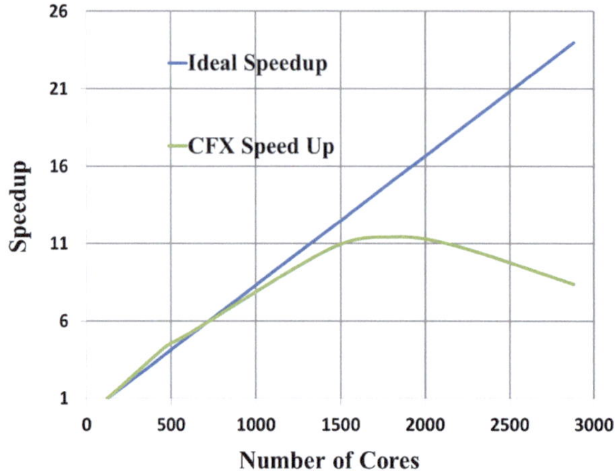

Fig. 6 Speedup measurements for simulations with *CFX* 16.1 in a fine unstructured grid composed of $83 \cdot 10^6$ elements and $24 \cdot 10^6$ nodes

6 Conclusions

To quantify the spatial discretization error, a Grid Convergence Index study was carried out for a natural convection flow simulation using the commercial *CFD* package *Ansys CFX* 16.1. This simulation is a part of the initial operating test *TH* 27 of the newly constructed two-room facility $THAI^+$. Five numerical grids with approximately $1.26 \cdot 10^6$, $3.02 \cdot 10^6$, $7.11 \cdot 10^6$, $16.85 \cdot 10^6$ and $39.73 \cdot 10^6$ elements were employed to assess the *GCI* values in 936 consistent points in $THAI^+$. The averaged *GCI* values for the grid set 3-2 ($7.11 \cdot 10^6$-$16.85 \cdot 10^6$) were 0.229, 0.513 and 0.018 for the temperature, relative humidity and pressure, respectively. These low *GCI* values are an indication of a good quality grid-independent solution for mesh 3 with $7.11 \cdot 10^6$ elements. The parallel performance of *CFX* 16.1 was also investigated in this work. The results of the scaling tests using a numerical grid with $83 \cdot 10^6$ elements and $24 \cdot 10^6$ nodes showed a super-linear speedup up to 710 cores. For higher core numbers, *CFX* speedup decreases. However, it remains always close to the ideal one. The maximum speedup was about 11.4/15 at 1800 cores. The computational time at this maximum point is equal to the physical time at a ratio of 44:1.

Future work on the grid convergence index may focus on the applicability of the *GCI* method in two-phase flows using the newly developed phase change model, presented in this publication. In addition, the parallel performance of *CFX* 16.1 using other partitioning methods besides the Multilevel Graph Partitioning Software *MeTiS* should be investigated, in order to determine the most appropriate method for containment simulations.

Acknowledgements This work was supported by the German Federal Ministry of Economic Affairs and Energy (BMWi) on the basis of a decision by the German Bundestag, project number 1501493.

References

1. Babić, M., Kljenak, I., Mavko, B.: Simulation of atmosphere mixing and stratification in the THAI experimental facility with a CFD code. In: International Conference Nuclear Energy for New Europe, Bled, Slovenia (2005)
2. Zirkel, A., Doebbener, G., Laurien,E.: CFD simulation of forced flow within the THAI model containment. In: 17th International Conference on Nuclear Engineering, Bruessels, Belgium (2008)
3. IAEA: Use of computational fluid dynamics codes for safety analysis of nuclear reactor systems. Summary report of a technical meeting, Pisa, 11–14 Nov 2002
4. Houkema, M., Siccama, N.B., Lycklama à, J.A., Nijeholt, Komen, E.: Validation of the CFX4 CFD code for containment thermal-hydraulics. Nucl. Eng. Des. **238**, 590–599 (2008)
5. Babić, M., Kljenak, I., Mavko, B.: Prediction of light gas distribution in experimental containment facilities using the CFX4 code. Nucl. Eng. Des. **238**, 538–550 (2008)
6. Zhang, J., Laurien, E.: 3D numerical simulation of flow with volume condensation in presence of non-condensable gases inside a PWR containment. In: Nagel, W.E., Kröner, D.H., Resch, M.M. (eds.) High Performance Computing in Science and Engineering'14, pp. 479–497. Springer, Cham (2015)
7. Mimouni, S., Lamy, J.-S., Lavieville, J., Guieu, S., Martin, M.: Modelling of sprays in containment applications with a CMFD code. Nucl. Eng. Des. **240**, 2260–2270 (2010)
8. Malet, J., Mimouni, S., Manzini, G., Xiao, J., Vyskocil, L., Siccama, N.B., Huhtanen, R.: Gas Entrainment by one single French PWR spray, SARNET-2 spray benchmark. Nucl. Eng. Des. **282**, 44–53 (2015)
9. Stewering, J., Schramm, B., Sonnenkalb, M.: Validation of CFD-models for natural convection, heat transfer and turbulence phenomena. In: Computational Fluid Dynamics (CFD) for Nuclear Reactor Safety Applications (NRS), Bethesda (2010)
10. Roache, P.J.: Verification and Validation in Computational Science and Engineering, pp. 446. Hermosa, Albuquerque (1998)
11. Celik, I.B., Ghia, U., Roache, P.J., Freitas, C.J., Colemann, H., Radd, P.E.: Procedure for estimation and reporting of uncertainty due to discretization in CFD applications. ASME. J. Fluids Eng. **130**, 078001–078001-4 (2008)
12. Krappel, T., Ruprecht, A., Riedelbauch, S.: Flow simulation of a francis turbine using the SAS turbulence model. In: Nagel, W.E., Kröner, D.H., Resch, M.M. (eds.) High Performance Computing in Science and Engineering'13, pp. 455–463. Springer, Cham (2013)
13. Ishii, M., Hibiki, T.: Thermo-Fluid Dynamics of Two-Phase Flow. Springer, New York (2006)
14. Crowe, C.T., Sommerfeld, M., Tsuji, Y.: Multiphase Flows with Droplets and Particles. CRC-Press, Boca Raton (1998)
15. Menter, F.R.: Two-equation eddy-viscosity turbulence models for engineering applications. AIAA J. **32**, 1598 (1994)
16. Ranz, W.E., Marshall, W.R.: Evaporation from drops, Parts I & II. Chem. Eng. Prog. **48**, 141–146, 173–180 (1952)
17. Poling, B.E., Prausnitz, J.M., O'Connell, J.P.: The Properties of Gases and Liquids. McGraw-Hill, New York (2001)
18. Karypis, G., Kumar, V.: A fast and high quality multilevel scheme for partitioning irregular graphs. SIAM J. Sci. Comput. **20**, 359–392 (1998)

19. Freitag, M., Schmidt, E., Gupta, S.: Specification GRS–Report for Double-Blind Simulations of THAI Test TH27: Initial Operation Test of $THAI^+$: Part 1 Natural Convection with Steam Injection and Condensation (2015)
20. Eça, L., Hoekstra, M.: A procedure for the estimation of the numerical uncertainty of CFD calculations based on grid refinement studies. J. Comput. Phys. **262**, 104–130 (2014)
21. Longest, P.W., Vinchurkar, S.,: Effects of mesh style and grid convergence on particle deposition in bifurcating airway models with comparisons to experimental data. Med. Eng. Phys. **29**, 350–366 (2007)
22. Mansour, A., Laurien, E.: Simulation of a Natural Convection Flow with Humid Air in a Two-Room Geometry. Computational Fluid Dynamics (CFD) for Nuclear Reactor Safety Applications (NRS), Boston, 12–15 Sep 2016

Simulations of Unsteady Aerodynamic Effects on Innovative Wind Turbine Concepts

Annette Fischer, Levin Klein, Thorsten Lutz, and Ewald Krämer

Abstract Unsteady aerodynamics of a novel two-bladed wind turbine and a three-bladed wind turbine equipped with an innovative load reduction concept are investigated in the present paper using Computational Fluid Dynamics (CFD). The load reduction concept consists of coupled leading and trailing edge flaps. Unsteady aerodynamic effects caused by flap deflection are shown and a suitable temporal discretization for efficient simulation of coupled leading and trailing edge flaps is found. The behavior of the two-bladed turbine regarding the aerodynamic loads under extremely yawed conditions is presented and associated to the local flow conditions. The results of the studies allow a better prediction and understanding of unsteady aerodynamics of wind turbines.

1 Introduction

In order to supply the needs of the population for more energy on the one hand and to follow the political demand to reduce emissions on the other hand, alternative sources of energy gain in importance. Thereby, wind energy has become one of the most important and promising regenerative sources of energy in the last few years. To improve the competitiveness of wind energy, the cost of energy (CoE) has to be reduced. This can be done among other by designing more effective and durable wind turbines or by reducing material costs. Load alleviation systems like for example flaps can reduce fatigue loads and consequently increase the life time of wind turbines and turbines with a two-bladed rotor can save money as the material and installation costs are lower.

To design wind turbines with load alleviation systems or two bladed rotors appropriate, a wide range of investigations has to be performed. Some can be done using engineering models like blade element momentum theory (BEMT), but other require high fidelity methods like CFD. Especially unsteady aerodynamic effects, caused for example by flap movement or yaw misalignment, need to be investigated

A. Fischer (✉) • L. Klein • T. Lutz • E. Krämer
Institute of Aerodynamics and Gas Dynamics, Pfaffenwaldring 21, 70569 Stuttgart, Germany
e-mail: fischer@iag.uni-stuttgart.de

W.E. Nagel et al. (eds.), *High Performance Computing in Science and Engineering '16*, DOI 10.1007/978-3-319-47066-5_36

appropriate as they occur frequently during the life time of a wind turbine. 2 % up to 10 % of their operating time, wind turbines are exposed to yaw misalignment [18] and load fluctuations, caused for example by tower shadow, shear or gusts, occur several times per revolution of the rotor.

In the present paper a three-bladed wind turbine equipped with mechanically coupled leading and trailing edge flaps as well as a two-blade rotor under yawed inflow are investigated. The three-bladed rotor equipped with load alleviation system is the NREL 5MW wind turbine with some minor modifications introduced within the KIC-OFFWINDTECH project [1, 2], whereas the two-bladed turbine is a prototype developed by Skywind. For both setups, a grid convergence study according to Celik [4] was performed in order to estimate the influence of the numerical grids on the solution. For the three-bladed rotor the temporal discretization of unsteady effects with a 1p excitation frequency, representing an atmospheric boundary layer, caused by the flaps was investigated. To better understand the influence of yawed inflow on a two-bladed rotor, an investigation of the unsteady load fluctuations caused by yawed inflow and a comparison to a reference case without yawed inflow was done.

2 Numerical Setup and Computational Details

2.1 Numerical Methods

The investigations presented in this paper were performed using the finite volume solver *FLOWer* [14] by *DLR (German Aerospace Center)* that has already been applied in several other wind energy projects [11, 20, 22, 25, 26] and is continuously developed by the present working group. *FLOWer* solves the unsteady Reynold-averaged Navier-Stokes equations (*URANS*) on block-structured grids. A second order central discretization scheme JST [9] is used for spatial discretization and the temporal discretization is realized with an implicit dual time stepping scheme [8]. For turbulence modeling the Menter SST model is applied. The overlapping grid technique *CHIMERA* [3] enables the use of independently generated grids. For the present studies the components of the turbines have been meshed individually with fully resolved boundary layer ensuring $y^+ = 1$ of the first cell. By applying rotationally periodic boundary conditions only one blade of a rotor can be simulated. This allows computationally efficient studies on the rotor aerodynamics of wind turbines under uniform inflow conditions.

2.2 Numerical Setups

Table 1 gives an overview of the turbines and their operating conditions used for the investigations in this paper.

Table 1 Characteristics of the investigated turbines

	Two-bladed turbine	Three-bladed turbine
Rotor diameter [m]	107	126
Tower height [m]	135.1	90
Cone angle [°]	2	2.5
RPM [-]	17	11.7
Inflow velocity [m/s]	11.4	11.3

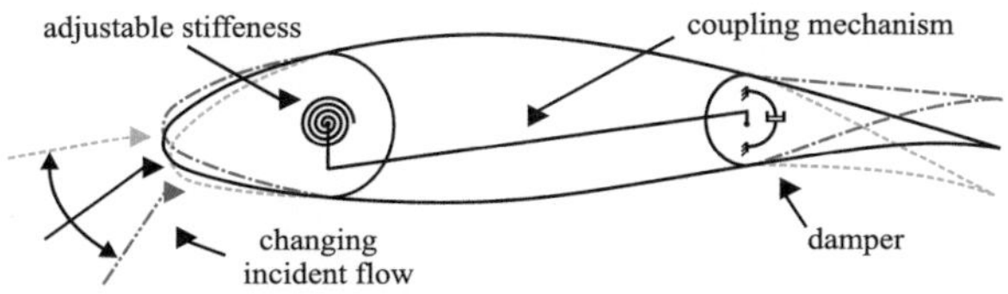

Fig. 1 "Adaptive camber mechanism" [15]

2.2.1 Three-Bladed Turbine

The present investigations were performed, using a one third model of the modified NREL 5 MW wind turbine. More information about the turbine can be found in [17]. For the present study, the turbine has been equipped with mechanically coupled leading and trailing edge flaps using a coupling ratio of three. This means that a deflection angle of $\gamma = -1°$ of the leading edge flap leads to a deflection angle of $\beta = 3°$of the trailing edge flap. This combined movement results in an increase of the camber of the airfoil (see Fig. 1). The concept, also known as adaptive camber concept, was developed by Lambie and Hufnagel at TU Darmstadt [15] as a passive load alleviation concept. For the present study, the flaps extend from 60 % to 80 % radius with the leading edge flap covering 20 % of the chord length and the trailing edge flap covering 30 % of the chord length.

In 2D, it was already investigated experimentally under steady inflow conditions [15] and dynamic inflow conditions [5, 13] and numerically under steady inflow conditions [7]. In the present paper prescribed flap deflections have been applied by grid deformation as described in [11].

The computational domain expands over a length of 2700 m and a radius of 720 m. It consists of four individual grids and 8.6 million cells of overall 16.5 million cells belong to the blade grid. All simulations were started with a steady computation in order to accelerate convergence, followed by an unsteady simulation.

2.2.2 Two-Bladed Turbine

The two-bladed turbine is the Skywind 3.4 MW prototype that has been subject of the German LARS project. It is embedded in a computational domain of

$1536 \times 1024 \times 832$ m (length $\times$ width $\times$ height). Six individual grids are placed in a Cartesian background grid which was refined using a hanging grid nodes technique. Altogether the computational setup consists of approx. 43.2 million cells. The first simulation was performed at uniform inflow conditions. To investigate the influence of yaw misalignment on the aerodynamics of the turbine, it has been rotated up to 30° around the vertically upward directed tower axis. In these simulations the time step corresponds to 2° azimuth. Additionally, to investigate the grid dependency, a one half model of the turbine has been generated. It consists of three independent grids for background, blade and hub. The background grid has the shape of a half cylinder with a length of 2700 m and a diameter of 1440 m.

3 FLOWer at HLRS

As an example, the two-bladed full model case is regarded. The problem size of 43.2 million cells has been computed on 1536 cpus. A simulation with 30 revolutions with 180 time steps per revolution and 50 inner iterations, as presented in the present paper, consumes approximately 42,000 cpuh. Since 2015, only minor changes in *FLOWer*, which do not affect the efficiency of the code, have been taken place. Therefore, the weak scaling test from [19] can be used as reference. Additionally, a strong scaling test has been performed. Both results are shown in Fig. 2. For these tests *FLOWer* was compiled with *ifort* on *Cray XC40*. For the weak scaling test the number of cells on each core was kept constant to 32^3. The strong scaling test was performed using 4096 times 32^3 cells. In both cases the number of cores was increased from 128 up to 4096 and for each run the time for 1000 iterations was taken and compared to the simulation on 128 cores. For up to 1024 cores both cases show the same efficiency. Using 4096 cores, *FLOWer* has an efficiency of 0.77 for weak scaling and 0.83 for strong scaling. This shows

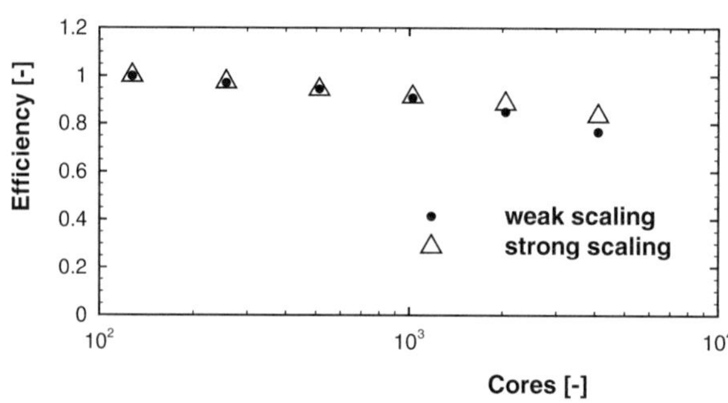

Fig. 2 "Efficiency of *FLOWer* on *Cray XC40* using *ifort* fortran compiler and a constant cell loading of 32^3 for each MPI process in case of weak scaling and 4096 times 32^3 cells in case of strong scaling"

Table 2 Number of cells for grid convergence studies

		Coarse	Medium	Fine
Two-bladed	Total	6,852,864	15,720,576	35,713,664
	Refinement factor	–	1.319	1.315
Three-bladed	Total	5,597,696	13,635,328	32,730,624
	Refinement factor	–	1.339	1.346

the suitability of *FLOWer* for the simulations regarded in this report and gives the opportunity for bigger cases.

4 Grid Convergence Study

4.1 Approach

For the three-bladed and the two-bladed turbine described in Sect. 2.2 grid convergence studies according to Celik [4] have been performed to evaluate the dependency of the CFD results on the grid resolution and to find suitable grids for further simulations. For the studies, three setups of different resolution have been generated for both turbines, using the one half model in the two-bladed case, see Table 2. As recommended by Celik, the grid refinement factor between two setups has been chosen to be larger than 1.3. For the investigation of the grid convergence of the three-bladed turbine the flaps have not been deflected and the grid had no refinement in the flap area.

4.2 Results of the Grid Convergence Study

Table 3 shows an extract of the results of the grid convergence studies. For both cases, two-bladed and three-bladed, the fine grid convergence index GCI_{fine}^{21} and the extrapolated error for the medium grid err_{ext}^{2} are listed. GCI_{fine}^{21} is an indicator for the grid dependency of the fine grid solution and err_{ext}^{2} represents the error of the medium grid solution to the extrapolated value.

Figure 3 shows the normalized sectional driving force with respect to the normalized radius. The three curves for the coarse, medium and fine setup are plotted as well as the extrapolated sectional driving force. Error bars indicate the local GCI_{fine}^{21}. In both cases the grid convergence is very good in the outer part of the rotor and worse in the inner part. In the root region there are big differences between the two cases. For the three-bladed turbine, the medium and fine grid solution lie very close and a low uncertainty is indicated. In contrast to this, the indicated uncertainty is very high for the two-bladed turbine as all setups overestimate the

Table 3 Selected results of the grid convergence studies.

		Power	Thrust	Driving force
Two-bladed	GCI_{fine}^{21}	0.05%	0.09%	3.60%
	err_{ext}^{2}	0.19%	0.28%	5.05%
Three-bladed	GCI_{fine}^{21}	0.02%	0.70%	0.26%
	err_{ext}^{2}	0.12%	0.82%	0.83%

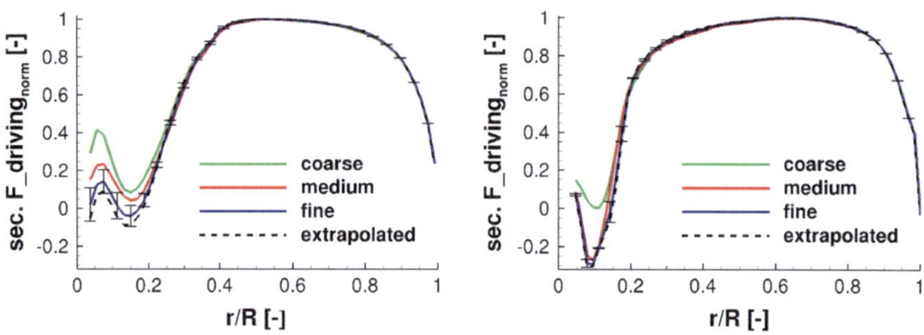

Fig. 3 Averaged sectional driving force for all three grids. *Left*: two-bladed, *right*: three-bladed

driving force close to the root compared to the extrapolated solution. This also explains the high GCI_{fine}^{21} in the integral driving force. Nevertheless, the influence on power is very low, as the sectional driving force is multiplied with the radius to get the moment. Overall the medium setup fulfills the contradictory requirements of little computational effort and high accuracy of the solution best.

5 Results

5.1 Influence of Temporal Discretization on the Simulation of Coupled Leading and Trailing Edge Flaps

Load reduction on wind turbines by dynamic flap deflection is associated with the occurrence of unsteady aerodynamic effects. Therefore, the temporal discretization of the flap deflection has to be investigated in order to ensure that unsteady aerodynamic effects are accurately resolved. The investigation of the temporal discretization was done for the 1p frequency, representing for example the impact of an atmospheric shear profile or a tilt angle of the rotor. In the middle of the flap, which corresponds to 70 % of the rotor radius the reduced frequency is k = 0.0338, where k is a function of the rotational frequency, the velocity and the chord length. According to Leishman [16], for $0 \leq k \leq 0.05$ a flow can be considered as quasi-steady and in most cases unsteady effects can be neglected. To prove this, 2D simulations of an airfoil extracted at the middle of the flap were performed. However, it turned out, that unsteady effects do occur (Fig. 4) and the shape of the

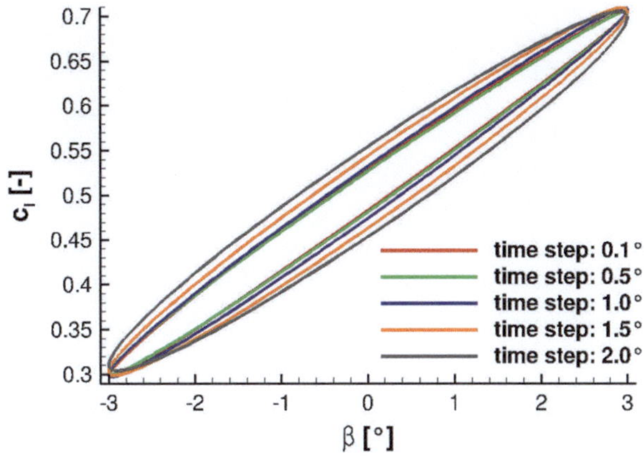

Fig. 4 c_l over β for different time steps. All calculations are performed with 30 inner iterations per time step

Table 4 Time steps per convective time unit

r/R	43 %	50 %	70 %	90 %	95 %
Time step: 0.5°	16.3	13.5	7.7	4.6	3.7
Time step: 1.0°	8.1	6.8	3.4	2.3	1.8
Time step: 2.0°	4.1	3.4	1.9	1.1	0.9

hysteresis curves depends on the time step. Consequently, those effects and their dependency on different factors were investigated.

If no complex separation occurs, 2D simulations are resolved at least with 50 to 100 time steps per convective time unit. Using 50 time steps per convective time unit at 70 % of the rotor radius of the NREL 5MW wind turbine would lead to approximately 4650 time steps for one revolution of the rotor which would not be feasible. In 3D simulations the time step is usually given in azimuth angle increments and typical values are between 1.0° and 3.0° per time step [19]. Table 4 shows the number of time steps per convective time unit for a selection of different time steps used in the 2D and 3D simulations.

In order to transfer the results from the 2D simulations to the 3D case, time steps normally used in full wind turbine simulations, corresponding to 0.1° up to 2.0°, were used for the present 2D investigations. A time step corresponding to 0.1° serves as reference in these simulations.

Figure 4 shows that the hysteresis curves for the different time steps converge to one solution if the time step is small enough. This effect is independent of the angle of attack and the flap angle as long as no separation occurs. It is also independent of the leading edge extension and only a minor dependency on the trailing edge extension was observed. Moreover, if the time step is small enough (in the present case $\leq 1°$), the number of inner iterations has only an insignificant influence on the predicted time shift. As additional time for grid deformation and Chimera search is needed at each time step, a larger time step with more inner iterations is preferable.

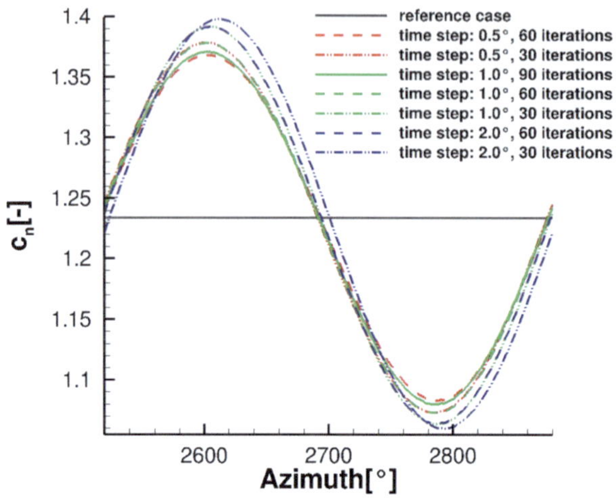

Fig. 5 c_n over the azimuth for different time steps and number of inner iterations extracted at r/R = 70 %

For the 3D simulations a steady computation with 45,000 iterations was performed, followed by four revolutions without flaps and four revolutions with flaps whereas only the last revolution was used for evaluation. Simulations with flap deflections of $\beta = \pm 3°$ and time steps corresponding to 0.5°, 1° and 2° azimuthal increment with 30, 60 and 90 inner iterations have been performed. The reference calculations without flaps were performed with a time step corresponding to 1.5° and 30 inner iterations.

In the 3D case, the choice of the time step has a stronger influence on the amplitude of the load coefficient as in the 2D case. Figure 5 shows the coefficient of the normal force c_n for the 3D case for different time steps and number of inner iterations.

The simulation with 0.5° time step and 30 inner iterations shows, as expected, the same results as the simulation with 1.0° time step and 60 inner iterations. Both simulations have 60 inner iterations per 1.0° azimuth. Regarding the computational costs, the case 1.0° and 60 inner iterations is preferable. Up to a certain time step size, the time shift of the solution is independent of the time step size and the number of inner iterations. Between 0.5° (30/60 inner iterations) and 1.0°(60/90 inner iterations) there is no difference in the time shift. However, if the number of inner iterations is too small (1.0°/30 and 2.0°/30) or the time step size is too big (2.0°), the time shift varies depending on these parameters. Therefore a time step of 2° or even larger or a combination with less than 60 inner iterations per 1.0° azimuthal increment is not recommendable.

The c_n amplitude for the different cases can be found in Fig. 6. It becomes obvious, that the amplitude reaches convergence with increasing number of inner iterations per time step. Therefore, a further increase of inner iterations per 1.0° as 90 is not worth regarding the benefit of a more accurate solution compared to the

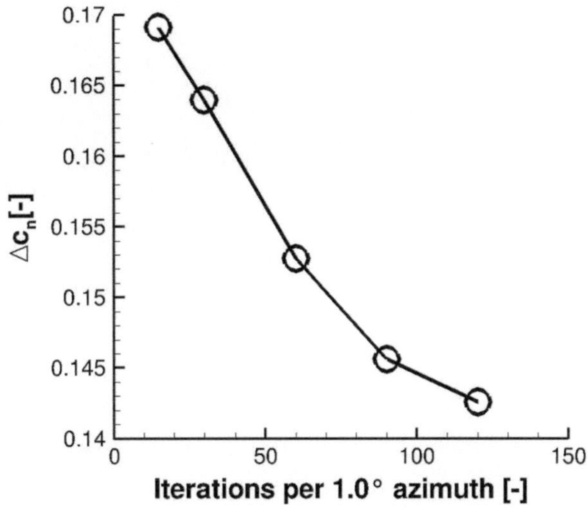

Fig. 6 c_n amplitude for different numbers of inner iterations per time step corresponding to 1.0°

simulation time. The same effects as described above can be seen for other force coefficients such as the tangential force coefficient c_t and the lift coefficient c_l.

A closer look to other blade sections shows, that the influence of the unsteady effects, caused by the 1p flap deflection, occurs all over the blade. Even in the root region, where the cylindrical part of the blade causes separation, the effects are superimposed. Moreover, the effects seen in Fig. 5 are the same over the whole radius. According to Table 4 the number of time steps per convective time unit for a time step according to 0.5° at 50 % radius is approximately twice the size it is for 70 % radius and approximately three times larger than at 90 %. Nonetheless, the differences between the solutions caused by the size of the time step and the number for inner iterations remain the same. This leads to the conclusion that no further significant benefit can be expected for the whole blade for a further increase of inner iterations or reduction of time step size. Therefore, the optimum combination of time step size and number of inner iterations regarding accuracy of the result and computational time for the present case with a 1p excitation frequency is 1.0° and 90 inner iterations.

5.2 Influence of Yawed Inflow on a Two-Bladed Turbine

In this section the full model simulation of the two-bladed turbine with 30° yaw is compared to the baseline 0° yaw case. Figure 7 shows the influence of the yawed inflow on the wake of the wind turbine. Compared to the baseline case the wake is

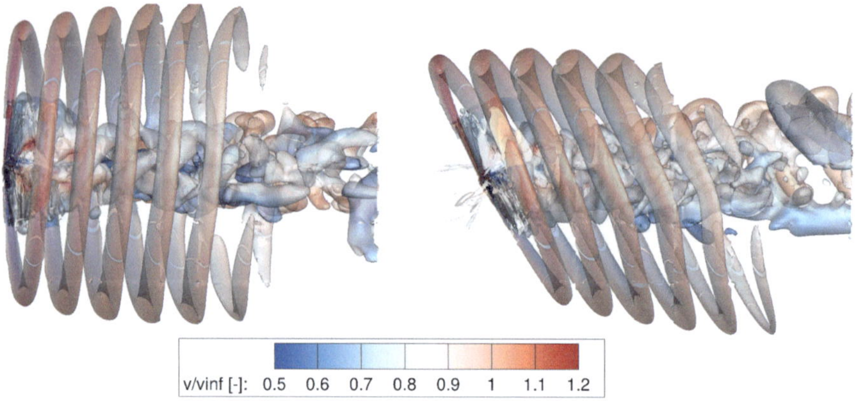

Fig. 7 Influence of yawed inflow on the wake of the two-bladed wind turbine. $\Lambda_2 = -2e^{-7}$ isosurfaces for vortex visualization. Contour levels indicate relative velocity magnitudes. View from top. *Left*: 0°, *right*: 30°

deflected towards the downwind side of the rotor. A common way to describe the decrease of power under yawed conditions is the cos^x function [18].

$$P = P_0 \cdot cos^x(\gamma) \tag{1}$$

Dahlberg [6] evaluated measurements from field as well as from wind tunnel and found exponents reaching from 1.88 to 5.14. In [21] a three-bladed turbine of approximately the same size as the two-bladed turbine has been investigated under yawed condition using *FLOWer*. An exponent of 2.38 was found there. The power of the investigated two-bladed turbine is reduced by 22.3 % under 30° yawed condition. This corresponds to an exponent of approximately $x = 1.7$ which is clearly lower than the exponents from literature stated above. The thrust and torque on one blade with respect to the azimuth are displayed in Fig. 8. The influence of the tower blockage can be seen clearly at 180° azimuth, where thrust is reduced by approx. 5 % and torque by approx. 10 % in the non-yawed case. Except of this local effect, both loads are almost constant over the azimuth for the non-yawed case. In the yawed case, both, thrust and torque are reduced. The maximum thrust is located between 90° and 120° azimuth, while the maximum torque is found at the upright position of the blade at 0° azimuth. Compared to thrust, the amplitude in torque is much higher but almost symmetrical. To understand the loads in more detail, the spanwise load distribution at 90° and 270° azimuth is displayed in Fig. 9. In the baseline case, the loads only differ close to the root caused by separation on the very thick airfoils. In the yawed case the sectional loads are lower or equal compared to the baseline case, except for the blade root region. Below 65 % relative radius the sectional loads are higher at 90° azimuth, above they are higher at 270° azimuth. Again the blade root region is neglected as there is a high fluctuation of the loads. At 90° azimuth the sectional thrust is almost the same in both cases for a relative radial position of 25 %

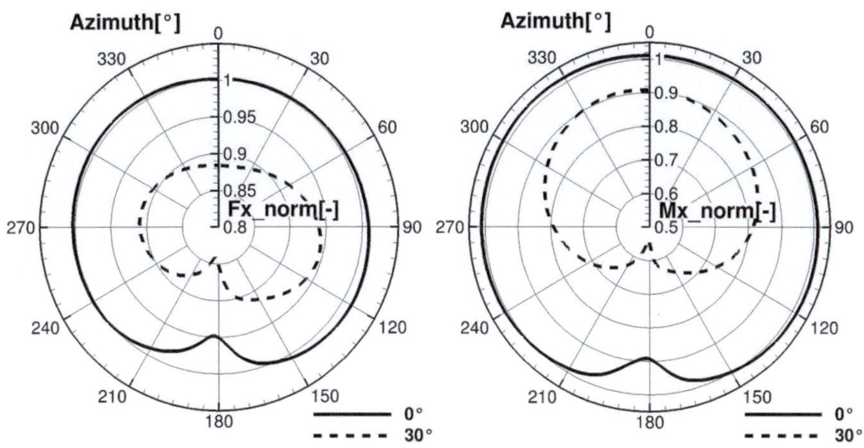

Fig. 8 Normalized blade thrust force and normalized blade torque for yawed and non-yawed inflow

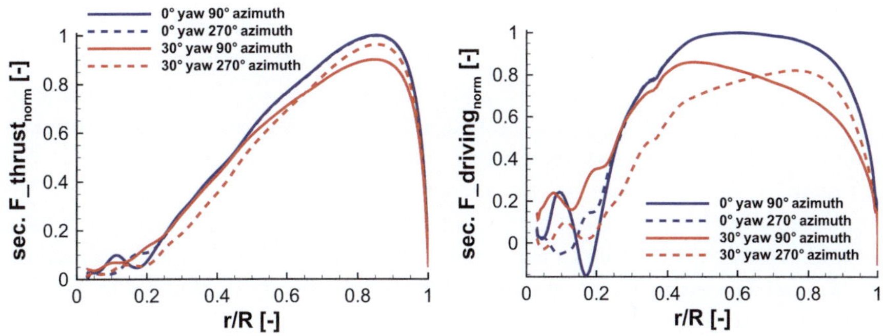

Fig. 9 Normalized sectional thrust force and normalized sectional driving force for yawed and non-yawed inflow

to 50 %, while the driving force only overlaps from 25 % to 30 % relative radius. Still, it is hard to explain the reasons for the observed load characteristics without knowledge of the local flow conditions at the blade. In BEMT based engineering codes for wind turbine aerodynamics, the local flow conditions are a direct outcome of the simulation as they are needed for the determination of the loads and for this purpose calculated iteratively every time step. In contrast to this, loads in CFD are a result of surface fluid interaction and the local flow at the blade is disturbed by the presence of the blade. To get local induction factor (a), local angle of attack (AoA) and local lift (c_l) and drag coefficient (c_d) from CFD results several methods exist, e.g. [10, 23, 24]. This is also helpful for comparison to BEMT results. The reduced axial velocity method from [10] gives very good results when applied on rotor only simulations with uniform inflow normal to the rotor plane as shown in [12]. Unfortunately, this method is only valid for steady flow states as the velocity is

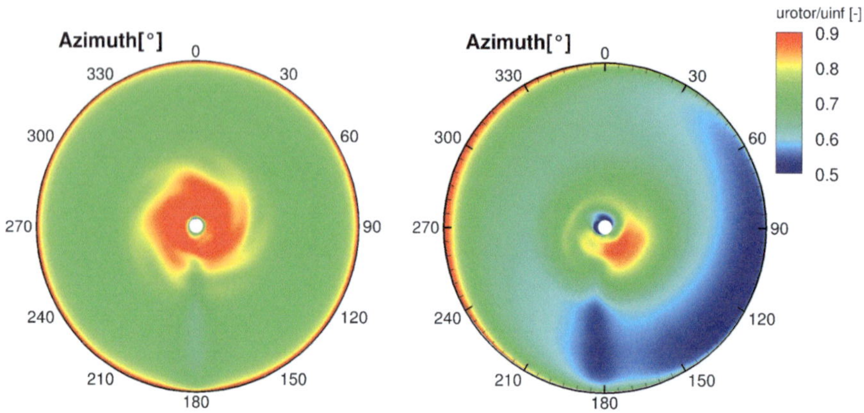

Fig. 10 Normalized axial velocity (in direction of the rotor axis) in the rotor plane. View from front in direction of the rotor axis. *Left*: 0°, *right*: 30°

averaged in circumferential direction. Because of this, for the presented study the reduced axial velocity method was adapted for full model simulations with steady inflow. It is based on the assumption, that the velocity in the rotor plane, averaged over $1/n_{blades}$ revolutions, is representative for the flow state and can be used to determine the local flow conditions.

Figure 10 shows the normalized averaged axial velocity in the rotor plane for both cases. For the non-yawed case it is almost constant over the azimuth, it only shows fluctuations in the inner part and an decrease of velocity at 180° caused by tower blockage. As it has to be expected the axial velocity is globally reduced in the yawed case. The strongest decrease can be found between 30° and 180° azimuth in the downwind side of the rotor with its maximum near the blade tip region. This gives a first hint for reduced loads at 90° in the outer region of the blade observed in Fig. 9.

Figure 11 shows the AoA and the relative local velocity normalized with the rotational velocity for two different spanwise positions for both cases. Compared to the non-yawed case, the average AoA is reduced in both spanwise positions but the amplitude is higher at the inner position caused by the lower rotational velocity. The curves are shifted to the upper right side in case of the 40 % radial position and to the upper left side in case of the 85 % radial position. Looking at the relative local velocity, one can see that it is dominated by the rotational velocity in the non-yawed case. At the inner position the local velocity is approx. 4 % higher than the rotational velocity while it is only 1 % at the outer position. In the yawed case the influence of the wind is much higher, the curves are shifted to the lower side of the rotor where the blades are moving against the wind, again the amplitude is higher at the inner position with the relative velocity ranging between 90 % and 117 %.

At 90° and 270° azimuth the local velocity magnitude is almost independent of yaw, the shapes of the sectional load graphs in Fig. 9 are only a result of the local

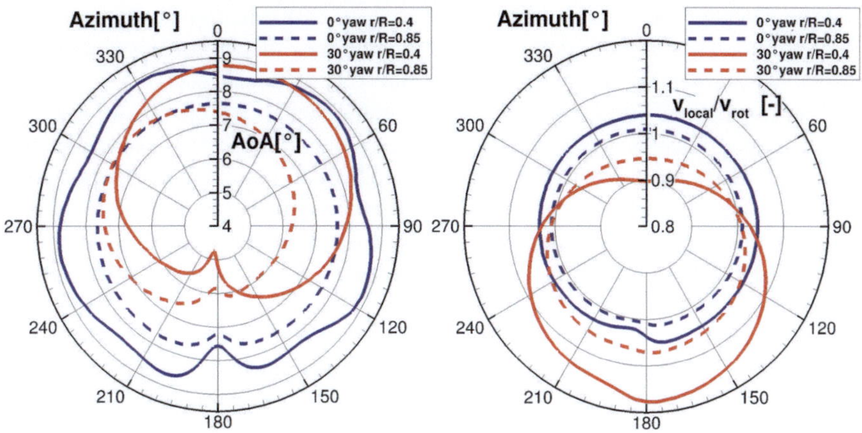

Fig. 11 Angle of attack and relative local velocity for yawed and non-yawed inflow

AoA. At a radial position of 40 % the AoA is higher at 90° azimuth, while at 85 % radius the AoA is higher at 270° azimuth which corresponds to the sectional forces in the diagrams.

For the integrated forces with respect to azimuth in Fig. 8 it can be concluded that torque is dominated by the local AoA, as the highest torque is found at 0° azimuth corresponding to the minimum in local velocity and to high AoAs at the regarded radial positions. Thrust is more dominated by the magnitude of the local velocity as the maximum is shifted to the lower half of the rotor. As observed above, the sectional thrust force in the inner part of the rotor mainly drives the integrated thrust force and thus the AoA in the inner part of the blade is responsible for the shift of the thrust force to the right side of the rotor.

6 Conclusion

The present article shows numerical investigations of innovative wind turbine applications with the CFD solver *FLOWer*. For both applications, which are in one case coupled leading and trailing edge flaps and in the other case a novel two-bladed wind turbine, grid convergence studies were performed in order to estimate the error caused by the grid resolution. Afterwards, the applications were investigated independently in order to gain a better understanding of the occurring unsteady aerodynamic effects and to improve their prediction as they can only be captured in detail by CFD.

For the coupled leading and trailing edge flaps an investigation of the temporal discretization was performed in order to evaluate the influence of the time step size and number of inner iterations on the numerical results. Prescribed flap deflections with a 1p frequency already result in unsteady aerodynamic effects and show dependency on the temporal discretization. A time step corresponding to 1.0° azimuth with 90 inner iterations was found to be efficient for this application, taking into account accuracy of the results and computational costs.

The two-bladed wind turbine was simulated under yawed conditions. The unsteady aerodynamic loads were evaluated in detail and it was found that the turbine shows lower decrease of power in yawed conditions compared to literature. A common approach for extraction of local flow conditions like local velocity and angle of attack from CFD results was adapted for full model simulations and the dependency of the aerodynamic loads to the azimuthal position could be associated to the local flow conditions.

The results of this studies help to increase the efficiency and accuracy of wind turbine CFD simulations and enable a better understanding of unsteady aerodynamics of wind turbines. These results can be used for the validation of engineering models which are usually applied for load prediction of wind turbines in industry.

Acknowledgements The authors gratefully acknowledge the *High Performance Computing Center Stuttgart* for providing computational resources within the project *WEALoads*. The studies presented in this article have been funded by the *Federal Ministry for Economic Affairs and Energy* and the *German Research Foundation (DFG)*.

References

1. Bekiropoulos, D., Lutz, T., Baltazar, J., Lehmkuhl, O., Glodic, N.: D2013-3.1: comparison of benchmark results from CFD-simulation. Deliverable report, KIC-OFFWINDTECH (2013)
2. Bekiropoulos, D., Rieß, R., Lutz, T., Krämer, E., Matha, D., Werner, M., Cheng, P.W.: Simulation of unsteady aerodynamic effects on floating offshore wind turbines. In: DEWEK (2012)
3. Benek, J.A., Steger, J.L., Dougherty, F.C., Buning, P.G.: Chimera. A Grid-Embedding Technique. Arnold Engineering Development Center Arnold Air Force Station, Tennessee Air Force Systems Command United States Air Force (1986)
4. Celik, I.B., Ghia, U., Roache, P.J., et al.: Procedure for estimation and reporting of uncertainty due to discretization in CFD applications. J. Fluids Eng.-Trans. ASME. **130**(7), 0780011–0780014 (2008)
5. Cordes, U., Hufnagel, K., Tropea, C., Kampers, G., Hölling, M., Peinke, J.: Experimental investigation of passive load reduction under dynamic inflow conditions. In: 33rd AIAA Applied Aerodynamics Conference, p. 3313 (2015)
6. Dahlberg, J., Montgomerie, B.: Research program of the Utgrunden demonstration offshore wind farm, final report part 2, wake effects and other loads. Swedish Defense Research Agency, FOI, pp. 2–17 (2005)
7. Fischer, A., Jost, E., Lutz, T., Krämer, E.: Numerical investigations of a passive load alleviation technique for wind turbines. In: 10th PhD Seminar on Wind Energy in Europe, Orléans, pp. 51–54, EAWE, 28–31 Oct 2014
8. Jameson, A.: Time dependent calculations using multigrid, with applications to unsteady flows past airfoils and wings. AIAA Paper, 1596:1991 (1991)
9. Jameson, A., Schmidt, W., Turkel, E., et al.: Numerical solutions of the euler equations by finite volume methods using Runge-Kutta time-stepping schemes. AIAA Paper, 1259:1981 (1981)

10. Johansen, J., Sørensen, N.N.: Aerofoil characteristics from 3D CFD rotor computations. Wind Energy **7**(4), 283–294 (2004)
11. Jost, E., Fischer, A., Lutz, T., Krämer, E.: Cfd studies of a 10 mw wind turbine equipped with active trailing edge flaps. In: 10th PhD Seminar on Wind Energy in Europe, Orléans, pp. 119–122, EAWE, 28–31 Oct 2014
12. Jost, E., Lutz, T., Krämer, E.: A parametric CFD study of morphing trailing edge flaps applied on a 10 mw offshore wind turbine. In: 13th Deep Sea Offshore Wind R&D Conference, EERA DeepWind'2016, Trondheim, 20–22 Jan 2016
13. Kampers, G., Peinke, J., Hölling, M., Cordes, U., Tropea, C.: Stochastic analysis of aerodynamic forces acting on a self-adaptive camber airfoil in turbulent inflow. In: 33rd AIAA Applied Aerodynamics Conference, p. 2427 (2015)
14. Kroll, N., Rossow, C.-C., Becker, K., Thiele, F.: The megaflow project. Aerosp. Sci. Technol. **4**(4), 223–237 (2000)
15. Lambie, B.: Aeroelastic investigation of a wind turbine airfoil with self-adaptive camber. PhD thesis, Technical University of Darmstadt (2011)
16. Leishman, J.G.: Principles of Helicopter Aerodynamics. Cambridge Aerospace Series, Cambridge, New York (2000)
17. Matha, D., Schuon, F., Lutz, T.: Baseline fowt definition v4. Deliverable report d3.1, KIC-OFFWINDTECH (2013)
18. Schepers, J.: Engineering models in wind energy aerodynamics. PhD thesis, TU Delft (2012)
19. Schulz, C., Fischer, A., Weihing, P., Lutz, T., Krämer, E.: Evaluation and control of loads on wind turbines under different operating conditions by means of CFD. In: High Performance Computing in Science and Engineering'15, pp. 463–478. Springer, Cham (2016)
20. Schulz, C., Klein, L., Weihing, P., Lutz, T., et al.: CFD studies on wind turbines in complex terrain under atmospheric inflow conditions. J. Phys. Conf. Ser. **524**, 012134 (2014). IOP Publishing
21. Schulz, C., Letzgus, P., Lutz, T., Krämer, E.: CFD study on the impact of yawed inflow on loads, power and near wake of a generic wind turbine. Wind Energy (to be published). doi:10.1002/we.2004
22. Schulz, C., Meister, K., Lutz, T., Krämer, E.: Investigations on the wake development of the Mexico rotor considering different inflow conditions. In: Contributions to the 19th STAB/DGLR Symposium, Munich, Germany 2014. Notes on Numerical Fluid Mechanics and Multidisciplinary Design. STAB. Springer, Nov 2014. Under review
23. Shen, W.Z., Hansen, M.O., Sørensen, J.N.: Determination of angle of attack (AOA) for rotating blades. In: Wind Energy, pp. 205–209. Springer (2007)
24. Shen, W.Z., Hansen, M.O., Sørensen, J.N.: Determination of the angle of attack on rotor blades. Wind Energy **12**(1), 91–98 (2009)
25. Weihing, P., Meister, K., Schulz, C., Lutz, T., et al.: CFD simulations on interference effects between offshore wind turbines. J. Phys. Conf. Ser. **524**, 012143 (2014). IOP Publishing
26. Weihing, P., Schulz, C., Lutz, T., Krämer, E.: CFD performance analyses of wind turbines operating in complex environments. In: Nagel, W.E., Kröner, D.H., Resch, M.M. (eds.) High Performance Computing in Science and Engineering'14, pp. 403–415. Springer, Cham (2015)

Part V
Transport and Climate

Christoph Kottmeier

In the field of "Transport and Climate", both the number and the CPU requirements of high-performance computing projects making use of the HLRS in Stuttgart and of the SSC in Karlsruhe have increased considerably in the last 2 years. Currently 11 projects are ongoing, mostly related to modelling large parts of the climate system. The topics cover a broad range of objectives as well as geographic regions. The CPU time requirements of such models strongly increase with higher and higher resolution, which are needed in oceanic and atmospheric models to resolve the small scales (turbulent, convective, mesoscale) of ambient flows. It is known from measurements that these highly energy-containing scales can strongly control larger-scale processes. Therefore it is important to represent their net effects adequately. This is done in coarsely resolved models by semi-empirical parametrizations. This is not fully satisfying, however, since such parametrizations can been hardly validated against measurements for the full parameter range of their application.

Therefore atmospheric and oceanic modellers go to higher resolution down to 1 km (and partly less), aiming that processes are directly simulated. Another general development is also reflected by the HLRS- and SSC-projects. More and more coupling between model submodules for, e.g., the atmosphere, the ocean, and ecosystems is realized. This also holds for nested models, where either 1-way coupling, and still rarely 2-way coupling is realized between a coarsely resolving model being applied for a large domain such as a global model and a limited area model that is run at high resolution. Developing such coupling tools requires substantial efforts in adaption and testing with many model test runs. The domain size of the limited area model can have strong effects on the outcome. Other critical

C. Kottmeier (✉)
Institut für Meteorologie und Klimaforschung, Karlsruher Institut für Technologie (KIT), Wolfgang-Gaede-Straße 1, 76131 Karlsruhe, Germany
e-mail: christoph.kottmeier@kit.edu

issues are due to the coincidence of slow processes in one of the coupled model systems (ocean and ice) and fast processes (atmosphere), which implies different time steps in numerical schemes.

Extensive testing on the model setups have to be done therefore, before long runs are performed, such as over decades up to centuries in climate modelling. Despite their high quality and very ambitious objectives it was decided not include the short interim reports from projects at an early stage into the review.

The projects chosen for oral presentation and for the HLRS report reflect very well the high importance of the HLRS and SSC computing facilities for highly visible research programmes in actual research.

The report on "Simulation of the rain belt of the West African Monsoon (WAM) in high resolution CCLM simulation (WASCAL-CCLM)" by IMK-IFU in Garmisch-Partenkirchen (KIT) addresses a vital problem for population in the semi-desert West African regions. The monsoon brings the only rain for potable water, agriculture, and ground water replenishment. It is highly important at which time the onset is and high intense in a certain year rainfall is. Both result from complex atmospheric interaction with surface processes and it a major problem to account for convection at high resolution. For HPC this means that mesh sizes have to be rather small and that the related HPC and storage requirements increase substantially.

Another project at SCC with focus on Australia is WA-AERO (Anthropogenic aerosol emissions and rainfall decline in South-West Australia) by IMK-IFU. It also addresses a desertication-threatened region, but in addition a open debate in climate science, namely the role of aerosols for cloud and precipitation formation.

In "High resolution climate projections using the WRF model on the HLRS (WRFCLIM)" a group from University of Hohenheim uses the Hazelhen of HLRS goes to km-scale resolution for smaller regions, but decadal simulation periods.

In the project LUCCi, the biogeophysical impacts of land surface on regional climate have been investigated for Central Vietnam (Vu Gia-Thu Bon basin) in using the regional climate model (RCM) Weather Research and Forecasting (WRF) Model. It is demonstrated that the replacement

of land surface due to an updated land-use/land cover data leads to significant changes of the biogeophysical properties of land surface, thereby altering the regional climate.

Aim of the project RUCACI is to investigate the impact of aerosol in high resolution climate runs for major parts of Africa. Aerosols, particularly mineral dust in West Africa, and their interactions with radiation and clouds represent one of the major uncertainties in our understanding of the climate system at regional scales. The online coupled comprehensive chemistry model system COSMO-ART, developed at KIT already showed in several case studies the potential of closing the gap between coarse global models and regionalized modelling. In order to apply it on decadal climate time scales the use of high performance computing becomes a necessity. In this study the effect of replacing the aerosol climatology usually used in regional climate simulations with COSMO-CLM by online calculated dust concentrations.

Simulation of the Rain Belt of the West African Monsoon (WAM) in High Resolution CCLM Simulation

Diarra Dieng, Gerhard Smiatek, Dominikus Heinzeller, and Harald Kunstmann

Abstract We present the results of our regional climate modeling experiments conducted on ForHLR1, using the consortium for small-scale modeling (COSMO) regional climate model CCLM over West Africa. This work is embedded in the context of the West African Science Service Center on Climate Change and Adapted Land Use (WASCAL) research project. We conduct nested runs at 50 and 12 km resolution driven by ERA-Interim data to assess the modeled location and intensity of the tropical rainbelt over West Africa for the period 1979–2013. The simulation period includes the years 1983 and 1999 with observed extreme anomalies (dry as well as wet). These anomalies are captured by our experiment: The model reproduces the observed zonal-mean variations in precipitation within the range of comparable regional climate model (RCM) studies, but reduces the dry bias in the Golf of Guinea and shows an increased accuracy for the driest years in general. Based on these encouraging results, we are currently extending our work towards historical climate runs and climate projections for an improved understanding of the different processes involved in the West Africa climate system and their role in generating extreme climatic conditions.

Keywords Regional climate modeling • COSMO-CLM • Tropical rainbelt • West Africa • WASCAL

D. Dieng (✉) • D. Heinzeller • H. Kunstmann
Institute of Meteorology and Climate Research (IMK-IFU), Karlsruhe Institute of Technology (KIT), Kreuzeckbahnstr.19, 82467, Garmisch-Partenkirchen, Germany

Institute of Geography, Chair for Regional Climate and Hydrology, University of Augsburg, 86135, Augsburg, Germany
e-mail: diarra.dieng@partner.kit.edu; dominikus.heinzeller@kit.edu; harald.kunstmann@kit.edu

G. Smiatek
Institute of Meteorology and Climate Research (IMK-IFU), Karlsruhe Institute of Technology (KIT), Kreuzeckbahnstr.19, 82467 Garmisch-Partenkirchen, Germany
e-mail: gerhard.smiatek@kit.edu

W.E. Nagel et al. (eds.), *High Performance Computing in Science and Engineering '16*, DOI 10.1007/978-3-319-47066-5_37

1 Introduction

Precipitation is one of the most difficult parameters to simulate and to compare to measurements, because its structure changes greatly over space and time [17]. Over West Africa, the problem in accurate simulations of rainfall lies in the complex interaction of several forcing processes with strong seasonal character and large inter-annual and decadal variability. For example, rainfall maxima occurring during the peak of the West African Monsoon (WAM) in June-July-August-September (JJAS) and the WAM system itself are a result of several atmospheric features, including the low-level monsoon flow, the mid tropospheric African Easterly Jet (AEJ), the African Easterly Waves (AEWs) and the upper level Tropical Easterly Jet (TEJ).

One pillar of the WASCAL (West African Science Service Center on Climate Change and Adapted Land Use) program is to improve the understanding and modeling capabilities for West Africa with a particular interest in climate variability and rainy season precipitation characteristics. Our investigation focuses on the reproduction of the spacial and temporal rainfall patterns over West Africa in a series of high-resolution CCLM climate model runs at 0.11° (approx. 12 km) resolution. The increased spatial resolution of our climate modeling experiments implies an improved representation of the topography, land use, vegetation and soil characteristics and is supposed to reduce the biases in current climate simulations. Our work extends existing CCLM studies investigating the rainfall characteristics [8, 15–17] and the WAM system [9] for the West African region. For the model evaluation, we explore the summer (June, July, August, September) rainfall variability, the latitudinal profile at four different selected locations centered on 15, 10, 0 and 10 °W, and the evolution of the monsoon rain band as simulated during the wet and the dry years.

2 Material and Methods

2.1 CCLM Model and Model Setup

In this context, long-term simulations are performed using the COSMO-CLM (Consortium for Small scale Modelling model in Climate Limited Area Model) model version 4.8 (cclm4.8_clm19). The non-hydrostatic CCLM model is based on the numerical weather prediction model COSMO developed by the Deutscher Wetterdienst (DWD) [3] and extended by the CCLM community. The here-presented CCLM runs apply land surface data from the ECOCLIMAP [12, 13] project, which provides vegetation characteristics such as leaf area index, plant cover, roughness length at monthly resolution to distinguish between the seasonal land cover characteristics [19]. The soil surface albedo is obtained from the MODIS (Moderate resolution Imaging Spectroradiometer) [11] data. Compared to the

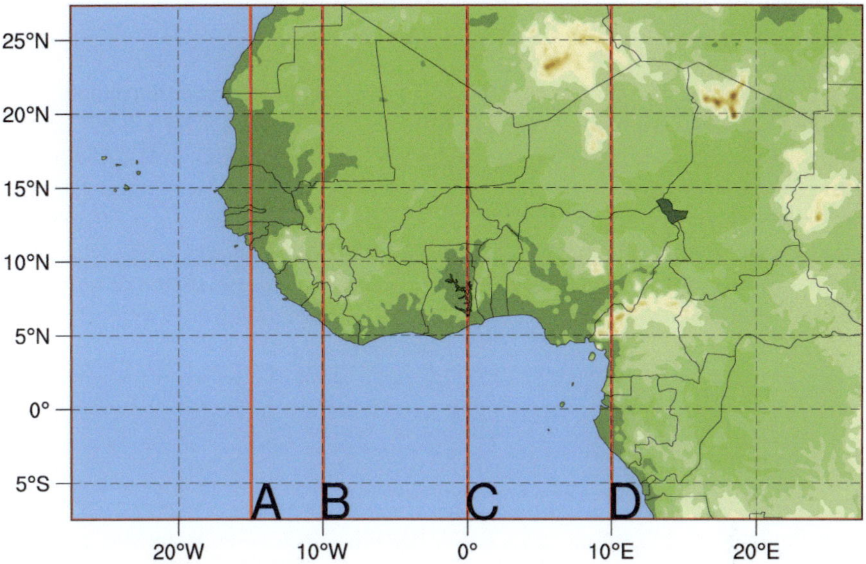

Fig. 1 CCLM simulation domain and topography for WASCAL at 12 km resolution runs. The *red lines* represent the different locations of the four longitudinal transects used in the study. *A*: lon = 15 °W, *B*: lon = 10 °W, *C*: lon = 0 °W and *D*: lon = 10 °E

standard CCLM setup, MODIS provides a more realistic information over deserts [10, 16]. The Runge-Kutta numerical integration scheme and the TKE advection scheme are selected and a vertical stratification of 40 verticals levels up to 10 hPa with additional layers close to the surface is employed. The model is driven with ERA-Interim boundary forcing data (Fig. 1).

2.2 *Investigation Area*

Earlier simulations performed by [1] pointed out that the choice of the simulation domain has an important impact on the modeled West African summer monsoon rainfall, because the ocean as well as the land-surface and the atmosphere are important drivers of the monsoon circulation and need to be taken into account. To capture the large scale atmospheric patterns in this region, we choose an extended high-resolution domain that covers entire West Africa, a fraction of the tropical Atlantic Ocean, the Cameroon and Fouta Djallon mountains, the Jos and Ethiopian plateaus, the Volta River basin as well as parts of the Sahara desert (−27.5 °W to 27.5 °E, −7.5 °N to 27.5 °N). This high-resolution domain (CCLM11) is nested in a lower-resolution domain (CCLM44) at 0.44° (approx. 50 km) resolution with a significantly larger geometrical extent (−37 °W to 48 °E, −18 °N to 36 °N), and forced by its 3-hourly lateral boundary conditions.

2.3 Observational Reference

The observational data for precipitation at monthly temporal resolution are obtained from the Tropical Rainfall Measuring Mission (TRMM) [7] with a 0.25° spatial resolution, the Global Precipitation Climatology Centre (GPCC) [18] 0.5° gridded rain gauge analysis, the Climate Research Unit (CRU) [14] 0.5° resolution data, and the Climate Hazards Group InfraRed Precipitation with Station data (CHIRPS) [4] gridded precipitation time series data. The monthly mean rainfall CHIRPS dataset is derived from a combination of satellite observations and rainfall station observations and provides gridded data at 0.05° (approx. 5 km) resolution from 1981 to the near present.

3 Results

The simulated mean rainfall and CRU, CHIRPS, GPCC, GPCP, TRMM observations for the rainy season June-July-August-September (JJAS) are depicted in Fig. 2. The CCLM11 spatial patterns are characterized by increasing rainfall amounts from both north and south, peaking at the monsoon rain band located between 2 and 15 °N. Largest rainfall amounts are simulated over the highlands of Guinea and Ethiopia and in the Cameroon mountains. In those areas, CCLM11 shows the largest absolute deviations from the observations of up to 3 mm/day. Relative biases, on the other hand, reach values of up to −60 % in the west of the modeled area with respect to CRU, CHIRPS, GPCC, GPCP, and up to −40 % with respect to TRMM. For the Sudan area with very little absolute precipitation, the relative biases exceed +60 %. In general, CCLM11 underestimates the observed values and extends the rain band further north than observed. As a consequence, the CCLM11 run shows a large wet bias in the north, where the 200 mm isohyet is located at about 17.5 °N (16 to 16.5 °N for the observations). At the same time, it exhibits a dry bias in the Golf of Guinea. An improvement is seen at the western coastline Africa and near Mont Cameroon regions, where the rainfall observed in CRU and GPCC is better reproduced in CCLM11.

In Fig. 3, we compare the observed and simulated longitudinal rainfall evolution in JJAS averaged along 10 °W and 10 °N over the period 1979–2013. The CCLM11 simulations underestimate the maximum precipitation in the region between 5 °N to about 12 °N in all months considered. The underestimation ranges from 80 to 100 mm, and the greatest value can be found in August. The reason for this shortcoming can be attributed to the limited ability of CCLM to fully transport the moisture from the ocean to the inland regions [16]. The simulated rainfall intensity decreases between 14 and 25 °N with a similar spatial pattern as in observations, but associated with a small bias in range of 5 mm. In summary, the CCLM11 simulations reveal some errors in capturing the position of the rainfall peak, shifting

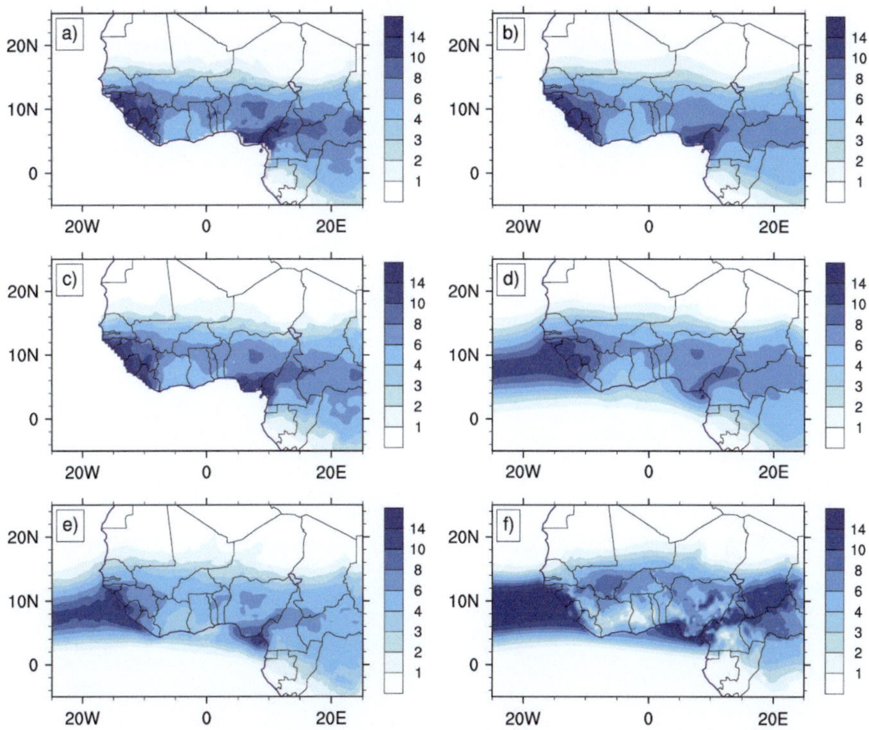

Fig. 2 Observed CRU (**a**), CHIRPS (**b**), GPCC (**c**), GPCP (**d**), TRMM (**e**) and simulated CCLM11 (**f**) JJAS rainfall distribution for the period 1979–2013

it 3° further north than observed. The precipitation bias varies substantially from June to August and ranges from 10 mm in June to over 90 mm in August.

The mean seasonal rainfall (mm/day) at the four selected longitudes for the period 1979–2013 is shown in Fig. 4. Close to the western continental coast (A), the simulated seasonal rainfall matches the observations well with peak values of up to 15 mm/day at the southern coastline. Further inland (B), the model underestimates the observed precipitation in the southern areas up to 12 °N, but matches the observations furthern northwards. For the longitudinal positions C and D, the model exhibits a northward shift of the rain band with a corresponding underestimation in the south. In both cases these deviations are a consequence of the orographic features, which are prominent in the south but less so in the north. Examples herefore are Lake Volta in Ghana and Jos Plateau in Nigeria, which are located between 5 and 10 °N along the transects C and D, respectively. In the transitional arid zone between 12 and 18 °N, the CCLM11 model run produces excessive rainfall amounts. This is more potentially be attributed to inaccurate time invariant data used in the model (e.g., land use and soil characteristics) as a result of poor observational coverage [6].

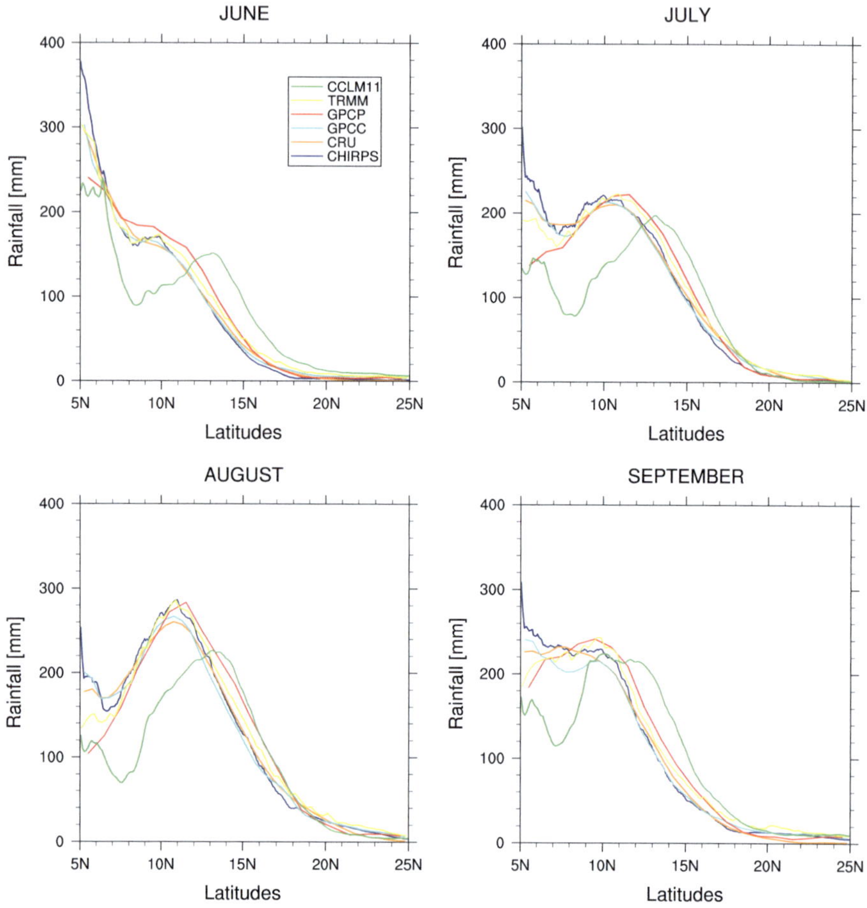

Fig. 3 Zonal average (10 °W–10 °E) of the observed and simulated mean rainfall (mm) for the months of June, July, August and September over the period 1979–2013

In Fig. 5, we compare the high-resolution CCLM11 to CRU, CHIRPS and GPCC to assess the movement of the monsoon rain band over West Africa during the driest (1983) and wettest year (1999) as identified in the Fig. 6. The annual cycle of the monsoon rainfall is characterized by three maxima. The first maximum occurs in May/June around 5–6° N with peak values of 200 mm/month of rainfall. This is followed by a retreat to 5 °N in September/October. Our CCLM11 model generally reproduces these features, but shows differences in the magnitude and the spatial extent of the simulated rainfall: In the driest year (left column), the modeled rainfall is generally underestimated with values up to 40 mm in June/July. Furthermore, it occurs one month too late as observed. For the wet year 1999 (right column), the simulated onset falls in April and occurs approximately one month earlier than observed. A quite good agreement is found for the location of the high

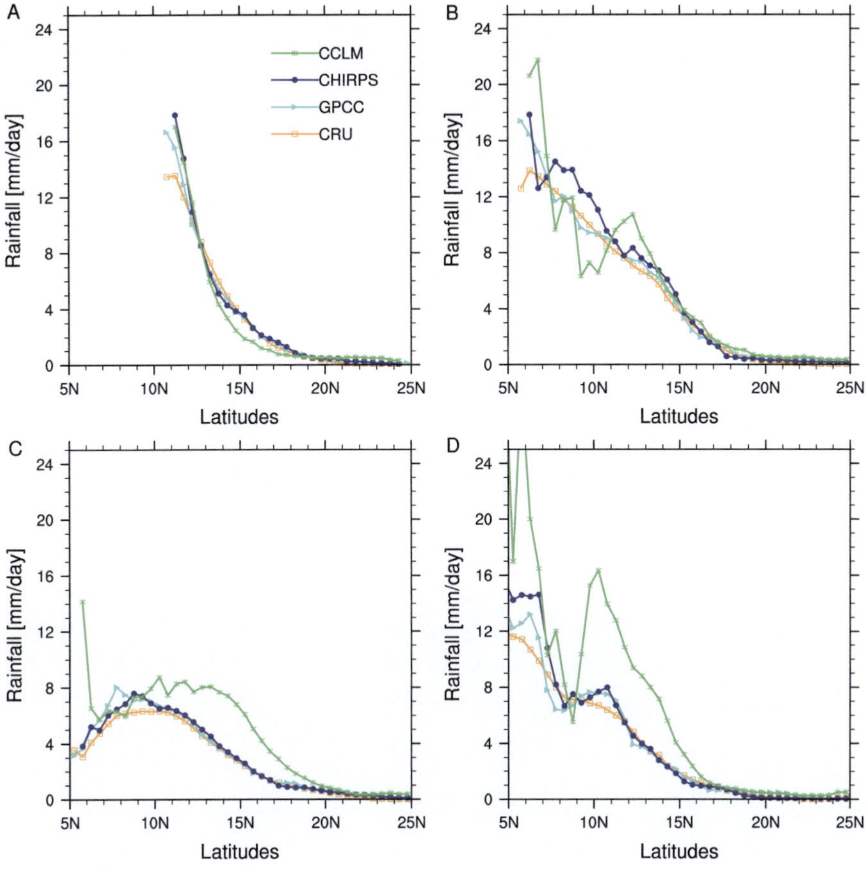

Fig. 4 Observed and simulated mean (June-July-August-September) rainfall (mm/day) for the period 1979–2013. Values averaged over 15 °W (**a**), 10 °W (**b**), 0 °W (**c**) and 10 °E (**d**)

precipitation intensity zone in August/September, however with an underestimation of the precipitation amount. In both years there is reasonable agreement with the observed end of the rainy saison. These results are in line with a previous study performed by [16] using CCLM, who concluded that CCLM underestimates the rainfall peak in the regions affected by the passage of the monsoon.

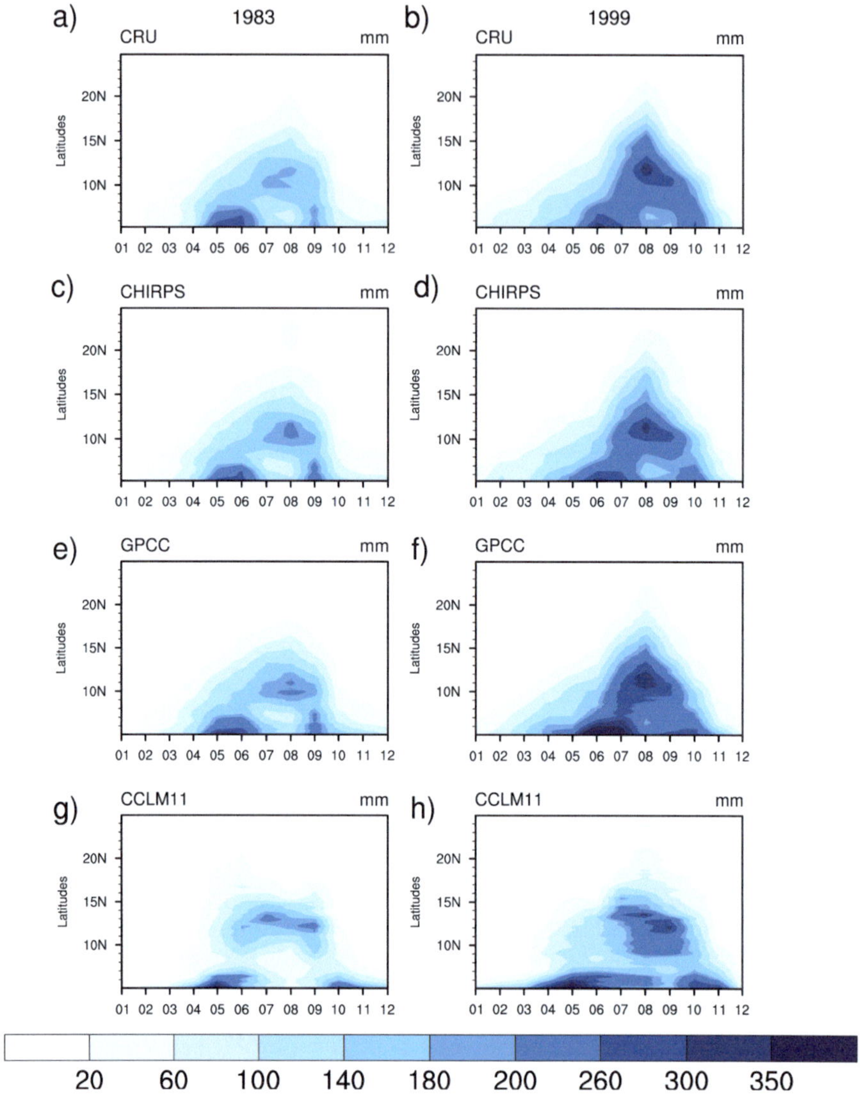

Fig. 5 Time-Latitude diagrams of monthly mean precipitation (in mm) averaged between 10° W and 10° E from CRU (**a**, **b**), CHIRPS (**c**, **d**), GPCC (**e**, **f**), CCLM11 (**g**, **h**) in the dry year 1983 (*left panel*), the wet year 1999 (*right panel*)

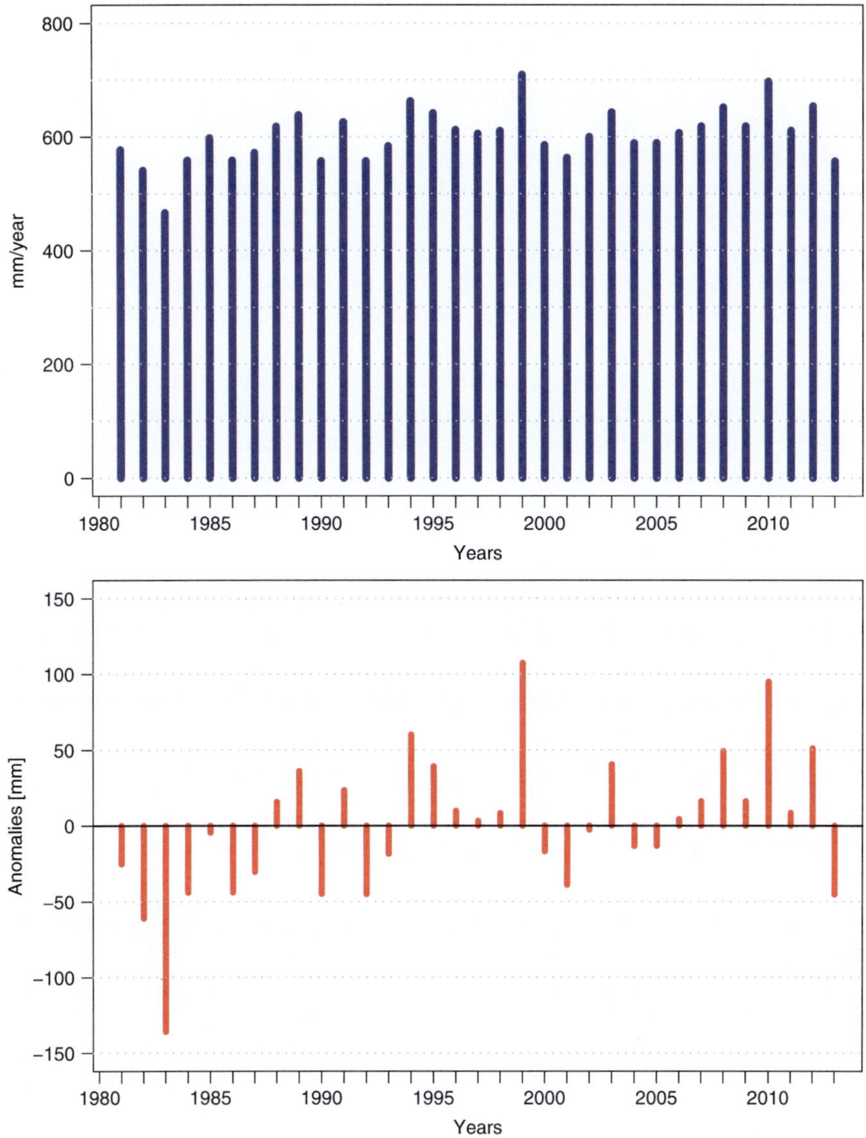

Fig. 6 Interannual variability of precipitation amount (*upper plot*) and precipitation anomalies (*bottom plot*) over West Africa from 1981 to 2010 for CHIRPS

4 Conclusion

Our results show that the high-resolution CCLM11 control run is able to reproduce the observed main climate characteristics, including the annual cycle of the West African Monsoon, within the range of comparable RCM evaluation studies. Despite the increased resolution, we find a northward shift of the monsoon rain band by about 3°, which results in a dry bias at the Coast of Guinea and a wet bias in the northern areas around 15 °N during the peak of the monsoon season in JJAS. The fact that the higher resolution of 12 km does not improve significantly the model results compared to existing lower-resolution experiments and our own CCLM44 simulation could be attributed to the fact that the monsoon precipitation over West Africa is dominated by convective rainfall [2]. At resolutions larger than 10 km, convective processes in the models are parameterized and therefore represented implicitly rather than calculated explicitly. We hypothesize that a convection-permitting resolution below 5–10 km will be able to address these deficiencies. However, the computational requirements forbid the application of such a high-resolution in long-term climate simulations.

A second aspect is the above-mentioned representation of time-invariant data sets in the model, such as surface and soil characteristics. Work is underway within the WASCAL program to address the poor data coverage on land and soil characteristics over West Africa using high-resolution satellite-composite products [5]. It is expected that integrating this high-resolution data in the physical parameterizations and the microphysics of the CCLM model will reduce the model bias.

5 Details on the Computation Setup

As proposed in our original application, we use a computational setup as summarized in Table 1 to conduct the high resolution simulations. Using 120 threads (6 nodes on ForHLR1), one simulation day for the at 0.11° resolution run requires 8.5 min real-time. The proposed entire 157 year simulation corresponds to 340 days real-time without overhead and queuing time. Figure 7 displays a scaling plot for the nested domain CCLM11 derived on ForHLR1 prior to our original application. To reduce the amount of data generated in our study, we make use of the compressed netCDF4 format and data thinning during the post-processing step directly on ForHLR1.

Table 1 Computational setup for nested model runs

Model run	Nodes/threads	Simulation type	CPUh/year	CPUh total
D2-0.11	6/120	ERA-Interim	6240	218,400
D2-0.11	6/120	ECHAM6-historical	6240	168,480
D2-0.11	6/120	ECHAM6-RCP4.5	6240	592,800
TOTAL				979,680

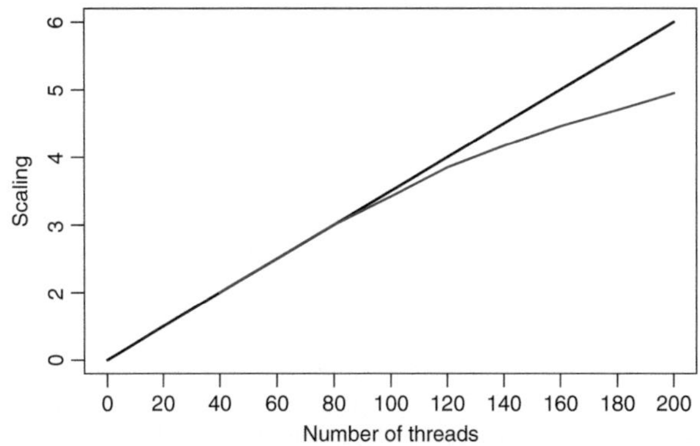

Fig. 7 Scaling plot for D2 Domain (*black line* indicates ideal scaling)

Acknowledgements This work was study funded by the German Federal Ministry of Science and Education (BMBF) within the WASCAL project. The authors thank the Steinbuch Centre for Computing (SCC) for providing access to the ForHLR I supercomputer.

References

1. Browne, N.A.K., Sylla, M.B.: Regional climate model sensitivity to domain size for the simulation of the West African Summer Monsoon Rainfall. Int. J. Geophys. **2012**, 17 (2012)
2. Diaconescu, E.P., Gachon, P., Scinocca, J., Laprise, R.: Evaluation of daily precipitation statistics and monsoon onset retreat over Western Sahel in multiple data sets. Clim. Dyn. **45**(5–6), 1325–1354 (2014)
3. Doms, G., Förstner, J., Heise, E., Herzog, H., Mironov, D., Raschendorfer, M., Reinhardt, T., Ritter, B., Schrodin, R., Schulz, J.-P., et al.: A description of the nonhydrostatic regional COSMO model. Part II: Physical Parameterization, p. 154 (2011)
4. Funk, C.C., Peterson, P.J., Landsfeld, M.F., Pedreros, D.H., Verdin, J.P., Rowland, J.D., Romero, B.E., Husak, G.J., Michaelsen, J.C., Verdin, A.P., et al.: A quasi-global precipitation time series for drought monitoring. U.S. Geolog. Surv. **832**(4), 4 (2014)
5. Gessner, U., Niklaus, M., Kuenzer, C., Dech, S.: Intercomparison of leaf area index products for a gradient of sub-humid to arid environments in West Africa. Remote Sens. **5**(3), 1235–1257 (2013)
6. Guillod, B.P., Davin, E.L., Kündig, C., Smiatek, G., Seneviratne, S.I.: Impact of soil map specifications for European climate simulations. Clim. Dyn. **40**(1–2), 123–141 (2013)

7. Huffman, G.J., Bolvin, D.T., Nelkin, E.J., Wolff, D.B., Adler, R.F., Gu, G., Hong, Y., Bowman, K.P., Stocker, E.F.: The TRMM multisatellite precipitation analysis TMPA: quasi-global, multiyear, combined-sensor precipitation estimates at fine scales. J. Hydrometeorol. **8**(1), 38–55 (2007)
8. Kaspar, F., Cubasch, U.: Simulation of East African precipitation patterns with the regional climate model CLM. Meteorol. Z. **17**(4), 511–517 (2008)
9. Kothe, S., Ahrens, B.: On the radiation budget in regional climate simulation for West Africa. J. Geophys. Res. Atmos. **115**(D23), 12 (2010)
10. Kotlarski, S., Keuler, K., Christensen, O.B., Colette, A., Déqué, M., Gobiet, A., Goergen, K., Jacob, D., Lüthi, D., van Meijgaard, E., et al.: Regional climate modeling on European scales: a joint standard evaluation of the EURO-CORDEX RCM ensemble. Geosci. Model Dev. **7**(4), 1297–1333 (2014)
11. Lawrence, P.J., Chase, T.N.: Representing a new MODIS consistent land surface in the Community Land Model (CLM 3.0). J. Geophys. Res. **112**(G1), 17 (2007)
12. Masson, V., Champeaux, J.L., Chauvin, F., Meriguet, C., Lacaze, R.: A global database of land surface parameters at 1 km resolution in meteorological and climate models. J. Clim. **16**(9), 1261–1282 (2003)
13. Masson, V., Champeaux, J.-L., Chauvin, F., Meriguet, C., Lacaze, R.: A global database of land surface parameters at 1 km resolution. Meteorol. Appl. **12**(1), 29–32 (2005)
14. Mitchell, T.D., Jones, P.D.: An improved method of constructing a database of monthly climate observations and associated highresolution grids. Int. J. Climatol. **25**(6), 693–712 (2005)
15. Nikulin, G., Jones, C., Giorgi, F., Asrar, G., Büchner, M., Cerezo-Mota, R., Christensen, O.B., Déqué, M., Fernandez, J., Hänsler, A., et al.: Precipitation climatology in an ensemble of CORDEX-Africa regional climate simulations. J. Clim. **25**(18), 6057–6078 (2012)
16. Panitz, H.-J., Dosio, A., Büchner, M., Lüthi, D., Keuler, K.: COSMO-CLM (CCLM) Climate simulations over CORDEX-Africa domain: Analysis of the ERA-interim driven simulations at 0.44° and 0.22° resolution. Clim. Dyn. **42**(11–12), 3015–3038 (2014)
17. Rockel, B., Geyer, B.: The performance of the regional climate model CLM in different climate regions, based on the example of precipitation. Meteorol. Z. **17**(4), 487–498 (2008)
18. Schneider, U., Becker, A., Meyer-Christoffer, A., Ziese, M., Rudolf, B.: Global precipitation analysis products of the GPCC. Deutscher Wetterdienst (2011)
19. Smiatek, G., Rockel, B., Schättler, U.: Time resolution data preprocessor for climate version of the COSMO Model COSMO-CLM. Meteorol. Z. **17**(4), 395–405 (2008)

Anthropogenic Aerosol Emissions and Rainfall Decline in South-West Australia

Dominikus Heinzeller, Wolfgang Junkermann, and Harald Kunstmann

Abstract It is commonly understood that the observed decline in precipitation in South-West Australia during the twentieth century is caused by anthropogenic factors. In our project *wa-aero* on ForHLR1, we focus on the role of rapidly rising aerosol emissions from anthropogenic sources in South-West Australia around 1970. An analysis of historical longterm rainfall data of the Bureau of Meteorology shows that South-West Australia as a whole experienced a gradual decline in precipitation over the twentieth century. However, on smaller scales and for the particular example of the Perth catchment area, a sudden drop in precipitation around 1970 is apparent. Modelling experiments at a convection-resolving resolution of 3.3 km using the Weather and Research Forecasting (WRF) model version 3.6.1 with the aerosol-aware Thompson-Eidhammer microphysics scheme are conducted for the period 1970–1974. A comparison of four runs with different prescribed aerosol emissions and without aerosol effects demonstrates that tripling the pre-1960s atmospheric CCN and IN concentrations, as suggested by airborne measurements, can suppress precipitation by 2–9 %, depending on the area and the season. An extended version of the results presented here was accepted for publication in the *Journal of Climate* in June 2016.

1 Introduction

Over the last century, South-West Australia experienced a substantial decrease in precipitation. This poses a great challenge for the isolated region around Perth and its hinterland, which relies on reliable water resources for living, for industry and

D. Heinzeller (✉) • H. Kunstmann
Institute of Meteorology and Climate Research (IMK-IFU), Karlsruhe Institute of Technology (KIT), Kreuzeckbahnstr.19, 82467 Garmisch-Partenkirchen, Germany

Department of Geography, Augsburg University, 86135 Augsburg, Germany
e-mail: heinzeller@kit.edu; harald.kunstmann@kit.edu

W. Junkermann
IMK-IFU, KIT, Institute of Meteorology and Climate Research, Karlsruhe Institute of Technology, Kreuzeckbahnstr.19, 82467 Garmisch-Partenkirchen, Germany
e-mail: wolfgang.junkermann@kit.edu

W.E. Nagel et al. (eds.), *High Performance Computing in Science and Engineering '16*, DOI 10.1007/978-3-319-47066-5_38

for agriculture. Naturally, the question arises to what extent this decline in rainfall is human-induced and how much global and local environmental changes contribute to it. This topic motivated numerous studies and measurement campaigns already in the 1970s and has been debated widely since then [4–6, 10, 11, 20].

Delworth et al. [10] used the global climate model GFDL (General Fluid Dynamics Laboratory) CM2.5 to analyse the causes of the decline in precipitation. In [11], they concluded that many aspects of the observed reduction in rainfall can be attributed to anthropogenic changes in levels of greenhouse gases and ozone in the atmosphere, whereas anthropogenic aerosols do not contribute significantly. This stands in contrast to numerous studies of the impact of aerosols on the build-up of clouds and precipitation through the formation of cloud particles and by exerting persistent radiative forcing on the climate system that disturbs dynamics [28]. Lee and Feingold [21] investigated aerosol effects on cloud field properties of convective clouds and concluded that aerosols do have substantial influence on the spatiotemporal distribution of convection and precipitation. The coarse resolution of 50 km of the global model used by [10] and the simplified treatment of aerosols therein may explain this discrepancy.

Bates et al. [4] reported from rainfall observations that the decrease in precipitation occurred in two distinctive steps around 1975 and 2000 rather than continuously. Likewise, they demonstrated a clear stepwise decrease in stream flow measurements at those times. Karoly [20] also stressed that the simulations of [10] underestimate the decline in rainfall and that their identified drivers for this decline are usually associated with changes to the Southern Hemisphere climate in summer, while the bulk of the precipitation in this region occurs in austral winter.

Changes in atmospheric circulation due to a constant rise in greenhouse gases and a constant depletion of ozone are large-scale features and usually induce a continuous change in precipitation on longer time scales than observed in parts of South-West Australia. Local changes to the environment, on the other hand, may have the potential to alter the local climate on very short time scales. With respect to the sudden drop in precipitation in the 1970s, several human-induced factors occurred just before or at this time:

1. The conversion of natural forest to agricultural land after World War II led to an almost complete deforestation of a 130,000 km^2 area by 1968, previously a biodiversity hotspot and now known as the "wheatbelt" [8, 25]. The deforestation had a strong impact on the aerosol concentration in this region through direct effects [13] and indirect effects [17, 24] with a time lag of about 15 years. Andrich and Imberger [3] compared coastal and inland rainfall and showed empirically that land clearing alone can account for 55–62 % of the observed decline in precipitation for the wheatbelt area south-east of the vermin fence. This is also supported by modelling experiments by [18], who showed that deforestation has been causing rainfall declines in this area, albeit their study was limited to two single events.
2. In 1966, the Muja Power Station was commissioned 22 km east of Collie (see Fig. 1). The coal power plant had a total output of 974 MW and as such was the

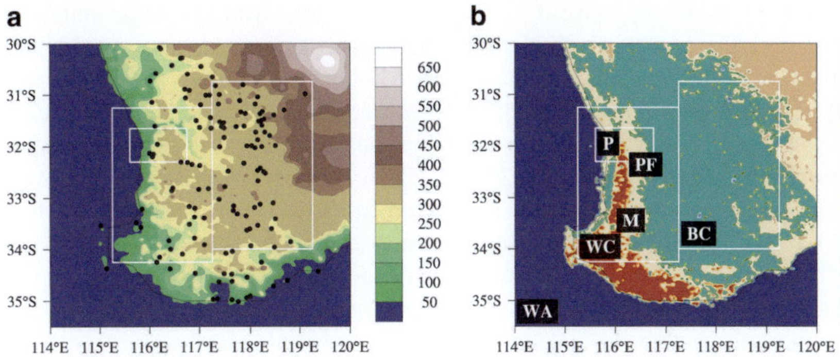

Fig. 1 Model topography (*left*, terrain height in m) and land-use classification (*right*) for a subset of the 3.3 km domain, labelled as West Australia (*WA*). Indicated are the three regions West Coast (*WC*), Perth/Freemantle (*PF*) and Back Country (*BC*) used in the analysis, as well as the location of Perth (*P*) and Muja Power (*M*). The *black dots* represent meteorological stations of the BOM [9] with a data availability of 90 % or more between 1920 and 2015. The dominant land-use categories are evergreen broadleaf forest (*red*), woody savannas (*ochre*) and croplands (*cyan*), which clearly mark the north-eastern border of the wheatbelt

largest source of aerosols in this region. Further power stations burning coal from the Collie mine were added in 1973 (Kwinana, varying several times between coal, gas and oil), in 1999 (Collie B) and in 2009 (Bluewaters). Additionally, the Kwinana refinery was continuously enlarged and eventually became the largest refinery in Australia. Airborne measurements taken during several flight campaigns in the 1970s led to an estimated total flux of 4×10^{19} particles per second from the Perth/Freemantle area, including the Collie region 150–200 km to the south-east, equivalent to a CCN production rate of 1×10^{19} particles per second [2]. This value is close to the total natural CCN production of all of Australia at that time [6].

Global circulation models suffer from a simplified treatment of aerosols and a relatively coarse resolution. Key components for the generation of rainfall such as convection and the interaction of cloud-condensating nuclei (CCN) and ice nuclei (IN) cannot be resolved and therefore are parameterised. For instance, in GFDL CM2.5, only direct aerosol effects are included implicitly in the model [10]. Regional climate models, on the other hand, have been taken to higher and higher resolution over recent years. New, sophisticated physics schemes have been added to explicitly treat the transport, growth and interaction of CCN and IN.

The Weather and Research Forecasting tool WRF [27] is widely used in numerical weather prediction and regional climate simulations. Since version 3.6, released in April 2014, the ARW (Advanced Research WRF) core of WRF contains an aerosol-aware microphysics option, the Thompson and Eidhammer scheme. Its main features are a fundamental, first order aerosol treatment and a direct coupling with radiation for aerosol indirect effects, which allows it to simulate the impact of aerosols on local weather and climate at a moderate increase in computational costs. In a first test of their new aerosol-aware scheme, Thompson and Eidhammer [29] confirmed that increased aerosol number concentrations result in larger numbers of cloud droplets of overall smaller size, leading to an increase in cloud albedo (first indirect effect) and delays in the development of precipitation (second indirect effect). However, as pointed out by [29], recent large-scale, high-resolution studies have shown that aerosol impacts on cloud systems interplay with the dynamics in a "naturally buffered" system [14, 26, 31]. These authors demonstrated that even large changes in aerosols resulted in surface precipitation differences of only a few percent overall, but also that stronger effects may occur locally.

In this study, we investigate the effect of the addition of aerosol physics on the simulated weather and climate in South-West Australia, and in particular address the question of whether the sudden increase in aerosols due to the emission from Muja Power Station can explain or at least contribute to the observed decrease in precipitation in the Perth/Freemantle area in the 1970s. We first revisit the historical rainfall observations of the Bureau of Meteorology to obtain a clear picture on the regional and temporal differences in rainfall decline over continental South-West Australia. In Sect. 2, we describe the observational data sets used and summarise the model configuration. In Sect. 3, we present and discuss the results obtained from our modelling experiments, while Sect. 4 is devoted to a summary and conclusions.

2 Methods

2.1 Observational Rainfall, Temperature and Pressure Data

We refer to the Bureau of Meteorology Daily Rainfall Climate Data [9] for the re-analysis of the decrease in precipitation over an extended period from 1920 to 2015. The starting date is chosen to guarantee a sufficiently large number of recording stations with high availability of data (90 % or more) for the entire period. Figure 1 displays all available stations in the area of study. A simple quality control is applied to the data to filter stations with zero or excessive annual precipitation.

In addition, we use gridded rainfall and near-surface air temperature data from the University of Delaware (UDEL) long term monthly means v3.01 [34] and from the Climate Research Unit (CRU) high-resolution time series data set v3.23 [15] at $0.5^\circ \times 0.5^\circ$ spatial resolution. Lastly, we include the HadSLP2 gridded global sea level pressure anomalies [1] at $5^\circ \times 5^\circ$ spatial resolution in our analysis.

2.2 Aerosol-Aware Regional Climate Model

We employ version 3.6.1 of the regional climate model WRF-ARW, released August 2014, to study the effect of aerosols and changes in their concentration on local weather and climate at very high resolution for a 4-year period from 1970 to 1974.

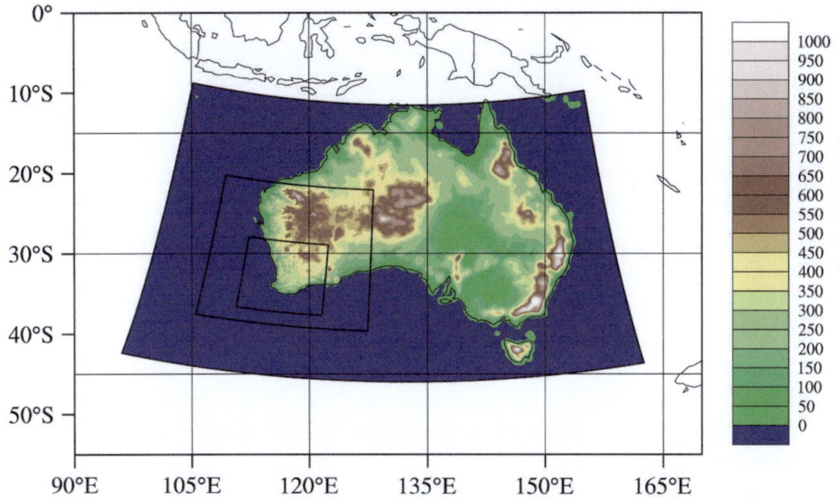

Fig. 2 Triple-nested domain configuration with 30, 10 and 3.3 km resolution. Lateral boundary conditions are taken from ERA40 re-analysis data at $1.0° \times 1.0°$ spatial resolution (110 km)

Lateral boundary conditions are supplied by ERA40 re-analysis data [30] at $1.0° \times 1.0°$ spatial resolution (110 km). For reasons explained below, a convection-resolving resolution of 3.3 km in a triple-nested approach is adopted (see Fig. 2). Due to computational constraints and the requirement to conduct multiple experiments with the innermost domain, we choose a one-way nesting approach, i.e. we switch off the feedback from the inner domains to the outer domains. Lateral boundary conditions from ERA40 to the 30 km domain are supplied every 6 h, and every 3 h for the nested domains using the ndown utility of WRF. Spectral nudging of temperature, wind and geopotential is applied above the planetary boundary layer to the outermost domain only to avoid a divergence of the regional model state from the forcing data set [22, 32]. We use the MODIS 21-class land-use table to describe the land surface properties, which dates back to 2001 and thus matches the conditions after the creation of the wheatbelt in South-West Australia in the 1960s. Several areas are selected for the analysis of the data, which are overlaid on the model topography and land-use map in Fig. 1: (South-)West Australia (WA), West Coast (WC), Perth/Freemantle (PF) and the Back Country (BC).

2.3 Aerosol-Aware Microphysics

The Thompson-Eidhammer aerosol-aware microphysics scheme was added in version 3.6 of WRF and allows the simulation of the effect of aerosols on local weather and precipitation for a moderate increase in computational costs by 16 %, compared to the standard Thompson microphysics scheme. In this implementation,

CCN are referred to as water-friendly aerosol particles and are created by summing sulphates, sea salts and organic carbon, while IN are referred to as ice-friendly aerosol particles and are created by summing five size bins of dust. Aerosols are treated in a fundamental, first order approach through activation of CCN and IN, depletion of aerosols (precipitation scavenging) and simplistic aerosol replenishment (surface emissions). The microphysics scheme is coupled directly to the RRTMG longwave (LW) and shortwave (SW) radiation schemes to in principle account for aerosol direct and indirect effects. It is important to note that in WRF version 3.6.1, this coupling is not complete: The calculation of the aerosol optical depth (AOD) is not informed by the new Thompson-Eidhammer scheme and thus assumes climatological aerosol concentrations (aerosol direct effect). However, the size of the aerosol particles, emitted by anthropogenic sources such as power plants and smelters in Australia and representing the bulk of the increase in aerosol concentration in the 1970s, has been measured for sizes between 5 and 100 nm [16]. While their exact size depends on the distance from the source and the available time for coagulation and growth [7, 16, 17], they are well below the range in which direct effects through scattering and absorption are important (>300 nm). Apart from the additional treatment of aerosols, physical consistency between the Thompson scheme and this new microphysics scheme is ensured. This allows us to assess the effect of the aerosol treatment or, more precisely, the aerosol indirect effects of small, anthropogenic aerosol particles, on the model results.

The aerosol scheme is not coupled to any cumulus scheme, which means that there is no depletion of aerosols by convective precipitation and no sub-grid scale aerosol activation. It is thus required to use a very high, convection-resolving horizontal resolution. Further, a relatively high vertical resolution and the activation of the new namelist variable `scalar_pblmix` are required to ensure that aerosols get mixed in the vertical by sub-grid turbulence. The specification of aerosols can be handled in two primary ways: (1) external data sets from climatology or other (chemistry) models, or (2) simplified vertical profiles prescribed in the model.

To fulfil the above requirements, we use a horizontal resolution of 3.3 km for the innermost domain. At this resolution, it can be assumed that convection is resolved at grid scale [23, 33]. To achieve a sufficiently high vertical resolution, in particular in the planetary boundary layer, we use 75 vertical levels with a lowermost level height of 25 m and 20 levels in the first 1000 m above surface. Such a small vertical grid spacing implies a reduction of the typical time step of 18 s (6 s per km horizontal resolution) to 4 s for model stability.

The initial aerosol concentrations are specified as simplified vertical profiles. The default vertical profile in WRFV3.6.1 depends on the terrain height and was designed to fit the continental U. S., for which the near-surface value is found to exist within an idealised boundary layer of approximately 200–1000 m, depending on starting elevation. An exponential decay of aerosol number from the higher numerical value in the boundary layer to the lower free tropospheric number is used to complete the vertical profile (Greg Thompson, private communication). This profile is adapted to describe different aerosol concentrations for South-West Australia: First, a standard vertical profile was created based on the airborne measurements and analysis of [5–7, 17], reflecting the clean environmental conditions prior to the commissioning of the Muja, Kwinana and Collie coal power plants. The initial profile is applied once at the starting time of the model integration and for every grid point, both over land and sea. During the model integration, the CCN and IN variables are advected and diffused exactly as other scalars (e.g. cloud ice number concentration). A simplified surface aerosol emission tendency is computed as a 2D field based on the horizontal grid spacing and starting aerosol number concentration for the CCN variable [29]. No surface emission tendency is applied for IN in this version of the code. The 2D tendency field is added each time step to the first model vertical level CCN value.

In this study, we address two questions. Firstly, we investigate the changes in the simulated weather and climate when aerosols are considered in the microphysics, using the initial aerosol profile presented above. Secondly, we study the impact of changes in the aerosol concentration from the clean environmental conditions to a polluted environment through modifications of the initial aerosol profile and the surface emission rates. In total, we conduct four model runs on the innermost domain for the period 1970–1974:

1. Standard run (wrf-std): In this configuration, we use the default Thompson microphysics scheme, which is coupled to the RRTMG LW/SW schemes, but does not treat aerosols explicitly. Because of the physical consistency between the Thompson and the Thompson-Eidhammer schemes, this run can be compared directly to the following runs to assess the impact of adding aerosol physics to the WRF model.
2. Aerosol run (wrf-aero): Here, we use the aerosol-aware Thompson-Eidhammer microphysics scheme with the initial aerosol profile and surface emission rates for South-West Australia. This run allows us to investigate the effect of aerosols (natural and anthropogenic) on the model without the contribution of the Muja Power Station or other large pollutants.
3. Aerosol boost run (wrf-aerox3): This configuration is identical to the aerosol run, but uses an initial aerosol profile trice as large for both CCN and IN, and accordingly, the CCN surface emission rate is also tripled. This run describes in a simple way the increase in aerosol concentrations due to the commissioning of the Muja Power Station and other sources of anthropogenic aerosols. The increase by a factor of three is motivated by differences in measured aerosol concentrations in the vicinity and far distance of the larger pollutants in the Muja/Collie area.
4. Muja Power run (wrf-muja): This configuration is identical to the aerosol run, but contains an additional source of anthropogenic aerosols injected into the model at the location of the power plant and with an emission rate as derived from observations. A total emission rate of $4.6 \times 10^8 \text{particles}/(\text{kg s})$ is added to the surface emissions in a circle sector with 20 km radius and 35° opening angle in direction north-east at the location of the Muja Power Station (116.26 °W, 33.34 °S). To account for the elevated emission from power plants at 250–400 m above ground, this additional source term is distributed evenly across the first 1500 m in height at every grid point in this sector.

We refer to previous work [12, 29] for a recommended WRF model configuration for this specific region and research question, which is summarised in Table 2 in the Appendix. It is important to remember that the differences between the standard Thompson scheme and the aerosol-aware Thompson-Eidhammer scheme are in the consideration of the aerosol *indirect* effects only: While the aerosol-aware runs compute size and number concentration of aerosols and thus cloud droplet numbers consistently, the standard run uses prescribed values for the cloud number droplets in the microphysics scheme. For both schemes, the aerosol *direct* effect is included through the calculation of the aerosol optical depth using climatological aerosol concentrations, which are independent of the initial aerosol profiles used for the different high-resolution model runs.

3 Results and Discussion

3.1 Long-Term Rainfall Trends in South-West Australia

Figure 3 displays the annual precipitation compiled from the [9] Daily Rainfall Climate Data (BOM hereafter) for the entire year, the wet season April to September, and the dry season October to March for the areas described in Fig. 1. Displayed are the mean of all stations with a data availability of 90 % or more and the stations with maximum and minimum rainfall over the entire period 1920–2015. Stations with minimum rainfall show no decrease over the entire period, while stations with maximum rainfall exhibit a dramatic decline in rainfall predominantly in the wet season. Independent of the area, the decrease in mean annual precipitation is observed entirely in the rainy season, while there is no change in the low amount of rainfall during the dry season. We would like to note here that the selection of stations used in the analysis matches that of [3] with the exception that the quality control flags of the station data and possible relocations of stations are not taken into account.

Stepwise and linear fits to the mean annual rainfall are displayed with their corresponding r^2 coefficients of determination. The data presented here and the fits to it at first glance contradict the current perception of a sudden, step-wise decrease in precipitation in the 1970s and at the beginning of the twenty-first century in South-West Australia [4]. A single, step-wise fit to the data results in a step of −246 mm/a (PF) to −42 mm/a (WA) around 1970 with a similar correlation coefficient as a continuous, linear decline by −430 mm/a (PF) to −71 mm/a (WA) from 1920 to 2015. For the three regions WA, WC and BC, the observations and our fits to them imply that a continuous decline in annual precipitation by 20 % between 1920 and 2015 matches the observations as well as a sudden drop by 10 % around 1970.

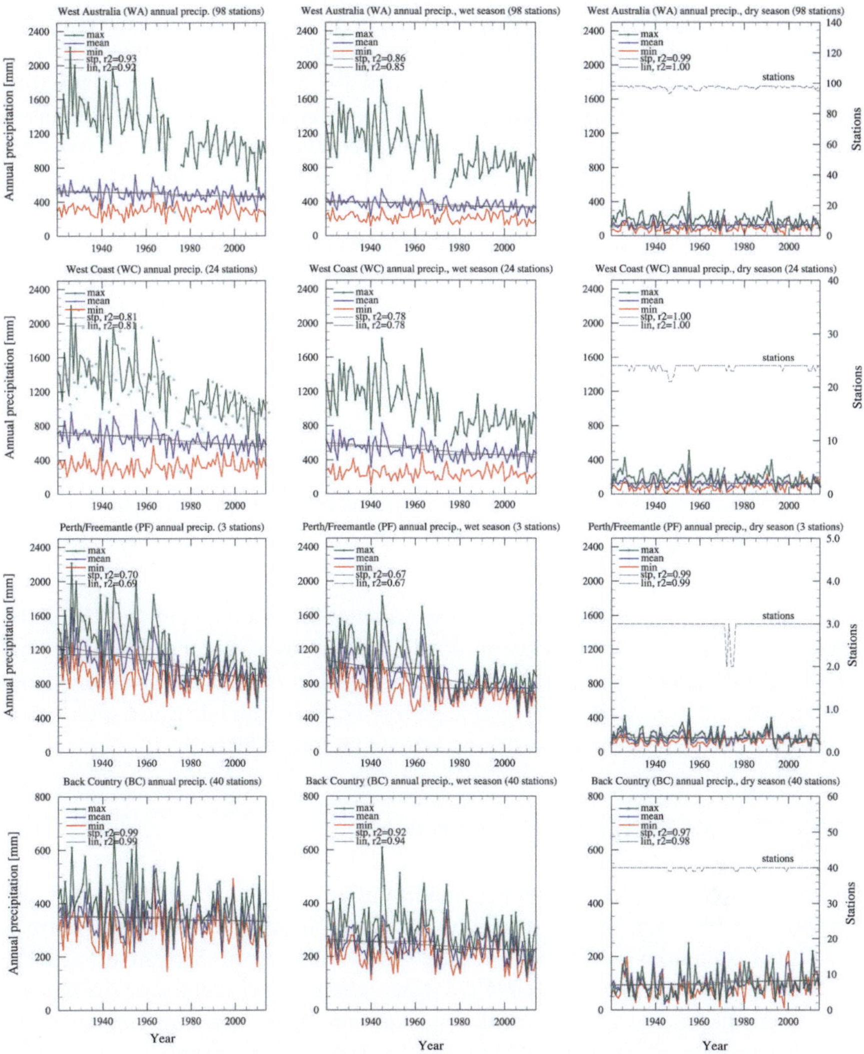

Fig. 3 Annual precipitation compiled from the [9] Daily Rainfall Climate Data for the areas described in Fig. 1 (*top* to *bottom*: WA, WC, PF, BC) and for the entire year (*left*), the wet season April to September (*middle*), and the dry season October to March (*right*). Displayed are the mean of all stations with a data availability of 90 % or more and the data of the station with the maximum/minimum rainfall over the entire period. Stepwise and linear fits to the mean annual rainfall are displayed with their r^2 coefficients

The extreme numbers for PF are a result of a very small number of three stations only that meet our criteria on data availability and for which two of the stations, 9010 Churchman Brook and 9031 Mundaring Weir, show a significant decrease in precipitation from 1500 mm/a to about 1000 mm/a for Churchman

Brook, and from 1100 mm/a to about 800 mm/a for Mundaring Weir, respectively. For these particular stations, the long-term records indeed resemble a sudden decrease in annual rainfall precipitation around 1970. It is interesting to note that both Churchman Brook and Mundaring Weir are located in the Perth catchment area at the Canning/Mundaring surface water storages and that the sudden drop in precipitation around 1970 for these stations fit perfectly with the observed dam levels displayed in [20], Fig. 1a.

Hence, our general findings of a continuous decline in precipitation for the WA, WC and BC areas do not contradict the sudden drop in observed river discharges reported by [4] and [20] for the Perth catchment area, for which small changes in circulation may have led to shifts in precipitation bands on a regional scale. In addition, anthropogenic factors such as irrigation and deforestation [3] may have influenced the dam water levels at that time.

3.2 Aerosol Effects in High-Resolution Regional Climate Modelling

Our findings in the previous section suggest that observational rainfall data can be interpreted as a continuous decline in precipitation or as a sudden decrease around 1970, depending on the area. In the following, we address the question of whether a sudden increase in small, anthropogenic particles can in principle cause such a drop in precipitation through first and second aerosol indirect effects.

3.2.1 CCN and IN Number Concentrations

Figure 4 displays the CCN number concentrations for the three different high-resolution model runs using aerosol-aware microphysics (wrf-aero, wrf-aerox3, wrf-muja) as contour-plot average for the wet season (April–September) and the dry season (October–March) 1970–1974. Near-surface wind vectors are overlaid on the contour plots. The CCN number concentration of the wrf-muja run, averaged over the area WA, exceeds that of the wrf-aerox3 run, i.e. three times the conditions prior to the commissioning of Muja Power and other large sources of anthropogenic aerosols. Despite being emitted by a tiny area around the location of Muja Power Station, the ultrafine aerosol particles are distributed widely and result in a higher CCN concentration along the West Coast and over the wheatbelt. While the annual average shows a symmetrical distribution around the emitting source (not displayed), the direction in which the ultrafine aerosol particles travel depends on the seasonality of the near-surface winds. In austral summer, the dominant near-surface wind direction is towards the north-west over land, while in austral winter the bulk of the CCN are pushed to the south-east of the area WA (Fig. 4). This suggests that the emissions from Muja Power are advected horizontally rather than mixed up into higher layers. The CCN number concentration varies with time around its

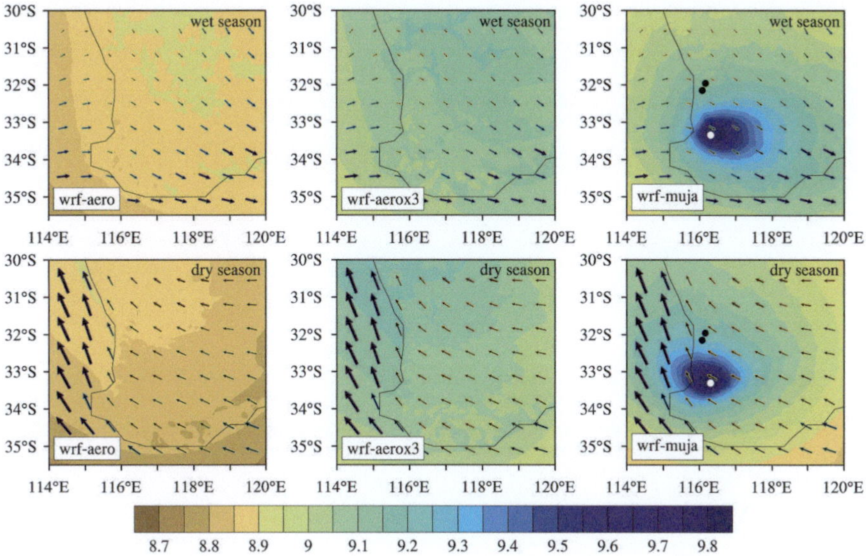

Fig. 4 CCN number concentrations [kg^{-1}] for the wet season and the dry season averages 1970–1974, summed up over the entire column at each grid point. Overlaid are 10 m surface winds. The *black dots* represent the positions of the BOM stations Churchman Brook (9010) and Mundaring Weir (9031), the *white dot* the location of the Muja Power station

initial values as a result of the continuous removal and replenishment through CCN activation, rain/snow/graupel collecting aerosols, cloud/rain evaporation and surface emissions.

3.2.2 Precipitation

The main focus of our investigation is the change in model precipitation when including aerosol physics in the microphysics schemes (i.e. the difference between wrf-std and the aerosol-aware runs), and when changing the aerosol concentration (i.e. the difference between wrf-aero and wrf-aerox3/wrf-muja). Figure 5 displays the mean monthly rainfall for the different model runs and for observational data from UDEL and BOM for the four regions of interest over land only, averaged over the entire period 1970–1974. The monthly precipitation is derived as the average of all stations in the region with a data availability of 90 % or more between 1970 and 1974 (BOM) or all grid points in the corresponding region (all others). Averaged over the entire region WA and the Back Country region BC, the gridded observations from UDEL and the station data from BOM agree well. For WC and even more so for PF, the UDEL observations show consistently smaller values than the BOM station data, which is a result of the small number of stations contributing to the BOM average and the "outlier" stations Churchman Brook and Mundaring Weir.

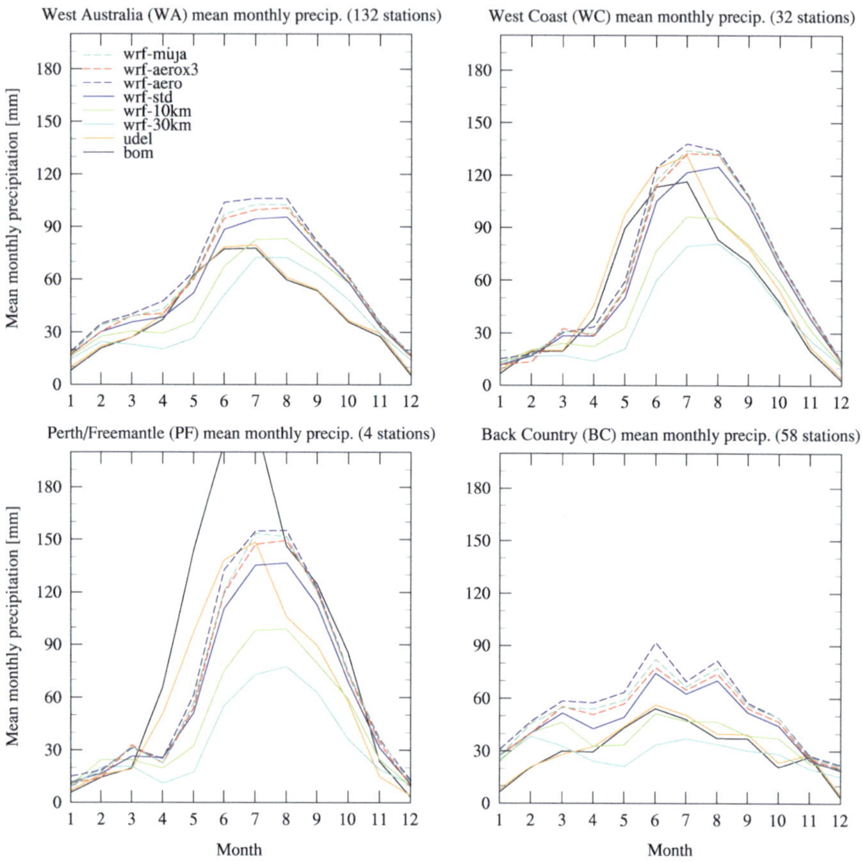

Fig. 5 Mean monthly precipitation 1970–1974 for the different model runs and observational data sets. For each region, the mean value is taken over all stations (bom) or over all grid points (all others), respectively

The monthly precipitation of the models is given by non-convective precipitation for the 3.3 km runs, where no cumulus scheme is employed, and by the sum of convective and non-convective precipitation for the 10 and 30 km runs. The amount of rainfall is clearly related to the horizontal resolution, with lower values for larger grid spacing. All 3.3 km models overpredict rainfall for WA/BC in both seasons and for WC/PF in the dry season, but match the WC/PF winter precipitation closely with mean biases of less than 10 mm/month. The coarser-resolution models provide better estimates of WA/BC precipitation for both season and for WC/PF in the dry season, but largely underestimate the WC/PF winter precipitation. Among the high-resolution runs, the non-aerosol-aware run wrf-std generates least precipitation, followed by wrf-aerox3, wrf-muja and wrf-aero with largest precipitation amounts, independent of the region and the season. In general, the high-resolution runs show large improvements in the spatial correlation for all regions during the wet season

Table 1 Accumulated precipitation [mm] at the end of the 4-year simulation period from 1 January 1970 to 1 January 1974 for the different model runs and observations from UDEL, for each of the regions over land only and split into wet season and dry season

Region	Season	UDEL	wrf-30 km	wrf-10 km	wrf-std	wrf-aero	wrf-aerox3	wrf-muja
Accumulated precipitation [mm]								
WA	Wet	1347	787	1075	1575	1758	1679	1737
	Dry	508	531	655	725	817	759	786
WC	Wet	2193	1226	1549	2207	2379	2321	2352
	Dry	535	383	492	562	633	598	610
PF	Wet	2342	1124	1560	2372	2603	2525	2587
	Dry	470	332	403	495	564	515	526
BC	Wet	940	675	918	1282	1435	1353	1412
	Dry	461	538	658	733	859	791	818

and for WA/BC during the dry season (not displayed). We speculate that this is due to the improved representation of the topography, i.e. the mountainous region in the north-east, in the 3.3 km models, while the coastal regions on average exhibit a smaller interannual variability due to the dominant wind direction from the south-east, i.e. over dry continental planes.

Table 1 compares the total amount of modelled precipitation, accumulated over the entire 4-year period 1970–1974, to the UDEL observations for each of the four regions and split into dry season and wet season. Averaged over the areas WA and BC, the coarse-resolution runs wrf-30 km and wrf-10 km exhibit smaller errors than the high-resolution runs. The coastal areas WC and PF are fit significantly better by the high-resolution runs, in particular for the wet season in austral winter. Among the four high-resolution runs, the standard run wrf-std with least precipitation performs best and in particular matches the observed WC and PF winter precipitation closely.

By comparing the accumulated precipitation values of the wrf-std and the wrf-aero runs, we find that the *addition* of aerosol physics with pre-1960s aerosol concentrations to the microphysics scheme leads to an increase in rainfall by 8.1 % (WC), 9.5 % (PF), 10.6 % (WA) and 12.1 % (BC). This effect is more pronounced for the dry season (between 11.2 % and 14.6 %) than for the wet season (between 7.2 % and 10.6 %). *Increasing* the aerosol concentration, however, reduces the amount of precipitation in all areas: Of the three aerosol runs, wrf-aero shows the largest values of accumulated precipitation and wrf-aerox3 (with 3 times larger CCN and IN concentrations) shows the smallest values: Compared to the pre-1960s run, precipitation is reduced by 3.1 % for WC, 4.0 % for PF, 5.4 % for WA, and 6.5 % for BC. Again, this effect is more pronounced for the dry season (between 5.5 % and 8.7 %) than for the wet season (between 2.4 % and 4.4 %). The wrf-muja run with additional CCN emissions from Muja Power, but otherwise standard aerosol concentrations from wrf-aero, lies in between.

It is nevertheless interesting to correlate the wrf-muja seasonal differences in decrease in precipitation for the four regions with the surface wind fields displayed in Fig. 4. For regions WA and WC, the ratios between the decrease in summer (dry season) and winter (wet season) precipitation relative to wrf-aero are 3.1 (3.7 %/1.2 %) and 3.3 (3.6 %/1.1 %), respectively. For region BC, where the bulk of the additional surface emissions from Muja Power are transported to in austral winter, we find a ratio of 2.9 (4.7 %/1.6 %). The opposite holds for region PF, which is most affected by the aerosol inflow from Muja Power in austral summer and for which we find a ratio of 11.0 (6.7 %/0.61 %). These numbers demonstrate the potential impact of aerosol emissions from a single source on local rainfall amounts.

4 Summary and Conclusions

In this study, we investigate the potential impact of changing aerosol concentrations on rainfall distribution and amount in South-West Australia. Prior to this, we revisit historical long-term observations of rainfall to differentiate between the nature of the processes contributing to the observed decline in precipitation in the twentieth century. On larger spatial scales for continental South-West Australia, we determine a continuous decline in precipitation rather than a sudden drop. This is in line with [20] and [11], who concluded that many aspects of the observed reduction in rainfall can be attributed to anthropogenic changes in levels of greenhouse gases and ozone in the atmosphere and changes to the transport and advection patterns over the Indian Ocean.

On smaller spatial scales and for the particular case of the Perth/Freemantle area, the observed decline in precipitation is too strong to be explained by changes of the large-scale atmospheric motion only. Further, we detect large differences between the individual stations and we identify in particular two stations (9010 Churchman Brook; 9031 Mundaring Weir) within the Perth water storage basin with a dramatic decline in annual precipitation by around 30 % between 1920 and 2015. More importantly, a significant share of this decrease falls into the 1970s. This coincides well with the observations of rainfall and streamflow measurements reported by [4] and [20]. We conclude that further processes of likely anthropogenic nature occurring on shorter time scales and smaller regional scales must have been involved. Candidates therefore are irrigation and deforestation [3] with subsequent release of ultrafine particles from salt lakes [17, 19], and anthropogenic aerosols emitted by coal power plants and smelters.

Here, we focus on the possible role of anthropogenic aerosols only and assume a constant land-use classification in our models, which dates back to 2001 and thus matches the conditions after the land clearing in South-West Australia. We consider in particular the emissions from large pollutants such as the Muja Power Station, commissioned in 1966 approximately 200 km south-east of the Perth catchment area, and the impact of a consistent treatment of aerosols on precipitation through first and second aerosol indirect effects using four different regional climate modelling experiments with a convection-permitting grid spacing of 3.3 km. We create pre- and post-industrial aerosol profiles of ultrafine and fine water-friendly aerosol particles (CCN) and ice-friendly aerosol particles (IN), based on air-borne measurements [5–7, 17].

First, we show that the emissions of ultrafine particles from Muja Power Station alone increase the amount of cloud-condensating nuclei (CCN) by a factor of three over the entire area and that this leads to a reduction in precipitation in the model along the West Coast, including the Perth/Freemantle area, as well as further inland. We further show that the emissions from Muja Power are only slowly mixed up into higher layers and follow the near-surface wind fields, which leads to a large inflow of aerosols into the Perth/Freemantle region in austral summer and into the Back Country region in austral winter. At the same time, precipitation around Perth is suppressed to a greater extent in austral summer than in austral winter compared to the average over the entire region of South-West Australia, whereas the opposite holds for the Back Country region. In a second experiment we increase both CCN and IN by a factor of three compared to pre-1960s levels. Tripling the pre-1960s aerosol concentrations of both CCN and IN (CCN only) corresponds to a decrease in annual precipitation between 3.1 % (1.7 %) for the West Coast and 6.5 % (2.8 %) for the Back Country. We also compare the pre-1960s aerosol run wrf-aero with the standard run wrf-std, which uses prescribed and constant cloud droplet numbers and CCN/IN number concentrations in the radiation and microphysics schemes. These values are significantly larger in the case of cloud droplets and IN number concentrations, and comparable in the case of CCN number concentrations. Relative to wrf-aero, the decrease in annual precipitation ranges from 8.1 % (WC) to 12.1 % (BC) for wrf-std and is thus stronger than for wrf-aerox3 and wrf-muja.

Our modelling results suggest that anthropogenic aerosol emissions can contribute to the observed sudden rainfall decline in the Perth/Freemantle area. While a decrease of around 10 % could account for the majority of the observed drop in the 1970s, it remains to be answered whether the combined CCN *and* IN emissions from Muja Power Station and from other large emitters in the area, for example the Kwinana Oil Refinery (commissioned in 1955), are sufficient. Further, the vast deforestation occurring in the same period can also lead to a decrease in precipitation. Hence, future work should try to disentangle the effects of deforestation and anthropogenic aerosols through a series of experiments with different land-use classifications or different aerosol concentrations and a combination of the two, based on detailed land-use maps and precise CCN and IN emissions from all major pollutants. We also suggest to repeat the experiments presented here for the East Coast of Australia, where no large-scale deforestation took place and the effect of rising anthropogenic aerosol emissions can be studied in isolation.

Acknowledgements The modelling experiments presented here required more than 2 Mio CPUh and were conducted on the Karlsruhe Institute of Technology Steinbruch Centre for Computing (KIT-SCC) ForHLR1 supercomputer. The authors acknowledge the European Centre for Medium-Range Weather Forecasts (ECMWF) for the dissemination of ERA40, the NOAA/OAR/ ESRL PSD, Boulder for providing the UDEL air temperature and precipitation data and the HadSLP2 sea level pressure data, the University of East Anglia, Climate Research Unit, for access to the CRU air temperature and precipitation data, and the Bureau of Meteorology, Australia, for the dissemination of the daily rainfall climate data. The authors are particularly grateful for the support of Greg Thompson (NCAR) in the design of the experiment and the setup of the WRF model.

Appendix

Table 2 WRF model configuration for the different domains at 30 km, 10 km and 3.33 km resolution and for the different types of high-resolution runs

Run	wa-30 km	wa-10 km	wa-std	wa-aero/ aerox3/muja
Microphysics	Thompson	Thompson	Thompson	Thompson-Eidhammer
Radiation	RRTMG LW/SW	RRTMG LW/SW	RRTMG LW/SW	RRTMG LW/SW
Cumulus	BMJ	BMJ	Off	Off
PBL	MYJ	MYJ	MYJ	MYJ
Surface layer	Janjic Eta	Janjic Eta	Janjic Eta	Janjic Eta
Land-surface	Noah LSM	Noah LSM	Noah LSM	Noah LSM
Scalar PBL mix	On	On	On	On
Grid FDDA	Above PBL	Off	Off	Off
o3input	2	2	2	2
aer_opt	1	1	1	1
Domain size	190 × 132 × 75	199 × 199 × 75	304 × 298 × 75	304 × 298 × 75
Time step	120 s	40 s	4 s	4 s
Rad. time step	30 m	10 m	3 m	3 m
Forcing interval	6 h	3 h	3 h	3 h
Computational setup				
HPC	IMK-IFU KEA	IMK-IFU KEA	KIT SCC ForHLR1	KIT SCC ForHLR1
Nodes	4	4	24	24
Total tasks	80	80	480	480

References

1. Allan, R., Ansell, T.: A new globally complete monthly historical gridded mean sea level pressure dataset (HadSLP2): 1850–2004. J. Clim. **19**, 5816–5842 (2006)
2. Andreae, M.O.: Correlation between cloud condensation nuclei concentration and aerosol optical thickness in remote and polluted regions. Atmos. Chem. Phys. Discus. **8**(3), 11293–11320 (2008)
3. Andrich, M.A., Imberger, J.: The effect of land clearing on rainfall and fresh water resources in Western Australia: a multi-functional sustainability analysis. Int. J. Sustain. Dev. World Ecol. **20**(6), 549–563 (2013)
4. Bates, B.C., Hope, P., Ryan, B., Smith, I., Charles, S.: Key findings from the Indian Ocean climate initiative and their impact on policy development in Australia. Clim. Change **89**(3–4), 339–354 (2008)
5. Bigg, E., Soubeyrand, S., Morris, C.: Persistent after-effects of heavy rain on concentrations of ice nuclei and rainfall suggest a biological cause. Atmos. Chem. Phys. **15**, 2313–2326 (2015)
6. Bigg, E., Turvey, D.: Sources of atmospheric particles over Australia. Atmos. Environ. **12**(8), 1643–1655 (1978)
7. Bigg, E.K.: Ice nucleus concentrations in remote areas. J. Atmos. Sci. **30**(6), 1153–1157 (1973)
8. Bradshaw, C.J.A.: Little left to lose: deforestation and forest degradation in Australia since European colonization. J. Plant Ecol. **5**(1), 109–120 (2012)
9. Bureau of Meteorology: Daily rainfall climate data: product code IDCJAC0009 (2015)
10. Delworth, T.L., Rosati, A., Anderson, W., Adcroft, A.J., Balaji, V., Benson, R., Dixon, K., Griffies, S.M., Lee, H.C., Pacanowski, R.C., Vecchi, G.A., Wittenberg, A.T., Zeng, F., Zhang, R.: Simulated climate and climate change in the GFDL CM2.5 high-resolution coupled climate model. J. Clim. **25**(8), 2755–2781 (2012)
11. Delworth, T.L., Zeng, F.: Regional rainfall decline in Australia attributed to anthropogenic greenhouse gases and ozone levels. Nat. Geosci. **7**, 583–587 (2014)
12. Fersch, B., Kunstmann, H.: Atmospheric and terrestrial water budgets: sensitivity and performance of configurations and global driving data for long term continental scale WRF simulations. Clim. Dyn. **42**(9–10), 2367–2396 (2013)
13. Gallagher, M.W., Nemitz, E., Dorsey, J.R., Fowler, D., Sutton, M.A., Flynn, M., Duyzer, J.H.: Measurements and parameterizations of small aerosol deposition velocities to grassland, arable crops, and forest: influence of surface roughness length on deposition. J. Geophys. Res. Atmos. **107**, AAC 8-1–AAC 8-10 (2002)
14. Grabowski, W.W., Morrison, H.: Indirect impact of atmospheric aerosols in idealized simulations of convective-radiative quasi equilibrium. Part II: Double-moment microphysics. J. Clim. **24**(7), 1897–1912 (2011)
15. Harris, I., Jones, P.D., Osborn, T.J., Lister, D.H.: Updated high-resolution grids of monthly climatic observations – the CRU TS3.10 Dataset. Int. J. Climatol. **34**, 623–642 (2014)
16. Junkermann, W., Hacker, J.M.: Ultrafine particles over Eastern Australia: an airborne survey. Tellus B **67**, 25308 (2015)
17. Junkermann, W., Hacker, J.M., Lyons, T., Nair, U.: Land use change suppresses precipitation. Atmos. Chem. Phys. Discus. **9**(3), 11481–11500 (2009)
18. Kala, J., Lyons, T., Nair, U.: Numerical simulations of the impacts of land-cover change on cold fronts in south-west Western Australia. Boun. Layer Meteorol. **138**, 121–138 (2010)
19. Kamilli, K.A., Ofner, J., Lendl, B., Schmitt-Kopplin, P., Held, A.: New particle formation above a simulated salt lake in aerosol chamber experiments. Environ. Chem. **12**(4), 489–503 (2015)
20. Karoly, D.J.: Climate change: human-induced rainfall changes. Nat. Geosci. **7**(8), 551–552 (2014)
21. Lee, S.S., Feingold, G.: Aerosol effects on the cloud-field properties of tropical convective clouds. Atmos. Chem. Phys. **13**(14), 6713–6726 (2013)

22. Miguez-Macho, G., Stenchikov, G.L., Robock, A.: Spectral nudging to eliminate the effects of domain position and geometry in regional climate model simulations. J. Geophys. Res. D Atmos. **109**, D13104 (2004)
23. Prein, A.F., Gobiet, A., Suklitsch, M., Truhetz, H., Awan, N.K., Keuler, K., Georgievski, G.: Added value of convection permitting seasonal simulations. Clim. Dyn. **41**(9–10), 2655–2677 (2013)
24. Ruprecht, J., Schofield, N.: Effects of partial deforestation on hydrology and salinity in high salt storage landscapes. II. Strip, soils and parkland clearing. J. Hydrol. **129**(1–4), 39–55 (1991)
25. Saunders, D.: Changes in the Avifauna of a region, district and remnant as a result of fragmentation of native vegetation: the wheatbelt of western Australia. A case study. Biol. Conserv. **50**(1–4), 99–135 (1989)
26. Seifert, A., Köhler, C., Beheng, K.D.: Aerosol-cloud-precipitation effects over Germany as simulated by a convective-scale numerical weather prediction model. Atmos. Chem. Phys. **12**(2), 709–725 (2012)
27. Skamarock, W., Klemp, J., Dudhi, J., Gill, D., Barker, D., Duda, M., Huang, X.-Y., Wang, W., Powers, J.: A description of the advanced research WRF version 3, NCAR/TN-475+STR. Technical report (2008)
28. Tao, W.-K., Chen, J.-P., Li, Z., Wang, C., Zhang, C.: Impact of aerosols on convective clouds and precipitation. Rev. Geophys. **50**, RG2001 (2012). doi:10.1029/2011RG000369
29. Thompson, G., Eidhammer, T.: A study of aerosol impacts on clouds and precipitation development in a large winter cyclone. J. Atmos. Sci. **71**, 3636–3658 (2014)
30. Uppala, S., Kållberg, P.W., Simmons, A.J., Andrae, U., Bechtold, V.D.C., Fiorino, M., Gibson, J.K., Haseler, J., Hernandez, A., Kelly, G.A., Li, X., Onogi, K., Saarinen, S., Sokka, N., Allan, R., Andersson, E., Arpe, K., Balmaseda, M.A., Beljaars, A.C.M., Berg, L.V.D., Bidlot, J., Bormann, N., Caires, S., Chevallier, F., Dethof, A., Dragosavac, M., Fisher, M., Fuentes, M., Hagemann, S., Hólm, E., Hoskins, B.J., Isaksen, L., Janssen, P.A.E.M., Jenne, R., Mcnally, A.P., Mahfouf, J.-F., Morcrette, J.-J., Rayner, N.A., Saunders, R.W., Simon, P., Sterl, A., Trenberth, K.E., Untch, A., Vasiljevic, D., Viterbo, P., Woollen, J.: The ERA-40 re-analysis. Q. J. R. Meteorol. Soc. **131**, 2961–3012 (2005)
31. van den Heever, S.C., Stephens, G.L., Wood, N.B.: Aerosol indirect effects on tropical convection characteristics under conditions of radiative-convective equilibrium. J. Atmos. Sci. **68**(4), 699–718 (2011)
32. von Storch, H., Langenberg, H., Feser, F.: A spectral nudging technique for dynamical downscaling purposes. Mon. Weather Rev. **128**, 3664–3673 (2000)
33. Weisman, M.L., Skamarock, W.C., Klemp, J.B.: The resolution dependence of explicitly modeled convective systems. Mon. Weather Rev. **125**(4), 527–548 (1997)
34. Willmott, C.J., Matsuura, K.: University of Delaware Terrestrial Air Temperature and Precipitation: Monthly and Annual Time Series (1950–1999) v3.01. http://www.esrl.noaa.gov/psd/data/gridded/data.UDel_AirT_Precip.html (2014). Accessed 6 Nov 2016

High-Resolution Climate Projections Using the WRF Model on the HLRS

Viktoria Mohr, Kirsten Warrach-Sagi, Thomas Schwitalla, Hans-Stefan Bauer, and Volker Wulfmeyer

Abstract The latest generation of climate projections for the twenty-first century are build on new emission scenarios based on Representative Concentration Pathways (RCPs). Within the world wide coordinated effort of the Coupled Model Intercomparison Project Phase 5 (CMIP5), their impact on climate is simulated with global general circulation models (GCMs) of the climate system with a spatial grid of 100–200 km resolution. High resolution information from a robust multi-model ensemble on possible ranges of future climate changes is essential for climate impact research and as background information for policy and economy. Within the Coordinated Regional Downscaling EXperiments (CORDEX), the global climate simulations are downscaled for most continental regions, e.g. a unique set of high resolution climate change simulations for Europe is currently established. This project contributes to this ensemble downscaling five GCM simulations from 1958 to 2100 with the Weather Research and Forecasting (WRF) model. The WRF simulations are currently performed with 0.44° and 0.11° resolution on the CRAY XC40 at the High Performance Computing Center Stuttgart (HLRS).

First results of the simulations on the 0.44° grid for the "historical" period from 1971–2000 and as comparison for two different future scenarios from 2071–2099 show an increase of the average temperature by 2–4 °C with respect to the chosen emission scenario, especially in the southeastern and northeastern part of Europe. In the future scenario where a moderate Greenhouse Gas emission increase is projected, the annual average precipitation in Germany is indicated to experience a decrease by 50–100 l/m^2. Considering the future scenario with a high projected emission increase, only marginal changes of the annual average precipitation are simulated.

V. Mohr (✉) • K. Warrach-Sagi • T. Schwitalla • H.-S. Bauer • V. Wulfmeyer
Institute of Physics and Meteorology, University of Hohenheim, Garbenstrasse 30, 70599 Stuttgart, Germany
e-mail: viktoria.mohr@uni-hohenheim.de; Kirsten.Warrach-Sagi@uni-hohenheim.de; thomas.schwitalla@uni-hohenheim.de; hans-stefan.bauer@uni-hohenheim.de; volker.wulfmeyer@uni-hohenheim.de

W.E. Nagel et al. (eds.), *High Performance Computing in Science and Engineering '16*, DOI 10.1007/978-3-319-47066-5_39

1 Introduction and Motivation

The projected increase of the anthropogenic emissions of CO_2 and other greenhouse gases within the next decades will have a considerable influence on the future climate and consequently on the society. Although climate change is a global issue, regionally the impact will be much more diverse. General circulation models (GCM) are currently the most advanced tools for simulating the response of the global climate system to increasing greenhouse gas concentrations. GCMs typically have a horizontal resolution of 100–200 km. However, to better understand also regional climate phenomena such as local extremes and in order to assess the effect of the expected climate change, scientists and end users like federal agencies and climate impact and adaptation researchers require projections on the regional scale with a higher horizontal resolution.

The Coupled Model Intercomparison Project Phase 5 (CMIP5) [1] provides a framework for coordinated global climate change experiments, where 20 different modelling groups performed global climate projections with their GCMs. CMIP5 contributed to the latest assessment report of the Intergovernmental Panel on climate change (IPCC). Those projections are based on Representative Concentration Pathways (RCPs) [2], representing four different possible greenhouse gas (GHG) concentration scenarios of the future climate. These scenarios are the RCP8.5, RCP6, RCP4.5 and RCP2.6 scenario. The number indicates the possible range in the change of radiative forcing (in W/m^2) by the year 2100 relative to pre-industrial values.

CORDEX (http://wcrp-cordex.ipsl.jussieu.fr), the COordinated Regional Downscaling EXperiment was established by the World Climate Research Programme (WCRP) in order to provide ensembles of regional climate simulations on a higher spatial resolution [3]. The task within CORDEX is to downscale the GCMs which contributed to the CMIP5 database with regional climate models to continental scale regions. For EURO-CORDEX, the European branch of the CORDEX initiative, simulations are done with 0.44° and 0.11° resolution (e.g. [4–6]). The higher resolution allows the simulation of smaller scale processes and feedback mechanisms and provides the results on a smaller spatial scale for end users.

The BMBF-funded (Federal Ministry for Education and Research) project ReKliEs-De (Regional Climate Ensembles Germany) (http://reklies.hlnug.de/) contributes to EURO-CORDEX by carrying out a certain number of regional climate projections. ReKliEs-De is a nationwide project for more accurate assessment of regional climate changes. The project aims to identify bandwidths and extreme values from the results of high-resolution regional climate projections for Germany and their preparation for Climate Impact Research and Policy Consulting. On this basis, more detailed studies on changes in the occurrence of extreme precipitation, drought or extreme heat can be carried out. In this new project the assessment of possible bandwidth and extreme expressions of these weather events shall be improved. This will provide more resilient statements and thus the results will be more usable for providing policy advice and the Climate Impact Research.

The DFG (Deutsche Forschungsgemeinschaft) funded Research Unit "FOR 1695 – Regional Climate Change" (klimawandel.uni-hohenheim.de) is an example for climate impact research at the University of Hohenheim. Within this project a climate simulation is further downscaled to a resolution of 3 km and below for Germany to study the interaction between agricultural landscapes and the atmosphere in a changing climate in Baden-Württemberg (SW Germany).

The scope of the WRFCLIM project at HLRS comprises of regional climate simulations, through dynamical downscaling of different GCM model outputs from 1958 to 2100 with the Weather Research and Forecasting (WRF) model [7], under the remit of the ReKliEs-De project and the Research Unit 1695. The simulations were started in Spring 2015 and are still ongoing. In the following a technical description of the simulations within WRFCLIM at HLRS is presented. A summary of the current analysis and some preliminary results are reported on the bases of the so far completed simulations.

2 WRF Simulations at HLRS

2.1 Description of Forcing Data

For the future climate projections, four different GCMs and two different RCP scenarios of the CMIP5 project, are applied as boundary forcing with the WRF model. The "historical" runs of the GCMs cover the period from 1850 to 2005. This period is forced by observed atmospheric composition changes of anthropogenic and natural sources. The "RCP" scenarios of the GCM's cover the period from 2006 to 2100. They represent mitigation scenarios that assume policy actions will be taken into account to achieve certain emission targets [1]. The numbers of the RCPs give a rough estimate of the range in the change of the radiative forcing by the year 2100 relative to the pre-industrial values. The forcing data we applied, the resolution of the GCMs, its scenarios and the chosen simulation period is presented in Table 1.

Table 1 Applied GCM data for the WRF simulations

GCM	Scenarios	Simulation period	GCM resolution
MPI-ESM-LR (Max Planck Institue-Earth System Model-Low Resolution)	Historical RCP8.5 RCP2.6	1958–2005 2006–2100 2006–2100	$1.8653° \times 1.875°$
MIROC5 (Model for Interdisciplinary Research on Climate)	Historical RCP8.5	1958–2005 2006–2100	$1.4008° \times 1.40625°$
HadGEM2-ES (Hadley Global Environment Model 2 – Earth System)	Historical RCP8.5	1958–2005 2006–2100	$1.25° \times 1.875°$
EC-EARTH (European Earth System Model)	Historical RCP8.5	1958–2005 2006–2100	$1.1215° \times 1.125°$

2.2 Technical Description

The simulations are performed with WRF model version 3.6.1, using the CRAY XC40 System at the HLRS. WRF is coupled with the land surface model NOAH [8] and applied with the following parameterizations: the Morrison two-moment microphysics scheme [9], the Yonsei University (YSU) planetary boundary layer scheme [10], the Kain-Fritsch-Eta convection scheme [11] and the radiation transport scheme CAM for longwave and shortwave radiation [12].

WRF was applied to create climate projections for Europe with the domain being specified by CORDEX (Fig. 1a). Within ReKliEs-De the focal region of the analyses is Germany and its contributing river catchment areas (Fig. 1b). The simulations are forced 6 hourly at the lateral boundaries with data, which is generally available at approx 1°–2° grid resolution (see Table 1). WRF is applied one-way nested in a nesting approach via 0.44°–0.11° resolution. WRF was compiled at HLRS with PGI 14.7 and applied in a hybrid configuration using MP and OpenMP to optimize the speed of the simulation. During the performance of the simulations a huge amount of output data emerges. As this amount cannot be stored as raw output, it will be extracted and minimized within the postprocessing process. Table 2 shows the technical details of the different simulations performed on hazelhen.

Within 24 h walltime, it was possible to simulate approximately 4 years of the 50 km domain projections and 1 year of the 12 km domain projections respectively. So far (status April 2016) we were able to downscale the majority of the GCM's from 1958–2100 (historical + RCP's) to the grid size of 50 km. The simulations on the 12 km grid are about to be finalized for the "historical" (1958–2005) projections within the next weeks.

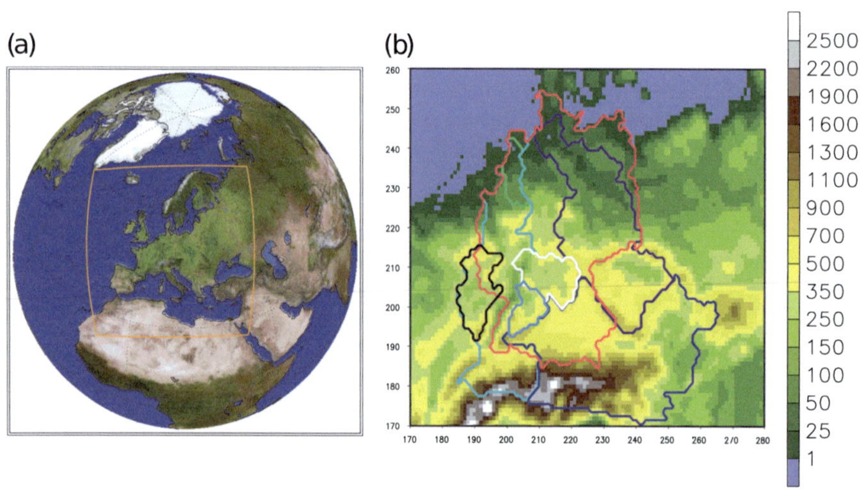

Fig. 1 EURO-CORDEX domain (**a**), and ReKliEs-De area of investigation with orography on 12 km resolution: Germany (*red*) and river catchments of Danube, Rhine, Elbe, Weser and Ems (colours) (**b**)

Table 2 Technical details of the WRF simulations from May 2015 to April 2016

Simulation	Nr. CPUs openMPI	Simulation period	Nr. of grid cells	Δt (s)	Status April 2016	Walltime	Nr. of simulations	Raw output size
EURO-CORDEX/ ReKIEs-De/ FOR 1696 0.44° historical	1536	01.01.1958–31.12.2005	129 × 139 × 50	180	Finalized	300 h	4	42 TB (210 GB/year)
EURO-CORDEX/ ReKIEs-De/ FOR 1696 0.44° RCP8.5/2.6	1536	01.01.2006–31.12.2100	129 × 139 × 50	180	Soon finalized	570 h	5	99 TB (210 GB/year)
EURO-CORDEX/ ReKIEs-De/ FOR 1696 0.11° historical	5400	01.01.1958–31.12.2005	452 × 460 × 50	60	Soon finalized	1200 h	4	700 TB (2.8 TB/year)
EURO-CORDEX/ ReKIEs-De/ FOR 1696 0.11° RCP8.5/2.6	5400	01.01.2006–31.12.2100	452 × 460 × 50	60	Expected end 2016	2280 h	5	1330 TB (2.8 TB/year)
FOR 1696 evaluation 0.44°	1536	01.01.1989–01.01.2014	129 × 139 × 50	180	Running; currently 1996	102 h	1	5,6 TB (210 GB/year)
FOR 1696 evaluation 0.11°	5400	01.01.1989–01.01.2014	452 × 460 × 50	60	Running; currently 1993	408 h	1	47,6 TB (2.8 TB/year)

3 Results

Since the model simulations are still running and the first results only became available recently, the following analyses is preliminary and shows only first results. Note that this section describes only exemplary the result of one GCM forcing one RCM. For a complete analyses of climate projections including bandwidth estimations it is essential to analyse the data from multi-model ensembles from different GCMs and RCMs as it will be done within ReKliEs-De.

3.1 Comparison of GCM and WRF: Temperature

To simulate the state of the atmosphere, a certain number of input variables are required by the WRF model. Among them, also the 3-D temperature field is needed at the lateral boundaries. Figure 2a displays the average temperature of the lowest model level from the coarse model resolution (~150 km) of the MPI- ESM-LR GCM for Europe. The temperature ranges from about 270 K (−3 °C) to 290 K (17 °C) from northern to southern Europe for the "historical" 30 year average 1971–2000. Figure 2b shows the near surface temperature simulated by the WRF model on 50 km resolution for the same period and area forced by the MPI-ESM-LR model (Fig. 2a). As mentioned above, only the EURO-CORDEX domain is simulated which is given on a rotated grid with lateral borders from around 25°N to 75°N

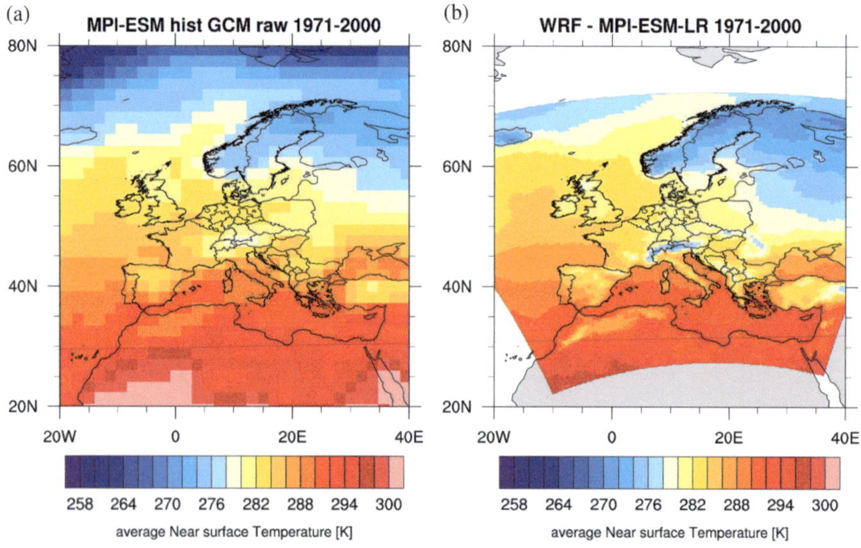

Fig. 2 Average near surface temperature from 1971-2000 from MPI-ESM-LR raw model output (**a**) and the downscaled simulation of MPI-ESM-LR from WRF (50 km) (**b**)

and 30°W to 50°E. In the downscaled simulation, the temperature ranges from around 268 K (−5 °C) in the north to 290 K (17 °C) in the southern part of the domain. The mountain elevations increase with model resolution, which explains that the temperature above Scandinavia and the Alpine region is lower, compared to the coarse model input. Due to the better orographic resolution especially of the Pyrenees, the alpine region and the scandinavian mountains in the WRF 50 km model, more details of the temperature pattern are indicated. This example represents pretty well, that GCMs are able to provide a basic representation of characteristics of the global climate on a large spatial scale. However, on the regional scale some important features are neglected due to a too coarse model resolution.

3.2 GCM-RCP and WRF-RCP Projections: Temperature

In Fig. 3, the simulated change of the average temperature is shown for the historical period 1971–2000 and the projection period 2071–2099 simulated with the MPI-ESM-LR for 2 GHG emission scenarios. The GCM raw model output is given by Fig. 3a, b whereas Fig. 3c, d depict the WRF downscaled simulations on 50 km grid resolution. Following the RCP2.6 on the coarse GCM grid (Fig. 3a), the temperature will increase by 0.5–1 °C in Europe. Considering the WRF simulation with the higher resolution, the temperature is increasing by 0.5 to ∼2 °C in the EURO-CORDEX domain. The coarse GCM model for RCP8.5 (Fig. 3b) gives a temperature increase of 1–3 °C whereas the downscaled projections on the 50 km grid reveal an increase of up to 4 °C in the southeastern part of Europe. The intensity of the simulated temperature increase for both scenarios is higher when considering the high resolution projections on 50 km resolution. Another distinctive feature of the RCM's (Fig. 3c, d) compared to the GCM's (Fig. 3a, b) is the indicated opposite sign of the warming showing a higher and in general a more intense increase of the average temperature in the eastern part of the domain than in the western part and over the Atlantic ocean.

3.3 WRF-RCP Projections: Precipitation

The annual average precipitation between 1971–2000 simulated by WRF on the 50 km grid and forced with MPI-ESM-LR data is presented in Fig. 4. In Germany the amount of annual precipitation varies from around 400 to 1000 l/year from the northeast to the southwest, which is in the magnitude range of observational data (see e.g. www.klimadiagramme.de). In the Alpine region in Germany, the precipitation can reach up to 1800 l/m^2. This demonstrates the general ability of WRF to simulate the amount of precipitation in Germany although WRF tends to overestimate precipitation in general [5].

The change of the annual mean precipitation from 1971–2000 to the future simulations for the period 2071–2099, according to the RCP projections is shown

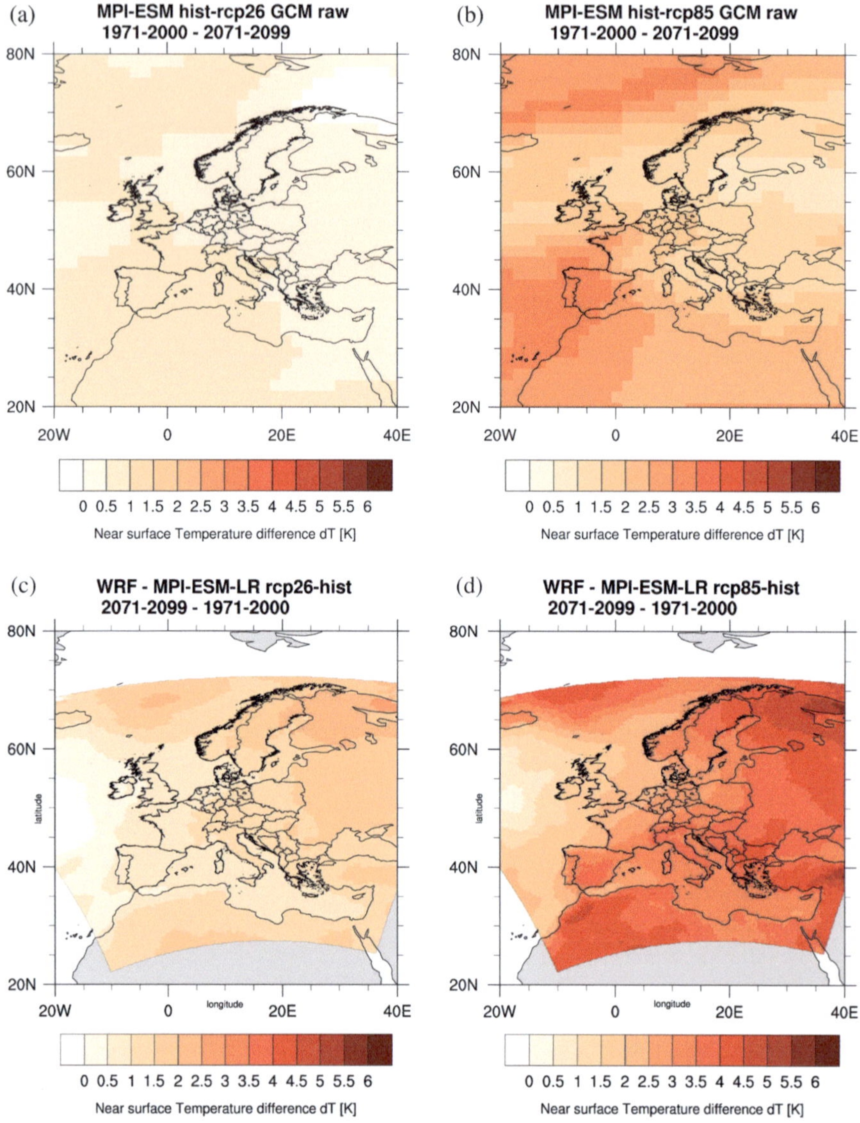

Fig. 3 Difference of near surface temperature from 1971–2000 and 2071–2099 of raw GCM output for RCP2.6 (**a**) and RCP8.5 (**b**). And average temperatures for the same period from the WRF model (50 km) forced by MPI-ESM-LR for RCP2.6 (**c**) and RCP8.5 (**d**)

in Fig. 5a for RCP2.6 and Fig. 5b for RCP8.5. The precipitation in Germany is projected to experience a slight decrease by about 50–100 l/year on the annual average regarding RCP2.6, a scenario with moderate increase of GHG emissions. For RCP8.5, where a rather strong increase of GHG emissions is assumed, the precipitation changes in Germany are projected to be moderate and smaller

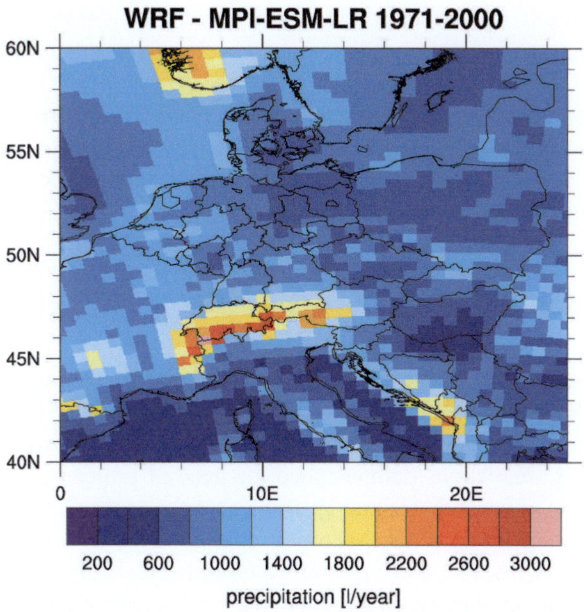

Fig. 4 Annual average precipitation from 1971–2000, simulated by WRF on the 50 km grid, forced with MPI-ESM-LR

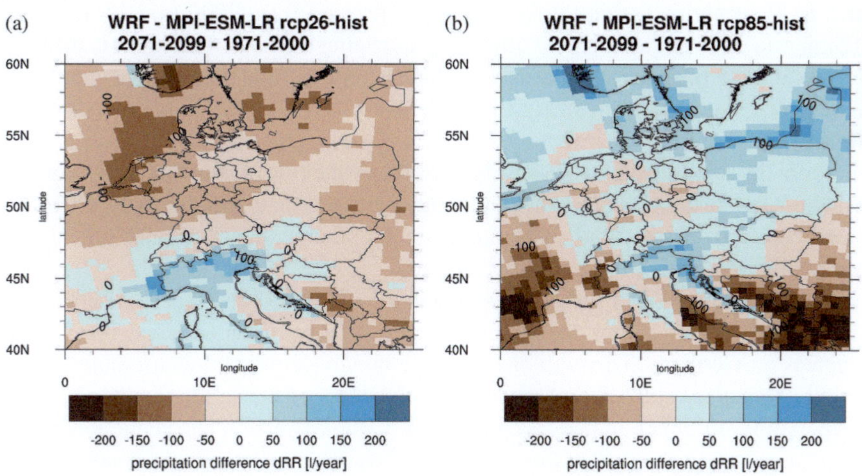

Fig. 5 Difference of the annual average precipitation from 1971–2000 and 2071–2099 simulated by WRF on 50 km grid. MPI-ESM-LR RCP2.6 forcing (**a**) and RCP8.5 forcing in (**b**)

compared to the RCP2.6 scenario varying from −50 to +50 l/year. Especially in the southeastern part of Europe a strong decrease of the annual precipitation is projected.

Although precipitation changes appear to be small for Germany on the annual average, on the seasonal scale the changes might be more intense and hence much more significant for the environment and the society than indicated by the annual average precipitation.

4 Conclusion and Outlook

The results highlighted some preliminary results of the downscaling of one GCM (MPI-ESM-LR) on the coarse model grid (∼150 km) with WRF to a refined grid (∼50 km) for a certain domain in Europe for 2 scenarios. The differences among GCM and RCM simulations demonstrated the need of projections using a higher spatial resolution in order to evaluate possible changes of the climate on a regional scale. The scope of the WRFCLIM project is to investigate the performance and benefit on a higher resolution of ∼12 km which is demanded by EURO-CORDEX. It is expected that simulations on a higher resolution improve their projection skill due to a better representation of the orographic effects. Especially metrics with a higher spatial variation will experience an improved representation e.g. like it was shown for the precipitation in[4] and [6].

High resolution simulations with WRF forced by 4 different GCM models and two RCP scenarios are currently beeing downscaled from 2000 to 2100 onto the 12 km grid. Due to variable and often long queueing times (2–7 days), it is difficult to predict the time the simulations are finished. The final projections should be realized by the end of the current year (2016) to fullfill ReKliEs-De objectives in time.

From the scientific point of view, besides the evaluation of the climate projections on the high resolution grid, also monthly and seasonal timescales need to be investigated in more detail as temperature or precipitation extremes are highly variable throughout the year within Germany.

The main objective of ReKliEs-De within the WRFCLIM project is the provision of "easy to use" special climate indices for end-users in order to assess the climate impacts on Germany. This will also be done by analyzing model ensembles, to be able to give also estimation of errors and robustness of the simulation results. The preparation of climate indices accompanied by the extraction and reduction of the simulation output and further investigations addressing the impact of climate change will be the main task within the next months

Acknowledgements This work is part of the ReKliEs-De project funded by the BMBF (Federal Ministry for Education and Research) and the Research Unit 1695 funded by the DFG (Deutsche Forschungsgemeinschaft). We are thankful for the support from the staff of the DKRZ (Deutsches Klimarechenzentrum), to be able to access GCM data. Computational Resources for the model simulations on the HLRS CRAY XC40 within WRFCLIM were kindly provided by HLRS. We would like to thank the staff for their great support.

References

1. Taylor, K.E., Stouffer, R.J., Meehl, G.A.: Bull. Am. Meteorol. Soc. **93**(4), 485 (2012)
2. Van Vuuren, D.P., Edmonds, J., Kainuma, M., Riahi, K., Thomson, A., Hibbard, K., Hurtt, G.C., Kram, T., Krey, V., Lamarque, J.F., et al.: Clim. Chang. **109**, 5 (2011)
3. Giorgi, F., Jones, C., Asrar, G.R., et al.: World Meteorol. Organ. (WMO) Bull. **58**(3), 175 (2009)
4. Warrach-Sagi, K., Schwitalla, T., Wulfmeyer, V., Bauer, H.S.: Clim. Dyn. **41**(3–4), 755 (2013)
5. Kotlarski, S., Keuler, K., Christensen, O.B., Colette, A., Déqué, M., Gobiet, A., Goergen, K., Jacob, D., Lüthi, D., van Meijgaard, E., Nikulin, G., Schär, C., Teichmann, C., Vautard, R., Warrach-Sagi, K., Wulfmeyer, V.: Geosci. Model Dev. **7**(4), 1297 (2014)
6. Prein, A., Gobiet, A., Truhetz, H., Keuler, K., Goergen, K., Teichmann, C., Maule, C.F., van Meijgaard, E., Déqué, M., Nikulin, G., et al.: Clim. Dyn. **46**(1–2), 383 (2016)
7. Skamarock, W.C., Klemp, J.B., Dudhia, J., Gill, D.O., Barker, D.M., Wang, W., Powers, J.G.: A description of the advanced research wrf version 2. Technical report, DTIC Document (2005)
8. Chen, F., Dudhia, J.: Mon. Weather Rev. **129**, 569 (2001)
9. Morrison, H., Thompson, G., Tatarskii, V.: Mon. Weather Rev. **137**, 991 (2009)
10. Hong, S.Y., Noh, Y., Dudhia, J.: Mon. Weather Rev. **134**, 2318 (2006)
11. Kain, J.S.: J. Appl. Meteorol. **43**, 170 (2004)
12. Collins, W.D., Rasch, P.J., Boville, B.A., Mc Caa, J.R., Williamson, D.L., Kiehl, J.T., Briegleb, B., Bitz, C., Lin, S.J., Zhang, M., Dai, Y.: Description of the NCAR community atmosphere model (cam 3.0), 226pp. NCAR technical Note NCAR/TN-464+STR, NCAR, Boulder (2004)

Biogeophysical Impacts of Land Surface on Regional Climate in Central Vietnam

Ngoc Bich Phuong Nguyen, Harald Kunstmann, Patrick Laux, and Johannes Cullmann

Abstract The biogeophysical impacts of land surface on regional climate have been investigated for the Vu Gia-Thu Bon basin in Central Vietnam using the regional climate model (RCM) Weather Research and Forecasting (WRF) Model. The replacement of land surface due to an updated land-use/land cover data leads to change the biogeophysical properties of land surface, thereby altering the regional climate. Results show that generally surface air temperatures increase by about 0.5 °C over the basin. Remarkable increases in surface air temperatures are about 2 °C appearing in cities. Annual precipitation decreases by about 800 mm over the Western basin and increases by about 1500 mm over the Southern basin. In general, if roughness length decreases (increases), horizontal wind speed and average maximum Convective Available Potential Energy (CAPE) increase (decrease) by about 1 ms^{-1} and 20 JKg^{-1}.

1 Introduction

Many investigations have shown that human-induced land surface change impacts strongly on climate at local and regional climate [2, 6, 12]. The impacts are a result of alterations in biogeophysical processes of land surface. In this study, we assess the biogeophysical impacts of land surface on the regional climate of the Vu Gia-Thu Bon (VGTB) basin of Central Vietnam.

N.B.P. Nguyen (✉)
IHP/HWRP Secretariat, Federal Institute of Hydrology, Am Mainzer Tor 1, 56058 Koblenz, Germany
e-mail: nn.bichphuong@gmail.com

H. Kunstmann • P. Laux
Karlsruhe Institute of Technology (KIT), Institute for Meteorology and Climate Research, Atmospheric Environmental Research (IMK-IFU), Kreuzeckbahnstrasse 19, 82467 Garmisch-Partenkirchen, Germany
e-mail: harald.kunstmann@kit.edu; patrick.laux@kit.edu

J. Cullmann
Climate and Water Department, World Meteorological Organization, 7bis, avenue de la Paix, CP No. 2300, CH-1211 Geneva 2, Switzerland
e-mail: dhwr@wmo.int

W.E. Nagel et al. (eds.), *High Performance Computing in Science and Engineering '16*, DOI 10.1007/978-3-319-47066-5_40

The land use/land cover (LULC) representation in regional climate models (RCMs) is crucial to simulate the interaction between the land and the atmosphere; however, LULC data is often inaccuracy due to out-of-date data and insufficient resolution [e.g. 7, 19]. Therefore, it is necessary to update the LULC data for more accurate in representing the interactions. In this study, we have used the RCM WRF model to simulate the interaction between the land and the atmosphere over the VGTB basin. The LULC representation in the WRF model has been updated using the LULC data acquired from Land Use and Climate Change interaction (LUCCi) project. The updated LULC data leads to alter the biogeophysical properties of land surface i.e. surface heat fluxes and roughness length, resulting in changes in the regional climate.

This study continues analysing the impacts of updated LULC data for the Vu Gia-Thu Bon basin in Central Vietnam in the previous study [14]. This study focuses the biogeophysical impacts of land surface on the regional climate.

2 Data and Methods

To assess the influences of LULC replacements on the regional climate, two ensembles were performed based on the default LULC data in the WRF model (referred to as WRF LULC-default) and the updated LULC data for the VGTB basin (referred to as the WRF LULC-LUCCi) with the boundary condition ERA-interim. The setting for the WRF model has followed the setting in [10, 14], which identified a suitable setup of WRF physical parametrizations for Southeast Asian. Each ensemble includes five simulations starting with 1-day lagged in January-2009 for generating slight disturbances in initial condition. The ensembles allow to examine the alteration of the regional climate owing to LULC replacements with measurable amount of dispersion arisen by a strong non-linearity of natural variability in the atmosphere [e.g. 8, 11]. To analyse the alteration of seasonal and annual variability due to the updated LULC data, the sets of simulations were conducted for 6 years from 2009 to 2015, in which the year 2009 was considered as a model spin-up time. In the spin-up time, the model achieves thermal and hydrological equilibrium between the land and the atmosphere; therefore, the spin-up time was disregarded in analysing results. The ensemble based on the WRF LULC-default is referred as Control and the ensemble based on the WRF LULC-LUCCi is referred to as CASE-2010.

Figure 1 shows the replacement of LULC due to the updated LULC data. Woodland areas replaced by mixed forest, cropland and grassland are about 31, 15 and 18 % area of the VGTB basin, respectively. About 12 % area is the replacement of cropland by mixed forest. Broad-leaf forest is replaced by grassland, accounting for about 5 % area. About 10 % area has no change and about 9 % area is other LULC replacements. In most of the Eastern basin, woodland is replaced by cropland. A large area of broad-leaf forest is replaced by grassland in the Western basin. In the Northern and Southern VGTB basin, cropland is mainly replaced by mixed forest.

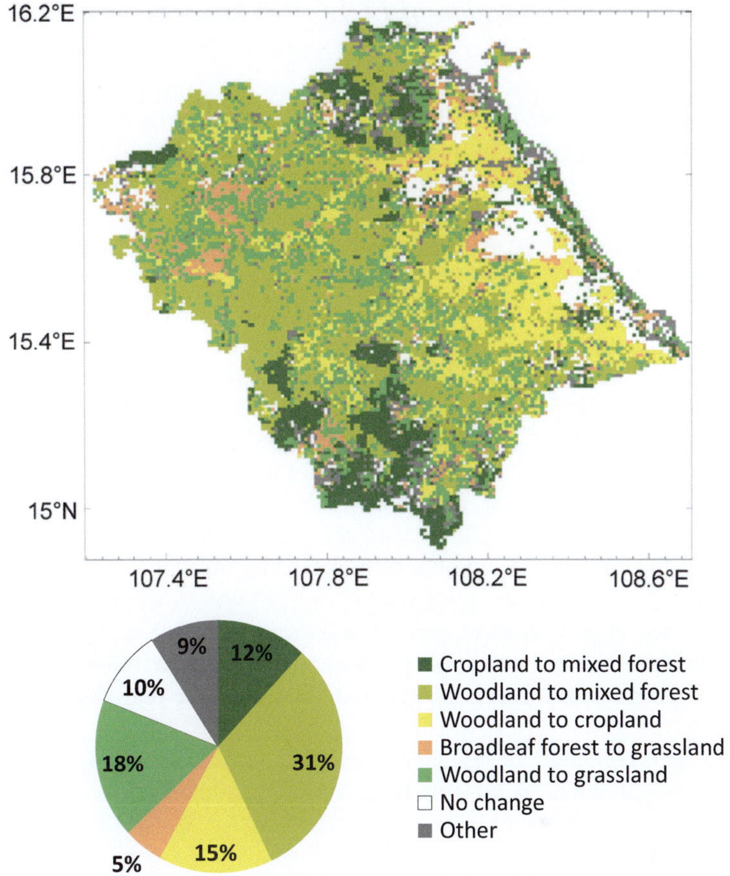

Fig. 1 Converted LULC types from WRF LULC-default to WRF LULC-LUCCi. Percentage changes for the VGTB basin

3 Results

The types of LULC replacement (e.g. forest by grassland, woodland by cropland) identify the alteration in physical properties of land surface, affecting on climate through changes in heat, moisture and momentum. Changes in heat and moisture fluxes are expressed by changes in albedo and partitioning in latent and sensible heat flux. The changes in momentum are exhibited by changes in roughness length. After updating LULC map for the VGTB basin in the WRF model, there are five main types of LULC replacements over the basin, i.e. the replacement of woodland by cropland, broad-leaf forest by grassland, cropland by mixed forest, woodland by mixed forest and woodland by grassland (Fig. 1).

3.1 Temporal Modification of LULC Physical Properties

The changes in monthly surface heat fluxes were analysed for all types of the LULC replacements (Fig. 2). Results were drawn by averaging all grid points representing LULC replacement type in the study area. In general, ground heat fluxes were two orders of magnitude less than latent and sensible heat fluxes. The modification of ground heat fluxes was nearly zero value. The changes in latent and sensible heat fluxes are in a range of −20 to 20 $\mathrm{Wm^{-2}}$ depending on characteristics of the LULC replacements.

Figure 2a shows that the woodland replaced by cropland was a reason for increased latent heat fluxes in dry season (March to August). The maximum increased latent heat flux was about 10 $\mathrm{Wm^{-2}}$ in June. The increased latent heat fluxes tended to increase evapotranspiration and then resulted in surface cooling. However, the replacement caused decreased sensible heat fluxes during most of year due to the increase in albedo and the decrease of roughness length. The maximum decrease in sensible heat flux was about 15 $\mathrm{Wm^{-2}}$ appearing in April and May. The reduction in sensible heat fluxes were the cause of decreased heat loss into the atmosphere, thereby suppressing the surface cooling. In rainy season (October to November), latent heat fluxes decreased by about 10 $\mathrm{Wm^{-2}}$, while a weak increase in sensible heat fluxes found in these months was about 2 $\mathrm{Wm^{-2}}$. The reduction of the heat fluxes tended to warm the surface in rainy season. In general, the total changes in the turbulent heat fluxes were reduced during the year, thereby warming the surface.

The larger reduction of roughness length and vegetation greenness when broad-leaf forest replaced by grassland has a remarkable effect on decreased

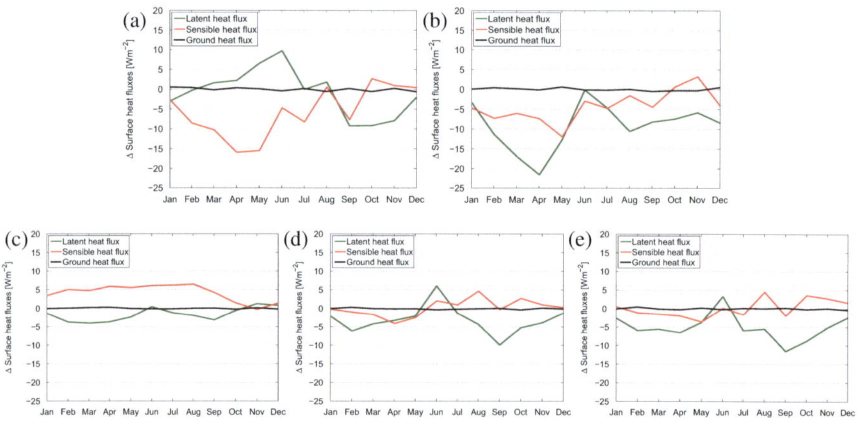

Fig. 2 Alteration of monthly heat fluxes based on LULC conversion types over the VGTB basin (WRF LULC-LUCCi *minus* WRF LULC-default). (**a**) *Δ* Woodland to cropland. (**b**) *Δ* Broadleaf forest to grassland. (**c**) *Δ* Cropland to mixed forest. (**d**) *Δ* Woodland to mixed forest. (**e**) *Δ* Woodland to grassland

evapotranspiration and increased albedo (Fig. 2b). Consequently, latent and sensible heat fluxes were reduced in the year, except a weak increase in sensible heat flux of about 3 Wm^{-2} in November. The maximum decrease in surface heat fluxes was about 30 Wm^{-2} observed in April. The decreases in total heat fluxes is the cause of warming surface and reducing convective cloud, indicating a decrease in precipitation.

Cropland replaced by mixed forest has an opposite changes in physical properties with the replacement of woodland by cropland or broad-leaf forest to grassland. Figure 2c indicates that sensible heat fluxes increased due to the decrease in albedo and warmer surface. Latent heat fluxes were reduced during the year; as a result, evapotranspiration was reduced, causing the surface warming. Latent heat fluxes decreased by about 3 Wm^{-2} in most of year, except no changes in June, November and December. Sensible heat fluxes increased by about 5 Wm^{-2} from January to September and fluctuated around zero value in November and December.

The replacement of woodland by mixed forest (Fig. 2d) and woodland by grassland (Fig. 2e) are the main LULC replacements over the VGTB basin, accounting for 31 % and 18 %, respectively. The alteration of heat fluxes due to these conversions are similar. Latent heat fluxes were reduced during the year except June with increases of about 5 Wm^{-2}. The maximum decrease in latent heat flux was about 10 Wm^{-2} in September. The decreased latent heat fluxes might result in a surface warming. A weak decrease in sensible heat fluxes was found from January to May. From June to December, sensible heat fluxes were in a range of –5 to 5 Wm^{-2}.

The changes in the surface heat fluxes shows a clear seasonality characterized by a dry season from February to August and rainy season from September to November. The alteration of surface heat fluxes may affect seasonal variability of climate variables.

3.2 Spatial Modification of LULC Physical Properties

Figure 3 shows the alteration of roughness length and albedo, as well as the changes in annual soil moisture and surface heat fluxes because of the updated LULC data in the VGTB basin. In the Eastern lowland basin, most of woodland was replaced by cropland, thereby decreasing roughness length and increasing albedo. The decrease in roughness length has a warming effect, while the increase in albedo tend to cool the surface owing to decreasing net absorb radiation. The transfer of sensible heat fluxes into the atmosphere was reduced by the cooling surface. While, latent heat fluxes had no changes in this LULC replacement. As a result, total turbulent heat fluxes decreased, causing less surface heat fluxes away from the surface, causing a warming effect. Sensible heat fluxes decreased by about 14 Wm^{-2} and surface soil moisture decreased by about 0.1 m^3m^{-3}.

In the Western basin, there is a large area of broad-leaf forest replaced by grassland similar to deforestation. Remarkable decreases in roughness length and evapotranspiration caused to warm the surface, while the increases in albedo had

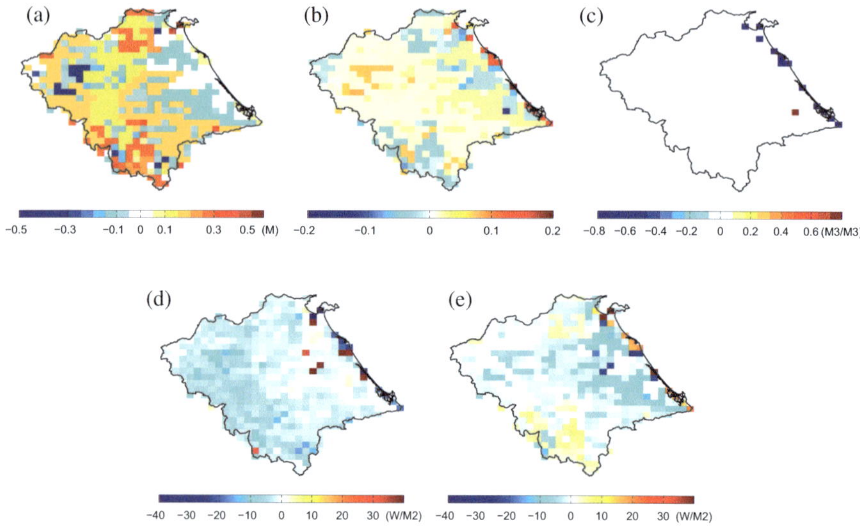

Fig. 3 Alteration of annual roughness length, albedo, soil moisture and heat fluxes over the VGTB basin due to updated LULC map (WRF LULC-LUCCi *minus* WRF LULC-default). (**a**) Δ Roughness length. (**b**) Δ Albedo. (**c**) Δ Soil moisture at surface. (**d**) Δ Latent heat flux. (**e**) Δ Sensible heat flux

the effect of cooling. The roughness length was remarkably reduced by 0.5 and the albedo increased about 0.09. The replacement led to decrease both latent and sensible heat fluxes. The decrease in latent heat fluxes were greater than the decrease in sensible heat fluxes about 5 Wm^{-2}. Sensible heat fluxes reduced by about 15 Wm^{-2}. The decreases in turbulent heat fluxes tended to warm the land surface and decrease precipitation (Fig. 5). There was a decrease in soil moisture of about 0.1 m^3m^{-3} observed over the area. This result is consistent with the surface warming and less precipitation due to deforestation [e.g. 5, 9, 16].

In the Northern and Southern basin, roughness length was increased due to the replacement of cropland by mixed forest, resulting in decreased albedo. Changes in albedo and roughness length tended to increase sensible heat fluxes, cooling the land surface. However, the replacement of cropland was the result of decreases in latent heat fluxes, consequence in less evapotranspiration and warming the surface. The modifications acted to increase slightly surface air temperatures as a result of the competition between the warming due to the decreased albedo and evapotranspiration, and the cooling due to the increased sensible heat fluxes.

Although the main LULC replacements in the Central basin and scattered over the highland of VGTB basin are woodland replaced by mixed forest (31 % area of the VGTB basin) and woodland replaced by grassland (18 % area of the VGTB basin), the replacement was no significant effects on the changes in albedo and surface heat fluxes. Consequently, there were no significant changes in surface air temperature. The albedo, soil moisture and sensible heat fluxes stayed in the same levels. There were weak decreases in latent heat fluxes in a range of 2–5 Wm^{-2}.

Some areas along the coastline were replaced by urban land. Although the extension of urban area was about 1 % area of the basin, the extension was the causes of remarkable increases in sensible heat fluxes and considerable decreases in latent heat fluxes. Consequently, surface air temperature was strongly affected.

4 Impacts of Updated LULC on Climate Variables

The LULC replacements over the VGTB basin result in the alteration of physical properties of land surface, thereby changes in the climate variables. The role of LULC on regional climate as a first-order climate forcing has been well-documented [e.g. 1, 3, 9].

4.1 Temporal Impacts of Updated LULC on Climate Variables

Figure 4 exhibits the alteration of climate variables due to the updated LULC data. Soil moisture at the first level (10 cm), which dominates the variability of the latent heat fluxes, experienced in decreases in a range of about 0.03 (4 %)–0.03 m^3m^{-3} (17 %) for all LULC replacement types in the VGTB basin. The replacement of broad-leaf forest by grassland caused the largest decrease in soil moisture because of the reduction in convective cloud. The decreased soil moisture in the dry months (April and May) was various among LULC replacement types, while the decreased soil moisture stayed in the same level with a decrease of about 15 % for most of LULC replacements. This evidence indicates the changes in soil moisture depending on LULC types and seasonal variability.

The changes in surface air temperature indicated a relation with the modification of surface heat fluxes (Fig. 4b). Overall, all replacements tended to increase surface air temperature of about 0.3 °C during the year The increased surface air temperature results from the decreases in net surface heat fluxes. In May, these replacements have less increases in surface air temperature than the other months due to the increased sensible heat fluxes. In October, surface air temperature increased in a range of 0.3–0.5 °C. The changes of surface air temperature depended on the type of LULC replacements. The maximum increase in surface temperature was 0.5 °C observed in October when replacing woodland by grassland.

Precipitation exhibited a decrease of about 50 mm in September and an increase of about 50 mm in October, and fluctuation in ±20 mm in the other months for all LULC replacements. From January to August, precipitation altered about 50 mm for all LULC replacements. However, precipitation altered about 20 % of the Control in precipitation amount for these months.

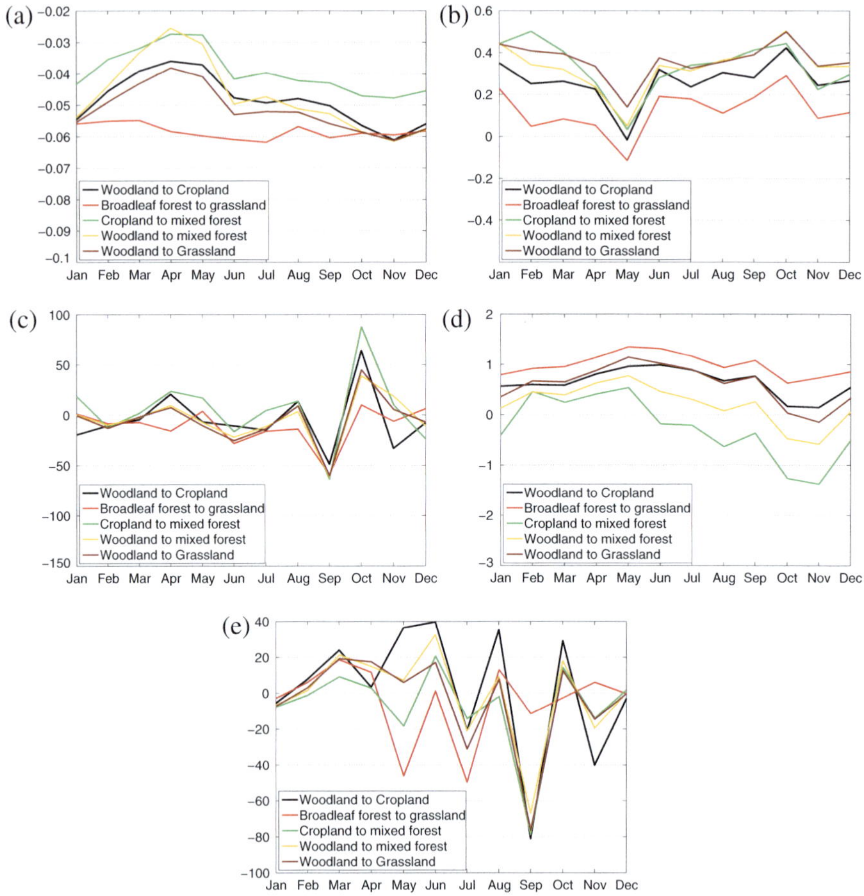

Fig. 4 Alteration of climate variables based on LULCC types over the VGTB basin due to the updated LULC map (WRF LULC-LUCCi *minus* WRF LULC-default). (**a**) Δ Soil moisture (m^3m^{-3}). (**b**) Δ Surface temperature (°C). (**c**) Δ Precipitation (mm). (**d**) Δ Wind speed (ms^{-1}). (**e**) Δ Maximum CAPE (JKg^{-1})

Figure 4d shows that the changes in wind speed were the result from the alteration of roughness length. The replacement of broad-leaf forest by grassland resulted in a strengthening of wind speed during the year in a range of 0–1.2 ms^{-1}, while the replacement of cropland by mixed forest caused decreases in wind speed of about 1 ms^{-1} from August to November.

The alteration of surface heat fluxes and surface energy balance tends to modification the structure and stability of the overlying troposphere, which can be exhibited through the convective available potential energy (CAPE). To provide insight into the genesis and intensity of convection, averaged maximum CAPE was calculated. Magnitude of the CAPE are much smaller than what would be expected for an individual storm due to regardless of weather conditions [17]. The changes in monthly mean of daily maximum CAPE are shown in Fig. 4e. All replacements led to increase in maximum CAPE in March about 20 JKg^{-1} (20 % of the Control). In the other months, the maximum CAPE were in a range of −80 to 40 JKg^{-1}. However, there was only change about 10 % of the Control.

4.2 Spatial Impacts of Updated LULC on Climate Variables

The consequences of the updated LULC data on climate variables are shown in Fig. 5. Overall, the updated LULC caused increases in surface air temperatures of about 0.8 °C over the VGTB basin due to the reduction in turbulent heat fluxes. The impacts of land forcing on the regional climate is dominated by the modification in turbulent heat fluxes. This result is consistent with previous studies, in which changes in albedo dominate the climate response in temperate regions while the

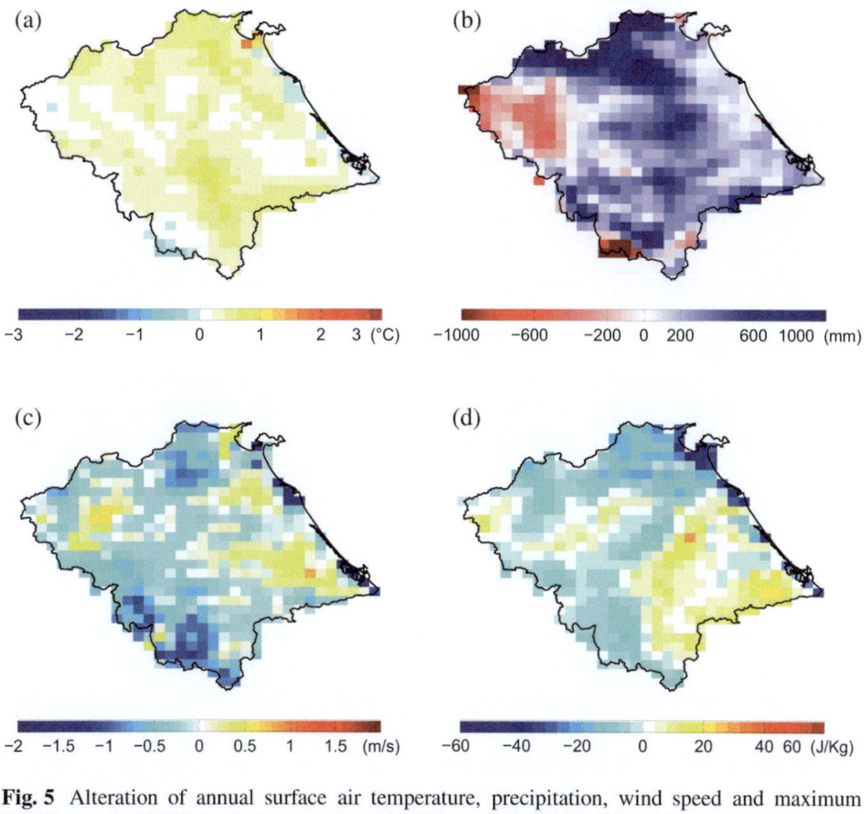

Fig. 5 Alteration of annual surface air temperature, precipitation, wind speed and maximum CAPE over the VGTB basin due to updated LULC map (WRF LULC-LUCCi *minus* WRF LULC-default). (**a**) Δ Surface air temperature. (**b**) Δ Precipitation

changes in evapotranspiration, heat fluxes and roughness length drive in the tropics [5, 9, 16].

In the Eastern basin, the decrease in surface heat fluxes due to decrease in roughness length tended to warm the surface, while the increase in albedo tended to cooling the surface. The net influences were a slight warming over the area. Surface air temperatures increased by about 0.3 °C. The alteration in precipitation was not directly impacted by the land forcing. Precipitation increased in a range of 50–200 mm. Wind speed shows the conjunction with the decreases in roughness length. The wind speed increased by about $0.5\,\mathrm{ms^{-1}}$ over the area. Convective available potential energy (CAPE) is the amount of energy a parcel of air would have if lifted a certain distance vertically through the atmosphere. The maximum CAPE in a day represents the unstable atmosphere in magnitude. The maximum CAPE increased by about $18\,\mathrm{JKg^{-1}}$.

In the Western basin, broad-leaf forest replaced by grassland and cropland resulted in a decrease of convective cloud due to reduction in surface heat fluxes. Precipitation decreased by about 600 mm over this area. The sensitivity of deforestation in tropical area to convective cloud and precipitation was mentioned in a large number of studies [e.g. 4, 18, 20]. These studies demonstrated that deforestation is the cause of a decrease in precipitation in a wide range from 1 to 20 % of the Control. The decreased precipitation depends on topographic effects and natural spatial variability of precipitation. [18] indicated that the east Asian summer monsoon is sensitive to deforestation in the Indo-China region. The replacement results in considerable increases in albedo and remarkable decreases in turbulent heat fluxes. Two drivers are in the opposite direction, leading a slight warming over the area. Wind speed and maximum CAPE were impacted by the reduction in roughness length. The wind speed increased in a range of 0.5–$0.8\,\mathrm{ms^{-1}}$ and maximum CAPE increased by about $10\,\mathrm{JKg^{-1}}$.

In the Northern and Southern basin, although the replacement of cropland by mixed forest led to increase sensible heat fluxes, surface air temperature increased slightly due to the decreases in latent heat fluxes and albedo. Some studies indicated that the replacement of cropland to forest lead to decrease near surface temperature and increase in latent heat fluxes [e.g. 13, 15]. However, water available for evapotranspiration is not efficiency during dry season, results in decreases in latent heat fluxes, thereby warming the land surface. At the Da Nang city, surface air temperature experienced an increase of about 1.5 °C because of the extended urban. Precipitation increased by about 1,000 mm over these areas, which may be associate with the increased surface air temperature and increased sensible heat fluxes into the atmosphere. Due to the increase in roughness length, wind speed and maximum CAPE were decreased by about $1.5\,\mathrm{ms^{-1}}$ and $20\,\mathrm{JKg^{-1}}$, respectively. The decreases in wind speed and maximum CAPE were a cause of increased sensible heat fluxes and surface temperature over these areas.

Table 1 Number of simulated months using the regional climate model WRF

Experiments	Perturbed runs	Months for perturbed runs	Total
WRF LULC-default	5	72	360
WRF LULC-LUCCi	5	72	360
			720 simulated months

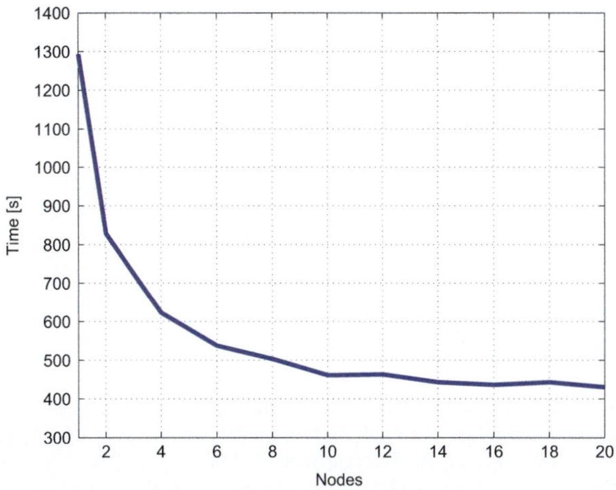

Fig. 6 Benchmark of the WRF simulating the 2 days at ForHLR

5 CPU Usage and Storage Capacities for This Study

To investigate the biophysical impacts of land surface on the regional climate in the VGTB basin in Central Vietnam. This study has used the RCM WRF model to dynamically downscale global climate information to the study area with the boundary condition ERA-interim. All simulations have been performed for the period from 2009 to 2014. The number of simulated months is presented in the Table 1.

According to the Fig. 6, simulating WRF with 10 nodes (200 CPUs) is reasonable. Therefore, the simulation were performed using 200 CPUs for each month. The computing time for the three domains for 1 month was 4 h which results in $4 \times 200 = 800$ CPUh per month. This means that for computing for 720 months $800 \times 720 = 576{,}000$ CPUh were consumed.

Acknowledgements This research is funded by the Federal Ministry of Education and Research (research project: Land Use and Climate Change Interactions in Central Vietnam (LUCCi), reference number 01LL0908C). The provision of CPU and storage capacities at Karlruhe Institute of Technology (KIT), Steinbuch Centre for Computing (SCC) and Karlruhe Institute of Technology (KIT), Institute of Meteorology and Climate Research (IMK-IFU) is highly acknowledged.

References

1. Bonan, G.B.: Forests and climate change: forcings, feedbacks, and the climate benefits of forests. Science **320**, 1444–1449 (2008)
2. Caldas, M.M., Goodin, D., Sherwood, S., Campos Krauer, J.M., Wisely, S.M.: Land-cover change in the Paraguayan Chaco: 2000–2011. J. Land Use Sci. **10**, 1–18 (2015)
3. Charney, J., Quirk, W.J., Chow, S.-H., Kornfield, J.: A comparative study of the effects of albedo change on drought in semi-arid regions. J. Atmos. Sci. **34**, 1366–1385 (1977)
4. Costa, M.H., Pires, G.F.: Effects of Amazon and Central Brazil deforestation scenarios on the duration of the dry season in the arc of deforestation. Int. J. Climatol. **30**, 1970–1979 (2010)
5. Davin, E.L., de Noblet-Ducoudré, N.: Climatic impact of global-scale deforestation: radiative versus nonradiative processes. J. Clim. **23**, 97–112 (2010)
6. Deng, X., Zhao, C., Yan, H.: Systematic modeling of impacts of land use and land cover changes on regional climate: a review. Adv. Meteorol. **2013**, 1–11 (2013)
7. Ezber, Y., Lutfi Sen, O., Kindap, T., Karaca, M.: Climatic effects of urbanization in Istanbul: a statistical and modeling analysis. Int. J. Clim. **27**, 667–679 (2007)
8. Giorgi, F., Bi, X.: A study of internal variability of a regional climate model. J. Geophys. Res. Atmos. (1984–2012) **105**, 29503–29521 (2000)
9. Kvalevåg, M.M., Myhre, G., Bonan, G., Levis, S.: Anthropogenic land cover changes in a GCM with surface albedo changes based on MODIS data. Int. J. Climatol. **30**, 2105–2117 (2010)
10. Laux, P., Lorenz, C., Thuc, T., Ribbe, L., Kunstmann, H., et al.: Setting up regional climate simulations for Southeast Asia. In: High Performance Computing in Science and Engineering, vol. 12, pp. 391–406. Springer, Berlin/New York (2013)
11. Lee, S.-J., Berbery, E.H.: Land cover change effects on the climate of the La Plata Basin. J. Hydrometeorol. **13**, 84–102 (2012)
12. Mahmood, R., Pielke, R.A., Hubbard, K.G., Niyogi, D., Dirmeyer, P.A., McAlpine, C., Carleton, A.M., Hale, R., Gameda, S., Beltrán-Przekurat, A., et al.: Land cover changes and their biogeophysical effects on climate. Int. J. Climatol. **34**, 929–953 (2014)
13. Nagendra, H., Southworth, J.: Reforesting landscapes: linking pattern and process, vol. 10. Springer, Dordrecht (2009)
14. Nguyen, N.B.P., Laux, P., Cullmann, J., Kunstmann, H.: High performance computing in science and engineering 15: Transactions of the High Performance Computing Center, Stuttgart (HLRS) 2015. In: Do We Have to Update the Land-Use/Land-Cover Data in RCM Simulations? A Case Study for the Vu Gia-Thu Bon River Basin of Central Vietnam, pp. 623–635. Springer, Berlin/New York (2016)
15. Pielke, R.A., Pitman, A., Niyogi, D., Mahmood, R., McAlpine, C., Hossain, F., Goldewijk, K.K., Nair, U., Betts, R., Fall, S., et al.: Land use/land cover changes and climate: modeling analysis and observational evidence. Wiley Interdiscip. Rev. Clim. Change **2**, 828–850 (2011)
16. Pongratz, J., Reick, C., Raddatz, T., Claussen, M.: Biogeophysical versus biogeochemical climate response to historical anthropogenic land cover change. Geophys. Res. Lett. **37**(8) (2010). doi:10.1029/2010GL043010
17. Riemann-Campe, K., Fraedrich, K., Lunkeit, F.: Global climatology of convective available potential energy (CAPE) and convective inhibition (CIN) in ERA-40 reanalysis. Atmos. Res. **93**, 534–545 (2009)
18. Sen, O.L., Wang, Y., Wang, B.: Impact of Indochina deforestation on the East Asian summer monsoon*. J. Clim. **17**, 1366–1380 (2004)
19. Sertel, E., Robock, A., Ormeci, C.: Impacts of land cover data quality on regional climate simulations. Int. J. Climatol. **30**, 1942–1953 (2010)
20. Wang, J., Chagnon, F.J., Williams, E.R., Betts, A.K., Renno, N.O., Machado, L.A., Bisht, G., Knox, R., Bras, R.L.: Impact of deforestation in the Amazon basin on cloud climatology. Proc. Natl. Acad. Sci. **106**, 3670–3674 (2009)

Reducing the Uncertainties of Climate Projections: High-Resolution Climate Modeling of Aerosol and Climate Interactions on the Regional Scale Using COSMO-ART: Interaction of Mineral Dust with Atmospheric Radiation over West-Africa

Bernhard Vogel, Hans-Juergen Panitz, and Heike Vogel

Abstract Aim of this project is to investigate the impact of aerosol in high resolution climate runs. At the moment aerosols and their interactions with radiation and clouds represent one of the major uncertainties in our understanding of the climate system as they can be described only roughly in coarse resolution global models. The online coupled comprehensive chemistry model system COSMO-ART already showed in several case studies the potential of closing this gap. In order to apply it on decadal climate time scales the use of high performance computing becomes a necessity. In this study we quantified the effect of replacing the aerosol climatology usually used in regional climate simulations with CLM by online calculated dust concentrations. The model domain covered the DEPARTURE region. Only radiation feedback was accounted for neglecting aerosol cloud interactions. Interactive dust improved the agreement of simulated precipitation in comparison with observations.

1 Motivation

Aerosols and their interactions with radiation and clouds represent one of the major uncertainties in our understanding of the climate system. While on the global scale a lot of modeling activity is going on to narrow this uncertainty there is not to much effort visible on the regional climate scale. On the other hand due to the coarse spatial resolution of global models the aerosol and micro physical processes and their interaction are only roughly described. The online coupled comprehensive chemistry model system COSMO-ART already showed in several case studies the potential of closing this gap. We used this model system to perform a sensitivity

B. Vogel (✉) • H.-J. Panitz • H. Vogel
Institute of Meteorology and Climate Research – Department Troposphere Research, Karlsruhe Institute of Technology (KIT), Kaiserstraße 12, 76131 Karlsruhe, Germany
e-mail: bernhard.vogel@kit.edu; hans-juergen.panitz@kit.edu; heike.vogel@kit.edu

W.E. Nagel et al. (eds.), *High Performance Computing in Science and Engineering '16*, DOI 10.1007/978-3-319-47066-5_41

study quantifying the effect of different mineral dust climatology available from literature and of online calculated dust and its feedback with radiation followed by altering precipitation over selected domains within Africa. This elucidates the relative importance of online coupled aerosol radiation interactions on the regional scale climate.

2 The Model System COSMO-ART

In order to quantify the feedback processes between aerosols and the state of the atmosphere on the continental to regional scale the fully online integrated model system COSMO-ART with two-way interactions between different atmospheric processes has been developed [1–3]. The operational weather forecast model COSMO of the Deutscher Wetterdienst [4] was extended to treat secondary aerosols as well as directly emitted components like soot, mineral dust, sea salt and biological material and their feedback with radiation and clouds. The gas phase chemistry module (RADMKA) is based on RADM2 and includes several improvements. We updated rate constants according to IUPAC, updated the mechanism concerning biogenic VOCs, made extensions for the hydrolysis of N2O5 and included new sources for HONO. The KPP mechanism can be used for a flexible modification of the chemical mechanism. COSMO-ART uses the modal approach to describe the size distribution. New particles can be formed by nucleation of sulfuric acid. The processes condensation, coagulation, sedimentation, and washout are taken into account. A volatility basic set approach is used to describe the secondary organic aerosol [5]. The thermodynamic module ISORROPIAII [6] is applied. Emissions of mineral dust are calculated at each grid point and each time step for three individual modes, depending on the simulated friction velocity and surface parameters [7]. Figure 1 shows the feedback processes that are realized in COSMO-ART.

3 Sensitivity Study

3.1 Model Set Up

We carried our three individual model runs for the model domain shown in Fig. 2. In two of them (Tanre and AeroCom) different dust climatology were used to calculate the radiative fluxes. In one simulation (ART) the online calculated dust concentrations were used to calculate the optical properties of the mineral dust. Aerosol cloud interactions were not accounted for in all three model runs. For each scenario a time period of 10 years was simulated.

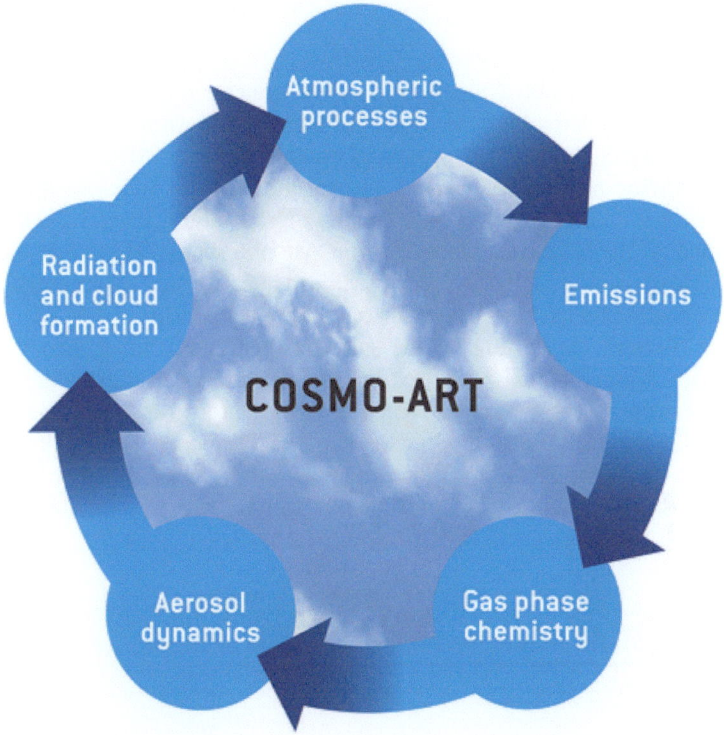

Fig. 1 Feedback processes realized in COSMO-ART

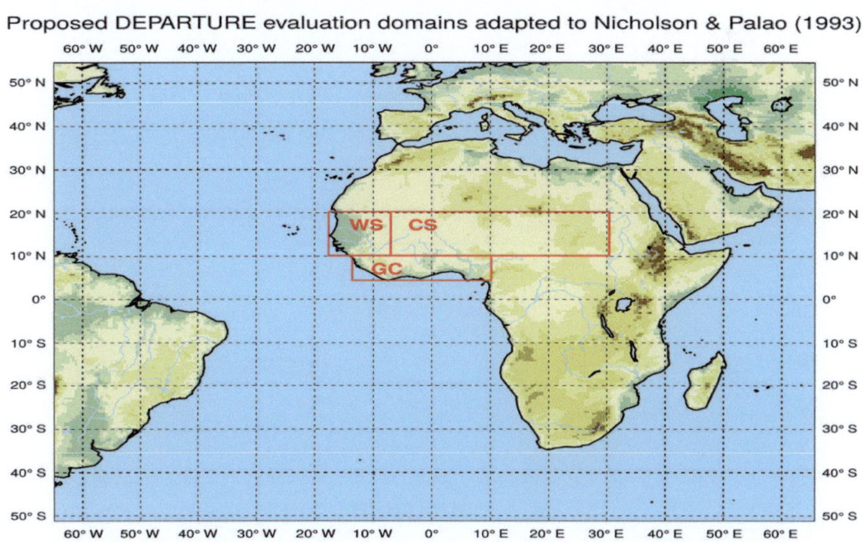

Fig. 2 Model domain. The *red boxes* indicate subdomains that were used for the evaluation of the model results. *CS* = Central Sahel, *WS* = Western Sahel, *GC* = Guinea coast

3.2 Model Results

Although we have enabled the interaction of mineral dust with radiation only and disabled the interaction with cloud formation we have to expect changes of precipitation which is a key meteorological quantity for West Africa. This is caused by the following process chain. Dust aerosol modifies the radiative fluxes. This induces temperature changes. Caused by these temperature changes the flow field is modified on the local scale as well as on synoptic scales. This alters the cloud formation and consequently the precipitation. Here, we will concentrate on the modifications of precipitation caused by the different scenarios.

Figure 3 shows the results for the sub domain Central Sahel. A comparison with observed precipitation (red curve) shows that all model results are in close agreement with the observations. When looking at the root mean square error it shows that the fully interactive model run (ART) gives better results than the scenarios with prescribed climatology. For the subdomain Guinea coast (Fig. 4) all scenarios are overestimating the precipitation. However, again the interactive

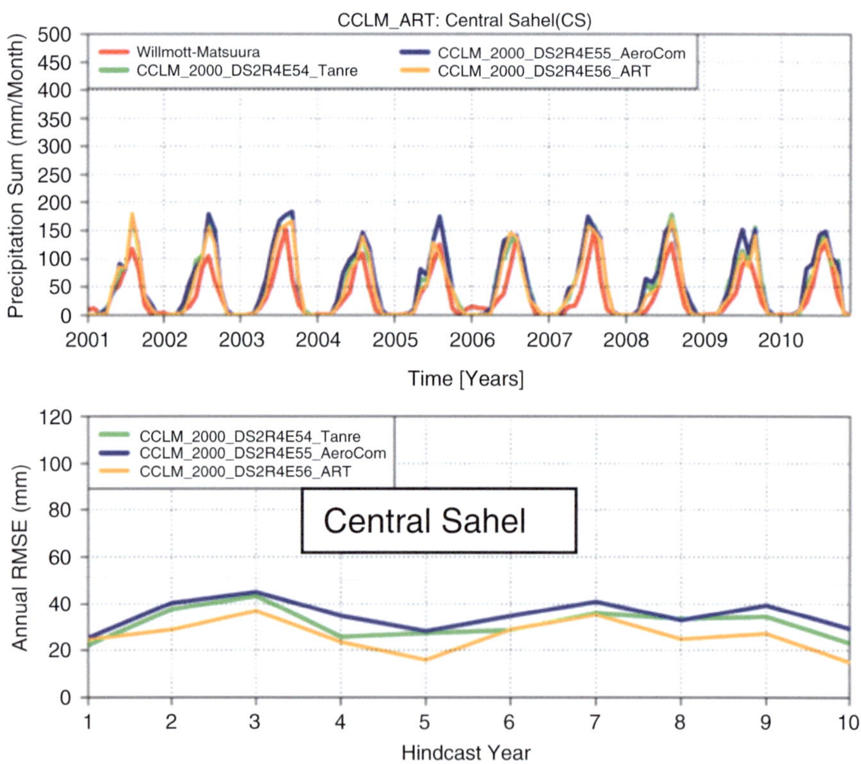

Fig. 3 Results for sub domain CS. *Top*: Simulated sum of precipitation per month for the different scenarios. The *red curve* shows the observation. *Bottom*: Annual root mean square error

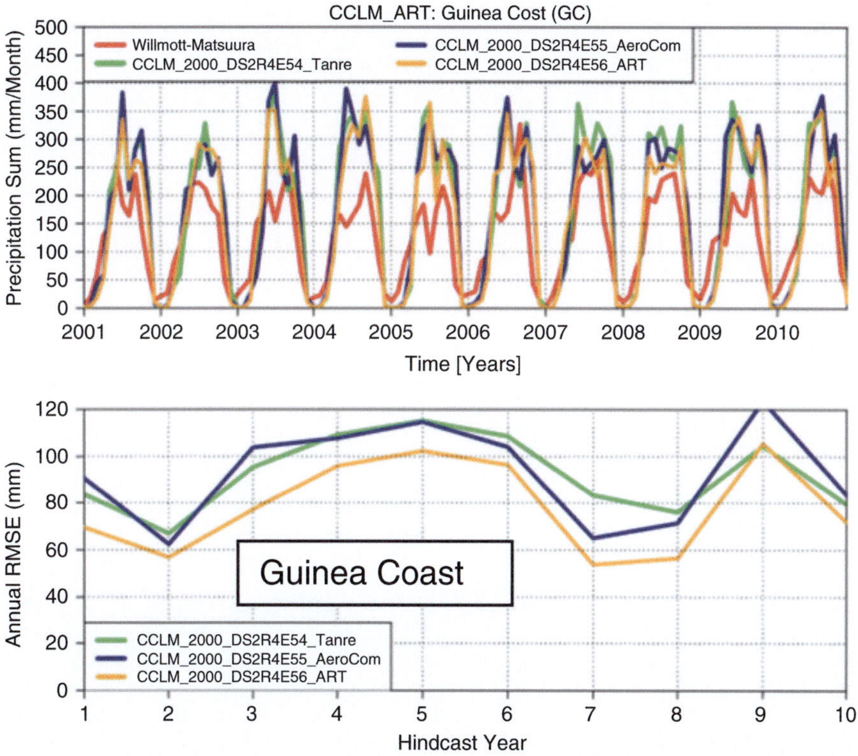

Fig. 4 Results for sub domain GC. *Top*: Simulated sum of precipitation per month for the different scenarios. The *red curve* shows the observation. *Bottom*: Annual root mean square error

scenario improves the agreement with observation for all years (bottom of Fig. 4). The results for sub domain West Sahel (Fig. 5) again show that all scenarios are in close agreement with the observations. When looking at the root mean square error the quality of the model results is comparable for all of the scenarios.

4 Resources

The simulations were carried out on a 275*207*35 grid with a time step of 240 s and a Runge-Kutta 3rd order time integration scheme. In each case 51 nodes were used. The simulation of one scenario with a prescribed mineral dust climatology required 1500 node hours on the CRAY XC40 'Hazel Hen'. A simulation with the interactive mineral dust required 5610 node hours. This is an increase by a factor of four. In addition to the increased computational costs a surplus of 1650 GB data storage in comparison with a pure COSMO run (with dust climatology) was needed.

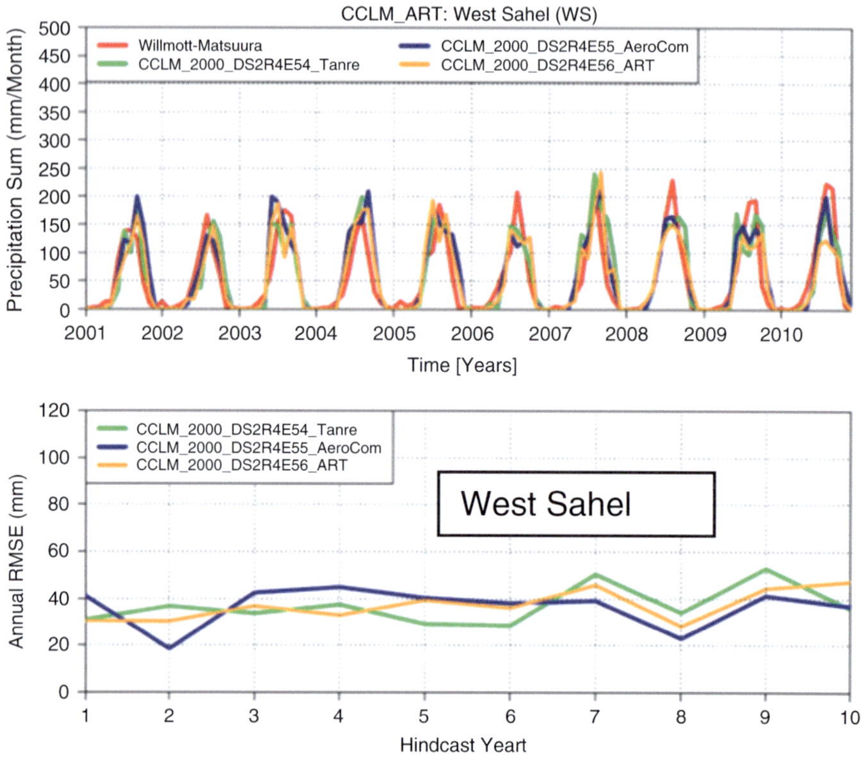

Fig. 5 Results for sub domain WS. *Top*: Simulated sum of precipitation per month for the different scenarios. The *red curve* shows the observation. *Bottom*: Annual root mean square error

5 Summary

Interactive dust improved the agreement of simulated precipitation in comparison with observations for some of the sub domains. The increase when taking into account the online calculated mineral dust and its optical properties dust emission is only a factor of four. This shows that high performance computing allows to carry our decadal regional climate simulations including interactive mineral dust. The results are taken as a starting point for further investigations.

References

1. Vogel, B., Vogel, H., Bumer, D., Bangert, M., Lundgren, K., Rinke, R., Stanelle, T.: The comprehensive model system COSMO-ART Radiative impact of aerosol on the state of the atmosphere on the regional scale. Atmos. Chem. Phys. **9**(22), 8661–8680 (2009). doi:10.5194/acp-9-8661-2009

2. Knote, C., et al.: Towards an online-coupled chemistry-climate model: evaluation of trace gases and aerosols in COSMO-ART. Geosci. Model Dev. **4**(4), 1077–1102 (2011). doi:10.5194/gmd-4-1077-2011
3. Bangert, M., et al.: Saharan dust event impacts on cloud formation and radiation over Western Europe. Atmos. Chem. Phys. **12**(9), 4045–4063 (2012). doi:10.5194/acp-12-4045-2012
4. Baldauf, M., Seifert, A., Foerstner, J., Majewski, D., Raschendorfer, M., Reinhardt, T.: Operational convective-scale numerical weather prediction with the COSMO model: description and sensitivities. Mon. Weather Rev. **139**(12), 3887–3905 (2011). doi:10.1175/MWR-D-10-05013.1
5. Athanasopoulou, E., et al.: Fire risk, atmospheric chemistry and radiative forcing assessment of wildfires in eastern Mediterranean. Atmos. Environ. **95**, 113–125 (2014). doi:10.1016/j.atmosenv.2014.05.077
6. Fountoukis, C., Nenes, A.: ISORROPIA II: a computationally efficient thermodynamic equilibrium model for ,Ca2+,Mg2+,NH4+,Na+,SO42,NO3,Cl,H2O aerosols. Atmos. Chem. Phys. **7**, 4639–4659 (2007). doi:10.5194/acp-7-4639-2007
7. Stanelle, T., Vogel, B., Vogel, H., Bumer, D., Kottmeier, C.: Feedback between dust particles and atmospheric processes over West Africa during dust episodes in March 2006 and June 2007. Atmos. Chem. Phys. **10**(22), 10771–10788 (2010). doi:10.5194/acp-10-10771-2010

Part VI
Miscellaneous Topics

Wolfgang Schröder

In the previous chapters topics such as fluid mechanics, structural mechanics, aerodynamics, thermodynamics, chemistry, combustion, and so forth have been addressed. In the following another degree of interdisciplinary research is emphasized. The articles clearly show the link between applied mathematics, fundamental physics, computer science, and the ability to develop certain models such that a closed mathematical description can be achieved which can be solved by highly sophisticated algorithms on up-to-date high performance computers. In other words, it is the collaboration of several scientific fields which on the one hand, defines the area of numerical simulations and on the other hand, determines the progress in fundamental and applied research. The subsequent papers, which represent an excerpt of various projects linked with the Höchstleistungszentrum Stuttgart (HLRS) and the Steinbuch Centre for Computing (SCC), will confirm that numerical simulations are not only used to compute some quantitative results but to corroborate basic physical models and to even develop new theories.

In the first contribution, the Chair of Thermodynamics and Energy Technology, University of Paderborn simulates transport properties for binary liquid mixtures using equilibrium molecular dynamics (EMD) together with the Green-Kubo formalism. The Maxwell-Stefan (MS), Fick- and self-diffusion coefficients, the shear viscosity, and the thermal conductivity were studied for all binary mixtures that can be formed out of methanol, ethanol, acetone, benzene, cyclohexane, toluene and carbon tetrachloride (CCl_4), with exception of the binary methanol + cyclohexane, which shows a miscibility gap under ambient conditions. These binary mixtures were selected because of the unusually good availability of experimental transport data such that a through assessment of the simulation data can be made. The binary liquid mixtures were studied over the whole composition range at ambient

W. Schröder (✉)
Institute of Aerodynamics, RWTH Aachen University, Wüllnerstr. 5a, 52062 Aachen, Germany
e-mail: office@aia.rwth-aachen.de

temperature and pressure, corresponding to a set of more than 200 transport data points. Each simulation point was calculated employing 4000 molecules with production runs of 10^7 time steps. The simulation results were compared, wherever possible, to experimental data and to a set of predictive equations.

The Institute of Applied Materials, Reliability of Components and Systems, Karlsruhe Institute of Technology deals with large-scale phase-field simulations. The combination of different chemical elements allows to obtain new and improved materials, as required for novel applications. Especially directionally solidified multicomponent eutectic alloys exhibit a wide range of patterns in the microstructure, which are correlated to the mechanical properties. The pattern formation during solidification depends on the chemical elements and the applied process parameters. Large-scale phase-field simulations are used to study the pattern formation of directional solidified ternary eutectics. Three different systems, starting from a model system towards the system Al-Ag-Cu are investigated, using three growth velocities. The three-dimensional simulation results are quantitatively compared and a broad variety of arising patterns for the studied systems is found. The results of the velocity variation follow the predictions from the analytic Jackson-Hunt approach.

The Geophysics Institute, Karlsruhe Institute of Technology presents applications of full waveform inversion (FWI) to field data. FWI is a powerful imaging technique which exploits the richness of seismic waveforms. It is further developed to obtain multi-parameter images at high resolution. Physical parameters are involved such as velocities and attenuation of seismic waves as well as mass density are involved. They are essential for a reliable petrophysical characterization of subsurface structures in hydrocarbon exploration, geotechnical applications and underground constructions. Referring to this, FWI is successfully applied to field datasets recorded in the Black Sea and in the shallow-water area of a river delta in the Atlantic Ocean. Detailed subsurface images are obtained containing rock formations which might be potential gas deposits. Additionally, synthetic studies are performed as preparatory steps to verify methodological improvements for further field-data applications. Resolution capabilities of FWI are demonstrated for imaging geological structures beneath salt bodies. Strategies to recover attenuation information from seismic data are investigated and a joint inversion of surface waves is performed to image the very shallow subsurface.

The Goethe Center for Scientific Computing, Goethe University Frankfurt has developed a massively parallel multigrid solver with level dependent smoothers. An issue that had not been fully addressed in previous studies is the difficulty to solve elliptic potential differential equations (PDE) on massively parallel computers in the presence of anisotropic coefficients or anisotropic elements in the underlying grid. While parallelism has been considered in previous studies, massively parallel systems as today's supercomputers with hundred thousands of computing cores did not exist at that time and thus no optimization regarding massive scalability has been performed. Recently, massively parallel multigrid has been described for the solution of elliptic PDEs for the special case with strong vertical anisotropies on structured grids. Considering the real world problem of drug discussion through the human skin, the former approaches have been extended to construct a method that employs geometric multigrid on massively parallel computers for problems with highly anisotropic elements using a combination of specialized refinement techniques and smoothers resulting in a robust and highly scalable solver for anisotropic problems. The special grid layout of the model problem thereby requires a solver which can handle anisotropies in all spatial directions on unstructured grids.

Molecular Simulation Study of Transport Properties for 20 Binary Liquid Mixtures and New Force Fields for Benzene, Toluene and CCl_4

Gabriela Guevara-Carrion, Tatjana Janzen, Y. Mauricio Muñoz-Muñoz, and Jadran Vrabec

1 Introduction

Nowadays, molecular modeling and simulation is being actively applied in physical, chemical and biological sciences as well as in engineering research and its importance will increase further in the future [31]. In the context of the chemical industry, molecular simulation has emerged as an alternative tool to estimate a wide variety of bulk phase thermodynamic property data, e.g., heat of formation, phase densities, transport coefficients, solubilities, rate constants, as well as to gain a deeper understanding of the subjacent molecular processes. Owing to the rapid increase in computing power and the development of new algorithms, the range of molecules that can be treated and the accuracy of the results is growing rapidly [18]. Traditionally, transport data have played a lesser role than other thermodynamic properties like vapor-liquid equilibria (VLE). Accurate experimental techniques for the measurement of transport properties were only developed around 1970, thus, the availability of such data is still low [52]. Furthermore, experimental measurements alone are not able to meet the demand for transport properties from the industry that may comprise several hundreds of data points for a single technical process [52]. On the other hand, classical theoretical methods are often incapable to accurately predict transport properties, especially when dealing with mixtures of liquids containing associating compounds.

G. Guevara-Carrion • T. Janzen • Y. Mauricio Muñoz-Muñoz • J. Vrabec (✉)
Lehrstuhl für Thermodynamik und Energietechnik (ThEt), Universität Paderborn,
Warburger Str. 100, 33098 Paderborn, Germany
e-mail: jadran.vrabec@upb.de.

W.E. Nagel et al. (eds.), *High Performance Computing in Science and Engineering '16*, DOI 10.1007/978-3-319-47066-5_42

This work reports on an extensive study on transport properties of 20 binary liquid mixtures using equilibrium molecular dynamics (EMD) together with the Green-Kubo formalism. In this context, the Maxwell-Stefan (MS), Fick- and self-diffusion coefficients, the shear viscosity and the thermal conductivity were studied for all binary mixtures that can be formed out of methanol, ethanol, acetone, benzene, cyclohexane, toluene and carbon tetrachloride (CCl_4), except for the binary methanol + cyclohexane, which shows a miscibility gap under ambient conditions. These binary mixtures were selected because of the unusually good availability of experimental transport data such that a thorough assessment of the simulation data can be made. The binary liquid mixtures were studied over the whole composition range at ambient temperature and pressure, corresponding to a set of more than 200 transport data points. Each simulation point was calculated employing 4000 molecules with production runs of 10^7 time steps. The simulation results were compared, wherever possible, to experimental data and to a set of predictive equations.

The quality of the simulation results is highly dependent on the quality of the force field that describes the molecular interactions. In fact, the prediction of transport properties is a challenging test for molecular force fields. E.g., this task was proposed for mixtures of the type water + short alcohol as a benchmark for water force fields [53]. In the present work, rigid, united-atom, non-polarizable force fields were used. The models for methanol, ethanol, acetone and cyclohexane were taken from previous work of our group [35, 46, 47, 56], whereas new force fields for benzene, toluene and CCl_4 are reported here. The force fields were developed starting from quantum chemical calculations with subsequent optimization of the site-site distances and force field parameters to experimental VLE and self-diffusion coefficient data.

2 New Force Fields

Force fields account for the intermolecular interactions, including hydrogen-bonding, by a set of Lennard-Jones (LJ) sites and superimposed point charges, point dipoles or point quadrupoles which may or may not coincide with the LJ site positions. The potential energy u_{kl} between two molecules k and l can be written as

$$u_{kl}\left(r_{klab}\right) = \sum_{a=1}^{S_k^{LJ}} \sum_{b=1}^{S_l^{LJ}} 4\epsilon_{klab} \left[\left(\frac{\sigma_{klab}}{r_{klab}} \right)^{12} - \left(\frac{\sigma_{klab}}{r_{klab}} \right)^{6} \right] + \sum_{c=1}^{S_k^e} \sum_{d=1}^{S_l^e} \frac{1}{4\pi\varepsilon_0} \left[\frac{q_{kc} q_{ld}}{r_{klcd}} \right] +$$

$$\sum_{c=1}^{S_k^e} \sum_{d=1}^{S_l^e} \frac{1}{4\pi\varepsilon_0} \left[\frac{\mu_{kc}\mu_{ld}}{r_{klcd}^3} \cdot f_1\left(\boldsymbol{\omega}_k, \boldsymbol{\omega}_l\right) + \frac{Q_{ld}\mu_{kc} + Q_{kc}\mu_{ld}}{r_{klcd}^4} \cdot f_2\left(\boldsymbol{\omega}_k, \boldsymbol{\omega}_l\right) \right] \quad (1)$$

where r_{klab}, ϵ_{klab}, σ_{klab} are the distance, the LJ energy parameter and the LJ size parameter, respectively, for the pair-wise interaction between LJ site a on the molecule k and the LJ site b on molecule l. The vacuum permittivity is ε_0, whereas q_{kc}, μ_{kc} and Q_{kc} denote the point charge magnitude, the dipole and the quadrupole moments of the electrostatic interaction site c on molecule k. The expression $f(\boldsymbol{\omega}_k, \boldsymbol{\omega}_l)$ stands for the dependence of the electrostatic interactions on the orientations $\boldsymbol{\omega}_k$ and $\boldsymbol{\omega}_l$ of the molecules k and l [21]. The summation limits S_k^{LJ} and S_k^e indicate the number of LJ and electrostatic sites, respectively. It should be noted that a point quadrupole can be approximated by three collinear point charges q, $-2q$ and q separated by l each, where $Q = 2ql^2$.

The force fields for benzene and toluene were obtained with the parameterization procedure proposed by Muñoz-Muñoz et al. [35]. Benzene was modeled by six LJ sites with superimposed point quadrupole sites. Internal bond angles of 120° and dihedral angles of 0° were kept constant so that all sites were located in a plane. The quadrupole moment of benzene was equally distributed among all LJ sites to avoid artifacts when mixtures with small molecules are considered. Initially, the quadrupole sites were located at the carbon positions and their value was set to $Q = Q_T/6$, where Q_T is the quadrupole moment magnitude of the benzene model by Bonnaud et al. [4]. The site-site distance between the LJ and quadrupole sites were modified in small steps as described in Ref. [35] until a suitable combination of parameters was obtained. Finally, the reduced units method by Merker et al. [34] was applied to obtain the definitive force field parameters.

Toluene was modeled on the basis of the benzene model with an additional LJ site, representing the methyl group, located a distance δ from the ring. The LJ parameters of the methyl site were taken from Schnabel et al. [46]. All site-site distances were optimized until accurate results of the thermodynamic properties were obtained following the above described procedures.

The critical point of the new benzene force field is located at $T = 561$ K, $\rho = 4.01$ mol l^{-1} and $p = 5.0$ MPa. The relative deviations with respect to experimental data [49] for critical temperature, critical density and critical pressure of benzene are −0.1 %, +2.9 % and +2.5 %, respectively. The predicted VLE properties exhibit an average deviation of 7.3 % for vapor pressure, 1.0 % for saturated liquid density and 2.4 % for enthalpy of vaporization in the regarded temperature range. In the temperature range between 280 and 333 K at ambient pressure, self-diffusion coefficient and shear viscosity obtained with the new benzene force field deviate on average by 5.7 % and 6.2 % from correlations of experimental data by Fischer and Weiss [15], respectively. The thermal conductivity deviates on average by 11 % from a correlation of experimental data [39, 45]. Figure 1 shows the calculated VLE and transport properties in comparison with the corresponding reference equations of state or experimental data. The critical temperature of the new toluene force field is 594 K, the critical density is 3.24 mol l^{-1} and the critical pressure is 4.4 MPa. The according relative deviations from experiment [26] are +0.5 %, +2.1 % and +6.4%, respectively. The VLE properties exhibit average deviations of 18.3 % for vapor pressure, 1.1 % for saturated liquid density and 3.7 % for enthalpy of vaporization in the studied temperature range. The ability of the new toluene force field to predict

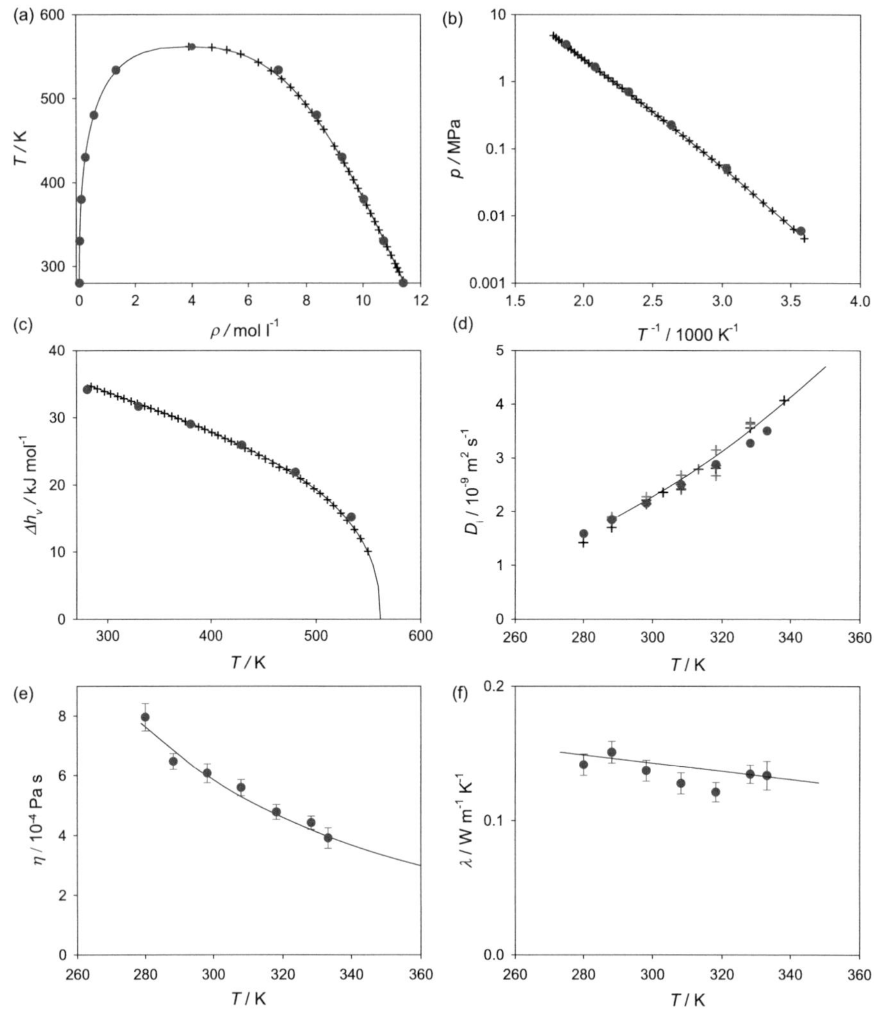

Fig. 1 Temperature dependence of the vapor-liquid equilibrium and transport properties of benzene. Simulation results (*blue solid circle*) for (**a**) saturated liquid density, (**b**) vapor pressure and (**c**) enthalpy of vaporization are compared with experimental data (*plus*) and an equation of state (*solid curve*) [49]. (**d**) Simulation results for the self-diffusion coefficient (*blue solid circle*) are compared with experimental data by Rathbun and Babb [41] (*black plus*), McCool and Wolf [32] (*blue plus*), Hiraoka [20] (*dark blue plus*), Graupner and Winter [17] (*red plus*), Falcone et al. [11] (*green plus*) and the correlation of experimental data by Fischer and Weiss [15] (*solid curve*). (**e**) Simulation results for the shear viscosity (*blue solid circle*) are compared with a correlation of experimental data [15] (*solid curve*). (**f**) Simulation results for the thermal conductivity (*blue solid circle*) are shown together with a correlation of experimental data [39] (*solid curve*)

self-diffusion coefficient, shear viscosity and thermal conductivity was tested in the temperature range between 273.15 and 350 K at ambient pressure. An average relative deviation (ARD) of 5 % was found for the self-diffusion coefficient when compared with experimental data [19, 24, 38, 50, 55]. Shear viscosity and thermal conductivity deviate both on average by approximately 10 % from correlations of experimental data [39], cf. Fig. 2.

CCl_4 was modeled by five LJ sites and five atom-centered point charges, i.e., one per atom. The magnitude of the point charges, obtained via quantum chemical calculations with the Møller-Plesset 2 method and the 6-31G* basis set, were taken from the NIST database [7]. The LJ parameters for the carbon atom, located in the center of the molecule, were taken from Merker et al. [33]. Thus, there were three parameters left to be determined, two LJ parameters for the four identical chlorine sites and the site-site distance between carbon and chlorine. These three parameters were iteratively optimized by carrying out molecular simulations of the VLE in the range of 300–525 K and of liquid CCl_4 at 298.15 K and comparing the results to experimental data of vapor pressure, saturated liquid density and self-diffusion coefficient. The critical point of the new CCl_4 force field is located at $T = 551\,\text{K}$, $\rho = 3.59\,\text{mol l}^{-1}$ and $p = 4.3\,\text{MPa}$, yielding relative deviations with respect to experimental data [5] of −1.0 %, −5.7 % and −0.8 %. The predicted VLE properties exhibit average deviations of 24.3 % for vapor pressure, 0.6 % for saturated liquid density and 13.9 % for enthalpy of vaporization in the studied temperature range. The transport properties of the CCl_4 model were calculated in the temperature range between 268 and 343 K at ambient pressure. Compared with correlations of experimental data an ARD of 9 % was found for the self-diffusion coefficient [6, 14, 19, 32, 41], 27 % for the shear viscosity [22, 29] and 19 % for the thermal conductivity [13, 25], cf. Fig. 3.

3 Methodology

Transport properties were sampled with EMD and the Green-Kubo formalism. The Green-Kubo expression for the self-diffusion coefficient D_i is related to the individual molecule velocity autocorrelation function

$$D_i = \frac{1}{3N_i} \int_0^\infty dt \left\langle \mathbf{v}_i^k(t) \cdot \mathbf{v}_i^k(0) \right\rangle. \tag{2}$$

Here, $\mathbf{v}_i^k(t)$ is the center of mass velocity vector of molecule k of component i at some time t and N_i is the number of molecules of component i. The brackets <...> denote the canonical (NVT) ensemble average. Equation (2) is an average over all N_i molecules in the ensemble because all contribute to the self-diffusion coefficient. The self-diffusion coefficient that describes the mobility of species i in a mixture is also termed intradiffusion coefficient.

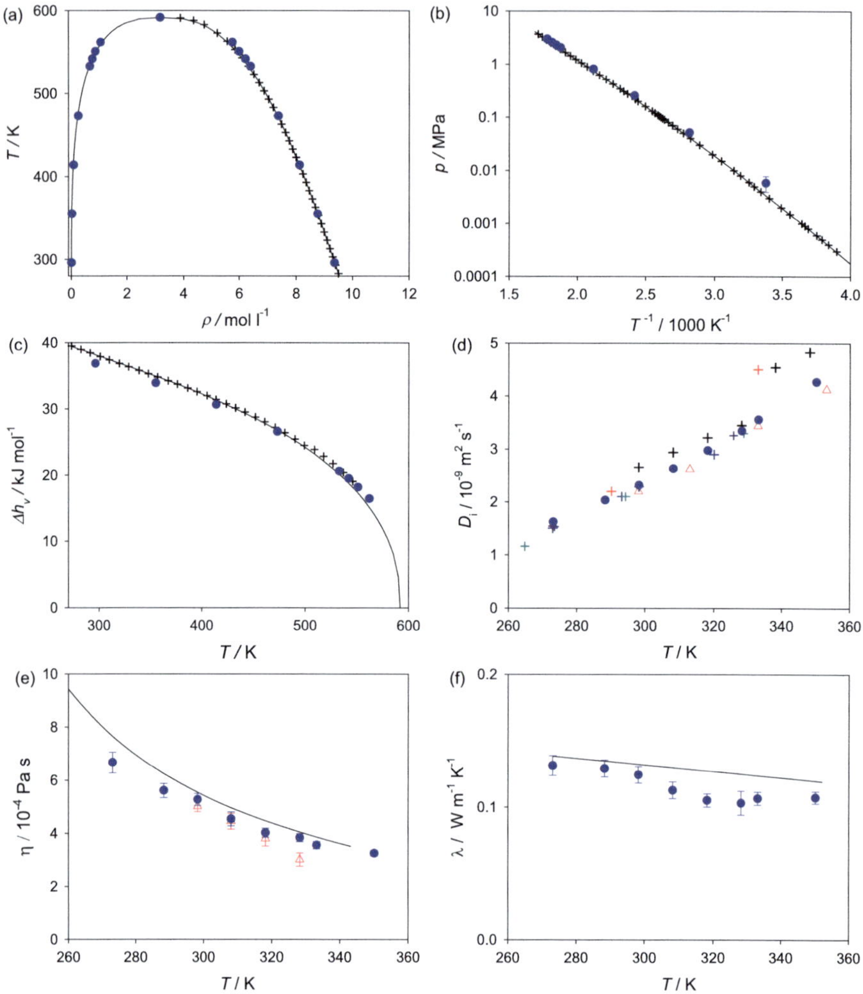

Fig. 2 Temperature dependence of the vapor-liquid equilibrium and transport properties of toluene. Simulation results (*blue solid circle*) for (**a**) saturated liquid density, (**b**) vapor pressure and (**c**) enthalpy of vaporization are compared with experimental data (*plus*) and an equation of state (*solid curve*) [26]. (**d**) Simulation results for the self-diffusion coefficient (*blue solid circle*) are compared with experimental data by Pickup and Blum [38] (*black plus*), Winfield [55] (*blue plus*), Trepădus et al. [50] (*red plus*), Krüger and Weiss [24] (*green plus*) and Harris et al. [19] (*dark blue plus*). The self-diffusion coefficient from the molecular model by Nieto-Draghi et al. [36, 37] (*red triangle*) is also shown. (**e**) Simulation results for the shear viscosity (*blue solid circle*) are compared with a correlation of experimental data [45] (*solid curve*) and the molecular model by Nieto-Draghi et al. [36, 37] (*red triangle*). (**f**) Simulation results for the thermal conductivity (*blue solid circle*) are shown together with a correlation of experimental data [39] (*solid curve*)

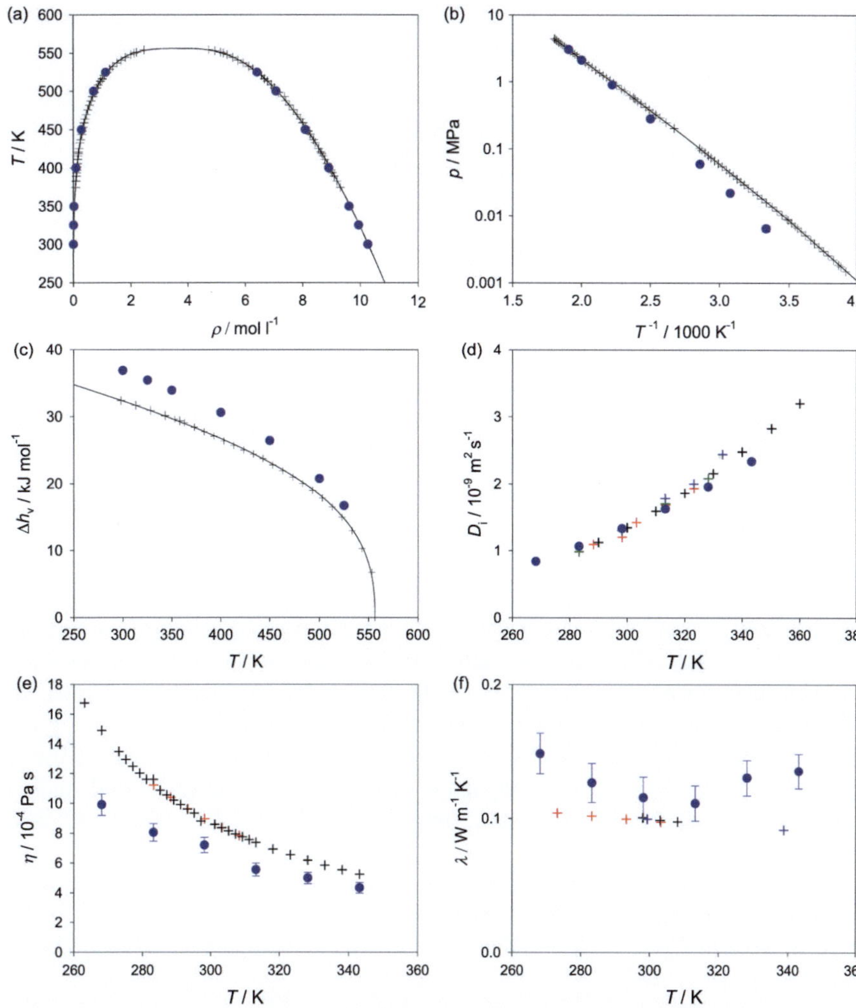

Fig. 3 Temperature dependence of the vapor-liquid equilibrium and transport properties of CCl_4. Simulation results (*blue solid circle*) for (**a**) saturated liquid density, (**b**) vapor pressure and (**c**) enthalpy of vaporization are compared with experimental data (*plus*) and an equation of state (*solid curve*). (**d**) Simulation results for the self-diffusion coefficient (*blue solid circle*) are compared with experimental data by Fischer and Weiss [15] (*black plus*), McCool and Wolf [32] (*green plus*), Collins and Mills [6] (*red plus*), Harris et al. [19] (*blue plus*) and Rathbun and Babb [41] (*dark blue plus*). (**e**) Simulation results for the shear viscosity (*blue solid circle*) are compared with experimental data by Luchinskii [29] (*black plus*) and Ikeuchi et al. [22] (*red plus*). (**f**) Simulation results for the thermal conductivity (*blue solid circle*) are shown together with experimental data by Rowley et al. [43] (*black plus*), Fischer [13] (*blue plus*) and Lei et al. [25] (*red plus*)

The MS diffusion coefficient $Đ_{ij}$ can be determined from the Onsager coefficients L_{ij} with the Green-Kubo expression [23]

$$L_{ij} = \frac{1}{3N} \int_0^\infty dt \left\langle \sum_{k=1}^{N_i} \mathbf{v}_i^k(0) \cdot \sum_{l=1}^{N_j} \mathbf{v}_j^l(t) \right\rangle, \tag{3}$$

where N is the total number of molecules. In this context, the MS diffusion coefficient for binary mixtures is given by [23]

$$Đ_{ij} = \frac{x_j}{x_i} L_{ii} + \frac{x_i}{x_j} L_{jj} - L_{ij} - L_{ji} \cdot \tag{4}$$

In this way, the MS diffusion coefficient can be sampled directly, but cannot be measured experimentally in the laboratory like the Fick diffusion coefficient D_{ij}. However, both diffusion coefficients are related by $D_{ij} = Đ_{ij} \cdot \Gamma$, where Γ is the thermodynamic factor given by

$$\Gamma = 1 + x_1 \left(\frac{\partial \ln \gamma_1}{\partial x_1} \right)_{T,p} = 1 + x_2 \left(\frac{\partial \ln \gamma_2}{\partial x_2} \right)_{T,p}, \tag{5}$$

where γ_i and x_i stand for the activity coefficient and molar fraction of component i, respectively. The MS diffusion coefficient can thus be transformed to the Fick diffusion coefficient and vice versa, if the thermodynamic factor is known. In this work, the thermodynamic factor was obtained from excess Gibbs energy G^{E} models fitted to experimental VLE data. Because the thermodynamic factor is sensitive to the underlying thermodynamic model, it was calculated for all studied mixtures employing three different G^{E} models, i.e., Wilson [54], NRTL [42] and UNIQUAC [1].

The shear viscosity η is associated with the autocorrelation function of the off-diagonal elements of the stress tensor J_p^{xy}

$$\eta = \frac{1}{Vk_BT} \int_0^\infty dt \left\langle J_p^{xy}(t) \cdot J_p^{xy}(0) \right\rangle, \tag{6}$$

where V stands for the volume. The component J_p^{xy} of the microscopic stress tensor $\mathbf{J}_p$ is given by [18]

$$J_p^{xy} = \sum_{k=1}^{N} m_k v_k^x v_k^y - \frac{1}{2} \sum_{k=1}^{N} \sum_{l \neq k}^{N} r_{kl}^x \frac{\partial u(r_{kl})}{\partial r_{kl}^y}. \tag{7}$$

Here, k and l denote different molecules of any species. The upper indices x and y stand for the spatial vector components, e.g., for velocity v_k^x or site-site distance r_{kl}^x. Equations (6) and (7) may directly be applied to mixtures. Five independent terms of the stress tensor J_p^{xy}, J_p^{xz}, J_p^{yz}, $(J_p^{xx} - J_p^{yy})/2$ and $(J_p^{yy} - J_p^{zz})/2$ were considered to improve statistics [2].

The thermal conductivity λ is given by the autocorrelation function of the elements of the microscopic heat flow J_q^x

$$\lambda = \frac{1}{Vk_B T^2} \int_0^\infty dt \left\langle J_q^x(t) \cdot J_q^x(0) \right\rangle. \tag{8}$$

In mixtures, energy and mass transport occur in a coupled manner, thus, the heat flow for a mixture of n components is given by [10]

$$\mathbf{J}_q = \frac{1}{2} \sum_{i=1}^{n} \sum_{k=1}^{N_i} \left[m_k^i \left(\mathbf{v}_k^i\right)^2 + \mathbf{w}_k^i \mathbf{I}_k^i \mathbf{w}_k^i + \sum_{j=1}^{n} \sum_{l \neq k}^{N_j} u\left(r_{kl}^{ij}\right) \right] \cdot \mathbf{v}_k^i$$

$$- \frac{1}{2} \sum_{i=1}^{n} \sum_{j=1}^{n} \sum_{k=1}^{N_i} \sum_{l \neq k}^{N_j} \mathbf{r}_{kl}^{ij} \cdot \left(\mathbf{v}_k^i \cdot \frac{\partial u\left(r_{kl}^{ij}\right)}{\partial \mathbf{r}_{kl}^{ij}} + \mathbf{w}_k^i \mathbf{\Gamma}_{kl}^{ij} \right) - \sum_{i=1}^{n} h_i \sum_{k=1}^{N_i} \mathbf{v}_k^i, \tag{9}$$

where $\mathbf{w}_k^i$ is the angular velocity vector of molecule k of component i and $\mathbf{I}_k^i$ its matrix of angular momentum of inertia. $u(r_{kl}^{ij})$ is the intermolecular potential energy and $\mathbf{\Gamma}_{kl}^{ij}$ is the torque due to the interaction between molecules k and l. The lower indices i and j denote the components of the mixture and h_i is the partial molar enthalpy of component i.

Molecular dynamics simulations were performed with the program *ms*2 [9, 16]. In a first step, a simulation in the isobaric-isothermal (*NpT*) ensemble was carried out to calculate the density and enthalpy at the desired temperature, pressure and composition. In the second step, a *NVT* ensemble simulation was performed at this temperature, density and composition to determine the transport properties. Newton's equations of motion were solved with a fifth-order Gear predictor-corrector numerical integrator. The temperature was controlled by velocity scaling. In all simulations, the integration time step was 0.877 fs. The simulations contained 4000 molecules and were carried out in a cubic volume with periodic boundary conditions where the cut-off radius was set to $r_c = 17.5$ Å. LJ long range interactions were considered using angle averaging [30]. Electrostatic long-range corrections were approximated by the reaction field technique with conducting boundary conditions ($\epsilon_{RF} = \infty$). Analogous *NVT* simulations with an extended cut-off radius that reached half of the edge length of the cubic simulation volume were employed to calculate the radial distribution function (RDF). Here, starting from well-equilibrated configurations, production runs had a duration between 5 and 10×10^4 time steps.

The simulations in the *NpT* ensemble were equilibrated over 1.2×10^5 time steps, followed by a production run over 5×10^5 time steps. In the *NVT* ensemble, the simulations were equilibrated over 3×10^5 time steps, followed by production runs of 10^7 time steps. The self- and MS diffusion coefficients, shear viscosity and thermal conductivity were calculated by Eqs. (2), (3), (4), (6) and (8) with up to 4×10^4 independent time origins of the autocorrelation functions. The sampling length of the autocorrelation functions was 17.5 ps for all mixtures. That extensive length of the autocorrelation functions was chosen such that long-time tail corrections were not necessary. The separation between the time origins was chosen so that all autocorrelation functions have decayed at least to $1/e$ of their normalized value to achieve their time independence [48]. Statistical uncertainties of the predicted values were estimated with a block averaging method [3].

4 Simulation Results

This report comprises a wide variety of liquid mixtures, from thermodynamically almost-ideal to highly non-ideal. Generally, mixtures behave ideally when the structure and thermodynamic properties of its net components are quite similar, such as methanol + ethanol or benzene + toluene. On the other hand, strongly non-ideal mixtures, which are thermodynamically more challenging, are in many cases a combination of highly polar with non-polar compounds, e.g., methanol + toluene, ethanol + benzene or acetone + cyclohexane.

For all 20 mixtures, the density, MS- and self-diffusion coefficient, the shear viscosity and the thermal conductivity were determined by molecular simulation for 11 composition points and compared with experimental data, if available. The Fick diffusion coefficient was calculated in all cases with the thermodynamic factor based on the Wilson G^E model [54], which gives the best fit of the experimental data. The resulting overall average relative deviation is 16 % for the set of 20 binary mixtures. In the same line, shear viscosity and thermal conductivity were predicted with an overall average deviation from experimental data of 8 % and 11 %, respectively. The self-diffusion coefficients are predicted with an overall relative deviation lower than 8.6 %. Figure 4 shows in more detail the average relative deviation from experiments of the present simulation results for Fick diffusion coefficient, shear viscosity and thermal conductivity. In general, a good agreement is found between experimental data and simulation results, with the exception of the binaries including CCl_4. Because of the large amount of data, graphical representations of the simulation results are given for a few selected mixtures that comprehend the different thermodynamic behavior present in the regarded set of mixtures.

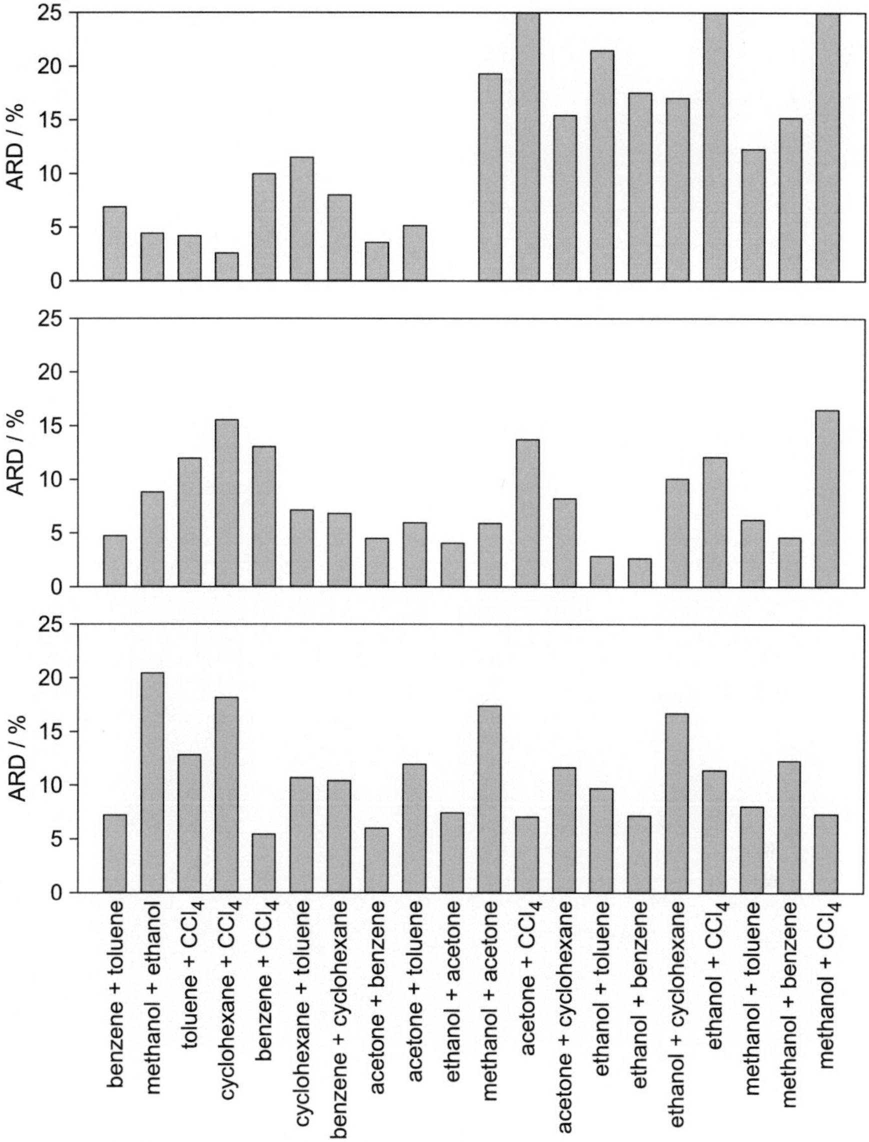

Fig. 4 Average relative deviation (ARD) of present simulation results for Fick diffusion coefficient (*top*), shear viscosity (*center*) and thermal conductivity (*bottom*) from the best polynomial fit of the available experimental data

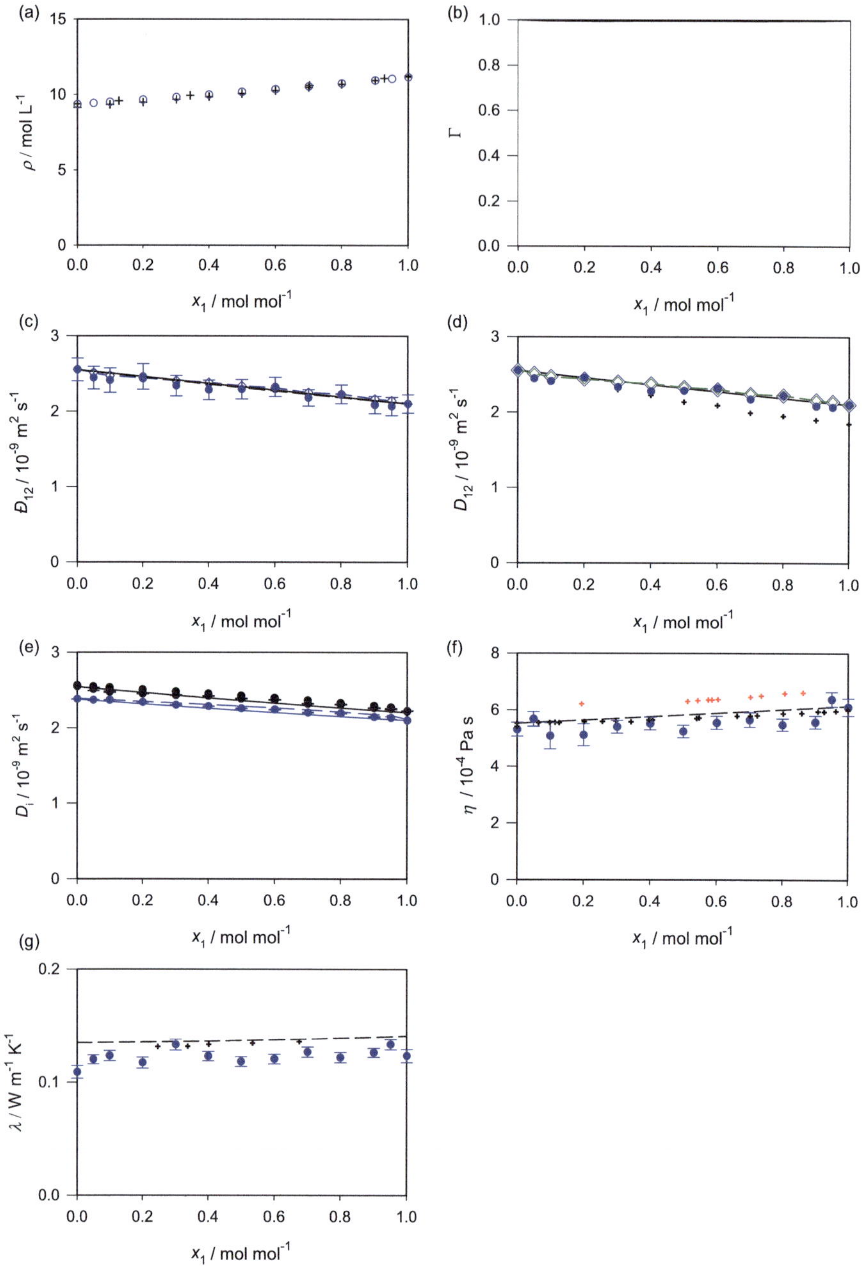

Fig. 5 Results for benzene (1) + toluene at 298.15 K and 0.1 MPa. (**a**) Simulation results for the density (*blue open circle*) are compared with experimental data. (**b**) Thermodynamic factor. (**c**) Simulation results for the Maxwell-Stefan diffusion coefficient (*blue solid circle*) are compared with the models by Darken [8] (*blue open circle*), Vignes [51] (*black dashed curve*), Li et al. [27] (*blue curve with diamonds*) and Zhou et al. [57] (*black solid curve*) based on present

4.1 Benzene + Toluene

Because the components of this mixture are chemically similar, it behaves nearly ideal. This fact can be clearly observed for the calculated thermodynamic factor, cf. Fig. 5b, which is approximately unity over the whole composition range. Per definition, an ideal mixture has a thermodynamic factor equal to unity. Figure 5 shows also the simulation results for the density, MS- and self-diffusion coefficient, shear viscosity and thermodynamic factor, as well as the calculated Fick diffusion coefficient in comparison with experimental data and some predictive equations from the literature. As can be observed, all properties are almost a linear function of the mole fraction. Thus, simple interpolative predictive equations are able to accurately predict the mixture properties with low deviations. Transport data reported here agree well with the experiments. The average relative deviations for Fick diffusion coefficient, shear viscosity and thermal conductivity are 6.9 %, 4.7 % and 7.9 %, respectively.

4.2 Acetone + Benzene

The thermodynamic factor of the mixture acetone + benzene deviates moderately from unity, and therefore from ideality, cf. Fig. 6b. A similar composition dependence of the thermodynamic factor can be observed for mixtures of acetone with toluene, methanol, and ethanol, as well as for those of cyclohexane with benzene and toluene. In general, the transport properties deviate to some extent from linearity. E.g., the composition dependence of the Fick- and self-diffusion coefficient as well as shear viscosity exhibit a small concave curvature. This can be observed in Fig. 6 compared with experimental data and some predictive equations from the literature. The simulation results agree well with the experimental values, with average relative deviations between 3 % and 6 %, performing better than the tested predictive equations.

Fig. 5 (continued) simulation data. (**d**) Simulation results for the Fick diffusion coefficient (*blue solid circle*) are compared with experimental data (*plus*). The models by Li et al. [27] (*blue curve with diamonds*), Zhou et al. [57] (*black solid curve*) and Zhu et al. [58] (*green curve with diamonds*) based on present simulation data are also shown. (**e**) Simulation results for the self-diffusion coefficients of benzene (*black solid circle*) and toluene (*blue solid circle*) are compared with the models by Li et al. [27] (*dashed curve*) and Liu et al. [28] (*solid curve*). (**f**) Simulation results for the shear viscosity (*blue solid circle*) are shown together with the viscosity of the ideal mixture (*dashed curve*) and experimental data (*plus*). (**g**) Simulation results for the thermal conductivity (*blue solid circle*) are compared with the predictions from the Filippov relation [12] (*dashed curve*) and experimental data (*plus*)

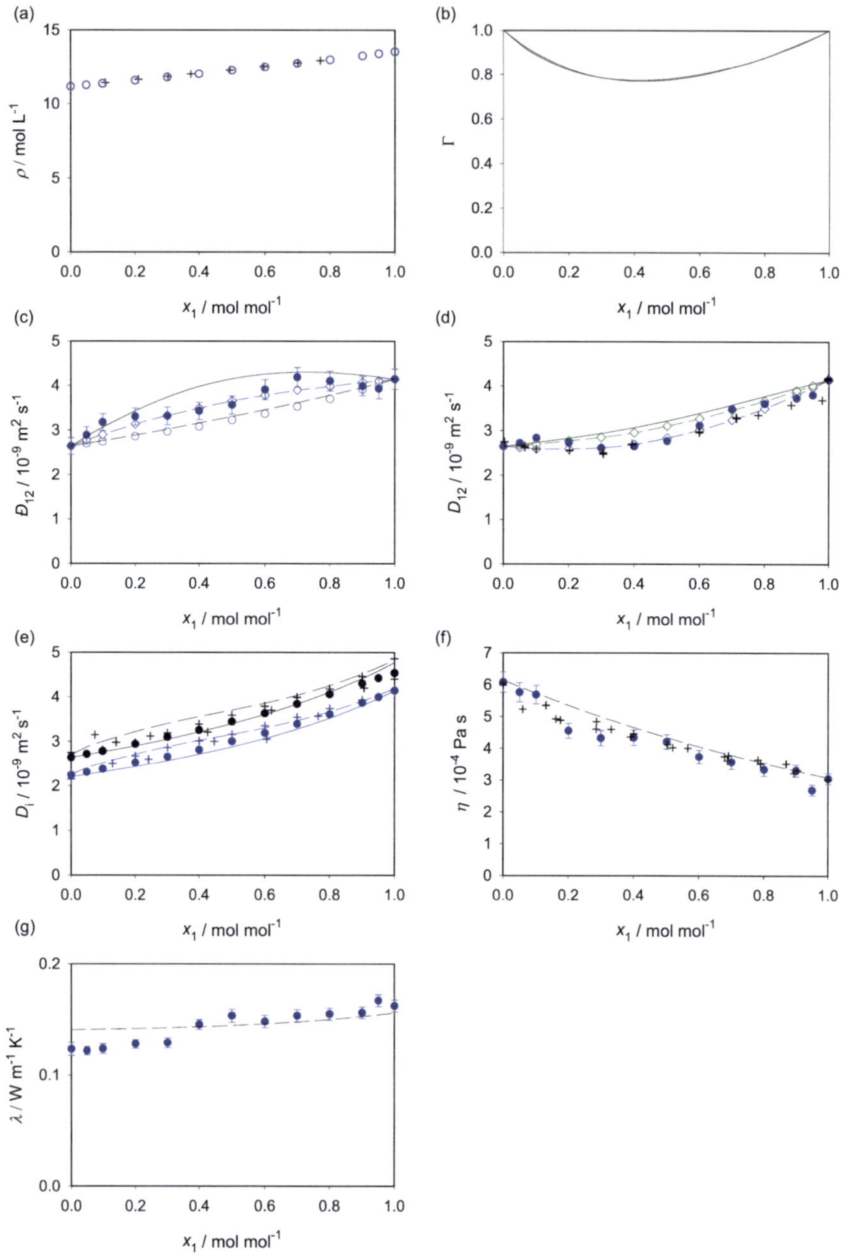

Fig. 6 Results for acetone (1) + benzene at 298.15 K and 0.1 MPa. (**a**) Simulation results for the density (*blue open circle*) are compared with experimental data. (**b**) Thermodynamic factor; the *shaded area* represents the range of the results of the three considered G^E models. (**c**) Simulation results for the Maxwell-Stefan diffusion coefficient (*blue solid circle*) are compared with the models by Darken [8] (*blue open circle*), Vignes [51] (*dashed curve*), Li et al. [27] (*blue curve with diamonds*) and Zhou et al. [57] (*solid curve*) based on present simulation data. (**d**) Simulation

4.3 *Ethanol + Cyclohexane*

Mixtures containing one alcohol usually show strong non-idealities for different thermodynamic and transport properties, e.g., MS, Fick, self-diffusion coefficients, shear viscosity or excess volume, as in the case of ethanol + cyclohexane. This behavior can be explained by association effects related to the presence of hydrogen-bonding. The thermodynamic factor of this mixture, shown in Fig. 7b, can reach values close to zero and exhibits a strong composition dependence. The shaded area, related to the chosen G^E model, is important so that the Fick diffusion coefficient can be calculated only within a relatively large uncertainty. Other binary mixtures showing a similar behavior are the ones of methanol and ethanol with benzene, toluene and CCl_4.

Figure 7 shows the predicted transport properties of ethanol + cyclohexane together with experimental data and selected predictive equations from the literature. It can be observed that shear viscosity, the Fick diffusion coefficient and self-diffusion coefficient of ethanol show a strong negative deviation from linearity with a minimum located around 0.2 mol mol^{-1} of ethanol. This sharp decrease of the diffusion coefficients at low alcohol concentration has been related to cluster formation due to solute self-association [44].

In order to study the microscopic structure responsible for this behavior RDF were sampled. The RDF $g_{A-B}(r)$ between like and unlike sites were calculated together with the running coordination number

$$N_{A-B}(r) = 4\pi\ \varrho \int_0^r r^2 g_{A-B}(r)\, dr, \qquad (10)$$

where r is the distance from the reference site and ϱ is the bulk number density of site B.

Three selected RDF for this mixture at different compositions are shown in Fig. 8. The double peak related to the hydrogen-bonding structure of ethanol g_{O-H} is enhanced as the ethanol concentration decreases. Thus, the nearest neighbor

Fig. 6 (continued) results for the Fick diffusion coefficient (*blue solid circle*) are compared with experimental data (*plus*). The models by Li et al. [27] (*blue curve with diamonds*), Zhou et al. [57] (*solid curve*) and Zhu et al. [58] (*green curve with diamonds*) based on present simulation data are also shown. (**e**) Simulation results for the self-diffusion coefficients of acetone (*black solid circle*) and benzene (*blue solid circle*) are compared with experimental data (*plus*) and the models by Li et al. [27] (*dashed curve*) and Liu et al. [28] (*solid curve*). (**f**) Simulation results for the shear viscosity (*blue solid circle*) are shown together with the viscosity of the ideal mixture (*dashed curve*) and experimental data (*plus*). (**g**) Simulation results for the thermal conductivity (*blue solid circle*) are compared with the predictions from the Filippov relation [12] (*dashed curve*)

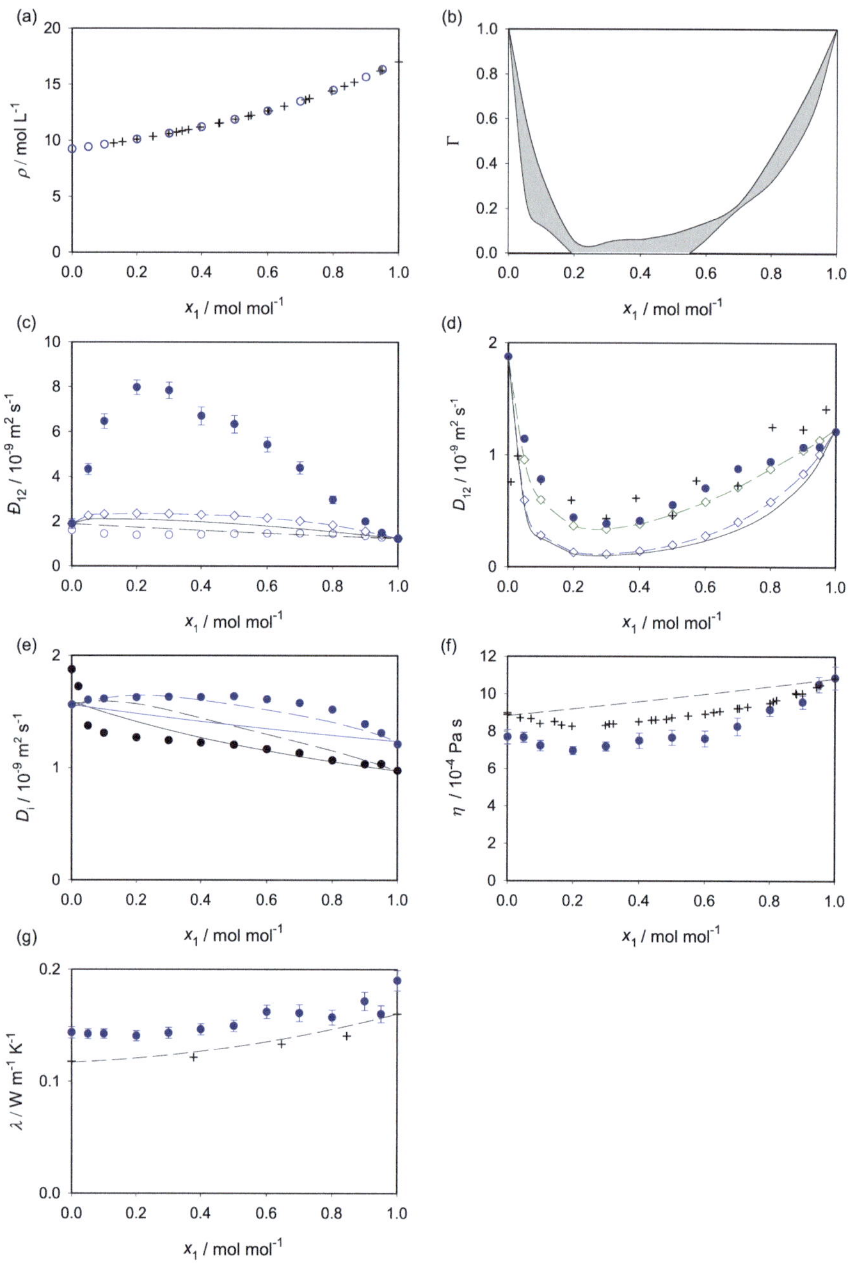

Fig. 7 Results for ethanol (1) + cyclohexane at 298.15 K and 0.1 MPa. (**a**) Simulation results for the density (*blue open circle*) are compared with experimental data. (**b**) Thermodynamic factor; the *shaded area* represents the range of the results of the three considered G^E models. (**c**) Simulation results for the Maxwell-Stefan diffusion coefficient (*blue solid circle*) are compared with the models by Darken [8] (*blue open circle*), Vignes [51] (*dashed curve*), Li et al. [27] (*blue curve with diamonds*) and Zhou et al. [57] (*solid curve*) based on present simulation data. (**d**) Simulation

hydrogen-bonding structure at low ethanol concentration appears to be more stable than that of the pure liquid. The insensitivity of the peak location on the composition suggests that the ethanol molecules conserve their nearest neighbors local environment, indicating the presence of clusters due to the strong ethanol self-association. The presence of clusters can also be inferred from the RDF between the like sites of the solvents, cf. Fig. 8. These findings are supported by the analysis of the simulation snapshots of these type of mixtures, which back up the conclusion that ethanol molecules are microsegregated [40] and explain the strong decrease of the Fick and self-diffusion coefficients at low ethanol composition. A comparison between the snapshots of three different mixtures from thermodynamically nearly ideal to highly non-ideal is given in Fig. 9.

5 Conclusion

Molecular dynamics simulation and the Green-Kubo formalism were employed to predict the transport properties of 20 binary liquid mixtures at ambient conditions. For this task, two new molecular models were developed in this work. These force fields are able to well reproduce the experimental data.

In this way a set of more than 200 state points was investigated. In general, a good agreement between simulation results and experimental values was found. The relative average deviations are almost always below 20 %, with the exception of the mixtures containing CCl_4. Further, the microscopic molecular structure responsible for the macroscopic behavior of the mixtures was analyzed employing radial distribution functions. It was found that thermodynamically challenging mixtures that include one hydrogen-bonding component tend to form clusters due to self-association at low alcohol mole fractions.

Fig. 7 (continued) results for the Fick diffusion coefficient (*blue solid circle*) are compared with experimental data (*plus*). The models by Li et al. [27] (*blue curve with diamonds*), Zhou et al. [57] (*solid curve*) and Zhu et al. [58] (*green curve with diamonds*) based on present simulation data are also shown. (**e**) Simulation results for the self-diffusion coefficients of ethanol (*black solid circle*) and cyclohexane (*blue solid circle*) are compared with the models by Li et al. [27] (*dashed curve*) and Liu et al. [28] (*solid curve*). (**f**) Simulation results for the shear viscosity (*blue solid circle*) are shown together with the viscosity of the ideal mixture (*dashed curve*) and experimental data (*plus*). (**g**) Simulation results for the thermal conductivity (*blue solid circle*) are compared with the predictions from the Filippov relation [12] (*dashed curve*) and experimental data (*plus*)

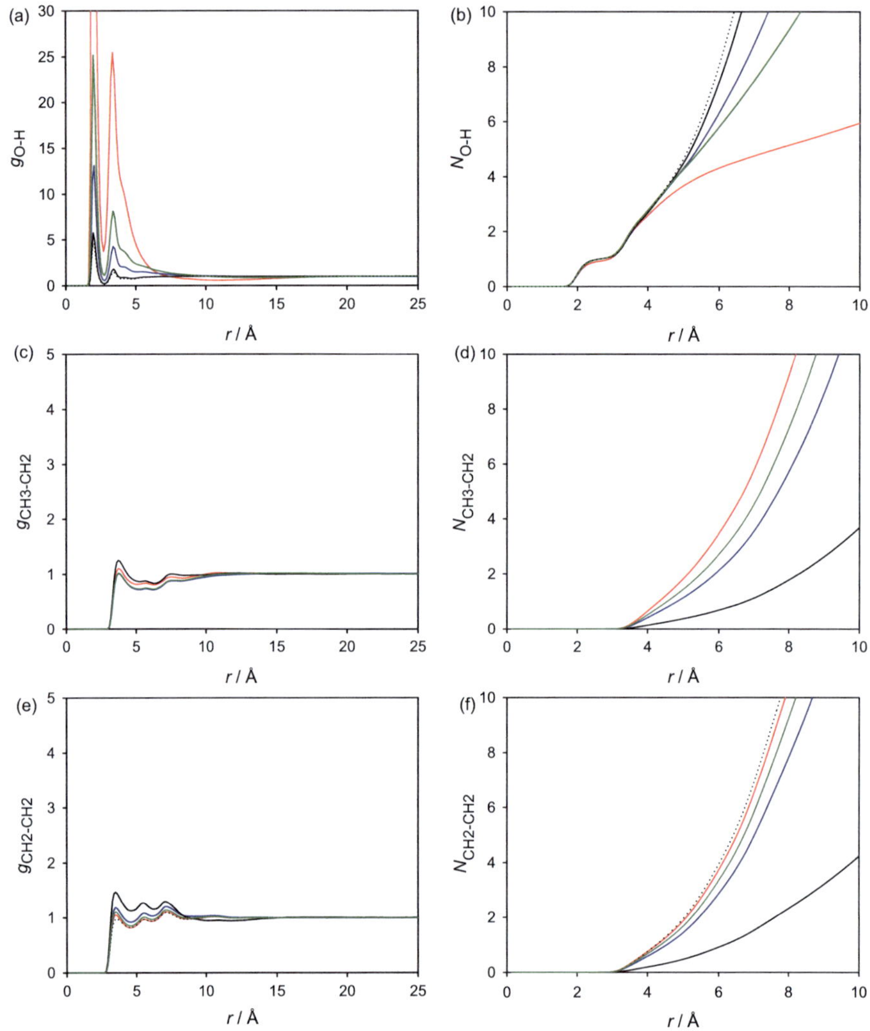

Fig. 8 Selected radial distribution functions (*left*) and the corresponding running coordination numbers (*right*) of ethanol (1) + cyclohexane at 298.15 K and 0.1 MPa between the oxygen and hydroxyl hydrogen sites of ethanol $g_{\text{O}-\text{H}}$ (**a, b**), the methylene and methyl sites of ethanol and cyclohexane $g_{\text{CH3}-\text{CH2}}$ (**c, d**) and the methyl sites of cyclohexane $g_{\text{CH2}-\text{CH2}}$ (**e, f**). Data for pure ethanol and cyclohexane (*black dotted curve*) as well as for the mixtures with $x_1 = 0.1$ (*red curve*), 0.3 (*green curve*), 0.5 (*blue curve*) and 0.9 mol mol^{-1} (*black curve*) are depicted

Fig. 9 Snapshots of toluene (1) + benzene (*top*), methanol (1) + acetone (*center*) and ethanol (1) + CCl_4 (*bottom*), at three mole fractions $x_1 = 0.1$ (*left*), 0.5 (*center*) and 0.9 mol mol^{-1} (*right*). At mole fractions of 0.1 and 0.9 the molecules of the solvent were suppressed to improve visibility. The methyl and methylene groups are shown in *orange*, the methine sites in *brown*, the oxygen atoms in *red* and the chlorine atoms in *green*

Acknowledgements We gratefully acknowledge support by Deutsche Forschungsgemeinschaft. This work was carried out under the auspices of the Boltzmann-Zuse Society (BZS) of Computational Molecular Engineering. The simulations were performed on the national supercomputer Hazel Hen at the High Performance Computing Center Stuttgart (HLRS) within the project MMHBF2.

References

1. Abrams, D.S., Prausnitz, J.M.: Statistical thermodynamics of liquid mixtures: a new expression for the excess Gibbs energy of partly or completely miscible systems. AIChE J. **21**, 116–128 (1975)
2. Alfe, D., Gillan, M.J.: First-principles calculation of transport coefficients. Phys. Rev. Lett. **81**, 5161–5164 (1988)
3. Allen, M.P., Tildesley, D.J.: Computer simulation of liquids. Clarendon Press, Oxford (1987)
4. Bonnaud, P., Nieto-Draghi, C., Ungerer, P.: Anisotropic united atom model including the electrostatic interactions of benzene. J. Phys. Chem. B **111**, 3730–3741 (2007)
5. Campbell, A., Chatterjee, R.: The critical constants and orthobaric densities of acetone, chloroform, benzene, and carbon tetrachloride. Can. J. Chem. **47**, 3893–3898 (1969)
6. Collings, A., Mills, R.: Temperature-dependence of self-diffusion for benzene and carbon tetrachloride. Trans. Faraday Soc. **66**, 2761–2766 (1970)
7. Computational Chemistry Comparison and Benchmark Data Base, Standard Reference Data Base No. 101. The National Institute of Standards and Technology. http://cccbdb.nist.gov/mulliken2.asp (2015)
8. Darken, L.S.: Diffusion, mobility and their interrelation through free energy in binary metallic systems. Trans. Am. Inst. Min. Met. Eng. **175**, 184–201 (1948)
9. Deublein, S., Eckl, B., Stoll, J., Lishchuk, S.V., Guevara-Carrion, G., Glass, C.W., Merker, T., Bernreuther, M., Hasse, H., Vrabec, J.: ms2: a molecular simulation tool for thermodynamic properties. Comput. Phys. Commun. **182**, 2350–2367 (2011)
10. Evans, D.J., Morris, G.P.: Statistical Mechanics of Nonequilibrium Liquids. Academic, London (1990)
11. Falcone, D.R., Douglass, D.C., McCall, D.W.: Self-diffusion in benzene. J. Phys. Chem. **71**, 2754–2755 (1967)
12. Filippov, L.P.: Teploprovodnost' rastvorov associirovannyh zhidkostej. Vest. Mosk. Univ., Ser. Fiz. Mat. Estestv. Nauk **10**, 67–69 (1955)
13. Fischer, S.: Experimentelle und theoretische Untersuchung des Einflusses der thermischen Strahlung auf die effektive Wärmeleitfähigkeit von Flüssigkeiten. Ph.D. thesis, Universität Siegen, Germany (1984)
14. Fischer, J.D.: Transporteigenschaften reiner Flüssigkeiten und binärer Mischungen mit unterschiedlichen Wechselwirkungsparametern. Ph.D. thesis, TH Darmstadt (1986)
15. Fischer, J., Weiss, A.: Transport properties of liquids. V. Self diffusion, viscosity, and mass density of ellipsoidal shaped molecules in the pure liquid phase. Ber. Bunsenges. Phys. Chem. **90**, 896–905 (1986)
16. Glass, C.W., Reiser, S., Rutkai, G., Deublein, S., Köster, A., Guevara-Carrion, G., Wafai, A., Horsch, M., Bernreuther, M., Windmann, T., Hasse, H., Vrabec, J.: ms2: a molecular simulation tool for thermodynamic properties, new version release. Comp. Phys. Commun. **185**, 3302–3306 (2014)
17. Graupner, K., Winter, E.R.S.: Some measurements of the self-diffusion coefficients of liquids. J. Chem. Soc. (Resumed) **1**, 1152–1150 (1952)
18. Gubbins, K.E., Quirke, N.: Introduction to Molecular Simulation and Industrial Applications: Methods, Examples and Prospects. Gordon and Breach Science Publishers, Amsterdam (1996)
19. Harris, K.R., Alexander, J.J., Goscinska, T., Malhotra, R., Woolf, L.A., Dymond, J.H.: Temperature and density dependence of the selfdiffusion coefficients of liquid n-octane and toluene. Mol. Phys. **78**, 235–248 (1993)
20. Hiraoka, H.: Self-diffusion of benzene under pressure. Bull. Chem. Soc. Jpn. **32**, 423–424 (1959)
21. Hirschfelder, J.O., Curtiss, C.F., Bird, R.B.: Molecular theory of gases and liquids. Wiley, New York (1954)

22. Ikeuchi, H., Kanakubo, M., Okuno, S., Sato, R., Fujita, K., Hamada, M., Shoda, N., Fukai, K., Okada, K., Kanazawa, H.: Densities and viscosities of tris(acetylacetonato)cobalt(III) complex solutions in various solvents. J. Solut. Chem. **39**, 1428–1453 (2010)
23. Krishna, R., van Baten, J.M.: The darken relation for multicomponent diffusion in liquid mixtures of linear alkanes: an investigation using molecular dynamics (MD) simulations. Ind. Eng. Chem. Res. **44**, 6939–6847 (2005)
24. Krüger, G., Weiss, R.: Diffusionskonstanten einiger organischer Flüssigkeiten. Z. Naturforsch. A **25**, 777–780 (1970)
25. Lei, Q.F., Lin R.-S., Ni, D.Y., Hou, Y.C.: Thermal conductivities of some organic solvents and their binary mixtures. J. Chem. Eng. Data **42**, 971–974 (1997)
26. Lemmon, E.W., Span, R.: Short fundamental equations of state for 20 industrial fluids. J. Chem. Eng. Data **51**, 785–850 (2006)
27. Li, J., Liu, H., Hu, Y.: A mutual-diffusion-coefficient model based on local composition. Fluid Phase Equilib. **187–188**, 193–208 (2001)
28. Liu, X., Schnell, S.K., Simon, J.M., Bedeaux, D., Kjelstrup, S., Bardow, A., Vlugt, T.J.H.: Fick diffusion coefficients of liquid mixtures directly obtained from equilibrium molecular dynamics. J. Phys. Chem. B **115**, 12921–12929 (2011)
29. Luchinskii, G.: Mechanical characteristics of Halogene anhydride's molecules. Zh. Obshch. Khim. **7**, 2116–2127 (1937)
30. Lustig, R.: Angle-average for the powers of the distance between two separated vectors. Mol. Phys. **65**, 175–179 (1988)
31. Maginn, E.J., Elliot, J.R.: Historical perspective and current outlook for molecular dynamics as a chemical engineering tool. Ind. Eng. Chem. Res. **49**, 3059–3078 (2010)
32. McCool, M.A., Collings, A.F., Woolf, L.A.: Pressure and temperature dependence of the self-diffusion of benzene. J. Chem. Soc. Faraday Trans. 1 **68**, 1489–1497 (1972)
33. Merker, T., Engin, C., Vrabec, J., Hasse, H.: Molecular model for carbon dioxide optimized to vapor-liquid equilibria. J. Chem. Phys. **132**, 234512 (2010)
34. Merker, T., Vrabec, J., Hasse, H.: Engineering molecular models: efficient parameterization procedure and cyclohexanol as case study. Soft Matter **10**, 3–25 (2012)
35. Muñoz-Muñoz, Y.M., Guevara-Carrion, G., Llano-Restrepo, M., Vrabec, J.: Lennard-Jones force field parameters for cyclic alkanes from cyclopropane to cyclohexane. Fluid Phase Equilib. **404**, 150–160 (2015)
36. Nieto-Draghi, C., Bonnaud, P., Ungerer, P.: Anisotropic united atom model including the electrostatic interactions of methylbenzenes. I. Thermodynamic and structural properties. J. Phys. Chem. C **111**, 15686–15699 (2007)
37. Nieto-Draghi, C., Bonnaud, P., Ungerer, P.: Anisotropic united atom model including the electrostatic interactions of methylbenzenes. II. Transport properties. J. Phys. Chem. C **111**, 15942–15951 (2007)
38. Pickup, S., Blum, F.D.: Self-diffusion of toluene in polystyrene solutions. Macromolecules **22**, 3961–3968 (1989)
39. Poling, B.E., Thomson, D.W., Friend, D.G., Rowley, R.L., Wilding, W.V.: Section 2. Physical and chemical data. In: Perry, R.H., Green, D.W. (eds.) Perry's Chemical Engineers' Handbook, 8th edn. McGraw-Hill, New York (2008)
40. Požar, M., Seguier, J.B., Guerche, J., Mazighi, R., Zoranić, L., Mijaković, M., Kežić-Lovrinčević, B., Sokolić, F., Perera, A.: Simple and complex disorder in binary mixtures with benzene as a common solvent. Phys. Chem. Chem. Phys. **17**, 9885–9898 (2015)
41. Rathbun, R., Babb, A.: Self-diffusion in liquids. III. Temperature dependence in pure liquids. J. Phys. Chem. **65**, 1072–1074 (1961)
42. Renon, H., Prausnitz, J.M.: Local compositions in thermodynamic excess functions for liquid mixtures. AIChE J. **14**, 135–144 (1968)
43. Rowley, R., White, G.: Thermal conductivities of ternary liquid mixtures. J. Chem. Eng. Data **32**, 63–69 (1987)
44. Rutten, P.W.M.: Diffusion in Liquids. Delft University Press, Delft (1992)

45. Santos, F.J.V., Nieto de Castro, C.A., Dymond, J.H., Dalaouti, N.K., Assael, M.J., Nagashima, A.: Standard reference data for the viscosity of toluene. J. Phys. Chem. Ref. Data **35**, 1–8 (2006)
46. Schnabel, T., Vrabec, J., Hasse, H.: Henry's law constants of methane, nitrogen, oxigen and carbon dioxide in ethanol from 273 to 498 K: prediction from molecular simulation. Fluid Phase Equilib. **233**, 134–143 (2005)
47. Schnabel, T., Srivastava, A., Vrabec, J., Hasse, H.: Hydrogen bonding of methanol in supercritical CO2: comparison between 1H-NMR spectroscopic data and molecular simulation results. J. Phys. Chem. B **111**, 9871–9878 (2007)
48. Schoen, M., Hoheisel, C.: The mutual diffusion coefficient D_12 in binary liquid model mixtures. Molecular dynamics calculations based on Lennard-Jones (12-6) potentials. Mol. Phys. **52**, 33–56 (1984)
49. Thol, M., Lemmon, E.W., Span, R.: Equation of state for benzene for temperatures from the melting Line up to 725 K with pressures up to 500 MPa. High Temp. High Press. **41**, 81–97 (2012)
50. Trepădus, V., Răpeanu, S., Pădureanu, I., Parfenov, V.A., Novikov, A.G.: Study of molecular rotations in some aromatic compounds by cold neutron scattering. J. Chem. Phys. **60**, 2832–2839 (1974)
51. Vignes, A.: Diffusion in binary solutions. Variation of diffusion coefficient with composition. Ind. Eng. Chem. Fundam. **5**, 189–199 (1966)
52. Wakeham, W.A.: Transport properties and industry. In: Letcher, T.M. (ed.) Chemical Thermodynamics for Industry. The Royal Society of Chemistry, London (2004)
53. Wensink, E.J.W., Hoffmann, A.C., van Maaren, P.J., van der Spoel, D.: Dynamic properties of water/alcohol mixtures studied by computer simulation. J. Chem. Phys. **119**, 7308–7317 (2003)
54. Wilson, G.M.: Vapor-liquid equilibrium. A new expression for the excess free energy of mixing. J. Am. Chem. Soc. **86**, 127–130 (1964)
55. Windfield, D.J.: Measurement of the apparent diffusion coefficient of toluene by quasielastic neutron scattering. J. Chem. Phys. **54**, 3643–3645 (1971)
56. Windmann, T., Linnemann, M., Vrabec, J.: Fluid phase behavior of nitrogen + acetone and oxygen + acetone by molecular simulation, experiment and the Peng-Robinson equation of state. J. Chem. Eng. Data **59**, 28–38 (2014)
57. Zhou, M., Yuan, X., Zhang, Y., Yu, K.T.: Local CompLocal composition based Maxwell–Stefan diffusivity model for binary liquid Systemsosition based Maxwell–Stefan diffusivity model for binary liquid systems. Ind. Eng. Chem. Res. **52**, 10845–10852 (2013)
58. Zhu, Q., Moggridge, G.D., D'Agostino, C.: A local composition model for the prediction of mutual diffusion coefficients in binary liquid mixtures from tracer diffusion coefficients. Chem. Eng. Sci. **132**, 250–258 (2015)

Large-Scale Phase-Field Simulations of Directional Solidified Ternary Eutectics Using High-Performance Computing

J. Hötzer, M. Kellner, P. Steinmetz, J. Dietze, and B. Nestler

Abstract The combination of different chemical elements allows to obtain new and improved materials, as required for novel applications. Especially directionally solidified multicomponent eutectic alloys exhibit a wide range of patterns in the microstructure, which are correlated to the mechanical properties. The pattern formation during solidification depends on the chemical elements and the applied process parameters. Large-scale phase-field simulations are used to study the pattern formation of directional solidified ternary eutectics. Three different systems, starting from a model system towards the system Al-Ag-Cu are investigated, using three growth velocities. The three-dimensional simulation results are quantitatively compared and a broad variety of arising patterns for the studied systems is found. The results of the velocity variation follow the predictions from the analytic Jackson-Hunt approach.

1 Introduction

The development of novel applications requires materials with defined properties. The macroscopic mechanical properties are influenced by the chemical composition as well as the evolving microstructure. Ternary eutectic alloys exhibit a wide range of microstructures. Especially during the directional solidification, parallel to the growth front, various patterns form. Despite several experimental works [5–10] to investigate the patterns formation in the ternary eutectic system Al-Ag-Cu, where various patterns are reported, the influence of physical and process parameters is

---

The authors "J. Hötzer, M. Kellner and P. Steinmetz" contributed equally.

J. Hötzer (✉) • M. Kellner • P. Steinmetz • J. Dietze • B. Nestler
Institute of Applied Materials, Reliability of Components and Systems (IAM-ZBS), Karlsruhe Institute of Technology (KIT), Haid-und-Neu-Str. 7, 76131 Karlsruhe, Germany

Institute of Materials and Processes, Hochschule Karlsruhe Technik und Wirtschaft, Moltkestr. 30, Karlsruhe, Germany
e-mail: johannes.hoetzer@kit.edu; michael.kellner@kit.edu; philipp.steinmetz@kit.edu; johannes.dietze@student.kit.edu; britta.nestler@kit.edu

W.E. Nagel et al. (eds.), *High Performance Computing in Science and Engineering '16*, DOI 10.1007/978-3-319-47066-5_43

not yet fully understood. To study for example the influence of gravity on the pattern formation, directional solidification experiments of Al-Ag-Cu are conducted on the international space station (ISS) [19]. For the simulative investigation of the pattern evolution we exploit large-scale phase-field simulations. A thermodynamic consistent phase-field model based on the Grand potential approach is applied [4, 13, 18]. Previous studies proved, that large-scale simulations are required to study the patterns formation for different systems in order to avoid effects of the domain boundary on the morphology [2, 12, 15, 21–24]. In systematic simulation studies, we analyze the influences of the equilibrium concentrations, the interface energies and the solidification velocity on the microstructure evolution. We start from an idealized symmetric ternary system and change it towards the real asymmetric system Al-Ag-Cu. To quantitatively classify and compare the arising patterns, a novel analysis method based on the second moment of inertia is introduced.

2 Methods

In the following, the phase-field model and the novel quantitative analysis method are presented.

2.1 *Phase-Field Method*

Based on a thermodynamic consistent Grand Potential functional [4, 18] and an Allen-Cahn approach the evolution equations for the four phase-fields ϕ_α (three solid and one liquid phase) are derived, to simulate the directional solidification. Together with the evolution equations for the $K = 3$ chemical potentials $\boldsymbol{\mu}$ and the analytic temperature T, a coupled set of partial differential equations is obtained to simulate the directional solidification of multicomponent and multiphase alloys. The evolution equations are formulated as:

$$\tau\epsilon\frac{\partial\phi_\alpha}{\partial t} = \underbrace{\begin{array}{c} -\epsilon T\left(\dfrac{\partial a(\boldsymbol{\phi},\nabla\boldsymbol{\phi})}{\partial\phi_\alpha} - \nabla\cdot\dfrac{\partial a(\boldsymbol{\phi},\nabla\boldsymbol{\phi})}{\partial\nabla\phi_\alpha}\right) \\ -\frac{1}{\epsilon}T\dfrac{\partial\omega(\boldsymbol{\phi})}{\partial\phi_\alpha} - \displaystyle\sum_{\beta=1}^{N}\psi_\beta(\boldsymbol{\mu},T)\dfrac{\partial h_\beta(\boldsymbol{\phi})}{\partial\phi_\alpha}\end{array}}_{:=-\delta\Psi/\delta\phi_\alpha} - \frac{1}{N}\sum_{\beta=1}^{N}{}^{\delta\Psi}/_{\delta\phi_\beta}, \tag{1}$$

$$\begin{aligned}\frac{\partial\boldsymbol{\mu}}{\partial t} = &\left[\sum_{\alpha=1}^{N}h_\alpha(\boldsymbol{\phi})\left(\frac{\partial\boldsymbol{c}_\alpha(\boldsymbol{\mu},T)}{\partial\boldsymbol{\mu}}\right)\right]^{-1}\Bigg(\nabla\cdot\Big(\boldsymbol{M}(\boldsymbol{\phi},\boldsymbol{\mu},T)\nabla\boldsymbol{\mu} - \boldsymbol{J}_{at}(\boldsymbol{\phi},\boldsymbol{\mu},T)\Big)\\ &-\sum_{\alpha=1}^{N}\boldsymbol{c}_\alpha(\boldsymbol{\mu},T)\frac{\partial h_\alpha(\boldsymbol{\phi})}{\partial t} - \sum_{\alpha=1}^{N}h_\alpha(\boldsymbol{\phi})\left(\frac{\partial\boldsymbol{c}_\alpha(\boldsymbol{\mu},T)}{\partial T}\right)\frac{\partial T}{\partial t}\Bigg).\end{aligned} \tag{2}$$

The differences of the Grand potentials ψ, acting as driving force for the phase transitions, can be derived from parabolic fits of the Gibbs energies, provided from thermodynamic CALPHAD databases [3]. The model is implemented in the massive parallel framework WALBERLA [2, 11]. A detailed description of the model is given in [13], the discretisation in [14] and applied computational optimizations and the scaling behaviour in [2].

2.2 Analysis the Second Moment of Inertia

In cross sections parallel to the growth front, rods and lamellae of the three solid phases arrange in different patterns during the directional solidification. To compare single rods and lamellae from experiments and simulations a quantitative method, based on the second moment of inertia, is presented. Adapting the method of [17] to quantify grains and combining it with the algorithm of [1], enables the automated analysis of the conducted phase-field simulations with periodic boundaries. The second moments of area are defined as

$$\hat{\mu}_{p,q} = \int_A x^p y^q dA \quad \forall\, p, q \in \{0, 1, 2\} \mid p + q = 2\,. \tag{3}$$

From this, the inertia tensor in the principle axis system of the form

$$\mathbb{J} = \begin{pmatrix} \hat{\mu}_{02} & \hat{\mu}_{11} \\ \hat{\mu}_{11} & \hat{\mu}_{20} \end{pmatrix} = \begin{pmatrix} \int_A y^2 dA & \int_A xydA \\ \int_A xydA & \int_A x^2 dA \end{pmatrix} \tag{4}$$

is derived. The first and second main invariances of $\mathbb{J}$ are expressed in dimensionless form by dividing it with the surface area. Subsequently, the dimensionless invariances are inverted, following [17], leading to

$$\hat{\omega}_1 = \frac{A^2}{\hat{\mu}_{02} + \hat{\mu}_{20}} \tag{5}$$

$$\hat{\omega}_2 = \frac{A^4}{\hat{\mu}_{02}\hat{\mu}_{20} - \hat{\mu}_{11}^2}\,. \tag{6}$$

After scaling $\hat{\omega}_1$ and $\hat{\omega}_2$ with the invariances of a circle, the operating numbers Ω_1 and Ω_2 can be derived, as presented in [17].

$$\Omega_1 = \frac{\hat{\omega}_1}{\hat{\omega}_{1,\text{circle}}} = \frac{A^2}{2\pi(\hat{\mu}_{02} + \hat{\mu}_{20})} \tag{7}$$

$$\Omega_2 = \frac{\hat{\omega}_2}{\hat{\omega}_{2,\text{circle}}} = \frac{A^4}{16\pi^2(\hat{\mu}_{02}\hat{\mu}_{20} - \hat{\mu}_{11}^2)} \tag{8}$$

The Ω_2 is related to its shape of the considered object and the Ω_1 value to its distortion. To determine the inertia tensor in a principle axis system, the barycenter of the considered structure has to be calculated. Due to the periodic boundary conditions in the simulations, the structures can expand over the domain boundaries. To calculate the barycenters of these structures, the algorithm of Bai and Breen [1] is applied. Initially, all cells with the coordinates x_i, y_i, belonging to the structure, are projected on a line with the normalized coordinate $s_i = {}^{x_i}/_{x_{max}}$ with $s_i \in [0, 1]$. This line is defined normal to the corresponding periodic boundaries and x_{max} is the width in this direction. Afterwards, the points on the line are mapped on a circle using the coordinate transformation of each point s_i to $\hat{x}_i$, $\hat{y}_i$ with

$$\hat{x}_i = \cos(2\pi s_i) \ , \tag{9}$$

$$\hat{y}_i = \sin(2\pi s_i) \ . \tag{10}$$

These coordinates of all N points are averaged by an arithmetic mean, separately as

$$\bar{\hat{x}} = \frac{1}{N}\sum_{i=1}^{N}\hat{x}_i \ , \tag{11}$$

$$\bar{\hat{y}} = \frac{1}{N}\sum_{i=1}^{N}\hat{y}_i \ . \tag{12}$$

The two average values are projected back in the original coordinate system using

$$\bar{x} = x_{max}\left(\frac{atan2(-\bar{\hat{y}}, -\bar{\hat{x}})}{2\pi} + \frac{1}{2}\right) \ . \tag{13}$$

This algorithm is repeated for each dimension with periodic boundaries and can also be used to calculate the inertia tensor in the principle axis system with the Steiner's theorem.

With this method, rods in the microstructure from simulations as well as experiments can be automatically analyzed and quantitatively compared, independent from the rod size. Also the method can be applied to investigate the shape evolution of a single rod during the solidification process. This allows to identify a stationary growth state and derive a criterion to stop the simulation. With this in-situ analysis the required computational time can be reduced.

3 Ternary Eutectic Directional Solidification

The influence of the process conditions on the pattern formation of ternary eutectics during directional solidification, is investigated with large-scale phase-field simulations. For this, the growth velocity for three different systems is systematically varied.

For the simulations a setup as following is applied: Starting from an initial Voronoi tessellation, the nuclei of the three solid phases grow coupled in direction of an imprinted temperature gradient. This gradient is pulled in a defined direction with the velocity v. To simulate an infinite domain, periodic boundaries are applied on the sides parallel to the growth direction and a Dirichlet boundary condition is used to model a infinite liquid domain above the solidification front.

The equilibrium concentrations of the three systems S_1, S_2 and S_3 are depicted in the ternary isotherm concentration diagram Fig. 1. The ternary eutectic point is described by the equilibrium concentration of the liquid. With these systems the transition from an ideal system (S_1), as investigated in [22], to a model of the real system Al-Ag-Cu (S_3) as studied in [13, 21, 24], is shown. In system S_1, the equilibrium concentrations of the solid phases α, β and γ are equally distributed around the liquid equilibrium concentration and the interface energies are set equal. For the systems S_2 and S_3 the arrangement of the solid equilibrium concentrations around the eutectic point are chosen similar to [13]. Equal interface energies, which

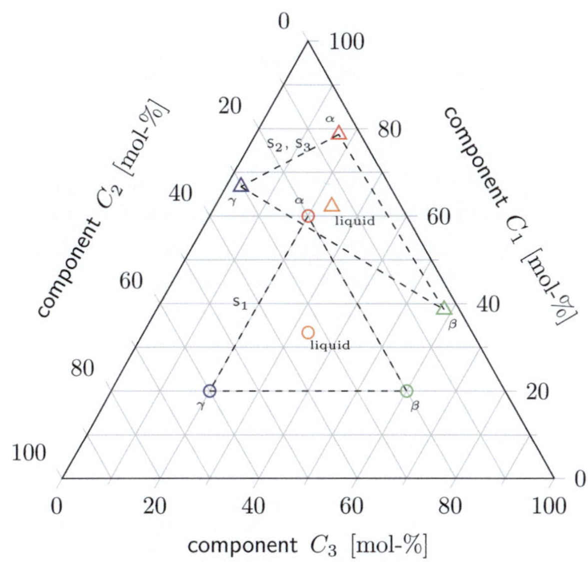

Fig. 1 Equilibrium concentrations of the solid phases and the liquid for system S_1 (marked with *circles*) and the systems S_2, S_3 (marked with *triangles*)

Table 1 Physical and numerical parameters for the conducted simulations

Parameter	Simulation value
D	5
T_0	0.91
∇T	10^{-4}
dx	1.0
dt	$3.2 \cdot 10^{-2}$
ε	4.0

are the same as in system S_1, are applied for system S_2, while system S_3 is modeled with the interface energies from [13, 22].

For these three systems, simulations with the velocities $v_1 = 1.74 \cdot 10^{-3}$, $v_2 = 2.76 \cdot 10^{-3}$ and $v_3 = 2.61 \cdot 10^{-3}$ are conducted. A selection of the common parameters for all simulations are summarized in Table 1.

The simulations are conducted with 3 million time steps in a domain of $800 \times 800 \times 250$ cells. Due to the applied moving window technique [2, 25], this corresponds to a growth height of approximately 6300 cells. Stationary growth is ensured for all presented simulations. To reduce the output data, only the surface meshes are stored. For each simulation, 13, 600 cores for 8 h were utilized, resulting in approximately 200 Gbyte of simulation data. In all simulation results, the phase α is marked in red, β in green and γ in blue. In the following, the simulation results are described and discussed.

3.1 System S_1

The simulation fronts after 3 million time steps for the three velocities v_1 to v_3 of system S_1 are depicted in Fig. 2. For all simulations different aligned hexagonal structures evolve, divided by contact zones, as discussed in previous work [22]. With an increasing velocity, the microstructure becomes finer. This is accompanied with an increase in both, the number of rods as well as the interface length. The observed refinement is in accordance with the analytical Jackson-Hunt approach [16]. Therefore, the characteristics of the two-dimensional Jackson-Hunt approach are also fulfilled for three-dimensional growth, as shown in [22] with phase-field simulations. This approach predicts, that for the minimum undercooling, the lamellar spacing λ and v are related as

$$\lambda^2 v = \text{constant} . \tag{14}$$

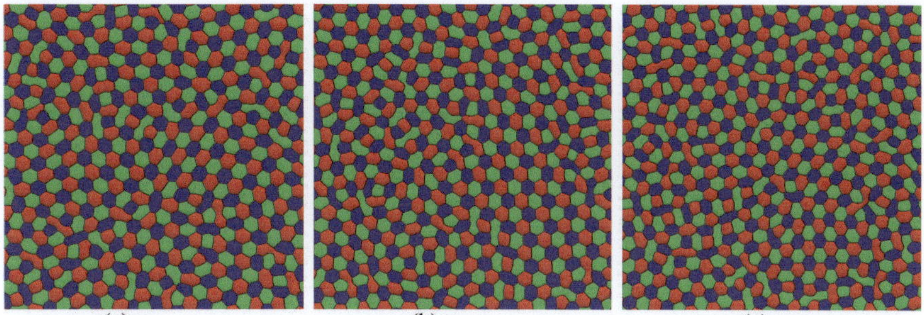

(a) v_1 (b) v_2 (c) v_3

Fig. 2 Solidification front of system S_1 after 3 million time steps for the three velocities $v_1 = 1.74 \cdot 10^{-3}$ (**a**), $v_2 = 2.76 \cdot 10^{-3}$ (**b**) and $v_3 = 2.61 \cdot 10^{-3}$ (**c**)

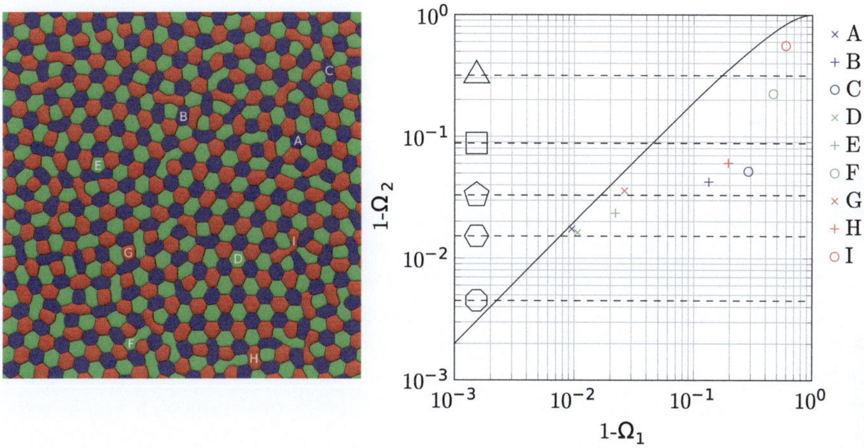

Fig. 3 In the *left pattern image*, selected rods from the simulated micrograph of system S_1 are marked with A-I for the velocity v_2 and on the *right side*, the position of the rods are plotted in an $1 - \Omega_1$ over $1 - \Omega_2$ diagram

The lamellar spacing λ describes the width of a repeating phase arrangement. Also the phase fractions remain constant for the three different velocities and similar patterns evolve. In Fig. 3 selected rods of system S_1 for the velocity v_2 are quantitatively compared. Therefor the method introduced in Sect. 2.2, using the second moment of inertia, is applied. In the diagram of Fig. 3, $1 - \Omega_2$ is plotted over $1 - \Omega_1$ in double-logarithmic scale, for a better comparison of the rods. The blue γ rods are depicted as A to C. The letters D to F identify three blue β rods and G to I red α rods. In the simulation various forms of rods can be quantitatively distinguished. A regular hexagonal form is visually as well quantitatively found for the rods A and D. The other rods exhibit different values of the operating numbers Ω. These different forms are indicated by the intersections of the curve, defined by $\Omega_2 = \Omega_1^2$, with the dashed horizontal lines in the diagram. This

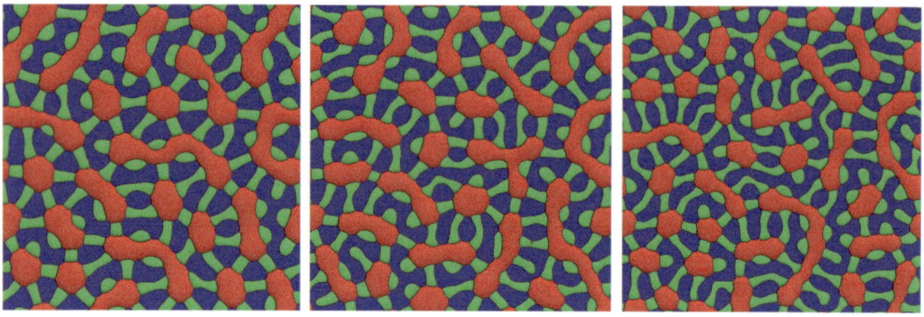

(a) v_1 (b) v_2 (c) v_3

Fig. 4 Solidification front of system S_2 after 3 million time steps for the three velocities v_1 (**a**), v_2 (**b**) and v_3 (**c**)

emphasizes the visual observation, that the rods do not align in a regular hexagonal structure.

3.2 System S_2

In the next step the parabolic free energies are shifted to those of the Al-Ag-Cu, as given in [13], but still equal interface energies are applied. The location and arrangement of the equilibrium concentrations are shown in the liquidus projection in Fig. 1.

In Fig. 4, the rods arrange in chain-like structures consisting of β and γ rods, embedded in a matrix phase α for all three velocities. As also observed in the simulations of [13], junctions and ring-like structures occur. This behavior is reported in experiments of directionally solidified ternary eutectics [5–10]. Due to the asymmetric arrangement of the equilibrium concentrations around the eutectic point, different phase fractions evolve for the three simulated growth velocities. This is caused by an adjustment of the front undercooling, depending on the growth velocity predicted in the analytical approach from Jackson and Hunt [16]. Therefore, different concentrations in the solid phases are established. Similar to system S_1 the structures become finer with an increase of the velocity.

3.3 System S_3

To accurately approximate the real material system, interface energies similar to [13, 21] are applied for system S_3. The solidification fronts at the end of the simulations are shown in Fig. 5 for the three different velocities v_1, v_2 and v_3.

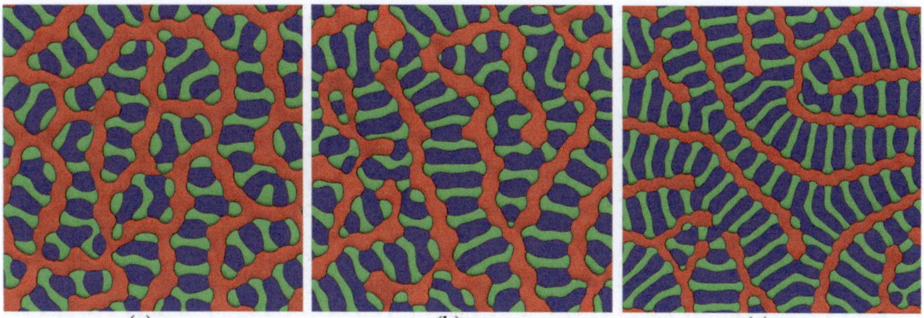

(a) v_1 (b) v_2 (c) v_3

Fig. 5 Solidification front of system S_3 after 3 million time steps for the three velocities v_1 (**a**), v_2 (**b**) and v_3 (**c**)

In the microstructure corresponding to the velocity v_1, island structures evolve as short chains [13]. For higher velocities the island structures merge and more aligned chains arise. Similar to the systems S_1 and S_2, the chain linkages become finer as analytically predicted [16]. With different growth velocities, the same tendencies in the change of the phase fractions can be seen in the systems S_2 and S_3. The chains in system S_3 are less branched, compared to system S_2. Due to the larger variety of patterns in system S_3, an influence of the interface energies on the pattern formation can be observed.

3.4 Comparison of the Three Systems with the Method of the Second Moment of Inertia

For a quantitative comparison of the arising patterns in the three systems S_1, S_2 and S_3 the previously introduced method of the second moment of inertia is applied. In Fig. 6 the probability of a Ω_1 and Ω_2 value for a certain rod shape is shown. On the left side, the probability of the shapes for the β rods and on the right side for the γ rods are depicted for the velocity v_1. For system S_1, a pronounced peak near $\Omega_1 = 0.9924$ and $\Omega_2 = 0.9848$ can be seen for both phases. These values correspond to an undistorted hexagon. For system S_2, no peak can be observed for the β rods and for γ, the peak is smaller than for system S_1. This trend continues for system S_3. The variance of the probabilities over Ω_1 and Ω_2 increase from system S_1 to system S_3. The introduced analysis with the second moment of inertia reflects the visual observation of different rod shapes in a quantitative manner.

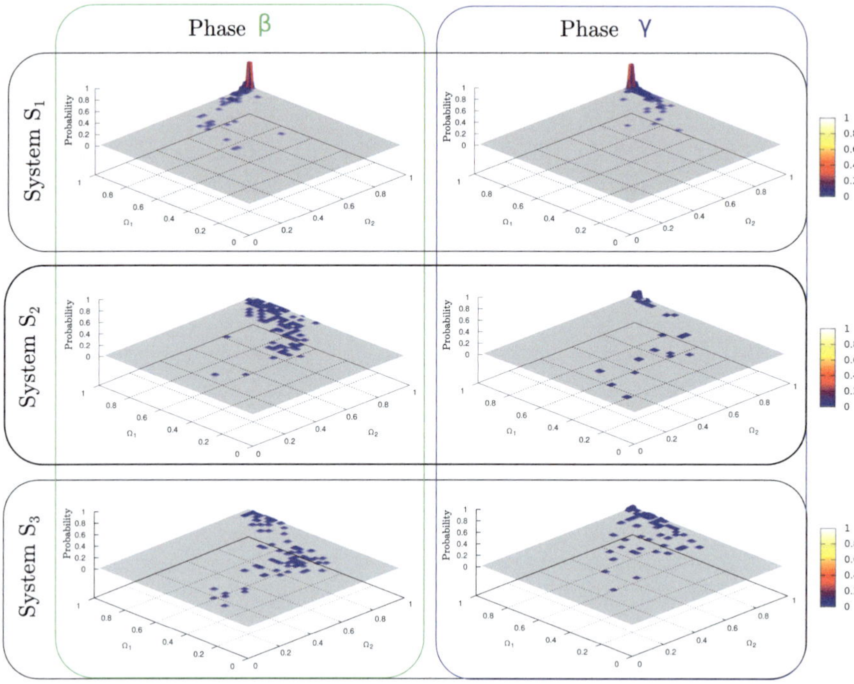

Fig. 6 Probability of a Ω_1 and Ω_2 value for a certain rod shape of the phases β (*left column*) and γ (*right column*) of the systems S_1, S_2 and S_3

4 Conclusion

In this work, we investigated the influence of process and physical parameters on the pattern formation of directionally solidified ternary eutectics. An idealized ternary eutectic alloy (system S_1) is systematically changed towards the real system of Al-Ag-Cu (system S_3). The pattern formation for all systems is investigated for three different velocities and quantitatively compared with the presented method of the second moment of inertia.

The conducted simulations follow the predictions of the analytical Jackson-Hunt approach, that higher velocities result in finer microstructures. The systems S_2 and S_3 refer to asymmetric arranged equilibrium concentration and we observe that the phase fractions change for the different velocities. This effect of the velocity on the phase fractions is also reported from experiments (Dennstedt, A.: 2016-03-21. Private communication). In our simulations, a variation of the interface energies leads to a better visual accordance with experimental micrographs. With the method of the second moment of inertia, it is quantitatively shown, that a larger deviation from the idealized system results in a higher variance of the rod shapes. We conclude that the analysis method is suited to classify single rod shapes as well as the total microstructure.

In experiments, various solidification processes take place at different length and time scales and a complex spatial interplay between the arising structures occurs. The coupled ,growth of primary dendrites with an interdendritic eutectic substructure is one of these processes [20] and will be focus of our forthcoming simulations and research. To further improve the prediction of solidification processes with simulations, additional physical phenomena have to be incorporated. Even for the actual generation of supercomupters these kinds of multiscale microstructures are challenging concerning the computational effort.

References

1. Bai, L., Breen, D.: Calculating center of mass in an unbounded 2d environment. J. Gr. GPU Game Tools **13**(4), 53–60 (2008)
2. Bauer, M., Hötzer, J., Steinmetz, P., Jainta, M., Berghoff, M., Schornbaum, F., Godenschwager, C., Köstler, H., Nestler, B., Rüde, U.: Massively parallel phase-field simulations for ternary eutectic directional solidification (2015). arXiv preprint arXiv:1506.01684, accepted at SuperComputing 2015
3. Choudhury, A., Kellner, M., Nestler, B.: A method for coupling the phase-field model based on a grand-potential formalism to thermodynamic databases. Curr Opin Solid State Mater Sci **19**(0), 287–300 (2015)
4. Choudhury, A., Nestler, B.: Grand-potential formulation for multicomponent phase transformations combined with thin-interface asymptotics of the double-obstacle potential. Phys. Rev. E 85:021602 (2012)
5. Dennstedt, A., Choudhury, A., Ratke, L., Nestler, B.: Microstructures in a ternary eutectic alloy: devising metrics based on neighbourhood relationships. In: IOP Conference Series: Materials Science and Engineering, Kazan (2014)
6. Dennstedt, A., Helfen, L., Steinmetz, P., Nestler, B., Ratke, L.: 3D synchrotron imaging of a directionally solidified ternary eutectic. Metall. Mater. Trans. A **47**, 981–984 (2015)
7. Dennstedt, A., Ratke, L.: Microstructures of directionally solidified Al-Ag-Cu ternary eutectics. Trans. Indian Inst. Met. **65**(6), 777–782 (2012)
8. Dennstedt, A., Ratke, L., Choudhury, A., Nestler, B.: New metallographic method for estimation of ordering and lattice parameter in ternary eutectic systems. Metall. Microstruct. Anal. **2**(3), 140–147 (2013)
9. Genau, A.L., Ratke, L.: Crystal orientation and morphology in Al–Ag–Cu ternary eutectic. IOP Conf. Ser.: Mater. Sci. Eng. **27**(1), 012032 (2012)
10. Genau, A., Ratke, L.: Morphological characterization of the Al-Ag-Cu ternary eutectic. Int. J. Mater. Res. **103**(4), 469–475 (2012)
11. Godenschwager, C., Schornbaum, F., Bauer, M., Köstler, H., Rüde, U.: A framework for hybrid parallel flow simulations with a trillion cells in complex geometries. In: Proceedings of SC13: International Conference for High Performance Computing, Networking, Storage and Analysis, p. 35. ACM, New York (2013)
12. Hötzer, J., Jainta, M., Steinmetz, P., Dennstedt, A., Nestler, B.: Die Vielfalt der Musterbildung in Metallen (2015)
13. Hötzer, J., Jainta, M., Steinmetz, P., Nestler, B., Dennstedt, A., Genau, A., Bauer, M., Köstler, H., Rüde, U.: Large scale phase-field simulations of directional ternary eutectic solidification. Acta Materialia **93**, 194–204 (2015)
14. Hötzer, J., Tschukin, O., Said, M.B., Berghoff, M., Jainta, M., Barthelemy, G., Smorchkov, N., Schneider, D., Selzer, M., Nestler, B.: Calibration of a multi-phase field model with quantitative angle measurement. J. Mater. Sci. **51**(4), 1788–1797 (2016)

15. Hötzer, J., Steinmetz, P., Jainta, M., Schulz, S., Kellner, M., Nestler, B., Genau, A., Dennstedt, A., Bauer, M., Köstler, H., Rüde, U.: Phase-field simulations of spiral growth during directional ternary eutectic solidification. Acta Materialia **106**, 249–259 (2016)
16. Jackson, K.A., Hunt, J.D.: Lamellar and rod eutectic growth. Aime Met. Soc. Trans. **236**, 1129–1142 (1966)
17. MacSleyne, J.P., Simmons, J.P., De Graef, M.: On the use of 2-d moment invariants for the automated classification of particle shapes. Acta Materialia **56**(3), 427–437 (2008)
18. Plapp, M.: Unified derivation of phase-field models for alloy solidification from a grand-potential functional. Phys. Rev. E **84**, 031601 (2011)
19. Rex, S.: ACCESS e.V., RWTH Aachen, SETA – Das Erstarrungsverhalten von mehrkomponentigen Legierungen, 2014-03-07. Accessed 25 Feb 2016
20. Rinaldi, M.D., Sharp, R.M., Flemings, M.C.: Growth of ternary composites from the melt: Part II. Metall. Trans. **3**(12), 3139–3148 (1972)
21. Steinmetz, P., Yabansu, Y.C., Hötzer, J., Jainta, M., Nestler, B., Kalidindi, S.R.: Analytics for microstructure datasets produced by phase-field simulations. Acta Materialia **103**, 192–203 (2016)
22. Steinmetz, P., Hötzer, J., Kellner, M., Dennstedt, A., Nestler, B.: Large-scale phase-field simulations of ternary eutectic microstructure evolution. Comput. Mater. Sci. **117**, 205–214 (2016)
23. Steinmetz, P., Hötzer, J., Nestler, B.: Charakterisierung mehrkomponentiger Materialstrukturen durch den Einsatz von Hchstleistungsrechnern und Data Mining Konzepte (2015)
24. Steinmetz, P., Kellner, M., Hötzer, J., Dennstedt, A., Nestler, B.: Phase-field study of the pattern formation in Al–Ag–Cu under the influence of the melt concentration. Comput. Mater. Sci. **121**, 6–13 (2016)
25. Vondrous, A., Selzer, M., Hötzer, J., Nestler, B.: Parallel computing for phase-field models. Int. J. High Perform. Comput. Appl. **28**(1), 61–72 (2014)

Seismic Applications of Full Waveform Inversion

A. Kurzmann, L. Gaßner, N. Thiel, M. Kunert, R. Shigapov, F. Wittkamp, T. Bohlen, and T. Metz

Abstract Full waveform inversion (FWI) is a powerful imaging technique which exploits the richness of seismic waveforms. We further developed FWI to obtain multi-parameter images at high resolution. Here, we involve physical parameters, such as velocities and attenuation of seismic waves as well as mass density, which are essential for a reliable petrophysical characterization of subsurface structures in hydrocarbon exploration, geotechnical applications and underground constructions. Referring to this, we successfully applied FWI to field datasets recorded in the Black Sea and in the shallow-water area of a river delta in the Atlantic Ocean. We obtained detailed subsurface images containing rock formations which might be potential gas deposits. Additionally, we performed synthetic studies as preparatory steps to verify methodological improvements for further field-data applications. Here, we demonstrate resolution capabilities of FWI for imaging geological structures beneath salt bodies, investigate strategies to recover attenuation information from seismic data and perform a joint inversion of surface waves to image the very shallow subsurface.

1 Introduction

In a time where natural resources are precious, the number of underground constructions is increasing. It is important to map earth's geological structures accurately by collecting seismic data and transforming them into subsurface images. We develop full waveform inversion (FWI) as a cutting-edge seismic inversion technique that accounts for the full information content of seismic recordings. Each echo from geological discontinuities is used to unscramble the subsurface. FWI retrieves multi-parameter models of the subsurface by solving the full wave equation. It allows to map structures on sub-wavelength scales. Thus, FWI helps to improve both petrophysical interpretation and geotechnical characterization of the subsurface.

A. Kurzmann (✉) • L. Gaßner • N. Thiel • M. Kunert • R. Shigapov • F. Wittkamp • T. Bohlen • T. Metz
Karlsruhe Institute of Technology (Geophysical Institute), Karlsruhe, Germany
e-mail: andre.kurzmann@kit.edu; thomas.bohlen@kit.edu; tilman.metz@kit.edu

W.E. Nagel et al. (eds.), *High Performance Computing in Science and Engineering '16*, DOI 10.1007/978-3-319-47066-5_44

First implementations of FWI were conducted in the 1980s in the time domain by [23] and [15] as well as in the frequency-domain in the 1990s by [19] (see [25] for a general FWI overview). Particularly due to huge improvements in high-performance computing, these FWI strategies have emerged as an efficient imaging tool. In our work we concentrate on the implementation of the time-domain FWI and its application to seismic field-data problems. It comprises two- and three-dimensional (e.g., [4]) modeling of viscoelastic wavefields and exploits – as a main advantage – straightforward and efficient parallelization by domain decomposition [2] and source parallelization [12] leading to a significant speedup on parallel computers.

Within the scope of the HPC project *KITFWT*, we present applications of FWI to field-data (Sects. 3.1, 3.2, and 3.3) as well as further methodological developments illustrated by synthetic examples (Sects. 3.4 and 3.5).

2 Methodology

2.1 Full Waveform Inversion

FWI aims to find the optimal subsurface model by iteratively minimizing the misfit function between recorded and synthetic seismic data. That is, by solving the "forward problem" this model has to explain the recorded seismic data. The iterative optimization scheme of FWI – combining "forward problem" and "inverse problem" – comprises several steps shown in Fig. 1.

In detail, the method is initialized by two main inputs. First, we choose 2D or 3D initial parameter models of the subsurface, such as seismic velocities v_P of compressional wave (P-wave) and v_S of shear wave (S-wave), mass density ρ as well as attenuation represented by quality factors Q_P and Q_S for both wave types. They are assigned to the starting model at the first FWI iteration. The initial model can be estimated from a-priori information or computed by conventional seismic imaging methods. Second, the recorded data is obtained from seismic measurements involving many source locations and receivers. Typical acquisitions are performed offshore by utilizing air guns as sources and hydrophones located at sea surface/sea floor or onshore with hammerblow sources and geophones.

Within the FWI framework, for each source of the acquisition geometry, seismic modeling is applied (solution of "forward problem", see [2, 24]). That is, using the initial source wavelet the wavefield is emitted by the source and forward-propagates across the medium. A time series of spatial wavefield volumes has to be stored in memory. Synthetic seismic data is obtained at the receivers and the difference of synthetic and recorded data is calculated – resulting in residuals. In order to improve the minimization of the misfit function, a comprehensive multi-stage workflow focusses on both different model scales by applying frequency filtering to the data (e.g., [3, 21]) or choosing subsets of the data.

For each source, the residual wavefield is back-propagated from the receivers to the source position. The cross-correlation of forward- and back-propagated wavefields yields source-specific steepest-descent gradients (solution of "inverse

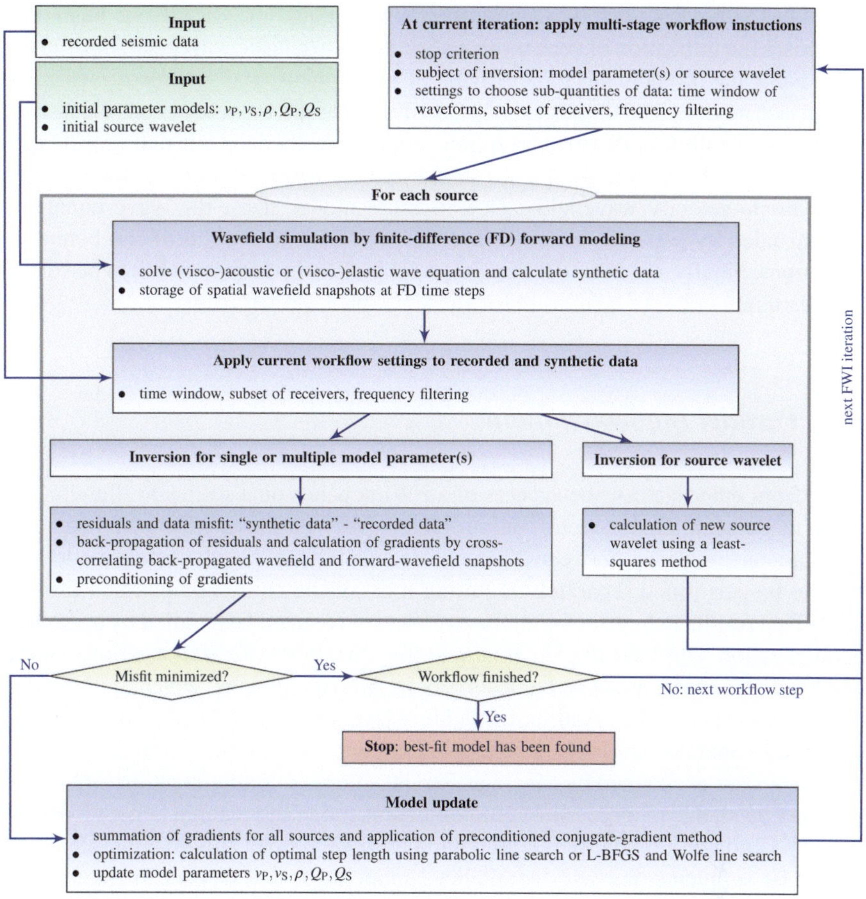

Fig. 1 General FWI scheme used for iterative improvement of physical model parameters of the subsurface by minimizing the misfit between modeled and recorded seismic data

problem", see [10, 15, 18, 23]). The computation of the global gradient for the entire acquisition geometry is given by the summation of all source-specific gradients. Subsequent optimization methods, such as preconditioned conjugate-gradient method and L-BFGS method are applied. The update of the model parameter(s) is the final step of a FWI iteration. The gradient has to be scaled by an optimal step length to get a proper model update. The estimation of the step length might require a significant amount of additional forward modelings. Furthermore, regarding the source wavelet, an equivalent inverse problem is given. The true wavelet is not known. The initial wavelet (a rough estimation or synthetic signal) is subject to FWI, too, and optimized during inversion by a least-squares method [7, 19].

Seismic modeling represents the fundamental part of FWI. In dependence of the field of application, the wave-propagation physics for an underlying subsurface model has to be described by an appropriate wave equation. On the one hand, that comprises the utilization of (visco-)acoustic or (visco-)elastic wave equation. On the other hand, the problem has to be solved for two-dimensional or three-dimensional subsurface models. The numerical implementation of the wave equations consists of a time-domain finite-difference (FD) time-stepping method in cartesian coordinates. In detail, the FD-scheme solves the stress-velocity formulation by utilizing stress and particle-velocity wavefields. Due to finite model sizes, the wave equations are expanded by perfectly matched layer terms (PML) to avoid artificial boundary reflections. Finally, at each FWI iteration a 2D or 3D wave equation has to be solved for a certain number of sources in forward- and back-propagation.

2.2 Parallel Implementation

Apart from other factors, the success of a FWI depends on a sufficient illumination of the model area. Thus, several source and receiver positions are necessary (reasonable numbers may vary between 20 and more than 500). For each source, modelings have to be performed separately requiring most of the entire computation time of FWI. That results in huge computational efforts, which can be handled by a massive parallelization. Our FWI implementation offers two types of parallelization. On the one hand, the model area can be decomposed into subdomains, which are assigned to all available cores [2]. Additional padding layers with half the size of the spatial differential operator are located around the model. At each time step these model boundaries are exchanged by Message-Passing-Interace communication (*MPI*). On the other hand, due to increasing communication, modelings cannot benefit from the decomposition of a model into a high number of very small subdomains. Hence, it should be supplemented with parallelized modelings with respect to the sources [11, 12]. The combination of domain decomposition and source parallelization results in nearly perfect speedup on supercomputers.

3 Results Obtained on FORHLR Phase I

3.1 Application of FWI to Image Submarine Gas Hydrate Deposits

3.1.1 Motivation

To study gas hydrate deposits in the Danube Deep Sea Fan in the Black Sea off the coast of Romania several geophysical experiments have been carried out including reflection and refraction seismic measurements. Within the SUGAR-III project (SUGAR – **SU**bmarine **GA**s Hydrate **R**esources) 15 ocean-bottom seismometer (OBS) stations equipped with pressure and particle velocity sensors have been deployed and have been covered by eight profiles of about 14 km length each. In this project, we apply FWI to data of a profile covering five stations to study subseafloor deposits of hydrated sediments which are possibly underlain by free gas.

3.1.2 FWI Setup and General Inversion Approach

For the acoustic FWI the data of the pressure sensors is utilized. As a starting model we use the resulting compressional-wave velocity model of a travel time tomography and a density model is derived by an empirical relation.

To account for the unknown source signature a correction filter is derived which matches the signature of the main events in the measured field data and in the modeled seismograms calculated for the starting model. This filter is inverted for at the beginning of each frequency stage by a waterlevel deconvolution and is then convolved with the original wavelet that has been used for the forward propagation. Another forward propagation is then performed with the corrected wavelet. Within one iteration the residuals of the measured and modeled data are minimized using the objective function suggested by [5]. Practically, this means the traces are normalized and, thus, a comparability of field and synthetic data is ensured. The gradients are spatially preconditioned to suppress undesired model updates, e.g. in the watercolumn and close to the OBS positions. Further preconditioning is applied to enhance updates in the deeper parts of the model.

3.1.3 Data Preparation

To prepare the field data for the inversion a data transformation is required to correct for 3D geometrical spreading effects when applying a 2D-FWI approach. This is accomplished by a convolution of each trace with $1/\sqrt{t}$ (t: traveltime) and a multiplication by $v\sqrt{2t}$ (v: velocity).

Examination of the field data showed strong ringing following the arrival of the direct waves which is possibly caused by receiver characteristics. To reduce influence of these dominant signals the timewindow including the direct wave and following simply reflected signals has been excluded from the inversion. We therefore only use refracted wave signals and multiple reflections.

3.1.4 Results

The inverted model of wave velocity v_P shows structures that are mainly horizontally orientated. In the central part of the profile high velocity anomalies become visible (see Fig. 2). The depth of the recovered anomalies coincides with the location of the BSR horizon. This, and a velocity increase of about 200 m/s indicates the presence of hydrated sediments. No indication for a gas layer is observed though.

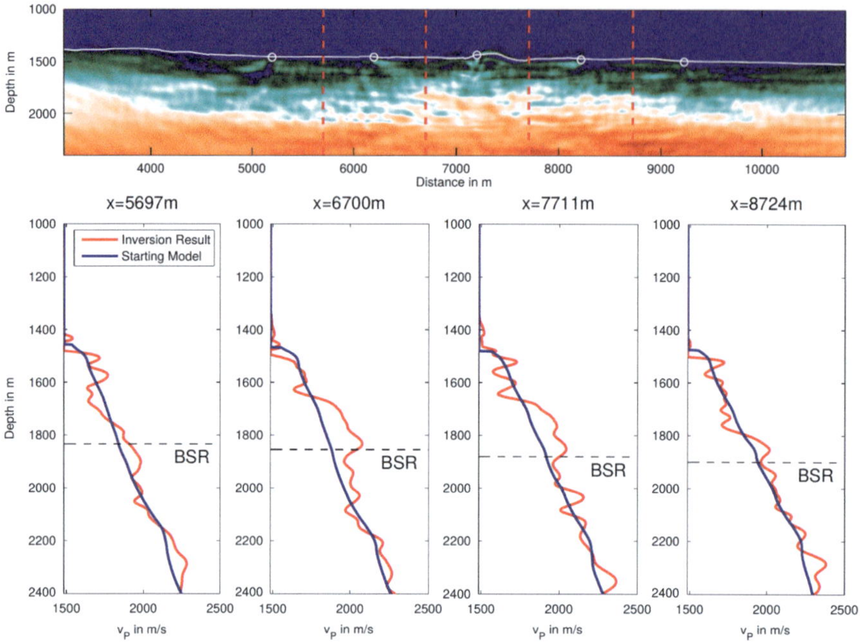

Fig. 2 *Top*: inverted v_P model, OBS locations (*white circles*), location of depth profiles shown below (*red dashed lines*); *bottom*: depth profiles of starting (*blue*) and final (*red*) v_P-model, seafloor (*white line*), depth of BSR horizon (*black dashed lines*)

A good overall match of the field data with the modeled seismograms for the final model could be achieved by the inversion (see Fig. 3). Also the direct wave phases that have not been considered for inversion could be fitted well.

3.1.5 Summary

We applied FWI to data recorded by ocean-bottom seismometers in order to identify hydrated sediments which are possibly underlain by free gas. Their presence is indicated by a bottom simulating reflector (BSR) which was tracked in the 3D reflection seismic data. To recover parameters of the subsurface we used an acoustic 2D FWI approach. The recovered model of compressional-wave velocity shows high-velocity anomalies in a depth that agrees well with the observed BSR horizon.

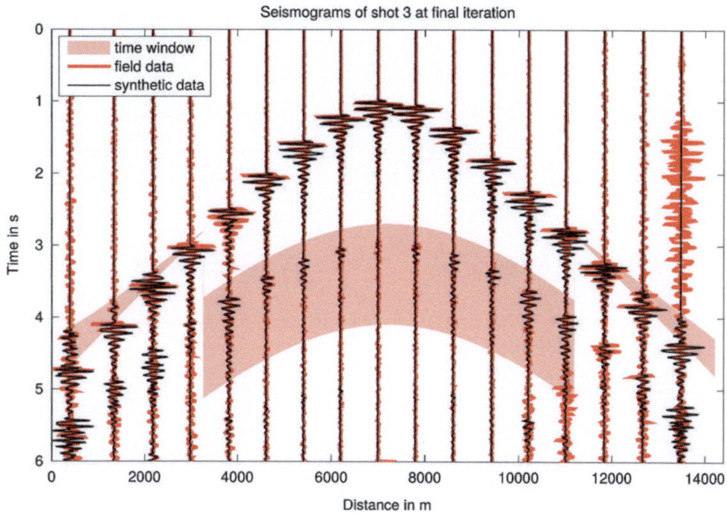

Fig. 3 Exemplary seismograms of OBS 3: measured field data (*red*), synthetic data for the final v_P-model (*black*); time window used for the inversion is marked by the *shaded red area*

3.2 Application of FWI to Marine Data Obtained in a River Delta

3.2.1 Motivation

The field-data application of FWI is still challenging and not common practise. 2D acoustic FWI is usually used to update a kinematic subsurface model and to improve the results of standard seismic imaging methods. In this work we apply 2D acoustic FWI to a marine seismic data set acquired in a river delta using ocean bottom cables (OBC). The aim of the seismic survey was to characterise a deep oil and gas deposit. Object of this work is an improved reconstruction of the near-surface region. Here, rising gases may stop at impenetrable sediments and accumulate to gas "pockets" reducing seismic velocities, which is a potential source for difficulties in standard imaging methods.

3.2.2 Prerequisites for FWI and Inversion Strategy

The field data were acquired in an OBC-geometry illustrated in Fig. 5a. Two-hundred and forty hydrophones were placed at the sea floor, whereas the source array (airguns) was dragged by a ship and triggered near the sea surface. Considering the solution of the 2D wave equation in forward modeling, we perform a 3D-to-2D transformation [17] of the recorded field data (example shown in Fig. 4 (left)).

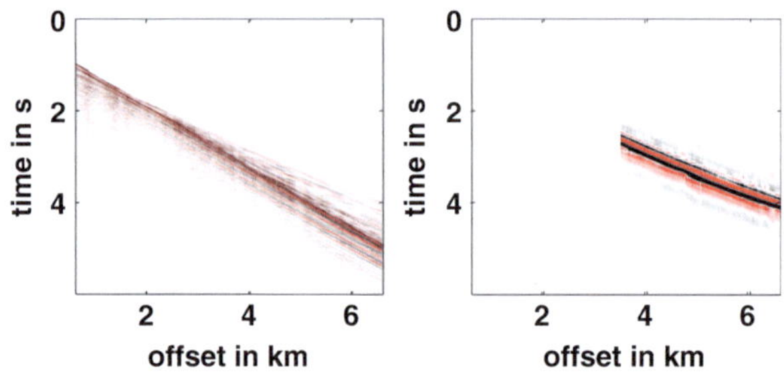

Fig. 4 *Left*: exemplary trace-normalized field data seismogram; *right*: windowed and filtered field data seismogram used in FWI; the seismograms belong to the source 13 located at $x = 2.5$ km (see Fig. 5a)

A suitable inversion strategy is essential for the success of FWI. It is represented by a comprehensive workflow consisting of several stages with particular instructions, such as stop criterions for each stage, frequency filtering as well as data windowing in time and offset (source-receiver distance). The existence of low frequencies is essential for a succesful FWI application. As the complexity of the misfit function increases with increasing frequencies, it is common procedure to start with low frequencies. After detecting a sufficient convergence of the data misfit at current stage, higher frequencies are added to recover structures at smaller scales [3]. In this work, we consider frequencies between 1.5 and 30 Hz. Furthermore, the windowing steps allow the focus on refracted waves, help to stabilize FWI and, thus, improve the resulting velocity model. As the first arrivals of the direct water wave do not carry informations about the subsurface, we focus on far offsets and ignore receivers with offset <3500 m (Fig. 4 (right)). The initial subsurface model has to contain large-scale structures representing the low frequencies, which are not available in the data. We use the result of a traveltime tomography as a v_P starting model shown in Fig. 5a. Due to the potential existence of gas accumulations and weakly consolidated rocks, we apply FWI in the acoustic approximation with consideration of attenuation as a modeling parameter. We only consider the FWI update of the v_P model.

3.2.3 Inversion Results

The final v_P model is shown in Fig. 5b showing a satisfactory resolution in the centre of the model (3 km $\leq x \leq$ 9 km and depths up to 1.2 km). Here, FWI recovered several geological structures which can be interpreted meaningfully. The remaining model areas are poorly illuminated. In shallow parts beneath the receiver array we identify layered structures including two types of significant low-velocity zones. On the one hand, rising gases may form accumulations (e.g., "gas pockets" close to the seafloor between $x = 4$ km and $x = 6$ km) at impenetrable layers reducing

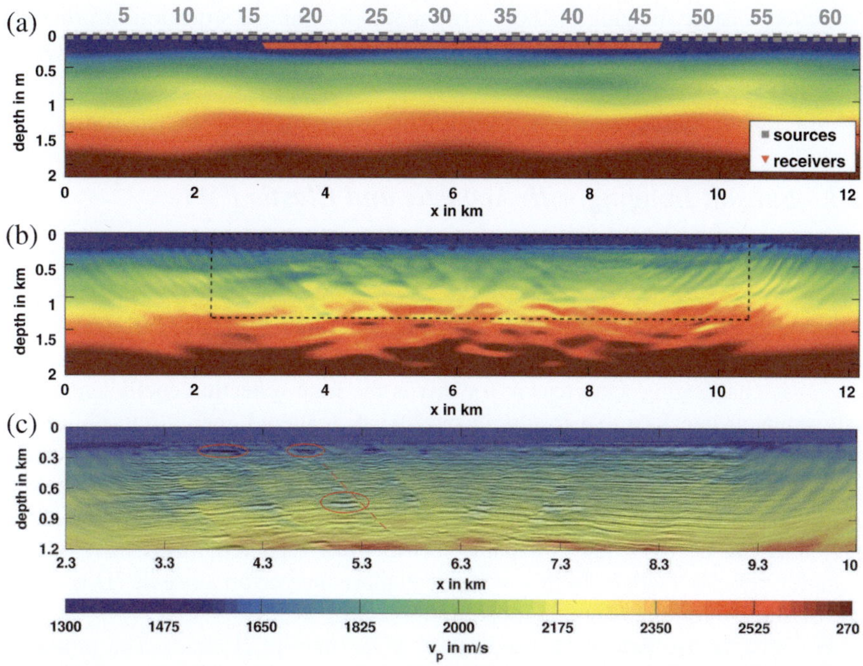

Fig. 5 (**a**) Starting v_P model based on a provided traveltime tomography. The coloured markers illustrate the acquisition geometry. Sources (*gray*) were triggered at the sea surface and hydrophones (*red*) were located at the sea floor. (**b**) Final v_P model obtained from FWI. The dashed rectangle represents the section in (**c**). (**c**) Overlay of final v_P model and the provided result of a reflection seismic imaging method. Examples of high similarity between both methods are highlighted by *red* markers

the seismic velocities. On the other hand, we find dipping structures in terms of geological fault zones causing both rising gases and accumulations.

We validated the quality of the velocity model by comparison with the result of a conventional reflection seismic imaging method (pre-stack depth migration), shown in Fig. 5c. Disregarding a different content of wavelengths or frequencies, respectively, they show a very good match, in particular fault zones and areas of strong seismic contrasts, such as gas accumulations.

3.2.4 Summary

In this work, we present a successful application of the acoustic FWI to marine seismic field data. Based on a simple starting model, small-scale structures, such as fault zones or other low-velocity areas with gas accumulations, could be recovered. Although we only considered a fraction of the data, FWI was able to obtain a velocity model which shows a high similarity to the result of a conventional reflection seismic imaging method. Both methods combine structural information of the subsurface and physical parameters, which is an important step for further petrophysical characterization.

3.3 Subsalt Imaging with Acoustic and Elastic FWI

3.3.1 Motivation

Salt bodies proved to be promising sites for the search for hydro carbonates. For classical imaging techniques the reconstruction of structures beneath or near salt bodies is challenging. One reason for this is the high reflection coefficient at the salt-sediment-interface that results in only weak scattered energy returning from the subsalt regions. Additional reasons are the complex shape of the salt bodies, trapped sediments in the salt body and a rugose surface. These characteristics lead to complex wavefields and regions with poor illumination. The solution for these problems can be FWI that is capable to use weak scattered waves travelling in complex velocity models. In this work we explore the performance of 2D acoustic and elastic FWI in time domain for subsalt imaging.

For this we use field data provided by PGS (marine 2D line). The profile is 265 km long with a total number of 5300 shots. The model is 15 km deep and for a better handling, the model was divided into three subpart. Only the right subpart will be shown in the following. The subpart has a size of 88.5×12 km, a grid distance of 12.5 m and a record length of 12 s for each of the 99 source points 804 receivers are used (moving streamer geometry).

3.3.2 Modeling

The initial velocity model (provided by PGS) includes a salt layer, a salt body and a velocity gradient as background (Fig. 7). The original acquisition geometry was extracted and used for the modeling. To validate the acquisition geometry and verify the suitability of the starting model. In Fig. 6 one acoustically modeled shot and the appropriated field data shot are displayed. The good match of the main events shows the sufficiency of the starting model.

3.3.3 Results of Resolution Study

The resolution in different parts of the model is influenced by the wave coverage and the velocity in the model. To estimate the resolution the true model is perturbed with a chequerboard pattern (Fig. 7). Each block has a size of 300 × 300 m and has a velocity perturbation of plus or minus 2 % of the true velocity. The starting model corresponds to the true model, however, without perturbations. The inversion

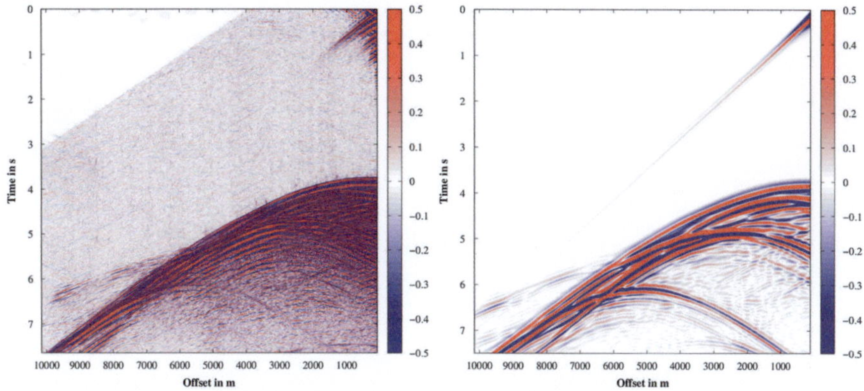

Fig. 6 Comparison of field data (**a**) and acoustically modeled data (**b**)

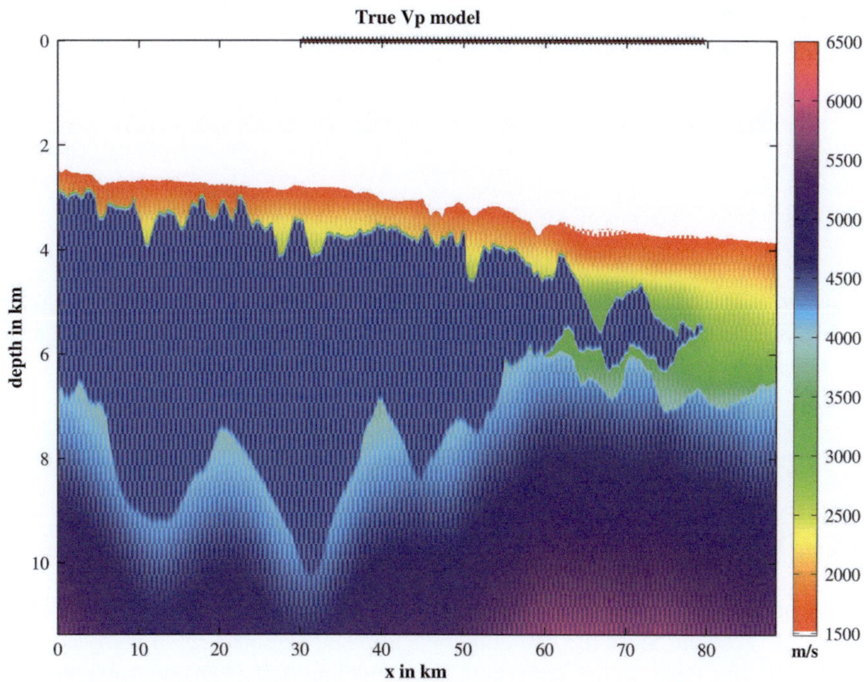

Fig. 7 Resolution study: perturbed model (true model)

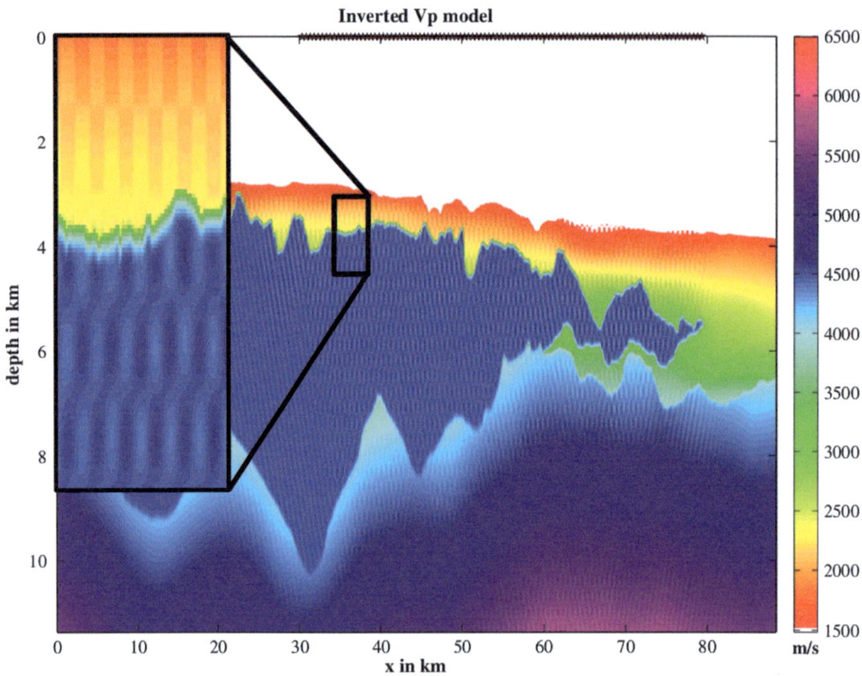

Fig. 8 Resolution study: inversion result

is performed purely acoustic in a frequency band of 3–10 Hz. The result is displayed in Fig. 8. The precise imaging of the chequerboards shows a good illumination in the target area (subsalt) for the given acquisition geometry.

3.4 Viscoacoustic FWI for Spatially Uncorrelated Problems

3.4.1 Motivation

Attenuation and dispersion of seismic waves play important role and need to be taken into account. Since the first numerical implementations [22] of viscoacoustic FWI until the most recent ones both the modeling and inversion were mostly developed in the frequency domain – exploiting its benefits, such as easy implementation of attenuation, computation of gradients without extra-cost. Time-domain FWI in attenuative media is less popular. The implementation of strictly constant Q within a wide frequency band is not so easy. Therefore, we have to consider a sum of relaxation mechanisms [14]. However, an advantage of time-domain implementations is efficient parallelizability. Our understanding of attenuation mechanisms and ability to get reliable Q estimates [9] are still limited. The problem is compounded by the fact that scattering effects mimic intrinsic attenuation. This opens the main question of this work: Can spatial distributions of velocity and attenuation be accurately recovered by applying FWI to synthetic marine reflection data? We investigate the applicability of time-domain viscoacoustic FWI and show numerical results using spatially uncorrelated models of velocity and attenuation [13].

3.4.2 Methodology

In time domain a general linear viscoacoustic equation of motion consists of a system of differential equations including stress-velocity formulation and relaxation mechanisms. Here, the conventional constant Q_P model, i.e., $Q_P(\omega) = const.$ with frequency ω, can be approximated by the generalized standard linear solid (GSLS) with a certain number of relaxation mechanisms [1]. The viscoacoustic medium is defined by ρ, the relaxed bulk modulus (containing v_P) and Q_P. While the forward problem comprises the solution of the viscoacoustic wave equation, the solution of the inverse problem involves the preconditioned conjugate-gradient method and independent parabolic line search with respect to the parameters v_P and Q_P.

3.4.3 Synthetic FWI Experiment

We apply viscoacoustic inversion to the 2D Marmousi model and investigate its impact on spatially uncorrelated models of v_P and Q_P. Density model, source wavelet and parameters in the water layer are assumed to be known. The acquisition geometry is a marine streamer at sea surface consisting of 32 explosive sources as well as a maximum number of 300 hydrophones. The true Q_P (Fig. 9d) model is derived from v_P (Fig. 9a) and turned upside down to avoid spatial correlation. The initial models for FWI can be found in Fig. 9b, e. Using the workflow framework of FWI, the sequential inversion of both parameters (i.e., first inverting for v_P, then both v_P and Q_P) recovers a satisfactory v_P model (Fig. 9c). The Q_P model shows significant artefacts (Fig. 9f). However, we can distinguish a quite good Q_P reconstruction in shallow areas and the unreliable footprint of v_P in deeper areas due to the cross-talk between both model parameters and insensitivity of seismic data to attenuation.

3.4.4 Summary

We implemented time-domain viscoacoustic FWI based on the GSLS. We tested its applicability on synthetic reflection marine data using the 2D Marmousi model. In contrast to the most of existing studies, we considered spatially uncorrelated models of v_P and Q_P. While v_P is recovered very well, Q_P is inverted satisfactorily only in shallow parts. We obtained an excellent fit of recorded and modeled data which

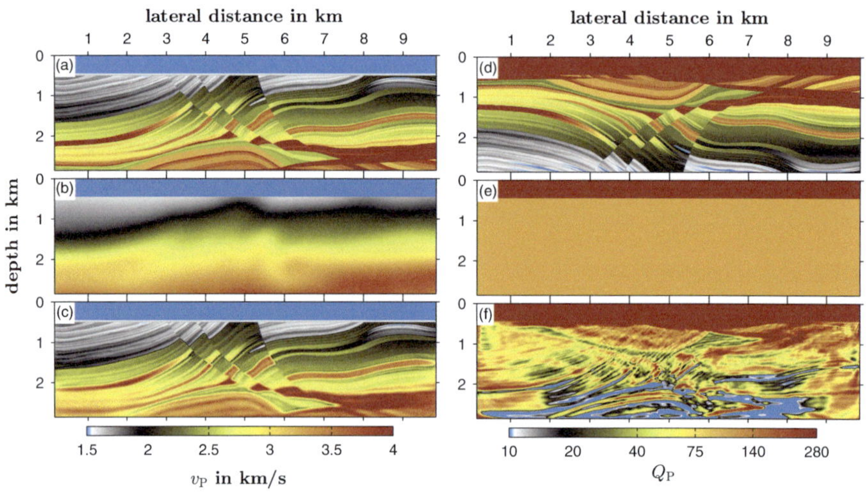

Fig. 9 v_P model: true (**a**), initial (**b**), final (**c**); Q_P model: true (**d**), initial (**e**), final (**f**)

can be interpreted either as low sensitivity of the synthetic data to deeper parts or a cross-talk effect where the Q_P-related data misfit is explained by the v_P model. In viscoacoustic inversion, the conventional assumption of correlated velocity and attenuation subsurface structures might induce an incorrect interpretation. Our results make clear that (a) further development of inversion strategies is necessary to extract the desired attenuation information from seismic data, and (b) the investigation of multiparameter inverse problems with spatially uncorrelated parameters has to be considered as a necessary step to verify the reliability of these strategies.

3.5 *Joint-FWI of Rayleigh and Love Waves in Shallow Seismics*

3.5.1 Motivation

Shallow-seismic Rayleigh and Love waves are attractive for geotechnical site investigations. They exhibit a high signal to noise ratio in field data recordings and have a high sensitivity to the S-wave velocity, an important geotechnical parameter to characterize the very shallow subsurface. In recent time full waveform inversion (FWI) has been successfully applied to reconstruct shallow 2-D shear wave velocity models using either Rayleigh waves (e.g., [8, 20]) or Love waves (e.g., [6, 16]). In most publications Rayleigh waves have been utilized. The aims of this synthetic study are (1) to compare the performance of individual waveform inversions of

Rayleigh or Love waves and (2) to explore the benefits of a simultaneous joint-inversion of both types of surfaces waves.

3.5.2 FWI Test Setting

In synthetic reconstruction experiments we explore the performance of individual and joint Rayleigh and Love wave FWI. We use a synthetic 2-D test model which emulates a realistic situation where a circular-shaped shallow small-scale low-velocity anomaly (trench) is embedded in a depth dependent background model. The models of S-wave velocity v_S, P-wave velocity v_P and density ρ are shown in Fig. 10 (left column). The sources and receivers are distributed over the model surface with a spacing of 5 and 0.8 m, respectively. The source excites frequencies between 0 and 20 Hz. During the multi-step inversion the frequency content increases gradually from 0 to 20 Hz. The multi-parameter inversion has three steps: (1) inversion of v_S only, (2) simultaneous inversion of v_P and v_S and (3) a final inversion of all elastic parameters. The used L-BFGS algorithm utilized the last 20 model and gradient differences.

3.5.3 Results

The results of the FWI reconstruction tests are shown in Fig. 10. The used 1-D starting models consist of a linear gradient up to a depth of 9 m. Below the true and the starting models are identical.

Individual Love wave FWI reconstructs the v_S model satisfactorily, also the shallow anomaly is reconstructed well. The ρ model is recovered surprisingly well, especially as normalized seismograms were used and the fact that the impact of density is mainly to the absolute wave amplitude as a function of offset. The objective function is reduced by four orders of magnitude and the velocity seismograms are fitted very well, thereby only a small residuum is remaining. The final Love wave FWI result of v_S and ρ appear as smoothed versions of the true models.

Individual Rayleigh wave FWI reconstructs the v_S model similarily well as the Love wave FWI. However, v_S and ρ suffer from vertically orientated artifacts underneath each source. These artifacts are most likely caused by the high amplitudes and specific radiation of Rayleigh waves below the source locations and are further enhanced in the gradients of v_S by wrong P-wave velocities in the vicinity of the sources. For the presented starting model the reconstruction of v_P was limited to the upper two meters. The inverted seismograms show small residuums, especially for the first arrivals. The objective function could be reduced by one order of magnitude.

The results of the **simultaneous joint FWI** are also affected by the source artifacts of the Rayleigh wave FWI. Nevertheless, the shallow trench is reconstructed in v_S successfully. However, the trench is not visible in v_P and only partly in the

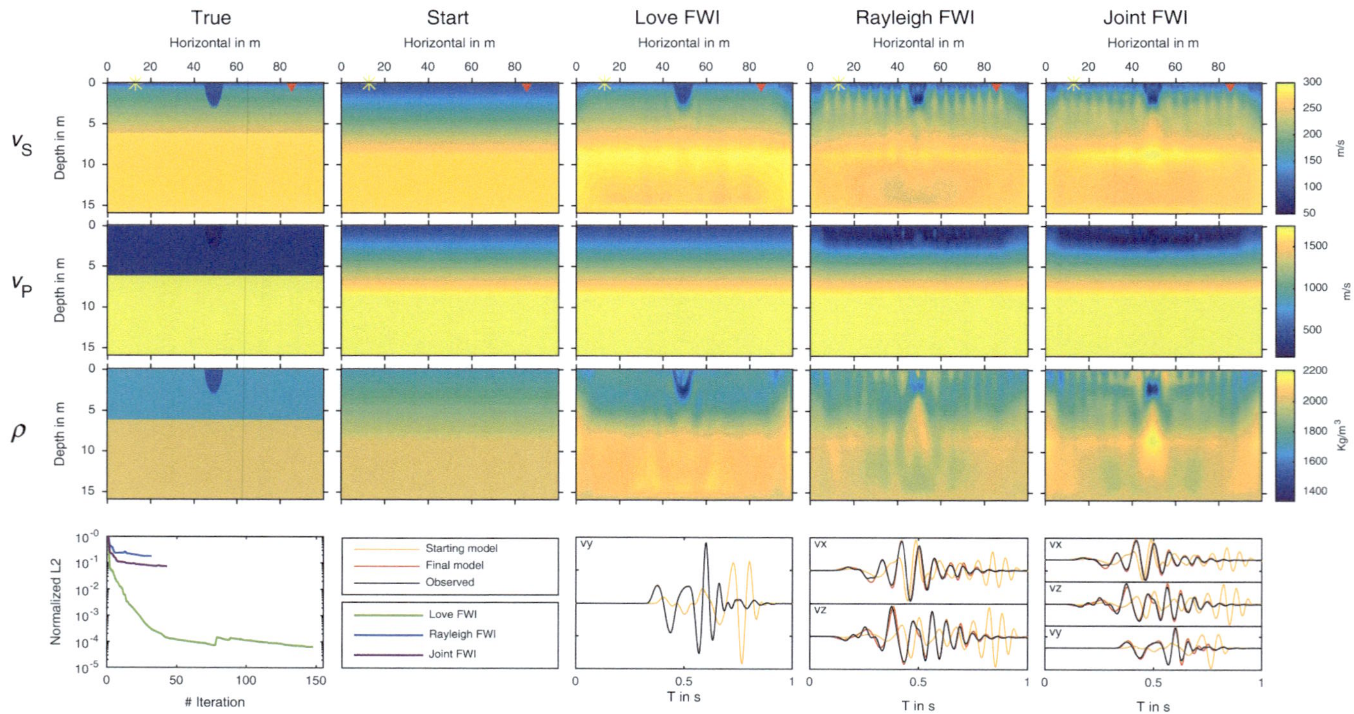

Fig. 10 Results of surface wave reconstruction tests. True and starting models are shown in column 1 and 2, respectively. The final FWI results for Love wave, Rayleigh wave and the joint FWI are shown in columns 3, 4 and 5, respectively. Below each FWI result, the seismograms for the true, starting and final models are given. The source (receiver) position corresponding to the seismograms is labeled by a yellow star (red triangle). The evolution of the L2-norm (lower left corner) is normalized to the maximal value in each case

inverted ρ model. The general appearance of the final v_P and ρ models is quite similar to individual Rayleigh wave FWI. The fit of the velocity seismograms v_x and v_z is slightly better than individual Rayleigh wave FWI.

3.5.4 Summary

We investigated the performance of individual waveform inversions of shallow-seismic Rayleigh or Love waves and studied the benefits of a simultaneous joint-inversion of both types of surfaces waves. We utilized a synthetic 2-D test model which emulates a realistic situation where a circular-shaped shallow low-velocity anomaly is embedded in a depth dependent background model. The true S-wave velocity model could be reconstructed by both the individual waveform inversion of Rayleigh and Love wave FWI as well as by the simultaneous joint-inversion of both wave types. The FWI of Rayleigh waves, however, suffers from artifacts below the source positions. The individual FWI of Love waves does not suffer from source artifacts and thus allows for an excellent final fit of both model and waveforms. In this case the single inversion of Love waves is thus superior to the individual inversion of Rayleigh waves and the join-inversion of Rayleigh and Love waves.

4 Computational Efforts of FWI on FORHLR Phase I

Based on the field-data example shown in Sect. 3.2, we estimated the resource consumption of one FWI application as follows:

- finite-difference discretization and 2D wave simulations:
  - spatial discretization: 12,160 × 2160 grid points (**26 million grid points**)
  - 60,000 time steps for each simulation
  - number of seismic sources: 50
  - total amount of **10,280 simulations** within whole FWI
- parallelization:
  - domain decomposition: 2D model is divided into 20 × 24 subdomains
  - source parallelization: simulations for 5 sources are computed at once
  - allocation of **2,400 CPU cores**
- resource consumption and computational performance of FWI framework:
  - number of iterations: 72
  - computation time for whole FWI job: 38.6 h (**92,640 core hours**)
  - memory consumption for wavefield storage: 835 MB/core (**total: 1.9 TB**)

Acknowledgements The scientific projects in this work are kindly supported by the sponsors of the Wave Inversion Technology (WIT) Consortium and funded by BWWi (grant number 03SX381C). We also gratefully acknowledge financial support by the Deutsche Forschungsgemeinschaft (DFG) through CRC 1173. The computations were performed on the computational resource "ForHLR Phase I" funded by the Ministry of Science, Research and the Arts Baden-Württemberg and DFG ("Deutsche Forschungsgemeinschaft").

References

1. Blanch, J., Robertsson, J., Symes, W.: Modeling of a constant Q: methodology and algorithm for an efficient and optimally inexpensive viscoelastic technique. Geophysics **60**(1), 176–184 (1995)
2. Bohlen, T.: Parallel 3-D viscoelastic finite difference seismic modeling. Comput. Geosci. **28**, 887–899 (2002)
3. Bunks, C., Saleck, F.M., Zaleski, S., Chavent, G.: Multiscale seismic waveform inversion. Geophysics **60**(5), 1457–1473 (1995)
4. Butzer, S., Kurzmann, A., Bohlen, T.: 3D elastic full-waveform inversion of small-scale heterogeneities in transmission geometry. Geophys. Prospect. **61**(6), 1238–1251 (2013)
5. Choi, Y., Alkhalifah, T.: Application of multi-source waveform inversion to marine streamer data using the global correlation norm. Geophys. Prospect. **60**(4), 748–758 (2012)
6. Dokter, E., Köhn, D., Wilken, D., Rabbel, W.: Application of elastic 2D waveform inversion to a near surface SH-wave dataset. In: 76th EAGE Conference and Exhibition (2014)
7. Forbriger, T.: Inversion flachseismischer Wellenfeldspektren. Dissertation, Stuttgart, University of Stuttgart (2001)
8. Groos, L., Schäfer, M., Forbriger, T., Bohlen, T.: The role of attenuation in 2D full-waveform inversion of shallow-seismic body and Rayleigh waves. Geophysics **79**(6), R247–R261 (2014)
9. Kamei, R., Pratt, R.G.: Inversion strategies for visco-acoustic waveform inversion. Geophys. J. Int. **194**(2), 859–884 (2013)
10. Köhn, D., De Nil, D., Kurzmann, A., Przebindowska, A., Bohlen, T.: On the influence of model parametrization in elastic full waveform tomography. Geophys. J. Int. **191**(1), 325–345 (2012)
11. Kurzmann, A.: Applications of 2D and 3D full waveform tomography in acoustic and viscoacoustic complex media. Dissertation, Karlsruhe Institute of Technology, Karlsruhe (2012)
12. Kurzmann, A., Köhn, D., Przebindowska, A., Nguyen, N., Bohlen, T.: 2D acoustic full waveform tomography: performance and optimization. In: 71st EAGE Conference and Technical Exhibition (2009)
13. Kurzmann, A., Shigapov, R., Bohlen, T.: Viscoacoustic full waveform inversion for spatially correlated and uncorrelated problems in reflection seismics. In: 77th EAGE Conference and Exhibition (2015)
14. Liu, H.-P., Anderson, D.L., Kanamori, H.: Velocity dispersion due to anelasticity; implications for seismology and mantle composition. Geophys. J. Int. **47**(1), 41–58 (1976)
15. Mora, P.: Nonlinear two-dimensional elastic inversion of multioffset seismic data. Geophysics **52**, 1211–1228 (1987)
16. Pan, Y., Xia, J., Xu, Y., Gao, L., Xu, Z.: Love-wave waveform inversion in time domain for shallow shear-wave velocity. Geophysics **81**(1), R1–R14 (2016)
17. Pica, A., Diet, J.P., Tarantola, A.: Nonlinear inversion of seismic reflection data in a laterally invariant medium. Geophysics **55**(3), 284–292 (1990)
18. Plessix, R.-E.: A review of the adjoint-state method for computing the gradient of a functional with geophysical applications. Geophys. J. Int. **167**(2), 495–503 (2006)
19. Pratt, R.: Seismic waveform inversion in the frequency domain, Part 1: theory and verification in a physical scale model. Geophysics **64**, 888–901 (1999)

20. Schäfer, M., Groos, L., Forbriger, T., Bohlen, T.: Line-source simulation for shallow-seismic data. Part 2: full-waveform inversion – a synthetic 2-D case study. Geophys. J. Int. **198**(3), 1405–1418 (2014)
21. Sirgue, L., Pratt, R.G.: Efficient waveform inversion and imaging: a strategy for selecting temporal frequencies. Geophysics **69**(1), 231–248 (2004)
22. Song, Z., Williamson, P., Pratt, R.: Frequency-domain acoustic-wave modeling and inversion of crosshole data: Part II – inversion method, synthetic experiments and real-data results. Geophysics **60**(3), 796–809 (1995)
23. Tarantola, A.: Inversion of seismic reflection data in the acoustic approximation. Geophysics **49**, 1259–1266 (1984)
24. Virieux, J.: P-SV wave propagation in heterogeneous media: velocity-stress finite-difference method. Geophysics **51**(4), 889–901 (1986)
25. Virieux, J., Operto, S.: An overview of full-waveform inversion in exploration geophysics. Geophysics **74**, WCC1–WCC26 (2009)

A Massively Parallel Multigrid Method with Level Dependent Smoothers for Problems with High Anisotropies

Sebastian Reiter, Andreas Vogel, Arne Nägel, and Gabriel Wittum

Abstract Anisotropic layers, as often seen in biological and geological domains, impose difficulties to several aspects of numerical simulations. In this article we examine how the highly scalable approach to massively parallel geometric multigrid solvers presented in Reiter et al. (Comput Vis Sci 16(4):151–164, 2013) can be extended to problem domains featuring such anisotropies. Considering the real world problem of drug diffusion through the human skin we combine hierarchically distributed multigrids, anisotropic refinement, and level dependent smoothing strategies to create a robust and highly scalable multigrid solver for anisotropic domains.

Keywords Multigrid • Parallelization • Anisotropy • Smoothing

1 Introduction

The development of algebraic solvers for discretizations of partial differential equations on massively parallel computers is an active research field. Multigrid methods [7] have been employed with great efficiency for elliptic PDEs on large super-computers [1, 3, 4, 6, 9, 10, 13–15, 17]. In [13] we demonstrated that the geometric multigrid solver of the software package UG4 [16] has nearly optimal weak scaling properties for up to 262,144 processes and more than 10^{10} unknowns. The study indicates that very good scalability should be achievable for even higher numbers of processes and unknowns with the given approach, once the required resources are available.

An issue that had not been fully addressed in previous studies are the difficulties to solve those equations on massively parallel computers in the presence of anisotropic coefficients or anisotropic elements in the underlying grid. Techniques exist to address those problems for general settings for lower process numbers, e.g., involving the use of anisotropic refinement to construct a specialized grid

S. Reiter (✉) • A. Vogel • A. Nägel • G. Wittum
G-CSC, Goethe-Universität Frankfurt, Kettenhofweg 139, 60325 Frankfurt (M.), Germany
e-mail: sebastian.reiter@gcsc.uni-frankfurt.de; andreas.vogel@gcsc.uni-frankfurt.de; arne.naegel@gcsc.uni-frankfurt.de; wittum@gcsc.uni-frankfurt.de

W.E. Nagel et al. (eds.), *High Performance Computing in Science and Engineering '16*, DOI 10.1007/978-3-319-47066-5_45

hierarchy [2]. While parallelism is considered in [2], massively parallel systems as todays supercomputers with hundred thousands of computing cores did not exist at that time and no optimizations regarding massive scalability have thus been performed. In [10] massively parallel multigrid is described for the solution of elliptic PDEs for the special case with strong vertical anisotropies on structured grids.

Considering the real world problem of drug diffusion through the human skin [11], we extended the methods described in [2] to construct a method that employs geometric multigrid on massively parallel computers for problems with highly anisotropic elements using a combination of specialized refinement techniques and smoothers resulting in a robust and highly scalable solver for anisotropic problems. The special grid layout of the model problem thereby requires a solver which can handle anisotropies in all spatial directions on unstructured grids.

2 Problem Description

The motivating biological question for the construction of our solver is the numerical simulation of substance transport through the human skin. Using simulations of such processes helps to estimate the risk assessment of chemical exposures and at the same time the need for in vitro and in vivo testing can be reduced. However, the special structure of the human skin imposes several numerical challenges due to the anisotropic geometry and physical coefficients varying by orders of magnitude, cf., e.g., [8, 11]. The uppermost part of the skin, called *stratum corneum (SC)*, consists of multiple layers of cells (*corneocytes*) which are connected by thin channels (*lipid layers*) (cf. Fig. 1). Since parameters affecting transport and diffusivity of a substance vary strongly in those subdomains, both have to be considered in a simulation. For benchmark purposes we solve a modified heat equation

$$\frac{\partial}{\partial t}(Ku) + \nabla \cdot (-\mathbf{D}K\nabla u) = 0$$

in a computational domain $\Omega = \Omega_{lip} \cup \Omega_{cor}$. Here, u corresponds to the chemical activity, $K = K(x)$ and $\mathbf{D} = \mathbf{D}(x)$ are spatially dependent partition and diffusion coefficients respectively. Concentrations are given by $c = Ku$ and may undergo discontinuities which reflect domain dependent variations in lipophilicity and hydrophilicity. For reasons of simplicity we used $K = 1$ and

$$D(x) = \begin{cases} 1 & \text{if } x \in \Omega_{lip} \\ 10^{-3} & \text{if } x \in \Omega_{cor}. \end{cases}$$

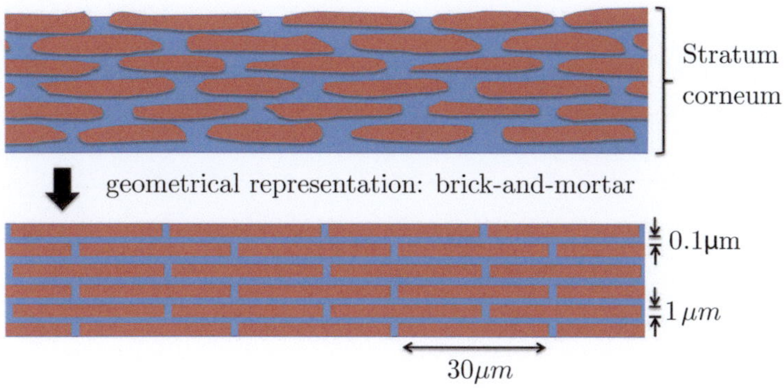

Fig. 1 Idealized model of the *stratum corneum*: *brick-and-mortar* grid

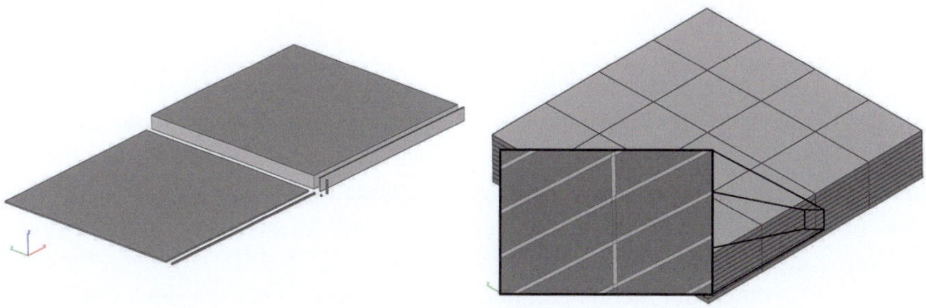

Fig. 2 *Left*: Individual hexahedral elements used for the *brick-and-mortar* FE-grid. *Right*: Anisotropic coarse grid of the 3d *brick-and-mortar* geometry (1280 elements)

For this study we focused on the idealized but still realistic *brick-and-mortar* domain. To this end, we used finite element grids consisting of highly anisotropic elements. This level of anisotropy is required to reduce the number of elements in the coarse grids, while still capturing the topology and morphology of the underlying domain.

For the considered geometry only hexahedral elements with varying degrees of anisotropy were used. The elements used to construct the coarse grid and an overview over the resulting FE-grid are depicted in Fig. 2. In the given example the highest aspect ratios are 1:15, however, the presented methods work for much higher aspect ratios, too.

The steady state solution of the regarded problem setup is depicted in Fig. 3.

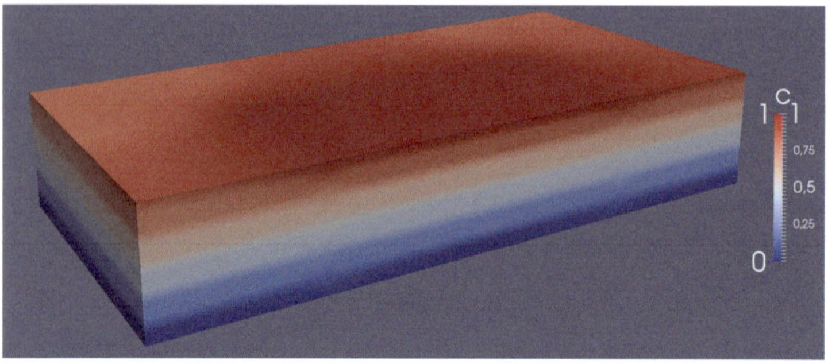

Fig. 3 Cut through the domain showing the steady state solution on level 5

3 Solver Setup

In [13] we employed a massively parallel geometric multigrid solver for the Poisson problem on grids with isotropic cells. While the rather simple Jacobi smoother used in those studies is fast and perfectly scalable, it is not suitable for anisotropic problems, since its smoothing properties deteriorate in this case. Iterative methods like the ILU method on the other hand are known to possess good smoothing properties also for highly anisotropic problems (cf. [2, 5, 18]). However, for most applicable methods an efficient parallel implementation is not feasible. The typical strategy for parallelization is then to employ those methods in Block-Jacobi-type fashion, i.e., on each process the more sophisticated smoother is executed locally, while interprocess couplings are treated using a Jacobi method.

While this setup works nicely for smaller process numbers, the iteration numbers typically increase with the number of processes being involved. As a second aspect, load imbalances can have a severe impact on the runtime of solver initializations and such load imbalances are typically given for massively parallel simulations on unstructured grids. This in particular holds true, when setup times do not grow like $O(n)$, as, e.g., for a threshold based ILUT.

In order to construct an optimal solver which can handle high anisotropies while still providing nearly optimal scalability, we combined the benefits of the Jacobi smoother for isotropic elements with the efficiency and robustness of the ILU smoothers for anisotropic problems. This combination is possible using a special refinement technique, which reduces the anisotropy of elements with each refinement until the resulting grid can be considered isotropic. The anisotropic and isotropic refinement rules used for the elements from Fig. 2 are depicted in Fig. 4.

In each refinement step only edges which are longer than a certain threshold are refined. Starting with a threshold of half of the length of the longest edge, the threshold is halved after each step so that shorter edges will be refined in the next iteration. For the shown brick-and-mortar geometry this technique leads to an isotropic grid after a certain number of refinements, depending on the highest aspect ratio.

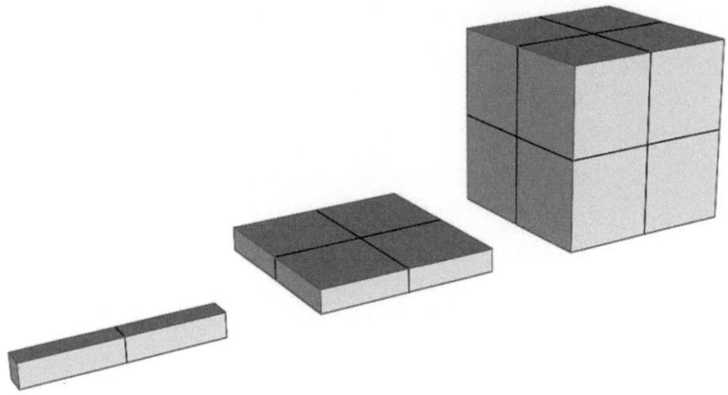

Fig. 4 Anisotropic refinement schemes for different shapes. *Black edges* are introduced during refinement

Table 1 Solver components: solvers on lower levels serve as base solvers for the preconditioners on higher levels

Levels	2	2–4	4-top
Solver	LU (exact)	CG	CG
Preconditioner	–	GMG	GMG
Smoother	–	ILU	Jacobi
Cycle	–	V (3,3)	V (3,3)

On the lower levels in which anisotropic elements are present we employ a multigrid method with robust ILU smoothing, whereas on higher levels, which contain nearly isotropic elements, a highly scalable multigrid method with Jacobi smoothing is used. This setup is even more justified considering the fact, that lower levels can be distributed to only a subset of the available processes, since they only contain a fraction of the elements of higher levels. We can thus make sure that complex ILU smoothing is only performed on a fixed number of processes, thus not interfering with the scalability on higher levels. To this end, we are using the hierarchical distribution approach described in [13]. Technically this setup is realized in the software package UG4 [16] by employing a multigrid method with ILU smoothing on lower levels as a base solver for the multigrid method with damped Jacobi smoothing ($omega = 0.5$) on higher levels. The prescribed tolerance for the intermediate coarse grid solver was a relative reduction by three orders of magnitude. Table 1 shows the different properties of the involved solver components.

4 Parallelization

Parallelization is performed using the approach detailed in [13]. Hierarchical distribution hereby plays an important role, since we apply the multigrid method with ILU smoothing on a smaller subset of processes only. The base solver for this lower ILU smoothed multigrid method may finally run on one process only.

Since anisotropic refinement may lead to load-imbalances, redistributions of the grid hierarchy may be necessary to improve those balances. To this end, we are using a fast parallel bisection strategy for distributed multigrid hierarchies as described in [12].

All communication between different processes is realized through the message passing interface *MPI*.

5 Results

A weak scaling study was performed on the Cray XC40 super computer *Hazel Hen* at the HLRS Stuttgart which features 7712 compute nodes, each with 128 GB of memory, 24 cores per node (virtually 48 through hyperthreading) and a peak performance of 7420 TFlops.

The study was performed by solving the aforementioned human skin brick-and-mortar model using the described solver setup. To allow for better comparability of the different runs, the number of outer CG-iterations was thereby fixed to 12, which resulted in a relative reduction of the defect by approximately 10^{-6} in all runs. The study starts on 1 process and for each subsequent run we refine the grid once more using regular refinement, thus increasing the number of elements by a factor of 8. At the same time we also increase the number of involved processes by a factor of 8 to guarantee a constant workload per process for all runs. Since we executed the parallel base-solver of the outer multigrid method on level 4, our study starts with level 5. Table 2 shows the number of unknowns and the run times of the different runs. The scaling behavior of assembly, solver initialization, and solving is also shown in Fig. 5.

Table 3 gives an overview over levelwise distribution qualities. The distribution quality q_l of a level l of the hierarchy is computed as

$$q_l := \frac{n_l^{total} - n_l^{max}}{n_l^{max} \cdot (P_l - 1)},$$

Table 2 Each line corresponds to an individual run. Recorded are the number of processes (*PEs*), the number of levels (*Levels*), the number of unknowns (*DoFs*), the run times of assembly (T_{ass}), solver initialization (T_{ini}), and solving (T_{sol})

PEs	Levels	DoFs	T_{ass} (s)	T_{ini} (s)	T_{sol} (s)
8	6	522,720	0.48	1.38	6.58
64	7	4,181,760	0.80	2.03	6.95
512	8	33,454,080	0.85	2.10	7.25
4096	9	267,632,640	0.87	2.10	7.15
32,768	10	2,141,061,120	0.86	2.13	7.60

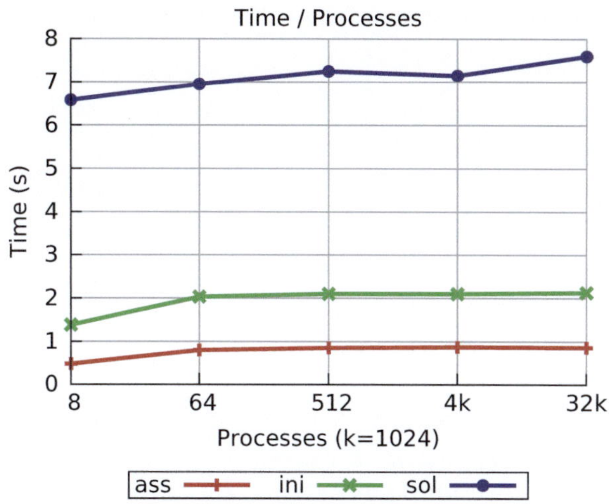

Fig. 5 Scaling of the run times of assembly (*ass*), solver initialization (*ini*), and solving (*sol*)

Table 3 Distribution qualities for each level of the multigrid hierarchy for the different runs

PE	0	1	2	3	4	5	6	7	8	9
8	1	1	1	1	1	1	–	–	–	–
64	1	1	0.92	0.93	0.94	0.99	0.99	–	–	–
512	1	1	1	0.91	0.95	0.94	0.94	0.94	–	–
4096	1	1	1	0.91	0.95	0.89	0.89	0.89	0.89	–
32,768	1	1	1	0.74	0.83	0.93	0.78	0.78	0.8	0.8

where $P_l > 1$ is the number of processes of the given process-hierarchy on level l, n_l^p is the number of elements in level l on process p, and

$$n_l^{total} := \sum_{p=1}^{P_l} n_l^p,$$

$$n_l^{max} := \max_{p=1,\ldots,P_l} n_l^p.$$

For $P_l = 1$ numerator and denominator both vanish and we define $q_l = 1$. q_l is thus in the range $[0, 1]$, where $q_l = 0$ means that all elements of level l are contained on one process only and $q_l = 1$ reflects an equal share of elements amongst all processes.

In Table 4 the number of processes used on each level for the individual runs are specified. They were chosen so that each level on each process would at least contain 32 elements (given an ideal load-balance). The first redistribution is performed for up to 256 processes. For each further redistribution the number of processes is multiplied by 64 and capped, if the maximum number of processes is reached.

Table 4 Number of processes used on each level for the individual runs

PE	0	1	2	3	4	5	6	7	8	9
8	1	1	8	8	8	8	–	–	–	–
64	1	1	64	64	64	64	64	–	–	–
512	1	1	1	256	256	512	512	512	–	–
4096	1	1	1	256	256	4096	4096	4096	4096	–
32,768	1	1	1	256	256	256	16,384	16,384	32,768	32,768

Both matrix assembly and solver initialization are performed process locally. The increase in run time T_{ass} and T_{ini} from 8 to 64 processes is related to the slight load-imbalance which can be observed for higher process numbers (cf. Table 3). Nevertheless, the scaling behavior of both assembly and initialization is very good and perfectly suited for large scale parallel runs.

The solver scalability is satisfactory as well. The run times T_{sol} increase slightly the more processes are involved and two effects are in play here: The distribution quality of the grid hierarchy deteriorates the larger the number of processes involved. This reflects the fact that some processes have more work to do than others in each program section due to the slight load imbalance. The slight imbalance is to be expected for an unstructured grid in which no special properties can be exploited for partitioning. However, for runs up to 256 processes the increasing parallelization of the intermediate base-solver on level 4 has a positive effect on total solver run times.

As demonstrated in [13], the underlying multigrid implementation in UG4 has nearly optimal scaling properties for perfectly balanced grids. The slightly worse scaling properties in the study at hand are thus likely to be linked to the observed load-imbalance. Nevertheless, given the complexity of the problem at hand we think that the achieved run-times are still convincing. The achieved results demonstrate the applicability of the presented approach to gain insight into complex biological processes through high-resolution numerical simulations on massively parallel computers.

Acknowledgements We thank the HLRS for the opportunity to use *Hazel Hen* and their kind support.

References

1. Baker, A.H., Falgout, R.D., Kolev, T.V., Yang, U.M.: Multigrid smoothers for ultra-parallel computing. SIAM J. Sci. Comput. **33**, 2864–2887 (2011)
2. Bastian, P., Wittum, G.: Adaptive multigrid methods: the UG concept. In: Adaptive Methods – Algorithms, Theory and Applications: Proceedings of the Ninth GAMM-Seminar, Kiel, 22–24 Jan 1993, pp. 17–37. Vieweg+Teubner Verlag, Wiesbaden (1994)
3. Bastian, P., Blatt, M., Scheichl, R.: Algebraic multigrid for discontinuous Galerkin discretizations of heterogeneous elliptic problems. Numer. Linear Algebra Appl. **19**(2), 367–388 (2012)

4. Bergen, B., Gradl, T., Rude, U., Hulsemann, F.: A massively parallel multigrid method for finite elements. Comput. Sci. Eng. **8**(6), 56–62 (2006)
5. Bramble, J., Zhang, X.: Uniform convergence of the multigrid v-cycle for an anisotropic problem. Math. Comput. **70**(234), 453–470 (2001)
6. Gmeiner, B., Köstler, H., Stürmer, M., Rüde, U.: Parallel multigrid on hierarchical hybrid grids: a performance study on current high performance computing clusters. Concurr. Comput.: Pract. Exp. **26**(1), 217–240 (2014)
7. Hackbusch, W.: Multi-grid Methods and Applications, vol. 4. Springer, Berlin/New York (1985)
8. Heisig, M., Lieckfeldt, R., Wittum, G., Mazurkevich, G., Lee, G.: Non steady-state descriptions of drug permeation through stratum corneum. I. The biphasic brick-and-mortar model. Pharm. Res. **13**(3), 421–426 (1996)
9. Heppner, I., Lampe, M., Nägel, A., Reiter, S., Rupp, M., Vogel, A., Wittum, G.: Software framework ug4: parallel multigrid on the hermit supercomputer. In: High Performance Computing in Science and Engineering 12, pp. 435–449. Springer, Berlin/London (2013)
10. Müller, E.H., Scheichl, R.: Massively parallel solvers for elliptic partial differential equations in numerical weather and climate prediction. Q. J. R. Meteorol. Soc. **140**(685), 2608–2624 (2014)
11. Nägel, A., Heisig, M., Wittum, G.: Detailed modeling of skin penetration–an overview. Adv. Drug Deliv. Rev. **65**(2), 191–207 (2013) Modeling the human skin barrier – towards a better understanding of dermal absorption
12. Reiter, S.: Effiziente Algorithmen und Datenstrukturen für die Realisierung von adaptiven, hierarchischen Gittern auf massiv parallelen Systemen. PhD thesis, Universität Frankfurt am Main (2014)
13. Reiter, S., Vogel, A., Heppner, I., Rupp, M., Wittum, G.: A massively parallel geometric multigrid solver on hierarchically distributed grids. Comput. Vis. Sci. **16**(4), 151–164 (2013)
14. Sampath, R.S., Biros, G.: A parallel geometric multigrid method for finite elements on octree meshes. SIAM J. Sci. Comput. **32**, 1361–1392 (2010)
15. Sundar, H., Biros, G., Burstedde, C., Rudi, J., Ghattas, O., Stadler, G.: Parallel geometric-algebraic multigrid on unstructured forests of octrees. In: Proceedings of the International Conference on High Performance Computing, Networking, Storage and Analysis, SC '12, pp. 43:1–43:11, Los Alamitos. IEEE Computer Society Press (2012)
16. Vogel, A., Reiter, S., Rupp, M., Nägel, A., Wittum, G.: UG 4: a novel flexible software system for simulating PDE based models on high performance computers. Comput. Vis. Sci. **16**(4), 165–179 (2013)
17. Williams, S., Lijewski, M., Almgren, A., Van Straalen, B., Carson, E., Knight, N., Demmel, J.: s-step Krylov subspace methods as bottom solvers for geometric multigrid. In: 28th International Parallel and Distributed Processing Symposium, pp. 1149–1158. IEEE, Piscataway (2014)
18. Wittum, G.: On the robustness of ILU smoothing. SIAM J. Sci. Stat. Comput. **10**(4), 699–717 (1989)

MIX
Papier aus verantwortungsvollen Quellen
Paper from responsible sources
FSC® C105338

If you have any concerns about our products,
you can contact us on
ProductSafety@springernature.com

In case Publisher is established outside the EU,
the EU authorized representative is:
Springer Nature Customer Service Center GmbH
Europaplatz 3, 69115 Heidelberg, Germany

Printed by Libri Plureos GmbH
in Hamburg, Germany